Lecture Notes in Artificial Intelligence 1831

Subseries of Lecture Notes in Computer Science
Edited by J. G. Carbonell and J. Siekmann

Lecture Notes in Computer Science
Edited by G. Goos, J. Hartmanis and J. van Leeuwen

T0216458

Springer
Berlin
Heidelberg
New York
Barcelona
Hong Kong
London
Milan
Paris
Singapore
Tokyo

David McAllester (Ed.)

Automated Deduction – CADE-17

17th International Conference on Automated Deduction
Pittsburgh, PA, USA, June 17-20, 2000
Proceedings

Springer

Series Editors

Jaime G. Carbonell,Carnegie Mellon University, Pittsburgh, PA, USA
Jörg Siekmann, University of Saarland, Saarbrücken, Germany

Volume Editor

David McAllester
AT&T Labs Research
180 Park Avenue, Florham Park, N.J., 07932-0971, USA
E-mail: dmac@research.att.com

Cataloging-in-Publication Data applied for

Die Deutsche Bibliothek - CIP-Einheitsaufnahme

Automated deduction : proceedings / CADE-17, 17th International
Conference on Automated Deduction, Pittsburgh, PA, USA, June 17 - 20,
2000. David McAllester (ed.). - Berlin ; Heidelberg ; New York ;
Barcelona ; Hong Kong ; London ; Milan ; Paris ; Singapore ; Tokyo :
Springer, 2000
    (Lecture notes in computer science ; Vol. 1831 : Lecture notes in
    artificial intelligence)
    ISBN 3-540-67664-3

CR Subject Classification (1998): I.2.3, F.4.1, F.3.1

ISBN 3-540-67664-3 Springer-Verlag Berlin Heidelberg New York

Typesetting: Camera-ready by author, data conversion by Christian Grosche, Hamburg
Printed on acid-free paper   SPIN 10721959   06/3142   5 4 3 2 1 0

Preface

For the past 25 years the CADE conference has been the major forum for the presentation of new results in automated deduction. This volume contains the papers and system descriptions selected for the 17th International Conference on Automated Deduction, CADE-17, held June 17-20, 2000, at Carnegie Mellon University, Pittsburgh, Pennsylvania (USA).

Fifty-three research papers and twenty system descriptions were submitted by researchers from fifteen countries. Each submission was reviewed by at least three reviewers. Twenty-four research papers and fifteen system descriptions were accepted. The accepted papers cover a variety of topics related to theorem proving and its applications such as proof carrying code, cryptographic protocol verification, model checking, cooperating decision procedures, program verification, and resolution theorem proving.

The program also included three invited lectures: "High-level verification using theorem proving and formalized mathematics" by John Harrison, "Scalable Knowledge Representation and Reasoning Systems" by Henry Kautz, and "Connecting Bits with Floating-Point Numbers: Model Checking and Theorem Proving in Practice" by Carl Seger. Abstracts or full papers of these talks are included in this volume. In addition to the accepted papers, system descriptions, and invited talks, this volume contains one page summaries of four tutorials and five workshops held in conjunction with CADE-17.

The CADE-17 ATP System Competition (CASC-17), held in conjunction with CADE-17, selected a winning system in each of four different automated theorem proving divisions. The competition was organized by Geoff Sutcliffe and Christian Suttner and was overseen by a panel consisting of Claude Kirchner, Don Loveland, and Jeff Pelletier. This was the fifth such competition held in conjunction with CADE. Since the contest was held during the conference the winners were unknown as of this printing and the results are not described here.

I would like to thank the members of the program committee and all the referees for their care and time in selecting the submitted papers. I would also like to give a special thanks to Bill McCune for setting up and maintaining the web site for the electronic program committee meeting.

April 2000   David McAllester

Conference Organization

Referees

Tamarah Arons
Clark Barrett
Peter Baumgartner
Gertrud Bauer
Adam Cichon
Witold Charatonik
Ingo Dahn
Anatoli Degtyarev
Grit Denker
Dan Dougherty
Alain Finkel
Dana Fisman
Laurent Fribourg
Juergen Giesl
Andy Gordon
Enrico Guinchiglia
Andrew Haas
Ian Horrocks
Joe Hurd
Paul Jackson

Robert B. Jones
Yonit Kesten
Gerwin Klein
Konstantin Korovin
Keiichirou Kusakari
Raya Leviatan
Ian Mackie
Rupak Majumdar
Fred Mang
Claude March
R. C. McDowell
D. Mery
Robin Milner
Marius Minea
Michael Norrish
Jens Otten
Nicolas Peltier
Leonor Prensa-Nieto
Alexandre Riazanov
C. Ringeissen

Sitvanit Ruah
Axel Schairer
Renate Schmidt
Philippe Schnoebele
Elad Shahar
Ofer Shtrichman
G. Sivakumar
Viorica Sofronie-Stokkermans
Dick Stearns
Frieder Stolzenburg
Jürgen Stuber
Yoshihito Toyama
Rakesh Verma
Bob Veroff
David von Oheimb
Uwe Waldmann
Christoph Weidenbac
Markus Wenzel
Benjamin Werner

Previous CADEs

CADE-1, Argonne National Laboratory, USA, 1974
CADE-2, Oberwolfach, Germany, 1976
CADE-3, MIT, USA, 1977
CADE-4, University of Texas at Austin, 1979
CADE-5, Les Arcs, France, 1980 (Springer LNCS 87)
CADE-6, Courant Institute, New York, 1982 (Springer LNCS 138)
CADE-7, Napa, California, 1984 (Springer LNCS 170)
CADE-8, University of Oxford, UK, 1986 (Springer LNCS 230)
CADE-9, Argonne National Laboratory, USA, 1988 (Springer LNCS 310)
CADE-10, Kaiserslautern, Germany, 1990 (Springer LNAI 449)
CADE-11, Saratoga Springs, New York, 1992 (Springer LNAI 607)
CADE-12, Nancy, France, 1994 (Springer LNAI 814)
CADE-13, Rutgers University, USA, 1996 (Springer LNAI 1104)
CADE-14, James Cook University, Australia, 1997 (Springer LNAI 1249)
CADE-15, Lindau, Germany, July 6-10, 1998. (Springer LNAI 1421)
CADE-16, Trento, Italy, July, 1999. (Springer LNAI 1632)

Table of Contents

Invited Talk:

High-Level Verification Using Theorem Proving and Formalized Mathematics
(Extended Abstract) 1
John Harrison

Session 1:

Machine Instruction Syntax and Semantics in Higher Order Logic 7
Neophytos G. Michael and Andrew W. Appel

Proof Generation in the Touchstone Theorem Prover 25
George C. Necula and Peter Lee

Wellfounded Schematic Definitions 45
Konrad Slind

Session 2:

Abstract Congruence Closure and Specializations 64
Leo Bachmair and Ashish Tiwari

A Framework for Cooperating Decision Procedures 79
Clark W. Barrett, David L. Dill, and Aaron Stump

Modular Reasoning in Isabelle 99
Florian Kammüller

An Infrastructure for Intertheory Reasoning 115
William M. Farmer

Session 3:

Gödel's Algorithm for Class Formation 132
Johan Gijsbertus Frederik Belinfante

Automated Proof Construction in Type Theory Using Resolution 148
Marc Bezem, Dimitri Hendriks, and Hans de Nivelle

System Description: TPS: A Theorem Proving System for Type Theory .. 164
Peter B. Andrews, Matthew Bishop, and Chad E. Brown

The Nuprl Open Logical Environment 170
Stuart F. Allen, Robert L. Constable, Rich Eaton, Christoph Kreitz,
and Lori Lorigo

System Description: ARA - An Automatic Theorem Prover for Relation
Algebras 177
Carsten Sinz

Invited Talk:

Scalable Knowledge Representation and Reasoning Systems 183
Henry Kautz

Session 4:

Efficient Minimal Model Generation Using Branching Lemmas 184
Ryuzo Hasegawa, Hiroshi Fujita, and Miyuki Koshimura

FDPLL — A First Order Davis–Putnam–Longeman–Loveland Procedure . 200
Peter Baumgartner

Rigid *E*-Unification Revisited 220
Ashish Tiwari, Leo Bachmair, and Harald Ruess

Invited Talk:

Connecting Bits with Floating-Point Numbers: Model Checking and
Theorem Proving in Practice 235
Carl-Johan Seger

Session 5:

Reducing Model Checking of the Many to the Few 236
E. Allen Emerson and Vineet Kahlon

Simulation Based Minimization 255
Doran Bustan and Orna Grumberg

Rewriting for Cryptographic Protocol Verification 271
Thomas Genet and Francis Klay

System Description: *SAT: A Platform for the Development of Modal
Decision Procedures 291
Enrico Giunchiglia and Armando Tacchella

System Description: DLP 297
Peter Patel-Schneider

Two Techniques to Improve Finite Model Search 302
Gilles Audemard, Belaid Benhamou, and Laurent Henocque

Session 6:

Eliminating Dummy Elimination 309
Jürgen Giesl and Aart Middeldorp

Extending Decision Procedures with Induction Schemes 324
Deepak Kapur and Mahadavan Subramaniam

Complete Monotonic Semantic Path Orderings 346
Cristina Borralleras, Maria Ferreira, and Albert Rubio

Session 7:

Stratified Resolution 365
Anatoli Degtyarev and Andrei Voronkov

Support Ordered Resolution 385
Bruce Spencer and Joseph D. Horton

System Description: IVY 401
William McCune and Olga Shumsky

System Description: SystemOnTPTP 406
Geoff Sutcliffe

System Description: PTTP+GLiDes: Semantically Guided PTTP 411
Marianne Brown and Geoff Sutcliffe

Session 8:

A Formalization of a Concurrent Object Calculus up to α-Conversion 417
Guillaume Gillard

A Resolution Decision Procedure for Fluted Logic 433
Renate A. Schmidt and Ullrich Hustadt

ZRes: The Old Davis–Putman Procedure Meets ZBDD ....·........... 449
Philippe Chatalic and Laurent Simon

System Description: MBASE, an Open Mathematical Knowledge Base 455
Andreas Franke and Michael Kohlhase

System Description: TRAMP: Transformation of Machine-Found Proofs
into ND-Proofs at the Assertion Level 460
Andreas Meier

Session 9:

On Unification for Bounded Distributive Lattices 465
Viorica Sofronie-Stokkermans

Reasoning with Individuals for the Description Logic $\mathcal{SHIQ}$ 482
Ian Horrocks, Ulrike Sattler, and Stephan Tobies

System Description: Embedding Verification into Microsoft Excel 497
Graham Collins and Louise A. Dennis

System Description: Interactive Proof Critics in XBarnacle 502
Mike Jackson and Helen Lowe

Tutorials:

Tutorial: Meta-logical Frameworks 507
Carsten Schürmann

Tutorial: Automated Deduction and Natural Language Understanding ... 509
Stephen Pulman

Tutorial: Using TPS for Higher-Order Theorem Proving and ETPS
for Teaching Logic 511
Peter B. Andrews and Chad E. Brown

Workshops:

Workshop: Model Computation - Principles, Algorithms, Applications 513
Peter Baumgartner, Chris Fermueller, Nicolas Peltier, and Hantao Zhang

Workshop: Automation of Proofs by Mathematical Induction 514
Carsten Schürmann

Workshop: Type-Theoretic Languages: Proof-Search and Semantics 515
Didier Galmiche

Workshop: Automated Deduction in Education 516
Erica Melis

Workshop: The Role of Automated Deduction in Mathematics 517
Simon Colton, Volker Sorge, and Ursula Martin

Author Index 519

High-Level Verification Using Theorem Proving and Formalized Mathematics
(Extended Abstract)

John Harrison

Intel Corporation, EY2-03
5200 NE Elam Young Parkway
Hillsboro, OR 97124, USA
johnh@ichips.intel.com

Abstract. Quite concrete problems in verification can throw up the need for a nontrivial body of formalized mathematics and draw on several special automated proof methods which can be soundly integrated into a general LCF-style theorem prover. We emphasize this point based on our own work on the formal verification in the HOL Light theorem prover of floating point algorithms.

1 Formalized Mathematics in Verification

Much of our PhD research [11] was devoted to developing formalized mathematics, in particular real analysis, with a view to its practical application in verification, and our current work in formally verifying floating point algorithms shows that this direction of research is quite justified.

First of all, it almost goes without saying that some basic facts about real numbers are useful. Admittedly, floating point verification *has* been successfully done in systems that do not support real numbers at all [16,17,19]. After all, floating point numbers in conventional formats are all rational (with denominators always a power of 2). Nevertheless, the whole point of floating point numbers is that they are approximations to reals, and the main standard governing floating point correctness [13] defines behavior in terms of real numbers. Without using real numbers it is already necessary to specify the square root function in an unnatural way, and for more complicated functions such as *sin* it seems hardly feasible to make good progress in specification or verification without using real numbers explicitly.

In fact, one needs a lot more than simple algebraic properties of the reals. Even to define the common transcendental functions and derive useful properties of them requires a reasonable body of analytical results about limits, power series, derivatives etc. In short, one needs a formalized version of a lot of elementary real analysis, an unusual mixture of the general and the special. A typical general result that is useful in verification is the following:

If a function f is differentiable with derivative f' in an interval $[a, b]$, then a sufficient condition for $f(x) \leq K$ throughout the interval is that $f(x) \leq K$ at the endpoints a, b and at all points of zero derivative.

D. McAllester (Ed.): CADE-17, LNAI 1831, pp. 1–6, 2000.

This theorem is used, for example, in finding a bound for the error incurred in approximating a transcendental function by a truncated power series. The formal HOL version of this theorem looks like this:

```
|- (!x. a <= x /\ x <= b ==> (f diffl (f' x)) x) /\
   f(a) <= K /\
   f(b) <= K /\
   (!x. a <= x /\ x <= b /\ (f'(x) = &0) ==> f(x) <= K)
   ==> (!x. a <= x /\ x <= b ==> f(x) <= K)
```

A typical concrete result is a series expansion for π [1]:

$$\pi = \sum_{n=0}^{\infty} \frac{1}{16^n}\left(\frac{4}{8n+1} - \frac{2}{8n+4} - \frac{1}{8n+5} - \frac{1}{8n+6}\right)$$

This allows us to approximate π arbitrarily closely by rational numbers. Doing so is important both for detailed analysis of trigonometric range reduction (reducing an argument x to a trigonometric function to r where $x = r + N\pi/2$) and to dispose of trivial side-conditions. For example, an algorithm might rely on the fact that $sin(x)$ is positive for some particular x, and we can verify this by confirming that $0 < x < \pi$ using an approximation of π. In HOL, the formal theorem is as follows:

```
|- (\n. inv(&16 pow n) * (&4 / &(8 * n + 1) - &2 / &(8 * n + 4) -
        &1 / &(8 * n + 5) - &1 / &(8 * n + 6)))) sums pi
```

The mathematics needed in floating-point verification is an unusual mixture of these general and special facts, and it's sometimes the kind that isn't widely found in textbooks. For example, an important result we use is the power series expansion for the cotangent function (for $x \neq 0$):

$$\cot(x) = \frac{1}{x} - \frac{1}{3}x - \frac{1}{45}x^3 - \frac{2}{945}x^5 - \cdots$$

To derive this straightforward-looking theorem, both getting a simple recurrence relation for the coefficients *and* a reasonably sharp bound on their size, is fairly non-trivial. A typical mathematics book either doesn't mention such a concrete result at all, or gives it without proof as part of a "cookbook" of well-known useful results. After some time browsing in a library, we eventually settled on formalizing a proof in Knopp's classic book on infinite series [14]. Formalizing this took several days of work, drawing extensively on existing analytical lemmas in HOL. A side-effect is that we derived a general result on harmonic sums, the simplest special cases of which are the well-known:

$$1 + 1/2^2 + 1/3^2 + 1/4^2 + \cdots = \pi^2/6$$

and

$$1 + 1/2^4 + 1/3^4 + 1/4^4 + \cdots = \pi^4/90$$

Knopp remarks

> It is not superfluous to realize all that was needed to obtain even the first of these elegant formulae.

We may add that it is even more surprising that such extensive mathematical developments are used simply to verify that a floating point tangent function satisfies a certain error bound. Of course, one also needs plenty of specialized facts about floating point arithmetic, e.g. important properties of rounding. These theories have also been developed in HOL Light [12] but we will not go into more detail here.

2 Proof in HOL Light

The theorem prover we are using in our work is HOL Light [8],[1] a version of the HOL prover [5]. HOL is a descendent of Edinburgh LCF [6] which first defined the 'LCF approach' that these systems take to formal proof. LCF provers explicitly generate proofs in terms of extremely low-level primitive inferences, in order to provide a high level of assurance that the proofs are valid. In HOL Light, as in most other LCF-style provers, the proofs (which can be very large) are not usually stored permanently, but the strict reduction to primitive inferences in maintained by the abstract type system of the interaction and implementation language, which for HOL Light is CAML Light [4,23]. The primitive inference rules of HOL Light, which implements a simply typed classical higher order logic, are very simple, and will be summarized below.

$$\frac{}{\vdash t = t} \text{ REFL}$$

$$\frac{\Gamma \vdash s = t \quad \Delta \vdash t = u}{\Gamma \cup \Delta \vdash s = u} \text{ TRANS}$$

$$\frac{\Gamma \vdash s = t \quad \Delta \vdash u = v}{\Gamma \cup \Delta \vdash s(u) = t(v)} \text{ MK\_COMB}$$

$$\frac{\Gamma \vdash s = t}{\Gamma \vdash (\lambda x. s) = (\lambda x. t)} \text{ ABS}$$

$$\frac{}{\vdash (\lambda x. t)x = t} \text{ BETA}$$

$$\frac{}{\{p\} \vdash p} \text{ ASSUME}$$

$$\frac{\Gamma \vdash p = q \quad \Delta \vdash p}{\Gamma \cup \Delta \vdash q} \text{ EQ\_MP}$$

---

[1] See http://www.cl.cam.ac.uk/users/jrh/hol-light/index.html

$$\frac{\Gamma \vdash p \quad \Delta \vdash q}{(\Gamma - \{q\}) \cup (\Delta - \{p\}) \vdash p = q} \quad \text{DEDUCT\_ANTISYM\_RULE}$$

$$\frac{\Gamma[x_1, \ldots, x_n] \vdash p[x_1, \ldots, x_n]}{\Gamma[t_1, \ldots, t_n] \vdash p[t_1, \ldots, t_n]} \quad \text{INST}$$

$$\frac{\Gamma[\alpha_1, \ldots, \alpha_n] \vdash p[\alpha_1, \ldots, \alpha_n]}{\Gamma[\gamma_1, \ldots, \gamma_n] \vdash p[\gamma_1, \ldots, \gamma_n]} \quad \text{INST\_TYPE}$$

In MK_COMB, the types must agree, e.g. $s : \sigma \to \tau$, $t : \sigma \to \tau$, $u : \sigma$ and $v : \sigma$. In ABS, we require that x is not a free variable in any of the assumptions Γ. In ASSUME, p must be of Boolean type, i.e. a proposition.

All theorems in HOL are deduced using just the above rules, starting from three *axioms*: Extensionality, Choice and Infinity. There are also definitional mechanisms allowing the introduction of new constants and types, but these are easily seen to be logically conservative and thus avoidable in principle.

CAML Light also serves as a programming medium allowing higher-level derived rules (e.g. to automate linear arithmetic, first order logic or reasoning in other special domains) to be programmed as reductions to primitive inferences, so that proofs can be partially automated. This is very useful in practice. In floating point proofs we make extensive use of quite intricate facts of linear arithmetic, such as:

```
|- x <= a /\ y <= b /\
   abs(x - y) < abs(x - a) /\ abs(x - y) < abs(x - b) /\
   (x <= b ==> abs(x - a) <= abs(x - b)) /\
   (y <= a ==> abs(y - b) <= abs(y - a))
   ==> (a = b)
```

Proving these by low-level primitive inferences can be tedious in the extreme, so it is immensely valuable to have the process automated. Similarly, we often use first order automation to avoid tedious low-level reasoning (e.g. chaining together many inequalities) or exploit symmetries via lemmas such as:

```
|- (!x y. P x y = P y x) /\
   (!x y. Q x ==> P x y)
   ==> !x y. Q x \/ Q y ==> P x y
```

Because these are all programmed as reductions to primitive inferences, we have the security of knowing that any errors in the derived rule cannot result in false "theorems" as long as the few primitive rules are sound. This can be especially important in verification of real industrial systems, since an error in a 'proof' can invalidate the entire result.

The basic LCF approach of exploiting traditional automated techniques [3, 15] or high-level methods of proof description [9] by reducing them to primitive inferences in a single core logic seems to us a very fruitful one. Of course, it has an efficiency penalty, but as we argue in [7], it is not usually too severe except in a few special cases. Nevertheless, there is still much more work to be done to make systems like HOL Light really usable by a nonspecialist. In our opinion, the most impressive system for formalizing abstract mathematics is Mizar [18, 22], and importing the strengths of that system into LCF-style provers is a popular topic of research [10, 21, 24, 26].

The first sustained attempt to actually formalize a body of mathematics (concepts and proofs) was *Principia Mathematica* [25]. This successfully derived a body of fundamental mathematics from a small logical system. However, the task of doing so was extraordinarily painstaking, and indeed Russell [20] remarked that his own intellect 'never quite recovered from the strain of writing it'. The correctness theorems we are producing in our work often involve tens or hundreds of millions of applications of primitive inference rules, and build from foundational results about the natural numbers up to nontrivial and highly concrete applied mathematics. Yet using HOL Light, which can bridge the abyss between simple primitive inferences and the demands of real applications, doing so is quite feasible.

References

1. D. Bailey, P. Borwein, and S. Plouffe. On the rapid computation of various polylogarithmic constants. *Mathematics of Computation*, 66:903–913, 1997.
2. Yves Bertot, Gilles Dowek, André Hirschowitz, Christine Paulin, and Laurent Théry, editors. *Theorem Proving in Higher Order Logics: 12th International Conference, TPHOLs'99*, volume 1690 of *Lecture Notes in Computer Science*, Nice, France, 1999. Springer-Verlag.
3. Richard John Boulton. Efficiency in a fully-expansive theorem prover. Technical Report 337, University of Cambridge Computer Laboratory, New Museums Site, Pembroke Street, Cambridge, CB2 3QG, UK, 1993. Author's PhD thesis.
4. Guy Cousineau and Michel Mauny. *The Functional Approach to Programming*. Cambridge University Press, 1998.
5. Michael J. C. Gordon and Thomas F. Melham. *Introduction to HOL: a theorem proving environment for higher order logic*. Cambridge University Press, 1993.
6. Michael J. C. Gordon, Robin Milner, and Christopher P. Wadsworth. *Edinburgh LCF: A Mechanised Logic of Computation*, volume 78 of *Lecture Notes in Computer Science*. Springer-Verlag, 1979.
7. John Harrison. Metatheory and reflection in theorem proving: A survey and critique. Technical Report CRC-053, SRI Cambridge, Millers Yard, Cambridge, UK, 1995. Available on the Web as
http://www.cl.cam.ac.uk/users/jrh/papers/reflect.dvi.gz.
8. John Harrison. HOL Light: A tutorial introduction. In Mandayam Srivas and Albert Camilleri, editors, *Proceedings of the First International Conference on Formal Methods in Computer-Aided Design (FMCAD'96)*, volume 1166 of *Lecture Notes in Computer Science*, pages 265–269. Springer-Verlag, 1996.
9. John Harrison. A Mizar mode for HOL. In Joakim von Wright, Jim Grundy, and John Harrison, editors, *Theorem Proving in Higher Order Logics: 9th International Conference, TPHOLs'96*, volume 1125 of *Lecture Notes in Computer Science*, pages 203–220, Turku, Finland, 1996. Springer-Verlag.

10. John Harrison. Proof style. In Eduardo Giménez and Christine Paulin-Mohring, editors, *Types for Proofs and Programs: International Workshop TYPES'96*, volume 1512 of *Lecture Notes in Computer Science*, pages 154–172, Aussois, France, 1996. Springer-Verlag.

11. John Harrison. *Theorem Proving with the Real Numbers*. Springer-Verlag, 1998. Revised version of author's PhD thesis.

12. John Harrison. A machine-checked theory of floating point arithmetic. In Bertot et al. [2], pages 113–130.

13. IEEE. Standard for binary floating point arithmetic. ANSI/IEEE Standard 754-1985, The Institute of Electrical and Electronic Engineers, Inc., 345 East 47th Street, New York, NY 10017, USA, 1985.

14. Konrad Knopp. *Theory and Application of Infinite Series*. Blackie and Son Ltd., 2nd edition, 1951.

15. Ramaya Kumar, Thomas Kropf, and Klaus Schneider. Integrating a first-order automatic prover in the HOL environment. In Myla Archer, Jeffrey J. Joyce, Karl N. Levitt, and Phillip J. Windley, editors, *Proceedings of the 1991 International Workshop on the HOL theorem proving system and its Applications*, pages 170–176, University of California at Davis, Davis CA, USA, 1991. IEEE Computer Society Press.

16. J Strother Moore, Tom Lynch, and Matt Kaufmann. A mechanically checked proof of the correctness of the kernel of the $AMD5_K86$ floating-point division program. *IEEE Transactions on Computers*, 47:913–926, 1998.

17. John O'Leary, Xudong Zhao, Rob Gerth, and Carl-Johan H. Seger. Formally verifying IEEE compliance of floating-point hardware. *Intel Technology Journal*, 1999-Q1:1–14, 1999. Available on the Web as http://developer.intel.com/technology/itj/q11999/articles/art_5.htm.

18. Piotr Rudnicki. An overview of the MIZAR project. Available on the Web as http://web.cs.ualberta.ca/~piotr/Mizar/MizarOverview.ps, 1992.

19. David Rusinoff. A mechanically checked proof of IEEE compliance of a register-transfer-level specification of the AMD-K7 floating-point multiplication, division, and square root instructions. *LMS Journal of Computation and Mathematics*, 1:148–200, 1998. Available on the Web via http://www.onr.com/user/russ/david/k7-div-sqrt.html.

20. Bertrand Russell. *The autobiography of Bertrand Russell*. Allen & Unwin, 1968.

21. Don Syme. Three tactic theorem proving. In Bertot et al. [2], pages 203–220.

22. Andrzej Trybulec. The Mizar-QC/6000 logic information language. *ALLC Bulletin (Association for Literary and Linguistic Computing)*, 6:136–140, 1978.

23. Pierre Weis and Xavier Leroy. *Le langage Caml*. InterEditions, 1993. See also the CAML Web page: http://pauillac.inria.fr/caml/.

24. Markus Wenzel. Isar - a generic intepretive approach to readable formal proof documents. In Bertot et al. [2], pages 167–183.

25. Alfred North Whitehead and Bertrand Russell. *Principia Mathematica (3 vols)*. Cambridge University Press, 1910.

26. Vincent Zammit. On the implementation of an extensible declarative proof language. In Bertot et al. [2], pages 185–202.

Machine Instruction Syntax and Semantics in Higher Order Logic

Neophytos G. Michael and Andrew W. Appel

Computer Science Department, Princeton University, 35 Olden Street
Princeton, NJ 08544, USA
nmichael@cs.princeton.edu
appel@cs.princeton.edu

Abstract. Proof-carrying code and other applications in computer security require machine-checkable proofs of properties of machine-language programs. These in turn require axioms about the opcode/operand encoding of machine instructions and the semantics of the encoded instructions. We show how to specify instruction encodings and semantics in higher-order logic, in a way that preserves the factoring of similar instructions in real machine architectures. We show how to automatically generate proofs of instruction decodings, global invariants from local invariants, Floyd-Hoare rules and predicate transformers, all from the specification of the instruction semantics. Our work is implemented in ML and Twelf, and all the theorems are checked in Twelf.

1 Introduction

The security problem for mobile code or for component software is this: an untrusted program (or program fragment) is to execute in a host environment (the *code consumer*), and we want to ensure that it will do no harm. Proof Carrying Code (PCC) [1] is a framework for solving this problem by providing such assurances to the host. In the PCC framework the code consumer advertises a *safety policy* which specifies the logic in which it will accept proofs, the regions of readable or writable addresses, and so on. The code producer must construct a proof that the machine-language program satisfies the safety policy; the proof might be generated using hints from the compiler that generated the code. This proof along with the code is communicated to the host environment and the host verifies it before executing the code. PCC has significant advantages over other approaches that address the same problem (such as software fault isolation [6] or byte code interpretation [7]): no performance penalty is taken since the code is run at native speeds, and the proofs are performed on native machine code so no unsoundness can be introduced in the translation (or compilation) from the proved program to the one that will actually execute. For well-chosen safety policies, the proofs can be generated completely automatically.

In Appel and Felty [5] we gave an overview of our PCC system and described how it differs from the approach taken by Necula [2]. Instead of building type-inference rules into the safety policy, we model types as defined predicates using

D. McAllester (Ed.): CADE-17, LNAI 1831, pp. 7–24, 2000.

the primitives of ordinary logic; we prove typing rules as lemmas, and show how to model a wide variety of type constructors. This way the PCC safety policy is independent of the code producer's programming language and type system. The machine description semantics are moved from the verification-condition generator to the safety policy. More specifically our safety policy consists of the following:

1. The logic: a fairly standard higher order logic[1] ($\mathcal{L}$) consisting of eight inference rules for the logic and twenty-nine for arithmetic (with addition and multiplication taken as primitives).
2. The machine code syntax and semantics: this is encoded as the definition of the **step** relation ($\mapsto$) that describes the syntax and semantics of the machine. **Step** formally captures the notion of a single instruction execution. These axioms also define the **decode** relation that completely specifies instruction opcodes and operands (machine syntax) for all legal machine-code instructions.
3. Safety constraints: these are statements[2] in $\mathcal{L}$ that describe general properties of the runtime system (such as readable and safe-to-jump memory locations). They may also contain typing judgments for the initial contents of the register bank.

The small size of the logic is one of the major advantages of our approach. It contains no inference rules on types and no Hoare-logic rules for instructions (thus avoiding all complications due to substitution). Since it is so small, the proof checker can be likewise small. Thus the trusted computing base (TCB) can be verified easily (either by hand or through other means). A small TCB is the essence of PCC.

To simplify the presentation of the following sections we will use the *toy* machine (from [5]), a word-addressed 16-bit CPU. Its instruction set is presented in figure 1. Our system currently works with two other machine architectures (Sparc and Mips) and when appropriate we will also use examples from these.

2   Overview

Our focus in this paper is twofold: concise axioms modeling machine architectures, and efficient proofs using those axioms.

---

[1] Our logic $\mathcal{L}$, is a sublogic of the Calculus of Constructions [11] and of the logic used in the HOL theorem prover [12], so our proofs can be checked in either Coq or HOL. Our current implementation uses Twelf [4].

[2] We offer a brief introduction to the syntax of our object logic: A metalogic (Twelf) type is a **type**, and an object-logic type is a **tp**. Object-logic types are constructed from **num** (the type of rationals), **form** (the type of formulas) and the **arrow** constructor. Object-level terms of type T have type (**tm** T) in the metalogic. Terms of type (**pf** A) are terms representing proofs of object formula A. The term **lam** $[x]\,F(x)$ is the object-logic function that maps x to $F(x)$ and **@** is the application operator for λ-terms. See Appel and Felty [5] for more details.

Instruction	Fields				Effect
add	0	d	$s1$	$s2$	$r_d := r_{s1} + r_{s2}$
addi	1	d	s	c	$r_d := r_s + \text{sign\_ext}(c)$
load	2	d	s	c	$r_d := m[r_s + \text{sign\_ext}(c)]$
store	3	$s1$	$s2$	c	$m[r_{s2} + \text{sign\_ext}(c)] := r_{s1}$
jump	4	d	s	c	$r_d := r_{pc}\,;\, r_{pc} := r_s + \text{sign\_ext}(c)$
bgt	5	$s1$	$s2$	c	if $r_{s1} > r_{s2}$ then $r_{pc} := r_{pc} + \text{sign\_ext}(c)$
beq	6	$s1$	$s2$	c	if $r_{s1} = r_{s2}$ then $r_{pc} := r_{pc} + \text{sign\_ext}(c)$

Fig. 1. The toy machine instruction set.

We will describe in detail our **step** relation and show how it succinctly captures the syntax and semantics of real machines. Since it is by far the largest piece of our safety policy we are of course concerned about its correctness. To this end we will show how parts of it can be automatically generated from existing systems. Here we tackle the syntax of machine instructions using machine descriptions from the New Jersey Machine Code Toolkit [8]. We also show how to automatically generate proofs of correspondence between machine code integers and statements involving the **decode** relation.

We will describe the engineering aspects of generating small proofs of safety. Program safety is proved using a coinduction theorem based on progress and preservation of an invariant. We construct invariant expressions whose size is linear in the number of program instructions, and structure the progress and preservation proofs so that – modulo the parts that will have to be built by our tactical theorem prover – they are linear in size. In building these invariants we need to use the weakest preconditions of instructions and we will show how to automatically generate lemmas for a Hoare logic of machine language from the **step** relation. Our safety proofs will be linear-sized trees of applications of these Hoare lemmas.

Figure 2 shows our system operating on a small program that computes the sum of a linked list of integers. The goal of the system is to prove that the initial machine configuration (**IMC**) is **safe**, in symbols the following theorem:

$$\text{IMC}(r_0, m_0) \rightarrow \text{safe}(r_0, m_0)$$

where

$$\text{IMC}(r, m) := m(100) = 8976 \land \cdots \land m(105) = 24859$$
$$\text{safe}(r, m) := \forall r', m'\ (r, m \mapsto^* r', m') \rightarrow \exists r'', m''(r', m' \mapsto r'', m'').$$

The **IMC** describes parts of memory at the moment the program will run (in this case only the part containing the program itself). The **step** relation $r, m \mapsto r', m'$ formally describes a single instruction execution, i.e. given a machine at state (r, m), after execution of the instruction found at $r(\text{pc})$, the machine will be at state (r', m'). The **safe** property states that no matter how far the execution

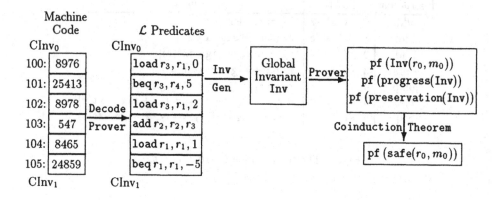

Fig. 2. Generating safety proofs.

proceeds, it never gets stuck, i.e. executes an illegal instruction or performs an illegal fetch.

The PCC system is presented with a list of machine code instructions (i.e. integers). The instruction stream is fed through the decode-prover whose job is to discover the instruction each integer represents, and to produce the symbolic representation of each instruction – which is a predicate that describes the instruction's semantics. The decode-prover also produces proofs of this correspondence. Following this, the predicates are fed into the invariant-generator which builds the global invariant to be used in the coinduction proof. Constructing invariants is not computable in general, so the prover requires hints in the form of local loop invariants decorating the targets of backward branches. Once the global invariant is built we must prove the three preconditions of the coinduction theorem[3] (see figure 3) in order to apply it. This is done by the prover and given the three proofs we apply the rule to finally establish $\mathtt{safe}(r_0, m_0)$.

$$\mathtt{progress(Inv)} := \forall r, m \; \mathtt{Inv}(r, m) \to \exists r', m' \, (r, m \mapsto r', m')$$

$$\mathtt{preservation(Inv)} := \forall r, m, r', m' \; \mathtt{Inv}(r, m) \land (r, m \mapsto r', m') \to \mathtt{Inv}(r', m')$$

$$\frac{\mathtt{Inv}(r, m) \qquad \mathtt{progress(Inv)} \qquad \mathtt{preservation(Inv)}}{\mathtt{safe}(r, m)}$$

Fig. 3. The coinduction theorem.

---

[3] A note on notation: in the interest of brevity we will sometimes use mathematical notation when presenting Twelf terms.

$$\text{upd}(f, d, x, f') := \forall z \text{ if } (z = d) \text{ then } f'(z) = x \text{ else } f'(z) = f(z)$$

$$\text{i\_add}(d, s_1, s_2)(r, m, r', m') := \exists sum \text{ plus\_mod16}(r(s_1), r(s_2), sum) \land$$
$$\text{upd}(r, d, sum, r') \land \text{no\_mem\_change}(m, m')$$

$$\text{i\_load}(d, s_1, c)(r, m, r', m') := \exists cext, addr$$
$$\text{sign\_ext}(3, c, cext) \land$$
$$\text{plus\_mod16}(r(s1), cext, addr) \land$$
$$\text{upd}(r, d, m(addr), r') \land$$
$$\text{readable}(addr) \land \text{no\_mem\_change}(m, m')$$

Fig. 4. Semantics of the add and load instruction of the toy machine.

3 Machine Semantics

In this section we show that the semantics of machine instructions can be easily and concisely expressed in higher order logic. We begin by explaining the idea using the toy machine, and then explore the problems in defining a semantic description of a real CPU.

Each instruction defines a relation between the machine state (registers, memory) before and after its execution. We treat both the memory and register bank as functions from integers to integers. Each instruction then becomes a predicate which takes (r, m, r', m') as input, and holds when the instruction can safely take state (r, m) to (r', m'). In figure 4 we show the terms expressing the semantics for the "add" and "load" instructions of the toy machine. The Twelf term i_add (what we will call a *constructor* in section 4) expects three arguments (d, s_1, s_2) and returns a predicate of type instr, defined as:

$$\text{instr} = \text{regs} \rightarrow \text{mem} \rightarrow \text{regs} \rightarrow \text{mem} \rightarrow \text{form}.$$

It is this predicate that we view as the semantics of the instruction. Thus for the add instruction, i_add(d, s_1, s_2) holds when for some integer sum, the following three equations hold:

$$sum = (r(s_1) + r(s_2)) \bmod 2^{16}$$
$$\forall x \text{ if } (x = d) \text{ then } r'(x) = sum \text{ else } r'(x) = r(x)$$
$$\forall x \, m(x) = m'(x).$$

The situation is similar for the semantics of the "load" instruction. But we wish to consider a program safe only if all of its memory accesses are within a specified region. Therefore our step relation admits only a subset of executable load instructions: those that load from readable addresses. The designer of the safety policy must provide axioms that define the readable predicate. In general the semantics of each instruction must enforce the proper conditions under which the instruction can be executed. For "add" there are no such conditions; we can always add two numbers.

Real hardware can be a lot more complex than our simplistic toy machine. On a modern CPU one has to deal with the issues of delayed branches, address alignment, stores and loads of different sizes, condition registers, sign extension, instructions with multiple effects, and ALU operations not directly expressible in our arithmetic, to mention just a few. We claim that all of these can be handled relatively easily with the right set of abstractions and definitions. Space restrictions only allow us to deal with a representative subset here. We will use the Sparc CPU in the presentation.

- Condition Registers: We model condition registers exactly as we model physical registers. We assign a number to each of them that is outside the range of representable register numbers and refer to them exactly the same way we refer to regular registers. Instructions that need to modify individual bits do so by the use of appropriate definitions (see the **bits** predicate below).
- Delayed Branches: In order to keep their deep pipelines filled, some modern CPUs have introduced the notion of a delayed branch. On such CPUs one (or more) of the instructions following a branch will be executed even if the branch is taken, before the CPU starts executing instructions from the target address. We will assume a single instruction delay slot (the solution can be easily generalized to a delay slot of n instructions). We introduce another register called the next program counter[4] (npc) which holds the address at which the pc will be next. In the semantics of a branch instruction, if the branch is to be taken we simply set $r(npc) = \texttt{target}$ and the **step** relation takes care of updating $r(pc)$ to $r(npc)$ at the appropriate time.
- Address Alignment: Machine addresses have to be properly aligned depending on the instruction that uses them. Using the $\texttt{bits}(r, l, v, w)$ predicate (which holds when the value in the binary representation of w between bits r and l equals v; in symbols $\texttt{bits}(r, l, v, w) \Leftrightarrow v = \lfloor \frac{w}{2^r} \rfloor \bmod 2^{r-l+1}$) we can easily express such constraints. In a load-word instruction for the Sparc for instance we would insist that $\texttt{bits}(0, 1, 0, address)$ holds.
- Stores/Loads: We chose to model memory by a function $m : \texttt{num} \rightarrow \texttt{num}$ that we define only on word-aligned addresses. This way we avoid the complications of modifying individual bytes in a word. When we wish to store a byte quantity, the entire word must be fetched from memory, the byte spliced into it, and then stored back in memory. For load we have a similar situation. With the appropriate definitions all these operations can be specified painlessly. With careful selection of predicates most of them can be shared between the load and store instructions. One such example is the predicate form_address below. It computes a word aligned address and offset from an unaligned one, and ensures that the original address was well aligned with

---

[4] This is in fact how the hardware manages delayed branches. Some machines make the npc register explicit in the specifications [10].

respect to the size of the value we are trying to load/store.

$$\texttt{form\_address}(u\_addr, alignment\_bit, addr, offset, size) :=$$
$$\texttt{bits}(0, alignment\_bit, offset, u\_addr) \land$$
$$\texttt{minus\_mod32}(u\_addr, offset, addr) \land \texttt{modulo}(offset, size, 0)$$

– Arithmetic Operations: Some of the arithmetic operations performed by modern CPUs are not directly expressible as functions in our logic. We cannot, for example, write the function that computes the bitwise "exclusive or" of two integers since our arithmetic primitives include only addition and multiplication and we have no recursion at the object level. Such operations are however, trivially expressed as relations (predicates). Here is for instance the xor predicate:

$$\texttt{xor}(a,b,c) = \forall i \, \exists x,y,r \, \texttt{bits}(i,i,x,a) \land \texttt{bits}(i,i,y,b) \land$$
$$(\texttt{if } x = y \texttt{ then } r = 0 \texttt{ else } r = 1) \land \texttt{bits}(i,i,r,c)$$

Factoring via Higher-order Predicates. Machine instruction sets are highly factored, both in syntax and semantics. Consider for instance the ALU operations of any modern RISC chip. The ALU takes its input from two registers (or a register and a constant) and produces the result in another. The only difference between instructions is the operation performed. Our use of higher order logic allows us to exploit such factoring very effectively. We find the commonalities in families of instructions (even between families as in the load/store case above), factor those out and reuse well-chosen definitions. Here is an example from the Sparc. The definition of i_aluxcc is reused to define 23 different instructions. Argument with_carry specifies whether the instruction operates with a "carry", modifies_icc specifies whether it modifies the integer condition codes, and func is the predicate describing the operation performed by the instruction.

```
alu_fun = num arrow num arrow num arrow form.

i_aluxcc : tm (form arrow form arrow alu_fun arrow alu_typ) =
 lam3 [with_carry : tm form][modifies_icc : tm form][func : tm alu_fun]
  lam3 [rs1][reg_imm][rd]
   lam4 [r][m][r'][m']
    (exists3 [v][v'][r'']
     (load_reg_imm @ r @ reg_imm @ v) and
     (compute_with_carry @ with_carry @ func @ r @ rs1 @ v @ v') and
     (compute_cc @ modifies_icc @ r @ r'' @ v') and
     (upd_reg @ r'' @ rd @ v' @ r') and
     (no_memory_change m m')).

i_AND   = i_aluxcc @ false @ false @ and_oper.
i_ANDcc = i_aluxcc @ false @ true  @ and_oper.
  "       "      "     "      "        "          -- 21 cases omitted.
```

Moreover we exploit commonality between machines. Many of our definitions that deal with the mechanics of splicing values into words, sign extension, and arithmetic operations, are shared between semantic descriptions of different machines. Higher-order predicates are useful in expressing this kind of sharing; note that the i_aluxcc predicate above is higher order.

4   The Decode Relation

On a von Neumann machine, each instruction is represented in memory by an integer. The decode relation makes this notion precise. It is a predicate of four arguments (m, w, i, s) stating that address w in memory m contains the encoding of instruction i that has size s. Modern microprocessors have hundreds of instructions and to construct this relation manually would be a daunting task. The observation that the information we wish to encode is very similar to the information used by an assembler/disassembler led us to look for an automatic way to generate the relation.

The New Jersey Machine Code Toolkit [8] helps programmers write applications that process machine code – assemblers, disassemblers, code generators, and so on. The toolkit lets programmers encode and decode machine instructions symbolically. It transforms symbolic manipulations into bit manipulations, guided by a specification that defines mappings between symbolic and binary representations of instructions. Of interest to us here is the specification language (called SLED) for encoding and decoding assembly-language representations of machine instructions [9]. It is a concise, elegant, and semantically well-founded language, a fact that has made the translation into logic fairly painless. In fact our translation into $\mathcal{L}$ can be viewed as a semantics for the language.

Before describing our encoding of SLED into $\mathcal{L}$ we offer a brief introduction to the language. In order to accommodate machines with non-uniform instruction sizes the toolkit works with streams of *tokens* instead of instructions. Each instruction consists of one or more tokens. Tokens are further partitioned into *fields* which are sequences of contiguous bits within a token. *Patterns* in SLED serve two purposes: firstly they are used to constrain the division of streams into tokens, and secondly to constrain the values of fields in those tokens. Patterns can be combined with various operators to produce new patterns. The toolkit is concerned with two representations of machine instructions: machine code and assembly language. *Constructors* are used to connect the two representations.

Figure 5 presents a SLED specification of the toy machine architecture. The first two lines specify the 16-bit token instr and its fields: op which occupies bits 12 to 15, rd which occupies bits 8 to 11, and so on. The next line specifies a list of patterns (add, ..., beq,) and for each one, it constrains the op field to have the value 0, ..., 6 respectively. Finally the constructors clause specifies the toy machine instructions. A special toolkit shortcut is used here: if no pattern is specified in the constructor definition then all the names used in the constructor must be either patterns or fields and their conjunction is taken to be the pattern that will be generated by the constructor. In the next subsections we show how to map fields, patterns, and constructors into higher-order logic.

```
fields of instr (16)
    op 12:15  rd 8:11  rs1 4:7  rs2 0:3  c 0:3

patterns [add addi load store jump bgt beq] is op = 0 to 6

constructors  add    rd, rs1, rs2
              addi   rd, rs1, c
              load   rd, rs1, c
              store  rd, rs1, c
              jump   rd, rs1, c
              bgt    rd, rs1, c
              beq    rd, rs1, c
```

Fig. 5. The SLED specification for the toy machine.

4.1 Mapping Fields into $\mathcal{L}$

The definition of the **bits** predicate (from section 3) makes it straightforward to map fields into $\mathcal{L}$. All that it takes is to supply the right and left bit specifiers of each field to this predicate. Since our definitions are curried, defining fields in $\mathcal{L}$ becomes very convenient and almost as terse as it is in SLED. For the toy machine the first two fields are translated as follows:

```
op = bits @ (const 12) @ (const 15).
rd = bits @ (const 8) @ (const 11).
```

The **op** predicate expects two integers as arguments (**v**, **word**), and it holds when **v** is equal to the integer between the 12th and 15th bit of **word**.

4.2 Mapping Patterns into $\mathcal{L}$

Patterns in SLED constrain both the division of streams into tokens and the values of the fields in those tokens. They are composed of constraints on fields. Patterns can be combined using various operators to form other patterns. The RISC machine descriptions we have considered so far contain only conjunction and disjunction operators, and those are the ones we currently translate. We expect no problems in translating the rest when we choose to deal with CISC machines. Conjunction is used to constrain multiple fields within a single token. When p and q are patterns, the pattern "p & q" matches if both p and q match. For example, in the SLED description for Sparc [8] we find:[5]

```
patterns
  [ TABLE_F2 CALL TABLE_F3 TABLE_F4 ] is op = {0 to 3}
```

---

[5] This is another example of the terseness of SLED. In the definitions of these patterns Ramsey [9] makes use of a SLED feature called *generating expressions*, which describe ranges of lists either explicitly or implicitly as shown in the example.

```
[ UNIMP Bicc SETHI FBfcc CBccc ]     is TABLE_F2 &  op2 = [0 2 4 6 7]
NOP                                  is SETHI &  rd = 0  &  imm22 = 0
```

In the first line `TABLE_F2` is defined as the pattern that wants the op field to equal zero, in the second line `TABLE_F2` is used in the definition of `SETHI` which is defined as the conjunction of patterns `TABLE_F2` and $op2 = 4$. Finally in the last line pattern `SETHI` is used in the definition of the `NOP` pattern.[6] Patterns of this kind are very easy to translate into $\mathcal{L}$. We make use of a higher level infix "and" operator defined as:

```
num_pred = num arrow form.
&& : tm num_pred -> tm num_pred -> tm num_pred =
[p1][p2] lam [w] (p1 @ w) and (p2 @ w).
```

Given `&&` it is now easy to deal with conjunctive patterns by simply "anding" together the different conjuncts after mapping each of them to an $\mathcal{L}$ predicate. The example above then becomes:

```
p_TABLE_F2 = op @ (const 0).
p_SETHI    = p_TABLE_F2 && (op2 @ (const 4)).
p_NOP      = p_SETHI && (rd @ (const 0)) && (imm22 @ (const 0)).
```

Disjunction in patterns is usually used to group patterns for related instructions. In the following example from the Sparc SLED we use disjunction to group the logical, shift, and arithmetic instructions into three groups, which are then disjunctively combined into a pattern that matches any ALU instruction.

```
patterns
   logical is AND | ANDcc | ANDN | ANDNcc | OR | ORcc | ORN | ORNcc | ...
   arith   is ADD | ADDcc | ADDX | ADDXcc | TADDcc | TADDccTV | ...
   shift   is SLL | SRL  | SRA
   alu     is logical | arith | shift
```

Disjunction patterns are mostly used as opcodes to constructors and we show how we deal with them in the next subsection.

4.3   Mapping Constructors into $\mathcal{L}$

A constructor maps a list of operands to a pattern which stands for the binary representation of an operand or an instruction. There are two kinds of constructors, typed and untyped. Typed constructors generate instruction operands and untyped constructors generate instructions. The following definition from the Sparc specification is an example of a typed constructor:

```
constructors   imode simm13! : reg_or_imm is  i = 1 & simm13
               rmode rs2     : reg_or_imm is  i = 0 & rs2
```

---

[6] A `NOP` on the Sparc is a `SETHI` on r_0 with value 0, and since r_0 is hardwired to zero it has no effect.

Each line in the definition of a constructor specifies the opcode, the operands, the constructor type, and matching pattern. Usually the opcode is the constructor's name (as in this case). Constructors generate disjoint sum types. In the above, imode : num → reg_or_imm is the canonical injection from num into the reg_or_imm type – likewise for rmode : num → reg_or_imm. The type is defined implicitly at first use. Each constructor is applicable when the pattern following the is keyword is satisfied.

The above constructor definition captures the following idiom: many Sparc instructions (such as add r1, reg_or_imm, r2) take either a register or a constant as one of their arguments. The hardware differentiates between the two instances by the value of bit 13 (field i) in the representation of the instruction. Depending on the value of i, either imode or rmode can be applied, giving in each case a reg_or_imm.

We translate a typed constructor into $\mathcal{L}$ as follows. We first create a new object-logic type for the constructor type. For each of the injective arrows (imode and rmode above) we create an injective Twelf term (c_imode and c_rmode), as well as a discriminator term (p_imode and p_rmode). Finally we generate a predicate that decides the type itself (p_reg_or_imm), i.e. a term that when given an object of that type and a word decides whether that word contains the given object. We show these terms for the example below:

```
reg_or_imm : tp
c_imode : num ⟶ reg_or_imm
c_rmode : num ⟶ reg_or_imm

p_imode(simm) := i(1) && simm13(simm)
p_rmode(s₂)   := i(0) && rs2(s₂)

p_reg_or_imm(regimm, word) :=
    (∃simm p_imode(simm, word) ∧ regimm = c_imode(simm)) ∨
    (∃s₂ p_rmode(s₂, word) ∧ regimm = c_rmode(s₂))
```

Untyped constructors represent the instructions themselves. Their translation into $\mathcal{L}$ is not much different from the typed case so we omit it.

Factoring via Higher-order Predicates. The extensive factoring present in the SLED specifications (through the wide use of "or" patterns) carries over to the translated higher-order logic terms. When translating a constructor that uses an "or" pattern as an opcode, we do not generate a unique term for each instruction but instead build just a single term that describes all of them. This way we preserve SLED's economy of syntax. Here is an example for the ALU instructions of the Sparc shown earlier. The constructor in the spec is the following:

```
constructors  alu rs1, reg_or_imm, rd
```

$$\begin{aligned}
\text{p\_instr}(word, i) &:= (\text{p\_add} \,\|_2\, \text{p\_addi} \,\|_2\, \text{p\_load} \,\|_2\, \text{p\_store} \,\|_2 \\
&\qquad \text{p\_jump} \,\|_2\, \text{p\_bgt} \,\|_2\, \text{p\_beq})(word, i) \\
\text{decode}(m, w, i, s) &:= (s = 1) \wedge \text{p\_instr}(m(w), i) \\
\text{step}(r, m, r', m') &:= \exists i, r'', size \; \text{decode}(m, r(pc), i, size) \wedge \\
&\qquad \text{upd}(r, pc, r(pc) + size, r'') \wedge \\
&\qquad i(r'', m, r', m')
\end{aligned}$$

Fig. 6. The decode and step relations for the toy machine.

and we generate the following two terms for it:

$$\begin{aligned}
&\text{p\_alu\_aux}(p\_i, i\_cons, s_1, regimm, s_2, word, i) := \\
&\quad (p\_i \;\&\&\; \text{rs1}(s_1) \;\&\&\; \text{p\_reg\_or\_imm}(regimm) \;\&\&\; \text{rs2}(s_2))(word) \wedge \\
&\quad i = i\_cons(s_1, regimm, s_2) \\
&\text{p\_alu}(word, i) := \exists s_1, rimm, s_2 \; (\text{p\_alu\_aux}(\text{p\_AND}, \text{i\_AND}) \qquad\qquad \|_5 \\
&\qquad\qquad\qquad\qquad\qquad\qquad \text{p\_alu\_aux}(\text{p\_ANDcc}, \text{i\_ANDcc}) \;\|_5
\end{aligned}$$

$$\vdots \quad \vdots \quad \vdots \quad \vdots \qquad - \text{35 cases omitted}$$

$$\text{p\_alu\_aux}(\text{p\_SRA}, \text{i\_SRA}))(s_1, rimm, s_2, word, i)$$

where p_AND is the opcode pattern, i_AND is the instruction constructor and likewise for the rest of them. Here again we make use of a higher level "or" ($\|_5$) operator to factor out the common arguments to the auxiliary predicate.

Our **decode-generator** is a 3200-line ML program that operates directly on SLED specifications. Since it generates a large portion of our safety policy it ought to be considered trusted code (along with the SLED specifications). We feel that this is a small enough program that can be thoroughly and convincingly debugged into correctness. Furthermore its output is human readable and only a constant factor bigger (between 2x and 3x) than the original SLED specification. Thus the output can easily be inspected and debugged directly. The program currently does not share any code with the New Jersey Machine Code Toolkit although the front-end code and some of the analysis that the two programs perform could be shared. We plan to investigate an integration of the two tools in the future.

4.4 The Decode and Step Relations

We are finally in a position to present the **decode** relation for the toy machine (see figure 6). After all the instruction predicates have been emitted, the **decode-generator** creates a predicate for the top-level token (i.e. **instr** in the case of the toy spec). This predicate is the disjunction of all the instruction predicates (modulo factoring as described above). Decode is then defined in terms of this predicate. Figure 6 also shows the step relation for the toy machine. It is a predicate mapping the machine state $(r, m) \mapsto (r', m')$ by requiring the existence of an instruction i, a register bank r'', and an integer $size$ such that location $r(pc)$ in memory m decodes to i, updating the register bank r with the next pc produces r'', and finally instruction i safely maps (r'', m) to (r', m'). Step models the meaning of a single instruction execution.

5 Machine Code Proofs

In this section we discuss some of the issues in generating the proofs used in the coinduction theorem (figure 3).

5.1 Hoare-Logic Predicates for Local Invariants

In the Floyd-Hoare logic one tries to establish statements of the form $P\{S\}Q$, where S is a program statement, and P, Q are logical formulae. $P\{S\}Q$ means that if P holds, and S executes to completion, then Q holds. The logic specifies a set of axioms and inference rules that allow the deduction of statements of this form. The assignment axiom for instance states: $\vdash P[E/V]\{V:=E\}P$. In our framework we have no such axioms or rules; nevertheless, our preservation statement (in figure 3) bears a striking resemblance to a Hoare judgment. What is stated there is in essence equivalent to:

$$Inv(r, m)\ \{(r, m) \mapsto (r', m')\}\ Inv(r', m') \tag{1}$$

i.e. if the invariant holds at (r, m), then it must hold at the new state (r', m') at which we were taken by the execution of some instruction (a single step). This similarity is of course no accident; we wish to exploit the well understood theory of Hoare logic in order to construct the weakest preconditions that will allow us to prove preservation.

Our invariant (as described in detail in previous work [5]) is in essence a disjunction of statements[7] of the form $r(pc) = n \land \mathtt{decode}(m, n, i, 1) \land I_n(r, m)$ where i is the instruction found at $m(n)$ and I_n is the local invariant at n. To make the situation more concrete assume that at $r(pc)$ we find instruction $\mathtt{add}(r_1, r_2, r_3)$ $(r_1 := r_2 + r_3)$, and that after completion of this instruction, we

---

[7] The invariant presented in Appel and Felty [5] could grow exponentially large for certain kinds of programs. By the use of appropriate higher-order definitions we have remedied this problem and now produce invariants that are always linear in the number of program instructions and in the size of the compiler-inserted loop invariants (see subsection 5.2). The structure of the new invariant is beyond the scope of this paper. The discussion in this section is equally applicable to either kind of invariant.

wish predicate $Q(r, m)$ to hold at the new state. The question now is what should I_n be in order to be able to prove equation 1, or equivalently the statement:

$$r(pc) = n \land \mathtt{decode}(m, n, i, 1) \land i = \mathtt{add}(r_1, r_2, r_3) \land I_n(r, m) \land \\ (r, m \mapsto r', m') \to Q(r', m'). \tag{2}$$

It is not difficult to see that one such I_n is the following: $Q(r, m)[(r_2 + r_3)/r_1]$, i.e. the formula we get after applying the assignment axiom of Hoare logic to the postcondition $Q(r, m)$. In building the invariant though, we do not wish to perform substitution of terms for two main reasons. Firstly, if we are not careful during substitution the local invariants could grow exponentially large.[8] The goal is to end up with small proofs of safety; an exponentially large theorem is unlikely to have a small proof. Secondly, our logic does not contain axioms that express term substitution; such axioms would render the proof checker more complex and would defeat our efforts for a small TCB. Instead we view substitution as a relation between terms and express the notion concisely by higher-order definitions. These definitions allow us to express $I_n(r, m)$ in terms of $Q(r, m)$ in such a way that the size of local invariants stays constant, and substitution is completely avoided (at this stage). We define predicate let_upd in terms of upd (introduced in figure 4) as follows:

$$\mathtt{let\_upd}(r, a, v, f) := \forall r' \; \mathtt{upd}(r, a, v, r') \to f(r').$$

Predicate let_upd specifies that for any function r' that updates r at a with value v, $f(r')$ must hold (we note that there is exactly one such r'; upd is deterministic).

Using this predicate we can succinctly express the weakest precondition for each of our instructions. Below we show the term for the add instruction; compare hx_add with the semantics of add shown in figure 4.

$$\mathtt{hx\_add}(d, s_1, s_2, post)(r, m) := \exists sum \; \mathtt{plus\_mod16}(r(s_1), r(s_2), sum) \land \\ \mathtt{let\_upd}(r, d, sum, \lambda r'.post(r', m))$$

The last argument to hx_add is the postcondition, and the return value is a predicate on (r, m) expressing the weakest precondition for the add. Our system currently generates all the predicate transformers (such as hx_add above) automatically for each instruction from the step relation of each machine. The program performing the translation is not part of the TCB; if there is a bug in it then we will simply fail to prove preservation.

In proving preservation we will have to prove a statement very similar to that in equation 2 for each instruction in our program (but see section 5.2). Such statements can be proved once and for all as lemmas and applied each time the corresponding instruction is encountered. The extensive use of such

---

[8] Consider for example the program $(r_2 := r_1 + r_1; r_3 := r_2 + r_2; r_4 := r_3 + r_3)$ with postcondition $Q(r_4)$. Its weakest precondition is $Q(((r_1 + r_1) + (r_1 + r_1)) + ((r_1 + r_1) + (r_1 + r_1)))$. The size of the argument to Q grows by a factor of two for each assignment.

lemmas will have a profound effect on the size of our safety proofs. We have currently proven such lemmas for all the instructions of the toy machine by hand. It is our intention to generate them and their proofs automatically from the **step** relation of each machine.

5.2 Domain Specific Proofs

Precondition strengthening (shown below) is another rule of Hoare logic.

$$\frac{P' \rightarrow P \qquad P\{S\}Q}{P'\{S\}Q} \tag{3}$$

It states that if $P\{S\}Q$ then one may replace P by a stronger predicate. This scenario occurs when we deal with program loops, as we explain next. Safety proofs for programs with loops require the use of loop invariants. Construction of loop invariants is not computable in general, so our theorem prover requires hints in the form of typing judgments at every location that is the target of a backward jump. At such locations though, our **invariant-generator** would have computed a local invariant I_n (this is the weakest precondition of the instruction – see subsection 5.1). We wish to replace I_n by H_n (the typing hint at that location) as the precondition of that instruction, but in order to be able to do that we must establish that $H_n \rightarrow I_n$. After that, a lemma application similar to rule 3 allows us to conclude $H_n\{S\}Q$. We are building a tactical theorem prover that understands the structure of types and is able to produce such proofs. The "linear size of proofs" discussed in this paper excludes the size of the strengthening proofs. These are not necessarily large but a description of their structure is beyond the scope of this paper.

5.3 Decode Proof-Generation

Proofs involving the **decode** relation can be hard to generate since the definition itself is quite involved. Our **decode-prover** (see figure 2) is a **Twelf** logic program that analyzes the machine-code stream and not only discovers which instruction each integer represents but also produces a proof of this fact. More concretely, if integer n represents instruction i, we get a proof of statement **instruction**(n, i) from which a proof of **decode**(m, w, i, s) follows trivially (given a proof that $n = m(w)$). The **decode-prover** for the toy machine is about 600 lines of **Twelf**, currently hand written. We plan to generate the **decode-prover** itself from the SLED specification of each machine. Note that the **decode-prover** is not part of the TCB; any bug in it will simply produce an invariant from which it will be impossible to show preservation.

6   Related Work

There has been a large amount of work in the area of proofs of machine language programs using both first order [14] and higher order logics [15][16]. Some of this work was focused on proving the correctness of the compiler or the code generator (see for instance [13]). For a historical survey see Calvert [18]. The practice of proving the Hoare rules as lemmas (see subsection 5.1 and 5.2) in an underlying logic is widespread among the program-verification community [15][16][17].

Two pieces of work are most related to ours: Wahab [15] is concerned with correctness (not just safety) of assembly language programs. He defines a flow-graph language expressive enough to describe sequential machine code programs (he deals with the Alpha AXP processor). Substitution is a primitive operator and the logic contains rules detailing term equality under substitution. He proves the Hoare-logic rules as theorems and uses abstraction in order to massage the code stream and get shorter correctness proofs. The translation from machine code to the flow-graph language does not go through a "decode" relation. Also the use of substitution as a primitive makes this approach unsuitable for our purposes since it complicates the TCB.

Boyer and Yu [14] formally specify a subset of the MC68020 microprocessor within the logic of the Boyer-Moore Theorem Prover [19], a quantifier-free first order logic with equality. Their specification of the step relation is similar to ours (they also include a decode relation) but in their approach these relations are functions. The theorem prover they use allows them to "run" the step function on concrete data (i.e. once the step function is specified they automatically have a simulator for the CPU). Their logic, albeit first-order, appears to be larger than ours mainly because of its wealth of arithmetic operators (decoding can be done directly from the specification). Also their machine descriptions are larger than ours; the subset of the 68020 machine description is about 128K bytes while our description of the Sparc is less than half that size. Admittedly, the Motorola chip is much more complex than the Sparc, but we suspect that most of the size difference is attributed to our extensive use of factoring facilitated by higher order logic.

7   Conclusion and Future Work

We have shown how higher-order logic can be used to succinctly describe the syntax and semantics of machine instructions, in a manner that preserves the natural factoring of each architecture. Our step relation formally captures the notion of a single instruction execution. It consists mainly of two pieces: (1) the decode relation that specifies the syntax of machine instructions, and (2) axioms describing the semantics of each instruction by predicates mapping machine states to machine states. The decode relation is generated automatically from existing compiler tools. Large parts of the safety proof involving decode can be generated completely automatically. We explained how to build Hoare-logic predicate transformers from our step relation in order to simplify the construction of the global invariant, and how lemmas can be used to minimize the size of safety proofs involving this invariant. The system is implemented in Twelf [4] and all theorems have been mechanically checked.

We are building a PCC system that will be used to generate safety proofs for many different architectures. Building all the pieces of figure 2 for each machine would be a daunting and unrewarding task. We instead intend to generate most of the prover components shown in figure 2 completely automatically. Since the `decode-prover` is in essence a machine-code disassembler, we intend to generate it directly from the `decode` relation of each machine or alternatively from each machine's SLED specification. Note that the `decode-prover` not only disassembles but also builds proofs involving `decode`. The `invariant-generator` is again machine-instruction dependent and can also be generated directly from `decode` (we already generate the predicate transformers expressing the weakest precondition for each instruction automatically from `step`). It is our intention to automatically generate the Hoare-logic lemmas (of subsection 5.1) along with their proofs from `step` since there will be a large number of them and their proofs tend to be rather long. The proof of `preservation` (see figure 3) requires an inversion lemma for `decode`. We have not proved this lemma for any machine yet, but we expect the proof to be mundane and long (linear in the size of the instruction set). Our plan is to generate these proofs from `decode`. Finally we are working on a tactical theorem prover that will fill in parts of the proofs involving compiler inserted invariants at locations of backward branches (see subsection 5.2).

References

1. George Necula. Proof Carrying Code. In *The 24th ACM SIGPLAN-SIGACT Symposium on Principles of Programming Languages,* pages 106-119, New York, January 1997. ACM Press.
2. George Ciprian Necula. Compiling with Proofs. PhD thesis, School of Computer Science, Carnegie Mellon University, Pittsburgh, PA, September 1998.
3. Frank Pfenning. Logic Programming in the LF logical framework. In Gérard and Gordon Plotkin, editors, Logical Frameworks, pages 149-181. Cambridge University Press, 1991.
4. Frank Pfenning and Carsten Schürmann. System description: Twelf - a meta-logical framework for deductive systems. In *the 16th International Conference on Automated Deduction.* Springer-Verlag, July 1999.
5. Andrew Appel and Amy Felty. A Semantic Model For Types and Machine Instructions for Proof-Carrying Code. In *the 27th ACM SIGPLAN-SIGACT Symposium on Principles of Programming Languages (POPL '00),* January 2000.
6. R. Wahbe, S. Lucco, T. Anderson, and S. Graham. Efficient software-based fault isolation. In *Proc. 14th ACM Symposium on Operating System Principles,* pages 203-216, New York, 1993. ACM Press.
7. Tim Lindholm and Frank Yellin. The Java Virtual Machine Specification. Addison Wesley, 1997.
8. Norman Ramsey, Mary Fernandez. The New Jersey Machine-Code Toolkit. In *Proceedings of the 1995 USENIX Technical Conference,* pages 289-302, New Orleans, LA, Han. 1995.
9. Norman Ramsey, Mary Fernandez. Specifying Representations of Machine Instructions. In *ACM Transactions on Programming Languages and Systems,* pages 492-524 Vol. 19, No. 3, May 1997.

10. SPARC International, Inc. The SPARC Architecture Manual v. 8, Prentice-Hall, Inc. 1992.
11. Thierry Coquand and Gérard Huet. The calculus of constructions. *Information and Computation*, 76(2/3), pages 95-120, February/March 1988.
12. M. J. C. Gordon and T. F. Melham (editors). Introduction to HOL: A theorem proving environment for higher order logic, Cambridge University Press, 1993.
13. R. Milner and R. Weyhrauch. Proving Compiler Correctness in a Mechanized Logic. In *Machine Intelligence*, 7:51-70, 1972.
14. Robert S. Boyer and Yuan Yu. Automated Correctness Proofs of Machine Code Programs for a Commercial Microprocessor. In *the 11th International Conference of Automated Deduction*, pages 416-430. Springer-Verlag, 1992.
15. M. Wahab. Verification and Abstraction of Flow-Graph Programs with Pointers and Computed Jumps. Technical Report, University of Warwick, Coventry, UK.
16. M. Gordon. A Mechanized Hoare Logic of State Transitions. In *A Classical Mind: Essays in Honour of C. A. R. Hoare*, pages 143-159. Edited by A. W. Roscoe (Prentice-Hall, 1994).
17. M. Gordon. Mechanizing Programming Logics in Higher Order Logic. In *Current Trends in Hardware Verification and Automated Theorem Proving*, pages 387-439. Edited by G. Birtwistle and P. A. Subrahmanyam (Springer-Verlag, 1989).
18. David William John Stringer-Calvert. Mechanical Verification of Compiler Correctness. Ph.D. thesis, University of York, 1998.
19. Robert S. Boyer and J Strother Moore. A Computational Logic Handbook. Academic Press 1988.

Proof Generation in the
Touchstone Theorem Prover

George C. Necula[1] and Peter Lee[2]

[1] University of California, Electrical Engineering and Computer Science Department
Berkeley, CA 94720, USA
necula@cs.berkeley.edu
[2] Carnegie Mellon University, School of Computer Science
Pittsburgh, PA 15213, USA
petel@cs.cmu.edu

Abstract. The ability of a theorem prover to generate explicit derivations for the theorems it proves has major benefits for the testing and maintenance of the prover. It also eliminates the need to trust the correctness of the prover at the expense of trusting a much simpler proof checker. However, it is not always obvious how to generate explicit proofs in a theorem prover that uses decision procedures whose operation does not directly model the axiomatization of the underlying theories. In this paper we describe the modifications that are necessary to support proof generation in a congruence-closure decision procedure for equality and in a Simplex-based decision procedure for linear arithmetic. Both of these decision procedures have been integrated using a modified Nelson-Oppen cooperation mechanism in the Touchstone theorem prover, which we use to produce proof-carrying code. Our experience with designing and implementing Touchstone is that proof generation has a relatively low cost in terms of design complexity and proving time and we conclude that the software-engineering benefits of proof generation clearly outweighs these costs.

1 Introduction

There are several reasons why a theorem prover ought to produce easily checkable derivations of the formulas it proves. First, that way the soundness of the theorem prover does not have to be trusted since it is reduced to the soundness of a much simpler proof checker. This allows theorem proving tasks to be delegated to anonymous or even untrusted parties, such as remote proving servers, without loss of confidence in the result. On the software-engineering side, the testing and maintenance of a proof-generating theorem prover can be simplified considerably at the cost of implementing a simple proof checker. Our initial motivation for developing such a theorem prover was to assist with the generation proof-carrying code [Nec97], in which an explicit proof of safety is attached to mobile code to allow a code receiver to verify easily the compliance of the code with a safety policy.

D. McAllester (Ed.): CADE-17, LNAI 1831, pp. 25–44, 2000.
© Springer-Verlag Berlin Heidelberg 2000

The complexity of proof generation in a theorem prover depends on the prover design. For example, a simple theorem prover can be written as an interpreter for a logic program consisting of a transcription of axioms and inference rules. In fact, the first implementation of proof-carrying code used the Elf [Pfe94] system to search for proofs when the logic was expressed as an LF signature, in the style described in [Pfe91]. For such a theorem prover it is a simple bookkeeping task to record the proof as the sequence of the inference rules used on the successful search path.

The problem is complicated somewhat in theorem provers based on decision procedures, such as PVS [ORS92] or Simplify [DLNS98], because of the indirect relationship between the decision algorithm, sometimes described in terms of graphs [Sho81] or matrices [Nel81], and the axiomatization of the theories involved. In this paper we describe an extension for proof generation of the Nelson-Oppen cooperating decision-procedures model and then we show how to implement proof generation in a congruence closure decision procedure for equality and in a Simplex-based decision procedure for linear arithmetic. We implemented these decision procedures along with a few others in the Touchstone theorem prover that we use in our proof-carrying code experiments. One noteworthy feature of our implementation is that proofs of intermediate subgoals are generated lazily and only if they turn out to be on the successful proof search path. With this optimization the overhead of proof generation is a 30% increase in the size of the prover source code and a 15% increase of proving time.

Proof generation or logging appears in various forms in other theorem provers as well. In LCF-style tactic-based provers (e.g. Isabelle [Pau94] and HOL [Gor85]) the lack of decision procedures allows a simple implementation of proof logging in the form of a trace of the successful proof search path. In theorem provers that do use decision procedures (e.g., PC-Nqthm [BM79], PVS [ORS92], Simplify [DLNS98]) most often the prover records only the user input and the invocations of the decision procedures, to allow batch-mode proof playback. This means that the implementations of the decision procedures must be trusted since they are also part of the proof checker. In addition to Touchstone, a select number of other theorem provers combine decision procedures for efficiency and proof generation for assurance. One of them is the Stanford Validity Checker [SD99], which uses a different set of decision procedures and the Shostak method for integrating decision procedures instead of the Nelson-Oppen method discussed here.

A more closely related result is Boulton's integration [Bou93,Bou95] of a fully-expansive implementation of the Nelson-Oppen method in the HOL theorem prover [Gor85]. While we used some of the same techniques as Boulton (such as the lazy generation of proof objects [Bou92]), our work is different from Boulton's in two respects. Boulton chooses to use a version of Fourier-Motzkin elimination for deciding linear arithmetic formulas, in order to simplify the task of generating proofs ([Bou93], page 80). We have opted for a more complex but apparently more efficient decision procedure based on the Simplex algorithm. A second difference is that Boulton uses a functional programming style in order to implement easily the required undo feature of decision procedures. We decided to use an imperative style and program the undo feature explicitly in order to have better control on the memory usage. As a result, we were able to implement the undo feature for the Simplex algorithm with a very small cost by not actually reverting the data structures to their original form but to another one that is equivalent. This, coupled with the modifications that are required to the linear-programming version of Simplex to make it usable in a Nelson-Oppen prover complicates the proof generation problem for Simplex. We show a solution to this problem in Section 3.2.

In addition to presenting the particular techniques that we use to generate proofs from decision procedures, a substantial part of this paper summarizes our experience in building and using the Touchstone theorem prover for producing proof-carrying code. We discuss both the additional programming-complexity cost of proof generation along with the benefits of proof generation for debugging and maintaining the prover. In fact, we show that the ability to generate proofs and thus to check easily each run of the prover allowed us to use aggressive implementation techniques in order to gain efficiency. Our measurements show that Touchstone appears to be faster than Boulton's implementation of the Nelson-Oppen strategy in the HOL theorem prover. In order to achieve this we had to adopt a more aggressive imperative programming style which led to a number of subtle design and programming errors that could have been avoided in a purely functional implementation. However this did not turn out to be a reliability problem because the proof checker quickly pointed out our programming errors.

Considering the complexity of implementing proof generation and the run-time cost of synthesizing proofs on one hand and, on the other hand, the number of design and implementation errors that were uncovered by proof checking during testing and maintenance along with the added value of the theorem prover as a proof-carrying code generator, we strongly advocate that theorem provers ought to generate easily checkable proofs.

2   Overview of the Touchstone Theorem Prover

Touchstone has a modular design based on a strategy for combining decision procedures first described by Nelson and Oppen [NO79]. The innovation in Touchstone lies in a modification of the Nelson-Oppen strategy to allow for proof-generating decision procedures and also in the techniques used to generate proofs in individual decision procedures. In this paper we discuss such techniques for the congruence closure decision procedure for equality and a Simplex-based decision procedure for linear arithmetic.

Touchstone handles the fragment of first-order logic shown in Figure 1, where the languages of literals L and expressions E can be extended with additional operators or function symbols. There are two motivations for restricting ourselves to such a small subset of first-order logic formulas. First, this fragment is

$$
\begin{array}{lll}
\text{Goals} & G ::= L \mid \top \mid G_1 \wedge G_2 \mid H \supset G \mid \forall x.G \\
\text{Hypotheses} & H ::= L \mid \top \mid H_1 \wedge H_2 \mid \mid H_1 \vee H_2 \mid \exists x.H \\
\text{Literals} & L ::= E_1 = E_2 \mid E_1 \neq E_2 \mid p(E_1, \ldots, E_n) \\
\text{Expressions} & E ::= n \mid E_1 + E_2 \mid f(E_1, \ldots, E_n) \mid \cdots
\end{array}
$$

Fig. 1. The syntax of formulas handled by Touchstone.

sufficient for expressing verification conditions for programs whose loop invariants and function pre/postconditions are themselves restricted to the language H of hypotheses. This is true in all applications to date of proof-carrying code where, in fact, we currently use only conjunctions of literals as hypotheses. Secondly, this fragment of intuitionistic logic has the convenient property that all inference rules are invertible and thus we can use a very simple yet complete inversion proof-search procedure without any disjunctive or existential choices. In essence, the hard part of the proving task in this fragment of logic lies with the decision procedures that handle goal literals. The prover can be extended to handle more logical connectives all the way to higher-order hereditary Harrop formulas following, for example, the strategies described in [Mil91] or [MNPS91].

A decision procedure for a given theory T in Touchstone knows how to decide whether a set of literals entails another literal, in the case when all of the literals involved contain only function symbols from T. Most decision procedures in Touchstone are implemented in terms of satisfiability procedures that can detect when a set of literals is unsatisfiable. In practice, goal formulas contain literals from multiple theories and, although necessary, it is not sufficient to have decision procedures for these isolated theories. Furthermore, combining decision procedures is not as straightforward as it might seem. To illustrate this point consider the theory $\mathbb{Q}$ of rational numbers with the free symbols $+, -, \geq$ and the numerals along with the usual axioms of rational arithmetic. Consider also the theory $\mathbb{E}$ with one uninterpreted unary function symbol "f". The satisfiability problems for each of these theories considered separately were solved long ago by Fourier for $\mathbb{Q}$ and by Ackermann for $\mathbb{E}$ [Ack54]. Consider now the following goal from the combined theory $\mathbb{Q} + \mathbb{E}^1$:

$$
f(f(x) - f(y)) \neq f(z) \wedge y \geq x \wedge x \geq y + z \wedge z \geq 0 \tag{1}
$$

Informally, to demonstrate that the above set of literals is not satisfiable, we would first use the two literals in the middle to infer in $\mathbb{Q}$ that "$0 \geq z$" and then the last literal to demonstrate that "$z = 0$" and hence also that "$x = y$". Then, we use the congruence rule of $\mathbb{E}$ to infer that "$f(x) = f(y)$". Then we move again in $\mathbb{Q}$ to prove that "$f(x) - f(y) = z$" and then back to $\mathbb{E}$ to prove that "$f(f(x) - f(y)) = f(z)$". This allows $\mathbb{E}$ to detect the contradiction with the first literal and to declare that the set of literals is not satisfiable. This example demonstrates that, in general, the decision procedures must interact in a non-obvious way to detect unsatisfiability.

---

[1] This example is taken from [Nel81].

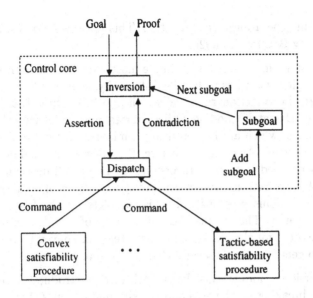

Fig. 2. The overall structure of the Touchstone theorem prover.

Nelson and Oppen show that it is enough for each decision procedure to broadcast only the contradictions and the *equalities between variables* that it discovers. This simple cooperation mechanism is shown in [Nel81] to be complete when the theories involved are convex. A theory is *not convex* if it has a set of literals that entails a proper disjunction of equalities between variables without entailing any single equality. For example, the theory $\mathbb{Z}$ of integer linear arithmetic is not convex since $y = z + 1 \wedge y \geq x \wedge x \geq z$ entails $x = y \vee x = z$.

The Nelson-Oppen architecture can be adapted to deal with non-convex theories by performing a case-split whenever a conjunction of literals entails a disjunction. Informally, the prover tries to guess which one of the disjuncts holds and asserts it to all decision procedures. If this does not lead to unsatisfiability then the next disjunct is tried. For this procedure to be correct there are additional technical requirements that the theories must satisfy as explained in [Nel81].

The structure and operation of Touchstone is shown in Figure 2. Input goals are first broken into literal goals and assertions by the "Inversion" module. Hypothesis literals and negated goal literals are asserted along with their proofs to the "Dispatch" module that essentially implements a broadcast medium between decision procedures. Each decision procedure receives either proved asserted literals or proved equality of variables discovered by other decision procedures. Decision procedures can also discover a contradiction, which is then propagated to the "Inversion" module. The "Subgoal" module is discussed later. As an optimization, equalities discovered and broadcasted by decision procedures are not accompanied by an actual proof but only by a token that identifies the originator decision procedure. Proofs are produced on demand only if the equality is actually used in generating a contradiction. This optimization is similar to the one described by Boulton [Bou92].

The "Inversion" module is fairly simple mostly due to the limited fragment of first-order logic that it has to handle. Informally, goals are first broken using the appropriate introduction rules. When the goal is a literal its negation is asserted to the "Dispatch" module. Other sources of assertions are the left-hand sides of assertions, which are broken using elimination rules into a sequence of literals to be asserted. As discussed before, the "Dispatch" module broadcasts the assertions received from the "Inversion" module to all decision procedures and expects them to return either a contradiction or a set of entailed equalities between variables. This proof procedure is complete for the fragment of logic that we consider here. The completeness at the level of literals is ensured by the correctness theorem for the Nelson-Oppen strategy for convex theories.

In order to generate proofs several changes have to be made:

- The "Inversion" module must keep track of the introduction rules that it uses while breaking up the goal and the elimination rules that it uses while breaking up the assertion.
- The "Dispatch" module and consequently all decision procedures must receive the asserted literals accompanied by a proof.
- A decision procedure that discovers an equality must also be able to produce a proof of that equality, possibly in terms of the proofs accompanying the assertions that it received previously.
- Furthermore, a decision procedure that discovers a contradiction must be able to exhibit a proof of falsehood.

All proof objects maintained in the system are represented as terms in the Edinburgh Logic Framework (LF) [HHP93], which conveniently allows higher-order proof representations, so we can use the simple LF type checker as a proof checker.

We decided to use an imperative implementation of the system so that each decision procedure can maintain state without having to pass it around as it would be necessary in a purely functional implementation. To ensure proper scoping of assertions and thus to maintain the soundness of the prover, the inversion module announces when decision procedures must "forget" certain assertions they have received. This is implemented by programming the "Inversion" module to issue pair of commands snapshot and undo. These commands are broadcast to decision procedures by the "Dispatch" module. The intended semantics of the undo operation is that all decision procedures should adjust their internal data structures so that they do not reflect assertions that were made after the matching snapshot operation. This ensures that assertions are properly retracted from the system at the time of the undo.

2.1   The Subgoal Module

Touchstone deviates from the Nelson-Oppen architecture described in [Nel81] by allowing the use of decision procedures based on tactics. A tactic-based decision procedure can, in addition to detecting contradictions and equalities between variables, announce subgoals that, if proved, would allow the decision procedure to detect a contradiction. Each such subgoal is announced to the "Subgoal" module along with a function that given a proof of the subgoal generates a proof of falsehood. We refer to this function as the proof transformer. The "Subgoal" module is used extensively in Touchstone for the implementation of various decision procedures for type checking, as explained in Section 4.

Touchstone attempts to prove tactic-generated subgoals only when the current proof is about to fail. If the "Inversion" module notices that no contradiction is announced after it asserts the negation of a goal literal then it queries the "Subgoal" module for a list of subgoals that were announced by tactic-based decision procedures. The "Inversion" considers these subgoals in turn and if any one of them is proved it can use the associated proof transformer to generate the desired contradiction.

An added benefit of the "Subgoal" module is that non-convex decision procedures can be incorporated quite naturally. If a decision procedure cannot discover an equality but it does discover a proper disjunction of equalities it is in a position to announce a subgoal consisting of a proper conjunction of disequalities. This will lead the "Inversion" to perform a case-split and to try to prove all disequalities independently, which is exactly the behavior desired in the presence of non-convex theories.

This completes the description of the modules responsible with the control of the decision procedures. In the next section we describe some general principles for implementing proof-generating decision procedures and then we examine in more detail two of the decision procedures of Touchstone.

3   Proof-Generating Decision Procedures

All decision procedures in Touchstone use the same internal representation of literals in the form of a global *expression directed acyclic graph*, or the E-DAG. The E-DAG contains a node for each unique subexpression of the goal formula. In addition to the global E-DAG, each decision procedure is free to maintain its own internal state.

Each decision procedure is required to implement at least three functions:

- The `assert` function that given a literal and its proof asserts the literal to the decision procedure. This function should return a list of equalities that were discovered along with their proofs, or a contradiction along with a proof of falsehood. As a side-effect this function can also announce subgoals along with their proof transformers to the "Subgoal" module.
- The `snapshot` and `undo` functions, as described above.

There are at least a couple of ways to implement the `snapshot` and `undo` operations. The simplest way is for each decision procedure to maintain a stack of the input assertions. Each time a new assertion arrives it is considered with

$$\frac{}{E = E}\ \text{eqid}\quad \frac{E_2 = E_1}{E_1 = E_2}\ \text{eqsym}\quad \frac{E_1 = E_2 \quad E_2 = E_3}{E_1 = E_3}\ \text{eqtr}$$

$$\frac{E_1 = E_2 \quad E_1 \neq E_2}{\bot}\ \text{falsei}\quad \frac{E_1 = E_1' \quad \cdots \quad E_n = E_n'}{\mathtt{f}(E_1, \ldots, E_n) = \mathtt{f}(E_1', \ldots, E_n')}\ \text{congr}$$

Fig. 3. The axioms of the theory $\mathbb{E}$ of equality.

respect to the assertions already memorized on the stack for the purpose of detecting equalities or announcing a contradiction. Each **snapshot** places a marker on the stack and each **undo** pops the stack up to the nearest marker. This simple strategy is typical of the decision procedures that use backward chaining. In Touchstone, the modular arithmetic and the typing decision procedures use this simple strategy.

Another strategy is used by the forward-chaining decision procedures. These typically maintain internal data structures that reflect the current assertions in an internal form. As new assertions arrive they are internalized in the data structure and equalities are propagated. To be usable in Touchstone, such decision procedures must be able to revert their data structures to the state at the matching **snapshot**. Decision procedures can implement the undo operation by using non-mutable data structures or by maintaining a list of the destructive operations performed on the state so that they can be undone. The latter strategy is used in Touchstone by the congruence closure and the Simplex decision procedures.

3.1 Proof Generation in the Congruence Closure Algorithm

A central theory in any implementation of the Nelson-Oppen architecture is the theory of equality. The free functions of the theory $\mathbb{E}$ are "$=$" and "$\neq$" along with any uninterpreted function symbols. The axioms of the theory are those shown in Figure 3. There is one congruence rule for each function symbol in the system.

The theory $\mathbb{E}$ was first shown decidable by Ackermann [Ack54] by reducing the problem to that of constructing the congruence closure of a relation on a graph. If $\mathcal{R}$ is an equivalence relation over a set of terms, we say that two terms $\mathtt{f}(t_1, \ldots, t_n)$ and $\mathtt{f}(t_1', \ldots, t_n')$ are *congruent* if t_i is related to t_i' by $\mathcal{R}$ for all $i = 1, \ldots, n$. The *congruence closure* of a relation $\mathcal{R}$ is the smallest extension of $\mathcal{R}$ that is both an equivalence relation and relates all its congruent terms. To see if a given equality "$t = u$" follows from a set of equalities $\mathcal{R}$, we first construct the congruence closure $\mathcal{R}'$ of $\mathcal{R}$ and then check to see if $t = u \in \mathcal{R}'$. This is the sense in which an algorithm for computing the congruence closure of a set of equalities can be at the base of a decision procedure for $\mathbb{E}$. The implementation of congruence closure in Touchstone is a proof-generating extension of that described in [Nel81].

In addition to the E-DAG, the congruence closure algorithm uses its own internal data structures to represent the congruence closure of the current equality assertions. Thus, a mapping root is maintained to map each subexpression to a representative for its equivalence class. A mapping forbid maps a class representative C to a set of nodes that are known to be distinct from C. Finally, the set of incoming equality and inequality assertions are stored on an undoStack along with their proofs. Additionally, whenever a congruence is discovered, the corresponding equality along with its proof is pushed on the undoStack.

For the purposes of discussing the proof generation strategy we do not need to see the whole implementation but just the invariant that it maintains. This invariant is shown below, with the notation $class(a)$ denoting the set of nodes with the same root as a:

C1. $\text{root}(a) = \text{root}(b)$ if and only if $a \equiv b$ or there exist $a_1, \ldots, a_{n+1}$ such that
- $a \equiv a_1$, $b \equiv a_{n+1}$, and
- $(a_i = a_{i+1}, \text{pf}_i) \in$ undoStack or $(a_{i+1} = a_i, \text{pf}_i) \in$ undoStack for all $i = 1, \ldots, n$

C2. $class(a) \cap \text{forbid}(\text{root}(b)) \neq \emptyset$ if and only if there exist $a' \in class(a)$ and $b' \in class(b)$ such that $(a' \neq b', \text{neqab}) \in$ undoStack.

Based on this invariant we can define a function $\text{prfEq}(a, b)$ that given two expressions with the same root produces a proof of their equality, and also a function $\text{mkEqContra}(a, b, eqab)$ that given two expressions as in Invariant C2 along with a proof of their equality, produces a proof of falsehood, as shown below.

$\text{prfEq}(a : \text{node}, b : \text{node}) =$ /* $\text{root}(a) \equiv \text{root}(b)$ */
   if $a \equiv b$ then **return** $\text{eqid}(a)$
   let $a_1, \ldots, a_{n+1}$ as in the invariant C1

$$pf'_i = \begin{cases} pf_i & \text{if } (a_i = a_{i+1}, pf_i) \in \text{undoStack} \\ \text{eqsym}(pf_i) & \text{if } (a_{i+1} = a_i, pf_i) \in \text{undoStack} \end{cases}$$

   return $\text{eqtr}(pf'_1, \text{eqtr}(pf'_2, \ldots, \text{eqtr}(pf'_{n-1}, pf'_n) \ldots))$

$\text{mkEqContra}(a, b, eqab) =$ /* $class(a) \cap \text{forbid}(\text{root}(b)) \neq \emptyset$ */
   let $a' \in class(a), b' \in class(b)$ such that $(a' \neq b', neqab) \in$ undoStack
   return $\text{falsei}(\text{eqtr}(\text{prfEq}(a', a), \text{eqtr}(eqab, \text{prfEq}(b, b'))), neqab)$

The decision procedure operates as follows. When an equality $a = b$ is asserted we first check the Invariant C2 condition to see if we detected a contradiction, in which case we use mkEqContra to generate the proof of falsehood required for the Contradiction exception. Otherwise we push the asserted equality along with its proof on the undoStack and we merge the classes of a and b updating the forbid sets accordingly to preserve the invariant. Finally, we check for newly introduced congruences. Any such congruence is both an equality to be announced to other decision procedures and an input to a recursive invocation of the merge procedure. For this latter step we use the congr rule to generate the appropriate proof.

When a disequality assertion $a \neq b$ along with its proof $neqab$ is encountered we first check whether a and b are equivalent. If they are then we announce a Contradiction with the proof falsei(prfEq(a, b), $neqab$). Otherwise we just add $\{(a \neq b, neqab)\}$ to the undoStack and we update the forbid sets of both a and b.

Note that congruence closure is a convex decision procedure and thus, it does not need to create case splits. Also, as proven in [NO80] and [Nel81] for similar implementations, the congruence closure algorithm is a sound and complete decision procedure for $\mathbb{E}$. The algorithmic complexity of the algorithm is determined by the method used to discover the congruent pairs of nodes. In Touchstone this is done with a simple algorithm of complexity $\mathcal{O}(n^2)$, where n is the number of nodes in the E-DAG. The complexity can be reduced to $\mathcal{O}(n \log n)$ using the more complex strategy described in [DST80]. The proof generation extensions do not affect the algorithmic complexity. For more implementation details the reader is invited to consult [Nel81,Nec98].

3.2   Proof-Generation in the Simplex Algorithm

Now we turn our attention to the decision procedure for the theory $\mathbb{Z}$ of integer numerals along with the operators $\{+, -, \geq, >\}$. As a notational convenience we also consider multiplication by integer numerals, which we write using the infix · operator. The decision problem for $\mathbb{Z}$ is essentially the problem of deciding whether one linear inequality is a consequence of several other inequalities. There are several decision procedures for $\mathbb{Z}$ in the literature. Some cover only special cases [AS80,Pra77,Sho81] while others attempt to solve the general case [Ble74,Nel81]. Here we will consider a general decision procedure based on the Simplex algorithm for linear programming, as described in [Nel81]. Like most decision procedures for $\mathbb{Z}$, Simplex is not complete since it works essentially with rational numbers. However, Simplex is powerful enough to handle the kinds of inequalities that typically arise in program verification. For space reasons, we discuss here only on the very basic properties of the Simplex algorithm used as a decision procedure and we focus on the modifications necessary for proof generation. The reader is invited to consult [Nec98,Nel81] for more details and a complete running example.

The internal data structure used by the Simplex algorithm is the *tableau*, which consists of a matrix of rational numbers q_{ij} with rows $i \in 1..r$ and columns $j \in 1..c$ along with a vector of rational numbers q_{i0}. All rows and columns in the tableau are owned by an expression in the E-DAG. We write $R(i)$ to denote the owner of row i and $C(j)$ to denote the owner of column j. The main property of the tableau is that each row owner can be expressed as a linear combination of column owners, as follows:

$$R(i) \simeq q_{i0} + \sum_{j=1}^{c} q_{ij} \cdot C(j) \quad (i = 1..r) \tag{2}$$

We use the $\simeq$ notation to denote equality between symbolic expressions modulo the rules of the commutative group of addition.

The Simplex tableau as described so far encodes only the linear relationships between the owning expressions. The actual encoding of the input inequality assertions is by means of *restrictions* on the values that owning expressions can take. There are two kinds of restrictions that Simplex must maintain. A row or a column can be either +*-restricted*, which means that the owning expression can take only values greater or equal to zero, or *-restricted*, in which case the owning expression can only be equal to zero.

To illustrate the operation of Simplex as a decision procedure for arithmetic consider the task of detecting the contradiction in the following set of literals:

$$\{1 - x \geq 0,\ 1 - y \geq 0,\ -2 + x + y \geq 0,\ -1 + x - y \geq 0\}$$

When Simplex asserts an inequality, it rewrites the inequality as $e \geq 0$ and it introduces a new row in the tableau owned by e. To simplify the notation we are going to use the names $s_1, \ldots, s_4$ for the left-hand sides of the our four inequations. After adding each row, Simplex attempts to mark it as +-restricted (since the input assertions comes with a proof that the owning expression is positive). But in doing so the tableau might become unsatisfiable, so Simplex first performs Gaussian elimination (called *pivoting* in Simplex terminology) to increase the entry in column 0 to a strictly positive value. If it is possible to bring the tableau in a form where each +-restricted row has a positive entry in column 0, then the tableau is satisfiable. One particular assignment that satisfies all +-restrictions is obtained by setting all variables or expressions owning the columns to zero.

In the process of adding the third inequality from above the tableau is as shown below in position (a) below. The row owned by s_3 cannot be marked as +-restricted because $q_{30} = -2 \not\geq 0$. To increase the value of q_{30} Simplex performs two pivot operations (eliminating x from s_1 and y from s_2) and brings the tableau in the state shown below in position (b).

	x	y			s_1^+	s_2^+			s_1^*	s_2^*			s_1^*	s_2^*
s_1^+	1	-1		x	1	-1		x	1	-1		x	1	-1
s_2^+	1		-1	y	1		-1	y	1		-1	y	1		-1
s_3	-2	1	1	s_3	0	-1	-1	s_3^*	0	-1	-1	s_3^*	0	-1	-1
												s_4	-1	-1	1
	(a)				(b)				(c)				(d)		

Now Simplex can safely mark the row s_3 as +-restricted and, if it were not for the need to detect all equalities between variables, it could proceed to process the fourth inequality. In order to detect easily all inequalities Simplex must maintain the tableau in a form in which all owners of rows and columns that are constrained to be equal to zero are marked as *-restricted. It turns out that in order to detect rows that must be made *-restricted it is sufficient to look for +-restricted rows that become maximized at 0. A row i is said to be *maximized* at q_{i0} when all its non-zero entries are either in *-restricted columns or are negative and are in +-restricted columns. One such row is s_3 in the tableau (b). Since

s_1 and s_2 are known to be positive and s_3 is a negative-linear combination of them, it follows that $s_3 \leq q_{30} = 0$. On the other hand after processing the third assertion we know that $s_3 \geq 0$ which leads Simplex to decide that s_3 is both $+$-restricted and maximized at 0, hence it must be equal to zero. Furthermore all non-zero entries in the row of s_3 must now be in $*$-restricted columns. Thus Simplex marks all of s_1, s_2, and s_3 as $*$-restricted rows, bringing the tableau in the state shown above in position (c). In this state Simplex notices that the rows of x and y differ only in $*$-restricted columns (which are known to be zero), hence it announces that x is equal to y.

Finally, when the fourth assertion is added, the tableau becomes as shown in position (d) above. Now Simplex notices that s_4 is maximized at -1 and consequently that it is impossible to increase the value of q_{40} by pivoting. Since Simplex also knows that $s_4 \geq 0$ (from the fourth assertion), it has discovered a contradiction.

In the rest of this section we discuss how one can extract proofs of variable equalities and contradictions from the Simplex tableau. Before that let us point out that the implementation of the undo feature in Simplex does not have to revert the tableau to the original state at the time of the matching **snapshot**. Instead, all it has to do is to remove the rows, columns and restrictions that were added since then. This is much less expensive than a full undo.

The operation of the Simplex algorithm maintains the following invariants:

S1. If $R(i)$ is restricted (either $+$-restricted or $*$-restricted) then there is a proof of $R(i) \geq 0$. We refer to this proof as $Proof(R(i))$. Similar for columns.

S2. If $R(i)$ is $+$-restricted then $q_i \geq 0$

S3. If $R(i)$ is $*$-restricted then $q_i = 0$ and $q_{ij} \neq 0$ implies $C(j)$ is $*$-restricted

S4. If $C(j)$ is $*$-restricted then there exists $*$-restricted $R(i)$ such that $q_{ij} < 0$ and $q_{ik} \leq 0$ for all $k > j$. We say that row i restricts column j.

Only Invariant S1 is introduced solely for the purpose of proof generation; the others are necessary for the correct operation of Simplex.

Simplex detects contradictions when it tries to add a $+$-restriction to a row i that is maximized at a negative value q_{i0}. The key ingredient of a proof of falsehood in this case is a proof that $R(i) \leq q_{i0}$. Simplex constructs this proof indirectly by constructing first a proof that $R(i) = q_{i0} + E$ and then a proof that $E \leq 0$. Furthermore, Simplex chooses E to be the expression constructed as the linear combination of column owners as specified by the entries in row i. If we ignore the $*$-restricted columns, then all of the non-zero entries in the maximized row i are negative and are in columns j that are $+$-restricted. For each such column, according to Invariant S1, there is a proof that $C(j) \geq 0$.

To construct these proofs Simplex uses the inference rules shown below:

$$\frac{R_i \geq 0 \quad R_i = q + E \quad E \leq 0}{\bot} \; \text{sfalse} \; (q < 0)$$

$$\frac{}{E_1 = E_2} \; \text{arith} \; (E_1 \simeq E_2) \qquad \frac{E_1 \leq 0 \quad E_2 \geq 0}{E_1 + q \cdot E_2 \leq 0} \; \text{geqadd} \; (q \geq 0)$$

mapRow$(i) = \Phi \leftarrow \emptyset$
    <u>foreach</u> $k = 1 .. c$ such that $q_{ik} < 0$ <u>do</u> $\Phi(C(k)) \leftarrow + q_{ik}$
    <u>foreach</u> $k = 1 .. c$ such that $q_{ik} > 0$ <u>do</u> $\Phi = \text{mapCol}(k, q_{ik}, \Phi)$
    <u>return</u> Φ

mapCol$(j, q, \Phi) = $ let row i be the restictor of j as in invariant S4 $(q_{ij} < 0)$
    $\Phi(R(i)) \leftarrow + q/q_{ij}$
    <u>foreach</u> $k \neq j$ such that $q_{ik} < 0$ <u>do</u> $\Phi(C(k)) \leftarrow + -q \cdot q_{ik}/q_{ij}$
    <u>foreach</u> k such that $q_{ik} > 0$ <u>do</u> $\Phi \leftarrow \text{mapCol}(k, -q \cdot q_{ik}/q_{ij}, \Phi)$
    <u>return</u> Φ

Fig. 4. Extracting coefficients from the Simplex tableau.

The `sfalse` rule is used to generate contradictions as explained above. Its first hypothesis is obtained directly from the proof of the incoming inequality $R(i) \geq 0$. The second hypothesis is constructed directly using the rule `arith` whose side-condition holds because of the main tableau representation invariant. Finally, the third hypothesis is constructed using repeated use of the `geqadd` rule for all the elements of the negative linear combination of restricted column owners, as read from the row i in the tableau. It is a known fact from linear algebra that a contradiction is entailed by a set of linear inequalities if and only if a false inequality involving only numerals can be constructed from a positive linear combination of the original inequalities. Thus these rules are not only necessary for Simplex but also sufficient for any linear arithmetic proof procedure.

The situation is somewhat more complicated due to the presence of *-restricted columns that might contain strictly positive entries in a maximized row (such as is the case in the row s_4 of tableau (d) shown before). To solve this complication we must also be able to express every *-restricted column as a negative linear combination of restricted owners. Simplex uses the two functions `mapRow` and `mapCol` shown in Figure 4 to construct negative linear combinations for a maximized row or a *-restricted column. In Figure 4 the notation Φ denotes a map from expressions to negative rational factors. All integer numerals n are represented as the numeral 1 with coefficient n. We use $\emptyset$ to denote the empty map. The operation $\Phi(E) \leftarrow + q$ updates the map Φ increasing the coefficient of E by q.

The reader is invited to verify using the Simplex invariants that if `mapRow` is invoked on a maximized row and if `mapCol` is invoked on a *-restricted column with a positive factor q then the resulting map Φ contains only restricted expressions with negative coefficients. The termination of `mapCol` is ensured by the Invariant S4.

The reader can verify that by running `mapRow(4)` on the tableau (d) we obtain that $s_4 \simeq -1 - 2 \cdot s_1 - 1 \cdot s_3$, which in turn says that the negation of the fourth inequality can be verified by multiplying the first equality by 2 and adding the third equality.

The Simplex equality proofs are generated using the inference rule seq shown below.

$$\frac{E_1 - E_2 = E \quad E \leq 0 \quad E_2 - E_1 = E' \quad E' \leq 0}{E_1 = E_2} \text{ seq}$$

Consider for example the case of two rows i_1 and i_2 whose entries are distinct only in *-restricted columns. We temporarily add to the tableau a new row r whose entries are $q_{i_1 j} - q_{i_2 j}$. Since row r is maximized at 0 we can use mapProof to produce the first two hypotheses of seq, just like we did for sfalse. Then we negate the entries in r and we use again mapProof to produce the last two hypotheses.

4   Lessons Learned while Building and Using Touchstone

In this section we describe our initial experience with building and using the Touchstone theorem prover. We programmed the theorem prover in the Standard ML of New Jersey dialect of ML. The whole project, including the control core along with the congruence closure and Simplex decision procedures and also with decision procedures for modular arithmetic and type checking, consists of around 11,000 lines of source code. Of these, about 3,000 are dedicated solely to proof representation, proof generation, proof optimization (such as local reduction and turning proofs by contradiction into simpler direct proofs) and proof checking.

The relatively small size of the proof generating component in Touchstone can be explained by the fact that most heuristics and optimizations that are required during proving are irrelevant to proof generation. Take for example the Simplex decision procedure. Its implementation has over 2000 lines of code, a large part of which encodes heuristics for selecting the best sequence of pivots. The proof-generating part of Simplex is a fairly straightforward reading of the tableau once the contradiction was found, as shown in Figure 4.

One important design feature of proof generation in Touchstone is that proofs are produced lazily, only when a contradiction is found. This is similar in spirit with the lazy techniques described by Boulton [Bou92]. We do not generate explicit proofs immediately when we discover and propagate an equality. Instead we only record which decision procedure discovered the equality and later, if needed, we ask that decision procedure to generate an explicit proof of the equality. This lazy approach to proof generation means that the time consumed for generating proofs is small when compared to the time required for proving since in most large proving tasks a large part of the time is spent exploring unsuccessful paths. Our experiments with producing type safety proofs for assembly language show that only about 15% of the time is spent on proof generation.

It was obvious from the beginning of the project that there will be a design and coding complexity cost to be paid for the proof-generating capability of the prover. We accepted this cost initially because we needed a mechanical way to build the proofs required for our proof-carrying code experiments. We did not anticipate how much this feature would actually simplify the building, testing and maintenance of the theorem prover. Indeed, we estimate that the ability to cross-check the operation of the prover during proof generation saved us many weeks or maybe months of testing and debugging.

One particular aspect of the theorem prover that led to many design and programming errors was that all decision procedures and the control core must be incremental and undoable. This is complicated by decision procedures that perform a pseudo-undo operation in the interest of efficiency. For example, the Simplex decision procedure does not revert the tableau to the exact state it was at the last snapshot operation but only to an equivalent state obtained simply by deleting some rows and columns. In the presence of such decision procedures the exact order of intermediate subgoals and discovered equalities depends on all previously processed subgoals. This defeats a common technique for isolating a bug by reducing the size of the goal in which it is manifested. It often happens that by eliminating seemingly unrelated subgoals the error disappears because the order in which entailed equalities is changed.

Proof-generation as a debugging mechanism continues to be valuable even as the prover matures. While we observe a decrease in the number of errors, we also observe a sharp increase in the average size of the proving goals that trigger an error. Indeed the size of these goals is now such that it would have been impractical to debug the prover just by manual inspection of a trace. Secondly, proof-generation gave us significant assistance with the upgrade and maintenance of the prover, as a broken invariant is promptly pointed out by the proof-generation infrastructure. We should also point out that for maximum assurance we checked proofs using the same small proof checker that we also use for proof-carrying code. However, we noticed that most prover errors surfaced as failures by the proof-generating infrastructure to produce a proof and only a very small number of bugs resulted in invalid proofs. A lesson that can be drawn here is that the software engineering advantages of proof-generation in theorem provers can be obtained by just going through the process of generating a proof without actually having to record the proof.

4.1   Using Touchstone to Build Proof-Carrying Code

The main motivation for building Touchstone was for use in a proof-carrying code system. A typical arrangement for the generation of proof-carrying code is shown in Figure 5. Note that on the right-hand side we have the untrusted components used to produce PCC while on the left-hand side we have the trusted infrastructure for checking PCC.

This figure applies to the particular case in which the safety policy consists of type safety in the context of a simple first-order type system with pointers and arrays. The process starts with a source program written in a type-safe subset of the C programming language. The source program is given to a certifying compiler that, in addition to producing optimized machine code, also generates function specifications and loop invariants based on types. We will return to the issue of modeling types in first-order logic shortly.

The code augmented with loop invariants is passed through a verification condition generator (VcGen) that produces a verification condition (VC). The

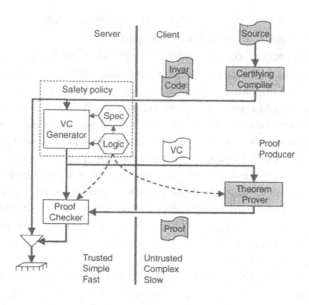

Fig. 5. A typical arrangement for building proof-carrying code.

VC is provable only if the code satisfies the loop invariants and the specifications and only if all memory operations are safe. To encode proof obligations for memory safety in a general way, VcGen emits formulas of the form "`saferd(E)`" to say that the memory address denoted by the symbolic expression E is readable, and "`safewr(E,E')`" to say that the value denoted by E' is writable at the address denoted by E.

The verification condition is passed then to Touchstone, which proves it and returns a proof encoded in a variant of the Edinburgh LF language. This allows an LF type checker on the receiver side to fully validate the proof with respect to the verification condition. The key idea behind proof-carrying code is that the whole process of producing a safe executable can be split into an complex and slow untrusted component on one side and a simple and fast trusted safety checker. It was therefore a key requirement that the proving and proof checking tasks be separated as shown in the picture. For more details on the system described here the reader is invited to consult [Nec98].

What remains to be discussed are the details of the logical theory that is used to model types and to derive memory safety. We use a theory of first-order types with constructors for types and a typing predicate. A few of the terms and formulas used along with three of the inference rules are shown below:

$$\frac{A : \mathtt{array}(T, L) \quad I \geq 0 \quad I < L}{A + 4 * I : \mathtt{ptr}(T)} \qquad \frac{A : \mathtt{ptr}(\mathtt{pair}(T_1, T_2))}{A + 4 : \mathtt{ptr}(T_1)} \qquad \frac{A : \mathtt{ptr}(T)}{\mathtt{saferd}(A)}$$

The first rule says that a pointer to an element of type T can be obtained by indexing in an array whose element type is T. In this rule the array type is dependent on the length of the array and the element size is considered to be 4 bytes. The second rule is used to reason about tuple destructors. Finally, the last rule is the only rule that introduces the saferd predicate, basically saying that in this safety policy readability of memory locations is dictated by types.

To handle this theory of types in Touchstone we make heavy use of the "Subgoal" module. We implemented a tactic that does backward chaining on the rules of the theory. A central element of the tactic is a heuristic that finds likely valid formulas of the form "$A : \text{ptr}(T)$" by looking at the form of A and at the current assertions (originating from function preconditions and loop invariants in this case). While the other elements of Touchstone have some completeness properties, the tactic for the theory of types need only be powerful enough to "understand" the code produced by our compiler. And since the compiler starts with an obviously well-typed source program the only difficulties can be introduced by optimizations. In fact, our compiler is very aggressive in optimizing array-bounds checks and hence the theorem prover must be able to prove itself all the arithmetic facts that the compiler has discovered and proved. As a result the Simplex satisfiability procedure is exercised quite heavily in this setting.

Our experiments with Touchstone in this setting have shown several interesting facts. This separation of tasks does indeed achieve a separation of complexity and running cost. In terms of code size the untrusted components are about four times larger than the trusted ones. In particular the Touchstone prover is four times larger than the proof checker. Furthermore, while Touchstone grows continuously as we incorporate more heuristics and better tactics we found that the proof-checking component has remained largely unchanged over a couple of years.

A sample of the experimental data that we collected is shown in Figure 6. This table shows, for a few programs, the sizes of the verification condition generated and of the associated proofs along with the time required for theorem proving and proof checking. The sizes are expressed in number of AST nodes while the timings are expressed in milliseconds. The measurements were performed on a DEC Alpha with a 21064 processor running at 175MHz. Notice that in these experiments proof checking is about an order of magnitude faster than theorem proving.

At this point we would like to point out that our imperative implementation of the Nelson-Oppen prover appears to be much faster than the functional implementation of Boulton in the HOL prover, as described in [Bou93,Bou95]. We ran Touchstone on the 11 examples shown on page 94 in [Bou93]. All of these examples are very small ranging from 9 to 43 AST nodes. While Boulton ran the measurements on a Sparcstation 2 with a 40Mhz processor we ran them on an Alpha with a 175Mhz. But even after multiplying the Touchstone running times by a factor of 5 to compensate for this difference it still results that Touchstone is faster by a factor ranging from 5 to 80 on these examples.

There are several reasons behind the better performance in Touchstone. One of them is that Touchstone handles only a small fragment of first-order logic and can thus use a very fast goal-directed proof procedure. HOL extended with the

Program	Lines	VC size (AST nodes)	Proving time (ms)	Proof size (AST nodes)	Check time (ms)
bcopy	16	82	25	64	4
edge	88	224	143	528	15
kmp	67	483	108	344	9
qsort	142	1444	127	1770	16
sharpen	153	420	257	477	23
simplex	303	7055	1272	3912	120
unpack	259	5759	1912	1750	92

Fig. 6. Experimental data collected using the Touchstone prover in the context of generating PCC for type safety.

Nelson-Oppen cooperative decision procedure on the other hand uses a more general proof procedure based on conversion to disjunctive normal form. Another possible reason for the disparity in the performance is that while HOL uses a satisfiability procedure for arithmetic based on Fourier-Motzkin variable elimination, Touchstone uses an efficient implementation of Simplex. Boulton explains that the Fourier-Motzkin elimination was chosen in HOL because of the simplicity of proof generation. We show that even an efficient version of Simplex can be made fully expansive, although almost surely at a larger programming cost.

Finally, we suspect that another reason that makes Touchstone faster than Boulton's implementation of the Nelson-Oppen strategy is the use of an imperative implementation style. By making very judicious use of memory and we observe very little garbage collection during proving. In contrast, Boulton's implementation uses a functional programming style leading to very elegant implementation of a crucial part of the prover: the undo mechanism. Instead our implementation of undo is quite a bit more complex and is responsible for many of the bugs that we discovered. In quite a few cases we forgot to undo certain changes thus leading to unsoundness. This did not turn out to be a big problem because the proof checking mechanisms quickly pointed out our errors.

Other times the undo procedure mistakenly removed too many assertions to some situations in which the prover was not able to prove predicates that it was intended to prove. We were helped in this situation by the fact that we were using Touchstone essentially to verify that a number of optimizations performed by our compiler preserve type safety. By design of both the compiler and the prover, every failed proof attempt points to either a compilation bug or a completeness bug in the prover. This is how we found a very large number of bugs in the compiler and some in the prover.

5 Conclusion

We describe in this paper an implementation of a Nelson-Oppen theorem prover enhanced with the ability to generate easily-checkable proof objects for all the predicates it proves. Our implementation, just like the one described by Nelson, was designed with efficiency in mind in order to handle verification condition of non trivial programs. This led us to use more complex algorithms and implementation techniques than a related functional implementation in the context of the HOL theorem prover.

The added complexity of our design seem to pay off in terms of efficiency. But it also led us to making many subtle design and programming errors that threatened both the soundness and the completeness of our prover. Fortunately, soundness was never in real danger because of Touchstone's proof generating ability, which enables us to use a simple proof checker to validate the correctness of each run.

As a general conclusion, we feel that the benefits of proof-generation in theorem provers clearly outweigh the additional cost of designing and implementing proof-generating decision procedures.

Acknowledgments

We would like to thank the anonymous referees for making many valuable suggestions that have improved this paper substantially.

References

Ack54.   Wilhelm Ackermann. *Solvable Cases of the Decision Problem.* Studies in Logic and the Foundations of Mathematics. North-Holland, Amsterdam, 1954.

AS80.   Bengt Aspvall and Yossi Shiloach. A polynomial time algorithm for solving systems of linear inequalities with two variables per inequality. *SIAM Journal on Computing*, 9(4):827–845, 1980.

Ble74.   W.W. Bledsoe. The Sup-Inf method in Presurger arithmetic. Technical report, University of Texas Math Dept., December 1974.

BM79.   Robert Boyer and J. Strother Moore. *A Computational Logic.* Academic Press, 1979.

Bou92.   Richard J. Boulton. A lazy approach to fully-expansive theorem proving. In *International Workshop on Higher Order Logic Theorem Proving and its Applications*, pages 19–38, Leuven, Belgium, September 1992. North-Holland. IFIP Transactions.

Bou93.   Richard John Boulton. *Efficiency in a Fully-Expansive Theorem Prover.* PhD thesis, University of Cambridge, December 1993.

Bou95.   Richard J. Boulton. Combining decision procedures in the HOL system. In *8th International Workshop on Higher Order Logic Theorem Proving and its Applications*, volume 971 of *Lecture Notes in Computer Science*, pages 75–89. Springer-Verlag, September 1995.

DLNS98. David L. Detlefs, K. Rustan M. Leino, Greg Nelson, and James B. Saxe. Extended static checking. SRC Research Report 159, Compaq Systems Research Center, 130 Lytton Ave., Palo Alto, December 1998.

DST80. Peter J. Downey, Ravi Sethi, and Robert E. Tarjan. Variations on the common subexpressions problem. *Journal of the ACM*, 27(4):758–771, 1980.

Gor85. Michael Gordon. HOL: A machine oriented formulation of higher-order logic. Technical Report 85, University of Cambridge, Computer Laboratory, July 1985.

HHP93. Robert Harper, Furio Honsell, and Gordon Plotkin. A framework for defining logics. *Journal of the Association for Computing Machinery*, 40(1):143–184, January 1993.

Mil91. Dale Miller. A logic programming language with lambda-abstraction, function variables, and simple unification. *Journal of Logic and Computation*, 1(4):497–536, September 1991.

MNPS91. Dale Miller, Gopalan Nadathur, Frank Pfenning, and Andre Scedrov. Uniform proofs as a foundation for logic programming. *Annals of Pure and Applied Logic*, 51:125–157, 1991.

Nec97. George C. Necula. Proof-carrying code. In *The 24th Annual ACM Symposium on Principles of Programming Languages*, pages 106–119. ACM, January 1997.

Nec98. George C. Necula. *Compiling with Proofs*. PhD thesis, Carnegie Mellon University, September 1998. Also available as CMU-CS-98-154.

Nel81. Greg Nelson. Techniques for program verification. Technical Report CSL-81-10, Xerox Palo Alto Research Center, 1981.

NO79. Greg Nelson and Derek Oppen. Simplification by cooperating decision procedures. *ACM Transactions on Programming Languages and Systems*, 1(2):245–257, October 1979.

NO80. Greg Nelson and Derek C. Oppen. Fast decision procedures based on congruence closure. *Journal of the Association for Computing Machinery*, 27(2):356–364, April 1980.

ORS92. S. Owre, J. M. Rushby, and N. Shankar. PVS: A prototype verification system. In Deepak Kapur, editor, *11th International Conference on Automated Deduction (CADE)*, volume 607 of *Lecture Notes in Artificial Intelligence*, pages 748–752, Saratoga, NY, June 1992. Springer-Verlag.

Pau94. L. C. Paulson. Isabelle: A generic theorem prover. *Lecture Notes in Computer Science*, 828:xvii + 321, 1994.

Pfe91. Frank Pfenning. Logic programming in the LF logical framework. In Gérard Huet and Gordon Plotkin, editors, *Logical Frameworks*, pages 149–181. Cambridge University Press, 1991.

Pfe94. Frank Pfenning. Elf: A meta-language for deductive systems (system description). In Alan Bundy, editor, *12th International Conference on Automated Deduction*, LNAI 814, pages 811–815, Nancy, France, June 26–July 1, 1994. Springer-Verlag.

Pra77. Vaughan R. Pratt. Two easy theories whose combination is hard. Unpublished manuscript, 1977.

SD99. Aaron Stump and David L. Dill. Generating proofs from a decision procedure. In A. Pnueli and P. Traverso, editors, *Proceedings of the FLoC Workshop on Run-Time Result Verification*, Trento, Italy, July 1999.

Sho81. Robert Shostak. Deciding linear inequalities by computing loop residues. *Journal of the ACM*, 28(4):769–779, October 1981.

Wellfounded Schematic Definitions

Konrad Slind

Cambridge University Computer Laboratory

Abstract. A *program scheme* looks like a recursive function definition, except that it has free variables 'on the right hand side'. As is well-known, equalities between schemes can capture powerful program transformations, *e.g.*, translation to tail-recursive form. In this paper, we present a simple and general way to define program schemes, based on a particular form of the wellfounded recursion theorem. Each program scheme specifies a schematic induction theorem, which is automatically derived by formal proof from the wellfounded induction theorem. We present a few examples of how formal program transformations are expressed and proved in our approach. The mechanization reported here has been incorporated into both the HOL and Isabelle/HOL systems.

Program schemes form the foundation of an interesting class of program development methodologies which advocate the incremental instantiation of abstract programs, preserving important properties all the while, until a suitable concrete program results. There has been a great deal of work on program transformation, for background see [6, 18, 33, 23, 26].

Although program transformation theories are being applied a lot informally, work on program transformation in mechanized proof assistants is not as abundant, in spite of the evident interest in using such systems as platforms for program development and transformation. One reason for this may be that, currently, such environments (*e.g.*, [13, 25, 22, 3]) tend to be based on logics of total functions and it is not clear how a program scheme can be regarded as a total function, since many schemes allow instantiations such that the resulting function is *not* total. In spite of this, we will describe a simple and general technique by which schemes may be defined such that totality is enforced.

1 Formal Basis

We work in a higher order logic commonly called HOL [13]; a description of the logic may be found in the Appendix. We adopt the common approach of using the native functions of the logic to represent programs; recursive programs are modelled with the use of a wellfounded recursion theorem. There are several equivalent definitions of wellfoundedness [28]; the following asserts that the relation $R : \alpha \to \alpha \to \text{bool}$ is wellfounded iff every non-empty set has an R-minimal element.

D. McAllester (Ed.): CADE-17, LNAI 1831, pp. 45–63, 2000.
© Springer-Verlag Berlin Heidelberg 2000

Definition 1 (Wellfoundedness).

$$\mathsf{WF}(R) \equiv \forall P.\ (\exists w.\ P\ w) \supset \exists min.\ P\ min \wedge \forall b.\ R\ b\ min \supset \neg P\ b.$$

From this definition, the following general induction and recursion theorems can be proved (the interested reader can find details in [30]):

Theorem 2 (Wellfounded Induction).

$$\mathsf{WF}(R) \supset (\forall x.\ (\forall y.\ R\ y\ x \supset P\ y) \supset P\ x) \supset \forall x.\ P\ x.$$

Theorem 3 (Wellfounded Recursion).

$$\forall f\ R\ M.\ (f = \mathsf{WFREC}\ R\ M) \supset \mathsf{WF}(R) \supset \forall x.\ f(x) = M\ (f\,|\,R, x)\ x.$$

$\mathsf{WFREC} : (\alpha \to \alpha \to \mathsf{bool}) \to ((\alpha \to \beta) \to (\alpha \to \beta)) \to \alpha \to \beta$ can be thought of as a 'controlled' fixpoint operator; since it is only used to prove Theorem 3, we omit its quite obfuscatory definition. Also used in the statement of Theorem 3 is a ternary operator that restricts a function to a certain set of values. [1]

Definition 4 (Restriction).

$$(f\,|\,R, y) \equiv \lambda x.\ \texttt{if}\ R\ x\ y\ \texttt{then}\ f\ x\ \texttt{else}\ \mathsf{Arb}.$$

Theorem 5. $R\ x\ y \supset (f\,|\,R, y)\ x = f\ x.$

2 The Technique

We shall present our approach with the hand derivation of an example; the automation of the technique will be taken up in Section 3. Consider the following description of the 'while' construct familiar from imperative programming:

$$\mathsf{While}\ s = \texttt{if}\ B\ s\ \texttt{then}\ \mathsf{While}\ (C\ s)\ \texttt{else}\ s.$$

This is a syntactic specification of a class of functions determined by the parameters B and C. To start the derivation, the description is translated into a functional:

$$\lambda While\ s.\ \texttt{if}\ B\ s\ \texttt{then}\ While\ (C\ s)\ \texttt{else}\ s. \qquad (1)$$

Instantiating M in the recursion theorem with (1) yields

$$\begin{array}{l} \left[\begin{array}{l} \mathsf{WF}\ R, \\ f = \mathsf{WFREC}\ R\ (\lambda While\ s.\ \texttt{if}\ B\ s\ \texttt{then}\ While\ (C\ s)\ \texttt{else}\ s) \end{array}\right] \\ \vdash \\ \hline \quad \forall x.\ f(x) = \texttt{if}\ B\ x\ \texttt{then}\ (f\,|\,R, x)\ (C\ x)\ \texttt{else}\ x. \end{array} \qquad (2)$$

---

[1] In set theory, or logics of partial functions, function restriction may result in a partial function. In a logic of total functions, such as HOL, a restriction of a function is still a total function, giving a fixed but arbitrary value when applied outside of the restriction.

By assuming $\forall s.\ B\ s \supset R\ (C\ s)\ s$ (we discuss the origin of this assumption in Section 3), it is possible to derive

$$
\begin{bmatrix}
\text{WF } R,\ \forall s.\ B\ s \supset R\ (C\ s)\ s, \\
f = \text{WFREC } R\ (\lambda While\ s.\ \texttt{if } B\ s\ \texttt{then } While\ (C\ s)\ \texttt{else } s)
\end{bmatrix}
\tag{3}
$$
$$\vdash$$
$$f\ x = \texttt{if } B\ x\ \texttt{then } f\ (C\ x)\ \texttt{else } x.$$

The assumptions WF R and $\forall s.\ B\ s \supset R\ (C\ s)\ s$ are the 'termination conditions' of (3). Now we apply the Principle of Constant Definition to define While. This is the central step in our method. The indefinite description operator (ε) is applied to *choose* a wellfounded relation R meeting the termination conditions. Notice also that the distinction between *parameters* (B and C) and *arguments* (s) is supported by the different binding sites in the definition: parameters are arguments to the definition itself, while the original argument s is a bound variable in the functional.

$$
\text{While} \equiv \lambda B\ C.\ \text{WFREC}(\varepsilon R.\ \text{WF } R \wedge \forall s.\ B\ s \supset R\ (C\ s)\ s)
$$
$$
(\lambda While\ s.\ \texttt{if } B\ s\ \texttt{then } While\ (C\ s)\ \texttt{else } s).
\tag{4}
$$

Eliminating (4) from the hypotheses of (3) yields

$$
\begin{bmatrix}
\text{WF } (\varepsilon R.\ \text{WF } R \wedge \forall s.\ B\ s \supset R\ (C\ s)\ s), \\
\forall s.\ B\ s \supset (\varepsilon R.\ \text{WF } R \wedge \forall s.\ B\ s \supset R\ (C\ s)\ s)\ (C\ s)\ s
\end{bmatrix}
$$
$$\vdash$$
While $B\ C\ s = \texttt{if } B\ s\ \texttt{then }$ While $B\ C\ (C\ s)\ \texttt{else } s.$

Finally, assuming $\text{WF}(R)$ and $\forall s.\ B\ s \supset R\ (C\ s)\ s$ and then applying the Select Axiom allows the conclusion

$$
\left[\,\text{WF } R, \forall s.\ B\ s \supset R\ (C\ s)\ s\,\right]
\tag{5}
$$
$$\vdash$$
While $B\ C\ s = \texttt{if } B\ s\ \texttt{then }$ While $B\ C\ (C\ s)\ \texttt{else } s.$

Remark 6. The derived equation (5) looks like a normal higher-order function; however, had we tried to define While as a higher order function in the standard manner, *i.e.*, with no parameters, then B, C and s would be treated as arguments—and thus bound in the functional—and the termination conditions would be equivalent to the proposition

$$
\exists R.\ \text{WF}(R) \wedge \forall B\ C\ s.\ R\ (B, C, C\ s)\ (B, C, s),
$$

which is not provable since C could be taken to be the identity function, but there is no wellfounded relation R such that $R\ x\ x$.

Remark 7. Our treatment of parameters is not specific to wellfounded recursion: it works for any fixpoint operator. In particular, for any fix satisfying the well-known equation $\text{fix}(M) = M \; (\text{fix}(M))$, it is merely a common subexpression elimination to get $\forall g. \; (g = \text{fix}(M)) \supset \forall x. \; g \; x = M \; g \; x$. By abstracting free variables $\mathcal{P}_1, \ldots, \mathcal{P}_k$ of M, this can be transformed to

$$\vdash \forall g. \; (g \, \mathcal{P}_1 \ldots \mathcal{P}_k = \text{fix}(M)) \supset \forall x. \; g \, \mathcal{P}_1 \ldots \mathcal{P}_k \, x = M \, (g \, \mathcal{P}_1 \ldots \mathcal{P}_k) \, x.$$

With hindsight, the treatment of parameters in inductive definition packages such as those reported in [21, 24, 15] can be seen as concrete applications of this theorem.

2.1 Induction

It is well known that the wellfounded relation used to prove termination for a function can also be used to derive an induction theorem, in which the induction predicate is assumed to hold for the arguments to recursive calls. For ML-style pattern-matching recursion equations of the form

$$f(pat_1) \equiv rhs_1[f(a_{11}), \ldots, f(a_{1k_1})]$$
$$\vdots \tag{6}$$
$$f(pat_n) \equiv rhs_n[f(a_{n1}), \ldots, f(a_{nk_n})],$$

an induction theorem of the following form (where $\Gamma(a_{ij})$ is the *context* of recursive call $f\,(a_{ij})$) can be derived by formal proof from Theorem 2:

$$
\left(
\forall
\left(
\begin{array}{cc}
(\forall(\Gamma(a_{11}) \supset P \, a_{11})) & \wedge \\
\vdots & \\
(\forall(\Gamma(a_{1k_1}) \supset P \, a_{1k_1})) & \wedge
\end{array}
\right) \supset P(pat_1)
\right) \wedge
$$

$$\vdots$$
$$\wedge$$

$$
\left(
\forall
\left(
\begin{array}{cc}
(\forall(\Gamma(a_{n1}) \supset P \, a_{n1})) & \wedge \\
\vdots & \\
(\forall(\Gamma(a_{nk_n}) \supset P \, a_{nk_n})) & \wedge
\end{array}
\right) \supset P(pat_n)
\right) \supset \forall v. \; P \; v.
$$

The assumptions to this theorem will be the termination conditions of f, as explained in [31], where the automatic derivation of such induction theorems is described. An earlier treatment of the derivation of induction for functions in a simpler object language is described in [5].

It might seem to be problematic to derive induction for program schemes since the termination relation is not known; however, an appropriate induction theorem can still be derived: all that is required is to *assume* that a suitable termination relation exists. We demonstrate the idea by deriving the following induction theorem for the While function:

$$
\begin{bmatrix}
\text{WF } R, \\
\forall s. \; B \, s \supset R \, (C \, s) \, s
\end{bmatrix}
\vdash \forall P. \; (\forall s. \; (B \, s \supset P \, (C \, s)) \supset P \, s) \supset \forall v. \; P \; v. \tag{7}
$$

The derivation begins by assuming the antecedent of (7) and the termination conditions of the definition.

1. $\forall s.\ (B(s) \supset P\,(C\,s)) \supset P\,s$ Assume
2. $\forall s.\ B(s) \supset R\,(C\,s)\,s$ Assume
3. $[2] \vdash B(s) \supset R\,(C\,s)\,s$ $\forall$-elim(2)
4. $\forall y.\ R\,y\,s \supset P\,s$ Assume
5. $[4] \vdash R\,(C\,s)\,s \supset P\,(C\,s)$ $\forall$-elim(4)
6. $[2, B\,s] \vdash R\,(C\,s)\,s$ Undisch(3)
7. $[4, 2, B\,s] \vdash P\,(C\,s)$ $\supset$-elim (5) (6)
8. $[4, 2] \vdash B\,s \supset P\,(C\,s)$ $\supset$-intro (7)
9. $[1, 4, 2] \vdash P\,s$ $\supset$-elim (1) (8)
10. $[1, 2] \vdash (\forall y.\ R\,y\,s \supset P\,s) \supset P\,s$ $\supset$-intro (9)
11. $[1, 2] \vdash \forall s.\ (\forall y.\ R\,y\,s \supset P\,s) \supset P\,s$ $\forall$-intro (10)

In step 11, the antecedent of the wellfounded induction theorem (Theorem 2) has been derived, and a few further obvious steps deliver (7), as desired.

Remark 8. If the semantics of a Hoare triple $\{P\}\,C\,\{Q\}$ are defined by

$$\text{Hoare } P\,C\,Q \equiv \forall s.\ P\,s \supset Q\,(C\,s)$$

then the following While rule for total correctness has an easy[2] proof by induction with (7):

$$\begin{bmatrix} \text{WF } R, \\ \forall s.\ B\,s \supset R\,(C\,s)\,s \end{bmatrix} \vdash \begin{array}{l} \text{Hoare } (\lambda s.\ P\,s \wedge B\,s)\ C\ P\ \supset \\ \text{Hoare } P\ (\text{While } B\ C)\ (\lambda s.\ P\,s \wedge \neg B\,s). \end{array}$$

3   Automation

A useful level of support for deriving program schemes can be supplied by generalizing and automating the steps taken in the While example. The particular interface we have implemented takes as input recursion equations of the form given in (6) and performs the following steps:

1. Translates the equations into a functional $\mathcal{F}$, using a pattern-matching translation based on those used in functional programming language implementations [2, 20].
2. Instantiates M in the recursion theorem with $\mathcal{F}$.
3. Extracts termination conditions $TC_1(R), \ldots TC_k(R)$ from the results of step 2, where R is a variable representing the wellfounded relation.
4. Computes the free variables $P_1 \ldots P_i$ of $\mathcal{F}$, then defines the constant denoting the desired function:

$$f \equiv \lambda P_1 \ldots P_i.\ \text{WFREC } (\varepsilon R.\text{WF}(R) \wedge TC_1(R) \wedge \ldots \wedge TC_k(R))\ \mathcal{F}.$$

---

[2] The proof takes four tactic applications in a current version [14] of the Hol98 proof assistant.

Two things are important here: (1) using the description operator to choose a suitable wellfounded relation meeting the results of step 3; and (2) making sure to separate parameters from arguments in the definition.

5. Eliminates the result of step 4 from the hypotheses of the result of step 2 (the instantiated recursion theorem).

6. Assumes each of $\mathsf{WF}(R), TC_1(R), \ldots TC_k(R)$ and then eliminates the description operator terms, via application of the Select Axiom. Now the desired termination constraints have been derived.

7. Derives the induction theorem from the termination conditions.

8. Returns the recursion equations and the induction theorem.

Fortunately, the algorithms of [30, 31] generalize naturally to support steps 1 to 8. The key to automation is step 3, in which termination conditions are automatically extracted. This is accomplished by use of a special contextual rewriter, which attempts to rewrite the instantiated recursion theorem (coming from step 2) with the conditional rewrite rule for function restriction (Theorem 5). In searching for matches for this rule, the rewriter is essentially searching for every recursive call site in the original equations. The rewriter uses its stock of contextual rules to gather and discard context Γ as it makes its search; when a recursive call site $(f \mid R, pat_i)\ (a_{ij})$ is found (in context $\Gamma(a_{ij})$), the termination condition $\Gamma(a_{ij}) \supset R\ a_{ij}\ pat_i$ is *captured* by performing a small proof which stores the termination condition on the assumptions. After the rewriting process terminates, one is left with a theorem, the conclusion of which is the desired recursion equations, and the assumptions of which are the termination conditions (from which the induction theorem can be derived). An important point about these manipulations is that they all take place by deductive steps in the object logic, so the results are sound. Detailed descriptions of the algorithms, including their extension to mutual recursion, nested recursion, and higher order recursion can be found in [32].

To extend these algorithms to program schemes is particularly simple: all that need be done is to take care never to quantify scheme variables in any of the derivations. If there are no scheme variables, the algorithms perform exactly as in [30, 31], so schemes are a smooth extension to the existing apparatus.

4   Formal Program Transformations

Program schemes are helpful for giving suitably abstract descriptions of classes of functions. A further application of schemes comes from using them as a basis for program transformation: instances of schemes may be identified, provided *applicability conditions* are satisfied. There are various ways of representing program transformations formally; we choose to represent them simply as theorems (specifically, as constrained recursion equations). Proving a program transformation typically involves an application of the induction theorem for one of the program schemes being equated.

Example 9. The following scheme expresses a class of *linear recursive* programs:

linRec$(x) \equiv$ if *Atomic x* then *A x* else *Join* (linRec (*Dest x*)) (*D x*).

Under certain conditions, instances of linRec are equal to corresponding instances of the following tail-recursive scheme, which uses an accumulating parameter:

accRec$(x, u) \equiv$ if *Atomic x* then *Join* (*A x*) *u*
$\qquad$ else accRec (*Dest x, Join* (*D x*) *u*).

Intuitively, the recursive calls of linRec must get 'stacked up' somehow, waiting for deeper recursive calls to return. In contrast, calls to accRec need not be stacked. If the combination function *Join* is associative, then the implicit bracketing of the stacked recursive calls can be replaced with a single data value that gets modified and passed at each recursive call. We now formalize this intuition. The result of defining linRec is (we omit the induction theorem):

$$\begin{bmatrix} \text{WF } R, \\ \forall x.\ \neg Atomic\ x \supset R\ (Dest\ x)\ x \end{bmatrix}$$
$\vdash$

linRec *D Dest Join A Atomic x* =
$\quad$ if *Atomic x* then *A x*
$\quad$ else *Join* (linRec *D Dest Join A Atomic* (*Dest x*)) (*D x*),

and that for accRec is (we conjoin the induction theorem to the recursion equation):

$$\begin{bmatrix} \text{WF } R, \\ \forall x.\ \neg Atomic\ x \supset R\ (Dest\ x)\ x \end{bmatrix}$$
$\vdash$
(accRec *D Dest A Join Atomic* (x, u) =
$\quad$ if *Atomic x* then *Join* (*A x*) *u*
$\quad$ else accRec *D Dest A Join Atomic* (*Dest x, Join* (*D x*) *u*))
$\wedge$
$\forall P.\ (\forall x\ u.\ (\neg Atomic\ x \supset P(Dest\ x, Join\ (D\ x)\ u)) \supset P(x, u)) \supset \forall v\ v_1.\ P(v, v_1)$.

The formal program transformation is then captured in the following theorem:

$$\begin{bmatrix} \text{WF } R, \\ \forall x.\ \neg Atomic\ x \supset R\ (Dest\ x)\ x, \\ \forall p\ q\ r.\ Join\ p\ (Join\ q\ r) = Join\ (Join\ p\ q)\ r \end{bmatrix}$$
$\vdash$

$\forall x\ u.\ Join$ (linRec *D Dest Join A Atomic x*) *u*
$$=$$
accRec *D Dest A Join Atomic* (x, u).

Proof. Apply the induction theorem for accRec, then expand the definitions of linRec and accRec. $\square$

Example 10. The following scheme for binary recursion uses the parameters *Left* and *Right* to break the input into two parts on which to recurse:

$$\mathsf{binRec}(x) \equiv \mathtt{if}\ Atomic\ x\ \mathtt{then}\ A\ x$$
$$\mathtt{else}\ Join\ (\mathsf{binRec}\ (Left\ x))\ (\mathsf{binRec}\ (Right\ x)).$$

The result of the definition is (omitting the induction theorem)

$$\left[\begin{array}{l} \mathsf{WF}\ R, \\ \forall x.\ \neg Atomic\ x \supset R\ (Right\ x)\ x, \\ \forall x.\ \neg Atomic\ x \supset R\ (Left\ x)\ x \end{array}\right]$$
$$\vdash$$

$$\mathsf{binRec}\ Right\ Left\ Join\ A\ Atomic\ x =$$
$$\quad \mathtt{if}\ Atomic\ x\ \mathtt{then}\ A\ x$$
$$\quad \mathtt{else}\ Join\ (\mathsf{binRec}\ Right\ Left\ Join\ A\ Atomic\ (Left\ x))$$
$$\qquad\qquad\quad (\mathsf{binRec}\ Right\ Left\ Join\ A\ Atomic\ (Right\ x)).$$

The example comes from Wand [33], who used paper and pencil, and has been treated in PVS [29]. In his development, Wand was interested in explaining how continuations give the programmer a representation of the runtime stack, and thus can act as a bridge in the transformation of non-tail-recursive functions to tail recursive ones. In our development, we will avoid the continuation-passing intermediate representation (although it is simple for us to handle) and transform to tail recursion in one step.

Now we present a general tail recursion scheme for lists. In the definition, the parameter $Dest : \alpha \to \alpha$ list breaks the head h of the work list $h :: t$ into a list of new work, which it prepends to t before continuing; hence, the tailRec scheme is quite general because the argument to the second tail call may increase in length by any finite amount. (Wand and Shankar only consider tail recursions in which the *Dest* parameter can produce two new pieces of work.)

$$\mathsf{tailRec}\ ([\,], v) \equiv v$$
$$\mathsf{tailRec}\ (h :: t, v) \equiv \mathtt{if}\ Atomic\ h\ \mathtt{then}\ \mathsf{tailRec}\ (t, Join\ v\ (A\ h))$$
$$\qquad\qquad\qquad\quad \mathtt{else}\ \mathsf{tailRec}\ (Dest\ h\ @\ t,\ v).$$

The result of this definition is (including the induction theorem)

$$\left[\begin{array}{l} \mathsf{WF}\ R, \\ \forall v\ t\ h.\ \neg Atomic\ h \supset R\ (Dest\ h\ @\ t, v)\ (h :: t, v), \\ \forall v\ t\ h.\ \ \ Atomic\ h \supset R\ (t, Join\ v\ (A\ h))\ (h :: t, v) \end{array}\right]$$
$$\vdash$$

$(\text{tailRec } Dest\ A\ Join\ Atomic\ ([\,],v) = v)\ \wedge$
$(\text{tailRec } Dest\ A\ Join\ Atomic\ (h::t,v) =$
    $\text{if } Atomic\ h$
    $\text{then tailRec } Dest\ A\ Join\ Atomic\ (t, Join\ v\ (A\ h))$
    $\text{else tailRec } Dest\ A\ Join\ Atomic\ (Dest\ h\ @\ t, v))$
$\wedge$

$$\left(\begin{array}{l} \forall P.\ (\forall v.\ P\ ([\,],v))\ \wedge \\ \quad (\forall h\ t\ v.\, \neg Atomic\ h \supset P\ (Dest\ h\ @\ t, v)\ \wedge \\ \qquad\qquad Atomic\ h \supset P\ (t, Join\ v\ (A\ h)) \supset P\ (h::t,v)) \\ \quad \supset \forall v\ v_1.\ P\ (v, v_1). \end{array} \right)$$

We intend to prove an equivalence between binRec and tailRec but the transformation seems to require the termination constraints for both binRec and tailRec to be satisfied. However, a bit of thought reveals that a useful fact about finite multisets can simplify matters, by allowing one constraint to be expressed in terms of the other.

Definition 11 (msetPred). *Let m be a finite multiset and $R : \alpha \to \alpha \to$ bool a relation on elements of m. The relation msetPred R is built by removing an x from m and replacing it with a finite multiset of elements, each of which is R-smaller than x.*

Theorem 12. $\text{WF}(R) \supset \text{WF}(\text{msetPred } R)$
Proof. *The classic(al) proof can be found in [8]; a recent constructive proof is described in [27, Chapter II].* $\square$

Now we show how the termination condition of tailRec can be reduced to the (simpler) one of binRec:

$\text{WF } R \wedge (\forall h\ y.\ \neg Atomic\ h \wedge \text{mem } y\ (Dest\ h) \supset R\ y\ h)$
    $\supset$
$\exists R'.\ \text{WF } R' \wedge$
    $(\forall h\ t\ v.\ \neg Atomic\ h \supset R'\ (Dest\ h\ @\ t,\ v)\ (h::t,v)) \wedge$
    $(\forall h\ t\ v.\ \ Atomic\ h \supset R'\ (t,\ Join\ v\ (A\ h))\ (h::t,v))$

Proof. Assume WF R and $\forall h\ y.\ \neg Atomic\ h \wedge \text{mem } y\ (Dest\ h) \supset R\ y\ h)$. R' is a relation on pairs. The witness for R' operates over the first projection of the pair, *i.e.*, over lists, and maps a list into a multiset of the list elements. Since R is wellfounded, msetPred over the multiset is wellfounded and thus the witness is wellfounded. The remaining two conjuncts are both true, the first by assumption and the definition of msetPred, and the second by the definition of msetPred, since no elements are being put back into the multiset. $\square$

With this reduction, one can state and prove the following general theorem relating binary recursion and tail recursion. The essential insight is that the work list l of tailRec represents a linearization of the binary tree of calls of binRec. Thus going from left to right through the work list, invoking binRec and accumulating the results, should deliver the same answer as executing tailRec on the work list.

We formalize this left-to-right pass by the auxiliary function rev_itlist.[3] Note how the *Dest* parameter of tailRec has been specialized with $\lambda x.\ [Left\ x,\ Right\ x]$.

$$\begin{bmatrix} \text{WF } R, \\ \forall x.\ \neg Atomic\ x \supset R\ (Left\ x)\ x \wedge R\ (Right\ x)\ x, \\ \forall p\ q\ r.\ Join\ (Join\ p\ q)\ r = Join\ p\ (Join\ q\ r) \end{bmatrix}$$
$$\vdash$$
$$\forall l\ v_0.$$
$$\quad \text{rev\_itlist}(\lambda tr\ v.\ Join\ v\ (\text{binRec}\ Right\ Left\ Join\ A\ Atomic\ tr))\ l\ v_0$$
$$=$$
$$\quad \text{tailRec}(\lambda x.[Left\ x,\ Right\ x])\ A\ Join\ Atomic\ (l, v_0)$$

Proof. Induct with the induction theorem for tailRec. The base case is straightforward; the step case is also essentially trivial, since it only involves using the induction hypotheses and rewriting with the definitions of rev_itlist, tailRec, and binRec. □

Finally, the desired program transformation

$$\begin{bmatrix} \text{WF } R, \\ \forall x.\ \neg Atomic\ x \supset R\ (Left\ x)\ x \wedge R\ (Right\ x)\ x, \\ \forall p\ q\ r.\ Join\ (Join\ p\ q)\ r = Join\ p\ (Join\ q\ r) \end{bmatrix}$$
$$\vdash$$
$$\forall x\ v_0.$$
$$\quad Join\ v_0\ (\text{binRec}\ Right\ Left\ Join\ A\ Atomic\ x)$$
$$=$$
$$\quad \text{tailRec}\ (\lambda x.\ [Left\ x,\ Right\ x])\ A\ Join\ Atomic\ ([x], v_0)$$

can be obtained by instantiating the work list l to comprise the initial item of work $[x]$, and then reducing the definition of rev_itlist away.

Example 13. Now we derive a program transformation originally presented by Bird [4], and later mechanized by Shankar [29]. Consider a datatype btree of binary trees with constructors

$$\text{LEAF} : \alpha\ \text{btree}$$
$$\text{NODE} : \alpha\ \text{btree} \to \alpha \to \alpha\ \text{btree} \to \alpha\ \text{btree}.$$

The so-called *catamorphism* (iterator) for this type is

$$\text{btreeRec LEAF } v\ f \equiv v$$
$$\text{btreeRec (NODE } t_1\ M\ t_2)\ v\ f \equiv f\ (\text{btreeRec } t_1\ v\ f)\ M\ (\text{btreeRec } t_2\ v\ f).$$

Most mechanizations of higher order logic automate such definitions; however, the so-called *anamorphism* (or unfold, or co-recursor) for this type has not been

---

[3] rev_itlist, also known as foldl to functional programmers, is defined as

$$\text{rev\_itlist } f\ []\ v \equiv v$$
$$\text{rev\_itlist } f\ (h :: t)\ v \equiv \text{rev\_itlist } f\ t\ (f\ h\ v).$$

straightforward to define in these systems. Understanding the following definition of unfold : $\alpha \rightarrow \beta$ btree may be eased by considering it as operating over an abstract datatype α which supports operations $More : \alpha \rightarrow$ bool and $Dest : \alpha \rightarrow \alpha * \beta * \alpha$.

$$
\begin{aligned}
&\text{unfold } x \equiv \texttt{if } More\ x \\
&\qquad\qquad \texttt{then let } (y_1, b, y_2) = Dest\ x \\
&\qquad\qquad\qquad \texttt{in} \\
&\qquad\qquad\qquad\qquad \text{NODE (unfold } y_1)\ b\ (\text{unfold } y_2) \\
&\qquad\qquad \texttt{else LEAF}.
\end{aligned}
$$

The automatically computed constraints attached to the definition are the following:

$$
\begin{bmatrix}
\text{WF } R, \\
\forall x\ y_1\ b\ y_2.\ More\ x \wedge ((y_1, b, y_2) = Dest\ x) \supset R\ y_2\ x, \\
\forall x\ y_1\ b\ y_2.\ More\ x \wedge ((y_1, b, y_2) = Dest\ x) \supset R\ y_1\ x.
\end{bmatrix}
$$

Notice that the mechanization is not currently smart enough to know that the two termination conditions share the same context. After some trivial manipulation to join the two termination conditions, the induction theorem for unfold is the following (omitting the hypotheses):

$$
\forall P.\ (\forall x.\ (\forall y_1\, b\, y_2. More\ x \wedge ((y_1, b, y_2) = Dest\ x) \supset P\ y_1 \wedge P\ y_2) \supset P\ x) \supset \forall v.\ P\ v. \tag{8}
$$

It is easy to generalize unfold to an arbitrary range type by replacing NODE and LEAF with parameters G and C:

$$
\begin{aligned}
&\text{fuse } x \equiv \texttt{if } More\ x \\
&\qquad\qquad \texttt{then let } (y_1, b, y_2) = Dest\ x \\
&\qquad\qquad\qquad \texttt{in} \\
&\qquad\qquad\qquad\qquad G\ (\text{fuse } y_1)\ b\ (\text{fuse } y_2) \\
&\qquad\qquad \texttt{else } C.
\end{aligned}
$$

The *fusion* theorem states that unfolding into a btree and then applying a structural recursive function to the result is equivalent to interweaving unfolding steps with the steps taken in the structural recursion. Thus two recursive passes over the data can be replaced by one:

$$
\begin{bmatrix}
\text{WF } R, \\
\forall x\ y_1\ b\ y_2.\ More\ x \wedge ((y_1, b, y2) = Dest\ x) \supset R\ y_1\ x\ \wedge R\ y_2\ x
\end{bmatrix}
$$
$$
\vdash
$$
$$
\forall x\ C\ G.\ \text{btreeRec (unfold } Dest\ More\ x)\ C\ G = \text{fuse } C\ Dest\ G\ More\ x.
$$

Proof. The proof is by induction using (8), followed by expanding the definitions of btreeRec, unfold, and fuse. $\square$

5   Related Work

The paper by Huet and Lang [18] is an important early milestone in the field of program transformation. They worked in the LCF system, using fixpoint induction to derive program transformations. Program schemes were not defined;

instead, transformations were represented via applications of the Y combinator, i.e., had the form *applicability conditions* $\supset$ Y $\mathcal{F}$ = Y $\mathcal{G}$, for functionals $\mathcal{F}$ and $\mathcal{G}$. An influential aspect of the work was the use of second order matching to automate the application of program transformations.

Work using PVS has represented program schemes and transformations by theories parameterized over the parameters of the scheme and having as proof obligations the applicability conditions of the transformation [29, 9, 10]. To apply the program transformation, the theory must be instantiated, and the corresponding concrete proof obligations proved.

In our technique—in contrast—the parameters of a scheme are arguments to the defined constant, and the proof obligations are constraints on the recursion equations and the induction theorem. Thus, theorems are used to represent both program schemes and program transformations. Instantiating a program transformation in our setting merely requires one to instantiate type variables and/or free term variables in a theorem. It remains to be seen if one representation is preferable to the other. In other ways, however, our approach seems to offer improved functionality:

1. Currently, our technique produces more general schemes, since termination conditions are phrased in terms of an arbitrary wellfounded relation, whereas termination relations in PVS are restricted to measure functions [22]. Similarly, a general induction theorem is automatically derived for each scheme in our setting, whereas the PVS user is limited to measure induction (or may alternatively derive a more general induction theorem 'by hand' from wellfounded induction).

2. Our technique is more convenient because it automatically generates—by deductive steps—termination conditions for schemes. Taking the example of unfold, one doesn't have to ponder the right constraints in our setting: they are delivered as part of the returned definition. In contrast, the definition of unfold in [29] requires expert knowledge of the PVS type system in order to phrase the right constraints on the *Dest* parameter. Since the termination conditions of a scheme constrain any program transformation that mentions the scheme, our approach should also ease the correct formulation of program transformations.

3. Our approach also works for mutually recursive schemes, which are not currently available in PVS.

The paper of Basin and Anderson [1] has much in common with our work: for example, both approaches represent schemes and transformations by HOL theorems (Basin and Anderson call these *rules*). Their work differs from ours by focusing on relations (they are interested in modelling logic programs) rather than recursive functions. They present two techniques: in the first, program schemes are not defined; instead, transformations are derived by wellfounded induction on the arguments of the specified recursive relations (the relations themselves are left as variables). In the second, a program scheme is represented by an inductively defined relation. The first approach suffers from lack of automation: termination constraints are not synthesized and induction theorems are not automatically derived. In contrast, their second approach requires no mention of wellfoundedness, and induction is automatically derived by the inductive definition package of Isabelle/HOL.

In [11], Farmer treats the definition of recursive functions in a logic of partial functions. Schematic functions are represented in a similar manner to our approach, but the automation issues we tackle have not been explored.

In the context of language design, Lewis *et al.* [19] use schemes to implement a degree of dynamic scoping in a statically scoped functional programming language. Their approach allows occurrences of a free variable, *e.g.*, $\mathcal{P}$, in the body of a program to be marked with special syntax, *e.g.*, ?$\mathcal{P}$. The program is then treated as being parameterized by all such variables. To instantiate $\mathcal{P}$ occurring in a program f by a ground value *val*, they employ a notation 'f with ?$\mathcal{P} = val$'. Although their work is phrased using operational semantics and ours is denotationally based, there are many similarities.

Finally, our approach gives a higher-order and fully formal account of the *steadfast transformation* idea of Flener *et al.* [12]. In contrast to their work, we need give no soundness proof since our transformations are generated by deductive steps in a sound logic.

6 Conclusions

We have shown how a very simple technique allows a smooth treatment of program schemes, their induction theorems, and program transformations. Although the ideas are presented in the HOL logic, they are broadly applicable: the only notable requirements are a recursion theorem of the right form, a basic definition principle for introducing abbreviations, and an indefinite description operator. We have also sketched how higher levels of automation may be achieved, based on the automatic extraction of termination conditions by contextual rewriting. A few standard examples have been covered and, in some cases, generalized.

We emphasize that transformations derived using our technique are sound. For any instantiation of the parameters of a scheme or transformation, the rules of deduction force the applicability constraints to be likewise instantiated, and those instantiations persist in the hypotheses until eliminated by deduction. An instantiated scheme or transformation with invalid constraints can of course be trivialized.

Future work should focus on the difficult problems involved in automating the application of program transformations. One potential benefit of our practice of always deriving induction theorems may be that, if the scheme and the induction theorem are treated as a unit during instantiation, the instantiated induction scheme will be available for reasoning about the instantiated program at each step in the instantiation chain.

The schematic definition facility we have presented has been implemented via simple extensions to the TFL [32] package: as a result, program schemes as described in this paper have been available in the public releases of both the Hol98 and Isabelle/HOL systems since summer 1999.

Acknowledgements

This research was carried out on EPSRC grant GR/L03071, and written up while the author was employed on ESPRIT Framework IV LTR 26241. Larry Paulson helped finalize the Isabelle/HOL instantiation.

References

1. Penny Anderson and David Basin. Program development schemata as derived rules. *Journal of Symbolic Computation*, 2000. To appear.
2. Lennart Augustsson. Compiling pattern matching. In J.P. Jouannnaud, editor, *Conference on Functional Programming Languages and Computer Architecture (LNCS 201)*, pages 368–381, Nancy, France, 1985.
3. Bruno Barras, Samuel Boutin, Cristina Cornes, Judicael Courant, Yann Coscoy, David Delahaye, Daniel de Rauglaudre, Jean-Christophe Filliatre, Eduardo Gimenez, Hugo Herbelin, Gerard Huet, Henri Laulhere, Cesar Munoz, Chetan Murthy, Catherine Parent-Vigouroux, Patrick Loiseleur, Christine Paulin-Mohring, Amokrane Saibi, and Benjamin Werner. *The Coq Proof Assistant Reference Manual*. INRIA, 6.3.1 edition, December 1999. Accessible at http://pauillac.inria.fr/coq/doc/main.html.
4. Richard Bird. Functional algorithm design. In B. Moeller, editor, *Mathematics of Program Construction, Third International Conference, (MPC'95)*, volume LNCS 947, pages 2–17, Kloster Irsee, Germany, July 17-21 1995.
5. Robert S. Boyer and J Strother Moore. *A Computational Logic*. Academic Press, 1979.
6. Rod Burstall and John Darlington. A transformation system for developing recursive programs. *Journal of the Association for Computing Machinery*, 24(1):44–67, January 1977.
7. Alonzo Church. A formulation of the Simple Theory of Types. *Journal of Symbolic Logic*, 5:56–68, 1940.
8. Nachum Dershowitz and Zohar Manna. Proving termination with multiset orderings. *CACM*, 22(8):465–476, 1979.
9. Axel Dold. Representing, verifying and applying software development steps using the PVS system. In V.S. Alagar and Maurice Nivat, editors, *Proceedings of the Fourth International Conference on Algebraic Methodology and Software Technology, AMAST'95, Montreal*, volume 936 of *Lecture Notes in Computer Science*, pages 431–435. Springer-Verlag, 1995.
10. Axel Dold. Software development in PVS using generic development steps. To appear in Springer LNCS, Proceedings of a Seminar on Generic Programming, April 1998.
11. William Farmer. Recursive definitions in IMPS. Available by anonymous FTP at ftp.harvard.edu, in directory imps/doc, file name recursive-definitions.dvi.gz, 1997.
12. P. Flener, K.-K. Lau, and M. Ornaghi. On correct program schemas. In N.E. Fuchs, editor, *Proceedings of LOPSTR'97 (LNCS 1463)*, pages 124–143. Springer-Verlag, 1998.
13. Mike Gordon and Tom Melham. *Introduction to HOL, a theorem proving environment for higher order logic*. Cambridge University Press, 1993.

14. Hardware Verification Group. *Hol98 User's Manual*. University of Cambridge, December 1999. Accessible at http://www.ftp.cl.cam.ac.uk/ftp/hvg/hol98.

15. John Harrison. Inductive definitions: automation and application. In E. Thomas Schubert, Phillip J. Windley, and James Alves-Foss, editors, *Proceedings of the 1995 International Workshop on Higher Order Logic theorem proving and its applications*, number 971 in LNCS, pages 200–213, Aspen Grove, Utah, 1995. Springer-Verlag.

16. John Harrison. HOL-Light: A tutorial introduction. In *Proceedings of the First International Conference on Formal Methods in Computer-Aided Design (FM-CAD'96)*, volume LNCS 1166, pages 265–269. Springer-Verlag, 1996.

17. John Harrison. *Theorem Proving with the Real Numbers*. CPHC/BCS Distinguished Dissertations. Springer, 1998.

18. Gerard Huet and Bernhard Lang. Proving and applying program transformations expressed with second-order patterns. *Acta Informatica*, 11:31–55, 1978.

19. Jeffery R. Lewis, Mark B. Shields, Erik Meijer, and John Launchbury. Implicit parameters: Dynamic scoping with static types. In Tom Reps, editor, *ACM Symposium on Principles of Programming Languages*, Boston, Massachusetss, USA, January 2000. ACM Press.

20. Luc Maranget. Two techniques for compiling lazy pattern matching. Technical Report 2385, INRIA, October 1994.

21. Tom Melham. A package for inductive relation definitions in HOL. In M. Archer, J. J. Joyce, K. N. Levitt, and P. J. Windley, editors, *Proceedings of the 1991 International Workshop on the HOL Theorem Proving System and its Applications*, pages 350–357. IEEE Computer Society Press, Davis, California, USA, August 1991.

22. S. Owre, J. M. Rushby, N. Shankar, and D.J. Stringer-Calvert. *PVS System Guide*. SRI Computer Science Laboratory, September 1998. Available at http://pvs.csl.sri.com/manuals.html.

23. Helmut A. Partsch. *Specification and Transformation of Programs: A Formal Approach to Software Development*. Texts and Monographs in Computer Science. Springer-Verlag, 1990.

24. Lawrence Paulson. A fixedpoint approach to implementing (co)inductive definitions. In Alan Bundy, editor, *12th International Conference. on Automated Deduction (CADE)*, volume LNAI 814, pages 148–161. Springer-Verlag, 1994. Revised version available at http://www.cl.cam.ac.uk/users/lcp/papers/recur.html under title 'A Fixedpoint Approach to (Co)inductive and Co(datatype) Definitions'.

25. Lawrence Paulson. *Isabelle : A Generic Theorem Prover*. Number 828 in LNCS. Springer-Verlag, 1994. Up-to-date reference manual can be found at http://www.cl.cam.ac.uk/Research/HVG/Isabelle/dist/.

26. Peter Pepper and Douglas R. Smith. A high-level derivation of global search algorithms (with constraint propagation). *Science of Computer Programming*, 28(2–3):247–271, April 1997.

27. Henrik Persson. *Type Theory and the Integrated Logic of Programs*. PhD thesis, Chalmers University of Technology, June 1999.

28. Piotr Rudnicki and Andrzej Trybulec. On equivalents of well-foundedness. *Journal of Automated Reasoning*, 23(3):197–234, 1999.

29. Natarajan Shankar. Steps towards mechanizing program transformations using PVS. In B. Moeller, editor, *Mathematics of Program Construction, Third Interna-*

tional Conference, (MPC'95), number 947 in Lecture Notes in Computer Science, pages 50–66, Kloster Irsee, Germany, July 17-21 1995.

30. Konrad Slind. Function definition in higher order logic. In *Theorem Proving in Higher Order Logics*, number 1125 in Lecture Notes in Computer Science, Abo, Finland, August 1996. Springer-Verlag.

31. Konrad Slind. Derivation and use of induction schemes in higher order logic. In *Theorem Proving in Higher Order Logics*, number 1275 in Lecture Notes in Computer Science, Murrary Hill, New Jersey, USA, August 1997. Springer-Verlag.

32. Konrad Slind. *Reasoning about Terminating Functional Programs*. PhD thesis, Institut für Informatik, Technische Universität München, 1999. Accessible at http://www.cl.cam.ac.uk/users/kxs/papers.

33. Mitchell Wand. Continuation-based program transformation strategies. *Journal of the ACM*, 1(27):164–180, January 1980.

Appendix

The HOL logic is a typed higher-order predicate calculus[13], derived from Church's Simple Theory of Types [7]. The HOL logic is classical and has a set theoretic semantics, in which types denote non-empty sets and the function space denotes total functions. Several mature mechanizations exist [14, 16, 25]. The HOL logic is built on the syntax of a lambda calculus with an ML-style polymorphic type system. The syntax is based on signatures for types (Ω) and terms (Σ_Ω). The type signature assigns arities to type operators, while the term signature delivers the types of constants.

Definition 14 (HOL Types). *The set of types is the least set closed under the following rules:*

type variable. *There is a countable set of type variables, which are represented with Greek letters, e.g., α, β, etc.*

compound type. *If c in Ω has arity n, and each of $ty_1, \ldots ty_n$ is a type, then $c(ty_1, \ldots ty_n)$ is a type.*

A type constant is represented by a 0-ary compound type. A large collection of types can be definitionally constructed in HOL, building on the initial types found in Ω: truth values (bool), function space (written $\alpha \to \beta$), and an infinite set of individuals (ind).

Terms are typed λ-calculus expressions built with respect to Σ_Ω. When we wish to show that a term M has type τ, the notation $M : \tau$ is used.

Definition 15 (HOL Terms). *The set of terms is the least set closed under the following rules:*

Variable. *if v is a string and ty is a type built from Ω then $v : ty$ is a term.*

Constant. *($c : ty$) is a term if $c : \tau$ is in Σ_Ω and ty is an instance of τ, i.e., there exists a substitution for type variables θ, such that each element of the range of θ is a type in Ω and $\theta(\tau) = ty$.*

Combination. *($M\ N$) is a term of type β if M is a term of type $\alpha \to \beta$ and N is a term of type α.*

Abstraction. *($\lambda v.\ M$) is a term of type $\alpha \to \beta$ if v is a variable of type α and M is a term of type β.*

Initially, Σ_Ω contains constants denoting equality ($=$), implication ($\supset$), and an indefinite description operator (ε). Types and terms form the basis of the *prelogic*, in which basic algorithmic manipulations on types and terms are defined: *e.g.*, the free variables of a type or term, α-convertibility, substitution, and β-conversion. For describing substitution, the notation $[M_1 \mapsto M_2]\,N$ is used to represent the term N where all free occurrences of M_1 have been replaced by M_2. Of course, M_1 and M_2 must have the same type in this operation. During substitution, every binding occurrence of a variable in N that would capture a free variable in M_2 is renamed to avoid the capture taking place.

Deductive system. In Figure 1, a useful set of inference rules is outlined, along with the axioms of the HOL logic. The derivable theorems are just those that can be generated by using the axioms and inference rules of Figure 1. More parsimonious presentations of this deductive system can be found in [13] or Appendix A of [17].

A theorem with hypotheses $P_1, \ldots, P_k$ and conclusion Q (all of type bool) is written $[P_1, \ldots, P_k] \vdash Q$. In the presentation of some rules, *e.g.*, $\lor$-elim, the following idiom is used: $\Gamma, P \vdash Q$. This denotes a theorem where P occurs as a hypothesis. A later reference to Γ then actually means $\Gamma - \{P\}$, *i.e.*, had P already been among the elements of Γ, it would now be removed.

Some rules, noted by use of the asterisk in Figure 1, have restrictions on their use or require special comment:

- $\forall$-intro. The rule application fails if x occurs free in Γ.
- $\exists$-intro. The rule application fails if N does not occur free in P. Moreover, only *some* designated occurrences of N need be replaced by x. The details of how occurrences are designated vary from implementation to implementation.
- $\exists$-elim. The rule application fails if the variable v occurs free in $\Gamma \cup \Delta \cup \{P, Q\}$.
- Abs. The rule application fails if v occurs free in Γ.
- tyInst. A substitution θ mapping type variables to types is applied to each hypothesis, and also to the conclusion.

An important feature of the HOL logic is $\varepsilon : (\alpha \to \text{bool}) \to \alpha$, Hilbert's *indefinite* description operator. A description term $\varepsilon x : \tau.\ P\ x$ is interpreted as follows: it delivers an arbitrary element e of type τ such that $P\ e$ holds. If there is no object that P holds of, then $\varepsilon x : \tau.\ P\ x$ denotes an arbitrary element of τ. This is summarized in the axiom $\vdash \forall P\ x.\ P\ x \supset P(\varepsilon x.\ P\ x)$.

Definition 16 (Arb). $\text{Arb} \equiv \varepsilon z : \alpha.\mathsf{T}$

The definition of Arb uses the Hilbert choice operator to denote an arbitrary but fixed value, for each type τ. Arb is fixed because T has no free variables; it is arbitrary because $\lambda v.\mathsf{T}$ holds for every element of τ. [4]

---

[4] F and T are the two constants of type bool denoting truth values in HOL.

⊃-intro	$\dfrac{\Gamma \vdash Q}{\Gamma - \{P\} \vdash P \supset Q}$	$\dfrac{\Gamma \vdash P \supset Q \quad \Delta \vdash P}{\Gamma \cup \Delta \vdash Q}$ ⊃-elim
∧-intro	$\dfrac{\Gamma \vdash P \quad \Delta \vdash Q}{\Gamma \cup \Delta \vdash P \wedge Q}$	$\dfrac{\Gamma \vdash P \wedge Q}{\Gamma \vdash P \quad \Gamma \vdash Q}$ ∧-elim
∨-intro	$\dfrac{\Gamma \vdash P}{\Gamma \vdash P \vee Q, \ \Gamma \vdash Q \vee P}$	$\dfrac{\Gamma_1 \vdash P \vee Q \qquad}{\Gamma_2, P \vdash M \quad \Gamma_3, Q \vdash M}{\Gamma_1 \cup \Gamma_2 \cup \Gamma_3 \vdash M}$ ∨-elim

⊃-intro $\dfrac{\Gamma \vdash Q}{\Gamma - \{P\} \vdash P \supset Q}$ $\dfrac{\Gamma \vdash P \supset Q \quad \Delta \vdash P}{\Gamma \cup \Delta \vdash Q}$ ⊃-elim

∧-intro $\dfrac{\Gamma \vdash P \quad \Delta \vdash Q}{\Gamma \cup \Delta \vdash P \wedge Q}$ $\dfrac{\Gamma \vdash P \wedge Q}{\Gamma \vdash P \quad \Gamma \vdash Q}$ ∧-elim

∨-intro $\dfrac{\Gamma \vdash P}{\Gamma \vdash P \vee Q, \ \Gamma \vdash Q \vee P}$ $\dfrac{\Gamma_1 \vdash P \vee Q \quad \Gamma_2, P \vdash M \quad \Gamma_3, Q \vdash M}{\Gamma_1 \cup \Gamma_2 \cup \Gamma_3 \vdash M}$ ∨-elim

∀-intro* $\dfrac{\Gamma \vdash P}{\Gamma \vdash \forall x.\ P}$ $\dfrac{\Gamma \vdash \forall x.\ P}{\Gamma \vdash [x \mapsto N]P}$ ∀-elim

∃-intro* $\dfrac{\Gamma \vdash P}{\Gamma \vdash \exists x.\ [N \mapsto x]P}$ $\dfrac{\Gamma \vdash \exists x.\ P \quad \Delta, [x \mapsto v]P \vdash Q}{\Gamma \cup \Delta \vdash Q}$ ∃-elim*

Assume $P \vdash P$ $\vdash M = M$ Refl

Sym $\dfrac{\Gamma \vdash M = N}{\Gamma \vdash N = M}$ $\dfrac{\Gamma \vdash M = N, \ \Delta \vdash N = P}{\Gamma \cup \Delta \vdash M = P}$ Trans

Comb $\dfrac{\Gamma \vdash M = N, \ \Delta \vdash P = Q}{\Gamma \cup \Delta \vdash M\ P = N\ Q}$ $\dfrac{\Gamma \vdash M = N}{\Gamma \vdash (\lambda v.M) = (\lambda v.N)}$ Abs*

tyInst* $\dfrac{\Gamma \vdash M}{\theta(\Gamma) \vdash \theta(M)}$ $\vdash (\lambda v.M)N = [v \mapsto N]M$ β-conv

---

Bool $\vdash P \vee \neg P$

Eta $\vdash (\lambda v.\ M\ v) = M$

Select $\vdash P\ x \supset P(\varepsilon x.\ P\ x)$

Infinity $\vdash \exists f : \text{ind} \to \text{ind}.\ (\forall x\ y.\ (f\ x = f\ y) \supset (x = y)) \wedge \exists y.\forall x.\ \neg(y = f\ x)$

Fig. 1. HOL deductive system

One of the most influential methodological developments in verification has been the adoption of *principles of definition* as logical prophylaxis, and implementations of HOL therefore tend to eschew the assertion of axioms.

Definition 17 (Principle of Constant Definition). *Given terms* $x : \tau$ *and* $M : \tau$ *in signature* Σ_Ω, *check that*

1. x *is a variable and the name of* x *is not the name of a constant in* Σ_Ω;
2. τ *is a type in* Σ_Ω;
3. M *is a term in* Σ_Ω *with no free variables; and*

4. Every type variable occurring in M occurs in τ.

If all these checks are passed, add a constant x : τ *to* Σ_Ω *and introduce an axiom* ⊢ x = M. □

Thus invocation of the Principle of definition, for suitable c and M, introduces c as an abbreviation for M. It is shown in [13] to be a sound means of extending the HOL logic. The notation c ≡ M is often used to show that a definition is being made. Derived definition principles, such as the one presented in this paper, reduce via deduction to application of the primitive Principle.

Abstract Congruence Closure and Specializations*

Leo Bachmair and Ashish Tiwari

Department of Computer Science
State University of New York
Stony Brook, NY 11794-4400, U.S.A
{leo,astiwari}@cs.sunysb.edu

Abstract. We use the uniform framework of *abstract congruence closure* to study the congruence closure algorithms described by Nelson and Oppen [9], Downey, Sethi and Tarjan [7] and Shostak [11]. The descriptions thus obtained abstract from certain implementation details while still allowing for comparison between these different algorithms. Experimental results are presented to illustrate the relative efficiency and explain differences in performance of these three algorithms. The transition rules for computation of abstract congruence closure are obtained from rules for *standard completion* enhanced with an *extension* rule that enlarges a given signature by new constants.

1 Introduction

Algorithms to compute "congruence closure" have typically been described in terms of directed acyclic graphs (dags) representing a set of terms, and a union-find data structure storing an equivalence relation on the vertices of this graph. In this paper, we abstractly describe some of these algorithms while still maintaining the "sharing" and "efficiency" offered by the data structures. This is achieved through the concept of an *abstract congruence closure*, c.f. [2, 3].

A key idea of abstract congruence closure is the use of new constants as names for subterms which yields a concise and simplified term representation. Consequently, complicated term orderings are no longer necessary or even applicable. There usually is a trade-off between the simplicity of terms thus obtained and the loss of term structure. In this paper, we get a middle ground where we keep the term structure as much as possible while still using extensions to obtain a simplified term representation. The paper also illustrates the use of an extended signature as a formalism to model and subsequently reason about data structures like the term dags, which are based on the idea of structure sharing.

In Section 2 we review the description of abstract congruence closure as a set of transition rules [2, 3]. The transition rules are derived from standard completion [1] enhanced with extension and suitably modified for the ground

---

* The research described in this paper was supported in part by the National Science Foundation under grant CCR-9902031.

D. McAllester (Ed.): CADE-17, LNAI 1831, pp. 64–78, 2000.
© Springer-Verlag Berlin Heidelberg 2000

case. Taking such an abstract view allows for a better understanding of the various graph-based congruence closure algorithms (Section 3), and also suggests new efficient procedures for constructing congruence closures (Section 4).

Preliminaries

Given a set $\Sigma = \cup_n \Sigma_n$ of function symbols and constants–called a *signature*–the set of (ground) terms $\mathcal{T}(\Sigma)$ over Σ is the smallest set containing Σ_0 and such that $f(t_1, \ldots, t_n) \in \mathcal{T}(\Sigma)$ whenever $f \in \Sigma_n$ and $t_i \in \mathcal{T}(\Sigma)$. The index n of the set Σ_n to which a function symbol f belongs is called the *arity* of the symbol f. Elements of arity 0 are called *constants*. A symbol $f \in \Sigma_k$ of arity k is also said to be a k-*ary* function symbol. The symbols $s, t, u, \ldots$ are used to denote terms in $\mathcal{T}(\Sigma)$; $f, g, \ldots$, function symbols. We write $t[s]$ to indicate that a term t contains s as a subterm and (ambiguously) denote by $t[u]$ the result of replacing a particular occurrence of s by u. A subterm of a term t is called *proper* if it is distinct from t.

An *equation* is a pair of terms, written as $s \approx t$. The *replacement* or *single-step rewrite relation*[1] $\to_E$ induced by a set of ground (or variable-free) equations E is defined by: $u[l] \to_E u[r]$ if, and only if, $l \approx r$ is in E. If $\to$ is a binary relation, then $\leftarrow$ denotes its inverse, $\leftrightarrow$ its symmetric closure, $\to^+$ its transitive closure and $\to^*$ its reflexive-transitive closure. Thus, $\leftrightarrow_E^*$ denotes the *congruence relation*[2], which is the same as the *equational theory* when E is ground, induced by a set E of ground equations. Equations are often called *rewrite rules*, and a set E a *rewrite system*, if one is interested particularly in the *rewrite relation* $\to_E^*$ rather than the equational theory $\leftrightarrow_E^*$.

If E is a set of equations, we write $E[s]$ to denote that the term s occurs as a subterm of some equation in E, and (ambiguously) use $E[t]$ to denote the set of equations obtained by replacing an occurrence of s in E by t.

A term t is in *normal form* with respect to a rewrite system R if there is no term t' such that $t \to_R t'$. We write $s \to_R^! t$ to indicate that t is a R-normal form of s. A rewrite system R is said to be (ground) *confluent* if every (ground) terms t has at most one normal form, i.e., if there exist s, s' such that $s \leftarrow_R^* t \to_R^* s'$, then, $s \to_R^* \circ \leftarrow_R^* s'$. It is *terminating* if there exists no infinite sequence $s_0 \to_R s_1 \to_R s_2 \cdots$ of terms. Rewrite systems that are (ground) confluent and terminating are called (ground) *convergent*.

2   Abstract Congruence Closure

We first review the concept of an abstract congruence closure [2, 3]. Let Σ be a signature and K be a set of constants disjoint from Σ. A D-*rule* (with respect to Σ and K) is a rewrite rule of the form $t \to c$ where t is a term from the set

---

[1] There is no difference between the replacement relation and the rewrite relation in the ground case.

[2] A congruence relation is a reflexive, symmetric and transitive relation on terms that is also a replacement relation.

$\mathcal{T}(\Sigma \cup K) - K$ and c is a constant in K[3]. A *C-rule* (with respect to K) is a rule $c \to d$, where c and d are constants in K. For example, if $\Sigma_0 = \{a, b, f\}$, and $E_0 = \{a \approx b, ffa \approx fb\}$ then $D_0 = \{a \to c_0, \ b \to c_1, \ ffa \to c_2, \ fb \to c_3\}$ is a set of D-rules over Σ_0 and $K_0 = \{c_0, c_1, c_2, c_3\}$. Original equations in E_0 can now be simplified using D_0 to give $C_0 = \{c_0 \approx c_1, c_2 \approx c_3\}$. The set $D_0 \cup C_0$ may be viewed as an alternative representation of E_0 over an extended signature. The equational theory presented by $D_0 \cup C_0$ is a conservative extension of the theory E_0. This reformulation of the equations E_0 in terms of an extended signature is (implicitly) present in all congruence closure algorithms, see Section 3.

A constant c in K is said to *represent* a term t in $\mathcal{T}(\Sigma \cup K)$ (via the rewrite system R) if $t \leftrightarrow^*_R c$. A term t is *represented* by R if it is represented by some constant in K via R. For example, the constant c_2 represents the term ffa via D_0.

Definition 1. *Let Σ be a signature and K be a set of constants disjoint from Σ. A ground rewrite system $R = D \cup C$ of D-rules and C-rules over the signature $\Sigma \cup K$ is said to be an* (abstract) congruence closure *(with respect to Σ and K) if (i) each constant $c \in K$ that is in normal form with respect to R, represents some term $t \in \mathcal{T}(\Sigma)$ via R, and (ii) R is ground convergent.*

If E is a set of ground equations over $\mathcal{T}(\Sigma \cup K)$ and in addition R is such that (iii) for all terms s and t in $\mathcal{T}(\Sigma)$, $s \leftrightarrow^*_E t$ if, and only if, $s \to^*_R \circ \leftarrow^*_R t$, then R will be called an (abstract) congruence closure *for E.*

Condition (i) essentially states that no superfluous constants are introduced; condition (ii) ensures that equivalent terms have the same representative; and condition (iii) implies that R is a conservative extension of the equational theory induced by E over $\mathcal{T}(\Sigma)$.

The rewrite system $R_0 = D_0 \cup \{c_0 \to c_1, \ c_2 \to c_3\}$ above is not a congruence closure for E_0, as it is not ground convergent. But we can transform R_0 into a suitable rewrite system, using a completion-like process described in more detail below, to obtain a congruence closure

$$R_1 = \{a \to c_1, \ b \to c_1, \ fc_1 \to c_3, \ fc_3 \to c_3, \ c_0 \to c_1, \ c_2 \to c_3\}.$$

Construction of Congruence Closures

We next present a general method for construction of congruence closures. Our description is fairly abstract, in terms of transition rules that manipulate triples (K, E, R), where K is the set of constants that extend the original fixed signature Σ, E is the set of ground equations (over $\Sigma \cup K$) yet to be processed, and R is the set of C-rules and D-rules that have been derived so far. Triples represent *states* in the process of constructing a congruence closure. Construction starts from *initial state* $(\emptyset, E, \emptyset)$, where E is a given set of ground equations.

---

[3] The definition of a D-rule is more general than the definition presented in [2, 3] as it allows for arbitrary non-constant terms on the left-hand side.

The transition rules can be derived from those for standard completion as described in [1], with some differences. In particular, (i) application of the transition rules is guaranteed to terminate, and (ii) a convergent system is constructed over an extended signature. The transition rules do *not* require any reduction ordering[4] on terms in $\mathcal{T}(\Sigma)$, but only only a simple ordering $\succ$ on terms in $\mathcal{T}(\Sigma \cup U)$[5] where U is an infinite set of constants from which new constants $K \subset U$ are chosen. In particular, if we assume $\succ_U$ is any ordering on the set U, then $\succ$ is defined as: $c \succ d$ if $c \succ_U d$ and $t \succ c$ if $t \rightarrow c$ is a D-rule. In this paper, the set $U = \{c_0, c_1, c_2, \ldots\}$, and we will assume $c_i \succ_U c_j$ iff $i < j$.

A key transition rule introduces new constants as names for subterms.

Extension:
$$\frac{(K, E[t], R)}{(K \cup \{c\}, E[c], R \cup \{t \rightarrow c\})}$$

where $t \rightarrow c$ is a D-rule, t is a term occurring in (some equation in) E, and $c \notin \Sigma \cup K$.

Following three rules are identical to the corresponding rules for standard completion.

Simplification:
$$\frac{(K, E[t], R \cup \{t \rightarrow c\})}{(K, E[c], R \cup \{t \rightarrow c\})}$$

where t occurs in some equation in E.

It is fairly easy to see that by repeated application of extension and simplification, any equation in E can be reduced to an equation that can be oriented by the ordering $\succ$.

Orientation:
$$\frac{(K \cup \{c\}, E \cup \{t \approx c\}, R)}{(K \cup \{c\}, E, R \cup \{t \rightarrow c\})}$$

if $t \succ c$.

Trivial equations may be deleted.

Deletion:
$$\frac{(K, E \cup \{t \approx t\}, R)}{(K, E, R)}$$

In the case of completion of ground equations, deduction steps can all be replaced by suitable simplification steps. In particular, most of the deduction steps can be described by collapse, and hence, the deduction rule considers only simple forms of overlap.

Deduction:
$$\frac{(K, E, R \cup \{t \rightarrow c, \ t \rightarrow d\})}{(K, E \cup \{c \approx d\}, R \cup \{t \rightarrow d\})}$$

---

[4] An *ordering* is any irreflexive and transitive relation on terms. A *reduction ordering* is an ordering that is also a well-founded replacement relation.

[5] Terms in $\mathcal{T}(\Sigma)$ are uncomparable by $\succ$.

In our case the usual side condition in the collapse rule, which refers to the *encompassment ordering*, can easily be stated in terms of the subterm relation.

$$\text{Collapse:} \quad \frac{(K, E, R \cup \{s[t] \to d,\ t \to c\})}{(K, E, R \cup \{s[c] \to d,\ t \to c\})}$$

if t is a proper subterm of s.

As in standard completion the simplification of right-hand sides of rules in R by other rules is optional and not necessary for correctness. The right-hand side term in any rule in R is always a constant.

$$\text{Composition:} \quad \frac{(K, E, R \cup \{t \to c,\ c \to d\})}{(K, E, R \cup \{t \to d,\ c \to d\})}$$

We use the symbol $\vdash$ to denote the one-step transition relation on states induced by the above transition rules. A *derivation* is a sequence of states $(K_0, E_0, R_0) \vdash (K_1, E_1, R_1) \vdash \cdots$.

Example 1. Consider the set of equations $E_0 = \{a \approx b,\ ffa \approx fb\}$. An abstract congruence closure for E_0 can be derived from $(K_0, E_0, R_0) = (\emptyset, E_0, \emptyset)$ as follows:

i	Constants K_i	Equations E_i	Rules R_i	Transition Rule
0	$\emptyset$	E_0	$\emptyset$	
1	$\{c_0\}$	$\{c_0 \approx b, ffa \approx fb\}$	$\{a \to c_0\}$	Ext
2	$\{c_0\}$	$\{ffa \approx fb\}$	$\{a \to c_0, b \to c_0\}$	Ori
3	$\{c_0\}$	$\{ffc_0 \approx fc_0\}$	$\{a \to c_0, b \to c_0\}$	Sim^2
4	$\{c_0, c_1\}$	$\{fc_1 \approx fc_0\}$	$R_3 \cup \{fc_0 \to c_1\}$	Ext
5	$\{c_0, c_1\}$	$\{fc_1 \approx c_1\}$	$R_3 \cup \{fc_0 \to c_1\}$	Sim
6	K_5	$\{\}$	$R_5 \cup \{fc_1 \to c_1\}$	Ori

The rewrite system R_6 is the required congruence closure.

The correctness of the transition rules presented here can be established in a way similar to the correctness of the transition rules for computing a congruence closure modulo associativity and commutativity [3]. The differences arise from the more general definition of D-rules, and the lack of any associative and commutative functions here.

The set of transition rules presented above are sound in the following sense: if $(K_0, E_0, R_0) \vdash (K_1, E_1, R_1)$, then, for all terms s and t in $\mathcal{T}(\Sigma \cup K_0)$, $s \leftrightarrow^*_{E_1 \cup R_1} t$ if and only if $s \leftrightarrow^*_{E_0 \cup R_0} t$. Additionally, let K_0 be a *finite* set of constants (disjoint from Σ), E_0 be a *finite* set of equations (over $\Sigma \cup K_0$), and R_0 be a *finite* set of D-rules and C-rules such that for every C-rule $c \to d \in R_0$, we have $c \succ_U d$. Then, any derivation starting from (K_0, E_0, R_0) is finite. If $(K_0, E_0, R_0) \vdash^* (K_m, E_m, R_m)$, then R_m is terminating. We call a state (K, E, R) *final* if no transition rule (except possibly composition) is applicable.

Theorem 1. *Let Σ be a signature and K_1 a finite set of constants disjoint from Σ. Let E_1 be a finite set of equations over $\Sigma \cup K_1$ and R_1 a finite set of D-rules and C-rules such that for every $c \in K_1$ represents some term $t \in \mathcal{T}(\Sigma)$ via $E_1 \cup R_1$, and $c \succ_U d$ for every C-rule $c \rightarrow d$ in R_1. If (K_n, E_n, R_n) is a final state such that $(K_1, E_1, R_1) \vdash^* (K_n, E_n, R_n)$, then $E_n = \emptyset$ and R_n is an abstract congruence closure for $E_1 \cup R_1$ (over Σ and K_1).*

3 Congruence Closure Strategies

The literature abounds with various implementations of congruence closure algorithms. We next describe the algorithms in [7], [9] and [11] as specific variants of our general abstract description. That is, we provide a description of these algorithms (modulo some implementation details) using abstract congruence closure transition rules.

Term directed acyclic graphs (dags) is a common data structure used to implement algorithms that work with terms over some signature—such as the congruence closure algorithm. In fact, many algorithms that have been described for congruence closure assume that the input is an equivalence relation on vertices of a given dag, and the desired output is an equivalence on the same dag that is defined by the congruence relation.

Figure 1 illustrate how a given term dag is (abstractly) represented using D-rules. The solid lines represent *subterm* edges, and the dashed lines represent a binary relation on the vertices. We have a D-rule corresponding to each vertex, and a C-rule for each dashed edge. Note that the D-rules corresponding to a conventional term dag representation are all of a special form $f(c_1, \ldots, c_k) \rightarrow c$, where $f \in \Sigma$ is a k-ary function symbol, and $c_1, \ldots, c_k, c$ are all new constants. Such rules will be called *simple* D-rules. The definition of D-rules given in Section 2 is more general, and allows for arbitrary terms on the left-hand sides. In a sense this corresponds to storing *contexts*, rather than just symbols from Σ, in each node (of the term dag). This is an attempt to keep as much of the term structure information as possible and still get advantages offered by a simplified term representation via extensions.

We need to specify a U set and an ordering $\succ_U$ on this set. Since elements of U serve only as names, we can choose U to be any countable set of symbols. An ordering $\succ_U$ need not be specified a-priori but can be defined on-the-fly as the derivation proceed. (The ordering has to be extended so that the irreflexivity and transitivity properties are preserved).

Traditional congruence closure algorithms also employ other data structures such as the following:

(i) Input dag: Starting from the state $(\emptyset, E_0, \emptyset)$, if we apply extension and simplification using strategy $(\mathbf{Ext} \circ \mathbf{Sim}^*)^*$ and making sure we create only simple D-rules, we finally get to a state (K_1, E_1, D_1) where all equations in E_1 are of the form $c \approx d$, for $c, d \in K_1$. The set D_1, then, represents the input dag and E_1 represents the (input) equivalence on the vertices of this dag. Note that due to eager simplification, we obtain representation of a dag with maximum possible

D-rules representing the term dag:

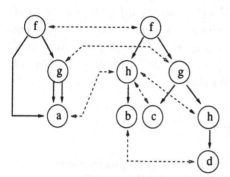

$$a \to c_1 \quad gc_1c_1 \to c_2 \quad fc_1c_2 \to c_3$$
$$b \to c_4 \quad hc_4 \to c_5 \quad c \to c_6$$
$$d \to c_7 \quad hc_7 \to c_8 \quad gc_6c_8 \to c_9$$
$$fc_5c_9 \to c_{10}$$

C-rules representing the relation on vertices:

$$c_1 \approx c_5 \quad c_2 \approx c_9 \quad c_3 \approx c_{10}$$
$$c_4 \approx c_7 \quad c_6 \approx c_5 \quad c_5 \approx c_8$$

Fig. 1. A term dag and a relation on its vertices

sharing. For example, if $E_0 = \{a \approx b, ffa \approx fb\}$, then $K_1 = \{c_0, c_1, c_2, c_3, c_4\}$, $E_1 = \{c_0 \approx c_1, c_3 \approx c_4\}$ and $R_1 = \{a \to c_0, b \to c_1, fc_0 \to c_2, fc_2 \to c_3, fc_1 \to c_4\}$.

(ii) Signature table: The *signature* table (indexed by vertices of the input dag) stores a *signature*[6] for some or all vertices. Clearly, the signatures are fully left-reduced D-rules.

(iii) Use table: The *use* table (also called predecessor list) is a mapping from the constant c to the set of all vertices whose signature contains c. This translates, in our presentation, to a method of indexing the set of D-rules.

(iv) Union Find: The union-find data structure that maintains equivalence classes on the set of vertices is represented by the set of C rules. If we apply orientation and simplification to the state (K_1, E_1, D_1) described above, using the strategy $(\mathbf{Ori} \circ \mathbf{Sim}^*)^*$, we obtain a state $(K_1, \emptyset, D_1 \cup C_1)$. The set C_1 is a representation of the Union-Find structure capturing the input equivalence on vertices. Continuing with the same example, C_1 would be the set $\{c_0 \to c_1, c_3 \to c_4\}$.

We note that, D-rules serve a two-fold purpose: they represent the input term dag, and also a signature table. We shall also note that Composition is used only implicitly in the various algorithms via path-compression on the union-find structure.

Shostak's Method

Shostak's congruence closure procedure was first described using *simple D-rules* and *C-rules* by Kapur [8]. We show here that Shostak's congruence closure procedure is a specific strategy over the general transition rules for abstract congruence closure presented here.

Shostak's congruence closure is *dynamic*: it can accept new equations after it has processed some equations, and can incrementally take care of the new

---

[6] The signature of a term $f(t_1, \ldots, t_k)$ is defined as $f(c_1, \ldots, c_k)$ where c_i is the name of the equivalence class containing term t_i.

equation. Its input state is $(\emptyset, E_0, \emptyset)$. Shostaks procedure can be described (at a fairly abstract level) as:

$$\textbf{Shos} = ((\textbf{Sim}^* \circ \textbf{Ext}^*)^* \circ (\textbf{Del} \cup \textbf{Ori}) \circ (\textbf{Col} \circ \textbf{Ded}^*)^*)^*$$

which is implemented as (i) pick an equation $s \approx t$ from the E-component, (ii) use simplification to normalize the term s to a term s' (iii) use extension to create *simple* D-rules for subterms of s' until s' reduces to a constant, say c, whence extension is no longer applicable. Perform steps (ii) and (iii) on the other term t as well to get a constant d. (iv) if c and d are identical then apply deletion (and continue with (i)), and if not, create a C-rule using orientation. (v) Once we have a new C-rule, perform all possible collapse step by this new rule, where each collapse step is followed by all the resulting deduction steps arising out of that collapse. The whole process is now repeated starting from step (i).

Shostak's procedure uses indexing based on the idea of the *use()* list. This *use()* based indexing is used to identify all possible collapse applications.

If the E-component of the state is empty while attempting to apply step (i), Shostak's procedure halts. It is fairly easy to observe that Shostak's procedure halts in a *final* state. Hence, Theorem 1 establishes that the R-component of Shostak's halting state contains a convergent system and is an abstract congruence closure.

Example 2. We use the set E_0 used in Example 1 of Section 2 to illustrate Shostak's method. We show some of the important intermediate steps of a Shostak derivation.

i	Constants K_i	Equations E_i	Rules R_i	Transition
0	$\emptyset$	E_0	$\emptyset$	
1	$\{c_0, c_1\}$	$\{ffa \approx fb\}$	$\{a \to c_0,\ b \to c_1,\ c_0 \to c_1\}$	$\textbf{Ext}^2 \circ \textbf{Ori}$
2	$\{c_0, c_1\}$	$\{ffc_1 \approx fb\}$	$\{a \to c_0,\ b \to c_1,\ c_0 \to c_1\}$	$\textbf{Sim}$
3	$\{c_0, \ldots, c_3\}$	$\{c_3 \approx fb\}$	$R_2 \cup \{fc_1 \to c_2,\ fc_2 \to c_3\}$	$\textbf{Ext}^2$
4	$\{c_0, \ldots, c_3\}$	$\{c_3 \approx c_2\}$	R_3	$\textbf{Sim}^2$
5	$\{c_0, \ldots, c_3\}$	$\emptyset$	$R_4 \cup \{c_3 \to c_2\}$	$\textbf{Ori}$

The Downey–Sethi–Tarjan Algorithm

The Downey, Sethi and Tarjan [7] procedures assumes that the input is a dag and an equivalence relation on its vertices, which, in our language, means that the starting state for this procedures is $(K_1, \emptyset, D_1 \cup C_1)$, where D_1 represents the input dag and C_1 represents the initial equivalence. It can be succinctly abstracted as:

$$\textbf{DST} = ((\textbf{Col} \circ (\textbf{Ded} \cup \{\epsilon\}))^* \circ (\textbf{Sim}^* \circ (\textbf{Del} \cup \textbf{Ori}))^*)^*$$

where ϵ is the null transition rule. This strategy is implemented as follows (i) if any collapse rule is applicable, it is applied and if, as a result any new deduction step is possible, it is done. This is repeated until no more collapse steps are possible. (ii) if no collapse steps are possible, then each C-equation in the E-component is picked up sequentially, fully-simplified (simplification) and then either deleted (deletion) or oriented (orientation).

Although the above description captures the essence of the Downey, Sethi and Tarjan procedure, a few implementation details need to be pointed out. Firstly, the Downey, Sethi and Tarjan procedure keeps the original dag (represented by D_1) intact[7], but changes signatures in a signature table. Hence, in the actual implementation described in [7], the $(\mathbf{Col} \circ (\mathbf{Ded} \cup \{\epsilon\}))^*$ strategy is applied by: (i) deleting all signatures that will be changed, i.e., deleting all D-rules which can be collapsed; (ii) computing new signatures using the *original* copy of the signatures stored in the form of the dag D_1; and, finally, (iii) inserting the newly computed signatures into the signature table and checking for possible deduction steps. Our description achieves the same end result, but, by doing fewer inferences.

Secondly, in the Downey, Sethi and Tarjan procedure, for efficiency, an equation $c \approx d$ is oriented to $c \to d$ if the c occurs fewer times than d in the signature table. This is done to minimize the number of collapse steps. Additionally, indexing based on the *use*() tables is used for efficiently implementing the specific strategy.

Let $(K_1, \emptyset, D_1 \cup C_1) \vdash^! (K_n, E_n, D_n \cup C_n)$ be a derivation using the DST strategy. Then, it is easily seen that the state $(K_n, E_n, D_n \cup C_n)$ is a final state, and hence the set $D_n \cup C_n$ is convergent, and also an abstract congruence closure. We remark here that D_n holds the information that is contained in the signature table, and C_n holds information in the union-find structure. The set C_n is usually considered the output of the Downey, Sethi and Tarjan procedure.

Example 3. We illustrate the Downey-Sethi-Tarjan algorithm by using the same set of equations E_0, used in Example 1 of Section 2. The start state is $(K_1, \emptyset, D_1 \cup C_1)$ where $K = \{c_0, \ldots, c_4\}$, $D_1 = \{a \to c_0,\ b \to c_1,\ fc_0 \to c_2,\ fc_2 \to c_3,\ fc_1 \to c_4\}$, and, $C_1 = \{c_0 \to c_1,\ c_3 \to c_4\}$.

i	Consts K_i	Eqns E_i	Rules R_i	Transition
1	K_1	$\emptyset$	$D_1 \cup C_1$	
2	K_1	$\emptyset$	$\{a \to c_0,\ b \to c_1,\ fc_1 \to c_2,$ $fc_2 \to c_3,\ fc_1 \to c_4\} \cup C_1$	Col
3	K_1	$\{c_2 \approx c_4\}$	R_2	Ded
4	K_1	$\emptyset$	$R_3 - \{fc_1 \to c_2\} \cup \{c_4 \to c_2\}$	Ori

Note that $c_4 \approx c_2$ was oriented in a way that no further collapses were needed thereafter.

The Nelson–Oppen Procedure

The Nelson-Oppen procedure is not exactly a completion procedure and it does *not* generate a congruence closure in our sense. The initial state of the Nelson-

---

[7] We could make a copy of the original D_1 rules and not change them, while keeping a separate copy as the signatures.

Oppen procedure is given by the tuple (K_1, E_1, D_1), where D_1 is the input dag, and E_1 represents an equivalence on vertices of this dag. The sets K_1 and D_1 remain unchanged in the Nelson-Oppen procedure. In particular, the inference rule used for deduction is different from the conventional deduction rule[8].

$$\text{NODeduction:} \quad \frac{(K, E, D \cup C)}{(K, E \cup \{c \approx d\}, D \cup C)}$$

if there exist two D-rules $f(c_1, \ldots, c_k) \to c$, and, $f(d_1, \ldots, d_k) \to d$ in the set D; and, $c_i \to'_C \circ \leftarrow'_C d_i$, for $i = 1, \ldots, k$.

The Nelson-Oppen procedure can now be (at a certain abstract level) represented as:

$$\mathbf{NO} = (\mathbf{Sim}^* \circ (\mathbf{Ori} \cup \mathbf{Del}) \circ \mathbf{NODed}^*)^*$$

which is applied in the following sense: (i) select a C-equation $c \approx d$ from the E-component, (ii) simplify the terms c and d using simplification steps until the terms can't be simplified any more, (iii) either delete, or orient the simplified C-equation, (iv) apply the NODeduction rule until there are no more non-redundant applications of this rule, (v) if the E-component is empty, then we stop, otherwise continue with step (i).

Certain details like the fact that newly added equations to the set E are chosen before the old ones in an application of orientation and indexing based on the $use()$ table, are abstracted away in this description.

Using the Nelson-Oppen strategy, assume we get a derivation (K_1, E_1, D_1)-$\vdash^*_{\mathbf{NO}} (K_n, E_n, D_n \cup C_n)$. One consequence of using a non-standard deduction rule, NODeduction, is that the resulting set $D_n \cup C_n = D_1 \cup C_n$ need not necessarily be convergent, although the the rewrite relation D_n / C_n [6] is convergent.

Example 4. Using the same set E_0 as equations, we illustrate the Nelson-Oppen procedure. The initial state is given by (K_1, E_1, D_1) where $K_1 = \{c_0, c_1, c_2, c_3, c_4\}$; $E_1 = \{c_0 \approx c_1, c_3 \approx c_4\}$; and, $D_1 = \{a \to c_0, b \to c_1, fc_0 \to c_2, fc_2 \to c_3, fc_1 \to c_4\}$.

i	Constants K_i	Equations E_i	Rules R_i	Transition
1	K_1	E_1	D_1	
2	K_1	$\{c_3 \approx c_4\}$	$D_1 \cup \{c_0 \to c_1\}$	Ori
3	K_1	$\{c_2 \approx c_4, c_3 \approx c_4\}$	R_2	NODed
4	K_1	$\{c_3 \approx c_4\}$	$R_2 \cup \{c_2 \to c_4\}$	Ori
5	K_1	$\emptyset$	$R_4 \cup \{c_3 \to c_4\}$	Ori

Consider deciding the equality $fa \approx ffb$. Even though $fa \leftrightarrow^*_{E_0} ffb$, the terms fa and ffb have distinct normal forms with respect to R_5. But terms in the original term universe have identical normal forms.

---

[8] This rule performs deduction modulo C-equations, i.e., we compute critical pairs between D-rules modulo the congruence induced by C-equations. Hence, the Nelson-Oppen procedure can be described as an *extended completion* [6] (or completion modulo C-equations) method over an extended signature.

4   Experimental Results

We have implemented five congruence closure algorithms, including those proposed by Nelson and Oppen (NO) [9], Downey, Sethi and Tarjan (DST) [7], and Shostak [11], and two algorithms based on completion—one with an indexing mechanism (IND) and the other without (COM). The implementations of the first three procedures are based on the representation of terms by directed acyclic graphs and the representation of equivalence classes by a union-find data structure. The completion procedure COM uses the following strategy:

$$((\text{Sim}^* \circ \text{Ext}^*)^* \circ (\text{Del} \cup \text{Ori}) \circ (\text{Com} \circ \text{Col})^* \circ \text{Ded}^*)^*.$$

The indexed variant IND uses a slightly different strategy

$$((\text{Sim}^* \circ \text{Ext}^*)^* \circ (\text{Del} \cup \text{Ori}) \circ (\text{Col} \circ \text{Com} \circ \text{Ded})^*)^*.$$

Indexing in the case of completion refers to the use of suitable data structures to efficiently identify which D-rules contain specified constants.

In a first set of experiments, we assume that the input is a set of equations presented as pairs of trees (representing terms). We added a preprocessing step to the NO and DST algorithms to convert the given input terms into a dag and initialize the other required data-structures. The other three algorithms interleave construction of a dag with deduction steps. The published descriptions DST and NO do not address the construction of a dag. Our implementation maintains the list of terms that have been represented in the dag in a hash table and creates a new node for each term not yet represented. We present below a sample of our results to illustrate some of the differences between the various algorithms.

The input set of equations E can be classified based on: (i) the size of the input and the number of equations, (ii) the number of equivalence classes on terms and subterms of E, and, (iii) the size of the use lists. The first set of examples are relatively simple and developed by hand to highlight strengths and weaknesses of the various algorithms. Example (a)[9] contains five equations that induce a single equivalence class. Example (b) is the same as (a), except that it contains five copies of all the equations. Example (c)[10] requires slightly larger use lists. Finally, example (d)[11] consists of equations that are oriented in the "wrong" way.

In Table 1 we compare the different algorithms by their total running time, *including* the preprocessing time. The times shown are the averages of several runs on a Sun Ultra workstation under similar load conditions. The time was computed using the *gettimeofday* system call.

---

[9] The equation set is $\{f^2(a) \approx a, f^{10}(a) \approx f^{15}(b), b \approx f^5(b), a \approx f^3(a), f^5(b) \approx b\}$.

[10] The equation set is $\{g(a,a,b) \approx f(a,b),\ gabb \approx fba,\ gaab \approx gbaa,\ gbab \approx gabb,\ gbba \approx gbab,\ gaaa \approx faa,\ a \approx c,\ c \approx d,\ d \approx e,\ b \approx c1,\ c1 \approx d1,\ d1 \approx e1\}$.

[11] The set is $\{g(f^i(a), h^{10}(b)) \approx g(a,b), i = \{1, \cdots, 25\}, h^{47}(b) \approx b, b \approx h^{29}(b), h(b) \approx c0, c0 \approx c1, c1 \approx c2, c2 \approx c3, c3 \approx c4, c4 \approx a, a \approx f(a)\}$.

	Eqns	Vert	Class	DST	NO	SHO	COM	IND
Ex.a	5	27	1	1.286	1.640	0.281	0.606	0.409
Ex.b	20	27	1	2.912	2.772	0.794	1.858	0.901
Ex.c	12	20	6	1.255	0.733	0.515	0.325	0.323
Ex.d	34	105	2	10.556	22.488	7.275	12.077	4.416

Table 1. Total running time (in milliseconds) for Examples $(a) - -(d)$. *Eqns* refers to the number of equations; *Vert* to the number of vertices in the initial dag; and *Class* to the number of equivalence classes induced on the dag.

Table 2 contains similar comparisons for considerably larger examples consisting of randomly generated equations over a specified signature. Again we show total running time, including preprocessing time[12].

	Eqns	Vert	Σ_0	Σ_1	Σ_2	d	Class	DST	NO	SHO	IND
Ex.1	10000	17604	2	0	2	3	7472	11.087	3.187	10.206	13.037
Ex.2	5000	4163	2	1	1	3	3	2.276	306.194	3.092	0.774
Ex.3	5000	7869	3	0	1	3	2745	2.439	1.357	3.521	3.989
Ex.4	6000	8885	3	0	1	3	9	3.551	1152.652	52.353	7.069
Ex.5	7000	9818	3	0	1	3	1	4.633	1682.815	47.755	5.471
Ex.6	5000	645	4	2	0	23	77	1.224	1.580	0.371	0.363
Ex.7	5000	1438	10	2	0	23	290	1.452	3.670	0.392	0.374

Table 2. Total running time (in seconds) for randomly generated equations. The columns Σ_i denote the number of function symbols of arity i in the signature and d denotes the maximum term depth.

In Table 3 we show the time for computing a congruence closure assuming terms are already represented by a dag. In other words, we do not include the time it takes to create a dag. Note that we include no comparison with Shostak's method, as the dynamic construction of a dag from given term equations is inherent in this procedure. However, a comparison with a suitable strategy (in which all extension steps are applied before any deduction steps) of IND is possible. We denote by IND* indexed completion based on a strategy that first constructs a dag. The examples are the same as in Table 2.

Several observations can be drawn from these results. First, the Nelson-Oppen procedure NO is competitive only when few deduction steps are performed and thus the number of equivalence classes is large. This is because it uses a non-standard deduction rule, which forces the procedure to unnecessarily repeat the same deductions many times over in a single execution. Not surprisingly, straight-forward completion without indexing is also inefficient when

---

[12] Times for COM are not included as indexing is indispensable for larger examples.

	DST	NO	IND*
Ex.1	0.919	0.296	0.076
Ex.2	0.309	319.112	1.971
Ex.3	0.241	0.166	0.030
Ex.4	0.776	1117.239	7.301

	DST	NO	IND*
Ex.5	0.958	1614.961	9.770
Ex.6	0.026	0.781	0.060
Ex.7	0.048	2.470	0.176

Table 3. Running time (in seconds) when input is in a dag form.

many deduction steps are necessary. Indexing is of course a standard technique employed in all practical implementations of completion.

The running time of the DST procedure critically depends on the size of the hash table that contains the signatures of all vertices. If the hash table size is large, enough potential deductions can be detected in (almost) constant time. If the hash table size is reduced, to say 100, then the running time increased by a factor of up to 50. A hash table with 1000 entries was sufficient for our examples (which contained fewer than 10000 vertices). Larger tables did not improve the running times.

Indexed Completion, DST and Shostak's method are roughly comparable in performance, though Shostak's algorithm has some drawbacks. For instance, equations are always oriented from left to right. In contrast, Indexed Completion always orients equations in a way so as to minimize the number of applications of the collapse rule, an idea that is implicit in Downey, Sethi and Tarjan's algorithm. Example (b) illustrates this fact. More crucially, the manipulation of the *use* lists in Shostak's method is done in a convoluted manner due to which redundant inferences may be done when searching for the correct non-redundant ones[13]. As a consequence, Shostak's algorithm performs poorly on instances where *use* lists are large and deduction steps are many such as in Examples (c), 4 and 5.

Finally, we note that the indexing used in our implementation of completion is simple—with every constant c we associate a list of D-rules that contain c as a subterm. On the other hand DST maintains at least two different ways of indexing the signatures, which makes it more efficient when the examples are large and deduction steps are plenty. On small examples, the overhead to maintain the data structures dominates. This also suggests that the use of more sophisticated indexing schemes for indexed completion might improve its performance.

5   Related Work and Conclusion

Kapur [8] considered the problem of casting Shostak's congruence closure [11] algorithm in the framework of ground completion on rewrite rules. Our work has been motivated by the goal of formalizing not just one, but several congruence closure algorithms, so as to be able to better compare and analyze them.

---

[13] The description in Section 3 accurately reflects the logical aspects of Shostak's algorithm, but does not provide details on data structures like the *use* lists.

We suggest that, abstractly, congruence closure can be *defined* as a ground convergent system; and that this definition does not restrict the applicability of congruence closure. The rule-based abstract description of the logical aspects of the various published congruence closure algorithms leads to a better understanding of these methods. It explains the observed behaviour of implementations and also allows one to identify weaknesses in specific algorithms. Additionally, using the abstract rules, we can also get efficient implementation of completion based congruence closure procedure—one can effectively utilize the theory of redundancy to figure out and eliminate inferences which are not necessary, and moreover also use knowledge about efficient indexing mechanisms.

The concept of an abstract congruence closure is also relevant for describing applications that use congruence closure algorithms. Some of these applications include efficient normalization by rewrite systems [4, 2], computing a complete set of rigid E-unifiers [13], and combination of decision procedures [11]. The notion of an abstract congruence closure is naturally extended to handle presence of associative-commutative operators, and this application is described in [3]. We believe that theories other than associativity and commutativity can also be incorporated with the inference rules for abstract congruence closure.

Congruence closure has also been used to construct a convergent set of ground rewrite rules in polynomial time by Snyder [12] and other works. Plaisted et. al. [10] gave a *direct* method, not based on using congruence closure, for completing a ground rewrite system in polynomial time. Hence our work completes the missing link, by showing that congruence closure is nothing but ground completion. In fact, the process of transforming a set of rewrite rules over an extended signature (representing an abstract congruence closure) into a convergent set of rewrite rules over the original signature can be easily described by additional transition rules [3]. Our approach is different from that of Snyder, and can be used to obtain a *more efficient* implementation partly because Snyder's algorithm needs *two* passes of the congruence closure algorithm, whereas we would need to compute the abstract congruence closure just once.

The concept of an abstract congruence closure as detailed here and the rules for computation open up new frontiers too. For example, the transition rules presented in Section 2 can be naturally implemented in MAUDE [5]. Moreover, specific strategies, such as the ones presented in Section 3 can be encoded easily too. This might provide a basis for automatically verifying the correctness of congruence closure algorithms[14].

Acknowledgements

We would like to thank the anonymous reviewers for their helpful comments.

---

[14] Personal communication with Manuel Clavel.

References

[1] L. Bachmair and N. Dershowitz. Equational inference, canonical proofs, and proof orderings. *J. ACM*, 41:236–276, 1994.

[2] L. Bachmair, C. Ramakrishnan, I. Ramakrishnan, and A. Tiwari. Normalization via rewrite closures. In P. Narendran and M. Rusinowitch, editors, *10th Int. Conf. on Rewriting Techniques and Applications*, pages 190–204, 1999. LNCS 1631.

[3] L. Bachmair, I. Ramakrishnan, A. Tiwari, and L. Vigneron. Congruence closure modulo associativity and commutativity. In H. Kirchner and C. Ringeissen, editors, *Frontiers of Combining Systems, 3rd Intl Workshop FroCoS 2000*, pages 245–259, 2000. LNAI 1794.

[4] L. P. Chew. *Normal forms in term rewriting systems*. PhD thesis, Purdue University, 1981.

[5] M. Clavel et al. *Maude: Specification and Programming in Rewriting Logic*. http://maude.csl.sri.com/manual/, SRI International, Menlo Park, CA, 1999.

[6] N. Dershowitz and J. P. Jouannaud. Rewrite systems. In J. van Leeuwen, editor, *Handbook of Theoretical Computer Science (Vol. B: Formal Models and Semantics)*, Amsterdam, 1990. North-Holland.

[7] P. J. Downey, R. Sethi, and R. E. Tarjan. Variations on the common subexpressions problem. *J. ACM*, 27(4):758–771, 1980.

[8] D. Kapur. Shostak's congruence closure as completion. In H. Comon, editor, *Proceedings of the 8th International Conference on Rewriting Techniques and Applications*, pages 23–37, 1997. Vol. 1232 of *Lecture Notes in Computer Science*, Springer, Berlin.

[9] G. Nelson and D. Oppen. Fast decision procedures based on congruence closure. *Journal of the Association for Computing Machinery*, 27(2):356–364, Apr. 1980.

[10] D. Plaisted and A. Sattler-Klein. Proof lengths for equational completion. *Information and Computation*, 125:154–170, 1996.

[11] R. E. Shostak. Deciding combinations of theories. *Journal of the ACM*, 21(7):583–585, 1984.

[12] W. Snyder. A fast algorithm for generating reduced ground rewriting systems from a set of ground equations. *Journal of Symbolic Computation*, 15(7), 1993.

[13] A. Tiwari, L. Bachmair, and H. Ruess. Rigid E-unification revisited. In D. McAllester, editor, *17th Intl Conf on Automated Deduction, CADE-17*, 2000.

A Framework for Cooperating Decision Procedures

Clark W. Barrett, David L. Dill, and Aaron Stump

Stanford University, Stanford, CA 94305, USA,
http://verify.stanford.edu

Abstract. We present a flexible framework for cooperating decision procedures. We describe the properties needed to ensure correctness and show how it can be applied to implement an efficient version of Nelson and Oppen's algorithm for combining decision procedures. We also show how a Shostak style decision procedure can be implemented in the framework in such a way that it can be integrated with the Nelson–Oppen method.

1 Introduction

Decision procedures for fragments of first-order or higher-order logic are potentially of great interest because of their versatility. Many practical problems can be reduced to problems in some decidable theory. The availability of robust decision procedures that can solve these problem within reasonable time and memory could save a great deal of effort that would otherwise go into implementing special cases of these procedures.

Indeed, there are several publicly distributed prototype implementations of decision procedures, such as Presburger arithmetic [15], and decidable combinations of quantifier-free first-order theories [2]. These and similar procedures have been used as components in applications, including interactive theorem provers [13,9], infinite-state model checkers [7,10,4], symbolic simulators [18], software specification checkers [14], and static program analyzers [8].

Nelson and Oppen [12] showed that satisfiability procedures for several theories that satisfy certain conditions can be combined into a single satisfiability procedure by propagating equalities. Many others have built upon this work, offering new proofs and applications [19,1].

Shostak [17,6,16] gave an alternative method for combining decision procedures. His method is applicable to a more restricted set of theories, but is reported to be more efficient and is the basis for combination methods found in SVC [2], PVS [13], and STeP [9,3]. An understanding of his algorithm has proven to be elusive.

Both STeP and PVS have at least some ability to combine the methods of Nelson and Oppen and Shostak [5,3], but not much detail has been given, and the methods used in PVS have never been published. As a result, there is still significant confusion about the relationship between these two methods and how to implement them efficiently and correctly.

D. McAllester (Ed.): CADE-17, LNAI 1831, pp. 79–98, 2000.

Our experience with SVC, a decision procedure for quantifier-free first-order logic based loosely on Shostak's method for combining cooperating decision procedures, has been both positive and negative. On the one hand, it has been implemented and is efficient and reliable enough to enable new capabilities in our research group and at a surprisingly large number of other sites. However, efforts to extend and modify SVC have revealed unnecessary constraints in the underlying theory, as well as gaps in our understanding of it.

This paper is an outcome of ongoing attempts to re-architect SVC to resolve these difficulties. We present an architecture for cooperating decision procedures that is simple yet flexible and show how the soundness, completeness, and termination of the combined decision procedure can be proved from a small list of clearly stated assumptions about the constituent theories. As an example of the application of this framework, we show how it can be used to implement and integrate the methods of Nelson and Oppen and Shostak. In so doing, we also describe an optimization applicable to the original Nelson and Oppen procedure and show how our framework simplifies the proof of correctness of Shostak's method. Due to the scope of this paper and space restrictions, many of the proofs have been abbreviated or omitted.

2   Definitions and Notation

Expressions in the framework are represented using the logical symbols **true**, **false**, and '=', an arbitrary number of variables, and non-logical symbols consisting of constants, and function and predicate symbols. We call **true** and **false** *constant formulas*. An *atomic formula* is either a constant formula, an equality between terms, or a predicate applied to terms. A *literal* is either an atomic formula or an equality between a non-constant atomic formula and **false**. Equality with **false** is used to represent negation. Formulas include atomic formulas, and are closed under the application of equality, conjunction and quantifiers. An expression is either a term or a formula. An expression is a *leaf* if it is a variable or constant. Otherwise, it is a *compound* expression, containing an operator applied to one or more children.

A *theory* is a set of first-order sentences. For the purposes of this paper, we assume that all theories include the axioms of equality. The *signature* of a theory is the set of function, predicate, and constant symbols appearing in those sentences. The *language* of a signature Σ is the set of all expressions whose function, predicate, and constant symbols come from Σ. Given a theory T with signature Σ, if ϕ is a sentence in the language of Σ, then we write $T \models \phi$ to mean that every model of T is also a model of ϕ. For a given model, M, an *interpretation* is a function which assigns an element of the domain of M to each variable. If Γ is a set of formulas and ϕ is a formula, then we write $\Gamma \models \phi$ to mean that for every model and interpretation satisfying each formula in Γ, the same model and interpretation satisfy ϕ. Finally, if Φ is a set of formulas, then $\Gamma \models \Phi$ indicates that $\Gamma \models \phi$ for each ϕ in Φ.

Expressions are represented using a directed acyclic graph (DAG) data structure such that any two expressions which are syntactically identical are uniquely represented by a single DAG. The following operations on expressions are supported.

Op(e)   the operator of e (just e itself if e is a leaf).
e[i]   the ith child of e, where e[1] is the first child.

If e_1 and e_2 are expressions, then we write $e_1 \equiv e_2$ to indicate that e_1 and e_2 are the same expression (syntactically identical). In contrast, $e_1 = e_2$ is simply intended to represent the expression formed by applying the equality operator to e_1 and e_2. Expressions can be annotated with various attributes. If a is an attribute, e.a is the value of that attribute for expression e. Initially, e.a $= \perp$ for each e and a, where $\perp$ is a special undefined value.

The following simple operations make use of an expression attribute called find to maintain equivalence classes of expressions. We assume that these are the only functions that reference the attribute. Note that when presenting pseudocode here and below, some required preconditions may be given next to the name and parameters of the function.

HasFind(a)   SetFind(a) {a.find = $\perp$ }
    RETURN a.find $\neq \perp$;   a.find := a;

Find(a) {HasFind(a)}   Union(a,b) {a.find $\equiv$ a $\wedge$ b.find $\equiv$ b }
    IF (a.find $\equiv$ a) THEN RETURN a;   a.find := b.find;
    ELSE RETURN Find(a.find);

In some similar algorithms, e.find is initially set to e, rather than $\perp$. The reason we don't do this is that it turns out to be convenient to use an initialized find attribute as a marker that the expression has been seen before. This not only simplifies the algorithm, but it also makes it easier to describe certain invariants about expressions.

The find attribute induces a relation $\sim$ on expressions: a $\sim$ b if and only if HasFind(a) $\wedge$ HasFind(b) $\wedge$ [Find(a)$\equiv$Find(b)]. For the set of all expressions whose find attributes have been set, this relation is an equivalence relation. The *find database*, denoted by $\mathcal{F}$, is defined as follows: a $= $ b $\in \mathcal{F}$ iff a $\sim$ b. The following facts will be used below.

Find Database Monotonicity. *If the preconditions for* SetFind *and* Union *are met, then if* $\mathcal{F}$ *is the find database at some previous time and* $\mathcal{F}'$ *is the find database now, then* $\mathcal{F} \subseteq \mathcal{F}'$.

Find Lemma. *If the preconditions for* Find, SetFind, *and* Union *hold, then* Find *always terminates.*

3   The Basic Framework

As mentioned above, the purpose of the framework presented in this paper is to combine satisfiability procedures for several first-order theories into a satisfiability procedure for their union. Suppose that $T_1, \ldots, T_n$ are n first-order theories, with signatures $\Sigma_1, \ldots \Sigma_n$. Let $T = \bigcup T_i$ and $\Sigma = \bigcup \Sigma_i$. The goal is to provide a framework for a satisfiability procedure which determines the satisfiability in T of a set of formulas in the language of Σ. Our approach follows that of Nelson and Oppen [12]. We assume that the intersection of any two signatures is empty and that each theory is *stably-infinite*. A theory T with signature Σ is called stably-infinite if any quantifier-free formula in the language of Σ is satisfiable in T only if it is satisfiable in an infinite model of T. We also assume that the theories are *convex*. A theory is convex if there is no conjunction of literals in the language of the theory which implies a disjunction of equalities without implying one of the equalities itself.

The interface to the framework from a client program consists of three methods: AddFormula, Satisfiable, and Simplify. Conceptually, AddFormula adds its argument (which must be a literal) to a set $\mathcal{A}$, called the *assumption history*. Simplify transforms an expression into a new expression which is equivalent modulo $T \cup \mathcal{A}$, and Satisfiable returns false if and only if $T \cup \mathcal{A} \models \mathbf{false}$. Since any quantifier-free formula can be converted to disjunctive normal form, after which each conjunction of literals can be checked separately for satisfiability, the restriction that the arguments to AddFormula be literals does not restrict the power of framework.

The framework includes sets of functions which are parameterized by theory. For example, if f is such a function, we denote by f_i the instance of f associated with theory T_i. If for some f and T_i, we do not explicitly define the instance f_i, it is assumed that a call to f_i does nothing. It is convenient to be able to call these functions based on the theory associated with some expression e. Expressions are associated with theories as follows. First, variables are partitioned among the theories arbitrarily. In some cases, one choice may be better than another, as discussed in Sec. 5.1 below. An expression in the language of Σ is associated with theory T_i if and only if it is a variable associated with T_i, its operator is a symbol in Σ_i, or it is an equality and its left side is associated with theory T_i. If an expression is associated with theory T_i, we call it an *i-expression*. We denote by $T(e)$ the index i, where e is an *i-expression*.

Figure 1 shows pseudocode for the basic framework. An input formula is first simplified it because it might already be known or reduce to something easier to handle. Simplification involves the recursive application of Find as well as certain rewrite rules. Assert calls Merge which merges two $\sim$-equivalence classes. Merge first calls Setup which ensures that the expressions are in an equivalence class.

There are four places in the framework in which *theory-specific* functionality can be introduced. TheorySetup, TheoryRewrite and PropagateEqualities are theory-parameterized functions. Also, each expression has a notify attribute containing a set of pairs $\langle f, d \rangle$, where f is a function and d is some data. Whenever Merge is called on an expression a = b, the find attribute of a changes to b, and $f(a = b, d)$ is called for each $\langle f, d \rangle \in$ a.notify. Typically, TheorySetup adds callback functions to the notify attribute of various expressions to guarantee that the theory's satisfiability procedure will be notified if one of those

```
AddFormula(e) { e is a literal }
  Assert(e);
  REPEAT
    done := true;
    FOREACH theory Tᵢ DO IF PropagateEqualitiesᵢ() THEN done := false;
  UNTIL done;

Assert(e) { e is a literal; T ∪ A ⊨ e }
  IF ¬Satisfiable() THEN RETURN;
  e' := Simplify(e);
  IF e' ≡ true THEN RETURN;
  IF Op(e') ≠ '=' THEN e' := (e' = true);
  Merge(e');

Merge(e) { Op(e) = '='; T ∪ A ⊨ e; see text for others }
  Setup(e[1]); Setup(e[2]);
  IF e[1] and e[2] are terms THEN TheorySetup_{T(e)}(e);
  Union(e[1],e[2]);
  FOREACH ⟨f,d⟩ ∈ e[1].notify DO f(e,d);

Setup(e)
  IF HasFind(e) THEN RETURN;
  FOREACH child c of e DO Setup(c);
  TheorySetup_{T(e)}(e);
  SetFind(e);

Simplify(e)
  IF HasFind(e) THEN RETURN Find(e);
  Replace each child c of e with Simplify(c);
  RETURN Rewrite(e);

Rewrite(e)
  IF HasFind(e) THEN RETURN Find(e);
  IF Op(e) = '=' THEN e' := RewriteEquality(e);
  ELSE e' := TheoryRewrite_{T(e)}(e);
  IF e ≢ e' THEN e' := Rewrite(e');
  RETURN e';

RewriteEquality(e)
  IF e[1] ≡ e[2] THEN RETURN true;
  IF one child of e is true THEN RETURN the other child;
  IF e[1] ≡ false THEN RETURN (e[2] = e[1]);
  RETURN e;

Satisfiable()
  RETURN true ≁ false;
```

Fig. 1. Basic Framework

expressions is merged with another expression. Finally, before returning from AddFormula, each theory may notify the framework of additional equalities it has deduced until each theory reports that there are no more equalities to propagate.

Theory-specific code is distinguished from the framework code shown in Fig. 1 and from *user code* which is the rest of the program. It may call functions in the framework, provided any required preconditions are met. Examples of theory-specific code for both Nelson–Oppen and Shostak style theories are given below, following a discussion of the abstract requirements which must be fulfilled by theory-specific code to ensure correctness.

4   Correctness of the Basic Framework

In order to prove correctness, we give a specification in terms of preconditions and postconditions and show that the framework meets the specification. Sometimes it is necessary to talk about the state of the program. Each run of a program is considered to be a sequence of *states*, where a state includes a value for each variable in the program and a location in the code.

4.1   Preconditions and Postconditions

The preconditions for each function in the framework except for Merge are shown in the pseudocode. In order to give the precondition for Merge, a few definitions are required.

A *path* from an expression e to a sub-expression s of e is a sequence of expressions $e_0, e_1, ..., e_n$ such that $e_0 \equiv e$, e_{i+1} is a child of e_i, and s is a child of e_n. A sub-expression s of an expression e is called a *highest find-initialized* sub-expression of e if HasFind(s) and there is a path from e to s such that for each expression e' on the path, $\neg$HasFind(e'). An expression e is called *find-reduced* if Find(s) $\equiv$ s for each highest find-initialized sub-expression s of e.

An expression e is called *merge-acceptable* if e is an equation and one of the following holds: e is a literal; $e[1]$ is **false** or an atomic predicate and $e[2] \equiv$ **true**; or $e[1] \equiv$ **true** and $e[2] \equiv$ **false**.

Merge Precondition.
Whenever Merge(e) is called, the following must hold.

1. e is merge-acceptable,
2. $e[1]$ and $e[2]$ are find-reduced,
3. $e[1] \not\equiv e[2]$, and
4. $T \cup \mathcal{A} \models e$.

In addition to the preconditions, the following postconditions must be satisfied by the parameterized functions.

TheoryRewrite Postcondition.
After `e' := TheoryRewrite(e)` or `e' := RewriteEquality(e)` is executed, the following must hold:

1. $\mathcal{F}$ is unchanged by the call,
2. if e is a literal, then e' is a literal,
3. if e is find-reduced, then `HasFind(e')` or e' is find-reduced, and
4. $T \cup \mathcal{F} \models e = e'$.

TheorySetup Postcondition.
After `TheorySetup` is executed, the find database is unchanged.

If all preconditions and postconditions hold for all functions called so far, we say that the program is in an *uncorrupted* state. Also, if **true** $\not\sim$ **false**, we say the program is in a *consistent* state. A few lemmas are required before proving that the preconditions and postconditions hold for the framework code.

Lemma 1. *If the program is in an uncorrupted state and* `Union(a,b)` *has been called, then since that call there have been no calls to* `Union` *where either argument was* a.

Proof. Once `Union(a,b)` is called, `a.find` $\not\equiv$ a and this remains true since it can never again be an argument to `SetFind` or `Union`.

Lemma 2 (Equality Find Lemma). *If* e $\equiv$ a = b *and the program is in an uncorrupted and consistent state whose location is not between the call to* `SetFind(e)` *and the next call to* `Union` *and* `HasFind(e)`, *then* a *and* b *are terms and* `Find(e)` $\equiv$ **false**.

Proof. Suppose `HasFind(e)`. Then `Setup(e)` was called. But by the definition of merge-acceptable, this can only happen if e[1] and e[2] are terms and `Merge(e = false)` was called, in which case `Union(e,false)` is called immediately afterwards. It is clear from the definition of merge-acceptable, that `Union` is never called with first argument **false** unless the second argument is **true**. Thus, if **true** $\not\sim$ **false**, it follows from Lemma 1 that `Find(e)` $\equiv$ **false**. $\square$

Lemma 3 (Literal Find Lemma). *If the program is in an uncorrupted state and* e *is a literal, then* `Find(e)` *is either* e, **true**, *or* **false**.

Proof. ¿From the previous lemma, it follows that if e is an equality, then `Find(e)` is either e, **true**, or **false**. A similar argument shows that the same is true for a predicate. $\square$

Lemma 4 (Simplify Lemma).
If the program is in an uncorrupted state after e' := `Simplify(e)` *is executed, then following are true:*

1. $\mathcal{F}$ *is unchanged by the call,*
2. *if* e *is a literal then* e' *is a literal,*
3. *if* e *is a literal or term, then* e' *is find-reduced, and*
4. $T \cup \mathcal{F} \models$ e = e'*.*

We must prove the following theorem. A similar theorem is required every time we introduce theory-specific code.

Theorem 1. *If the program is in an uncorrupted state located in the framework code, then the next state is also uncorrupted.*

Proof.
Find Precondition: Find is called in two places by the framework. In each case, we check the precondition before calling it.
SetFind Precondition: SetFind(e) is only called from Setup(e) which returns if HasFind(e). Otherwise, Setup performs a depth-first traversal of the expression and calls SetFind. It follows from the TheorySetup Postcondition and the fact that expressions are acyclic that the precondition is satisfied.
Union Precondition: Union(a,b) is only called if Merge(a = b) is called first. By the Merge precondition, a and b are find-reduced. It is easy to see that after Setup(a) and Setup(b) are called, Find(a) $\equiv$ a and Find(b) $\equiv$ b.
AddFormula Precondition: We assume that AddFormula is only called with literals.
Assert Precondition: Assert(e) is only called from AddFormula. In this case, e $\in \mathcal{A}$, so it follows that $T \cup \mathcal{A} \models$ e.
Merge Precondition: Merge(e') is called from Assert(e). We know that e is a literal, so by the Simplify Lemma, Simplify(e) is a literal and is find-reduced. It follows that e' is merge-acceptable and e'[1] and e'[2] are find-reduced and unequal. ¿From the Simplify Lemma, we can conclude that $T \cup \mathcal{F} \models$ e = e'. It follows from the soundness property (described next) that $T \cup \mathcal{A} \models$ e = e'. We know that $T \cup \mathcal{A} \models$ e, so it follows that $T \cup \mathcal{A} \models$ e'.
TheoryRewrite Postcondition: It is straight-forward to check that each of the requirements hold for RewriteEquality.

$\square$

4.2 Soundness

The satisfiability procedure is *sound* if whenever the program state is inconsistent, $T \cup \mathcal{A} \models$ **false**. Soundness depends on the invariance of the following property.

Soundness Property. $T \cup \mathcal{A} \models \mathcal{F}$*.*

Lemma 5. *If the program is in an uncorrupted state, then the soundness property holds.*

Proof. Initially, the find database is empty. New formulas are added in two places. The first is in `Setup`, when `SetFind` is called. This preserves the soundness property since it only adds a reflexive formula to $\mathcal{F}$. The other is in `Merge(e)`, when `Union(e[1],e[2])` is called. This adds the formula e to $\mathcal{F}$, but we know that $T \cup \mathcal{A} \models e$ by the Merge Precondition. It also results in the addition of any formulas which can be deduced using transitivity and symmetry, but these are also entailed because T includes equality.   □

Theorem 2. *If the program is in an uncorrupted state, then the satisfiability procedure is sound.*

Proof. Suppose `Satisfiable` returns false. This means that **true** $\sim$ **false**. It follows from the previous lemma that $T \cup \mathcal{A} \models$ **true** = **false**, so $T \cup \mathcal{A} \models$ **false**.   □

4.3   Completeness

The satisfiability procedure is *complete* if $T \cup \mathcal{A}$ is satisfiable whenever the program is in a consistent state in the user code.

We define the *merge database*, denoted $\mathcal{M}$, as the set of all expressions e such that there has been a call to `Merge(e)`. In order to describe the property which must hold for completeness, we first introduce a few definitions, adapted from [19].

Recall that an expression in the language of Σ is an i-expression if it is a variable associated with T_i, its operator is a symbol in Σ_i, or it is an equality and its left side is an i-expression. A sub-expression of e is called an i-*leaf* if it is a variable or a j-expression, with $j \neq i$, and every expression along some path from e is an i-expression. An i-leaf is an i-*alien* if it is not an i-expression. An i-expression in which every i-leaf is a variable is called *pure* (or i-*pure*).

With each term t which is not a variable, we associate a fresh variable $v(t)$. We define $v(t)$ to be t when t is a variable. For some expression or set of expressions S, we define $\gamma_i(S)$ by replacing all of the i-alien terms t in S by $v(t)$[1] so that every expression in $\gamma_i(S)$ is i-pure. We denote by $\gamma_0(S)$ the set obtained from S by replacing all maximal terms (i.e. terms without any superterms) t by $v(t)$. Let Θ be the set of all equations $t = v(t)$, where t is a sub-term of some formula in $\mathcal{M}$. It is easy to see that $T \cup \mathcal{M}$ is satisfiable iff $T \cup \mathcal{M} \cup \Theta$ is satisfiable.

Let $\mathcal{M}_i = \{ e \mid e \in \mathcal{M} \wedge e \text{ is an } i\text{-expression} \}$. Define Θ_i similarly. Notice that $(\mathcal{M} \cup \Theta)$ is logically equivalent to $\bigcup \gamma_i(\mathcal{M}_i \cup \Theta_i)$, since each can be transformed into the other by repeated substitutions.

---

[1] Since expressions are DAG's, we must be careful about what is meant by *replacing* a sub-expression. The intended meaning here and throughout is that the expression is considered as a tree, and only occurrences of the term which qualify for replacement in the tree are replaced. This means that some occurrences may not be replaced at all, and the resulting DAG may look significantly different as a result.

We define V, the set of *shared terms* as the set of all terms t such that $v(\mathrm{t})$ appears in at least two distinct sets $\gamma_i(\mathcal{M}_i \cup \Theta_i), 1 \leq i \leq n$. Let $E(V) = \{\mathrm{a} = \mathrm{b} \mid \mathrm{a}, \mathrm{b} \in V \wedge \mathrm{a} \sim \mathrm{b}\}$, and let $D(V) = \{\mathrm{a} \neq \mathrm{b} \mid \mathrm{a}, \mathrm{b} \in V \wedge \mathrm{a} \not\sim \mathrm{b}\}$. For a set of expressions S, an arrangement $\pi(S)$ is a set such that for every two expressions a and b in S, exactly one of a = b or a $\neq$ b is in $\pi(S)$. We denote by $\pi(V)$ the arrangement $E(V) \cup D(V)$ of V determined by $\sim$. Now we can state the property required for completeness.

Completeness Property. If the program is in a consistent state in the user code, then $T_i \cup \gamma_i(\mathcal{M}_i \cup \pi(V))$ is satisfiable.

The following lemmas are needed before proving completeness.

Lemma 6. *If the program is in an uncorrupted state, then* $T \cup \mathcal{M} \models \mathcal{F}$

Proof. Every formula in $\mathcal{F}$ is either in $\mathcal{M}$ or can be derived from formulas in $\mathcal{M}$ using reflexivity, symmetry, and transitivity of equality. $\qquad\square$

Lemma 7. *If the program is in an uncorrupted and consistent state in the user code, then* $T \cup \mathcal{M} \models \mathcal{A}$.

Proof. Suppose $e \in \mathcal{A}$. Then we know that `Assert(e)` was called at some time previously. We can conclude by monotonicity of the find database that true $\not\sim$ false at the time of that call. Thus, e' := `Simplify(e)` was executed. By the Simplify Lemma, if $\mathcal{F}_1$ was the find database at the time of the call, $T \cup \mathcal{F}_1 \models$ e = e'. Now, if $e' \equiv$ true, then $T \cup \mathcal{F}_1 \models$ e and so by monotonicity and Lemma 6, $T \cup \mathcal{M} \models$ e. Otherwise, `Merge` is called. Let x be the argument to `Merge`. It is easy to see that $T \cup \mathcal{F}_1 \models$ e = x. But x $\in \mathcal{M}$, so $T \cup \mathcal{M} \models$ x. It then follows easily by monotonicity and Lemma 6 that $T \cup \mathcal{M} \models$ e. $\qquad\square$

The following theorem is from [19].

Theorem 3. *Let T_1 and T_2 be two stably-infinite, signature-disjoint theories and let ϕ_1 be a set of formulas in the language of T_1 and ϕ_2 a set of formulas in the language of T_2. Let v be the set of their shared variables and let $\pi(v)$ be an arrangement of v. If $\phi_i \wedge \pi(v)$ is satisfiable in T_i for $i = 1, 2$, then $\phi_1 \wedge \phi_2$ is satisfiable in $T_1 \cup T_2$.*

Theorem 4. *If the procedure always maintains an uncorrupted state and the completeness property holds for each theory, then the procedure is complete.*

Proof. Suppose that for a consistent state in the user code, $T_i \cup \gamma_i(\mathcal{M}_i \cup \pi(V))$ is satisfiable for each i. This implies that $T_i \cup \gamma_i(\mathcal{M}_i \cup \Theta_i \cup \pi(V))$ is satisfiable (since each equation in Θ_i simply defines a new variable), which is logically equivalent (by applying substitutions from Θ_i) to $T_i \cup \gamma_i(\mathcal{M}_i \cup \Theta_i) \cup \gamma_0(\pi(V))$. Now, each set $\gamma_i(\mathcal{M}_i \cup \Theta_i)$ is a set of formulas in the language of T_i, and $\gamma_0(\pi(V))$ is an arrangement of the variables shared among these sets, so we can conclude by repeated application of Theorem 3 that $\bigcup \gamma_i(\mathcal{M}_i \cup \Theta_i)$ is satisfiable in T. But $\bigcup \gamma_i(\mathcal{M}_i \cup \Theta_i)$ is equivalent to $\mathcal{M} \cup \Theta$ which is satisfiable in T iff $T \cup \mathcal{M}$ is satisfiable. Finally, by Lemma 7, $T \cup \mathcal{M} \models \mathcal{A}$. Thus we can conclude that $T \cup \mathcal{A}$ is satisfiable. $\qquad\square$

4.4 Termination

We must show that each function in the framework terminates. The following requirements guarantee this.

Termination Requirements.

1. The preconditions for `Find`, `SetFind`, and `Union` always hold.
2. For each i-expression e, $\text{TheoryRewrite}_i(\text{e})$ terminates.
3. If s is a sequence of expressions in which the next member of the sequence e' is formed from the previous member e by calling $\text{TheoryRewrite}_i(\text{e})$, then beyond some element of the sequence, all the expressions are identical.
4. For each i-expression e, $\text{TheorySetup}_i(\text{e})$ terminates.
5. After `Union(a,b)` is called,
   (a) No new entries are added to `a.notify`.
   (b) Each call to each funtion in `a.notify` terminates.
6. For each theory T_i, $\text{PropagateEqualities}_i$ terminates and after calling $\text{PropagateEqualities}_i$ some finite number of times, it will always return false.

Theorem 5. *If the termination requirements hold, then each function in the framework terminates.*

Proof. The first condition guarantees that `Find` terminates, from which it follows that `Satisfiable` terminates. The next two ensure that `Rewrite` terminates. It then follows easily that `Simplify` must terminate. The next few conditions are sufficient to ensure that `Setup` and `Merge` terminate, from which it follows that `Assert` terminates. This, together with the last condition allows us to conclude that `AddFormula` terminates.   □

It is not hard to see that without any theory-specific code, these requirements hold.

5 Examples Using the Framework

In this section we will give two examples to show how the framework can accommodate different kinds of theory-specific code.

5.1 Nelson–Oppen Theories

A Nelson–Oppen style satisfiability procedure for a theory T_i must be able to determine the satisfiability of a set of formulas in the language of Σ_i as well as which equalities between variables are entailed by that of formulas [12]. We present a method for integrating such theories which is flexible and efficient.

Suppose we have a Nelson–Oppen style satisfiability procedure which treats alien terms as variables with the following methods:

```
AddFormula_i        Adds a new formula to the set A_i.
Satisfiable_i       True iff T_i ∪ γ_i(A_i) is satisfiable.
AddTermToPropagate_i Adds a term to the set Δ_i.
GetEqualities_i     Returns the largest set of equalities E_i between terms
                    in Δ_i such that T_i ∪ γ_i(A_i) ⊨ γ_i(E_i).
```

A new expression attribute, **shared** is used to keep track of which terms are relevant to more than one theory. Each theory is given an index, i, and the **shared** attribute is set to i if the term is used by theory i. If more than one theory uses the term, the **shared** attribute is set to 0. This is encapsulated in the **SetShared** and **IsShared** methods shown below.

```
SetShared(e,i)                              IsShared(e)
  IF e.shared = ⊥ THEN e.shared := i;         RETURN e.shared = 0;
  ELSE IF e.shared ≠ i THEN e.shared := 0;
  AddTermToPropagate_i(e);
```

Figure 2 shows the theory-specific code needed to add a theory T_i with a satisfiability procedure as described above. We will refer to a theory implemented in this way as a *Nelson–Oppen* theory. Each i-expression is passed to **TheorySetup_i**. **TheorySetup_i** marks these terms and their alien children as used by T_i. It also ensures that **Notify_i** will be called if any of these expressions are merged with something else. When **Notify_i** is called, the formula is passed along to the satisfiability procedure for T_i. These steps correspond to the decomposition into pure formulas in other implementations (but without the introduction of additional variables). **PropagateEqualities_i** asserts any equations between shared terms that have been deduced by the satisfiability procedure for T_i. This corresponds to the equality propagation step in other methods. It is sufficient to propagate equalities between shared variables, a fact also noted in [19].

We also introduce a new optimization. Not all theories need to know about all equalities between shared terms. A theory is only notified of an equality if the left side of that equality is a term that it has seen before. In order to guarantee that this results in fewer propagations, we have to ensure that whenever an equality between two terms is in M, if one of the terms is not shared, then the left term is not shared. We can easily do this by modifying **RewriteEquality** to put non-shared terms on the left. However, this is not necessary for correctness, a fact which allows the integration of Shostak-style satisfiability procedures which require a different implementation of **RewriteEquality** as described in Sec. 5.2 below.

A final optimization is to associate variable with theories in such a way as to to avoid causing terms to be shared unnecessarily. For example, if $x = t$ is a formula in M and x is a variable and t is an i-term, it is desirable for x to be an i-term as well (otherwise, t immediately becomes a shared term). In our implementation, expressions are type-checked and each type is associated with

```
TheorySetup_i (e)
  FOREACH i-alien child a of e DO BEGIN
    a.notify := a.notify ∪ { ⟨Notify_i, ∅⟩ };
    SetShared(a,i);
  END
  e.notify := e.notify ∪ { ⟨Notify_i, ∅⟩ };
  IF e is a term THEN SetShared(e,i);

TheoryRewrite_i (e)
  RETURN e;

PropagateEqualities_i ()
  propagate := false;
  IF Satisfiable() BEGIN
    IF ¬ Satisfiable_i() THEN Merge(true = false));
    ELSE FOREACH x = y ∈ GetEqualities_i DO
      IF IsShared(x) AND IsShared(y) AND x ≁ y THEN BEGIN
        propagate := true;
        Assert(x = y));
      END
  END
  RETURN propagate;

Notify_i (e)
  IF e[1] is an i-alien term THEN BEGIN
    x := Find(e[2]);
    x.notify := x.notify ∪ { ⟨Notify_i, ∅⟩ };
    e := (e[1] = x);
  END
  AddFormula_i (e);
```

Fig. 2. Code for implementing a Nelson–Oppen theory T_i.

a theory. Thus, we can easily guarantee this by associating x with the theory associated with its type.

Correctness. The proof of the following theorem is similar to that given for the framework code and is omitted.

Theorem 6. *If the program is in an uncorrupted state located in the theory-specific code for a Nelson–Oppen theory, then the next state is also uncorrupted.*

To show that the completeness property holds, we must show that if the program is in a consistent state in the user code, then $T_i \cup \gamma_i(\mathcal{M}_i \cup \pi(V))$ is satisfiable. This requires the following invariant to hold for each theory T_i.

Shared Term Requirement. There has been a call to SetShared(e,i) if $v(e)$ appears in $\gamma_i(\mathcal{M}_i \cup \Theta_i)$.

Lemma 8. *If T_i is a Nelson–Oppen theory, then the shared term requirement holds for T_i.*

Corollary 1. *If T_i is a Nelson–Oppen theory, and $v(t)$ appears in $\gamma_i(\mathcal{M}_i \cup \Theta_i)$, then $t \in \Delta_i$.*

Let $\Delta_i' = \Delta_i \cup \{x \mid x$ is a term and $t = x \in \mathcal{A}_i$ for some term $t\}$.

Lemma 9. *If T_i is a Nelson–Oppen theory and the program is in an uncorrupted state in the user code and $x = y \in \mathcal{M}$, where $x \in \Delta_i'$, then $x = z \in \mathcal{A}_i$, where $z \equiv \text{Find}(y)$ at some previous time.*

Proof. Suppose $x \in \Delta_i$. Then `SetShared` was called. It is easy to see from the code that at the time it was called, `Notify`$_i$ was added to `x.notify`. If on the other hand, $x \notin \Delta_i$, then $t = x \in \mathcal{A}_i$ for some t which is not an i-term. But then, when $t = x$ was added to $\mathcal{A}_i$, `Notify`$_i$ was added to `x.notify`. In each case, `Notify`$_i(x = y)$ will be called after `Merge(x = y)` is called, so that $x = \text{Find}(y)$ is added to $\mathcal{A}_i$. □

Lemma 10. *If T_i is a Nelson–Oppen theory and the program is in an uncorrupted state in the user code and $x \sim y$, where $x, y \in \Delta_i'$, then $T_i \cup \gamma_i(\mathcal{A}_i) \models \gamma_i(x = y)$.*

Proof. We can show by the previous lemma that since $\text{Find}(x) \equiv \text{Find}(y)$, there is a chain of equalities in $\mathcal{A}_i$ linking x to y. □

Let $D_i = \{a \neq b \mid a, b \in (\Delta_i \cap V)\}$, and let $D_i' = \{a \neq b \mid a, b \in (\Delta_i' \cap V)\}$.

Lemma 11. *If T_i is a Nelson–Oppen theory and the program is in an uncorrupted and consistent state in the user code, then $T_i \cup \gamma_i(\mathcal{A}_i \cup D_i)$ is satisfiable.*

Proof. No single disequality $x \neq y \in D_i$ can be inconsistent because if it were, that would mean $T_i \cup \gamma_i(\mathcal{A}_i) \models \gamma_i(x = y)$. But if this is the case, since `PropagateEqualities`$_i$ terminated, it must be the case that $x \sim y$. Since no single equality $x = y$ is entailed, it follows from the convexity of T_i, that no disjunction of equalities can be entailed. □

Lemma 12. *If T_i is a Nelson–Oppen theory and the program is in an uncorrupted and consistent state in the user code, then $T_i \cup \gamma_i(\mathcal{A}_i \cup D_i')$ is satisfiable.*

Proof. If $t_1' \neq t_2' \in D_i'$, we can find (by the definition of Δ_i') some t_1 and t_2 such that $t_1 \neq t_2 \in D_i$ and $\mathcal{A}_i \models (t_1 = t_1' \wedge t_2 = t_2')$. The result follows by the previous lemma. □

Theorem 7. *If each theory satisfies the shared term requirement and the program is in an uncorrupted and consistent state in the user code, then if T_i is a Nelson–Oppen theory, the completeness property holds for T_i.*

Proof. It is not hard to show that if $v(x) \in \gamma_i(\mathcal{A}_i \cup \Theta_i)$, then $x \in \Delta_i'$. It then follows that an interpretation satisfying $T_i \cup \gamma_i(\mathcal{A}_i \cup D_i')$ can be modified to also satisfy $\gamma_i(\pi(V))$. □

Termination. The only termination condition that is non-trivial is the last one. The following requirement is sufficient to fulfill this condition.

Nelson–Oppen Termination Requirement

Suppose that before a call to `Assert` from `PropagateEqualities`$_i$, n is the number of equivalence classes in $\sim$ containing at least one term $t \in V$. Then, either the state following the call to `Assert` is inconsistent or if m is the number of equivalence classes in $\sim$ containing at least one term $t \in V$ after returning from `Assert`, $m < n$.

If every theory is a Nelson–Oppen theory, it is not hard to see that this requirement holds. This is because each call to `Assert` merges the equivalence classes of two shared variables without creating any new ones.

5.2   Adding Shostak Theories

Suppose we have a theory T_i with no predicate symbols which provides two functions, σ and ω which we refer to as the canonizer and solver respectively. Note that if we have more than one such theory, we can often combine the canonizers and solvers to form a canonizer and solver for the combined theory, as described in [17][2]. The functions σ and ω have the following properties. σ is a canonizer for T_i if

1. $T_i \models \gamma_i(a = b)$ iff $\sigma(a) \equiv \sigma(b)$
2. $\sigma(\sigma(t)) \equiv \sigma(t)$ for all terms t.
3. $\gamma_i(\sigma(t))$ contains only variables occurring in $\gamma_i(t)$.
4. $\sigma(t) \equiv t$ if t is a variable or not an i-term.
5. If $\sigma(t)$ is a compound i-term, then $\sigma(x) = x$ for each child x of $\sigma(t)$.

ω is a solver[3] for T_i if

1. If $T_i \models \gamma_i(x \neq y)$ then $\omega(x = y) \equiv \textbf{false}$.
2. Otherwise, $\omega(x = y) \equiv a = b$ where a and b are terms,
3. $T_i \models (x = y) \leftrightarrow (a = b)$,
4. $\gamma_i(a)$ is a variable and does not appear in $\gamma_i(b)$,
5. neither $\gamma_i(a)$ nor $\gamma_i(b)$ contain variables not occurring in $\gamma_i(x = y)$,
6. $\omega(a = b) \equiv a = b$ and $\sigma(b) \equiv b$.

We call such a theory a Shostak theory. The code in Fig. 3 shows the additional code needed to integrate a Shostak theory.

---

[2] Although it has been claimed that solvers can always be combined to form a solver for the combined theory [6, 17], this is not always possible, as pointed out in [11]

[3] Shostak allows the solved form to be more general. To simplify the presentation, we assume the solver returns a single, logically equivalent, equation.

```
RewriteEquality(e)
  IF e[1] ≡ e[2] THEN RETURN true;
  IF one child of e is true THEN RETURN the other child;
  IF e[1] ≡ false THEN RETURN (e[2] = e[1]);
  IF e[1] is a term THEN RETURN ω(e);
  RETURN e;

TheorySetupᵢ(e)
  FOREACH a which is an i-leaf in e DO BEGIN
    IF Op(e) = '=' THEN a.notify := a.notify ∪ {⟨UpdateDisequality,e⟩};
    ELSE a.notify := a.notify ∪ {⟨UpdateShostak,e⟩};
    SetShared(a,i);
  END
  IF e is a term THEN SetShared(e,i);

TheoryRewriteᵢ(e)
  RETURN σ(e);

PropagateEqualitiesᵢ()
  RETURN false;

UpdateDisequality(x,y)
  IF ¬Satisfiable() ∨ ¬HasFind(y) THEN RETURN;
  Replace each i-leaf c in y with Find(c);
  y' := Rewrite(y);
  IF y' ≢ false THEN Merge(y' = false);

UpdateShostak(x,y)
  IF Find(y) ≡ y THEN BEGIN
    Replace each i-leaf c in y with Find(c) to get y';
    Merge(y = σ(y'));
  END
```

Fig. 3. Code for implementing a Shostak theory T_i.

Correctness. It is not hard to show that this code satisfies the preconditions and requirements of the framework.

Theorem 8. *If the program is in an uncorrupted state located in the theory-specific code for a Shostak theory, then the next state is also uncorrupted.*

Included in the Shostak code are the calls to SetShared necessary to allow this theory to be integrated with Nelson–Oppen theories. We have not included the code typically included for handling uninterpreted functions. This is because our approach allows us to consider uninterpreted functions as belonging to a separate Nelson–Oppen theory. Though we do not show how in this paper, any simple congruence closure algorithm can be integrated as a Nelson–Oppen theory. Omitting details related to uninterpreted functions simplifies the presentation and proof. We have also included code for handling disequalities, which Shostak's original procedure does not handle directly. We will give some intuition for how this works after making a few definitions.

Let $\Delta_i = \{t \mid t$ is an i-leaf in some expression $e \in \mathcal{M}\}$. Let $\mathcal{E} = \{a = b \mid a \in \Delta_i \wedge b \equiv \text{Find}(a)\}$. For an expression e, define $\tau(e)$ to be the expression obtained from e by replacing each i-leaf x in e by $\text{Find}(x)$. Shostak's method works by ensuring that $\text{Find}(t) \equiv \sigma(\tau(t))$. This together with the properties of the solver ensure that the set $\mathcal{E}$ is equivalent to a substitution, meaning it is easily satisfiable. These are the key ideas of the completeness argument.

Lemma 13. *If the program is in an uncorrupted and consistent state which is not inside of a call to* **Merge**, *then for each term* t *such that* $\text{HasFind}(t)$, $\text{Find}(t) \equiv \sigma(\tau(t))$. *Also, if* $\text{Find}(t) \equiv t$, *then* $\tau(t) \equiv t$.

Proof. When SetFind is first called on an expression e, the **Merge** preconditions together with the solver and canonizer guarantee that $e \equiv \sigma(\tau(e))$. Then, whenever an i-leaf is merged, UpdateShostak is called to preserve the invariant. $\square$

Lemma 14. *If the program is in an uncorrupted and consistent state in the user code, and* T_i *is a Shostak theory, then* $T_i \cup \gamma_i(\mathcal{E})$ *is satisfiable.*

Proof. Let M be a model of T_i, and let $x \in \Delta_i$. If $\text{Find}(x) \equiv x$, then assign $v(x)$ an arbitrary value. Otherwise, assign $v(x)$ the same value as $\gamma_i(\text{Find}(x))$. By the above lemma, this assignment satisfies $\gamma_i(\mathcal{E})$. $\square$

Lemma 15. *If the program is in an uncorrupted and consistent state in the user code and* T_i *is a Shostak theory, then* $T_i \cup \gamma_i(\mathcal{M}_i)$ *is satisfiable.*

Proof. Suppose $e \in \mathcal{M}_i$. Clearly $e[1] \sim e[2]$. If e is an equality between terms, it follows from Lemma 13 that $\sigma(\tau(e[1])) \equiv \sigma(\tau(e[2]))$. By properties of σ, it follows that $T_i \models \tau(e[1]) = \tau(e[2])$. Then, by the definition of $\mathcal{E}$, it follows that $T_i \cup \mathcal{E} \models e[1] = e[2]$ and hence $T_i \cup \gamma_i(\mathcal{E}) \models \gamma_i(e[1] = e[2])$. Suppose on the other hand that e is the literal $(x = y) = \textbf{false}$, and suppose that $T_i \cup \gamma_i(\mathcal{E}) \models \gamma_i(x = y)$. The same argument as above in reverse shows that $\text{Find}(x) \equiv \text{Find}(y)$. The UpdateDisequality code ensures that in this case **true** will get merged with **false**, contradicting the assumption that the state is consistent. Thus, $T_i \cup \gamma_i(\mathcal{E}) \not\models \gamma_i(x = y)$. Since T_i is convex, it follows that $T_i \cup \gamma_i(\mathcal{E} \cup \mathcal{M}_i)$ is satisfiable. $\square$

Theorem 9. *If the program is in an uncorrupted and consistent state in the user code and* T_i *is a Shostak theory, then the completeness property holds for* T_i.

Proof. The above lemma shows that $T_i \cup \gamma_i(\mathcal{E} \cup \mathcal{M}_i)$ is satisfiable. Suppose a and b are shared terms. If $a \sim b$, a similar argument to that given above shows that $T_i \cup \gamma_i(\mathcal{E}) \models \gamma_i(a = b)$. If, on the other hand $a \not\sim b$, it follows easily that $T_i \cup \gamma_i(\mathcal{E}) \not\models \gamma_i(a = b)$. Since each equality in $\gamma_i(\mathcal{M}_i \cup \pi(V))$ is entailed by $T_i \cup \gamma_i(\mathcal{E})$ and none of the disequalities are, it follows by convexity that $T_i \cup \gamma_i(\mathcal{M}_i \cup \pi(V))$ is satisfiable. $\square$

Termination. The idempotency of the solver and canonizer are sufficient to guarantee termination of rewrites. For each expression e, it is not hard to show that something is added to $e.\mathtt{notify}$ only if $\mathtt{Find(e)} \equiv e$. Consider the functions called by $\mathtt{Merge}$ which are $\mathtt{UpdateDisequality}$ and $\mathtt{UpdateShostak}$. Both of them call $\mathtt{Merge}$ recursively. Each of them reduce the value of some measure of the program state. For $\mathtt{UpdateDisequality}$, the measure is the number of equality expressions e such that $\mathtt{HasFind(e)}$ and $\omega(\tau(e)) \not\equiv \mathbf{false}$. For $\mathtt{UpdateShostak}$, the measure is the number of expressions e such that $\mathtt{Find(e)} \equiv e$ and $\mathtt{Find(c)} \not\equiv c$ for some i-leaf c of e. With some effort, it can be verified that none of the functions in the theory-specific code presented thus far which can be called after $\mathtt{Union}$ increase either of these measures. The other termination conditions are trivial.

Finally, in order to combine Shostak and Nelson–Oppen, the Shostak code must not break the Nelson–Oppen Termination Requirement. Any new call to $\mathtt{Merge}$ has the potential to "create" new shared terms by causing a new term to show up in $\mathcal{M}_i$ for some i. A careful analysis shows that if $\mathtt{Assert(x = y)}$ is called from the Nelson–Oppen code, any resulting call to $\mathtt{Merge}$ does not increase the number of equivalence classes containing shared terms. Lemma 13 ensures that by the time $\mathtt{Assert}$ has returned, $x \sim y$, so the number of equivalence classes containing shared terms decreases as required.

6   Conclusion

We have presented a framework for combining decision procedures for disjoint first-order theories, and shown how it can be used to implement and integrate Nelson–Oppen and Shostak style decision procedures.

This work has shed considerable light on the individual methods as well as on what is required to combine them. We discovered that a more restricted set of equalities can be propagated in the Nelson–Oppen framework without losing completeness. Also, by separating the uninterpreted functions from the Shostak method, the code is simpler and easier to verify.

We are working on an extension of the framework which would handle non-convex theories and more general Shostak solvers. In future work, we hope also to be able to relax the requirements that the theories be disjoint and stably-infinite. We also plan to complete and distribute a new version of SVC based on these results.

Acknowledgments

We would like to thank Natarajan Shankar at SRI for helpful discussions and insight into Shostak's decision procedure. This work was partially supported by the National Science Foundation Grant MIPS-9806889 and NASA contract NASI-98139. The third author is supported by a National Science Foundation Graduate Fellowship.

References

1. F. Baader and C. Tinelli. A new approach for combining decision procedures for the word problem, and its connection to the Nelson–Oppen combination method. In W. McCune, editor, *14th International Conference on Computer Aided Deduction*, Lecture Notes in Computer Science, pages 19–33. Springer-Verlag, 1997.
2. Clark Barrett, David Dill, and Jeremy Levitt. Validity checking for combinations of theories with equality. In M. Srivas and A. Camilleri, editors, *Formal Methods In Computer-Aided Design*, volume 1166 of *Lecture Notes in Computer Science*, pages 187–201. Springer-Verlag, 1996.
3. N. Bjorner. *Integrating Decision Procedures for Temporal Verification*. PhD thesis, Stanford University, 1999.
4. Michael A. Colon and Tomas E. Uribe. Generating finite-state abstractions of reactive systems using decision procedures. In *International Conference on Computer-Aided Verification*, volume 1427 of *Lecture Notes in Computer Science*, pages 293–304. Springer-Verlag, 1998.
5. D. Cyrluk. Private communication. 1999.
6. D. Cyrluk, P. Lincoln, and N. Shankar. On Shostak's Decision Procedure for Combinations of Theories. In M. McRobbie and J. Slaney, editors, *13th International Conference on Computer Aided Deduction*, volume 1104 of *Lecture Notes in Computer Science*, pages 463–477. Springer-Verlag, 1996.
7. Satyaki Das, David L. Dill, and Seungjoon Park. Experience with predicate abstraction. In *11th International Conference on Computer-Aided Verification*, pages 160–172. Springer-Verlag, July 1999. Trento, Italy.
8. David L. Detlefs, K. Rustan M. Leino, Greg Nelson, , and James B. Saxe. Extended static checking. Technical Report 159, Compaq SRC, 1998.
9. Z. Manna et al. STeP: Deductive-Algorithmic Verification of Reactive and Realtime Systems. In *8th International Conference on Computer-Aided Verification*, volume 1102 of *Lecture Notes in Computer Science*, pages 415–418. Springer-Verlag, 1996.
10. H.Saidi and N.Shankar. Abstract and model check while you prove. In *Proceedings of the 11th Conference on Computer-Aided Verification*. Springer-Verlag, July 1999. Trento, Italy.
11. J. Levitt. *Formal Verification Techniques for Digital Systems*. PhD thesis, Stanford University, 1999.
12. G. Nelson and D. Oppen. Simplification by cooperating decision procedures. *ACM Transactions on Programming Languages and Systems*, 1(2):245–57, 1979.
13. S. Owre, J. Rushby, and N. Shankar. PVS: A Prototype Verification System. In D. Kapur, editor, *11th International Conference on Automated Deduction*, volume 607 of *Lecture Notes in Artificial Intelligence*, pages 748–752. Springer-Verlag, 1992.
14. David Y.W. Park, Jens U. Skakkebæk, Mats P.E. Heimdahl, Barbara J. Czerny, and David L. Dill. Checking properties of safety critical specifications using efficient decision procedures. In *FMSP'98: Second Workshop on Formal Methods in Software Practice*, pages 34–43, March 1998.
15. William Pugh. The omega test: a fast and practical integer programming algorithm for dependence analysis. In *Communications of the ACM*, volume 8, pages 102–114, August 1992.
16. H. Ruess and N. Shankar. Deconstructing Shostak. In *17th International Conference on Computer Aided Deduction*, 2000.

17. R. Shostak. Deciding combinations of theories. *Journal of the Association for Computing Machinery*, 31(1):1–12, 1984.
18. J. Su, D. Dill, and J. Skakkebæk. Formally verifying data and control with weak reachability invariants. In *Formal Method In Computer-Aided Design*, 1998.
19. C. Tinelli and M. Harandi. A new Correctness Proof of the Nelson–Oppen Combination Procedure. In F. Baader and K. Schulz, editors, *1st International Workshop on Frontiers of Combining Systems (FroCoS'96)*, volume 3 of *Applied Logic Series*. Kluwer Academic Publishers, 1996.

Modular Reasoning in Isabelle

Florian Kammüller

GMD First, 12489 Berlin, Germany
florian@first.gmd.de

Abstract. The concept of locales for Isabelle enables local definition
and assumption for interactive mechanical proofs. Furthermore, dependent types are constructed in Isabelle/HOL for first class representation
of structure. These two concepts are introduced briefly. Although each
of them has proved useful in itself, their real power lies in combination.
This paper illustrates by examples from abstract algebra how this combination works and argues that it enables modular reasoning.

1 Motivation

Modules for theorem provers are a means for organizing theories of applications.
Generic interactive theorem provers like PVS [OSRSC98], IMPS [FGT93], and
HOL [GM93] define their applications as object logics. Modules are used to
maintain and structure these object logics. Being a classical software engineering
concept for re-usability and structuring, modules are the obvious method for
organizing formalizations of theorem provers.

Apart from just organizing big theories, advanced modular features — like parameterization and instantiation — give rise to use modules to represent (mathematical) structure logically. For example, the abstract algebraic structure of
groups is represented by a module in the following fashion (cf. [OSRSC98]).

```
Module Group [G: TYPE, o : G -> G -> G, inv: G -> G, e: G]
  ∀ x: G. x o e = x
  ∀ x: G. x o (inv x) = e
  ∀ x, y, z: G. x o (y o z) = (x o y) o z
```

The abstract character of groups is modeled in systems like PVS by using
(generic) sorts or explicit parameters to model the contents of the group. Reasoning about properties of group elements and the operation o is possible inside such
a theory. The parameterization enables the instantiation of the group theory to
actual groups. The abstractly derived results can thus be reused by an instantiation. This is also what we think of as modular reasoning; reasoning where the
abstraction and structuring of modules becomes part of the proof process.

However, an adequate way of reasoning is not possible in this setting. For
example, we must consider the class of all groups to enable reasoning about
general properties which hold, say, only for finite groups. This class of all groups
cannot be defined here because the theory level is separate from the reasoning
level. There are more examples, like quotients of groups forming groups again; the

D. McAllester (Ed.): CADE-17, LNAI 1831, pp. 99–114, 2000.
© Springer-Verlag Berlin Heidelberg 2000

problem is always the same. Since modules are not first class citizens, we cannot use the structure defined by a module in any formula. Hence, formalizations using modules to represent mathematical structure are not *adequate*; we can only reason about a restricted set of aspects of the (mathematical) world.

In rich type theories there is the concept of dependent types. Systems like Coq [D+93] and LEGO [LP92] implement such type theories. If the hierarchies of the type theory are rich enough then dependent types are first class citizens. Usually, type theories do not have advanced module concepts as they are known in interactive theorem provers, like PVS and IMPS. However, it is well known that dependent types may be used to represent modules (e.g. [Mac86]).

We verified by case studies (e.g. [KP99]) that a module system where the modules are first class citizens is actually necessary for an *adequate* representation of (mathematical) structures in the logic of a theorem prover. Yet, it turns out that we sometimes need just some form of local scope and not a first class representation. We need *locality*, i.e. the possibility to declare concepts whose scope is limited or temporary. Locality and adequacy are separate concerns that do not coincide generally. We propose to use separate devices, i.e. locales [KWP99] and dependent types [Kam99b]. We have designed and implemented them for Isabelle. In this paper, we show that in combination they realize modular reasoning.

In Section 2.1 we shortly introduce the concept of locales for Isabelle. The way we represent dependent types in Isabelle/HOL is sketched in Section 2.2. The introduction to these topics has been presented elsewhere and goes only as far as needed for the understanding of the following. In Section 3 we present various case studies. They illustrate the use of locales and dependent types and validate that the combination of these concepts enables modular reasoning.

2   Prerequisites and Concepts

Isabelle is a higher order logic theorem prover [Pau94]. It is generic, that is, it can be instantiated to form theorem provers for a wide range of logics. These can be made known to the prover by defining theories that contain sort and type declarations, constants, and related definitions and rules. The most popular object logics are Isabelle/HOL and Isabelle/ZF. A powerful parser supports intelligible syntactic abbreviations for user-defined constants.

Definitions, rules, and other declarations that are contained in an Isabelle theory are visible whenever that theory is loaded into an Isabelle session. All theories on which the current theory is built are also visible. All entities contained in a current theory stay visible for any other theory that uses the current one. Thus, theory rules and definitions are not suited for formalizing concepts that are of only local significance in certain contexts or proofs.

Isabelle theories form hierarchies. However, theories do not have any parameters or other advanced features typical for modules in theorem provers. That is, Isabelle did not have a module concept prior to the developments presented in the current section.

2.1   Locales

Locales [KWP99] declare a context of fixed variables, local assumptions and local definitions. Inside this context, theorems can be proved that may depend on the assumptions and definitions and the fixed variables are treated like constants. The result will then depend on the locale assumptions, while the definitions of a locale are eliminated.

The definition of a locale is static, i.e. it resides in a theory. Nevertheless, there is a dynamic aspect of locales corresponding to the interactive side of Isabelle. Locales are by default inactive. If the current theory context of an Isabelle session contains a theory that entails locales, they can be invoked. The list of currently active locales is called *scope*. The process of activating them is called *opening*; the reverse is *closing*.

Locales can be defined in a nested style, i.e. a new locale can be defined as the extension of an existing one. Locales realize a form of polymorphism with binding of type variables not normally possible in Isabelle (see Section 3.2).

Theorems proved in the scope of a locale may be exported to the surrounding context. The exporting device for locales dissolves the contextual structure of a locale. Locale definitions become expanded, locale assumptions attached as individual assumptions, and locale constants transformed into variables that may be instantiated freely. That is, exporting reflects a locale to Isabelle's meta-logic. Although they do not have a first class representation, locales have at least a meta-logical explanation. In Section 3.4 we will see that this is crucial for the sound combination of locales and dependent types.

Locales are part of the official distribution since Isabelle version 98-1. They can be used in all of Isabelle's object logics — not just Isabelle/HOL — and have been used already in many applications apart from the ones presented here.

2.2   Dependent Types as First Class Modules

In rich type theories, e.g. UTT [Bai98], groups can be represented as

$$\Sigma\, G : set.\ \Sigma\, e : G.\ \Sigma \circ : map_2\ G\ G\ G.\ \Sigma\, ^{-1} : map\ G\ G.\ \text{group\_axioms}$$

where group_axioms abbreviates the usual rules for groups, corresponding to the body of a module for groups. The elements G, e, $\circ$ and $^{-1}$ correspond to the parameters of a module and occur in group_axioms. Since this Σ-type can be considered as a term in a higher type universe, we can use it in other formulas. Hence, this modular formalization of groups is adequate.

Naïvly, a Σ-type may be understood as the Cartesian product $A \times B$ and a Π-type as a function type $A \to B$, but with the B having a "slot" of type A, i.e. being parameterized over an element of A. The latter part of this intuition gives rise to use these type constructors to model the parameterization, and hence the abstraction, of modules.

Isabelle's higher order logic does not have dependent types. For the first class representation of abstract algebraic structures we construct an embedding

of Σ-types and Π-types as typed sets into Isabelle/HOL using set-theoretic definitions [Kam99b]. Since sets in Isabelle/HOL are constants, algebraic structures
become first class citizens. Moreover, abstract structures that use other structures as parameters may be modeled as well. We call such structures higher order
structures. An example are group homomorphisms, i.e. maps from the carrier of
a group G to the carrier of a group H that respect operations.

$$Hom \equiv \Sigma \ G \in Group. \ \Sigma \ H \in Group.$$
$$\{\Phi \mid \Phi \in G.\langle cr \rangle \to H.\langle cr \rangle \land$$
$$(\forall x, y \in G.\langle cr \rangle. \ \Phi(G.\langle f \rangle \ x \ y) = H.\langle f \rangle \ \Phi(x) \ \Phi(y))\}$$

The postfix tags, like $.\langle f \rangle$ are field descriptors of the components of a structure.
In general, Σ is used as a constructor for higher order structures. In some cases,
however, a higher order structure is uniquely constructed, as for example the factorization of a group by one of its subgroups (see Section 3.3). In those cases, we
use the Π-type. We define a set-typed λ-notation that enables the construction
of functions of a Π-type.

3   Locales + Dependent Types = Modules

The main idea of this work is that a combination of the concepts of locales
and dependent types enables adequate representation and convenient proof with
modular structures. To validate this hypothesis we present various case studies
pointing out the improvements that are gained through the combination of the
two concepts. Some basic formalizations in Section 3.1, explain the use and
interaction of the two concepts. In Section 3.2, we reconsider the case study of
Sylow's theorem [KP99] that mainly illustrates the necessity of locales. Then, in
Section 3.3, we discuss in detail the quotient of a group that clearly proves the
need of the first class property of the dependent type representation. However,
it illustrates as well how the additional use of locales enhances the reasoning.
After summarizing other examples and analyzing the improvements we show in
Section 3.4 how operations on structures may be performed with the combined
use of locales and dependent types.

3.1   Formalization of Group Theory

Groups and Subgroups. The class of groups is defined as a set over a record
type with four elements: the carrier, the binary operation, the inverse and the
unit element that constitute a group. They can be referred to using the projections G.<cr>, G.<f>, G.<inv>, and G.<e> for some group G. Since the class of
all groups is a set, it is a first class citizen. We can write G $\in$ Group as a logical
formula to express "G is a group". Hence, all group properties can be derived
from the definition.

In the definition of the subgroup property we can use an elegant approach
which reads informally: *a subset H of G is a subgroup if it is a group with G's
operations.* Only since groups are first class citizens, we can describe subgroups

in that way. A Σ-structure is used to model the subgroup relation. The $(\!|$ $|\!)$ enclosed quadruple constructs the subgroup as an element of the record type of the elements of Group. Our λ-notation enables the restriction of the group operations to the subset H.

```
Σ G ∈ Group. {H | H ⊆ (G.<cr>) ∧
   (| carrier = H, bin_op = λ x ∈ H. λ y ∈ H. (G.<f>) x y,
      inverse = λ x ∈ H. (G.<inv>) x, unit = (G.<e>) |) ∈ Group}
```

The convenient syntax H <<= G, for H is a subgroup of G, may be used to abbreviate (G, H) ∈ subgroup.

In addition to the first class representation of groups and subgroups, we define a locale group to provide a local proof context for group related proofs. The fixes, assumes, and defines parts introduce the constants with their polymorphic types, the assumptions and definitions of the locale[1]. The 'a is a polymorphic type variable (see Section 3.2).

```
locale group =
  fixes
    G         :: "'a grouptype"
    e         :: "'a"
    binop     :: "'a => 'a => 'a"    (infixr "#" 80)
    inv       :: "'a => 'a"          ("i (_)"     [90]91)
  assumes
    Group_G   "G ∈ Group"
  defines
    e_def     "e == (G.<e>)"
    binop_def "x # y == (G.<f>) x y"
    inv_def   "i x == (G.<inv>) x"
```

This locale is attached to the theory file for groups. Prior to starting the proofs concerning groups, we open this locale and can subsequently use the syntax and the local assumption G ∈ Group throughout all proofs for groups. This improves the readability of the derivations as well as it reduces the length of the proofs. For example, instead of

```
[| G ∈ Group; x ∈ (G.<cr>); (G.<f>) x x = x |] ==> x = (G.<e>)
```

we can state this theorem now as

```
[| x ∈ (G.<cr>); x # x = x |] ==> x = e
```

Subgoals of the form G ∈ Group that would normally be created in proofs are not there any more because they are now matched by the corresponding locale rule. All group related proofs share this assumption. Thus, the use of a locale rule reduces the length of the proofs.

---

[1] We omitted an abbreviation for G.<cr> to contrast it from the group G.

Cosets. To enable the proof of Sylow's theorem and further results from group theory we define left and right cosets of a group, a product, and an inverse operation for subsets of groups. We create a separate theory for cosets named Coset containing their definitions.

```
r_coset G H a ≡ (λ x. (G.<f>) x a) '' H
l_coset G a H ≡ (λ x. (G.<f>) a x) '' H
set_r_cos G H ≡ r_coset G H '' (G.<cr>)
set_inv G H  ≡ (λ x. (G.<inv>) x) '' H
```

Cosets immediately give rise to the definition of a special class of subgroups, the so-called *normal subgroups* of a group.

```
Normal ≡  Σ G ∈ Group.
   {H | H <<= G  ∧  (∀ x ∈ (G.<cr>). r_coset G H x = l_coset G x H)}
```

We define the convenient syntax H <| G for (G, H) ∈ Normal. As is apparent from the definition, normal subgroups are a special case of subgroups of a group where left and right cosets coincide. This is not necessarily the case in non-Abelian groups.

Since the notion of cosets, e.g. r_coset G H a, depends on the binary operation of the group, they have the additional parameter G. The mathematical notation is Ha. We want have to at least a notation like H #> a.

Locales give us this support. We define a locale for the use of cosets to enable convenient syntax for cosets and products. This locale is defined as an extension of the locale for groups. In the scope of the locale coset, we can omit the group parameter G that is necessary for an adequate formalization and we can define local infix syntax. That is, we can write H #> a instead of r_coset G H a, I(H) instead of set_inv G H, H1 <#> H2 instead of set_prod G H1 H2, and {* H *} for set_r_cos G H. Logically, the short forms refer to the adequate definitions as may be revealed in theorems by export (see Section 2.1).

The theorems we derive about cosets and the set product of groups are needed as a calculational basis for Lagrange's theorem used in Sylow's proof (see Section 4.1) and in the theorems involving the quotient of a group (see Section 3.3). The binary operation of groups is lifted to the level of subsets of a group. We derive algebraic rules relating the coset operators <# and #> with the product operation for subsets <#>. For example, the theorem

```
set_prod G (r_coset G H x) (r_coset G H y) = r_coset G H ((G.<f>) x y)
```

can be written in the scope of the locale coset as

```
(H #> x) <#> (H #> y) = H #> (x # y)
```

The advantage is considerable, especially if we consider that the syntax is not only important when we type in a goal for the first time, but we are confronted with it in each proof step. Hence, the syntactical improvements are crucial for a good interaction with the proof assistant.

3.2   Sylow's Theorem

Sylow's theorem gives criteria for the existence of subgroups of prime power order in finite groups.

Theorem 1. *If G is a group, p a prime and p^α divides the order of G then G contains a subgroup of order p^α.*

In the first mechanization of the theorem [KP99], here referred to as the *ad hoc* version, we were forced to abuse the theory mechanism to achieve readable syntax for the main proof. We declared the local constants and definitions as Isabelle constants and definitions. To model local rules, we used axioms, i.e. Isabelle rules. This works, but contradicts the meaning of axioms and definitions in a theory (*cf.* Section 2).

Locales offer the ideal support for this procedure and the mechanization is methodically sound. In the theory of cosets, we define a locale for the proof of Sylow's theorem. The natural number constants we had to define in [KP99] as constants of an Isabelle theory become now locale constants. The names we use as abbreviations for larger formulas like the set $\mathcal{M} \equiv \{S \subseteq G_{cr} \mid card(S) = p^\alpha\}$ also become added as locale constants. So, the **fixes** section of the locale **sylow** is

```
locale sylow = coset +
  fixes
    p, a, m   :: "nat"
    calM      :: "'a set set"
    RelM      :: "('a set * 'a set)set"
```

The following **defines** section introduces the local definitions of the set $\mathcal{M}$ and the relation $\sim$ on $\mathcal{M}$ (here calM and RelM).

```
defines
  calM_def "calM == {s | s ⊆ (G.<cr>) ∧ card(s) = (p ^ a)}"
  RelM_def "RelM == {(N1,N2) | (N1,N2) ∈ calM × calM
                      ∧ (∃ g ∈ (G.<cr>). N1 = (N2 #> g) )}"
```

Note that the previous definitions depend on the locale constants p, a, and m (and G from locale **group**). We can abbreviate in a convenient way using locale constants without being forced to parameterize the definitions, i.e. without locales we would have to write **calM G p a m** and **RelM G p a m**. Furthermore, without locales the definitions of **calM** and **RelM** would have to be theory level definitions — visible everywhere — whereas now they are just local.

Finally, we add the locale assumptions to the locale **sylow**. Here, we can state all assumption that are local for the 52 theorems of the Sylow proof. In the mechanization of the proof without locales in [KP99] all these merely local assumptions had to become rules of the theory for Sylow.

```
assumes
  Group_G   "G ∈ Group"
  prime_p   "p ∈ prime"
  card_G    "order(G) = (p ^ a) * m"
  finite_G  "finite (G.<cr>)"
```

The locale **sylow** can subsequently be opened to provide just the right context to conduct the proof of Sylow's theorem in the way we discovered in the *ad hoc* approach [KP99] to be appropriate, but now we can define this context soundly. In the earlier mechanization of the theorem, we abused the theory mechanisms of constants, rules and definitions to that end. Apart from the fact that it is meaningless to define local entities globally, we could not use a polymorphic type 'a for the base type of groups. Using polymorphism would have led to inconsistencies, as we would have assumed the Sylow premises for all groups. Hence, we had to use a fixed type, whereby the theorem was not generally applicable to groups. Also, we restricted the definition of local proof contexts to the scope as outlined above, i.e. to entities that are used in the theorem. Having locales, we can extend the encapsulation of local proof context much further than in the *ad hoc* mechanization and closer to the way the paper proof operates.

More Encapsulation. Now, we may soundly use the locale mechanism for *any* merely locally relevant definition. In particular we can define the abbreviation

```
H == {g | g ∈ (G.<cr>)  ∧  M1 #> g = M1}
```

for the main object of concern, the Sylow subgroup that is constructed in the proof. Naturally, we refrained from using a definition for this set before because in the global theorem it is not visible at all, i.e. it is a temporary definition. But, by adding the above line to a new locale, after introducing a suitably typed locale constant in the **fixes** part, the proofs for Sylow's theorem improve a lot. A further measure taken now is to define in the new locale the two assumptions that are visible in most of the 52 theorems of the proof of Sylow's theorem. Summarizing, a locale for the central part of Sylow's proof is given by:

```
locale sylow_central = sylow +
  fixes
    H  :: "'a set"
    M  :: "'a set set"
    M1 :: "'a set"
  assumes
    M_ass "M ∈ calM / RelM ∧
           ¬(p ^ ((max-n r. p ^ r | m)+ 1) | card(M))"
    M1_ass "M1 ∈ M"
  defines
    H_def "H == {g | g ∈ (G.<cr>)  ∧  M1 #> g = M1}"
```

We open this locale after the first few lemmas when we arrive at theorems that use the locale assumptions and definitions. Subsequently, we assume that the locales **group** and **coset** are open. Henceforth, the conjectures become shorter and more readable than in the *ad hoc* version. For example,

```
[| M ∈ calM / RelM
   ∧ ¬(p ^ ((max-n r. p ^ r | m)+ 1) | card(M));
   M1 ∈ M; x ∈ {g | g ∈ (G.<cr>)  ∧  M1 #> g = M1};
   xa ∈ {g. g ∈ (G.<cr>)  ∧  M1 #> g = M1} |]
   ==> x # xa ∈ {g | g ∈ (G.<cr>)  ∧  M1 #> g = M1}
```

can now be stated as

```
[| x ∈ H; xa ∈ H |] ==> x # xa ∈ H
```

Figure 1 illustrates how the scoping for Sylow's proof works. Apart from the

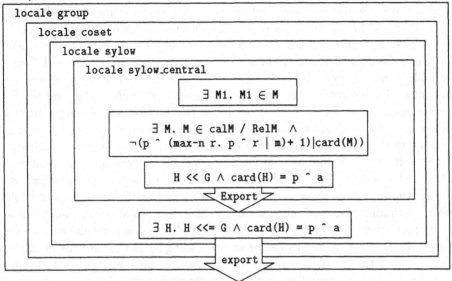

Fig. 1. Sylow's theorem at different levels of nested locales

main theorem, only two other theorems need to be exported from the innermost locale **sylow_central**. These two theorems prove existence of witnesses for the locale assumptions **M_ass** and **M1_ass** and are used to cancel the latter assumptions from the main theorem at the level of locale **sylow**. When we finally export the main theorem from the context of locale **sylow** using the generally normalizing function **export**, we achieve the desired form of Sylow's theorem which is independent from any local definitions and assumptions. The theorem stands alone as a global theorem of the Isabelle theory **Coset**, and is hence applicable to any group. This becomes visible in the resulting theorem by the question marks indicating schematic variables.

There is another feature of locales that we used and that is particularly decisive for the combination of locales and dependent types: a slightly changed polymorphism.

Adapted Polymorphism. The declarations of locale constants may use polymorphism, as seen in most of the examples so far, but this is different to the one usual in Isabelle. Usually, Isabelle's polymorphic declarations are completely independent of each other, e.g. if the same type variable 'a is used in two declarations, these constants may be still instantiated to different types. In locales, we enrich the expressiveness of polymorphic definitions by extending the scope of the polymorphic variable names over all constant declarations of a locale. This changes the usual polymorphism of Isabelle, in that equal names imply the same variable. That is, polymorphic variables are fixed by the variable names, e.g. 'a, inside the locale. Although the locale as an entity can still be instantiated to arbitrary constant types of appropriate sort, the instantiation is implicitly forced to be the same for all constants of a locale that use the same variable name.

This restriction only holds if the same names are used. Naturally we preserve the same freedom of expressiveness that was there before: if we use different variable names in polymorphic declarations of locale constants, they can be instantiated independently. An example is the use of two different groups for the construction of the direct product of groups (see Section 3.3). However, in the Sylow case study — and in abstract algebraic applications in general — this is exactly what we need: we want to constrain different constructors to the same type, while we still want to stay abstract, i.e. use polymorphic declarations. Most of the locales defined for the examples of this paper, e.g. groups, use one polymorphic type variable 'a in different locale constant declarations, while referring to constituents of one structure. That is, they are abstract, but the same type. This "connected" form of polymorphic declaration reflects the connection that is there in the dependent type corresponding to the locale. For example, the fact that the constituents of a Group element are ranging over the same base type, needs to be reflected to the polymorphic type 'a in the constant declarations of locale groups.

3.3   Quotient of a Group

If a group is factorized by one of its normal subgroups then the quotient together with the induced operations on the cosets is again a group. This is a quite standard result of group theory, but it is challenging because it contains a self-reference: a structure constructed from a group shall be a group again. The quotient of a group illustrates the need for structures as dependent types, and hence first class citizens. In addition, we will analyze to what extent locales can be helpful.

For this proof we define a new theory that builds on the theory of cosets. The factorization of a group by one of its normal subgroups is given by the set of cosets. The operations on the cosets are described by the group operations lifted to the level of cosets, i.e. the binary operation is given by the product of cosets, the inverse operation is given by the inverse coset, and the factor of the quotient serves as a unit element, a normal subgroup H. To describe this construction formally, we use our typed λ-notation (see Section 2.2).

```
FactGroup ≡
  λ G ∈ Group. λ H ∈ Normal ↓ G.
  ( carrier = set_r_cos G H,
    bin_op = λ X ∈ set_r_cos G H.
```

```
        λ Y ∈ set_r_cos G H. set_prod G X Y,
  inverse = λ X ∈ set_r_cos G H. set_inv G X,
  unit = H |)
```

We define the theory syntax G **Mod** H for the quotient **FactGroup** G H. To enhance the readability of the construction and thereby the proofs about it, we employ locales. We cannot use any nicer syntax in the above definition of the quotient because in the body of the λ-term above, the terms G and H are parameters. Hence, they have to stay flexible. However, using locales we can fix a group G and a normal subgroup H in G for the local proof context.

```
locale factgroup = coset +
  fixes
    F :: "('a set) grouptype"
    H :: "('a set)"
  assumes
    H_ass "H <| G"
  defines
    F_def "F == FactGroup G H"
```

By defining this locale as an extension of the locale **coset**, we incorporate all the syntactical abbreviations we defined for cosets and operations on cosets in Section 3.1. In addition, we have the group G already as a fixed local constant. The additional definition of the quotient as F lets us derive in the scope of this locale[2]

```
F = (| carrier = {* H *},
       bin_op = (λ X ∈ {* H *}. λ Y ∈ {* H *}. X <#> Y),
       inverse = (λ X ∈ {* H *}. I(X)),
       unit = H |)
```

The derivation is an application of Isabelle's simplifier to the corresponding definitions, and the reduction rules for λ. By the additional use of the locale properties of fixing and local definition, we achieve a readable syntax in a local scope.

   With these preparations, we can prove that this quotient is again a group, which is trivially stated as F ∈ **Group** in the scope of the locale. The proof is straightforward. By backward resolution with the introduction rule for the group property **GroupI** we can reduce it to six subgoals that can be solved by repeatedly applying previously derived results about cosets and the operations on them. Note that here the initial application of **GroupI** illustrates the advantage we gain through the normalization performed by export. Although we proved the rule **GroupI** for the fixed group G we can now apply it again to the group F which is even constructed with that same G.

   By exporting the result that F is a group we get the general formula

```
[| ?G ∈ Group; ?H <| ?G |]  ==> ?G Mod ?H ∈ Group
```

---

[2] Opening factgroup automatically opens coset and group.

In an earlier version of this experiment, we did not employ locales. The statement of the conjecture was even without locales not such a problem; it corresponded to the above formula. But, in the proof of the group property, where all the definitions of the lifted operations have to be employed, we were formerly exposed to formulas that were hard to read.

As a further illustration of the concept of higher order structures, we consider the proof that the constructed λ-term FactGroup is an element of a suitable Π-set.

```
FactGroup ∈ (Π G ∈ Group. (Normal ↓ G) -> Group)
```

This membership statement is equivalent to the structural proposition that the quotient of a group is a function mapping a group and a normal subgroup of this group to another group. We call this kind of theorem a *structural proposition* because membership in a structure (a set) entails the proposition we just proved, i.e. that the quotient is a group. If we interpret the sets that are our structures as types, then we see how the Curry-Howard isomorphism of propositions-as-types [How80] is embodied in a statement like above. In contrast to type theory, we do not need to state this isomorphism as a paradigm — it is inherent because we use sets: from the above we can derive the logical proposition.

Further Examples and Analysis. Other theorems we mechanized [Kam99a] but cannot present here due to space limitations are:

- the direct product of two groups is again a group
- the set of bijections with the appropriate operations of composition of bijections, inverse bijection, and identical bijection forms a group
- the automorphisms of a ring form a group
- the full version of Tarski's fixed point theorem, i.e. the classical theorem with the addition that the set of all fixed points of the continuous function f is itself a complete lattice.

As with the quotient of groups, we first performed the mechanization without the use of locales[3]. In comparison, we could reduce the size of the proofs by 50% using locales. Although in the latter version some savings are due to polishing the proofs by improving the applications of automatic simplification tactics, a larger portion is due to locales. Furthermore, the streamlining of the proofs was made much easier because of the greatly improved comprehensibility. Where we were lost before in grasping huge complicated terms, and thus sometimes misled from the optimum solution, the natural representation achieved by locales leads the way now. Locales allow us to use the same local definition and assumptions as in a module. At the same time the structures, like groups and rings, are first class citizens, whereby we achieve adequacy. Through the combination with locales the higher complexity of the adequate formalizations is balanced out.

Hence, the combination of locales and dependent types adds up to modular reasoning. Since locales are reflected into the meta-logic (see Section 2.1), this

---

[3] At the time the implementation was not capable of dealing with nested locales.

combination does not introduce inconsistencies and enables reuse of locales as we will see in the following section.

3.4   Operations on Modules

Through the embedding of structures as Σ-types and Π-types, we achieve first class representations of modules. Thereby, we are able to use structures in formulas. Moreover, we illustrate in the present section that we can express general operations on structures such as forgetful functors. We show how the substructure of a ring that is an Abelian group can be revealed. For this example we first have to explain how we formalized rings. Using extension of record types, we can build the base type for rings on the base type for groups **grouptype**.

```
record 'a ringtype = 'a grouptype +
   Rmult    :: "['a, 'a] => 'a"
```

Thereby, we inherit the components of groups and can form rings by just extending the latter by the second operation **Rmult**[4]. We add the syntax R.<m> for the additional element **Rmult** of a ring to adapt the notation for rings to the syntax of the group projections.

To isolate the group contained in a ring we can use an element of a Π-set. This λ-function represents a forgetful functor. It "forgets" some structure.

```
group_of :: "'a ringtype => 'a grouptype"
"group_of == λ R ∈ Ring.
   (| carrier = (R.<cr>), bin_op = (R.<f>),
       inverse = (R.<inv>), unit = (R.<e>)  |)"
```

Thereby, we are able to refer to the substructure of the ring that forms an Abelian group using the forgetful functor **group_of**. We can derive the theorem[5]

```
R ∈ Ring ==> group_of R ∈ AbelianGroup          [R_Abel]
```

This enables a better structuring and decomposition of proofs. In particular, we can use this functor when we employ locales for ring related proofs. Then we want to use the encapsulation already provided for groups by the locale **group**. To achieve this we define the locale for rings as an extension.

```
locale ring = group +
   fixes
      R    :: "'a ringtype"
      rmult :: "['a, 'a] => 'a" (infixr "**" 80)
   assumes
      Ring_R "R ∈ Ring"
   defines
      rmult_def "x ** y == (R.<m>) x y"
      R_id_G    "G == group_of R"
```

---

[4] Note that we formalize rings without 1. Mathematical textbooks sometimes use the notion of rings for rings with a 1 for convenience.

[5] **AbelianGroup** is the structure of Abelian, i.e. commutative, groups. Their definition is a simple extension from the one of groups by the additional commutativity.

Note that we are able to use the locale constant G again in a locale definition, i.e. R_id_G. This is sound because we have not defined G yet. If one gives a constant an inconsistent definition, then one will be unable to instantiate results proved in the locale. This way of reusing the local proof context of groups for the superstructure of rings illustrates the flexibility of locales as well as the ease of integration with the mechanization of structures given by Σ and Π.

Theorems that are proved in the setup of the locale ring using group results will have in the exported form the assumptions

    [| R ∈ Ring; group_of R ∈ Group ... |] ==> ...

But, as an implication of the theorem R_Abel, we can easily derive

    R ∈ Ring ==> group_of R ∈ Group

Thus, the second premise can be cancelled. Although we have to do a final proof step to cancel the additional premise, this shows the advantage of locales being reflected onto premises of the global representations of theorems: it is impossible to introduce unsoundness. A definition of a locale constant that is not consistent with its properties stated by locale rules would be not applicable. Since locale assumptions and definitions are explained through meta-assumptions, the resulting theorem would carry the inconsistent assumptions implicitly. We see that the nested structure of locales is consistent with the logical structure because locales are reflected to the meta-logic. Thereby reuse of the locale of groups is possible.

4   Conclusion

4.1   Related Work

The proof of the theorem of Lagrange has been performed with the Boyer Moore Prover [Yu90]. E. Gunter formalized group theory in HOL [Gun89]. In the higher order logic theorem prover IMPS [FGT93] some portion of abstract algebra including Lagrange is proved. Mizar's [Try93] library of formalized mathematics contains probably more abstract algebra theorems than any other system. However, to our knowledge we were the first to mechanically prove Sylow's first theorem. Since it uses Lagrange's theorem, we had to prove this first. In contrast to the formalization as seen in [Yu90] the form of Lagrange that we need for Sylow's theorem is not just the one stating that the order of the subgroup divides the order of the group but instead gives the precise representation of the group's order as the product of order of the subgroup and the *index* of this subgroup in G. Since we have a first class representation of groups, we can express this order equation and can use general results about finite sets to reduce it to simpler theorems about cosets. Hence, compared to [Yu90] our proof of Lagrange is simpler.

Locales implement a sectioning device similar to that in AUTOMATH [dB80] or Coq [Dow90]. In contrast to this kind of sections, locales are defined statically. Also, optional pretty printing syntax is part of the concept. The HOL system [GM93] has a concept of abstract theories based on Gunter's experiments with abstract algebra [Gun89,Gun90] in parts comparable to locales.

4.2 Discussion

Modules for theorem provers can be considered as a means to represent abstract structures. In that case modules need to be first class citizens to enable adequacy. Another aspect of modules is the locality they provide by their scoping, which is useful — if sometimes not necessary — for shortening and hence readability of formulas and proofs.

The embedding of dependent types combines the expressiveness of type theories with the convenience and power of higher order logic theorem proving. Although the dependent types are only modeled as typed sets of Isabelle/HOL we get the "expressive advantage". In contrast to earlier mechanizations of dependent types in higher order logic [JM93] our embedding is relatively light-weight as it is based on a simple set-theoretic embedding. At the same time the Π and Σ-types are strong enough to express higher-level modular notions, like mappings between parameterized structures.

Locales are a general concept for locality. They are not restricted to any particular object logic of Isabelle, i.e. they can be used for any kind of reasoning. Although they are not first class citizens, there is the export device that reflects locales to meta-logical assumptions, thereby explaining them in terms of Isabelle's meta-logic.

Locality and adequacy are separate aspects that may sometimes coincide, but in general they should be treated individually. We have illustrated this by showing how the devices of locales for locality and dependent types for adequacy add up to support modular reasoning. The presented case studies contain as well aspects that have to be expressed adequately by first class modules given by dependent types as ones that needed the structuring and syntactic support of locales. Moreover, we have shown that the separation of the concepts does not hinder their smooth combination. Where the first class representations become too complicated, locales can be used to reduce them. Moreover, in intricate combinations like the forgetful functor in Section 3.4 we have seen that the reflection of locales to the meta-logic preserves consistency and enables reuse.

Hence, instead of using one powerful but inadequate concept for modular reasoning, like classical modules, we think that locales combined with dependent types are appropriate. The separation is tailored for Isabelle, yet it is applicable to other theorem provers. Since the difference between adequacy and locality is a general theoretical issue, the conceptual design of a combination of two devices for the support of modular reasoning presented in this paper is of more general interest.

References

[Bai98]   A. Bailey. *The Machine-Checked Literate Formalisation of Algebra in Type Theory*. PhD thesis, University of Manchester, 1998.

[D+93]   Gilles Dowek et al. The Coq proof assistant user's guide. Technical Report 154, INRIA-Rocquencourt, 1993.

[dB80]   N. G. de Bruijn. A Survey of the Project AUTOMATH. In Seldin and Hindley [SH80], pages 579–606.

[Dow90]   G. Dowek. Naming and Scoping in a Mathematical Vernacular. Technical Report 1283, INRIA, Rocquencourt, 1990.

[FGT93]   W. M. Farmer, J. D. Guttman, and F. J. Thayer. IMPS: an Interactive Mathematical Proof System. *Journal of Automated Reasoning*, 11:213–248, 1993.

[GM93]   M. J. C. Gordon and T. F. Melham, editors. *Introduction to HOL, a Theorem Proving Environment for Higher Order Logic*. Cambridge University Press, 1993.

[Gun89]   E. L. Gunter. Doing Algebra in Simple Type Theory. Technical Report MS-CIS-89-38, Dep. of Computer and Information Science, University of Pennsylvania, 1989.

[Gun90]   E. L. Gunter. The Implementation and Use of Abstract Theories in HOL. In *Third HOL Users Meeting*, Aarhus University, 1990.

[How80]   W.A. Howard. The formulae-as-types notion of construction. In Seldin and Hindley [SH80], pages 479–490.

[JM93]   B. Jacobs and T. F. Melham. Translating Dependent Type Theory into Higher Order Logic. In M. Bezem and J. F. Groote, editors, *Typed Lambda Calculi and Applications*, number 664 in LNCS. Springer, 1993.

[Kam99a]   F. Kammüller. *Modular Reasoning in Isabelle*. PhD thesis, University of Cambridge, 1999. Technical Report 470.

[Kam99b]   F. Kammüller. Modular Structures as Dependent Types in Isabelle. In *Types for Proofs and Programs*, volume 1657 of *LNCS*. Springer, 1999.

[KP99]   F. Kammüller and L. C. Paulson. A Formal Proof of Sylow's First Theorem – An Experiment in Abstract Algebra with Isabelle HOL. *Journal of Automated Reasoning*, 23(3-4):235–264, 1999.

[KWP99]   F. Kammüller, M. Wenzel, and L. C. Paulson. Locales – a Sectioning Concept for Isabelle. In *Theorem Proving in Higher Order Logics, TPHOLs'99*, volume 1690 of *LNCS*. Springer, 1999.

[LP92]   Z. Luo and R. Pollack. Lego proof development system: User's manual. Technical Report ECS-LFCS-92-211, University of Edinburgh, 1992.

[Mac86]   D. B. MacQueen. Using Dependant Types to Express Modular Structures. In *Proc. 13th ACM Symp. Principles Programming Languages*. ACM Press, 1986.

[OSRSC98]   S. Owre, N. Shankar, J.M. Rushby, and D.W.J. Stringer-Calvert. PVS Language Reference. Part of the PVS Manual. Available on the Web as http://www.csl.sri.com/pvsweb/manuals.html, September 1998.

[Pau94]   L. C. Paulson. *Isabelle: A Generic Theorem Prover*, volume 828 of *LNCS*. Springer, 1994.

[SH80]   J.P. Seldin and J.R. Hindley, editors. *To H. B. Curry: Essays on Combinatory Logic*, Academic Press Limited, 1980.

[Try93]   A. Trybulec. Some Features of the Mizar Language. 1993. Available from Mizar user's group.

[Yu90]   Y. Yu. Computer Proofs in Group Theory. *Journal of Automated Reasoning*, 6:251–286, 1990.

An Infrastructure for Intertheory Reasoning

William M. Farmer

Department of Computing and Software
McMaster University, 1280 Main Street West
Hamilton, Ontario L8S 4L7, Canada
wmfarmer@mcmaster.ca

Abstract. The little theories method, in which mathematical reasoning is distributed across a network of theories, is a powerful technique for describing and analyzing complex systems. This paper presents an infrastructure for intertheory reasoning that can support applications of the little theories method. The infrastructure includes machinery to store theories and theory interpretations, to store known theorems of a theory with the theory, and to make definitions in a theory by extending the theory "in place". The infrastructure is an extension of the intertheory infrastructure employed in the IMPS Interactive Mathematical Proof System.

1 Introduction

Mathematical reasoning is always performed within some context, which includes vocabulary and notation for expressing concepts and assertions, and axioms and inference rules for proving conjectures. In informal mathematical reasoning, the context is almost entirely implicit. In fact, substantial mathematical training is often needed to "see" the context.

The situation is quite different in formal mathematics performed in logical systems often with the aid of computers. The context is formalized as a mathematical structure. The favored mathematical structure for this purpose is an axiomatic theory within a formal logic. An axiomatic theory, or *theory* for short, consists of a formal language and a set of axioms expressed in the language. It is a specification of a set of objects: the language provides names for the objects and the axioms constrain what properties the objects have.

Sophisticated mathematical reasoning usually involves several related but different mathematical contexts. There are two main ways of dealing with a multitude of contexts using theories. The *big theory method* is to choose a highly expressive theory—often based on set theory or type theory—that can represent many different contexts. Each context that arises is represented in the theory or in an extension of the theory. Contexts are related to each other in the theory itself.

An alternate approach is the *little theories method* in which separate contexts are represented by separate theories. Structural relationships between contexts are represented as interpretations between theories (see [4, 19]). Interpretations

D. McAllester (Ed.): CADE-17, LNAI 1831, pp. 115–131, 2000.

serve as conduits for passing information (e.g., definitions and theorems) from abstract theories to more concrete theories, or indeed to other equally abstract theories. As a result, the big theory is replaced with a network of theories—which can include both small compact theories and large powerful theories. The little theories approach has been used in both mathematics and computer science (see [10] for references). In [10] we argue that the little theories method offers important advantages for mechanized mathematics. Many of these advantages have been demonstrated by the IMPS Interactive Mathematical Proof System [9, 11] which supports the little theories method.

A mechanized mathematics system based on the little theories method requires a different infrastructure than one based on the big theory method. In the big theory method all reasoning is performed within a single theory, while in the little theories method there is both intertheory and intratheory reasoning. This paper presents an infrastructure for intertheory reasoning that can be employed in several kinds of mechanized mathematics systems including theorem provers, software specification and verification systems, computer algebra systems, and electronic mathematics libraries. The infrastructure is closely related to the intertheory infrastructure used in IMPS, but it includes some capabilities which are not provided by the IMPS intertheory infrastructure.

The little theories method is a major element in the design of several software specification systems including EHDM [18], IOTA [16], KIDS [20], OBJ3 [12], and Specware [21]. The intertheory infrastructures of these systems are mainly for constructing theories and linking them together into a network. They do not support the rich interplay of making definitions, proving theorems, and "transporting" definitions and theorems from one theory to another needed for developing and exploring theories within a network.

The Ergo [17] theorem proving system is another theorem proving system besides IMPS that directly supports the little theories method.[1] Its infrastructure for intertheory reasoning provides full support for *constructing theories* from other theories via inclusion and interpretation but only partial support for *developing theories* by making definitions and proving theorems. In Ergo, theory interpretation is static: theorems from the source theory of an interpretation can be transported to the target theory of the interpretation only when the interpretation is created [14]. Theory interpretation is dynamic in the intertheory infrastructure of this paper (and of IMPS).

The rest of the paper is organized as follows. The underlying logic of the intertheory infrastructure is given in section 2. Section 3 discusses the design requirements for the infrastructure. The infrastructure itself is presented in section 4. Finally, some applications of the infrastructure are described in section 5.

---

[1] Many theorem proving systems indirectly support the little theories methods by allowing a network of theories to be formalized within a big theory.

2   The Underlying Logic

The intertheory infrastructure presented in this paper assumes an underlying logic. Many formal systems, including first-order logic and Zermelo-Fraenkel set theory, could serve as the underlying logic. For the sake of convenience and precision, we have chosen a specific underlying logic for the infrastructure rather than treating the underlying logic as a parameter. Our choice is Church's simple theory of types [3], denoted in this paper by **C**.

The underlying logics of many theorem proving systems are based on **C**. For example, the underlying logic of IMPS (and its intertheory infrastructure) is a version of **C** called LUTINS [5, 6, 8]. Unlike **C**, LUTINS admits undefined terms, partial functions, and subtypes. By virtue of its support for partial functions and subtypes, many theory interpretations can be expressed more directly in LUTINS than in **C** [8].

We will give now a brief presentation of **C**. The missing details can be filled in by consulting Church's original paper [3] or one of the logic textbooks, such as [1], which contains a full presentation of **C**. We will also define a number of logical notions in the context of **C** including the notions of a theory and an interpretation.

2.1   Syntax of C

The *types* of **C** are defined inductively as follows:

1. ι is a type (which denotes the type of individuals).
2. $*$ is a type (which denotes the type of truth values).
3. If α and β are types, then $(\alpha \to \beta)$ is a type (which denotes the type of total functions that map values of type α to values of type β).

Let $\mathcal{T}$ denote the set of types of **C**.

A *tagged symbol* is a symbol tagged with a member of $\mathcal{T}$. A tagged symbol whose symbol is a and whose tag is α is written as a_α. Let $\mathcal{V}$ be a set of tagged symbols called *variables* such that, for each $\alpha \in \mathcal{T}$, the set of members of $\mathcal{V}$ tagged with α is countably infinite. A *constant* is a tagged symbol c_α such that $c_\alpha \notin \mathcal{V}$.

A *language* L of **C** is a set of constants. (In the following, let a "language" mean a "language of **C**".) An *expression of type α* of L is a finite sequence of symbols defined inductively as follows:

1. Each $a_\alpha \in \mathcal{V} \cup L$ is an expression of type α.
2. If F is an expression of type $\alpha \to \beta$ and A is an expression of type α, then $F(A)$ is an expression of type β.
3. If $x_\alpha \in \mathcal{V}$ and E is an expression of type β, then $(\lambda x_\alpha . E)$ is an expression of type $\alpha \to \beta$.
4. If A and B are expressions of type α, then $(A = B)$ is an expression of type $*$.

5. If A and B are expressions of type $*$, then $\neg A$, $(A \supset B)$, $(A \wedge B)$, and $(A \vee B)$ are expressions of type $*$.

6. If $x_\alpha \in \mathcal{V}$ and E is an expression of type $*$, then $(\forall x_\alpha . E)$ and $(\exists x_\alpha . E)$ are expressions of type $*$.

Expressions of type α are denoted by $A_\alpha, B_\alpha, C_\alpha$, etc. Let $\mathcal{E}_L$ denote the set of expressions of L. "Free variable", "closed expression", and similar notions are defined in the obvious way. Let $\mathcal{S}_L$ denote the set of *sentences* of L, i.e., the set of closed expressions of type $*$ of L.

2.2   Semantics of C

For each language L, there is a set $\mathcal{M}_L$ of *models* and a relation $\models$ between models and sentences of L. $M \models A_*$ is read as "M is a model of A_*". Let L be a language, $A_* \in \mathcal{S}_L$, $\Gamma \subseteq \mathcal{S}_L$, and $M \in \mathcal{M}_L$. M is a *model of* Γ, written $M \models \Gamma$, if $M \models B_*$ for all $B_* \in \Gamma$. Γ *logically implies* A_*, written $\Gamma \models A_*$, if every model of Γ is a model of A_*.

2.3   Theories

A *theory* of C is a pair $T = (L, \Gamma)$ where L is a language and $\Gamma \subseteq \mathcal{S}_L$. Γ serves as the set of axioms of T. (In the following, let a "theory" mean a "theory of C".) A_* is a *(semantic) theorem* of T, written $T \models A_*$, if $\Gamma \models A_*$. T is *consistent* if some sentence of L is not a theorem of T. A theory $T' = (L', \Gamma')$ is an *extension* of T, written $T \leq T'$, if $L \subseteq L'$ and $\Gamma \subseteq \Gamma'$. T' is a *conservative extension* of T, written $T \trianglelefteq T'$, if $T \leq T'$ and, for all $A_* \in \mathcal{S}_L$, if $T' \models A_*$, then $T \models A_*$.

The following lemma about theory extensions is easy to prove.

Lemma 1. *Let T_1, T_2, and T_3 be theories.*

1. *If $T_1 \leq T_2 \leq T_3$, then $T_1 \leq T_3$.*
2. *If $T_1 \trianglelefteq T_2 \trianglelefteq T_3$, then $T_1 \trianglelefteq T_3$.*
3. *If $T_1 \leq T_2 \leq T_3$ and $T_1 \trianglelefteq T_3$, then $T_1 \trianglelefteq T_2$.*
4. *If $T_1 \trianglelefteq T_2$ and T_1 is consistent, then T_2 is consistent.*

2.4   Interpretations

Let $T = (L, \Gamma)$ and $T' = (L', \Gamma')$ be theories, and let $\Phi = (\gamma, \mu, \nu)$ where $\gamma \in \mathcal{T}$ and $\mu : \mathcal{V} \to \mathcal{V}$ and $\nu : L \to \mathcal{E}_{L'}$ are total functions.

For $\alpha \in \mathcal{T}$, $\Phi(\alpha)$ is defined inductively as follows:

1. $\Phi(\iota) = \gamma$.
2. $\Phi(*) = *$.
3. If $\alpha, \beta \in \mathcal{T}$, then $\Phi(\alpha \to \beta) = \Phi(\alpha) \to \Phi(\beta)$.

Φ is a *translation from L to L'* if:

1. For all $x_\alpha \in \mathcal{V}$, $\mu(x_\alpha)$ is of type $\Phi(\alpha)$.
2. For all $c_\alpha \in L$, $\nu(c_\alpha)$ is of type $\Phi(\alpha)$.

Suppose Φ is a translation from L to L'. For $E_\alpha \in \mathcal{E}_L$, $\Phi(E_\alpha)$ is the member of $\mathcal{E}_{L'}$ defined inductively as follows:

1. If $E_\alpha \in \mathcal{V}$, then $\Phi(E_\alpha) = \mu(E_\alpha)$.
2. If $E_\alpha \in L$, then $\Phi(E_\alpha) = \nu(E_\alpha)$.
3. $\Phi(F_{\alpha \to \beta}(A_\alpha)) = \Phi(F_{\alpha \to \beta})(\Phi(A_\alpha))$.
4. $\Phi(\lambda x_\alpha . E_\beta) = (\lambda \Phi(x_\alpha) . \Phi(E_\beta))$.
5. $\Phi(A_\alpha = B_\alpha) = (\Phi(A_\alpha) = \Phi(B_\alpha))$
6. $\Phi(\neg E_*) = \neg \Phi(E_*)$.
7. $\Phi(A_* \,\square\, B_*) = (\Phi(A_*) \,\square\, \Phi(B_*))$ where $\square \in \{\supset, \wedge, \vee\}$.
8. $\Phi(\square x_\alpha . E_*) = (\square \Phi(x_\alpha) . \Phi(E_*))$ where $\square \in \{\forall, \exists\}$.

Φ is an *interpretation of T in T'* if it is a translation from L to L' that maps theorems to theorems, i.e., for all $A_* \in \mathcal{S}_L$, if $T \models A_*$, then $T' \models \Phi(A_*)$.

Theorem 1 (Relative Consistency). *Suppose Φ be an interpretation of T in T' and T' is consistent. Then T is consistent.*

Proof. Assume $\Phi = \langle \gamma, \mu, \nu \rangle$ is an interpretation of T in T', T' is consistent, and T is inconsistent. Then $F_* = (\exists x_\iota . \neg(x_\iota = x_\iota))$ is a theorem of T, and so $\Phi(F_*) = (\exists \mu(x_\iota) . \neg(\mu(x_\iota) = \mu(x_\iota)))$ is a theorem of T', which contradicts the consistency of T'. $\square$

The next theorem gives a sufficient condition for a translation to be an interpretation.

Theorem 2 (Interpretation Theorem). *Suppose Φ is a translation from L to L' and, for all $A_* \in \Gamma$, $T' \models \Phi(A_*)$. Then Φ is an interpretation of T in T'.*

Proof. The proof is similar to the proof of Theorem 12.4 in [6]. $\square$

3   Design Requirements

At a minimum, an infrastructure for intertheory reasoning should provide the capabilities to store theories and interpretations and to record theorems as they are discovered. We present in this section a "naive" intertheory infrastructure with just these capabilities. We then show that the naive infrastructure lacks several important capabilities. From these results we formulate the requirements that an intertheory infrastructure should satisfy.

3.1   A Naive Intertheory Infrastructure

We present now a naive intertheory infrastructure. In this design, the state of the infrastructure is a set of *infrastructure objects*. The infrastructure state is initially the empty set. It is changed by the application of *infrastructure operations* which add new objects to the state or modify objects already in the state. There are three kinds of infrastructure objects for storing theories, theorems, and interpretations, respectively, and there are four infrastructure operations for creating the three kinds of objects and for "installing" theorems in theories.

Infrastructure objects are denoted by boldface letters. The three infrastructure objects are defined simultaneously as follows:

1. A *theory object* is a tuple $\mathbf{T} = (n, L, \Gamma, \Sigma)$ where n is a string, L is a language, $\Gamma \subseteq \mathcal{S}_L$, and Σ is a set of theorem objects. n is called the *name* of $\mathbf{T}$ and is denoted by $[\mathbf{T}]$. (L, Γ) is called the *theory* of $\mathbf{T}$ and is denoted by $\mathrm{thy}(\mathbf{T})$.
2. A *theorem object* is a tuple $\mathbf{A} = ([\mathbf{T}], A_*, J)$ where $\mathbf{T} = (n, L, \Gamma, \Sigma)$ is a theory object, $A_* \in \mathcal{S}_L$, and J is a justification[2] that $\mathrm{thy}(\mathbf{T}) \models A_*$.
3. An *interpretation object* is a tuple $\mathbf{I} = ([\mathbf{T}], [\mathbf{T}'], \Phi, J)$ where $\mathbf{T}$ and $\mathbf{T}'$ are theory objects, Φ is a translation, and J is a justification that Φ is an interpretation of $\mathrm{thy}(\mathbf{T})$ in $\mathrm{thy}(\mathbf{T}')$.

Let $\mathbf{S}$ denote the infrastructure state. The four infrastructure operations are defined as follows:

1. Given a string n, a language L, and $\Gamma \subseteq \mathcal{S}_L$ as input, if, for all theory objects $\mathbf{T}' = (n', L', \Gamma', \Sigma') \in \mathbf{S}$, $n \neq n'$ and $\mathrm{thy}(\mathbf{T}) \neq \mathrm{thy}(\mathbf{T}')$, then create-thy-obj adds the theory object $(n, L, \Gamma, \emptyset)$ to $\mathbf{S}$; otherwise, the operation fails.
2. Given a theory object $\mathbf{T} \in \mathbf{S}$, a sentence A_*, and a justification J as input, if $\mathbf{A} = ([\mathbf{T}], A_*, J)$ is a theorem object, then create-thm-obj adds $\mathbf{A}$ to $\mathbf{S}$; otherwise, the operation fails.
3. Given two theory objects $\mathbf{T}, \mathbf{T}' \in \mathbf{S}$, a translation Φ, and a justification J as input, if $\mathbf{I} = ([\mathbf{T}], [\mathbf{T}'], \Phi, J)$ is an interpretation object, then create-int-obj adds $\mathbf{I}$ to $\mathbf{S}$; otherwise, the operation fails.
4. Given a theorem object $\mathbf{A} = ([\mathbf{T}], A_*, J) \in \mathbf{S}$ and a theory object $\mathbf{T}' = (n', L', \Gamma', \Sigma') \in \mathbf{S}$ as input, if $\mathrm{thy}(\mathbf{T}) \leq \mathrm{thy}(\mathbf{T}')$, then install-thm-obj replaces $\mathbf{T}'$ in $\mathbf{S}$ with the theory object $(n', L', \Gamma', \Sigma' \cup \{\mathbf{A}\})$; otherwise, the operation fails.

---

[2] The notion of a justification is not specified. It could, for example, be a formal proof.

3.2 Missing Capabilities

The naive infrastructure is missing four important capabilities:

A. Definitions. Suppose we would like to make a definition that the constant $\textsf{is\_zero}_{\iota \to *}$ is the predicate $(\lambda x_\iota . x_\iota = 0_\iota)$ in a theory T stored in a theory object $\mathbf{T} = (n, L, \Gamma, \Sigma) \in \mathbf{S}$. The naive infrastructure offers only one way to do this: create the extension $T' = (L', \Gamma')$ of T, where $L' = L \cup \{\textsf{is\_zero}_{\iota \to *}\}$ and

$$\Gamma' = \Gamma \cup \{\textsf{is\_zero}_{\iota \to *} = (\lambda x_\iota . x_\iota = 0_\iota)\},$$

and then store T' in a new theory object $\mathbf{T}'$ by invoking create-thy-obj. If $\textsf{is\_zero}_{\iota \to *}$ is not in L, T and T' can be regarded as the same theory since $T \trianglelefteq T'$ and $\textsf{is\_zero}_{\iota \to *}$ can be "eliminated" from any expression of L' by replacing every occurrence of it with $(\lambda x_\iota . x_\iota = 0_\iota)$.

Definitions are made all the time in mathematics, and thus, implementing definitions in this way will lead to an explosion of theory objects storing theories that are essentially the same. A better way of implementing definitions would be to extend T to T' "in place" by replacing T in $\mathbf{T}$ with T'. The resulting object would still be a theory object because every theorem of T is also a theorem of T'.

This approach, however, would introduce a new problem. If an interpretation object $\mathbf{I} = ([\mathbf{T}], [\mathbf{T}'], \Phi, J) \in \mathbf{S}$ and thy($\mathbf{T}$) is extended in place by making a definition $c_\alpha = E_\alpha$, then the interpretation Φ would no longer be an interpretation of T in T' because $\Phi(c_\alpha)$ would not be defined.

There are three basic solutions to this problem. The first one is to automatically extend Φ to an interpretation of T in T' by defining $\Phi(c_\alpha) = \Phi(E_\alpha)$. However, this solution has the disadvantage that, when an expression of T containing c_α is translated to an expression of T' via the extended Φ, the expression of T will be expanded into a possibly much bigger expression of T'.

The second solution is to automatically transport the definition $c_\alpha = E_\alpha$ from T to a T' via Φ by making a new definition of the form $d_\beta = \Phi(E_\alpha)$ in T' and defining $\Phi(c_\alpha) = d_\beta$. The implementation of this solution would require care because, when two similar theories are both interpreted in a third theory, common definitions in the source theories may be transported multiple times to the target theory, resulting in definitions in the target theory that define different constants in exactly the same way.

The final solution is to let the user extend Φ by hand whenever it is necessary. This solution is more flexible than the first two solutions, but it would impose a heavy burden on the user. Our experience in developing IMPS suggests that the best solution would be some combination of these three basic solutions.

B. Profiles. Suppose we would like to make a "definition" that the constant $\textsf{a\_non\_zero}_\iota$ has a value not equal to 0_ι in a theory T stored in a theory object $\mathbf{T} = (n, L, \Gamma, \Sigma) \in \mathbf{S}$. That is, we would like to add a new constant $\textsf{a\_non\_zero}_\iota$ to L whose value is specified, but not necessarily uniquely determined, by the sentence $\neg(\mathsf{a\_non\_zero}_\iota = 0_\iota)$. More precisely, let $T' = (L', \Gamma')$ where $L' = L \cup \{\mathsf{a\_non\_zero}_\iota\}$ and

$$\Gamma' = \Gamma \cup \{\neg(\mathsf{a\_non\_zero}_\iota = 0_\iota)\}.$$

If $\mathsf{a\_non\_zero}_\iota$ is not in L and the sentence $(\exists\, x_\iota \,.\, \neg(x_\iota = 0_\iota))$ is a theorem of T, then $T \trianglelefteq T'$.

We call definitions of this kind *profiles*.[3] A profile introduces a finite number of new constants that satisfy a given property. Like ordinary definitions, profiles produce conservative extensions, but unlike ordinary definitions, the constants introduced by a profile cannot generally be eliminated. A profile can be viewed as a generalization of a definition since any definition can be expressed as a profile.

Profiles are very useful for introducing new machinery into a theory. For example, a profile can be used to introduce a collection of objects plus a set of operations on the objects—what is called an "algebra" in mathematics and an "abstract datatype" in computer science. The new machinery will not compromise the original machinery of T because the resulting extension T' of T will be conservative. Since T' is a conservative extension of T, any reasoning performed in T could just as well have been performed in T'. Thus the availability of T' normally makes T obsolete.

Making profiles in the naive infrastructure leads to theory objects which store obsolete theories. The way of implementing definitions by extending theories in place would work just as well for profiles. As with definitions, extending theories in place could cause some interpretations to break. A combination of the second and third basic solutions to the problem given above for definitions could be used for profiles. The first basic solution is not applicable because profiles do not generally have the eliminability property of definitions.

C. Theory Extensions. Suppose that $\mathbf{S}$ contains two theory objects $\mathbf{T}$ and $\mathbf{T}'$ with $\mathrm{thy}(\mathbf{T}) \leq \mathrm{thy}(\mathbf{T}')$. In most cases (but not all), one would want every theorem object installed in $\mathbf{T}$ to also be installed in $\mathbf{T}'$. The naive infrastructure does not have this capability. That is, there is no support for having theorem objects installed in a theory object to automatically be installed in preselected extensions of the theory object. An intertheory infrastructure should guarantee that, for each theory object $\mathbf{T}$ and each preselected extension $\mathbf{T}'$ of $\mathbf{T}$, every theorem, definition, and profile installed in $\mathbf{T}$ is also installed in $\mathbf{T}'$.

D. Theory Copies. The naive infrastructure does not allow the infrastructure state to contain two theory objects storing the same theory. As a consequence, it is not possible to add a copy of a theory object to the infrastructure state. We will see in section 5 that creating copies of a theory object is a useful modularization technique.

---

[3] Profiles are called *constant specifications* in [13] and *constraints* in [15].

3.3   Requirements

Our analysis of the naive intertheory infrastructure suggests that the intertheory infrastructure should satisfy the following requirements:

R1 *The infrastructure enables theories and interpretations to be stored.*

R2 *Known theorems of a theory can be stored with the theory.*

R3 *Definitions can be made in a theory by extending the theory in place.*

R4 *Profiles can be made in a theory by extending the theory in place.*

R5 *Theorems, definitions, and profiles installed in a theory are automatically installed in certain preselected extensions of the theory.*

R6 *An interpretation of T_1 in T_2 can be extended in place to an interpretation of T_1' in T_2' if T_i is extended to T_i' by definitions or profiles for $i = 1, 2$.*

R7 *A copy of a stored theory can be created and then developed independently from the original theory.*

The naive infrastructure satisfies only requirements **R1** and **R2**. The IMPS intertheory infrastructure satisfies all of the requirements except **R4** and **R7**.

4   The Intertheory Infrastructure

This section presents an intertheory infrastructure that satisfies all seven requirements in section 3.3. It is the same as the naive infrastructure except that the infrastructure objects and operations are different. That is, the infrastructure state is a set of infrastructure objects, is initially the empty set, and is changed by the application of infrastructure operations which add new objects to the state or modify objects already in the state. As in the naive infrastructure, let **S** denote the infrastructure state.

4.1   Objects

There are five kinds of infrastructure objects. The first four are defined simultaneously as follows:

1. A *theory object* is a tuple $\mathbf{T} = (n, L_0, \Gamma_0, L, \Gamma, \Delta, \sigma, \mathcal{N})$ where:
   (a) n is a string called the *name* of $\mathbf{T}$. It is denoted by $[\mathbf{T}]$.
   (b) L_0 and L are languages such that $L_0 \subseteq L$. L_0 and L are called the *base language* and the *current language* of $\mathbf{T}$, respectively.
   (c) $\Gamma_0 \subseteq \mathcal{S}_{L_0}$ and $\Gamma \subseteq \mathcal{S}_L$ with $\Gamma_0 \subseteq \Gamma$. The members of Γ_0 and Γ are called the *base axioms* and the *current axioms* of $\mathbf{T}$, respectively.
   (d) $\Gamma \subseteq \Delta \subseteq \{A_* \in \mathcal{S}_L : \Gamma \models A_*\}$. The members of Δ are called the *known theorems* of $\mathbf{T}$, and Δ is denoted by $\mathrm{thms}(\mathbf{T})$.

(e) σ is a finite sequence of theorem, definition, and profile objects called the *event history* of $\mathbf{T}$.

(f) $\mathcal{N}$ is a set of names of theory objects called the *principal subtheories* of $\mathbf{T}$. For each $[\mathbf{T}'] \in \mathcal{N}$ with $\mathbf{T}' = (n', L_0', \Gamma_0', L', \Gamma', \Delta', \sigma', \mathcal{N}')$, $L_0' \subseteq L_0$, $\Gamma_0' \subseteq \Gamma_0$, $L' \subseteq L$, $\Gamma' \subseteq \Gamma$, $\Delta' \subseteq \Delta$, and σ' is a subsequence of σ.

The *base theory* of $\mathbf{T}$ is the theory (L_0, Γ_0) and the *current theory* of $\mathbf{T}$, written thy$(\mathbf{T})$, is the theory (L, Γ).

2. A *theorem object* is a tuple $\mathbf{A} = ([\mathbf{T}], A_*, J)$ where:
   (a) $\mathbf{T}$ is a theory object with thy$(\mathbf{T}) = (L, \Gamma)$.
   (b) $A_* \in \mathcal{S}_L$. A_* is called the *theorem* of $\mathbf{A}$.
   (c) J is a justification that $\Gamma \models A_*$.

3. A *definition object* is a tuple $\mathbf{D} = ([\mathbf{T}], c_\alpha, E_\alpha, J)$ where:
   (a) $\mathbf{T}$ is a theory object with thy$(\mathbf{T}) = (L, \Gamma)$.
   (b) c_α is a constant not in L.
   (c) $E_\alpha \in \mathcal{E}_L$. $c_\alpha = E_\alpha$ is called the *defining axiom* of $\mathbf{D}$.
   (d) J is a justification that $\Gamma \models O_*$ where O_* is $(\exists x_\alpha \,.\, x_\alpha = E_\alpha)$ and x_α does not occur in E_α.[4] O_* is called the *obligation* of $\mathbf{D}$.

4. A *profile object* is a tuple $\mathbf{P} = ([\mathbf{T}], \mathcal{C}, E_\beta, J)$ where:
   (a) $\mathbf{T}$ is a theory object with thy$(\mathbf{T}) = (L, \Gamma)$.
   (b) $\mathcal{C} = \{c_{\alpha_1}^1, \ldots, c_{\alpha_m}^m\}$ is a set of constants not in L.
   (c) $E_\beta = (\lambda x_{\alpha_1}^1 \cdots \lambda x_{\alpha_m}^m \,.\, B_*)$ where $x_{\alpha_1}^1, \ldots, x_{\alpha_m}^m$ are distinct variables. $E_\beta(c_{\alpha_1}^1) \cdots (c_{\alpha_m}^m)$ is called the *profiling axiom* of $\mathbf{P}$.
   (d) J is a justification that $\Gamma \models O_*$ where O_* is $(\exists x_{\alpha_1}^1 \cdots \exists x_{\alpha_m}^m \,.\, B_*)$. O_* is called the *obligation* of $\mathbf{P}$.

An *event object* is a theorem, definition, or profile object.

Let $\mathbf{T} \leq \mathbf{T}'$ mean thy$(\mathbf{T}) \leq$ thy$(\mathbf{T}')$ and $\mathbf{T} \trianglelefteq \mathbf{T}'$ mean thy$(\mathbf{T}) \trianglelefteq$ thy$(\mathbf{T}')$. $\mathbf{T}$ is a *structural subtheory* of $\mathbf{T}'$ if one of the following is true:

1. $\mathbf{T} = \mathbf{T}'$.
2. $\mathbf{T}$ is a structural subtheory of a principal subtheory of $\mathbf{T}'$.

$\mathbf{T}$ is a *structural supertheory* of $\mathbf{T}'$ if $\mathbf{T}'$ is a structural subtheory of $\mathbf{T}$.

For a theory object $\mathbf{T} = (n, L_0, \Gamma_0, L, \Gamma, \Delta, \sigma, \mathcal{N})$ and an event object e whose justification is correct, $\mathbf{T}[e]$ is the theory object defined as follows:

1. Let e be a theorem object $([\mathbf{T}'], A_*, J)$. If $\mathbf{T}' \leq \mathbf{T}$, then

$$\mathbf{T}[e] = (n, L_0, \Gamma_0, L, \Gamma, \Delta \cup \{A_*\}, \sigma^{\smallfrown}\langle e \rangle, \mathcal{N});$$

otherwise, $\mathbf{T}[e]$ is undefined.

2. Let e be a definition object $([\mathbf{T}'], c_\alpha, E_\alpha, J)$. If $\mathbf{T}' \leq \mathbf{T}$ and $c_\alpha \notin L$, then

$$\mathbf{T}[e] = (n, L_0, \Gamma_0, L \cup \{c_\alpha\}, \Gamma \cup \{A_*\}, \Delta \cup \{A_*\}, \sigma^{\smallfrown}\langle e \rangle, \mathcal{N})$$

where A_* is the defining axiom of e; otherwise, $\mathbf{T}[e]$ is undefined.

---

[4] In $\mathbf{C}$, $\Gamma \models (\exists x_\alpha \,.\, x_\alpha = E_\alpha)$ always holds and so no justification is needed, but in other logics such as LUTINS a justification is needed since $\Gamma \models (\exists x_\alpha \,.\, x_\alpha = E_\alpha)$ will not hold if E_α is undefined.

3. Let e be a profile object $([\mathbf{T}'], C, E_\beta, J)$. If $\mathbf{T}' \leq \mathbf{T}$ and $C \cap L = \emptyset$, then

$$\mathbf{T}[e] = (n, L_0, \Gamma_0, L \cup C, \Gamma \cup \{A_*\}, \Delta \cup \{A_*\}, \sigma\,\widehat{\,}\,\langle e \rangle, \mathcal{N})$$

where A_* is the profiling axiom of e; otherwise, $\mathbf{T}[e]$ is undefined.

An event history σ is *correct* if the justification in each member of σ is correct. For a correct event history σ, $\mathbf{T}[\sigma]$ is defined by:

1. Let $\sigma = \langle \rangle$. Then $\mathbf{T}[\sigma] = \mathbf{T}$.
2. Let $\sigma = \sigma'\,\widehat{\,}\,\langle e \rangle$. If $(\mathbf{T}[\sigma'])[e]$ is defined, then $\mathbf{T}[\sigma] = (\mathbf{T}[\sigma'])[e]$; otherwise, $\mathbf{T}[\sigma]$ is undefined.

Let the *base* of $\mathbf{T}$, written $\mathsf{base}(\mathbf{T})$, be the theory object

$$(n\_\mathsf{base}, L_0, \Gamma_0, L_0, \Gamma_0, \Gamma_0, \langle \rangle, \emptyset).$$

$\mathbf{T}$ is *proper* if the following conditions are satisfied:

1. Its event history σ is correct.
2. $\mathsf{thy}(\mathbf{T}) = \mathsf{thy}(\mathsf{base}(\mathbf{T})[\sigma])$.
3. $\mathsf{thms}(\mathbf{T}) = \mathsf{thms}(\mathsf{base}(\mathbf{T})[\sigma])$.

Lemma 2. *If T is a proper theory object, then A_* is a known theorem of T iff A_* is a base axiom of T or a theorem, defining axiom, or profiling axiom of an event object in the event history of T.*

Proof. Follows immediately from the definitions above.

Theorem 3. *If T is a proper theory object, then $\mathsf{base}(T) \trianglelefteq T$.*

Proof. Since $\mathbf{T}$ is proper, the event history σ of $\mathbf{T}$ is correct and $\mathsf{thy}(\mathbf{T}) = \mathsf{thy}(\mathsf{base}(\mathbf{T})[\sigma])$. We will show $\mathsf{base}(\mathbf{T}) \trianglelefteq \mathbf{T}$ by induction on $|\sigma|$, the length of σ.

Basis. Assume $|\sigma| = 0$. Then $\mathsf{thy}(\mathbf{T}) = \mathsf{thy}(\mathsf{base}(\mathbf{T}))$ and so $\mathsf{base}(\mathbf{T}) \trianglelefteq \mathbf{T}$ is obviously true.

Induction step. Assume $|\sigma| > 0$. Suppose $\sigma = \sigma'\,\widehat{\,}\,\langle e \rangle$. By the induction hypothesis, $\mathsf{base}(\mathbf{T}) \trianglelefteq \mathsf{base}(\mathbf{T})[\sigma']$. We claim $\mathsf{base}(\mathbf{T})[\sigma'] \trianglelefteq (\mathsf{base}(\mathbf{T})[\sigma'])[e]$. If e is a theorem object, then clearly $\mathsf{thy}(\mathsf{base}(\mathbf{T})[\sigma']) = \mathsf{thy}((\mathsf{base}(\mathbf{T})[\sigma'])[e])$ and so $\mathsf{base}(\mathbf{T})[\sigma'] \trianglelefteq (\mathsf{base}(\mathbf{T})[\sigma'])[e]$. If e is a definition or profile object, then $\mathsf{base}(\mathbf{T})[\sigma'] \trianglelefteq (\mathsf{base}(\mathbf{T})[\sigma'])[e]$ by the justification of e. Therefore, $\mathsf{base}(\mathbf{T}) \trianglelefteq \mathbf{T}$ follows by part (2) of Lemma 1. $\square$

We will now define the fifth and last infrastructure object: An *interpretation object* is a tuple $\mathbf{I} = ([\mathbf{T}], [\mathbf{T}'], \Phi, J)$ where:

1. $\mathbf{T}$ is a theory object called the *source theory* of $\mathbf{I}$.
2. $\mathbf{T}'$ is a theory object called the *target theory* of $\mathbf{I}$.
3. Φ is a translation.
4. J is a justification that Φ is an interpretation of $\mathsf{thy}(\mathsf{base}(\mathbf{T})[\sigma])$ in $\mathsf{thy}(\mathsf{base}(\mathbf{T}')[\sigma'])$ where σ and σ' are initial segments of the event histories of $\mathbf{T}$ and $\mathbf{T}'$, respectively.

4.2   Operations

The infrastructure design includes ten operations.

There are operations for creating the infrastructure objects:

1. Given a string n, a language L, a set Γ of sentences, and theory objects $\mathbf{T}^i = (n^i, L_0^i, \Gamma_0^i, L^i, \Gamma^i, \Delta^i, \sigma^i, \mathcal{N}^i) \in \mathbf{S}$ for $i = 1, \ldots, m$ as input, let
   (a) $L_0' = L_0^1 \cup \cdots \cup L_0^m$.
   (b) $\Gamma_0' = \Gamma_0^1 \cup \cdots \cup \Gamma_0^m$.
   (c) $L' = L^1 \cup \cdots \cup L^m$.
   (d) $\Gamma' = \Gamma^1 \cup \cdots \cup \Gamma^m$.
   (e) $\Delta' = \Delta^1 \cup \cdots \cup \Delta^m$.
   (f) $\sigma' = \sigma^1 {}^\smallfrown \cdots {}^\smallfrown \sigma^m$.
   If

$$\mathbf{T} = (n, L \cup L_0', \Gamma \cup \Gamma_0', L \cup L', \Gamma \cup \Gamma', \Gamma \cup \Delta', \sigma', \{[\mathbf{T}_1], \ldots, [\mathbf{T}_m]\})$$

   is a theory object and $n \neq [\mathbf{T}']$ for any theory object $\mathbf{T}' \in \mathbf{S}$, then create-thy-obj adds $\mathbf{T}$ to $\mathbf{S}$; otherwise, the operation fails.

2. Given a theory object $\mathbf{T} \in \mathbf{S}$, a sentence A_*, and a justification J as input, if $\mathbf{A} = ([\mathbf{T}], A_*, J)$ is a theorem object, then create-thm-obj adds $\mathbf{A}$ to $\mathbf{S}$; otherwise, the operation fails.

3. Given a theory object $\mathbf{T} \in \mathbf{S}$, a constant c_α, an expression E_α, and a justification J as input, if $\mathbf{D} = ([\mathbf{T}], c_\alpha, E_\alpha, J)$ is a definition object, then create-def-obj adds $\mathbf{D}$ to $\mathbf{S}$; otherwise, the operation fails.

4. Given a theory object $\mathbf{T} \in \mathbf{S}$, a set $\mathcal{C}$ of constants, an expression E_β, and a justification J as input, if $\mathbf{P} = ([\mathbf{T}], \mathcal{C}, E_\beta, J)$ is a profile object, then create-pro-obj adds $\mathbf{P}$ to $\mathbf{S}$; otherwise, the operation fails.

5. Given two theory objects $\mathbf{T}, \mathbf{T}' \in \mathbf{S}$, a translation Φ, and a justification J as input, if $\mathbf{I} = ([\mathbf{T}], [\mathbf{T}'], \Phi, J)$ is an interpretation object, then create-int-obj adds $\mathbf{I}$ to $\mathbf{S}$; otherwise, the operation fails.

There are operations for installing theorem, definition, and profile objects in theory objects:

1. Given a theorem object $\mathbf{A} = ([\mathbf{T}_0], A_*, J) \in \mathbf{S}$ and a theory object $\mathbf{T}_1 \in \mathbf{S}$, if $\mathbf{T}_0 \leq \mathbf{T}_1$, then install-thm-obj replaces every structural supertheory $\mathbf{T}$ of $\mathbf{T}_1$ in $\mathbf{S}$ with $\mathbf{T}[\mathbf{A}]$; otherwise, the operation fails.

2. Given a definition object $\mathbf{D} = ([\mathbf{T}_0], c_\alpha, E_\alpha, J) \in \mathbf{S}$ and a theory object $\mathbf{T}_1 \in \mathbf{S}$, if $\mathbf{T}_0 \leq \mathbf{T}_1$ and $\mathbf{T}[\mathbf{D}]$ is defined for every structural supertheory $\mathbf{T}$ of $\mathbf{T}_1$ in $\mathbf{S}$, then install-def-obj replaces every structural supertheory $\mathbf{T}$ of $\mathbf{T}_1$ in $\mathbf{S}$ with $\mathbf{T}[\mathbf{D}]$; otherwise, the operation fails.

3. Given a profile object $\mathbf{P} = ([\mathbf{T}_0], \mathcal{C}, E_\beta, J) \in \mathbf{S}$ and a theory object $\mathbf{T}_1 \in \mathbf{S}$, if $\mathbf{T}_0 \leq \mathbf{T}_1$ and $\mathbf{T}[\mathbf{P}]$ is defined for every structural supertheory $\mathbf{T}$ of $\mathbf{T}_1$ in $\mathbf{S}$, then install-pro-obj replaces every structural supertheory $\mathbf{T}$ of $\mathbf{T}_1$ in $\mathbf{S}$ with $\mathbf{T}[\mathbf{P}]$; otherwise, the operation fails.

There are operations to extend an interpretation object and to copy a theory object:

1. Given an interpretation object $\mathbf{I} = ([\mathbf{T}], [\mathbf{T}'], \Phi, J) \in \mathbf{S}$, a translation Φ', and a justification J' as input, if Φ' extends Φ and $\mathbf{I}' = ([\mathbf{T}], [\mathbf{T}'], \Phi', J')$ is an interpretation object, then extend-int replaces $\mathbf{I}$ in $\mathbf{S}$ with $\mathbf{I}'$; otherwise, the operation fails.

2. Given a string n and a theory object

$$\mathbf{T} = (n', L_0, \Gamma_0, L, \Gamma, \Delta, \sigma, \mathcal{N}) \in \mathbf{S}$$

as input, if $n \neq [\mathbf{T}']$ for any theory object $\mathbf{T}' \in \mathbf{S}$, then create-thy-copy adds the theory object

$$\mathbf{T}' = (n, L_0, \Gamma_0, L, \Gamma, \Delta, \sigma, \mathcal{N})$$

to $\mathbf{S}$; otherwise, the operation fails.

The infrastructure operations guarantee that the following theorem holds:

Theorem 4. *If the justification of every event object in S is correct, then:*

1. *Every object in S is a well-defined theory, theorem, definition, profile, or interpretation object.*
2. *Every theory object in S is proper.*
3. *Distinct theory objects in S have distinct names.*

Some Remarks about the Intertheory Infrastructure:

1. Theory and interpretation objects are modifiable, but event objects are not.
2. The event history of a theory object records how the theory object is constructed from its base theory.
3. The theory stored in a theory object $\mathbf{T}$ extends all the theories stored in the principal subtheories of $\mathbf{T}$.
4. Theorem, definition, and profile objects installed in a theory $\mathbf{T}$ in $\mathbf{S}$ are automatically installed in every structural supertheory of $\mathbf{T}$ in $\mathbf{S}$.
5. The infrastructure allows definitions and profiles to be made in a theory object $\mathbf{T}$ both by modifying $\mathbf{T}$ using install-def-obj and install-prof-obj and by creating an extension of $\mathbf{T}$ using create-thy-obj.
6. By Theorem 2, if Φ is a translation from thy($\mathbf{T}$) to thy($\mathbf{T}'$) which maps the base axioms of $\mathbf{T}$ to known theorems of $\mathbf{T}'$, then Φ is an interpretation of thy($\mathbf{T}$) in thy($\mathbf{T}'$).
7. The interpretation stored in an interpretation object is allowed to be incomplete. It can be extended as needed using extend-int.

5   Some Applications

5.1   Theory Development System

The intertheory infrastructure provides a strong foundation on which to build a system for developing axiomatic theories. The infrastructure operations enable theories and interpretations to be created and extended. Many additional operations can be built on top of the ten infrastructure operations. Examples include operations for transporting theorems, definitions, and profiles from one theory to another and for instantiating theories.

Given a theorem object $\mathbf{A} = ([\mathbf{T_0}], A_*, J_0)$ installed in $\mathbf{T} \in \mathbf{S}$ and an interpretation object $\mathbf{I} = ([\mathbf{T}], [\mathbf{T'}], \Phi, J) \in \mathbf{S}$ as input, the operation transport-thm-obj would invoke create-thm-obj and install-thm-obj to create a new theorem object $([\mathbf{T'}], \Phi(A_*), J')$ and install it in $\mathbf{T'}$. The justification J' would be formed from J_0 and J.

Given a constant d_β, a definition object $\mathbf{D} = ([\mathbf{T_0}], c_\alpha, E_\alpha, J)$ installed in $\mathbf{T} \in \mathbf{S}$, and an interpretation object $\mathbf{I} = ([\mathbf{T}], [\mathbf{T'}], \Phi, J) \in \mathbf{S}$ as input, if $\Phi(\alpha) = \beta$ and d_β is not in the current language of $\mathbf{T'}$, the operation transport-def-obj would invoke create-def-obj and install-def-obj to create a new definition object $([\mathbf{T'}], d_\beta, \Phi(E_\alpha), J')$ and install it in $\mathbf{T'}$; otherwise, the operation fails. The justification J' would be formed from J_0 and J. An operation transport-pro-obj could be defined similarly.

Given theory objects $\mathbf{T}, \mathbf{T'} \in \mathbf{S}$ and an interpretation object $\mathbf{I} = ([\mathbf{T_0}], [\mathbf{T'_0}], \Phi, J) \in \mathbf{S}$ as input, if $\mathbf{T_0} \leq \mathbf{T}$ and $\mathbf{T'_0} \leq \mathbf{T'}$, the operation instantiate-thy would invoke create-thy-obj to create a new theory object $\mathbf{T''}$ and create-int-obj to create a new interpretation object $\mathbf{I'} = ([\mathbf{T}], [\mathbf{T''}], \Phi', J)$ such that:

- $\mathbf{T''}$ is an extension of $\mathbf{T'}$ obtained by "instantiating" $\mathbf{T_0}$ in $\mathbf{T}$ with $\mathbf{T'}$. How $\mathbf{T'}$ is cemented to the part of $\mathbf{T}$ outside of $\mathbf{T_0}$ is determined by Φ. The constants of $\mathbf{T}$ which are not in $\mathbf{T_0}$ may need to be renamed and retagged to avoid conflicts with the constants in $\mathbf{T'}$.
- Φ' is an interpretation of thy($\mathbf{T}$) in thy($\mathbf{T''}$) which extends Φ.

For further details, see [7].

This notion of theory instantiation is closely related to the notion of theory instantiation proposed by Burstall and Goguen [2]; in both approaches a theory is instantiated via an interpretation. However, in our approach, any theory can be instantiated with respect to any of its subtheories. In the Burstall-Goguen approach, only "parameterized theories" can be instantiated and only with respect to the explicit parameter of the parameterized theory.

5.2 Foundational Theory Development System

A theory development system is *foundational* if every theory developed in the system is consistent relative to one or more "foundational" theories which are known or regarded to be consistent. Since the operations for installing theorems, definitions, and profiles in a theory always produce conservative extensions of the original theory by Theorem 3, these operations preserve consistency. Therefore, a foundational theory development system can be implemented on top of the infrastructure design by simply using a new operation for creating theory objects that is successful only when the theory stored in the object is consistent relative to one of the foundational theories.

The new operation can be defined as follows. Suppose T^* is a foundational theory. Given a string n, a language L, a set Γ of sentences, theory objects $T_1, \ldots, T_m \in S$, a translation Φ, and a justification J as input, if J is a justification that Φ is an interpretation of $T = (L, \Gamma)$ in $\text{thy}(T^*)$, the new operation would invoke create-thy-obj on $(n, L, \Gamma, \{[T_1], \ldots, [T_m]\})$ to create a theory object T and then invoke create-int-obj on $([T], [T^*], \Phi, J)$ to create an interpretation object I; otherwise the operation fails. If the operation is successful and J is correct, then $\text{thy}(T)$ would be consistent relative to $\text{thy}(T^*)$ by Theorem 1.

5.3 Encapsulated Theory Development

Proving a theorem in a theory may require introducing several definitions and proving several lemmas in the theory that would not be useful after the theorem is proved. Such "local" definitions and lemmas would become logical clutter in the theory. One strategy for handling this kind of clutter is to encapsulate local development in a auxiliary theory so that it can be separated from the development of the main theory. The infrastructure design makes this encapsulation possible.

Suppose that one would like to prove a theorem in a theory stored in theory object T using some local definitions and lemmas. One could use create-thy-copy to create a copy T' of T and create-int-obj to create a interpretation object I storing the identity interpretation of $\text{thy}(T')$ in $\text{thy}(T)$. Next the needed local definitions and lemmas could be installed as definition and theorem objects in T'. Then the theorem could be proved and installed as a theorem object in T'. Finally, the theorem could be transported back to T using the interpretation stored in I. The whole local development needed to prove the theorem would reside in T' completely outside of the development of T.

A different way to encapsulate local theory development is used in the ACL2 theorem prover [15].

5.4 Sequent-Style Proof System

A goal-oriented sequent-style proof system can be built on top of the intertheory infrastructure. A *sequent* would have the form $T \rightarrow A_*$ where T is a theory object called the *context* and A_* is a sentence in the current language of T

called the *assertion*. The system would include the usual inference rules of a sequent-style proof system plus rules to:

- Install a theorem, definition, or profile into the context of a sequent.
- Transport a theorem, definition, or profile from a theory object to the context of a sequent.

Some of the proof rules, such as the deduction rule, would add or remove axioms from the context of a sequent, thereby defining new theory objects. The proof rules for the rules of universal generalization and existential generalization would be implemented by installing a profile in the context of a sequent.

A sentence A_* in the current language of a theory object T would be proved as follows. create-thy-copy would be used to create a copy T' of T and create-int-obj would be used to create a interpretation object I storing the identity interpretation of thy(T') in thy(T). Then the sequent $T' \rightarrow A_*$ would be proved, possibly with the help of local or imported definitions and lemmas. The contexts created in the course of the proof would be distinct supertheories of T'. A theorem or definition installed in a context appearing in some part of the proof would be available wherever else the context appeared in the proof.

When the proof is finished, A_* would be installed as a theorem object in T'. The theorem could be then transported back to T using the interpretation stored in I. The theory objects needed for the proof—T' and its supertheories—would be separated from T and the other theory objects in S.

Acknowledgments

Many of the ideas in this paper originated in the design and implementation of IMPS done jointly by Dr. Joshua Guttman, Dr. Javier Thayer, and the author.

References

1. P. B. Andrews. *An Introduction to Mathematical Logic and Type Theory: To Truth through Proof.* Academic Press, 1986.
2. R. Burstall and J. Goguen. The semantics of Clear, a specification language. In *Advanced Course on Abstract Software Specifications*, volume 86 of *Lecture Notes in Computer Science*, pages 292–332. Springer-Verlag, 1980.
3. A. Church. A formulation of the simple theory of types. *Journal of Symbolic Logic*, 5:56–68, 1940.
4. H. B. Enderton. *A Mathematical Introduction to Logic.* Academic Press, 1972.
5. W. M. Farmer. A partial functions version of Church's simple theory of types. *Journal of Symbolic Logic*, 55:1269–91, 1990.
6. W. M. Farmer. A simple type theory with partial functions and subtypes. *Annals of Pure and Applied Logic*, 64:211–240, 1993.
7. W. M. Farmer. A general method for safely overwriting theories in mechanized mathematics systems. Technical report, The MITRE Corporation, 1994.
8. W. M. Farmer. Theory interpretation in simple type theory. In J. Heering et al., editor, *Higher-Order Algebra, Logic, and Term Rewriting*, volume 816 of *Lecture Notes in Computer Science*, pages 96–123. Springer-Verlag, 1994.

9. W. M. Farmer, J. D. Guttman, and F. J. Thayer Fábrega. IMPS: An updated system description. In M. McRobbie and J. Slaney, editors, *Automated Deduction—CADE-13*, volume 1104 of *Lecture Notes in Computer Science*, pages 298–302. Springer-Verlag, 1996.

10. W. M. Farmer, J. D. Guttman, and F. J. Thayer. Little theories. In D. Kapur, editor, *Automated Deduction—CADE-11*, volume 607 of *Lecture Notes in Computer Science*, pages 567–581. Springer-Verlag, 1992.

11. W. M. Farmer, J. D. Guttman, and F. J. Thayer. IMPS: An Interactive Mathematical Proof System. *Journal of Automated Reasoning*, 11:213–248, 1993.

12. J. A. Goguen and T. Winkler. Introducing OBJ3. Technical Report SRI-CSL-99-9, SRI International, August 1988.

13. M. J. C. Gordon and T. F. Melham. *Introduction to HOL: A Theorem Proving Environment for Higher Order Logic*. Cambridge University Press, 1993.

14. N. Hamilton, R. Nickson, O. Traynor, and M. Utting. Interpretation and instantiation of theories for reasoning about formal specifications. In M. Patel, editor, *Proceedings of the Twentieth Australasian Computer Science Conference*, volume 19 of *Australian Computer Science Communications*, pages 37–45, 1997.

15. M. Kaufmann and J S. Moore. Structured theory development for a mechanized logic, 1999. Available at
http://www.cs.utexas.edu/users/moore/publications/acl2-papers.html,

16. R. Nakajima and T. Yuasa, editors. *The IOTA Programming System*, volume 160 of *Lecture Notes in Computer Science*. Springer-Verlag, 1982.

17. R. Nickson, O. Traynor, and M. Utting. Cogito ergo sum—providing structured theorem prover support for specification formalisms. In K. Ramamohanarao, editor, *Proceedings of the Nineteenth Australasian Computer Science Conference*, volume 18 of *Australian Computer Science Communications*, pages 149–158, 1997.

18. J. Rushby, F. von Henke, and S. Owre. An introduction to formal specification and verification using EHDM. Technical Report SRI-CSL-91-02, SRI International, 1991.

19. J. R. Shoenfield. *Mathematical Logic*. Addison-Wesley, 1967.

20. D. Smith. KIDS: A knowledge-based software development system. In M. Lowry and R. McCartney, editors, *Automating Software Design*, pages 483–514. MIT Press, 1991.

21. Y. Srinivas and R. Jullig. Specware: Formal support for composing software. In *Proceedings of the Conference on Mathematics of Program Construction*, 1995.

Gödel's Algorithm for Class Formation

Johan Gijsbertus Frederik Belinfante

Georgia Institute of Technology, Atlanta, GA 30332-0160 (U.S.A.)
belinfan@math.gatech.edu

Abstract. A computer implementation of Gödel's algorithm for class formation in Mathematica™ is useful for automated reasoning in set theory. The original intent was to forge a convenient preprocessing tool to help prepare input files for McCune's automated reasoning program Otter. The program is also valuable for discovering new theorems. Some applications are described, especially to the definition of functions. A brief extract from the program is included in an appendix.

1 Introduction

Robert Boyer et al. (1986) proposed clauses capturing the essence of Gödel's finite axiomatization of the von Neumann-Bernays theory of sets and classes. Their work was simplified significantly by Art Quaife (1992a and 1992b). About four hundred theorems of elementary set theory were proved using McCune's automated reasoning program Otter. A certain degree of success has been achieved recently (1999a and 1999b) in extending Quaife's work. Some elementary theorems of ordinal number theory were proved, based on Isbell's definition (1960) of ordinal number, which does not require the axiom of regularity to be assumed.

An admitted disadvantage of Gödel's formalism is the absence of the usual class formation {x | p(x)} notation. Replacing the axiom schema for class formation are a small number of axioms for certain basic class constructions. Definitions of classes must be expressed in terms of two basic classes, the universal class V and the membership relation E, and seven other basic class constructors: the unary constructors complement, domain, flip and rotate, and the binary constructors pairset, cart, intersection. Gödel also included an axiom for inverse, but it can be deduced from the others.

2 A Brief Description of the GOEDEL Program

As a replacement for the axiom schema for class formation, Kurt Gödel (1940) proved a fundamental Class Existence Metatheorem Schema for class formation. His proof of this metatheorem is constructive; a recursive algorithm for converting customary definitions of classes using class formation to expressions built out of the primitive constructors is presented, together with a proof of termination. An implementation of Gödel's algorithm in Mathematica™ was created (1996)

D. McAllester (Ed.): CADE-17, LNAI 1831, pp. 132–147, 2000.

to help prepare input files for proofs in set theory using McCune's automated reasoning program Otter.

The likelihood of success in proving theorems using programs like Otter depends critically on the simplicity of the definitions used and the brevity of the statements of the theorems to be proved. To mitigate the effects of combinatorial explosion, one typically sets a weight limit to exclude complicated expressions from being considered. Although combinatorial explosion can not be prevented, the idea is to snatch a proof quickly before the explosion gets well under way.

Because one needs compact definitions for practical applications, and because the output of Gödel's original algorithm is typically extremely complicated, a large number of simplification rules were added to the Mathematica implementation of Gödel's algorithm. With the addition of simplification rules, Gödel's proof of termination no longer applies. No assurance can be given that the added simplification rules will not cause looping to occur, but we have tested the program on a suite of several thousand examples, and it appears that it can be used as a practical tool to help formulate definitions and to simplify the statements of theorems. The GOEDEL program contains no mechanism for carrying out deductions, but it does sometimes manage to prove statements by simplifying them to True.

Much of the complexity of Gödel's original algorithm stems from his use of Kuratowski's definition for ordered pairs. The Mathematica implementation does not assume any particular contruction of ordered pairs, but instead includes additional rules to deal with ordered pairs. The self-membership rule in the original algorithm was modified because in our work on ordinal numbers the axiom of regularity is not assumed.

The stripped down version of the GOEDEL program presented in the Appendix omits many membership rules for defined constructors as well as most of the simplification rules. The modified Gödel's algorithm is presented as a series of definitions for a Mathematica function class[x,p]. The first argument x, assumed to be the name of a set, must be either an atomic symbol, or an expression of the form pair[u, v] where u and v in turn are either atomic symbols or pairs, and so on. It should be noted that Gödel did not allow both u and v to be pairs, but this unnecessary limitation has been removed to make the formalism more flexible. The second argument p is some statement which can involve the variables that appear in x, as well as other variables that may represent arbitrary classes (not just sets). The statement can contain quantifiers, but the quantified variables must be sets. The Gödel algorithm does not apply to statements containing quantifiers over proper classes. The quantifiers forall and exists used in the GOEDEL program are explicitly restricted to set variables.

A few simple examples will be presented to illustrate how the GOEDEL program is used. For convenience, Mathematica style notation will be employed, which does not quite conform to the notational requirements of Otter. For example, Mathematica permits one to define intersection to be an associative and commutative function of any number of variables. For brevity we write a —→ b to mean that Mathematica input a produces Mathematica output b for some version of the GOEDEL program.

The functions FIRST and SECOND which project out the first and second components of an ordered pair, respectively, can be specified as the classes

$$\text{class}[\text{pair}[\text{pair}[x, y], z], \text{equal}[z, x]] \longrightarrow \text{FIRST},$$
$$\text{class}[\text{pair}[\text{pair}[x, y], z], \text{equal}[z, y]] \longrightarrow \text{SECOND}.$$

Examples which involve quantifiers include the domain and range of a relation:

$$\text{class}[x, \text{exists}[y, \text{member}[\text{pair}[x, y], z]]] \longrightarrow \text{domain}[z],$$
$$\text{class}[y, \text{exists}[x, \text{member}[\text{pair}[x, y], z]]] \longrightarrow \text{range}[z].$$

It is implicitly assumed that all quantified variables refer to sets, but the free variable z here can stand for any class.

3   Eliminating Flip and Rotate

Gödel's algorithm uses two special constructors flip[x] and rotate[x] which produce ternary relations. The ternary relation flip[x] is

$$\text{class}[\text{pair}[\text{pair}[u, v], w], \text{member}[\text{pair}[\text{pair}[v, u], w], x]]$$

while rotate[x] is

$$\text{class}[\text{pair}[\text{pair}[u, v], w], \text{member}[\text{pair}[\text{pair}[v, w], u], x]].$$

Because these functors are not widely used in mathematics, it may be of interest to note that they could be eliminated in favor of more familiar ones. One can rewrite flip[x] as composite[x, SWAP], where SWAP = flip[Id] is the relation

$$\text{class}[\text{pair}[\text{pair}[u, v], \text{pair}[x, y]], \text{and}[\text{equal}[u, y], \text{equal}[v, x]]] \longrightarrow \text{SWAP}.$$

Note that the functions that project out the first and second members of an ordered pair are related by SECOND = flip[FIRST] and FIRST = flip[SECOND].

The general formula for rotate[x] is more complicated, but Gödel's algorithm actually only involves the special case where x is a Cartesian product. In this special case one has the simple formula,

$$\text{rotate}[\text{cart}[x_-, y_-]] := \text{composite}[x, \text{SECOND}, \text{id}[\text{cart}[y, V]]].$$

Using these formulas, the constructors flip and rotate could be completely eliminated from Gödel's algorithm, as well as from Gödel's axioms for class theory, if one instead takes as primitives the constructors composite, inverse, FIRST and SECOND. We have done so in the abbreviated version of the GOEDEL program listed in the Appendix. The function SWAP mentioned above, for example, could be defined in terms of these new primitives as

```
intersection[composite[inverse[FIRST],SECOND],
            composite[inverse[SECOND],FIRST]] := SWAP.
```

4   Equational Set Theory without Variables

The simplification rules in the GOEDEL program can be used not only to simplify descriptions of classes, but can also be induced to simplify statements. Given any statement p, one can form the class class[w, p] where w is any variable that does not occur in the statement p. This class is the universal class V if p is true, and is the empty class when p is false. One can form a new statement equivalent to the original one by the definition

$$\text{assert}[p\_] := \text{Module}[\{w = \text{Unique}[]\}, \text{equal}[V, \text{class}[w, p]]]$$

The occurrence of class causes Gödel's algorithm to be invoked, the meaning of the statement p to be interpreted, and the simplification rules in the GOEDEL program to be applied. While there can be no assurance the transformed statement will actually be simpler than the statement one started with, in practice it often is. For instance, the input

$$\text{assert}[\text{equal}[\text{composite}[\text{cross}[x, x], \text{DUP}], \text{composite}[\text{DUP}, x]]]$$

produces the statement FUNCTION[composite[Id, x]] as output. To improve the readability of the output, in the current version of the GOEDEL program, rules have been added which may convert the equations obtained with assert back to simpler non-equational statements.

Since some theorem provers are limited to equational statements, it is of interest to reformulate set theory in equational terms. Alfred Tarski and Steven Givant (1987) have shown that all statements of set theory can be reformulated as equations without variables, somewhat reminiscent of combinatory logic. But whereas combinatory logic uses function-like objects as primitives, their calculus is based on the theory of relations. It has recently been proposed by Omodeo and Formisano (1998) that this formalism be used to recast set theory in a form accessible to purely equational automated reasoning programs. It is interesting to note that the assert mechanism in the GOEDEL program achieves the same objective. Any statement is converted by assert into an equation of the form equal[V, x]. If one prefers, one may also write this equation in the equivalent form equal[0, complement[x]].

Another consequence of the assert process is that one can always convert negative statements into positive ones. For example, the negative statement not[equal[0, x]] is converted by assert into the equivalent positive statement equal[V, image[V, x]]. Thus it appears that at least in set theory it does not make too much sense to make a big distinction between positive and negative literals, because one can always convert the one into the other. Also, one can always convert a clause with several literals into a unit clause; the clause or[equal[0, x], equal[0, y]], for example, is equivalent to the unit clause

$$\text{equal}[0, \text{intersection}[\text{image}[V, x], \text{image}[V, y]]].$$

The class image[V, x] which appears in these expressions is equal to the empty set if x is empty, and is equal to the universal class V if x is not empty. This class is quite useful for reformulating conditional statements as unconditional ones. Many equations in set theory hold only for sets and not for proper classes. For example, the union of the singleton of a class x is x when x is a set, but is the empty set otherwise. This rule can be written as a single equation which applies to both cases as follows:

$$U[\text{singleton}[x_-]] := \text{intersection}[x, \text{image}[V, \text{singleton}[x]]]$$

(This is in fact one of the thousands of rules in the GOEDEL program.) Although such unconditional statements initially appear to be more complex than the conditional statements that they replace, experience both with Otter and with the GOEDEL program indicates that the unconditional statements are in fact preferable. In Otter the unconditional rule can often be added to the demodulator list. In Mathematica, an unconditional simplification rule generally works faster than a conditional one.

When **assert** is applied to a statement containing quantifiers, the statement is converted to a logically equivalent equation without quantifiers. All quantified variables are eliminated. What happens is that the quantifiers are neatly built into equivalent set-theoretic constructs like **domain** and **composite**. For example, the axiom of regularity is usually formulated using quantifiers as:

$$\text{implies}[\text{not}[\text{equal}[x, 0]], \text{exists}[u, \text{and}[\text{member}[u, x], \text{disjoint}[u, x]]]].$$

When **assert** is applied, this statement is automatically converted into the equivalent quantifier-free statement

$$\text{or}[\text{equal}[0, x], \text{not}[\text{subclass}[x, \text{complement}[P[\text{complement}[x]]]]]].$$

In this case the quantifier was hidden in the introduced power class functor. Replacing x by its complement, one obtains the following neat reformulation of the axiom of regularity:

$$\text{implies}[\text{subclass}[P[x], x], \text{equal}[x, V]].$$

That is, the axiom of regularity says that the universal class is the only class which contains its own power class. When the axiom of regularity is not assumed, there may be other classes with this property. In particular, the Russell class RUSSELL = complement[fix[E]] has this property, a fact that is useful in the Otter proofs in ordinal number theory.

This reformulation of the axiom of regularity has the advantage over the original one in that its clausification does not introduce new Skolem functions.

5   Functions, Vertical Sections, and Cancellation Machines

The process of eliminating variables and hiding quantifiers is facilitated by having available a supply of standard functions corresponding to the primitive constructors, as well as important derived constructors. For example, Quaife introduced the function SUCC corresponding to the constructor

$$\text{succ}[x_-] := \text{union}[x, \text{singleton}[x]]$$

so that the statement that the set **omega** of natural numbers is closed under the successor operation could be written in the compact variable-free form as the condition subclass[image[SUCC, omega], omega]. This is just one of the many techniques that Quaife exploited to reduce the plethora of Skolem functions that had appeared in the earlier work of Robert Boyer, et al.

Replacing the function symbols of first order logic by bonafide set-theoretic functions not only helps to eliminate Skolem functions, but also improves the readability of the statements of theorems. A standard way to obtain definitions for most of these functions is in terms of a basic constructor VERTSECT, enabling one to introduce a lambda calculus for defining functions by specifying the result obtained when they are applied to an input. The basic idea is not limited to functions; any relation can be specified by giving a formula for its vertical sections. The vertical sections of a relation **z** are the family of classes

$$\text{image}[z, \text{singleton}[x]] = \text{class}[y, \text{member}[\text{pair}[x, y], z]].$$

One is naturally led to introduce the function which assigns these vertical sections:

$$\text{VERTSECT}[z] == \text{class}[\text{pair}[x, y], \text{equal}[y, \text{image}[z, \text{singleton}[x]]]]$$

(Formisano and Omodeo (1998) call this function $\nabla(z)$.) Gödel's algorithm converts this formula to the expression

```
VERTSECT[z] == composite[Id,intersection[
               complement[composite[E,complement[z]]],
               complement[composite[complement[E],z]]]].
```

Of course, for many relations **z** the vertical sections need not be sets. The domain of VERTSECT[z] in general is the class of all sets **x** for which image[z, singleton[x]] is also a set. We call a relation *thin* when all vertical sections are sets. In addition to functions, there are many important relations, such as **inverse[E]** and **inverse[S]**, that are thin.

Using Otter, we have proved many facts about VERTSECT, making it unnecessary to repeat such work for individual functions.

When **f** is a function, image[f, singleton[x]] is a singleton, and one can select the element in that singleton by applying either the sum class operation U, as Quaife does, or by applying the unary intersection operation A defined by

$$\text{class}[u, \text{forall}[v, \text{implies}[\text{member}[v, x], \text{member}[u, v]]]] \longrightarrow A[x]$$

or equivalently,

$$\text{complement}[\text{image}[\text{complement}[\text{inverse}[E]], x]] \longrightarrow A[x].$$

The difference between using U and A only affects the case that **x** is a proper class. Nevertheless, using A instead of U in the definition of application has many practical advantages.

For example, one can use VERTSECT to obtain a formula for any function from a formula for its application A[image[f, singleton[x]]]. This can be done neatly in the GOEDEL program by introducing the Mathematica definition

```
lambda[x_,e_]:=
Module[{y=Unique[]},VERTSECT[class[pair[x,y],member[y,e]]]]
```

This Mathematica function lambda satisfies:

$$\text{FUNCTION}[f] := \text{True}; \quad \text{lambda}[x, A[\text{image}[f, \text{singleton}[x]]]] \longrightarrow f,$$

It should be noted that nothing like this works when one replaces A by U because U does not distinguish between 0 and singleton(0), whereas A does. For the constant function $f := \text{cart}[x, \text{singleton}[0]]$, for example, one has $U[\text{image}[f, \text{singleton}[y]]] \longrightarrow 0$, whereas

$$A[\text{image}[f, \text{singleton}[y]]] \longrightarrow$$
$$\text{complement}[\text{image}[V, \text{intersection}[x, \text{singleton}[y]]]].$$

Because the formula for $U[\text{image}[f, \text{singleton}[y]]]$ has lost all information about the domain x of the function f, one cannot reconstruct f from this formula, but one can reconstruct f from the formula for $A[\text{image}[f, \text{singleton}[y]]]$.

As examples of definitions obtained using lambda we mention the function SINGLETON which takes any set to its singleton,

$$\text{lambda}[x, \text{singleton}[x]] \longrightarrow \text{VERTSECT}[\text{Id}],$$

and the function POWER which takes any set to its power set,

$$\text{lambda}[x, P[x]] \longrightarrow \text{VERTSECT}[\text{inverse}[S]].$$

The function VERTSECT[x] itself satisfies

$$\text{lambda}[w, \text{image}[x, \text{singleton}[w]]] \longrightarrow \text{VERTSECT}[x].$$

In addition to VERTSECT, it is also convenient to introduce a related constructor IMAGE, defined by

$$\text{VERTSECT}[\text{composite}[x_, \text{inverse}[E]]] := \text{IMAGE}[x].$$

The function IMAGE[x] satisfies

$$\text{lambda}[u, \text{image}[x, u]] \longrightarrow \text{IMAGE}[x].$$

The definition IMAGE[inverse[E]] := BIGCUP of the function BIGCUP which corresponds to the constructor U[x] was one of the first applications found for IMAGE. The function IMAGE[inverse[S]] is the hereditary closure operator, which takes any set x to its hereditary closure image[inverse[S],x]. This function is closely related to the POWER function mentioned earlier. The functions IMAGE[FIRST] and IMAGE[SECOND] take x to its domain and range, respectively, while IMAGE[SWAP] takes x to its inverse. The function IMAGE[cross[u,v]] takes x to composite[v,x,inverse[u]]. For example, the function that corresponds to the constructor flip is IMAGE[cross[SWAP,Id]].

The constructor **IMAGE** is not a functor in the category theory sense. The function **IMAGE[x]** does not in general preserve composites, but only when the right hand factor is thin:

$$\text{domain}[\text{VERTSECT}[t]] := V;$$
$$\text{IMAGE}[\text{composite}[x, t]] \longrightarrow \text{composite}[\text{IMAGE}[x], \text{IMAGE}[t]].$$

IMAGE preserves the global identity function: **IMAGE[Id]** $\longrightarrow$ **Id**; but in general **IMAGE[id[x]]** is not an identity function. It is nonetheless a useful function:

$$\text{lambda}[w, \text{intersection}[x, w]] \longrightarrow \text{IMAGE}[\text{id}[x]].$$

An important application of **VERTSECT** is to provide a mechanism for recovering a function **f** from a formula for **composite[inverse[E],f]**. One can use **VERTSECT** to cancel factors of **inverse[E]**; for example, the Mathematica input

```
FUNCTION[f1] := True; FUNCTION[f2] := True;
domain[f1] := V; domain[f2] := V;
Map[VERTSECT, composite[inverse[E],f1]==composite[inverse[E],f2]]
```

produces the output **f1 == f2**. When the assumption about the domains of the functions are omitted, the results are slightly more complicated, but one nonetheless can obtain a formula for each function in terms of the other.

It is possible to use **VERTSECT** to construct other such cancellation machines which cancel factors of **S**, **inverse[S]** or **DISJOINT**. These machines were found to be quite useful in our investigations of the binary functions which correspond to the constructors **intersection**, **cart**, **union** and so forth.

6   Binary Functions and Proof by Rotation

Binary functions such as **CART**, **CAP**, **CUP**, corresponding to the constructors **cart**, **intersection**, **union**, are important for obtaining variable-free expressions in many applications. To apply the **lambda** formalism to these functions, it is convenient to introduce the abbreviations

```
first[x_] := A[domain[singleton[x]]];
second[x_] := A[range[singleton[x]]].
```

One then has

$$\text{lambda}[x, \text{intersection}[\text{first}[x], \text{second}[x]]] \longrightarrow \text{CAP},$$
$$\text{lambda}[x, \text{union}[\text{first}[x], \text{second}[x]]] \longrightarrow \text{CUP},$$
$$\text{lambda}[x, \text{cart}[\text{first}[x], \text{second}[x]]] \longrightarrow \text{CART},$$

(It should be noted that **first** and **second** here are technically different from the rather similar constructors **1st** and **2nd** introduced by Quaife.)

Although Gödel's **rotate** functor can be completely eliminated, nevertheless it does in fact have many desirable properties. For example, the **rotate** functor preserves unions, intersections and relative complements, whereas composite preserves only unions. In the study of binary functions, the rotate constructor has turned out to be extremely useful. Often we can take one equation for binary functions and rotate it to obtain another.

The SYMDIF function corresponding to the symmetric difference operation is rotation invariant. Schroeder's transposition theorem can be given a succinct variable-free formulation as the statement that the relation

$$\text{composite[DISJOINT, COMPOSE]}$$

is rotation invariant, where DISJOINT is class[pair[x, y], disjoint[x, y]], and COMPOSE is the binary function corresponding to composite.

We mention three applications of these binary functions for defining classes. The class of all transitive relations can be specified as:

$$\text{class}[x, \text{subclass}[\text{composite}[x, x], x]] \longrightarrow \text{fix}[\text{composite}[S, \text{COMPOSE}, \text{DUP}]].$$

The class of all disjoint collections, specified as the input

```
class[z,forall[x,y, implies[and[member[x,z],member[y,z]],
                           or[equal[x,y],disjoint[x,y]]]]]
```

produces

$$\text{fix}[\text{image}[\text{inverse}[\text{CART}], \text{P}[\text{union}[\text{DISJOINT}, \text{Id}]]]]$$

as output. The class of all topologies, input as

```
class[t,and[subclass[image[BIGCUP,P[t]],t],
            subclass[image[CAP,cart[t,t]],t]]]
```

produces the output

```
intersection[
    complement[fix[composite[complement[E],BIGCUP,inverse[S]]]],
    fix[composite[S,IMAGE[CAP],CART,DUP]]].
```

7   Conclusion

Proving theorems in set theory with a first order theorem prover such as Otter is greatly facilitated by the use of a companion program GOEDEL which permits one to automatically translate from the notations commonly used in mathematics to the special language needed for the Gödel theory of classes.

Having an arsenal of set-theoretic functions that correspond to the function symbols of first order logic proves to be useful for systematically eliminating existential quantifiers and thereby avoiding the Skolem functions produced when formulas with existential quantifiers are converted to clause form. Although the main focus in this talk was on the use of the GOEDEL program to help find convenient definitions for all these functions, the GOEDEL program also permits one to discover useful formulas that these functions satisfy. By adding these formulas as new simplification rules, the program has grown increasingly powerful over the years.

The GOEDEL program currently contains well over three thousand simplification rules, many of which have been proved valid using Otter. The simplification rules can be used not only to simpify definitions, but also to simplify statements. This power to simplify statements has led to the discovery of many new formulas, especially new demodulators. Experience with Otter indicates that searches for proofs are dramatically improved by the presence of demodulators even when they are not directly used in the proof of a theorem because they help to combat combinatorial explosion.

Appendix. An Extract from the GOEDEL Program

```
Print[":Package Title: GOEDEL.M      2000 January 13 at 6:45 a.m. "];
(*
:Context: Goedel'
:Mathematica Version: 3.0         :Author: Johan G. F. Belinfante
:Summary: The GOEDEL program implements Goedel's algorithm for class
 formation, modified to avoid assuming the axiom of regularity, and
 Kuratowski's construction of ordered pairs.
:Sources: <description of algorithm, information for experts>
 Kurt Goedel, 1939 monograph  on consistency of the axiom of choice and
 the generalized continuum hypothesis, pp. 9-14.

:Warnings:  <description of global effects, incompatibilities>
 0 is used to represent the empty set.
 E is used to represent the membership relation.
:Limitations:   <special cases not handled, known problems>
 The simplification rules are not confluent; termination is not assured.
 There is no user control over the order that simplification rules are applied.
 This stripped down version of GOEDEL51.A23 lacks 95% of the simplification
 rules needed to produce good output. Mathematica's builtin Tracing commands
 are the only mechanism for discovering what rules were actually applied.
 :Examples: Sample files are available for various test suites.
*)
BeginPackage["Goedel'"]

and::usage = "and[x,y,...] is conjunction"

assert::usage = "assert[p] produces a statement equivalent to p by applying Goedel's
 algorthm to class[w,p]. Applying assert repeatedly sometimes simplifies a statement."

cart::usage = "cart[x,y] is the cartesian product of classes x and y."

class::usage = "class[x,p] applies Goedel's algorthm to the class of all sets x
 satisfying the condition p. The variable x may be atomic, or of the form pair[u,v],
 where u and v in turn can be pairs, etc."

complement::usage = "complement[x] is the class of all sets that do not belong to x"

composite::usage = "composite[x,y,...]  composite of x,y, ... "

domain::usage = "domain[x] is the domain of x"

E::usage = "E is the membership relation"

equal::usage = "equal[x,y] is the statement that the classes x and y are equal"
```

```
exists::usage = "exists[x,y,...,p] means there are sets x,y, ... such that p"

FIRST::usage = "FIRST is the function which takes pair[x,y] to y"

forall::usage = "forall[x,y,..., p] means that p holds for all sets x,y,..."

Id::usage = "Id is the identity relation"
id::usage = "id[x] is the restriction of the identity relation to x"

image::usage = "image[x,y] is the image of the class y under x"

intersection::usage = "intersection[x,y,...] is the intersection of classes x,y,..."

inverse::usage = "the relation inverse[x] is the inverse of x"

LeftPairV::usage = "LeftPairV is the function that takes x to pair[V,x]"

member::usage = "member[x,y] is the statement that x belongs to y"

not::usage = "not[p] represents the negation of p"

or::usage = "or[x,y,...] is the inclusive or"

P::usage = "the power class P[x] is the class of all subsets of x"

pair::usage = "pair[x,y] is the ordered pair of x and y."

range::usage = "range[x] is the range of x"

RightPairV::usage = "RightPairV is the function that takes x to pair[x,V]"

S::usage = "S is the subset relation"

SECOND::usage = "SECOND is the function that maps pair[x,y] to y"

singleton::usage = "singleton[x] has no member except x; it is 0 if x is not a set"

subclass::usage = "subclass[x,y] is the statement that x is contained in y"

U::usage = "the sum class U[x] is the union of all sets belonging to x"

union::usage = "union[x,y,...]  is the union of the classes x,y,... "

V::usage = "the universal class"

Begin["'Private'"]    (* begin the private context *)
(* definitions of auxiliary functions not exported *)
varlist[u_] := {u} /; AtomQ[u]
varlist[pair[u_,v_]] := Union[varlist[u],varlist[v]]

(* Is the expression x free of all variables which occur in y? *)
allfreeQ[x_,y_] := Apply[And,Map[FreeQ[x,#]&,varlist[y]]]

(* definitions of exported functions *)
(* Rules that must be assigned before attributes are set. *)
and[p_] := p
or[p_] := p

Attributes[and] := {Flat, Orderless, OneIdentity}
Attributes[or] := {Flat, Orderless, OneIdentity}

composite[x_] := composite[Id,x]
intersection[x_] := x
union[x_] := x

Attributes[composite] := {Flat,OneIdentity}
Attributes[intersection] := {Flat, Orderless, OneIdentity}
Attributes[union] := {Flat, Orderless, OneIdentity}
```

```
not[True]  := False                          (* Truth Table *)
not[False] := True


(* abbreviation for multiple quantifiers *)
exists[x_,y__,p_] := exists[x,exists[y,p]]

(* elimination rule for universal quantifiers *)
forall[x__,y_] := not[exists[x,not[y]]]

(* basic rules for membership *)
member[u_,0] := False

(* Added to avoid assuming axiom of regularity. Goedel assumes member[x,x] = 0. *)
class[w_,member[x_,x_]] :=
    Module[{y=Unique[]},class[w, exists[y,and[member[x,y],equal[x,y]]]]]

class[z_,member[w_,cart[x_,y_]]] := Module[{u = Unique[],v = Unique[]},
    class[z,exists[u,v,and[equal[pair[u,v],w],member[u,x],member[v,y]]]]]

member[pair[u_,v_],cart[x_,y_]] := and[member[u,x],member[v,y]]
member[u_,complement[x_]] := and[member[u,V],not[member[u,x]]]

class[w_,member[z_,composite[x_,y_]]] := Module[{t=Unique[],u=Unique[],v=Unique[]},
    class[w,exists[t,u,v,and[equal[z,pair[u,v]],and[member[pair[u,t],y],
    member[pair[t,v],x]]]]]]

class[z_,member[w_,cross[x_,y_]]] :=
    Module[{u1=Unique[],u2=Unique[],v1=Unique[],v2=Unique[]},
    class[z,exists[u1,u2,v1,v2,and[equal[pair[pair[u1,u2],pair[v1,v2]],w],
                      member[pair[u1,v1],x],member[pair[u2,v2],y]]]]]

(* Goedel's definition 1.5 *)
class[w_,member[u_,domain[x_]]] := Module[{v = Unique[]},class[w,exists[v,
                and[member[u,V],member[pair[u,v],x]]]]]

class[z_,member[w_,E]] := Module[{u = Unique[],v = Unique[]},class[z,exists[u,v,
                and[equal[pair[u,v],w],member[u,v]]]]]

class[w_,member[x_,FIRST]] := Module[{u = Unique[],v = Unique[]},class[w,
                exists[u,v,equal[pair[pair[u,v],u],x]]]]

class[z_,member[w_,Id]] := Module[{u = Unique[]},class[z,exists[u,equal[w,pair[u,u]]]]]

class[z_,member[w_,id[x_]]] := Module[{u = Unique[]},class[z,exists[u,
                and[member[u,x],equal[w,pair[u,u]]]]]]

class[w_,member[v_,image[z_,x_]]] := Module[{u = Unique[]},class[w,exists[u,
                and[member[v,V],member[u,x],member[pair[u,v],z]]]]]

member[u_,intersection[x_,y_]] := and[member[u,x],member[u,y]]

class[x_,member[w_,inverse[z_]]] := Module[{u = Unique[],v = Unique[]},
    class[x,exists[u,v,and[equal[pair[u,v],w],member[pair[v,u],z]]]]]

class[x_,member[w_,LeftPairV]] := Module[{u = Unique[],v = Unique[]},class[x,
                exists[u,v,and[equal[pair[u,v],w],equal[v,pair[V,u]]]]]]

member[x_,P[y_]] := and[member[x,V],subclass[x,y]]

class[u_,member[v_,pair[x_,y_]]] := Module[{z=Unique[]},class[u,exists[z,
                and[equal[pair[x,y],z],member[v,z]]]]]

class[w_,member[v_,range[z_]]] := Module[{u = Unique[]},class[w,exists[u,
                and[member[v,V],member[pair[u,v],z]]]]]

class[x_,member[w_,RightPairV]] := Module[{u = Unique[],v = Unique[]},class[x,
                exists[u,v,and[equal[pair[u,v],w],equal[v,pair[u,V]]]]]]
```

```
class[w_,member[x_,S]] := Module[{u = Unique[],v = Unique[]},class[w,exists[u,v,
    and[equal[pair[u,v],x],subclass[u,v]]]]]

class[w_,member[x_,SECOND]] := Module[{u = Unique[],v = Unique[]},class[w,
    exists[u,v,equal[pair[pair[u,v],v],x]]]]

member[u_,singleton[x_]] := and[equal[u,x],member[u,V]]

member[u_,union[x_,y_]] := or[member[u,x],member[u,y]]

class[w_,member[x_,U[z_]]] := Module[{y = Unique[]},class[w,exists[y,
    and[member[x,y],member[y,z]]]]]

class[w_,subclass[x_,y_]] := Module[{u = Unique[]},class[w,
    forall[u,or[not[member[u,x]], member[u,y]]]]]

class[x_,False]:=0
class[x_,True]:=V /; AtomQ[x]

class[pair[u_,v_],True] := cart[class[u,True],class[v,True]]

class[u_,member[u_,x_]] := x /; And[FreeQ[x,u],AtomQ[u]]

(* axiom B.1    membership relation *)
class[pair[u_,v_],member[u_,v_]] := E /; And[AtomQ[u],AtomQ[v]]

(* axiom B.2    intersection *)
class[x_,and[p_,q_]] := intersection[class[x,p],class[x,q]]
class[x_,or[p_,q_]] := union[class[x,p],class[x,q]]

(* axiom B.3    complement *)
class[x_,not[p_]] := intersection[complement[class[x,p]],class[x,True]]

(* axiom B.4 domain and Goedel's equation 2.8 on page 9 *)
class[x_,exists[y_,p_]] := domain[class[pair[x,y],p]]

(* axiom B.5    cartesian product *)
class[pair[u_,v_],member[u_,x_]] :=
        cart[x,V] /; And[FreeQ[x,u],FreeQ[x,v],AtomQ[u],AtomQ[v]]

(* axiom B.6    inverse *)
class[pair[u_,v_],member[v_,u_]] := inverse[E] /; And[AtomQ[u],AtomQ[v]]

(* an interpretation of Goedel's equation 2.41 on page 9 *)
class[pair[u_,v_],p_] := cart[class[u,p],class[v,True]] /; allfreeQ[p,v]

(* an interpretation of Goedel's equation 2.7 on page 9 *)
class[pair[u_,v_],p_] := cart[class[u,True],class[v,p]] /; allfreeQ[p,u]

(* Four rules to replace the rotation rules on Goedel's page 9: *)

class[pair[pair[u_,v_],w_],p_] := composite[class[pair[v,w],p],SECOND,
                     id[cart[class[u,True],V]]] /; allfreeQ[p,u]

class[pair[pair[u_,v_],w_],p_] := composite[class[pair[u,w],p],FIRST,
                     id[cart[V,class[v,True]]]] /; allfreeQ[p,v]

class[pair[w_,pair[u_,v_]],p_] := composite[id[cart[class[u,True],V]],
                     inverse[SECOND],class[pair[w,v],p]] /; allfreeQ[p,u]

class[pair[w_,pair[u_,v_]],p_] := composite[id[cart[V,class[v,True]]],
                     inverse[FIRST],class[pair[w,u],p]] /; allfreeQ[p,v]

(* special maneuver on page 10 of Goedel's monograph *)
class[u_,member[x_,y_]] := Module[{v = Unique[]},
    class[u,exists[v,and[equal[x,v],member[v,y]]]]] /; FreeQ[varlist[u],x]
```

```
(* new rules for equality *)
equal[x_,x_] := True


class[pair[u_,v_],equal[u_,v_]] := Id /; And[AtomQ[u],AtomQ[v]]
class[pair[u_,v_],equal[v_,u_]] := Id /; And[AtomQ[u],AtomQ[v]]

class[x_,equal[x_,y_]] :=
    intersection[singleton[y],class[x,True]] /; allfreeQ[y,x]

(* Goedel's Axiom A.3 of Coextension. *)
class[w_,equal[x_,y_]] :=
    intersection[class[w,subclass[x,y]], class[w,subclass[y,x]]] /;
    And[Or[Not[MemberQ[varlist[w],x]], Not[MemberQ[varlist[w],y]]],
        Not[SameQ[Head[x],pair]], Not[SameQ[Head[y],pair]]]

class[x_,equal[y_,x_]] := intersection[singleton[y],class[x,True]] /; allfreeQ[y,x]

equal[pair[x_,y_],0] := False
equal[pair[x_,y_],V] := False

(* equality of pairs *)
equal[pair[u_,v_],pair[x_,y_]] := and[equal[singleton[u],singleton[x]],
                                      equal[singleton[v],singleton[y]]]

class[w_,equal[singleton[u_],singleton[v_]]] := class[w,equal[u,v]] /;
                   MemberQ[varlist[w],u] || MemberQ[varlist[w],v] ||
                   member[u,V] || member[v,V]

(* flip equations involving a single pair to put pair on the left *)
equal[x_,y_pair] := equal[y,x]

(* rules that apply when x or y is known not to be a set *)
pair[x_,y_] := pair[V,y] /; Not[V === x] && not[member[x,V]]
pair[x_,y_] := pair[x,V] /; Not[V === y] && not[member[y,V]]

(* rule that applies when z does not occur in varlist[u] or when z occurs in x or y. *)
class[u_,equal[pair[x_,y_],z_]] := Module[{v=Unique[]},
        class[u,exists[v,and[equal[pair[x,y],v],equal[v,z]]]]] /;
        Not[MemberQ[varlist[u],z]] || Not[FreeQ[{x,y},z]]

(* rule that applies when z does occur in varlist[w] and z does not occur in either x or y.
   This rule only applies when x and y are known to be sets. *)
class[w_,equal[pair[x_,y_],z_]] := Module[{u=Unique[],v=Unique[]},
        class[(w/.z->pair[u,v]),and[equal[x,u],equal[y,v]]]] /;
        And[MemberQ[varlist[w],z],FreeQ[{x,y},z],
            Or[member[x,V],MemberQ[varlist[w],x]],
            Or[member[y,V],MemberQ[varlist[w],y]]]

(* rule that applies when one does not know whether or not x is a set *)
class[u_,equal[pair[x_,y_],z_]] := Module[{v=Unique[]},
    class[u,or[and[not[member[x,V]],equal[pair[V,y],z]],
    exists[v,and[equal[x,v],equal[pair[v,y],z]]]]]] /;
    Not[MemberQ[varlist[u],x]] && UnsameQ[V,x] && Not[member[x,V]] === True]

(* rule that applies when one does not know whether or not y is a set *)
class[u_,equal[pair[x_,y_],z_]] := Module[{v=Unique[]},
    class[u,or[and[not[member[y,V]],equal[pair[x,V],z]],
    exists[v,and[equal[y,v],equal[pair[x,v],z]]]]]] /;
    Not[MemberQ[varlist[u],y]] && UnsameQ[V,y] && Not[member[y,V]] === True]

class[pair[u_,v_],equal[pair[V,u_],v_]] := LeftPairV
class[pair[u_,v_],equal[pair[u_,V],v_]] := RightPairV

class[pair[u_,v_],equal[pair[V,v_],u_]] := inverse[LeftPairV]
class[pair[u_,v_],equal[pair[v_,V],u_]] := inverse[RightPairV]

image[inverse[RightPairV],x_] := 0 /; composite[Id,x] == x
image[inverse[LeftPairV],x_] := 0 /; composite[Id,x] == x
```

```
class[w_,equal[pair[V,y_],z_]] := Module[{v = Unique[]},
        class[w,or[and[not[member[y,V]],equal[pair[V,V],z]],
        and[member[y,V],exists[v,and[equal[pair[V,v],z],equal[v,y]]]]]]] /;
        Not[allfreeQ[y,w]]

class[w_,equal[pair[x_,V],z_]] := Module[{v = Unique[]},
        class[w,or[and[not[member[x,V]],equal[pair[V,V],z]],
        and[member[x,V],exists[v,and[equal[pair[v,V],z],equal[v,x]]]]]]] /;
        Not[allfreeQ[x,w]]

class[w_,equal[pair[V,V],w_]] := singleton[pair[V,V]]

class[w_,equal[pair[V,V],x_]] := Module[{v = Unique[]},
        class[w,exists[v,and[equal[pair[V,V],v],equal[v,x]]]]] /;
        Not[MemberQ[x,varlist[w]]]

(* assertions *)
assert[p_] := Module[{w = Unique[]}, equal[V,class[w,p]]]

(* a few simplification rules *)
cart[x_,0] := 0
cart[0,x_] := 0

complement[0] := V
complement[complement[x_]] := x
complement[union[x_,y_]] := intersection[complement[x],complement[y]]
complement[V] := 0

composite[x_,cart[y_,z_]] := cart[y,image[x,z]]
composite[cart[x_,y_],z_] := cart[image[inverse[z],x],y]
composite[Id,x_,y_] := composite[x,y]
composite[x_,Id] := composite[Id,x]
composite[Id,Id] := Id

domain[cart[x_,y_]] := intersection[x,image[V,y]]
domain[composite[x_,y_]] := image[inverse[y],domain[x]]
domain[Id] := V
domain[id[x_]] := x

id[V] := Id

image[0,x_] := 0
image[x_,0] := 0
image[composite[x_,y_],z_] := image[x,image[y,z]]
image[Id,x_] := x
image[id[x_],y_] := intersection[x,y]

intersection[cart[x_,y_],z_] := composite[id[y],z,id[x]]
intersection[V,x_] := x

inverse[0] := 0
inverse[cart[x_,y_]] := cart[y,x]
inverse[complement[x_]] := composite[Id,complement[inverse[x]]]
inverse[composite[x_,y_]] := composite[inverse[y],inverse[x]]
inverse[Id] := Id
inverse[inverse[x_]] := composite[Id,x]

range[Id] := V
union[0,x_] := x
End[ ]          (* end the private context *)

Protect[ and, assert, cart, class, complement, composite, domain, E,
equal, exists, FIRST, forall, Id, id, image, intersection, inverse,
LeftPairV, member, not, or, P, pair, range, RightPairV, S, SECOND,
singleton, subclass, U, union, V ]

EndPackage[ ]  (* end the package context *)
```

References

Belinfante, J. G. F., On a Modification of Gödel's Algorithm for Class Formation, Association for Automated Reasoning News Letter, No. 34 (1996) pp. 10–15.

Belinfante, J. G. F., On Quaife's Development of Class Theory, Association for Automated Reasoning Newsletter, No. 37 (1997) pp. 5–9.

Belinfante, J. G. F., Computer Proofs in Gödel's Class Theory with Equational Definitions for Composite and Cross, Journal of Automated Reasoning, vol. 22 (1999) pp. 311–339.

Belinfante, J. G. F., On Computer-Assisted Proofs in Ordinal Number Theory, Journal of Automated Reasoning, vol. 22 (1999), pp. 341–378.

Bernays, P., Axiomatic Set Theory, North Holland Publishing Co., Amsterdam. First edition: 1958. Second edition: 1968. Republished in 1991 by Dover Publications, New York.

Boyer, R., Lusk, E., McCune, W., Overbeek, R., Stickel M. and Wos, L., Set Theory in First Order Logic: Clauses for Gödel's Axioms, Journal of Automated Reasoning, volume 2 (1986), pages 287–327.

Formisano, A. and Omodeo, E. G., An Equational Re-Engineering of Set Theories, presented at the FTP'98 International Workshop on First Order Theorem Proving (November 23–25, 1998).

Gödel, K., The Consistency of the Axiom of Choice and of the Generalized Continuum Hypothesis with the Axioms of Set Theory, Princeton University Press, Princeton, 1940.

Isbell, J. R., A Definition of Ordinal Numbers, The American Mathematical Monthly vol. 67 (1960), pp. 51–52.

McCune, W. W., Otter 3.0 Reference Manual and Guide, Argonne National Laboratory Report ANL–94/6, Argonne National Laboratory, Argonne, IL, January 1994.

Megill, N. D., Metamath: A Computer Language for Pure Mathematics, 1997.

Noël, P. A. J., Experimenting with Isabelle in ZF set theory, Journal of Automated Reasoning, vol. 10 (1993), pp. 15–58.

Paulson, L. C., and Grąbczewski, K., Mechanizing Set Theory, Journal of Automated Reasoning, vol. 17, pp. 291–323 (1996).

Quaife, A., Automated Deduction in von Neumann-Bernays-Gödel Set Theory, Journal of Automated Reasoning, vol. 8 (1992), pp. 91–147.

Quaife, A., Automated Development of Fundamental Mathematical Theories, Ph.D. thesis, Univ. of California at Berkeley, Kluwer Acad. Publishers, Dordrecht, 1992.

Tarski, A., and Givant, S., A Formalization of Set Theory without Variables, American Mathematical Society Colloquium Publications, volume 41, Providence, Rhode Island, 1987.

Wos, L., Automated Reasoning: 33 Basic Research Problems, Prentice Hall, Englewood Cliffs, NJ, 1988.

Wos, L., The problem of finding an inference rule for set theory, Journal of Automated Reasoning, vol. 5 (1989), pp. 93–95.

Wos, L., Overbeek, R., Lusk, E. and Boyle, J., Automated Reasoning: Introduction and Applications, Second Edition, McGraw Hill, New York, 1992.

Wolfram, S., The Mathematica™ Book, Wolfram Media Inc., Champaign, Illinois, 1996.

Automated Proof Construction in Type Theory Using Resolution

Marc Bezem[1], Dimitri Hendriks[2], and Hans de Nivelle[3]

[1] Utrecht University, Department of Philosophy
bezem@phil.uu.nl
[2] Utrecht University, Department of Philosophy
hendriks@phil.uu.nl
[3] Max Planck Institute
nivelle@mpi-sb.mpg.de

Abstract. We provide techniques to integrate resolution logic with equality in type theory. The results may be rendered as follows.

- A clausification procedure in type theory, equipped with a correctness proof, all encoded using higher-order primitive recursion.
- A novel representation of clauses in minimal logic such that the λ-representation of resolution proofs is linear in the size of the premisses.
- A translation of resolution proofs into lambda terms, yielding a verification procedure for those proofs.
- The power of resolution theorem provers becomes available in interactive proof construction systems based on type theory.

1 Introduction

Type theory (= typed Lambda Calculus) offers a powerful formalism for formalizing mathematics. Strong points are: the logical foundation, the fact that proofs are first-class citizens, and the generality which naturally facilitates extensions, such as inductive types. Type theory captures definitions, reasoning and computation at various levels in an integrated way. In a type-theoretical system, formalized mathematical statements are represented by types, and their proofs are represented by λ-terms. The problem whether π is a proof of statement A reduces to checking whether the term π has type A. Computation is based on a simple notion of rewriting. The level of detail is such that the well-formedness of definitions and the correctness of derivations can automatically be verified.

However, there are also weak points. It is exactly the appraised expressivity and the level of detail that makes automation at the same time necessary and difficult. Automated deduction appears to be mostly successful in weak systems, such as propositional logic and predicate logic, systems that fall short to formalize a larger body of mathematics. Apart from the problem of the expressivity of these systems, only a minor part of the theorems that can be expressed can actually be proved automatically. Therefore it is necessary to combine automated

D. McAllester (Ed.): CADE-17, LNAI 1831, pp. 148–163, 2000.
© Springer-Verlag Berlin Heidelberg 2000

theorem proving with interactive theorem proving. Recently a number of proposals in this direction have been made. In [MS99] Otter is combined with the Boyer-Moore theorem prover. (A verified program rechecks proofs generated by Otter.) In [Hur99] Gandalf is linked to HOL. (The translation generates scripts to be run by the HOL-system.) In [ST95], proofs are translated into Martin-Löf's type theory, for the Horn clause fragment of first-order logic. In the Omega system [Hua96,Omega] various theorem provers have been linked to a natural deduction proof checker. The purpose there is to automatically generate proofs from so called *proof plans*. Our approach is different in that we generate complete proof objects for both the clausification and the refutation part.

Resolution theorem provers, such as Bliksem [Blk], are powerful, but have the drawback that they work with normal forms of formulae, so-called clausal forms. Clauses are (universally closed) disjunctions of literals, and a literal is either an atom or a negated atom. The clausal form of a formula is essentially its Skolem-conjunctive normal form, which need not be exactly logically equivalent to the original formula. This makes resolution proofs hard to read and understand, and makes the interactive navigation of the theorem prover through the search space very difficult. Moreover, optimized implementations of proof procedures are error-prone (cf. recent CASC disqualifications).

In type theory, the proof generation capabilities suffer from the small granularity of the inference steps and the corresponding astronomic size of the search space. Typically, one hyperresolution step requires a few dozens of inference steps in type theory. In order to make the formalisation of a large body of mathematics feasible, the level of automation of interactive proof construction systems such as Coq [Coq98], based on type theory, has to be improved.

We propose the following proof procedure. Identify a non-trivial step in a Coq session that amounts to a first-order tautology. Export this tautology to Bliksem, and delegate the proof search to the Bliksem inference engine. Convert the resolution proof to type theoretic format and import the result back in Coq. We stress the fact that the above procedure is as secure as Coq. Hypothetical errors (e.g. the clausification procedure not producing clauses, possible errors in the resolution theorem prover or the erroneous formulation of the lambda terms corresponding to its proofs) are irrelevant because the resulting proofs are type-checked by Coq. The security could be made independent of Coq by using another type-checker.

Most of the necessary meta-theory is already known. The negation normal form transformation can be axiomatized by classical logic. The prenex and conjunctive normal form transformations require that the domain is non-empty. Skolemization can be axiomatized by so-called Skolem axioms, which can be viewed as specific instances of the Axiom of Choice. Higher-order logic is particularly suited for this axiomatization: we get logical equivalence modulo classical logic plus the Axiom of Choice, instead of awkward invariants as equiconsistency or equisatisfiability in the first-order case.

By adapting a result of Kleene, Skolem functions and –axioms could be eliminated from resolution proofs, which would allow us to obtain directly a proof of the original formula (cf. [Pfe84]), but currently we still make use of the Axiom of Choice.

The paper is organized as follows. In Section 2 we set out a two-level approach and define a deep embedding to represent first-order logic. Section 3 describes a uniform clausification procedure. We explain how resolution proofs are translated into λ-terms in Sections 4 and 5. Finally, the outlined constructions are demonstrated in Section 6.

2   A Two-Level Approach

The basic sorts in Coq are $*^p$ and $*^s$. An object M of type $*^p$ is a logical proposition and denotes the class of proofs of M. Objects of type $*^s$ are usual sets such as the set of natural numbers, lists etc. In type theory, the typing relation is expressed by $t : T$, to be interpreted as 't belongs to set T' when $T : *^s$, and as 't is a proof of proposition T' when $T : *^p$. As usual, $\rightarrow$ associates to the right; $\rightarrow$ is used for logical implication as well as for function spaces. Furthermore, well-typed application is denoted by $(M\ N)$ and associates to the left. Scopes of bound variables are always extended to the right as far as possible. We use the notation

$$T : *^s := \begin{cases} constructor_0 : \cdots \rightarrow T \\ \quad\vdots \qquad\quad \vdots \\ constructor_n : \cdots \rightarrow T \end{cases}$$

to define the inductive set T, that is: the smallest set of objects that is freely generated by $constructor_0, \ldots, constructor_n$. Moreover, we use

$$\lambda t : T. \text{Cases } t \text{ of } \begin{cases} pattern_0 \Rightarrow rhs_0 \\ \quad\vdots \quad\ \vdots \ \ \vdots \\ pattern_m \Rightarrow rhs_m \end{cases}$$

for the exhaustive case analysis on t in the inductive type T. If t matches $pattern_i$, it is replaced by the right-hand side rhs_i.

We choose for a deep embedding in adopting a two-level approach for the treatment of arbitrary first-order languages. The idea is to represent first-order formulae as objects in an inductive set $o : *^s$, accompanied by an interpretation function E that maps these objects into $*^p$.[1] The next paragraphs explain why we distinguish a higher (*meta-*, *logical*) level $*^p$ and a lower (*object-*, *computational*) level o.

The universe $*^p$ includes higher-order propositions; in fact it encompasses full impredicative type theory. As such, it is too large for our purposes. Given a suitable signature, any first-order proposition $\varphi : *^p$ will have a formal counterpart $p : o$ such that φ equals $(E\ p)$, the interpretation of p. Thus, the first-order

---

[1] Both o as well as E depend on a fixed but arbitrary signature.

fragment of $*^p$ can be identified as the collection of interpretations of objects in o.

Secondly, Coq supplies only limited computational power on $*^p$, whereas o, as every inductive set, is equipped with the powerful computational device of higher-order primitive recursion. This enables the syntactical manipulation of object-level propositions.

Reflection of object-level propositions is used for the proof construction of first-order formulae in $*^p$, in the following way. Let $\varphi : *^p$ be a first-order proposition. Then there is some $\dot\varphi : o$ such that $(E\ \dot\varphi)$ is convertible with φ.[2] Moreover, suppose we have proved

$$\forall p : o.\ (E\ (T\ p)) \to (E\ p)$$

for some function $T : o \to o$. Then, to prove φ it suffices to prove $(E\ (T\ \dot\varphi))$. Matters are presented schematically in Figure 1. In Section 3 we discuss a concrete function T, for which we have proved the above. For this T, proofs of $(E\ (T\ \dot\varphi))$ will be generated automatically, as will be described in Sections 4 and 5.

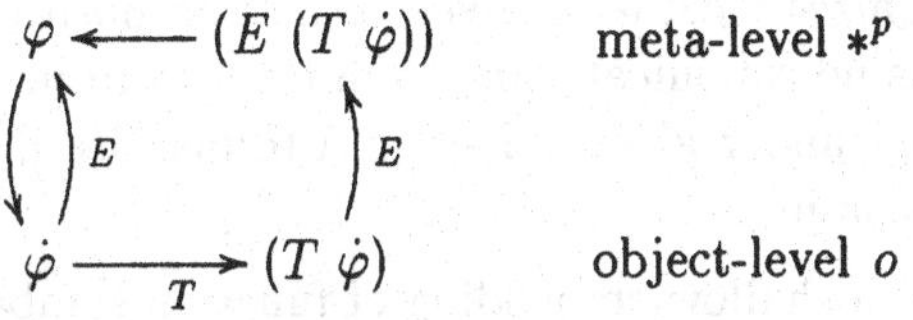

Fig. 1. Schematic overview of the general procedure. The proof of the implication from $(E\ (T\ \dot\varphi))$ to φ can be generated uniformly in $\dot\varphi$.

Object-Level Propositions and the Reflection Operation

In Coq, we have constructed a general framework to represent first-order languages with multiple sorts. Bliksem is (as yet) one-sorted, so we describe the setup for one-sorted signatures only.

Assume a domain of discourse $\sigma : *^s$. Suppose we have relation symbols $R_0, \ldots, R_k$ typed $\sigma^{e_0} \to *^p, \ldots, \sigma^{e_k} \to *^p$ respectively. Here $e_0, \ldots, e_k$ are natural numbers and σ^n is the Cartesian product of n copies of σ, that is:

$$\sigma^0 = unit \qquad \sigma^1 = \sigma \qquad \sigma^{n+2} = \sigma \times \sigma^{n+1}$$

The set $unit$ is a singleton with sole inhabitant tt.

---

[2] The mapping ˙ is a syntax-based translation outside Coq.

Let L be the non-empty[3] list of arities $[e_0, \ldots, e_k]$. We define o, the set of objects representing propositions, inductively by:

$$
o : *^s := \left\{
\begin{array}{l}
rel : \Pi i : (index\ L).\, \sigma^{(select\ L\ i)} \to o \\
\dot{\neg} : o \to o \\
\dot{\to}, \dot{\vee}, \dot{\wedge} : o \to o \to o \\
\dot{\forall}, \dot{\exists} : (\sigma \to o) \to o
\end{array}
\right.
$$

We use the dot-notation $\dot{}$ to distinguish the object-level constructors from Coq's predefined connectives. Connectives are written infix. The function $select$ is of type $\Pi L : (nelist\ nat).(index\ L) \to nat$, where $(index\ L)$ computes the set $\{0, \ldots, k\}$. We have $(select\ L\ i) =_{\beta\delta\iota} e_i$. Thus, an atomic proposition is of form $(rel\ i\ t)$, with t an argument tuple in σ^{e_i}.

The constructors $\dot{\forall}, \dot{\exists}$ map propositional functions of type $\sigma \to o$ to propositions of type o. This representation has the advantage that binding and predication are handled by λ-abstraction and λ-application. On the object-level, existential quantification of x over p (of type o, possibly containing occurrences of x) is written as $(\dot{\exists}\ (\lambda x : \sigma.\, p))$. Although this representation suffices for our purposes, it causes some well-known difficulties. E.g. we cannot write a boolean function which recognizes whether a given formal proposition is in prenex normal form. As there is no canonical choice of a fresh term in σ, it is not possible to recursively descend under abstractions in λ-terms. See [NM98, Sections 8.3, 9.2] for a further discussion.

For our purposes, a shallow embedding of function symbols is sufficient. We have not defined an inductive set $term$ representing the first-order terms in σ like we have defined o representing the first-order fragment of $*^p$. Instead, 'meta-level' terms of type σ are taken as arguments of object-level predicates. Due to this shallow embedding, we cannot check whether certain variables have occurrences in a given term. Because of that, e.g., distributing universal quantifiers over conjuncts can yield dummy abstractions. These problems could be overcome by using de Bruijn-indices (see [dB72]) for a deep embedding of terms in Coq.

The interpretation function E is a canonical homomorphism recursively defined as follows.

$$
E : o \to *^p := \lambda p : o.\, \text{Cases } p \text{ of} \left\{
\begin{array}{l}
(rel\ i\ t) \Rightarrow (R_i\ t) \\
\dot{\neg} p_0 \Rightarrow \neg(E\ p_0) \\
p_1 \dot{\to} p_2 \Rightarrow (E\ p_1) \to (E\ p_2) \\
p_1 \dot{\vee} p_2 \Rightarrow (E\ p_1) \vee (E\ p_2) \\
p_1 \dot{\wedge} p_2 \Rightarrow (E\ p_1) \wedge (E\ p_2) \\
(\dot{\forall}\ p_0) \Rightarrow \forall x : \sigma.\, (E\ (p_0\ x)) \\
(\dot{\exists}\ p_0) \Rightarrow \exists x : \sigma.\, (E\ (p_0\ x))
\end{array}
\right.
$$

---

[3] We require the signature to contain at least one relation symbol.

In the above definitions of o, its constructors and of E, the dependency on the signature has been suppressed. In fact we have:

$$o : *^s \to (nelist\ nat) \to *^s$$
$$rel : \Pi\sigma : *^s.\ \Pi L : (nelist\ nat).\ \Pi i : (index\ L).\ \sigma^{(select\ L\ i)} \to (o\ \sigma\ L)$$
$$E : \Pi\sigma : *^s.\ \Pi L : (nelist\ nat).\ (\Pi i : (index\ L).\ \sigma^{(select\ L\ i)} \to *^p) \to (o\ \sigma\ L) \to *^p$$

In the next section, we fix an arbitrary signature and mention the above dependencies implicitly only.

3 Clausification and Correctness

We describe the transformation to *minimal clausal form* (see Section 4), which is realized on both levels. On the object-level, we define an algorithm $mcf : o \to o$ that converts object-level propositions into their clausal form. On the meta-level, clausification is realized by a term mcf_{prf}, which transforms a proof of $(E\ (mcf\ p))$ into a proof of $(E\ p)$.

The algorithm mcf consists of the subsequent application of the following functions: $nnf, pnf, cnf, sklm, duqc, impf$ standing for transformations to negation, prenex and conjunctive normal form, Skolemization, distribution of universal quantifiers over conjuncts and transformation to implicational form, respectively. As an illustration, we describe the functions nnf and $sklm$.

Concerning negation normal form, a recursive call like

$$(nnf\ \neg(A \wedge B)) = (nnf\ \neg A) \vee (nnf\ \neg B)$$

is not primitive recursive, since $\neg A$ and $\neg B$ are not subformulae of $\neg(A \wedge B)$. Such a call requires general recursion. Coq's computational mechanism is higher-order primitive recursion, which is weaker than general recursion but ensures universal termination.

The function $nnf : o \to pol \to o$ defined below[4], makes use of the so-called polarity of an input formula. Polarities are: $pol : *^s := \begin{cases} \oplus : pol \\ \ominus : pol \end{cases}$.

---

[4] For $Q = \forall, \exists$, we write $Qx : \sigma.\ p$ instead of $(Q\ (\lambda x : \sigma.\ p))$.

$nnf : o \to pol \to o := \lambda p{:}o.\, \lambda a{:}pol.$ Cases $p\ a$ of

$$
\begin{cases}
(rel\ i\ t) \oplus \Rightarrow (rel\ i\ t) \\
(rel\ i\ t) \ominus \Rightarrow \neg(rel\ i\ t) \\
\neg p_0 \oplus \Rightarrow (nnf\ p_0\ \ominus) \\
\neg p_0 \ominus \Rightarrow (nnf\ p_0\ \oplus) \\
p_1 \dot\to p_2 \oplus \Rightarrow (nnf\ p_1\ \ominus) \dot\vee (nnf\ p_2\ \oplus) \\
p_1 \dot\to p_2 \ominus \Rightarrow (nnf\ p_1\ \oplus) \dot\wedge (nnf\ p_2\ \ominus) \\
p_1 \dot\vee p_2 \oplus \Rightarrow (nnf\ p_1\ \oplus) \dot\vee (nnf\ p_2\ \oplus) \\
p_1 \dot\vee p_2 \ominus \Rightarrow (nnf\ p_1\ \ominus) \dot\wedge (nnf\ p_2\ \ominus) \\
p_1 \dot\wedge p_2 \oplus \Rightarrow (nnf\ p_1\ \oplus) \dot\wedge (nnf\ p_2\ \oplus) \\
p_1 \dot\wedge p_2 \ominus \Rightarrow (nnf\ p_1\ \ominus) \dot\vee (nnf\ p_2\ \ominus) \\
(\dot\forall\ p_0) \oplus \Rightarrow \dot\forall x : \sigma.\,(nnf\ (p_0\ x)\ \oplus) \\
(\dot\forall\ p_0) \ominus \Rightarrow \dot\exists x : \sigma.\,(nnf\ (p_0\ x)\ \ominus) \\
(\dot\exists\ p_0) \oplus \Rightarrow \dot\exists x : \sigma.\,(nnf\ (p_0\ x)\ \oplus) \\
(\dot\exists\ p_0) \ominus \Rightarrow \dot\forall x : \sigma.\,(nnf\ (p_0\ x)\ \ominus)
\end{cases}
$$

We have proved the following lemma.

$$EM \to \forall p{:}o.\,((E\ p) \leftrightarrow (E\ (nnf\ p\ \oplus))) \wedge (\neg(E\ p) \leftrightarrow (E\ (nnf\ p\ \ominus)))$$

Where EM is the principle of excluded middle, defined in such a way that it affects the first-order fragment only.

$$EM : *^p := \forall p{:}o.\,(E\ p) \vee \neg(E\ p)$$

Skolemization of a formula means the removal of all existential quantifiers and the replacement of the variables that were bound by the removed existential quantifiers, by new terms, that is, Skolem functions applied to the universally quantified variables whose quantifier had the existential quantifier in its scope. Instead of quantifying each of the Skolem functions, we introduce an index type $skolT$, which may be viewed as a family of Skolem functions.

$$skolT : *^s := nat \to nat \to \Pi n{:}nat.\,\sigma^n \to \sigma$$

A Skolem function, then, is a term $(f\ i\ j\ n) : \sigma^n \to \sigma$, with $f : skolT$ and $i, j, n : nat$. Here, i and j are indices that distinguish the family members. If the output of nnf yields a conjunction, the remaining clausification steps are performed separately on the conjuncts. (This yields a significant speed-up in performance.) Index i denotes the position of the conjunct, j denotes the number of the replaced existentially quantified variable in that conjunct. The function $sklm$ is defined as follows.

$$
\begin{aligned}
&sklm : skolT \to nat \to nat \to \Pi n{:}nat.\,\sigma^n \to o \to o := \\
&\quad \lambda f{:}skolT.\, \lambda i, j, n{:}nat.\, \lambda t{:}\sigma^n.\, \lambda p{:}o. \\
&\quad \text{Cases } p \text{ of} \\
&\qquad \begin{cases}
(\dot\forall\ p_0) \Rightarrow \dot\forall x.\,(sklm\ f\ i\ j\ (n+1)\ (insert\ x\ n\ t)\ (p_0\ x)) \\
(\dot\exists\ p_0) \Rightarrow (sklm\ f\ i\ (j+1)\ n\ t\ (p_0\ (f\ i\ j\ n\ t))) \\
p' \Rightarrow p'
\end{cases}
\end{aligned}
$$

Given a variable $x : \sigma$, an arity $n : nat$, a tuple $t : \sigma^n$, the term $(insert\ x\ n\ t)$ adds x at the end of t, resulting in a tuple of type σ^{n+1}. Thus, if p is a universal statement, the quantified variable is added at the end of the so far constructed tuple t of universally quantified variables. In case p matches $(\exists\ p_0)$, the term $(f\ i\ j\ n\ t)$ is substituted for the existentially quantified variable (the 'hole' in p_0) and index j is incremented. The third case, p', exhausts the five remaining cases. As we force input formulae to be in prenex normal form (via the definition of mcf), nothing remains to be done.

We proved the following lemma.

$$\sigma \to AC \to \forall i : nat.\ \forall p : o.\ (E\ p) \to \exists f : skolT.\ (E\ (sklm\ f\ i\ 0\ 0\ tt\ p))$$

Here, $\sigma \to \cdots$ expresses the condition that σ is non-empty. AC is the Axiom of Choice, which allows us to form Skolem functions.

$$AC : *^p := \forall \alpha : \sigma \to skolT \to o.$$
$$(\forall x : \sigma.\ \exists f : skolT.\ (E\ (\alpha\ x\ f)))$$
$$\to \exists F : \sigma \to skolT.\ \forall x : \sigma.\ (E\ (\alpha\ x\ (F\ x)))$$

Reconsider Figure 1 and substitute mcf for T. We have proved that for all objects $p : o$ the interpretation of the result of applying mcf to p implies the interpretation of p. Thus, given a suitable signature, from any first-order formula $\varphi : *^p$, we can construct the classical equivalent $(E\ (mcf\ \hat{\varphi})) \in MCF$. The term mcf_{prf} makes clausification effective on the meta-level.

$$mcf_{prf} : EM \to AC \to \sigma \to \forall p : o.\ (E\ (mcf\ p)) \to (E\ p)$$

Given inhabitants $em : EM$ and $ac : AC$, an element $s : \sigma$, a proposition $p : o$ and a proof $\rho : (E\ (mcf\ p))$, the term $(mcf_{prf}\ em\ ac\ s\ p\ \rho)$ is a proof of $(E\ p)$. The term $(E\ (mcf\ p)) : *^p$ computes a format $C_1 \to \cdots \to C_n \to \bot$. Here $C_1, \ldots, C_n : *^p$ are universally closed clauses that will be exported to Bliksem, which constructs the λ-term ρ representing a resolution refutation of these clauses (see Sections 4 and 5). Finally, ρ is type-checked in Coq. Section 6 demonstrates the outlined constructions.

The complete Coq-script generating the correctness proof of the clausification algorithm comprises $\pm$ 65 pages. It is available at the following URL.

www.phil.uu.nl/~hendriks/claus.tar.gz

4   Minimal Resolution Logic

There exist many representations of clauses and corresponding formulations of resolution rules. The traditional form of a clause is a disjunction of literals, that is, of atoms and negated atoms. Another form which is often used is that of a sequent, that is, the implication of a disjunction of atoms by a conjunction of atoms.

Here we propose yet another representation of clauses, as far as we know not used before. There are three main considerations.

- A structural requirement is that the representation of clauses is closed under the operations involved, such as instantiation and resolution.
- The Curry-Howard correspondence is most direct between minimal logic ($\rightarrow$,$\forall$) and a typed lambda calculus with product types (with $\rightarrow$ as a special, non-dependent, case of Π). Conjunction and disjunction in the logic require either extra type-forming primitives and extra terms to inhabit these, or impredicative encodings.
- The λ-representation of resolution proofs should preferably be linear in the size of the premisses.

These considerations have led us to represent a clause like:

$$L_1 \vee \cdots \vee L_p$$

by the following classically equivalent implication in minimal logic:

$$\overline{L}_1 \rightarrow \cdots \rightarrow \overline{L}_p \rightarrow \bot$$

Here $\overline{L}_i$ is the complement of L_i in the classical sense (i.e. double negations are removed). If C is the disjunctive form of a clause, then we denote its implicational form by $[C]$. As usual, these expressions are implicitly or explicitly universally closed.

A resolution refutation of given clauses $C_1, \ldots, C_n$ proves their inconsistency, and can be taken as a proof of the following implication in minimal logic:

$$C_1 \rightarrow \cdots \rightarrow C_n \rightarrow \bot$$

The logic is called minimal as we use no particular properties of $\bot$. We are now ready for the definition of the syntax of minimal resolution logic.

Definition 1. *Let $\forall \vec{x}. \phi$ denote the universal closure of ϕ. Let Atom be the set of atomic propositions. We define the sets Literal, Clause and MCF of, respectively, literals, clauses and minimal clausal forms by the following abstract syntax.*

$$
\begin{aligned}
Literal &::= Atom \mid Atom \rightarrow \bot \\
Clause &::= \bot \mid Literal \rightarrow Clause \\
MCF &::= \bot \mid (\forall \vec{x}. \ Clause) \rightarrow MCF
\end{aligned}
$$

Next we elaborate the familiar inference rules for factoring, permuting and weakening clauses, as well as the binary resolution rule.

Factoring, Permutation, Weakening

Let C and D be clauses, such that C subsumes D propositionally, that is, any literal in C also occurs in D. Let $A_1, \ldots, A_p, B_1, \ldots, B_q$ be literals ($p, q \geq 0$) and write

$$[C] = A_1 \rightarrow \cdots \rightarrow A_p \rightarrow \bot$$

and

$$[D] = B_1 \to \cdots \to B_q \to \bot$$

assuming that for every $1 \leq i \leq p$ there is $1 \leq j \leq q$ such that $A_i = B_j$.

A proof of $[C] \to [D]$ is the following λ-term:

$$\lambda c : [C]. \, \lambda b_1 : B_1 \ldots \lambda b_q : B_q. \, (c \; \pi_1 \; \ldots \; \pi_p)$$

with $\pi_i = b_j$, where j is such that $B_j = A_i$.

Binary Resolution

In the traditional form of the binary resolution rule for disjunctive clauses we have premisses C_1 and C_2, containing one or more occurrences of a literal L and of $\overline{L}$, respectively. The conclusion of the rule, the resolvent, is then a clause D consisting of all literals of C_1 different from L joined with all literals of C_2 different from $\overline{L}$. This rule is completely symmetric with respect to C_1 and C_2.

For clauses in implicational form there is a slight asymmetry in the formulation of binary resolution. Let $A_1, \ldots, A_p, B_1 \ldots, B_q$ be literals ($p, q \geq 0$) and write

$$[C_1] = A_1 \to \cdots \to A_p \to \bot,$$

with one or more occurrences of the negated atom $A \to \bot$ among the A_i and

$$[C_2] = B_1 \to \cdots \to B_q \to \bot,$$

with one or more occurrences of the atom A among the B_j. Write the resolvent D as

$$[D] = D_1 \to \cdots \to D_r \to \bot$$

consisting of all literals of C_1 different from $A \to \bot$ joined with all literals of C_2 different from A. A proof of $[C_1] \to [C_2] \to [D]$ is the following λ-term:

$$\lambda c_1 : [C_1]. \, \lambda c_2 : [C_2]. \, \lambda d_1 : D_1 \ldots \lambda d_r : D_r. \, (c_1 \; \pi_1 \; \ldots \; \pi_p)$$

For $1 \leq i \leq p$, π_i is defined as follows. If $A_i \neq (A \to \bot)$, then $\pi_i = d_k$, where k is such that $D_k = A_i$. If $A_i = A \to \bot$, then we put

$$\pi_i = \lambda a : A. \, (c_2 \; \rho_1 \; \ldots \; \rho_q),$$

with ρ_j ($1 \leq j \leq q$) defined as follows. If $B_j \neq A$, then $\rho_j = d_k$, where k is such that $D_k = B_j$. If $B_j = A$, then $\rho_j = a$. It is easily verified that $\pi_i : (A \to \bot)$ in this case.

If $(A \to \bot)$ occurs more than once among the A_i, then $(c_1 \; \pi_1 \; \ldots \; \pi_p)$ need not be linear. This can be avoided by factoring timely. Even without factoring, a linear proof term is possible: by taking the following β-expansion of $(c_1 \; \pi_1 \; \ldots \; \pi_p)$ (with a' replacing copies of proofs of $(A \to \bot)$):

$$(\lambda a' : A \to \bot. \, (c_1 \; \pi_1 \; \ldots \; a' \; \ldots \; a' \; \ldots \; \pi_p))(\lambda a : A. \, (c_2 \; \rho_1 \; \ldots \; \rho_q))$$

This remark applies to the rules in the next subsections as well.

Paramodulation

Paramodulation combines equational reasoning with resolution. For equational reasoning we use the inductive equality of Coq. In order to simplify matters, we assume a fixed domain of discourse σ, and denote equality of $s_1, s_2 \in \sigma$ by $s_1 \approx s_2$.

Coq supplies us with the following terms:

$$eqrefl : \forall s : \sigma. \, (s \approx s)$$
$$eqsubst : \forall s : \sigma. \forall P : \sigma \to *^p . \, (P \, s) \to \forall t : \sigma. \, (s \approx t) \to (P \, t)$$
$$eqsym : \forall s_1, s_2 : \sigma. \, (s_1 \approx s_2) \to (s_2 \approx s_1)$$

As an example we define $eqsym$ from $eqsubst, eqrefl$:

$$\lambda s_1, s_2 : \sigma. \, \lambda h : (s_1 \approx s_2). \, (eqsubst \; s_1 \; (\lambda s : \sigma. \, (s \approx s_1))) \; (eqrefl \; s_1) \; s_2 \; h)$$

Paramodulation for disjunctive clauses is the rule with premiss C_1 containing the equality literal $t_1 \approx t_2$ and premiss C_2 containing literal $A[t_1]$. The conclusion is then a clause D containing all literals of C_1 different from $t_1 \approx t_2$, joined with C_2 with $A[t_2]$ instead of $A[t_1]$.

Let $A_1, \ldots, A_p, B_1 \ldots, B_q$ be literals ($p, q \geq 0$) and write

$$[C_1] = A_1 \to \cdots \to A_p \to \bot,$$

with one or more occurrences of the equality atom $t_1 \approx t_2 \to \bot$ among the A_i, and

$$[C_2] = B_1 \to \cdots \to B_q \to \bot,$$

with one or more occurrences of the atom $A[t_1]$ among the B_j. Write the conclusion D as

$$[D] = D_1 \to \cdots \to D_r \to \bot$$

and let l be such that $D_l = A[t_2]$. A proof of $[C_1] \to [C_2] \to [D]$ can be obtained as follows:

$$\lambda c_1 : [C_1]. \, \lambda c_2 : [C_2]. \, \lambda d_1 : D_1 \ldots \lambda d_r : D_r. \, (c_1 \; \pi_1 \; \ldots \; \pi_p)$$

If $A_i \neq (t_1 \approx t_2 \to \bot)$, then $\pi_i = d_k$, where k is such that $D_k = A_i$. If $A_i = (t_1 \approx t_2 \to \bot)$, then we want again that $\pi_i : A_i$ and therefore put

$$\pi_i = \lambda e : (t_1 \approx t_2). \, (c_2 \; \rho_1 \; \ldots \; \rho_q).$$

If $B_j \neq A[t_1]$, then $\rho_j = d_k$, where k is such that $D_k = B_j$. If $B_j = A[t_1]$, then we also want that $\rho_j : B_j$ and put (with $d_l : D_l$)

$$\rho_j = (eqsubst \; t_2 \; (\lambda s : \sigma. \, A[s]) \; d_l \; t_1 \; (eqsym \; t_1 \; t_2 \; e))$$

The term ρ_j has type $A[t_1]$ in the context $e : (t_1 \approx t_2)$. The term ρ_j contains an occurrence of $eqsym$ because of the fact that the equality $t_1 \approx t_2$ comes in the wrong direction for proving $A[t_1]$ from $A[t_2]$. With this definition of ρ_j, the term π_i has indeed type $A_i = (t_1 \approx t_2 \to \bot)$.

As an alternative, it is possible to expand the proof of $eqsym$ in the proof of the paramodulation step.

Equality Factoring

Equality factoring for disjunctive clauses is the rule with premiss C containing equality literals $t_1 \approx t_2$ and $t_1 \approx t_3$, and conclusion D which is identical to C but for the replacement of $t_1 \approx t_3$ by $t_2 \not\approx t_3$. The soundness of this rule relies on $t_2 \approx t_3 \vee t_2 \not\approx t_3$.

Let $A_1, \ldots, A_p, B_1 \ldots, B_q$ be literals ($p, q \geq 0$) and write

$$[C] = A_1 \to \cdots \to A_p \to \perp,$$

with equality literals $t_1 \approx t_2 \to \perp$ and $t_1 \approx t_3 \to \perp$ among the A_i. Write the conclusion D as

$$[D] = B_1 \to \cdots \to B_q \to \perp$$

with $B_{j'} = (t_1 \approx t_2 \to \perp)$ and $B_{j''} = (t_2 \approx t_3)$. We get a proof of $[C] \to [D]$ from

$$\lambda c : [C]. \, \lambda b_1 : B_1 \ldots \lambda b_q : B_q. \, (c \; \pi_1 \; \ldots \; \pi_p).$$

If $A_i \neq (t_1 \approx t_3 \to \perp)$, then $\pi_i = b_j$, where j is such that $B_j = A_i$. For $A_i = (t_1 \approx t_3 \to \perp)$, we put

$$\pi_i = (eqsubst \; t_2 \; (\lambda s : \sigma. \, (t_1 \approx s \to \perp)) \; b_{j'} \; t_3 \; b_{j''}).$$

The type of π_i is indeed $t_1 \approx t_3 \to \perp$.

Note that the equality factoring rule is constructive in the implicational translation, whereas its disjunctive counterpart relies on the decidability of $\approx$. This phenomenon is well-known from the double negation translation.

Positive and Negative Equality Swapping

The positive equality swapping rule for disjunctive clauses simply swaps an atom $t_1 \approx t_2$ into $t_2 \approx t_1$, whereas the negative rule swaps the negated atom. Both versions are obviously sound, given the symmetry of $\approx$.

We give the translation for the positive case first and will then sketch the simpler negative case. Let C be the premiss and D the conclusion and write

$$[C] = A_1 \to \cdots \to A_p \to \perp,$$

with some of the A_i equal to $t_1 \approx t_2 \to \perp$, and

$$[D] = B_1 \to \cdots \to B_q \to \perp.$$

Let j' be such that $B_{j'} = (t_2 \approx t_1 \to \perp)$. The following term is a proof of $[C] \to [D]$.

$$\lambda c : [C]. \, \lambda b_1 : B_1 \ldots \lambda b_q : B_q. \, (c \; \pi_1 \; \ldots \; \pi_p)$$

If $A_i \neq (t_1 \approx t_2 \to \perp)$, then $\pi_i = b_j$, where j is such that $B_j = A_i$. Otherwise

$$\pi_i = \lambda e : (t_1 \approx t_2). \, (b_{j'} \; (eqsym \; t_1 \; t_2 \; e))$$

such that also $\pi_i : (t_1 \approx t_2 \to \perp) = A_i$.

In the negative case the literals $t_1 \approx t_2$ in question are not negated, and we change the above definition of π_i into

$$\pi_i = (eqsym\ t_2\ t_1\ b_{j'}).$$

In this case we have $b_{j'} : (t_2 \approx t_1)$ so that $\pi_i : (t_1 \approx t_2) = A_i$ also in the negative case.

Equality Reflexivity Rule

The equality reflexivity rule simply cancels a negative equality literal of the form $t \not\approx t$ in a disjunctive clause. We write once more the premiss

$$[C] = A_1 \to \cdots \to A_p \to \perp,$$

with some of the A_i equal to $t \approx t$, and the conclusion

$$[D] = B_1 \to \cdots \to B_q \to \perp.$$

The following term is a proof of $[C] \to [D]$:

$$\lambda c : [C].\ \lambda b_1 : B_1 \ldots \lambda b_q : B_q.\ (c\ \pi_1\ \ldots\ \pi_p).$$

If $A_i \neq (t \approx t)$, then $\pi_i = b_j$, where j is such that $B_j = A_i$. Otherwise $\pi_i = (eqrefl\ t)$.

5 Lifting to Predicate Logic

Until now we have only considered inference rules without quantifications. In this section we explain how to lift the resolution rule to predicate logic. Lifting the other rules is very similar.

Recall that we must assume that the domain is not empty. Proof terms below may contain a variable $s : \sigma$ as free variable. By abstraction $\lambda s : \sigma$ we will close all proof terms. This extra step is necessary since $\forall s : \sigma.\ \perp$ does not imply $\perp$ when the domain σ is empty. This is to be compared to $\Box\perp$ being true in a blind world in modal logic.

Consider the following clauses

$$C_1 = \forall x_1, \ldots, x_p : \sigma.\ [A_1 \vee R_1]$$

and

$$C_2 = \forall y_1, \ldots, y_q : \sigma.\ [\neg A_2 \vee R_2]$$

and their resolvent

$$R = \forall z_1, \ldots, z_r : \sigma.\ [R_1\theta_1 \vee R_2\theta_2]$$

Here θ_1 and θ_2 are substitutions such that $A_1\theta_1 = A_2\theta_2$ and $z_1, \ldots, z_r$ are all variables that actually occur in the resolvent, that is, in $R_1\theta_1 \vee R_2\theta_2$ after application of θ_1, θ_2. It may be the case that $x_i \theta_1$ and/or $y_j \theta_2$ contain other variables than $z_1, \ldots, z_r$; these are understood to be replaced by the variable $s : \sigma$ (see above). It may be the case that θ_1, θ_2 do not represent a most general unifier. For soundness this is no problem at all, but even completeness is not at stake since the resolvent is not affected. The reason for this subtlety is that the proof terms involved must not contain undeclared variables.

Using the methods of the previous sections we can produce a proof π that has the type

$$[A_1 \vee R_1]\theta_1 \to [\neg A_2 \vee R_2]\theta_2 \to [R_1\theta_1 \vee R_2\theta_2].$$

A proof of $C_1 \to C_2 \to R$ is obtained as follows:

$$\lambda c_1 : C_1. \lambda c_2 : C_2. \lambda z_1 \ldots z_r : \sigma. (\pi \; (c_1 \; (x_1\theta_1)) \ldots (x_p\theta_1)) \; (c_2 \; (y_1\theta_2) \ldots (y_q\theta_2)))$$

We finish this section by showing how to assemble a λ-term for an entire resolution refutation from the proof terms justifying the individual steps. Consider a Hilbert-style resolution derivation $C_1, \ldots, C_m, C_{m+1}, \ldots, C_n$ with premisses $c_1 : C_1, \ldots, c_m : C_m$. Starting from n and going downward, we will define by recursion for every $m \leq k \leq n$ a term π_k such that

$$\pi_k[c_{m+1}, \ldots, c_k] : C_n$$

in the context extended with $c_{m+1} : C_{m+1}, \ldots, c_k : C_k$. For $k = n$ we can simply take $\pi_n = c_n$. Now assume π_{k+1} has been constructed for some $k \geq m$. The proof π_k is more difficult than π_{k+1} since π_k cannot use the assumption $c_{k+1} : C_{k+1}$. However, C_{k+1} is a resolvent, say of C_i and C_j for some $i, j \leq k$. Let ρ be the proof of $C_i \to C_j \to C_{k+1}$. Now define

$$\pi_k[c_{m+1}, \ldots, c_k] = (\lambda x : C_{k+1}.\pi_{k+1}[c_{m+1}, \ldots, c_k, x])(\rho \; c_i \; c_j) : C_n$$

The downward recursion yields a proof $\pi_m : C_n$ which is linear in the size of the original Hilbert-style resolution derivation. Observe that a forward recursion from m to n would yield the normal form of π_m, which could be exponential.

6   Example

Let P be a property of natural numbers such that P holds for n if and only if P does not hold for any number greater than n. Does this sound paradoxical? It is contradictory. We have $P(n)$ if and only if $\neg P(n+1), \neg P(n+2), \neg P(n+3), \ldots$, which implies $\neg P(n+2), \neg P(n+3), \ldots$, so $P(n+1)$. It follows that $\neg P(n)$ for all n. However, $\neg P(0)$ implies $P(n)$ for some n, contradiction.

A closer analysis of this argument shows that the essence is not arithmetical, but relies on the fact that $<$ is transitive and serial. The argument is also valid in a finite structure, say $0 < 1 < 2 < 2$. This qualifies for a small refutation problem, which we formalize in Coq. Type dependencies are given explicitly.

Thus, *nat* is the domain of discourse. We declare a unary relation P and a binary relation $<$.

$$P : nat \to *^P$$
$$< : nat \times nat \to *^P$$

Let $L : (nelist\ nat) := [1,2]$ be the corresponding list of arities. The relations are packaged by *Rel*.

$$Rel : \Pi i : (index\ L).\, nat^{(select\ L\ i)} \to *^P := \lambda i : (index\ L).\, \text{Cases } i \text{ of} \begin{cases} 0 \Rightarrow P \\ 1 \Rightarrow < \end{cases}$$

For instance, $(E\ nat\ L\ Rel\ (rel\ nat\ L\ 1\ (2,0))) =_{\beta\delta\iota} (Rel\ 1\ (2,0)) =_{\beta\delta\iota} 2 < 0$. It is convenient to represent the relations $P, <$ as object-level constants.

$$\dot{P} : nat \to (o\ nat\ L) \qquad := (rel\ nat\ L\ 0)$$
$$\dot{<} : nat \times nat \to (o\ nat\ L) := (rel\ nat\ L\ 1)$$

Let us construct the formal propositions *trans* and *serial*, stating that $\dot{<}$ is serial and transitive. (We write $n \dot{<} m$ instead of $(\dot{<}\ (n,m))$.)

$$trans : (o\ nat\ L) := \dot{\forall}x,y,z : nat.\,(x \dot{<} y \wedge y \dot{<} z) \dot{\to} x \dot{<} z$$
$$serial : (o\ nat\ L) := \dot{\forall}x : nat.\,\dot{\exists}y : nat.\,x \dot{<} y$$

We define *foo*.

$$foo : (o\ nat\ L) := \dot{\forall}x : nat.\,(\dot{P}\ x) \dot{\leftrightarrow} (\dot{\forall}y : nat.\,x \dot{<} y \dot{\to} \dot{\neg}(\dot{P}\ y))$$

Furthermore, we define *taut* on the object-level, representing the example informally stated at the beginning of this section. (If the latter is denoted by φ, then $taut = \dot{\varphi}$.)

$$taut : (o\ nat\ L) := (trans \dot{\wedge} serial) \dot{\to} \dot{\neg} foo$$

Interpreting *taut*, that is $\beta\delta\iota$-normalizing $(E\ nat\ L\ Rel\ taut)$, results in '*taut* without dots'.

We declare $em : (EM\ nat\ L\ Rel)$, $ac : (AC\ nat\ L\ Rel)$ and use 0 to witness the non-emptiness of *nat*. We reduce the goal $(E\ nat\ L\ Rel\ taut)$, using the result of Section 3, to the goal $(E\ nat\ L\ Rel\ (mcf\ nat\ L\ taut))$. If we prove this latter goal, say by a term ρ, then

$$(mcf_{prf}\ nat\ L\ Rel\ em\ ac\ 0\ taut\ \rho) : (E\ nat\ L\ Rel\ taut)$$

We normalize the new goal:

$$(E\ nat\ L\ Rel\ (mcf\ nat\ L\ taut)) =_{\beta\delta\iota}$$
$$\forall f : (skolT\ nat).$$
$$(\forall x,y,z : nat.\,x < y \to y < z \to (x < z \to \bot) \to \bot)$$
$$\to (\forall x : nat.\,(x < (f\ 1\ 0\ 1\ x) \to \bot) \to \bot)$$
$$\to (\forall x : nat.\,(x < (f\ 2\ 0\ 1\ x) \to \bot) \to ((P\ x) \to \bot) \to \bot)$$
$$\to (\forall x : nat.\,((P\ (f\ 2\ 0\ 1\ x)) \to \bot) \to ((P\ x) \to \bot) \to \bot)$$
$$\to (\forall x,y : nat.\,(P\ x) \to x < y \to (P\ y) \to \bot)$$
$$\to \bot$$

This is the minimal clausal form of the original goal. We refrained from exhibiting its proof ρ for reasons of space. The Coq-script generating ρ can be found in example.v in the tar file mentioned at the end of Section 3.

References

[Blk]   Bliksem is available at URL: www.mpi-sb.mpg.de/~bliksem.

[dB72]   N. de Bruijn. Lambda calculus notation with nameless dummies, a tool for automatic formula manipulation. *Indagationes Mathematicae* 34, pages 381–392, 1972.

[Coq98] B. Barras, S. Boutin, C. Cornes, J. Courant, J.-C. Filliâtre, E. Giménez, H. Herbelin, G. Huet, C. Muñez, C. Murthy, C. Parent, C. Paulin-Mohring, A. Saïbi, B. Werner. The Coq Proof Assistant Reference Manual, version 6.2.4. INRIA, 1998. Available at:
ftp.inria.fr/INRIA/coq/V6.2.4/doc/Reference-Manual.ps.

[Hen98] D. Hendriks. Clausification of First-Order Formulae, Representation & Correctness in Type Theory. Master's Thesis, Utrecht University, 1998. URL:
www.phil.uu.nl/~hendriks/thesis.ps.gz.

[Hua96] X. Huang. Translating machine-generated resolution proofs into ND-proofs at the assertion level. In *Proceedings of PRICAI-96*, pages 399–410, 1996.

[Hur99] J. Hurd. Integrating Gandalf and HOL. In *Proceedings TPHOL's 99*, number LNCS 1690, pages 311–321. Springer Verlag, 1999.

[MS99]   W. McCune and O. Shumsky. IVY: A preprocessor and proof checker for first-order logic. Preprint ANL/MCS-P775-0899, Argonne National Laboratory, Argonne IL, 1999.

[NM98]   G. Nadathur and D. Miller. Higher-order logic programming. In D. Gabbay e.a. (eds.) *Handbook of logic in artificial intelligence*, Vol. 5, pp. 499–590. Clarendon Press, Oxford, 1998.

[Omega] Omega can be found on www.ags.uni-sb.de/~omega/.

[Pfe84] F. Pfenning. Analytic and non-analytic proofs. In *Proceedings CADE 7*, number LNCS 170, pages 394–413. Springer Verlag, 1984.

[ST95]   J. Smith and T. Tammet. Optimized encodings of fragments of type theory in first-order logic. In *Proceedings Types 95*, number LNCS 1158, pages 265–287. Springer Verlag, 1995.

System Description: TPS:
A Theorem Proving System for Type Theory

Peter B. Andrews[1], Matthew Bishop[2], and Chad E. Brown[1]

[1] Department of Mathematical Sciences, Carnegie Mellon University
Pittsburgh, PA 15213, USA
Peter.Andrews@cmu.edu, cebrown@andrew.cmu.edu
[2] Department of Computer Science, King's College London, Strand
London WC2R 2LS, England
bishopm@dcs.kcl.ac.uk

1 Introduction

This is a brief update on the TPS automated theorem proving system for classical type theory, which was described in [3]. Manuals and information about obtaining TPS can be found at http://gtps.math.cmu.edu/tps.html.

In Section 2 we discuss some examples of theorems which TPS can now prove automatically, and in Section 3 we discuss an example which illustrates one of the many challenges of theorem proving in higher-order logic. We first provide a brief summary of the key features of TPS.

TPS uses Church's type theory [8] (typed λ-calculus) as its logical language. Wffs are displayed on the screen and in printed proofs in the notation of this system of symbolic logic.

One can use TPS in automatic, semi-automatic, or interactive mode to construct proofs in natural deduction style, and a mixture of these modes of operation is most useful for significant applications. Our current research is focused primarily on increasing the power of the purely automatic search procedures, since these are useful in speeding up the construction of proofs even if many of the key ideas must be supplied interactively.

When searching for a proof of a theorem, TPS first tries to find an *expansion proof* [11], of which an important component is a *mating* [1] (otherwise known as a *spanning set of connections* [4]). Various search procedures are implemented in TPS, most notably those described in [6], [5], [10], and [9]. The method of *dual instantiation* of definitions discussed in [7] is also implemented in TPS. Once an expansion proof has been found, it is translated into a natural deduction proof by the methods of [13] and [14].

Many aspects of the behavior of TPS can be varied by changing the settings of flags. These flags provide a convenient facility for exploring various aspects of the problem of searching for proofs, and are essential in setting bounds for the many dimensions of proof search in higher-order logic.

* This material is based upon work supported by the National Science Foundation under grant CCR-9732312.

D. McAllester (Ed.): CADE-17, LNAI 1831, pp. 164–169, 2000.

2 New Theorems

In the notation used by TPS, o is the type of truth values, ι is the type of individuals, and $(\alpha\beta)$ (which some authors prefer to write as $(\beta \to \alpha)$) is the type of functions from elements of type β to elements of type α. An entity of type $(o\alpha)$ is regarded as a set of elements of type α, and $f_{o\alpha}x_\alpha$ can be interpreted as meaning that x_α is in $f_{o\alpha}$. $\gamma\beta\alpha$ is an abbreviation for $((\gamma\beta)\alpha)$.

A dot stands for a left bracket whose mate is as far to the right as is consistent with the pairing of brackets already present.

The following theorems were all proven completely automatically by TPS once the flags were set. All the timings quoted below represent the internal runtime, excluding garbage-collect time, used by TPS to find an expansion proof and translate it into a natural deduction proof on a Tangent workstation with a Pentium III processor and 512 megabytes of RAM using Allegro Common Lisp 5.0 for Linux. The numbers are useful only for their approximate magnitudes. They may not represent optimal settings of the flags, and the times required to prove these theorems will probably increase as ways are found to move more of the burden of setting flags from users to TPS.

We start with several theorems concerned with various formulations of the Axiom of Choice.[1] We first list these formulations. In [15] these are presented as statements of axiomatic set theory; in a type-theoretic context their variables must be given types, and they take the form of axiom schemas. Logical relations between these formulations of the Axiom of Choice are then complicated by the need to have appropriate relations between the types which are involved.

AC1(β) from [15] : $\forall s_{o(o\beta)}.\forall X_{o\beta}[sX \supset \exists y_\beta Xy] \supset \exists f_{\beta(o\beta)}\forall X.sX \supset X.fX$
If s is a set of non-empty sets, there is a function f such that for every $x \in s$, $f(x) \in x$.

AC3(β, α) from [15] : $\forall r_{(o\beta)\alpha}\exists g_{\beta\alpha}\forall x_\alpha.\exists y_\beta rxy \supset rx.gx$
For every function r, there is a function g such that for every x, if x is in the domain of r and $r(x) \neq \emptyset$, then $g(x) \in r(x)$. (In a set-theoretic context it may be assumed that the values of r are sets, but in a type-theoretic context r must be given a type compatible with this assumption.)

AC17(α) from [15] : $\forall g_{o\alpha(\alpha(o\alpha))}.\forall h_{\alpha(o\alpha)}\exists u_\alpha[gh]u \supset \exists f_{\alpha(o\alpha)}gf.f.gf$
If s is a set (which we represent as the set of all elements of type α), t is the collection of all non-empty subsets of s, F is the set of all functions (which must have type $(\alpha(o\alpha))$) from t to s, and g is a function from F to t, then there is an $f \in F$ such that $f(g(f)) \in g(f)$.

AC(α) from [2] : $\exists f_{\alpha(o\alpha)}\forall X_{o\alpha}.\exists t_\alpha Xt \supset X.fX$
There is a universal choice function f (for elements of type α) such that if X is any non-empty set (whose elements are of type α), then $fX \in X$

THM532: AC1$(\beta) \supset$ AC3(β, α)		*(3.96 seconds)*
THM533: AC3$(\alpha, o\alpha) \supset$ AC1(α)		*(1.92 seconds)*
THM560: AC3$(\alpha, o\alpha) \equiv$ AC1(α)		*(19.26 seconds)*

---

[1] See [12] for a discussion of interactive proofs of many similar theorems.

THM534: $AC1(\alpha) \supset AC17(\alpha)$ *(4.71 seconds)*

THM541: $AC(\alpha) \equiv AC1(\alpha)$ *(5.00 seconds)*

THM531E: FINITE-SET $C_{o\alpha} \wedge B_{o\alpha} \subseteq C \supset$ FINITE-SET B

(21.50 seconds)

THM531E says that a subset of a finite set is finite. FINITE-SET is defined as $[\lambda X_{o\alpha} \forall P_{o(o\alpha)}.\forall E_{o\alpha}[\sim \exists t_{\alpha} Et \supset PE] \wedge \forall Y_{o\alpha} \forall x_{\alpha} \forall Z_{o\alpha}[[PY \wedge .Z \subseteq .Y + x] \supset PZ] \supset PX]$, which is one of several ways one can define finiteness inductively. $Y_{o\alpha} + x_{\alpha}$ is defined as $[\lambda t_{\alpha}.Y_{o\alpha} t \vee t = x_{\alpha}]$, which is another notation for $Y_{o\alpha} \cup \{x_{\alpha}\}$.

THM196B: $\sim [a_{\iota} = b_{\iota}] \supset \sim \forall j_{\iota\iota} \forall k_{\iota\iota}.$ITERATE+ $j[k \circ j] \supset$ ITERATE+ jk

(1.41 seconds)

ITERATE+ is defined as $\lambda f_{\alpha\alpha} \lambda g_{\alpha\alpha} \forall p_{o(\alpha\alpha)}.pf \wedge \forall j_{\alpha\alpha}[pj \supset p.f \circ j] \supset pg$, and so ITERATE+ fg means that g is an iterate of f — i.e., a function of the form $f \circ ... \circ f$. The symbol $\circ$ denotes the composition of functions, and is defined as $\lambda f_{\alpha\beta} \lambda g_{\beta\gamma} \lambda x_{\gamma} f.gx$. The theorem refutes the conjecture that if $k \circ j$ is an iterate of j, then k must be an iterate of j. Of course, the conjecture is trivially true if there is just one individual, so the theorem depends on the assumption that there are two distinct individuals. TPS proves the theorem by constructing the simple counterexample where k is the identity function and j is the constant function whose value is b. The proof consists of a verification that this is indeed a counterexample.

THM563: CLOS-SYS1 $.\lambda W_{o\beta}.W \, 0_{\beta} \wedge \forall x_{\beta} \forall y_{\beta}[W y \wedge x \leq y \supset W x] \wedge$ $\forall x \forall y \forall z_{\beta}.W x \wedge W y \wedge JOIN \, x \, y \, z \supset W z$ *(49.2 minutes)*

THM563 states that the collection of sets $W_{o\beta}$ which contain an element 0, are downward closed with respect to a binary relation $\leq$, and are closed with respect to a tertiary relation $JOIN$, is a closure system.

CLOS-SYS1 is defined as $\lambda CL_{o(o\beta)} \forall S_{o(o\beta)}.S \subseteq CL \supset CL.\bigcap S$, and $\bigcap$ is defined as $\lambda S_{o(o\beta)} \lambda x_{\beta} \forall W_{o\beta}.S W \supset W x$, so a closure system is a collection of sets closed under arbitrary intersections.

THM563 is very general, but we can illustrate it with the following special case. Let β be the type of finite binary trees. We use 0 as a name for the tree with a single node. We can define a partial ordering $\leq$ on this type by saying a tree x is less than a tree y if we can replace the leaves of x by some trees to obtain y. Thus, 0 is the smallest member of β. Given trees x and y, there is a tree called $[x \vee y]$ which is the $lub\{x, y\}$ such that $JOIN \, x \, y \, [x \vee y]$. We can represent each infinite binary tree by a certain set $W_{o\beta}$ of finite binary trees, which approximate the infinite tree in the sense illustrated below, where the finite trees approximate the infinite tree on the right.

(We can regard finite binary trees as special cases of infinite binary trees. A finite binary tree x_{β} when considered as an infinite binary tree is the set $\{y_{\beta} | y \leq x\}$, an object of type $(o\beta)$.) A set $W_{o\beta}$ that represents a tree contains the tree 0, is downward closed, and is closed with respect to joins. In the case of this example, THM563 shows that the set of infinite trees constitutes a closure system, and therefore forms a complete lattice under the subset ordering.

$\quad$ **X5204:** $\#f_{\alpha\beta}[\bigcup w_{o(o\beta)}] = \bigcup .\#[\#f]w$ $\hfill$ *(33.16 seconds)*
$\#$ is defined as $\lambda f_{\alpha\beta}\lambda x_{o\beta}\lambda z_\alpha \exists t_\beta.xt \wedge z = ft$, and so $\#f_{\alpha\beta}x_{o\beta}$ is the image of the set $x_{o\beta}$ under the function $f_{\alpha\beta}$. This is a polymorphic definition, and the instances of $\#$ in the theorem have appropriate types attached to them. $\bigcup$ is defined as $\lambda D_{o(o\alpha)}\lambda x_\alpha \exists S_{o\alpha}.DS \wedge Sx$; hence $\bigcup D_{o(o\alpha)}$ is the union of the collection $D_{o(o\alpha)}$ of sets.

$\quad$ **X5311A-EXT:** $\forall y_\alpha[\iota[= y] = y] \wedge \forall p_{o\alpha}\forall q_{o\alpha}[\forall x_\alpha[px \equiv qx] \supset \forall r_{o(o\alpha)}.rp \supset$
$rq] \supset \forall p.\Sigma^{11}p \supset p.\iota p$ $\hfill$ *(3.0 minutes)*
Σ^{11} is defined as $\lambda p_{o\alpha}\exists y_\alpha.py \wedge \forall z_\alpha.pz \supset y = z$, and represents the property of being a one-element set, This is essentially theorem 5311 from [2]. In order to prove it one needs axioms of descriptions and extensionality, so they are made antecedents of the main implication. The theorem says that if p is a one-element set, then the description operator ι maps p to the unique entity which is in p.

3   A Challenge

We conclude with a discussion of an example which poses a significant challenge for TPS and other theorem provers for higher-order logic and set theory. Cantor's theorem for sets says that if U is any set and W is its power set, then W has larger cardinality than U. This is usually expressed by saying that there is no surjection from U onto W. If one takes the members of U as the set of individuals, the theorem can be expressed simply by the wff $\sim \exists g_{o\iota\iota}\forall f_{o\iota}\exists j_\iota.gj = f$, which we called X5304 in [2] and [3]. TPS has been able to prove this for many years. However, one can also express the fact that W has larger cardinality than U by saying that there is no injection from W into U, which we formalize as follows:

$\quad$ **X5309:** $\sim \exists h_{\iota(o\iota)}\forall p_{o\iota}\forall q_{o\iota}.hp = hq \supset p = q$ $\hfill$ *(not proven)*
We call this the Injective Cantor Theorem.

$\quad$ Here is an informal proof of this theorem. Suppose there is a function $h :$ $W \to U$ such that (1) h is injective. Let (2) $D = \{ht \,|\, t \in W$ and $ht \notin t\}$. Note that (3) $D \in W$. Now suppose that (4) $hD \in D$. Then (by 2) there is a set t such that (5) $t \in W$ and (6) $ht \notin t$ and (7) $hD = ht$. Therefore (8) $D = t$ (by 1, 7), so (9) $hD \notin D$ (by 6, 8). This argument (4-9) shows that (10) $hD \notin D$. Thus (11) $hD \in D$ (by 2, 3, 10). This contradiction shows that there can be no such h.

$\quad$ It is easy to prove parts of this argument automatically. Define IDIAG to be $[\lambda h_{\iota(o\iota)}\#h.\lambda s_{o\iota}. \sim s.hs]$. Then [IDIAG h] represents the set D of the informal argument above. TPS can automatically prove the following theorems, from which X5309 follows trivially:

THM143B: $\forall h_{\iota(o\iota)}.\forall p_{o\iota}\forall q_{o\iota}[hp = hq \supset p = q] \supset {\sim} \text{IDIAG } h.h.\text{IDIAG } h$
(3.35 seconds)
THM144B: $\forall h_{\iota(o\iota)}.\text{IDIAG } h.h.\text{IDIAG } h$ *(0.47 seconds)*
However, a completely automatic proof of X5309 seems well beyond the present capabilities of Tps . The expansion proof which corresponds to the argument above involves instantiating a quantifier on a set variable with a wff which contains another quantifier on a set variable, which must also be instantiated with a wff which contains a quantifier. We may say that such an expansion proof has quantificational depth 3. Thus far Tps has found expansion proofs only of quantificational depth ≤ 2.

We may define the quantificational depth of a theorem to be the minimum of the quantificational depths of its expansion proofs. (Thus all theorems of first-order logic have quantificational depth at most 1). Research on methods of proving theorems which are deep in this sense should stimulate significant progress in higher-order theorem proving.[2]

References

1. Peter B. Andrews. Theorem Proving via General Matings. *Journal of the ACM*, 28:193–214, 1981.
2. Peter B. Andrews. *An Introduction to Mathematical Logic and Type Theory: To Truth Through Proof*. Academic Press, 1986.
3. Peter B. Andrews, Matthew Bishop, Sunil Issar, Dan Nesmith, Frank Pfenning, and Hongwei Xi. TPS: A Theorem Proving System for Classical Type Theory. *Journal of Automated Reasoning*, 16:321–353, 1996.
4. Wolfgang Bibel. *Automated Theorem Proving*. Vieweg, Braunschweig, second edition, 1987.
5. Matthew Bishop. A Breadth-First Strategy for Mating Search. In Harald Ganzinger, editor, *Proceedings of the 16th International Conference on Automated Deduction*, volume 1632 of *Lecture Notes in Artificial Intelligence*, pages 359–373, Trento, Italy, 1999. Springer-Verlag.
6. Matthew Bishop. *Mating Search Without Path Enumeration*. PhD thesis, Department of Mathematical Sciences, Carnegie Mellon University, April 1999. Department of Mathematical Sciences Research Report No. 99–223. Available at http://gtps.math.cmu.edu/tps.html.
7. Matthew Bishop and Peter B. Andrews. Selectively Instantiating Definitions. In Claude Kirchner and Hélène Kirchner, editors, *Proceedings of the 15th International Conference on Automated Deduction*, volume 1421 of *Lecture Notes in Artificial Intelligence*, pages 365–380, Lindau, Germany, 1998. Springer-Verlag.
8. Alonzo Church. A Formulation of the Simple Theory of Types. *Journal of Symbolic Logic*, 5:56–68, 1940.
9. Sunil Issar. Path-Focused Duplication: A Search Procedure for General Matings. In *AAAI-90. Proceedings of the Eighth National Conference on Artificial Intelligence*, volume 1, pages 221–226. AAAI Press/The MIT Press, 1990.

---

[2] We are not aware of a *proof* that X5309 or any other theorem has depth greater than 2, but we conjecture that there is no upper bound on the quantificational depths of theorems of type theory.

10. Sunil Issar. *Operational Issues in Automated Theorem Proving Using Matings.* PhD thesis, Carnegie Mellon University, 1991. 147 pp.
11. Dale A. Miller. A Compact Representation of Proofs. *Studia Logica,* 46(4):347–370, 1987.
12. Lawrence Paulson and Krzystztof Grabczewski. Mechanising Set Theory: Cardinal Arithmetic and the Axiom of Choice. *Journal of Automated Reasoning,* 17:291–323, 1996.
13. Frank Pfenning. *Proof Transformations in Higher-Order Logic.* PhD thesis, Carnegie Mellon University, 1987. 156 pp.
14. Frank Pfenning and Dan Nesmith. Presenting Intuitive Deductions via Symmetric Simplification. In M. E. Stickel, editor, *Proceedings of the 10th International Conference on Automated Deduction,* volume 449 of *Lecture Notes in Artificial Intelligence,* pages 336–350, Kaiserslautern, Germany, 1990. Springer-Verlag.
15. Herman Rubin and Jean E. Rubin. *Equivalents of the Axiom of Choice, II.* North-Holland, 1985.

The Nuprl Open Logical Environment

Stuart F. Allen, Robert L. Constable, Rich Eaton,
Christoph Kreitz, and Lori Lorigo*

Department of Computer Science, Cornell-University, Ithaca, NY 14853-7501
{sfa,rc,eaton,kreitz,lolorigo}@cs.cornell.edu

Abstract. The Nuprl system is a framework for reasoning about mathematics and programming. Over the years its design has been substantially
improved to meet the demands of large-scale applications. Nuprl LPE,
the newest release, features an open, distributed architecture centered
around a flexible knowledge base and supports the cooperation of independent formal tools. This paper gives a brief overview of the system
and the objectives that are addressed by its new architecture.

1 Introduction

The Nuprl proof development system [C$^+$86] is a framework for the development
of formalized mathematical knowledge as well as for the synthesis, verification,
and optimization of software. The original system was based on a significant
extension of Martin-Löf's intuitionistic Type Theory [ML84], which includes
formalizations of the fundamental concepts of mathematics, data types, and
programming. The system itself supports interactive and tactic-based reasoning, decision procedures, evaluation of programs, language extensions through
user-defined concepts, and an extendable library of verified knowledge from
various domains. Since its first release in 1984 it has been used in increasingly large applications in mathematics and programming, such as verifications of a logic synthesis tool [AL93] and of the SCI cache coherency protocol
[How96] as well as the verification and optimization of group communication
systems [KHH98,Kre99,L$^+$99].

Over the years it has turned out that the rapidly growing demands for formal
knowledge and tools cannot be met by a single closed system anymore. Automatic tools such as decision procedures, fully automatic theorem provers, proof
planners, rewrite engines, model checkers, and computer algebra systems have
been very successful in their respective areas, but have limited application domains. Proof assistants like Nuprl, Isabelle [Pau90], HOL [GM93], PVS [O$^+$96],
and Ωmega [B$^+$97] are more general but at a lesser degree of automation. Each
of these systems has accumulated a substantial amount of formalized knowledge
in its respective formalism, but no system contains all the currently available
formal knowledge. A variety of user interfaces have been developed for these
systems each with its own strengths and weaknesses.

* Part of this work was supported by DARPA grant F 30620-98-2-0198

D. McAllester (Ed.): CADE-17, LNAI 1831, pp. 170–176, 2000.
© Springer-Verlag Berlin Heidelberg 2000

These observations led to an entirely new design of the Nuprl system that shall replace the monolithic architecture of current theorem proving environments. The Nuprl LPE (logical programming environment) is an *open*, distributed architecture that integrates all its key subsystems as *independent* components and, by using a flexible knowledge base as its central component, supports the interoperability of current proof technology.

In the following we shall briefly discuss the key issues that shall be addressed by Nuprl LPE and describe its architecture as well as the available components.

2 Design Objectives

Besides preserving and expanding the strengths of the existing Nuprl system, the new design of the Nuprl LPE is based on the following objectives.

Interoperability: The Nuprl LPE shall provide a platform for the cooperation of proof systems and a common knowledge base that makes formal theories available to the individual systems. Special support for computational logics shall be offered, but other logics shall be accommodated as well.

Optimization and Productivity: To optimize software reuse and system maintenance, the key components of the Nuprl LPE have to be independently operating programs that communicate using a protocol. This will increase the system's productivity, as several inference engines can be run in parallel or even off-line while the user continues to work on other proof goals.

Accountability: As the Nuprl LPE shall accomodate a variety of logics, there cannot be an absolute notion of correctness anymore. Instead, the extent to which one may rely upon formalized knowledge in the library must be accounted for. Justifications for the validity of proofs depend upon what rules and axioms are admitted and on the reliability of the inference engines employed. The design has to make sure that such information can be easily exposed to determine which proofs are valid in a particular logic.

Information Preservation: The system has to make sure that knowledge cannot be destroyed or corrupted if a user erroneously overwrites a proof or if the system crashes before a proof could be saved. The system must guarantee that such information can always be recovered.

Large Scale Object Management: The system should use abstract object references rather than traditional naming schemes. This is invaluable for merging mass libraries, where name collisions are inevitable, and also for performing context-specific tasks.

3 The Nuprl LPE Architecture

Figure 1 illustrates the distributed open architecture of Nuprl LPE. The system is organized as a collection of communicating processes that are centered around a common knowledge base, called the *library*. The library contains definitions, theorems, inference rules, meta-level code (e.g. tactics), and structure objects

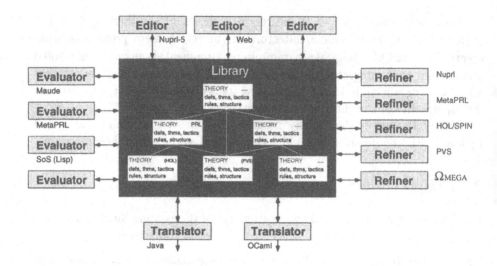

Fig. 1. Nuprl LPE distributed open architecture

that can be used to provide a modular structure for the library's contents. Inference engines (*refiners*), user interfaces (*editors*), rewrite engines (*evaluators*), and *translators* are started as independent processes that can connect to the library at any time.

The library can communicate with arbitrarily many other processes. This allows the user to connect *several refiners* and evaluators simultaneously, e.g. the Nuprl and MetaPRL [Met] refiners, major systems like HOL, PVS, Ωmega, or SPECWARE [SJ95], decision procedures, first-order provers, Mathematica [Wol88], and the Maude [C+99b] rewrite engine, and to have them cooperate through the library, which stores the formal knowledge required by these tools. It is also possible to run different refiners in parallel on the same proof goal or several instances of the same refiner on different proof goals.

Providing *several editors* enables several users to work in parallel on the same formal theory while using their favorite interface. At the same time external users can access the system through the Web without having to restart the whole system, as one would have to do in monolithic architectures.

Translators between the formal knowledge stored in the library and, for instance, programming languages like Java or Ocaml [Kre97,KHH98] allow the formal reasoning tools to supplement real-world software from various domains and thus provide a logical programming environment for the respective languages.

The Nuprl LPE provides special support for computational and constructive logics but it can accommodate other logics equally well. Its open architecture makes it possible that different systems with different formalisms and representation structures cooperate through a common knowledge base that can store all these informations. It is obvious that translations between different formalisms need to be developed to make such a cooperation possible and that several theoretical issues need to be addressed for each of them (see e.g. [How96,FH97]).

But the Nuprl LPE provides the necessary infrastructure for these translations. They only have to operate on the formal knowledge stored in the library and can be provided as independent external processes that are invoked as necessary. Translations can also be used in a transitive fashion, e.g implementing the translation between Maude and Nuprl [C+99a] automatically gives us access to all formalisms that have been translated into Maude.

In the current standard configuration, which we call Nuprl 5, the system essentially provides an extended functionality of the Nuprl 4 system [Jac94]. It consists of the library, the Nuprl 5 editor, and the Nuprl 5 refiner. The library contains all the definitions, theorems, inference rules, and tactics of Nuprl 4 as well as the structure objects that emulate the Nuprl 4 system architecture. The Nuprl 5 editor is capable of interpreting these structure objects while displaying and editing proofs and terms. The Nuprl 5 refiner is able to interpret inference rules and the ML code of the tactics. In the following we will describe the individual components more specifically.

The Knowledge Base. The knowledge base is based on a transaction model for entering and modifying objects. Changes to objects, e.g. the effects of editor commands or inference steps, are immediately committed to the library. This makes sure that knowledge doesn't get lost in case of a system failure, which could happen in systems that keep newly developed knowledge in memory until it is explicitly saved to disk. The knowledge base also provides the option to undo changes, redo transactions, or to have several processes view or work on the same object – essentially following the same protocols as databases.

However, changes do not overwrite an object but instead create a new version. The previous version is preserved until it is explicitly destroyed in a garbage collection process. A version control mechanism allows the user to recover previous versions of an object. This protects user data from being corrupted or destroyed erroneously and enables a user to create several proofs of the same theorem.

To account for the validity of library objects, the knowledge base supports *dependency tracking*. For this purpose a variety of information is stored together with an object, e.g. the logical rules and theorems on which it depends, the exact version of the refiners that were used to prove it, timestamps, etc. This information will help a user to validate theorems that rely on knowledge created by several systems, provided that the conditions for *hybrid validity* wrt. the underlying logics are well understood and stored in the library. For instance, a theorem referring to lemmata from Nuprl (constructive type theory) and HOL (classical higher order logic) would be marked as constructively valid if the HOL theorems involve only decidable predicates.

Apart from abstract links from terms to library objects the library does not impose any predefined structure. All visible structure, e.g. the directory structure as observed by the Nuprl 5 editor, is generated by structure objects that are explicitly present in the library. This allows exploiting the structure of the library and modifying it without having to change the representation of already stored knowledge. Structure, as we understand it, is only a matter of external presentation, not of internal representation.

For the same reason, there is a separation between objects and the names that users choose to denote them. Technically, the visible name is just a display version of the internal name, which makes it possible to chose the same name for different objects without creating internal name clashes and to disambiguate the display as needed.

The absence of a predefined library structure is a prerequisite for integrating formal knowledge from other systems besides Nuprl without requiring these systems to change their representation structure. The "only" thing that needs to be done is to emulate this structure in Nuprl's knowledge base.

User Interfaces. The main user interface of Nuprl 5 is the Nuprl 5 *navigator*. Its communication with the knowledge base is based on sending and receiving abstract terms. While displaying and editing terms it interprets the corresponding structure objects and displays them as directories, theorems, definitions, proofs, or mathematical expressions, sometimes opening new windows for this purpose. For the user, it provides the functionality of a structure editor: the user can mark subterms and edit slots in the displayed term and then cause the navigator to send the result back to the library, which processes the result while the user may continue to work with the editor.

The meaning of the abstract terms received by the knowledge base is determined by the structure objects already present. It could be interpreted as a command to store the term, to execute a tactic which subsequently calls one or several refiners, to open a proof editor window, or to send another term to be displayed. Obviously, this process involves a lot of management information in the terms being sent that is usually not shown to the user. However, the user has the right to edit (almost) all structure objects as well and thus customize the appearance of the information presented by the editor.

The Nuprl 5 editor is capable of interpreting objects as commands. A very convenient feature resulting from that is a hyperlink mechanism where clicking on that term causes the corresponding object to be raised, but more general ML expressions can be executed as well. This enables the user to trace back definitions of logical expressions and tactics or to customize the editor by adding buttons for common commands.

In addition to the Nuprl 5 navigator, Nuprl 5 provides emulations of the editors used in the previous release of Nuprl in order to ensure upward compatibility, as well as valuable extensions for facilitating proof browsing, merging, replaying and accounting. There is also a web front end [Nau98] that allows external users to browse the Nuprl library without having to install the whole system.

Inference Engines. The Nuprl 5 inference engine refines proof goals by executing ML code that may include references to library objects, particularly to the inference rules and tactics stored in the knowledge base. It applies the code to a given proof goal that it receives as an abstract term and returns the resulting list of subgoals back to the library. In the process it may invoke decision procedures and proof checkers. Based on the validations given in the rule objects it can also extract programs from proofs and evaluate them. The inference mechanism is fairly straightforward and compatible with the one in Nuprl 4.

As an alternative one may invoke the **MetaPRL** refiner [Met], a modularized version of **Nuprl**'s inference engine implemented in **OCaml**, that can run up to 100 times faster due to improvements in rewriting and evaluation. The communication between **Nuprl LPE** and **MetaPRL** utililizes the MathBus design [Mat].

We are currently working on connecting a variety of external refiners such as a constructive first-order theorem prover [K+00], the **HOL** system (via **Maude** [C+99a]), **Mathematica**, and **Isabelle** [Nau99]. We will also emulate the refiner of **Nuprl** 3 in order to be able to restore older theories that had not been migrated during the transition to **Nuprl** 4.

4   Progress and Availability

Nuprl LPE is the result of more than 25 years of experience with mathematical proof assistants. We have completed the implementation of the basic **Nuprl LPE** system, which provides a platform for the cooperation of a variety of proof systems through its open distributed architecture.

Using the **Nuprl LPE** infrastructure we have implemented the **Nuprl** 5 system consisting of the **Nuprl** 5 editor, refiner, evaluator, and the **Nuprl** 5 core library. The latter contains the terms and rules of the **Nuprl** type theory as well as the standard theories and tactics, which were migrated from **Nuprl** 4 to **Nuprl** 5. Besides the **Nuprl** 5 refiner a user may also invoke the **MetaPRL** refiner.

In addition to that we have migrated most of the user-defined theories from **Nuprl** 4 to **Nuprl** 5 and are currently testing the behavior of the new system under large scale applications like the verification of communication systems. Experience shows that the separation of library, editor, and refiner makes **Nuprl** 5 more efficient than its predecessor **Nuprl** 4, because it can take advantage of multiprocessor machines or a network of computers and run several competing refiners to solve a goal. We also observed an increased productivity of the system's users, who can now work on other tasks while a refiner solves a complex goal.

The completion of the basic **Nuprl LPE** system enables us to increase the system's capabilities by adding new editors, refiners, evaluators, or translators to the system. The open distributed architecture opens the door for a variety of research topics which can be turned into practically useful components as soon as their theoretical background has been explored.

Nuprl LPE is written mostly in Common Lisp, but uses some extensions that require Lucid or Allegro Lisp. An executable copy running under Linux is available at

http://www.cs.cornell.edu/Info/Projects/NuPrl/nuprl5/index.html.

References

[AL93]   M. Aagaard and M. Leeser. Verifying a logic synthesis tool in Nuprl. *CAV-93*, LNCS 663, pp. 72–83. Springer, 1993.

[B+97]   C. Benzmüller et al. *Ω*mega: Towards a mathematical assistant. *CADE-14*, LNAI 1249, pp. 252–256. Springer, 1997.

[C⁺86]   R. Constable et al. *Implementing Mathematics with the NuPRL proof development system*. Prentice Hall, 1986.

[C⁺99a]   M. Clavel, F. Duran, S. Eker, J. Meseguer, M.-O. Stehr Maude as a formal meta-tool. *FM'99*, LNCS 1709, pp. 1684–1703, Springer, 1999.

[C⁺99b]   M. Clavel, F. Duran, S. Eker, P. Lincoln, N. Marti-Oliet, J. Meseguer, J.F. Quesada.   The Maude system.   *RTA'99*, LNCS 1631, pp. 240–243, Springer, 1999.

[FH97]   Amy Felty and Douglas Howe. Hybrid Interactive Theorem Proving using Nuprl and HOL. *CADE-14*, LNAI 1249, pp. 351–365. Springer, 1997.

[GM93]   M. Gordon and T. Melham. *Introduction to HOL: a theorem proving environment for higher-order logic*. Cambridge University Press, 1993.

[How96]   D. Howe. Importing mathematics from HOL into NuPRL. *Theorem Proving in Higher Order Logics*, LNCS 1125, pp. 267–282. Springer, 1996.

[Jac94]   P. Jackson. *The Nuprl Proof Development System, Version 4.2*. Cornell University. Department of Computer Science, 1994.

[KHH98]   C. Kreitz, M. Hayden, J. Hickey. A proof environment for the development of group communication systems. *CADE-15*, LNAI 1421, pp. 317–332. Springer, 1998.

[Kre97]   C. Kreitz. Formal reasoning about communication systems I: Embedding ML into type theory. Technical Report TR97-1637, Cornell University. Department of Computer Science, 1997.

[Kre99]   C. Kreitz. Automated fast-track reconfiguration of group communication systems. *TACAS-99*, LNCS 1579, pp. 104–118. Springer, 1999.

[K⁺00]   C. Kreitz, J. Otten, S. Schmitt, B. Pientka. Matrix-based Constructive Theorem Proving. *Intellectics and Computational Logic*. Kluwer, 2000.

[L⁺99]   X. Liu, C. Kreitz, R. van Renesse, J. Hickey, M. Hayden, K. Birman, R. Constable. Building reliable, high-performance communication systems from components. *SOSP'99*, Operating Systems Review 34(5):80-92, 1999.

[Mat]   The MathBus Term Structure.
          http://www.cs.cornell.edu/simlab/papers/mathbus/mathTerm.htm.

[Met]   Metaprl home page.
          http://ensemble01.cs.cornell.edu:12000/cvsweb/meta-prl.

[ML84]   P. Martin-Löf. *Intuitionistic Type Theory*, Bibliopolis, 1984.

[Nau98]   P. Naumov. Publishing formal mathematics on the web. Technical Report TR98-1689, Cornell University. Department of Computer Science, 1998.

[Nau99]   P. Naumov. Importing Isabelle formal mathematics into Nuprl. Technical Report TR99-1734, Cornell University. Department of Computer Science, 1999.

[O⁺96]   S. Owre et al. PVS: Combining specification, proof checking and model checking. *CAV-96*, LNCS 1102, pp. 411–414, Springer, 1996.

[Pau90]   Lawrence C. Paulson. Isabelle: The next 700 theorem provers. *Logic and Computer Science*, pp. 361–386. Academic Press, 1990.

[SJ95]   Y. V. Srinivas and Richard Jüllig. SPECWARE: Formal Support for composing software. In *Mathematics of Program Construction*, 1995.

[Wol88]   S. Wolfram. *Mathematica: A System for Doing Mathematics by Computer*. Addison Wesley, 1988.

System Description: ARA – An Automatic Theorem Prover for Relation Algebras*

Carsten Sinz

Symbolic Computation Group, WSI for Computer Science
Universität Tübingen, D-72076 Tübingen, Germany
sinz@informatik.uni-tuebingen.de

Abstract. ARA is an automatic theorem prover for various kinds of relation algebras. It is based on Gordeev's Reduction Predicate Calculi for n-variable logic (RPC_n) which allow first-order finite variable proofs. Employing results from Tarski/Givant and Maddux we can prove validity in the theories of simple semi-associative relation algebras, relation algebras and representable relation algebras using the calculi RPC_3, RPC_4 and RPC_ω. ARA, our implementation in Haskell, offers different reduction strategies for RPC_n, and a set of simplifications preserving n-variable provability.

1 Introduction

Relations are an indispensable ingredient in many areas of computer science, such as graph theory, relational databases, logic programming, and semantics of computer programs, to name just a few. So relation algebras – which are extensions of Boolean algebras – form the basis of many theoretical investigations. As they can also be approached from a logic point of view, an application of ATP methods promises to be beneficial.

We follow the lines of Tarski [TG87], Maddux [Mad83] and Gordeev [Gor95] by converting equations from various relation algebraic theories to finite variable first-order logic sentences. We are then able to apply Gordeev's n-variable calculi RPC_n to the transformed formulae.

Our implementation ARA is a prover for the RPC_n calculi with a front-end to convert relation algebraic propositions to 3-variable first-order sentences. It implements a fully automatic proof procedure, various reduction strategies, and some simplification rules, to prove theorems in the theories SSA^1, RA, and RRA.

2 Theoretical Foundations

Gordeev's Reduction Predicate Calculi. Gordeev developed a cut free formalization of predicate logic without equality using only finitely many distinct

---

* This work was partially supported by DFG under grant Ku 966/4-1.
[1] This "simple" variant of semi-associative relation algebra (SA) includes the identity axiom $A \odot \overset{\circ}{1} = A$ only for literals A.

D. McAllester (Ed.): CADE-17, LNAI 1831, pp. 177–182, 2000.
© Springer-Verlag Berlin Heidelberg 2000

variables. In [Gor95] the reduction predicate calculi RPC_n were introduced. These term rewriting systems reduce valid formulae of n-variable logic to true. n-variable logic comprises only formulae containing no more than n distinct variables, the same restriction applies to RPC_n. Moreover, all RPC_n formulae are supposed to be in negation normal form.

Formulae provable in RPC_n are exactly those provable in the standard modus ponens calculus, or, alternatively, the sequent calculus with cut rule, using at most n distinct variables ([Gor95], Theorem 3.1 and Corollary). The RPC_n rewriting systems consist of the following rules:[2]

$$(R1) \qquad A \vee \top \longrightarrow \top$$
$$(R2) \qquad L \vee \neg L \longrightarrow \top$$
$$(R3) \qquad A \wedge \top \longrightarrow A$$
$$(R4) \quad A \vee (B \wedge C) \longrightarrow (A \vee B) \wedge (A \vee C)$$
$$(R5) \qquad \exists x A \longrightarrow \exists x A \vee A[x/t]$$
$$(R6) \qquad \forall x A \vee B \longrightarrow \forall x A \vee B \vee \forall y (\forall y B \vee A[x-y][x/y])$$
$$(R6') \qquad \forall x A \longrightarrow \forall x A \vee A[-x] \vee \forall y (A[x-y][x/y])$$

Here, L denotes an arbitrary literal, A, B and C are formulae, x and y are individual variables, and t is any term. A term may be a variable or a constant, function symbols do not occur in RPC_n. $F[x/t]$ denotes substitution of x by t, where $F[x/t]$ does not introduce new variables. $F[-x]$ and $F[x-y]$ are variable elimination operators (see [Gor95] for a rigorous definition), where $F[-x]$ deletes all free occurrences of x from F by replacing the respective (positive and negative) literals contained in F by falsum ($\bot$). The binary elimination operator $F[x-y]$ is defined by $F[x-y] = F[-y]$ if $x \neq y$, and $F[x-y] = F$ otherwise. Note, that the variable elimination operators are – just like substitution – metaoperators on formulae, and not part of the formal language of the logic itself.

A formula F is provable in RPC_n ($RPC_n \vdash F$) iff it can be reduced to true, i.e., iff $F \overset{*}{\longrightarrow}_{RPC} \top$. We call a formula sequence $(F_0, \ldots, F_n)$ a reduction chain if $F_i \longrightarrow_{RPC} F_{i+1}$. A reduction strategy is a computable function that extends a reduction chain by one additional formula. As all essential rules of RPC_n are of the form $F \longrightarrow F \vee G$, and thus no "wrong" reductions are possible, strategies can be used instead of a search procedure. This also means that no backtracking is needed and the RPC_n calculi are confluent on the set of valid formulae of n-variable logic.

ARA implements various reduction strategies rather than unrestricted breadth-first search or iterative deepening.

Translation from Relation Algebra to Predicate Logic. In order to prove formulae in the theory of relation algebra, we follow the idea of [TG87] and transform sentences of relation algebra to 3-variable first-order sentences. The transformation τ_{xyz} is straightforward, where x and y denote the predicate's arguments and z is a free variable. For brevity, we give only part of the definition

---

[2] $\vee$ and $\wedge$ are supposed to be AC-operators.

of τ_{xyz}, for predicate symbols, relative and absolute product ($\odot$ and $\cdot$):

$$\tau_{xyz}(R) \quad = x R y$$
$$\tau_{xyz}(\Phi \odot \Psi) = \exists z(\tau_{xzy}(\Phi) \wedge \tau_{zyx}(\Psi))$$
$$\tau_{xyz}(\Phi \cdot \Psi) \quad = \tau_{xyz}(\Phi) \wedge \tau_{xyz}(\Psi)$$

Here, Φ and Ψ stand for arbitrary relation algebraic expressions. A sentence $\Phi = \Psi$ from relation algebra is then translated by τ into an equivalence expression of first-order logic, a relational inclusion $\Phi \leq \Psi$ into an implication:

$$\tau(\Phi = \Psi) = \forall x \forall y(\tau_{xyz}(\Phi) \leftrightarrow \tau_{xyz}(\Psi))$$
$$\tau(\Phi \leq \Psi) = \forall x \forall y(\tau_{xyz}(\Phi) \rightarrow \tau_{xyz}(\Psi))$$

We can simulate proofs in various theories of relation algebra by using the following tight link between n-variable logic and relation algebras proved by Maddux (see, e.g., [Mad83]):

1. A sentence is valid in every *semi-associative relation algebra (SA)* iff its translation can be proved in 3-variable logic.
2. A sentence is valid in every *relation algebra (RA)* iff its translation can be proved in 4-variable logic.
3. A sentence is valid in every *representable relation algebra (RRA)* iff its translation can be proved in ω-variable logic.

We have to restrict SA in case of 3-variable logic to its simple variant SSA, as RPC_n – as well as the more familiar Hilbert-Bernays first-order formalism – contains only the simple Leibniz law. Compared to the generalized Leibniz law used in Tarski's and Maddux's formalisms, this simple schema is finitely (i.e., as an axiom) representable in RPC_n. The corresponding refinement is not necessary in case of RA and RRA, as the generalized Leibniz law for 3-variable formulae is deducible from its simple form in RPC_4, and thus in RPC_ω (see, e.g., [Gor99]).

3   Implementation

The ARA prover is a Haskell implementation of the RPC_n calculi with a front end to transform relation algebraic formulae to first-order logic. It offers different reduction strategies and a set of additional simplification rules. Most of these simplification rules preserve n-variable provability and can thus be used for SSA- and RA-proofs.

Input Language. The ARA system is capable of proving theorems of the form $\{E_1, \ldots E_n\} \vdash E$, where E and E_i are relation algebraic equations or inclusions.[3] Each equation in turn may use the connectives for relative and absolute sum

---

[3] Instead of an equation E a formula F may also be used, which is interpreted as the equation $F = 1$, where 1 stands for the absolute unit predicate.

and product, negation, conversion, and arbitrary predicates, among them the absolute and relative unit and the absolute zero as predefined predicates.

The translation of a relational conjecture of the form $\{E_1, \ldots E_n\} \vdash E$ to first-order logic is done in accordance with Tarski's deduction theorem for $\mathcal{L}^\times$ ([TG87], 3.3):

$$\tau(\{E_1, \ldots E_n\} \vdash E) = \tau(E_1) \wedge \ldots \wedge \tau(E_n) \rightarrow \tau(E)$$

To give an impression of how the actual input looks like, we show the representation of Dedekind's rule $(Q \odot R) \cdot S \leq (Q \cdot (S \odot R^\smile)) \odot (R \cdot (Q^\smile \odot S))$ as a conjecture for ARA:

```
|- (Q@R)*S < (Q*(S@R^))@(R*(Q^@S));
```

Literal and Reduction Tracking. To guarantee completeness of the deterministic proof search introduced by reduction strategies, we employ the technique of reduction tracking in our implementation. The idea is as follows: While successively constructing the reduction chain, record the first appearance of each reduction possibility[4] and track the changes performed on it. A strategy is complete, if each reduction possibility is eventually considered.

Literal tracking is used by the LP reduction strategy described later. It keeps track of the positions of certain literal occurrences during part of the proof.

Reduction Strategies. We implemented a trivial strategy based on the reduction tracking idea described above and several variants of a literal tracking strategy. The latter select a pair of complementary literals (and therefore are called LP strategies) that can be disposed of by a series of reduction steps. In order to find such pairs, an equation system is set up (similar to unification) that is solvable iff there is a reduction sequence that moves the literals (or one of their descendants) into a common disjunction. Then, RPC reductions are selected according to the equation system to make the literal pair vanish.

Additional Simplification Rules. To improve proof search behavior we added some simple rules and strategies to the RPC_n calculus that preserve n-provability:

1. Give priority to shortening rules $(R1)$, $(R2)$ and $(R3)$.
2. Remove quantifiers that bind no variables.
3. Minimize quantifier scopes.
4. Subgoal generation: To prove $F \wedge G$ prove first F, and then G.
5. Additional $\forall$-rule: $\forall x A \vee B \longrightarrow \forall y (A[x/y] \vee B)$ if $y \notin \mathrm{Fr}(A) \cup \mathrm{Fr}(B)$, which is used with priority over $(R6)$ and $(R6')$.
6. Delete pure literals.
7. Replace $F \vee \tilde{F}$ resp. $F \wedge \tilde{F}$ by F, if $\tilde{F}$ is a bound renaming of F.

---

[4] A reduction possibility consists of a position in the formula and, in case of the RPC-rules $(R5)$, $(R6)$ and $(R6')$, an additional reduction term.

Simplification rule 3 is applied only initially, before the actual proof search starts. Rules 5 and 7 are used to keep formula sizes smaller during proof search, where rule 5 is a special form of ($R6$) that has the purpose to accelerate introduction of so far unused variables.

Moreover, there are two additional simplification rules that, however, may change n-provability: (1) partial Skolemization to remove ∀-quantifiers and (2) replacement of free variables by new constants.

4 Experimental Results

We made some experiments with our implementation on a Sun Enterprise 450 Server running at 400 MHz. The Glasgow Haskell Compiler, version 4.04, was used to translate our source files. In Table 1 the results of our tests are summarized. The problem class directly corresponds to the number of variables used for the proof, as indicated at the end of Section 2.

Table 1. ARA run times for some relation algebra problems.

problem	source	class	strat.	proofs	steps	time
3.2(v)	[TG87]	SSA	LI	2	46	130
3.2(vi)	[TG87]	SSA	LI	2	41	100
3.2(xvii)	[TG87]	SSA	LI	3	22	40
3.1(iii)(ϵ)	[TG87]	RA	AI	6	104	200
3.2(xix)	[TG87]	RA	LA	3	25	50
Thm 2.7	[CT51]	RA	AI	1	12	40
Thm 2.11	[CT51]	RA	AI	1	19	50
Cor 2.19	[CT51]	RA	AI	1	77	75170
Dedekind	[DG98]	RA	AI	1	37	90
Cor 2.19	[CT51]	RRA	AI	1	38	140

In the last three columns the following information is given: the number of proofs that the problem consists of, the total number of RPC-reductions[5], and the total proof time in milliseconds. The first letter in the strategy column is "L" for the normal LP reduction strategy and "A" for the LP strategy with priority for the additional ∀-simplification rule. The second letter corresponds to the selection of disjunctive subformulae, i.e., A in rule ($R4$) and B in rule ($R6$). The strategy indicated by letter "I" selects a minimal suitable disjunction, "A" a maximal one.

The proof of Corollary 2.19 from [CT51] reveals an unexpectedly long runtime, which is reduced considerably by allowing more variables for the proof and thus switching to RRA.

---

[5] Only reductions with one of the rules ($R4$), ($R5$), ($R6$) and ($R6'$) are considered.

5 Conclusion and Future Work

By using ARA, many small and medium-sized theorems in various relation algebras could be proved. Formulae not containing the relative unit predicate can be handled quite efficiently, other formulae suffer from the fact that neither the RPC_n calculi nor the ARA prover offer a special treatment of equality.

Compared with other implementations [HBS94, vOG97, DG98], the most obvious differences are the automatic proof procedure using reduction strategies and the translation to the RPC_n calculi. The RALF system [HBS94] offers no automatic proof search, but has particular strengths in proof presentation. RALL [vOG97] is based on HOL and Isabelle, and thus is able to deal with higher order constructs. It also offers an experimental automatic mode using Isabelle's tactics. δRA is a Display Logic calculus for relation algebra, and its implementation [DG98] is based on Isabelle's metalogic. It also offers an automatic mode using Isabelle's tactics.

ARA can also be used to generate proofs in ordinary first-order logic and in restricted variable logics. As ARA is the initial implementation of a new calculus, we expect that further progress is very well possible. Implementation of new strategies or built-in equality may be viable directions for improvement.

Availability. The ARA system is available as source and binary distribution from `www-sr.informatik.uni-tuebingen.de/~sinz/ARA`.

References

[CT51]   L. H. Chin and A. Tarski. Distributive and modular laws in the arithmetic of relation algebras. *University of California Publications in Mathematics, New Series*, 1(9):341–384, 1951.

[DG98]   J. Dawson and R. Goré. A mechanized proof system for relation algebra using display logic. In *JELIA '98*, LNAI 1489, pages 264–278. Springer, 1998.

[Gor95]   L. Gordeev. Cut free formalization of logic with finitely many variables, part I. In *CSL '94*, LNCS 933, pages 136–150. Springer, 1995.

[Gor99]   L. Gordeev. Variable compactness in 1-order logic. *Logic Journal of the IGPL*, 7(3):327–357, 1999.

[HBS94]   C. Hattensperger, R. Berghammer, and G. Schmidt. RALF - a relation-algebraic formula manipulation system and proof checker. In *AMAST'93*, Workshops in Computing, pages 405–406. Springer, 1994.

[Mad83]   R. Maddux. A sequent calculus for relation algebras. *Annals of Pure and Applied Logic*, 25:73–101, 1983.

[TG87]   A. Tarski and S. Givant. *A Formalization of Set Theory without Variables*, volume 41 of *Colloquium Publications*. American Mathematical Society, 1987.

[vOG97]   D. von Oheimb and T. Gritzner. RALL: Machine-supported proofs for relation algebra. In *Automated Deduction - CADE-14*, LNAI 1249, pages 380–394. Springer, 1997.

Scalable Knowledge Representation and Reasoning Systems

Henry Kautz

AT&T Labs-Research
180 Park Ave
Florham Park NJ 07974, USA
kautz@research.att.com

Abstract. Traditional work in knowledge representation (KR) aimed to create practical reasoning systems by designing new representations languages and specialized inference algorithms. In recent years, however, an alternative approach based on compiling combinatorial reasoning problems into a common propositional form, and then applying general, highly-efficient search engines has shown dramatic progress. Some domains can be compiled to a tractable form, so that run-time problem-solving can be performed in worst-case polynomial time. But there are limits to tractable compilation techniques, so in other domains one must compile instead to a minimal combinatorial "core". The talk will describe how both problem specifications and control knowledge can be compiled together and then solved by new randomized search and inference algorithms.

D. McAllester (Ed.): CADE-17, LNAI 1831, pp. 183–183, 2000.
© Springer-Verlag Berlin Heidelberg 2000

Efficient Minimal Model Generation Using Branching Lemmas

Ryuzo Hasegawa, Hiroshi Fujita, and Miyuki Koshimura

Graduate School of Information Science and Electrical Engineering
Kyushu University, Kasuga-shi, Fukuoka 816-8580, Japan
{hasegawa,fujita,koshi}@ar.is.kyushu-u.ac.jp
http://ss104.is.kyushu-u.ac.jp/

Abstract. An efficient method for minimal model generation is presented. The method employs *branching assumptions and lemmas* so as to prune branches that lead to nonminimal models, and to reduce minimality tests on obtained models. This method is applicable to other approaches such as Bry's *complement splitting* and *constrained search* or Niemelä's *groundedness test*, and greatly improves their efficiency. We implemented MM-MGTP based on the method. Experimental results with MM-MGTP show a remarkable speedup compared to MM-SATCHMO.

1 Introduction

The notion of minimal models is important in a wide range of areas such as logic programming, deductive databases, software verification, and hypothetical reasoning. Some applications in such areas would actually need to generate Herbrand minimal models of a given set of first-order clauses.

Although the conventional tableaux and the Davis-Putnam methods can construct all minimal models, they may also generate nonminimal models that are redundant and thus would cause inefficiency. In general, in order to ensure that a model M is minimal, it is necessary to check if M is not *subsumed* by any other model. We call it a minimality test on M. Since minimality tests on obtained models become still more expensive as the number of models increases, it is important to avoid the generation of nonminimal models.

Recently two typical approaches in the tableaux framework have been reported. Bry and Yahya [1] presented a sound and complete procedure for generating minimal models and implemented MM-SATCHMO [2] in Prolog. The procedure rejects nonminimal models by means of *complement splitting* and *constrained search*. Niemelä also presented a propositional tableaux calculus for minimal model reasoning [8], where he introduced the *groundedness test* which substitutes for constrained searches. However, both approaches have the following problems: they perform unnecessary minimality tests on such models that are assured to be minimal through a simple analysis of a proof tree, and they cannot completely prune all redundant branches that lead to nonminimal models.

To solve these problems, we propose a new method that employs *branching lemmas*. It is applicable to the above approaches to enhance their ability.

D. McAllester (Ed.): CADE-17, LNAI 1831, pp. 184–199, 2000.
© Springer-Verlag Berlin Heidelberg 2000

Branching lemmas provide an efficient way of applying factorization [6] to minimal model generation, and the use of them is justified by the notion of *proof commitment*. Consider model generation with complement splitting. If no proof commitment occurs between a newly generated model M and any other model M' that has been obtained (no branch extended below a node labeled with a literal in M' is closed by the negation of a literal in M), M is guaranteed to be minimal, thus a minimality test on M can be omitted. In addition, by pruning many branches that result in nonminimal models, the search space will be greatly reduced. The above things can be achieved with branching lemmas.

We implemented the method on a Java version of MGTP [3,4] into which the functions of CMGTP [9] are already incorporated. We call this system MM-MGTP. It is applicable to first-order clauses as well as MM-SATCHMO. Experimental results show remarkable speedup compared to MM-SATCHMO.

This paper is organized as follows: in Section 2 the basic procedure of MGTP is outlined, while in Section 3 key techniques for minimal model generation are described. Then in Section 4 we define the branching lemma and explain how it works for minimal model generation. Section 5 refers to the features of MM-MGTP, and Section 6 proves the soundness and completeness of minimal model generation with complement splitting and branching lemmas. in Section 7 we compare experimental results obtained by running MM-MGTP and MM-SATCHMO, then discuss related work in Section 8.

2   Outline of MGTP

Throughout this paper, a clause is represented in implication form: $A_1 \wedge \ldots \wedge A_m \to B_1 \vee \ldots \vee B_n$ where $A_i(1 \leq i \leq m)$ and $B_j(1 \leq j \leq n)$ are literals; the left hand side of $\to$ is said to be the *antecedent*; and the right hand side of $\to$ the *consequent*. A clause is said to be *positive* if its antecedent is *true* $(m = 0)$, and *negative* if its consequent is *false* $(n = 0)$. A clause for $n \leq 1$ is called a *Horn* clause, otherwise a clause for $n > 1$ is called a *non-Horn* clause. A clause is said to be *range-restricted* if every variable in the consequent of the clause appears in the antecedent, and *violated* under a set M of ground literals if it holds that $\forall i(1 \leq i \leq m)A_i\sigma \in M \wedge \forall j(1 \leq j \leq n)B_j\sigma \notin M$ with some substitution σ.

A sequential algorithm of the MGTP procedure mg is sketched in Fig. 1. Given a set S of clauses, mg tries to construct a model by extending the current model candidate M so as to satisfy violated clauses under M (*model extension*). This process forms a proof tree called an *MG-tree*. In Fig. 1, operations to construct a *model tree* T consisting of only models are added to the original procedure, for use in a model checking type MM-MGTP to be explained later. The function $mg0$ takes, as an initial input, the consequents of positive Horn and non-Horn clauses, an empty model candidate M and a null model tree T, and returns SAT/UNSAT as a proof result. It works as follows:

(1) As long as the unit buffer U is not empty, $mg0$ picks up a unit literal u from U, and extends a model candidate M with u (*Horn extension*). $T' \oplus u$ means that u is attached to the leaf of T'. Then, the conjunctive matching

$U_0 \leftarrow$ positive Horn clauses; $D_0 \leftarrow$ positive non-Horn clauses;
$T \leftarrow \phi$; $Ans \leftarrow mg0(U_0, D_0, \emptyset, T)$;

function $mg0(U, D, M, \mathbf{var}\ T)$ { $T' \leftarrow \phi$;
  while $(U \neq \emptyset)$ { $U \leftarrow U \setminus \{u \in U\}$; $\cdots (1)$
    if $(u \notin M)$ { $M \leftarrow M \cup \{u\}$; $T' \oplus u$; $CJM(u, M)$;
      if (M is rejected) return UNSAT; }
    if $(U = \emptyset)$ { $Simp\&Subsump(D, M)$; $\cdots (2)$
      if (M is rejected) return UNSAT; } }
  if $(D \neq \emptyset)$ { $d \leftarrow (L_1 \vee \ldots \vee L_n) \in D$; $D \leftarrow D \setminus \{d\}$; $\cdots (3)$
    $A \leftarrow$ UNSAT;
    for $j \leftarrow 1$ to n { $T_j \leftarrow \phi$; $A \leftarrow A \circ mg0(U \cup \{L_j\}, D, M, T_j)$; }
    $T \oplus T' \oplus \langle T_1, \ldots, T_n \rangle$; return A; }
  else { $T \oplus T'$; return SAT; } $\cdots (4)$

Fig. 1. The MGTP procedure mg

$$\frac{L \in M \quad \neg L \in M}{\bot}$$

$$\frac{L(\neg L) \in M \quad \neg L(L) \vee C \in D}{C}$$

Fig. 2. Unit refutation   **Fig. 3.** Disjunction simplification

procedure $CJM(u, M)$ is invoked to search for clauses whose antecedents are satisfied by M and u. If such nonnegative clauses are found, their consequents are added to U or the disjunction buffer D according to the form of a consequent. When the antecedent of a negative clause is satisfied by $M \cup \{u\}$ in $CJM(u, M)$, or the unit refutation rule shown in Fig. 2 applies to $M \cup \{u\}$, $mg0$ rejects M and returns UNSAT (*model rejection*).

(2) When U becomes empty, the procedure $Simp\&Subsump(D, M)$ is invoked to apply the disjunction simplification rule shown in Fig. 3 and to perform subsumption tests on D against M. If a singleton disjunction is derived as a consequence of disjunction simplification, it is moved from D to U. When an empty clause is derived, $mg0$ rejects M and returns UNSAT.

(3) If D is not empty, $mg0$ picks up a disjunction d from D and recursively calls $mg0$ to expand M with each disjunct $L_j \in d$ (*non-Horn extension*). $A \circ B$ returns SAT if either A or B is SAT, otherwise returns UNSAT. $T' \oplus \langle T_1 \ldots T_n \rangle$ means that each sub model tree T_j of L_j is attached to the leaf of T', where $\langle \phi, \phi \rangle = \phi$ and $T \oplus \phi = T$.

(4) When both U and D become empty, $mg0$ returns SAT.

The nodes of an MG-tree except the root node are all labeled with literals used for model extension. A branch or a path from the root to a leaf corresponds to a model candidate. Failed branches are those closed by model rejection, and are marked with × at their leaves. A branch is a success branch if it ends with a node at which model extension cannot be performed any more. Figure 4 gives an

$$S1 =$$
$$\left\{ \begin{array}{l} \rightarrow a \vee b \vee c. \\ a \rightarrow b. \\ c \rightarrow . \end{array} \right\}$$

$$\begin{array}{ccc} & a & b & c \\ & | & \odot & \times \\ & b & & \\ & \otimes & & \end{array}$$

$$(B_{11} \wedge \ldots \wedge B_{1k_1}) \vee \ldots \vee (B_{n1} \wedge \ldots \wedge B_{nk_n})$$

B_{11}	$\ldots$	B_{n1}
$\vdots$	$\vdots$	$\vdots$
B_{1k_1}	$\ldots$	B_{nk_n}

Fig. 4. $S1$ and its MG-tree   **Fig. 5.** Splitting rule

MG-tree for the clause set $S1$. Here two models $\{a, b\}$ and $\{b\}$ are obtained, while a model candidate $\{c\}$ is rejected. $\{a, b\}$ is nonminimal since it is subsumed by $\{b\}$. We say that a model M subsumes M' if $M \subseteq M'$. The mark $\odot(\otimes)$ placed at a leaf on a success branch indicates that the model corresponding to the branch is minimal (respectively nonminimal).

MGTP allows an extended clause of the form $Ante \rightarrow (B_{11} \wedge \ldots \wedge B_{1k_1}) \vee \ldots \vee (B_{n1} \wedge \ldots \wedge B_{nk_n})$ as in [5]. The clause implies that model extension with it is performed according to the splitting rule shown in Fig. 5

Major operations in MGTP, such as conjunctive matching, subsumption testing, unit refutation, and disjunction simplification, comprise a membership test to check if a literal L belongs to a model candidate M. So speeding up the test is the key to achieving a good performance. For this, we introduced a facility called an *Activation-cell* (A-cell) [4]. It retains a boolean flag to indicate whether a literal L is in the current model candidate M under construction (active) or not (inactive). On the other hand, all occurrences of L are uniquely represented as a single object in the system (no copies are made), and the object has an ac field to refer to an A-cell. So, whether $L \in M$ or not is determined by merely checking the ac field of L. By using the A-cell facility, every major operation in MGTP can be performed in $\mathcal{O}(1)$ w.r.t. the size of the model candidate.

3   Minimal Model Generation

The first clause $\rightarrow a \vee b \vee c$ in $S1$ is equivalent to an extended clause $\rightarrow (a \wedge \neg b \wedge \neg c) \vee (b \wedge \neg c) \vee c$. By applying the splitting rule in Fig. 5 to the extended clause, the nonminimal model $\{a, b\}$ of $S1$ can be pruned since the unit refutation rule applies to b and $\neg b$. The added $\neg b$ and $\neg c$ are called *branching assumptions* and they are denoted by $[\neg b]$ and $[\neg c]$, respectively. In general, non-Horn extension with a disjunction $L_1 \vee L_2 \vee \ldots \vee L_n$ is actually performed using an augmented one $(L_1 \wedge [\neg L_2] \wedge \ldots \wedge [\neg L_n]) \vee (L_2 \wedge [\neg L_3] \wedge \ldots \wedge [\neg L_n]) \vee \ldots \vee L_n$, which exactly corresponds to an application of the *complement splitting* rule [1].

Complement splitting guarantees that the leftmost model in an MG-tree is always minimal, as proven by Bry and Yahya [1]. However, all other models generated to the right of it are not necessarily minimal. For instance, given the clause set $S2$ in Fig. 6, we obtain a minimal model $\{a\}$ on the leftmost branch, while obtaining a nonminimal model $\{b, a\}$ on the rightmost branch.

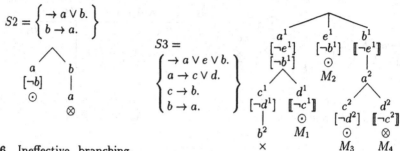

$$S2 = \left\{ \begin{array}{l} \to a \vee b. \\ b \to a. \end{array} \right\}$$

Fig. 6. Ineffective branching
assumption

$$S3 = \left\{ \begin{array}{l} \to a \vee e \vee b. \\ a \to c \vee d. \\ c \to b. \\ b \to a. \end{array} \right\}$$

Fig. 7. An MG-tree with branching lemmas

In order to ensure that every model obtained is minimal, MM-SATCHMO employs *constrained search* based on *model constraints* [1] as follows. When a minimal model $\{L_1, \ldots, L_m\}$ is found, a model constraint, i.e., a new negative clause $L_1 \wedge \ldots \wedge L_m \to$, is added to the given clause set. For instance, in Fig. 6, a negative clause $a \to$ is added to $S2$ when the minimal model $\{a\}$ is obtained. The negative clause forces the nonminimal model $\{b, a\}$ to be rejected.

However, this method needs to maintain negative clauses being added dynamically, the number of which might increase significantly. Moreover, it may bring rather heavy overhead due to conjunctive matching on the negative clauses, which is performed every time a model candidate is extended.

To alleviate the above memory consumption problem, Niemelä's approach [8] seems to be promising. His method works as follows. Whenever a model $M = \{L_1, \ldots, L_m\}$ is obtained, it is tested whether $\forall L \in M \ S \cup \overline{M} \models L$ holds or not, where S is the given clause set and $\overline{M} = \{\neg L' \mid L' \notin M\}$. This test, called the groundedness test, is nothing but reconstructing a new tableaux with a temporarily augmented clause set $S_M = S \cup \overline{M} \cup \{L_1 \wedge \ldots \wedge L_m \to\}$. If S_M is unsatisfiable, then it is concluded that M is minimal, otherwise nonminimal.

4   Branching Lemma

If branching assumptions are added symmetrically, inference with them becomes unsound. For instance, consider the clause set $S2'$ obtained by adding a clause $a \to b$ to $S2$ in Fig. 6. If $\neg a$ is added to the disjunct b, no models are obtained for $S2'$, although a minimal model $\{a, b\}$ does exist. However, for $S2$, $\neg a$ can be added to b to reject the model $\{b, a\}$, because the proof below a does not depend on that of b, that is, there is no mutual proof commitment between the two branches. In this situation, we can use $\neg a$ as a unit lemma in the proof below b.

Definition 1. *Let L_i be a disjunct in $L_1 \vee \ldots \vee L_n$ used for non-Horn extension. The disjunct L_i is called a* committing *disjunct, if a branch expanded below L_i is closed by the branching assumption $[\neg L_j]$ of some right sibling disjunct L_j ($i+1 \leq j \leq n$). On the other hand, every right sibling disjunct L_k ($i + 1 \leq k \leq n$) is called a* committed *disjunct from L_i.*

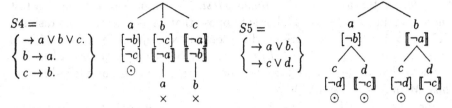

Fig. 8. Pruning by branching lemmas   Fig. 9. Omitting minimality test

Definition 2. *If L_i in $L_1 \vee \ldots \vee L_n$ used for non-Horn extension is not a committing disjunct, we add $\neg L_i$ to every right sibling disjunct L_j $(i + 1 \leq j \leq n)$. Such $\neg L_i$ is called a* branching lemma *and is denoted by $[\![\neg L_i]\!]$.*

For example, in Fig. 7, a^1 is a committing disjunct since the assumption $[\neg b^1]$ is used to close the leftmost branch expanded below a^1. Here, a superscript is added to a literal to identify an occurrence of the identical literal. e^1, b^1 are committed disjuncts since they are committed from a^1. Branching lemmas $[\![\neg c^1]\!]$, $[\![\neg e^1]\!]$, and $[\![\neg c^2]\!]$ are generated from non-committing disjuncts c^1, e^1, and c^2, respectively, whereas $[\![\neg a^1]\!]$ cannot be generated from the committing disjunct a^1.

Definition 3. *Let M be a model obtained in an MG-tree. If it contains a committed disjunct L_j in $L_1 \vee \ldots \vee L_n$ used for non-Horn extension, each committing disjunct L_i appearing as a left sibling of L_j is said to be a* committing disjunct relevant to M. *M is said to be a* safe *model if it contains no committed disjuncts. Otherwise, M is said to be a* warned *model.*

With branching lemmas, it is possible to prune branches that would lead to nonminimal models as shown in Fig. 8. In addition to this, branching lemmas have a great effect of reducing minimality tests as described below.

Omitting a Minimality Test. If an obtained model M is safe, M is assured to be minimal so that no minimality test is required. Intuitively, this is justified as follows. If $L_j \in M$ is a disjunct in some disjunction $L_1 \vee \ldots \vee L_n$, it cannot be a committed disjunct by Definition 3. For each left sibling disjunct $L_k (1 \leq k \leq j - 1)$, a model M' containing L_k, if any, satisfies the following: $L_j \notin M'$ under branching assumption $[\neg L_j]$, and $L_k \notin M$ under branching lemma $[\![\neg L_k]\!]$. Thus, M is not subsumed by M'.

For instance, in Fig. 9, all obtained models of $S5$ are assured to be minimal without performing any minimality test, since they are safe.

Restricting the Range of a Minimality Test. On the other hand, if an obtained model M is warned, it is necessary to perform a minimality test on M against models that have been obtained. However, minimality tests should be performed only against such models that contain committing disjuncts relevant to M. We call this a *restricted minimality test*. The reason why a minimality test is necessary in this case is as follows: Suppose that L_k is a committing disjunct relevant to M, L_j is the corresponding committed disjunct in M and M' is a model containing L_k. Although $L_j \notin M'$ under branching assumption $[\neg L_j]$, it may hold that $L_k \in M$ since the branching lemma $[\![\neg L_k]\!]$ is not allowed for M. Thus, M may be subsumed by M'.

For example, in Fig. 7, model M_1 is safe because it contains no committed disjunct. Thus, a minimality test on M_1 can be omitted. Models M_2, M_3, M_4 are warned because they contain the committed disjunct e^1 or b^1. Hence, they require minimality tests. Since a^1 is the committing disjunct relevant to each of them, minimality tests on them are performed[1] only against M_1 containing a^1.

5   Implementation of MM-MGTP

We have implemented two types of a minimal model generation prover MM-MGTP: *model checking* and *model re-computing*. The former is based on Bry and Yahya's method and the latter on Niemelä's method.

Model Checking MM-MGTP. Although a model checking MM-MGTP is similar to MM-SATCHMO, the way of treating model constraints differs somewhat. Instead of dynamically adding model constraints (negative clauses) to the given clause set, MM-MGTP retains them in the form of a model tree T. Thus, the constrained search for minimal models in MM-SATCHMO is replaced by a *model tree traversal* for minimality testing. For this, whenever a warned model M is obtained at (4) in Fig. 1, *mg* invokes the attached procedure *mchk*.

Consider Fig. 7 again. When the proof of a^1 has completed, the node N_{a^1} labeled with a^1 in T is marked as having a committing disjunct, and a pointer to the A-cell allocated for its parent (root) is assigned to a *com* field of the corresponding committed literals e^1, b^1. By this, e^1, b^1 are recognized to be committed just by checking their *com* fields, and branches below e^1, b^1 are identified to be warned. Hence, when a model M_3 is generated, *mchk* first finds the committed disjunct b^1 in M_3. Then, finding b^1's left sibling node N_{a^1}, *mchk* traverses down paths below N_{a^1} searching for a minimal model that subsumes M_3.

During the traversal of a path in T, each node on the path is examined whether a literal L labeling the node belongs to the current model M (active) or not by checking the *ac* field of L. If L is active, it means that $L \in M$, otherwise $L \notin M$. For the latter, *mchk* quits traversing the path immediately and searches for another one. If *mchk* reaches an active leaf, it means that M is subsumed by the minimal model on the traversed path, and thus M is nonminimal.

Here, we employ *early pruning* as follows. If the current model $M = \{L_1, \ldots, L_i, \ldots, L_m\}$ is subsumed by the previous model M' such that $M' \subseteq \{L_1, \ldots, L_i\}$, we can prune the branches below L_i. Although *mchk* is invoked after M has been

---

[1] For further refinement, a minimality test on M that contains no committing disjunct relevant to it can be omitted. This is the case for M_2.

generated, our method is more efficient than using model constraints, since it performs a minimality test not every time model extension with $L_j \in M$ occurs, but only once when a warned model M is obtained.

Model Re-computing MM-MGTP. A model re-computing MM-MGTP can also be implemented easily. In this version, model tree operations are removed from mg, and the re-computation procedure $rcmp$ for minimality testing is attached at (4) in Fig. 1. $rcmp$ is the same as mg except that some routines are modified for restarting the execution. It basically performs groundedness tests in the same way as Niemelä's: whenever a *warned* model $M = \{L_1, \ldots, L_m\}$ is obtained, mg invokes $rcmp$ to restart model generation for $S \cup \overline{M}$ with a negative clause $C_M = L_1 \wedge \ldots \wedge L_m \to$ being added temporarily. If $rcmp$ returns UNSAT, then M is assured to be minimal. Otherwise, a model M' satisfying $M' \subset M$ should be found, and then M will be rejected since it turns out to be nonminimal. Note that no model M'' satisfying $M'' \not\subseteq M$ will be generated by $rcmp$ because of the constraints $\overline{M}$ and C_M. Note also that those minimal models which subsume M, if any, must be found to the left of M in the MG-tree, due to complement splitting.

The slight difference with Niemelä's is that in place of C_M above, we use a *shortened negative clause* $L_{k_1} \wedge \ldots \wedge L_{k_r} \to$, consisting of committed disjuncts in M. It is obtained by removing from C_M uncommitted literals $L_u \in M$, i.e., those not committed from their left siblings. For instance in Fig. 7, when model $M_4 = \{b^1, a^2, d^2\}$ is obtained, a shortened negative clause $b^1 \to$ will be created instead of $b^1 \wedge a^2 \wedge d^2 \to$, since only b^1 is the committed disjunct in M_4.

The use of shortened negative clauses corresponds to the restricted minimality test and enables it to avoid groundedness tests on such uncommitted literals L_u. The validity of using shortened negative clauses is given in Theorem 4.

6   Soundness and Completeness

In this section, we present some results on soundness and completeness of the MM-MGTP procedure. First, we show that model generation with factorization is complete for generating minimal models. This implies that the MM-MGTP procedure is also complete because the use of branching assumptions and lemmas can be viewed as an application of factorization. Second, we give a necessary condition for a generated model to be nonminimal. The restricted minimality test keeps minimal model soundness because it is performed whenever the condition is satisfied. Last, we prove that using shortened negative clauses for a groundedness test guarantees the minimality of generated models.

Theorem 1. *Let T be a proof tree of a set S of clauses, N_1 and N_2 be sibling nodes in T, L_i a literal labeling N_i and T_i a subproof tree below $N_i(i = 1, 2)$, as shown in Fig. 10(a). If N_2 has a descendant node N_3 labeled with L_1, then for each model M through a subproof tree T_3 below N_3, there exists a model M' through T_1 such that $M' \subseteq M$ (Fig. 10(b)).*

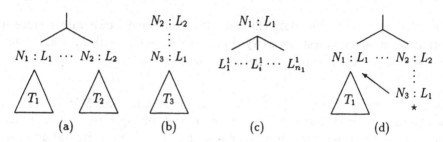

Fig. 10. Proof trees explaining Theorem 1, 2 and Definition 5

Proof. We define the sequence of literals $s_1, s_2, \ldots$, constituting M' through T_1 by induction. Let I be a set of literals on the path P ending with N_1. If N_1 is a leaf $(T_1 = \phi)$, P must be a success path and $M' = I \subseteq M$, because M is a model through T_3 and $L_1 \in M$. Otherwise, there is a clause $C_1 = \Gamma_1 \to L_1^1 \vee \ldots \vee L_{n_1}^1$ violated under I and used for model extension at N_1 (Fig. 10(c)). Then, $M \models \Gamma_1$ and $M \models L_1^1 \vee \ldots \vee L_{n_1}^1$ since M is a model. So, there exists a node labeled with s_1 such that $s_1 \in \{L_1^1, \ldots, L_{n_1}^1\}$ and $s_1 \in M$. Suppose that we have defined the first n literals of M' in T_1 to be $s_1, \ldots, s_n$, by traveling down the successor nodes whose labels belong to M. If the node labeled with s_n is a leaf, we are done. Otherwise, we may continue our definition of M'. The sequence ends with a label of a leaf finitely or continues forever. In either case, there exists a model M' containing literals in the sequence such that $M' = I \cup \{s_1, s_2, \ldots\} \subseteq M$. $\square$

The above is a fundamental theorem for proving the minimal model completeness of model generation with factorization. We define our factorization essentially in the same means as tableau factorization [6]. To avoid a circular argument, a factorization dependency relation is arranged on a proof tree.

Definition 4 (Factorization Dependency Relation). *A factorization dependency relation on a proof tree is a strict partial ordering $\prec$ relating sibling nodes in the proof tree. A relation $N_1 \prec N_2$ means that searching for minimal models below N_2 is committed to that below N_1.*

Definition 5 (Factorization). *Given a proof tree T and a factorization dependency relation $\prec$ on T, first select a node N_3 labeled with literal L_1 and another node N_1 labeled with the same literal L_1 such that (1) N_3 is a descendant of a node N_2 which is a sibling of N_1, and (2) $N_2 \not\prec N_1$. Next, close the branch extended to N_3 (denoted by $\star$) and modify $\prec$ by adding a relation $N_1 \prec N_2$, then forming the transitive closure of the relation. The symbol $\star$ means that the proof of N_3 is committed to that of N_1. The situation is depicted in Fig. 10(d).*

Corollary 1. *Let S be a set of clauses. If a minimal model M of S is built by model generation, then M is also built by model generation with factorization.*

Proof. Immediately from Theorem 1. $\square$

The model generation procedure is minimal model complete (in the sense that it generates all minimal models) for range-restricted clauses [1]. This implies the minimal model completeness of model generation with factorization.

Corollary 2. *(Minimal model completeness of model generation with factorization) Let S be a satisfiable set of range-restricted clauses and T a proof tree by model generation with factorization. If M is a minimal model of S, then M is found in T.*

We consider model generation with branching assumptions and lemmas as arranging factorization dependency relation on sibling nodes $N_1, \ldots, N_m$ labeled with $L_1, \ldots, L_m$, respectively, as follows: $N_j \prec N_i$ for all $j(i < j \leq m)$ if L_i is a committing disjunct, while $N_i \prec N_j$ if $[\![\neg L_i]\!]$ is used below N_j. This consideration leads to the minimal model completeness of the MM-MGTP procedure.

Corollary 3 (Minimal Model Completeness of MM-MGTP). *Let S be a satisfiable set of range-restricted clauses and T a proof tree by model generation with branching assumptions and branching lemmas. If M is a minimal model of S, then M is found in T.*

Although model generation with factorization can suppress the generation of nonminimal models, it may still generate them. In order to make the procedure sound, that is, to make it generate minimal models only, we need a minimality test on an obtained model. The following theorem gives a necessary condition for a generated model to be nonminimal.

Theorem 2. *Let S be a set of clauses and T a proof tree of S by model generation with factorization. Let N_1 and N_2 be sibling nodes in T, T_i a subproof tree below N_i and M_i a model through $T_i(i = 1, 2)$. If $N_2 \not\prec N_1$, then $M_1 \not\subseteq M_2$.*

Proof. Suppose that N_i is labeled with a literal $L_i(i = 1, 2)$ (Fig. 10(a)). It follows from $N_2 \not\prec N_1$ that (1) $N_1 \prec N_2$ or (2) there is no $\prec$-relation between N_1 and N_2. If (1) holds, $L_1 \notin M_2$ because every node labeled with L_1 in T_2 has been factorized with N_1. On the other hand, $L_1 \in M_1$. Therefore, $M_1 \not\subseteq M_2$. If (2) holds, $L_1 \notin M_2$ because there is no node labeled with L_1 in T_2. Therefore, $M_1 \not\subseteq M_2$. □

Theorem 2 says that (1) if $N_2 \not\prec N_1$, no minimality test on M_2 against M_1 is required, otherwise (2) if $N_2 \prec N_1$, we need to check the minimality of M_2 against M_1.

In MM-MGTP based on a depth-first-left-first search, omitting a minimality test on a safe model is justified by the above (1), while the restricted minimality test on a warned model is justified by (2). If N_1 were a left sibling of N_2 such that $N_1 \prec N_2$, e.g., a branching lemma $[\![\neg L_1]\!]$ is used below N_2, a minimality test on M_1 against M_2 will be required according to Theorem 2. However it is unnecessary in MM-MGTP since it always holds that $M_2 \not\subseteq M_1$ as follows.

Theorem 3. *Let T be a proof-tree by a depth-first-left-first search version of model generation with factorization and M_1 a model found in T. If a model M_2 is found to the right of M_1 in T, then $M_2 \not\subseteq M_1$.*

Proof. Let P_{M_i} be a path corresponding to M_i $(i = 1, 2)$. Then, there are sibling nodes N_1 on P_{M_1} and N_2 on P_{M_2}. Let L_i be a label of N_i $(i = 1, 2)$. Now assume that $M_2 \subseteq M_1$. This implies $L_2 \in M_1$. Then, there is a node N_{L_2} labeled with L_2 on P_{M_1}. However, N_{L_2} can be factorized with N_2 in the depth-first-left-first search. This contradicts that M_1 is found in T. Therefore, $M_2 \not\subseteq M_1$.   □

Corollary 4 (Minimal Model Soundness of MM-MGTP). *Let S be a satisfiable set of range-restricted clauses and T a proof tree by model generation with branching assumptions, branching lemmas, and restricted minimality tests. If M is a model found in T, then M is a minimal model of S.*

The following theorem says that a shortened negative clause for the groundedness test guarantees the minimality of generated models.

Definition 6. *Let S be a set of clauses, T a proof tree of S by model generation with factorization and M a model found in T. For each literal $L \in M$, N_L denotes a node labeled with L on the path of M. Let $M_p \subseteq M$ be a set satisfying the following condition: for every $L \in M_p$, there exists a node N such that $N_L \prec N$. Note that M_p is a set of committed disjuncts in M. C_{M_p} denotes a shortened negative clause of the form $L_1 \wedge \ldots \wedge L_m \rightarrow$, where $L_i \in M_p (i = 1, \ldots, m)$.*

Theorem 4. *Let S be a set of clauses. M is a minimal model of S if and only if $S_{M_p} = S \cup \overline{M} \cup \{C_{M_p}{}^2\}$ is unsatisfiable, where $\overline{M} = \{L' \rightarrow \; | \; L' \notin M\}$.*

Proof. (Only-if part) Let M' be a model of S. There are three cases according to the relationship between M and M': (1) $M' \setminus M \neq \emptyset$, (2) $M' = M$, or (3) $M' \subset M$. If (1) holds, M' is rejected by a negative clause in $\overline{M}$. If (2) holds, M' is rejected by the shortened negative clause C_{M_p}. The case (3) conflicts with the assumption that M is minimal. Now that no model of S is a model of S_{M_p}. Therefore, S_{M_p} is unsatisfiable.   □

Proof. (If part) Let T be a proof tree of S by model generation with factorization. Suppose that M is not minimal. Then there exists a model M' of S found in T such that $M' \subseteq M$. Let P_M, P'_M be the paths corresponding to M, M', respectively. Then, there are sibling nodes N and N' in T such that N is on P_M and N' on P'_M. Let L, L' be a label of N, N', respectively. In case of $N \prec N'$, M' conflicts neither with C_{M_p} because $L \notin M'$ nor with $\overline{M}$ because $M' \subseteq M$. Thus, M' is a model of S_{M_p}. This contradicts that S_{M_p} is unsatisfiable. In case of $N \not\prec N'$, since a node labeled with L' cannot appear on P_M, both $L' \notin M$ and $L' \in M'$ hold. This contradicts that $M' \subseteq M$. Therefore, M is minimal.   □

7   Experimental Results

This section compares experimental results on MM-MGTP with those on MM-SATCHMO and MGTP. Regarding MM-MGTP, we also compare four versions: model re-computing with/without branching lemmas (Rcmp+BL/Rcmp),

---

[2] If $M_p = \emptyset$, $L_1 \wedge \ldots \wedge L_m \rightarrow$ becomes an empty clause $\rightarrow$ which denotes contradiction. In this case, we conclude that M is minimal without a groundedness test.

and model checking with/without branching lemmas (Mchk+BL/Mchk). MM-MGTP and MGTP are implemented in Java, while MM-SATCHMO in ECL^{i}PS^e Prolog. All experiments were performed on a Sun Ultra10 (333 MHz,128 MB). Table 1 shows the results. The examples used are as follows.

ex1. $S_n = \{\to a_k \vee b_k \vee c_k \vee d_k \vee e_k \vee f_k \vee g_k \vee h_k \vee i_k \vee j_k \mid 1 \leq k \leq n\}$

This problem is taken from the benchmark examples for MM-SATCHMO. The MG-tree for ex1 is a balanced tree of branching factor 10, and every generated model is minimal. Since every success branch contains no committed disjunct, i.e., the corresponding model is safe, no minimality test is required if branching lemmas are used.

ex2. $S_n = \{a_{i-1} \to a_i \vee b_i \vee c_i, \ b_i \to a_i, \ c_i \to b_i \mid 2 \leq i \leq n\} \cup \{\to a_1\}$

The MG-tree for ex2 becomes a right-heavy unbalanced tree. Only the leftmost branch gives a minimal model, which subsumes all other models to the right. With branching lemmas, these nonminimal models can be rejected.

ex3. $T_1 = \{\to a_1 \vee b_1, \ a_1 \to b_1, \ b_1 \to a_2 \vee b_2, \ a_2 \to b_2 \vee d_1\}$
$T_2 = \{b_2 \to a_3 \vee b_3, \ a_3 \to a_2 \vee c_2, \ a_3 \wedge a_2 \to b_3 \vee d_2, \ a_3 \wedge c_2 \to b_3 \vee d_2\}$
$T_j = \{b_j \to a_{j+1} \vee b_{j+1}, \ a_{j+1} \to a_j \vee c_j, \ c_j \to a_{j-1} \vee c_{j-1},$
$\qquad a_{j+1} \wedge a_2 \to b_{j+1} \vee d_j, \ a_{j+1} \wedge c_2 \to b_{j+1} \vee d_j\} \ (j \geq 3)$
$S_n = \bigcup_{i=1}^{n} T_i$

The MG-tree for ex3 is a right-heavy unbalanced tree as for ex2. Since every success branch contains committed disjuncts, minimality tests are inevitable. However, none of the obtained models is rejected by the minimality test.

ex4. $S_a = \{\to a_i \vee b_i \vee c_i \vee d_i \vee e_i \mid 1 \leq i \leq 4\} \cup \{a_3 \to a_2, \ a_4 \to a_3, \ a_1 \to a_4\}$

ex5. $S_{abcd} = S_a \cup \{b_3 \to b_2, \ b_4 \to b_3, \ b_1 \to b_4\} \cup \{c_3 \to c_2, \ c_4 \to c_3, \ c_1 \to c_4\}$
$\qquad \cup \{d_3 \to d_2, \ d_4 \to d_3, \ d_1 \to d_4\}$

ex4 and ex5 are taken from the paper [8]. No nonminimal models can be rejected without using branching lemmas.

syn9-1. An example taken from the TPTP library [11], which is unsatisfiable.

Channel. A *channel-routing* problem [12] in which constraint propagation with negative literals plays an essential role to prune the search space. One can obtain only minimal models with MGTP. The last two first-order examples are used to estimate the overhead of minimality testing in MM-MGTP.

Table 1. Performance comparison

Problem	Rcmp+BL	Mchk+BL	Rcmp	Mchk	MM-SAT	MGTP
ex1 (N=5)	0.271 100000 0	0.520 100000 0	2.315 100000 0	0.957 100000 0	8869.950 100000 0	0.199 100000 0
ex1 (N=7)	34.150 10000000 0	OM (>144) − −	324.178 10000000 0	OM (>115) − −	OM (>40523) − −	19.817 10000000 0
ex2 (N=14)	0.001 1 26	0.001 1 26	82.112 1 1594322	16.403 1 1594322	1107.360 1 1594323	9.013 1594323 0
ex3 (N=16)	19.816 65536 1	5.076 65536 1	19.550 65536 1	5.106 65536 1	OM (>2798) − −	589.651 86093442 0
ex3 (N=18)	98.200 262144 1	26.483 262144 1	95.436 262144 1	26.103 262144 1	OM (>1629) − −	5596.270 774840978 0
ex4	0.002 341 96	0.002 341 96	0.009 341 160	0.003 341 160	0.3 341 284	0.004 501 0
ex5	0.001 17 84	0.001 17 84	0.002 17 88	0.001 17 88	0.25 17 608	0.001 129 0
syn9-1	0.105 0 19683	0.109 0 19683	0.101 0 19683	0.092 0 19683	TO (>61200) − −	0.088 0 19683
channel	4.016 51922 78	4.064 51922 78	46.166 51922 78	4.517 51922 78	NA − −	3.702 51922 78

top: time (sec), middle: No. of models, bottom: No. of failed branches.
MM-SAT: MM-SATCHMO, OM: Out of memory, TO: Time out,
NA: Not available due to lack of constraint handling

MM-MGTP vs. MM-SATCHMO. Since MM-SATCHMO aborted execution very often due to memory overflow, we consider the problems that MM-SATCHMO could solve. A great advantage of MM-MGTP is seen for ex1 that does not need any minimality test and ex2 in which branching lemmas have high pruning effects. The fastest version of MM-MGTP achieves a speedup of 33,000 and 1,100,000 for ex1 and ex2, respectively, compared to MM-SATCHMO. Even for small problems like ex4 and ex5, MM-MGTP is more than one hundred times faster than MM-SATCHMO. In addition, it is reported that the Niemelä's system takes less than 2 and 0.5 seconds for ex4 and ex5, respectively [8].

MM-MGTP vs. MGTP. Compared to MGTP, proving time for MM-MGTP is much shortened as the number of nonminimal models rejected by branching lemmas and minimality tests increases. In particular, ex2 and ex3 exhibit a great effect of minimal model generation with branching lemmas. Although ex1, syn9-1, and channel are problems such that no nonminimal model is created, very little overhead is observed for MM-MGTP that employs branching lemmas, because minimality tests can be omitted.

Rcmp vs. Mchk. Proving time for Rcmp increases from 2 to 5 times that for Mchk because of re-computation overhead, for propositional problems ex1 (except N=7) through ex5, that do not require a term memory [10]. For the first-order problem channel that requires the term memory, Rcmp is about 10 times slower than Mchk. This is because the overhead of term memory access is large, besides Rcmp doubles the access frequencies when performing groundedness tests.

Next, look at the branching lemma effect. For ex1 and channel, since minimality tests can be omitted with branching lemmas, Rcmp+BL and Mchk+BL obtain 8.5 to 11.5- and 1.84 to 1.1-fold speedup, respectively, compared to versions without branching lemmas. Although the speedup ratio is rather small for Mchk+BL, it proves that Mchk based on model tree traversal is very efficient. ex2 is a typical example to demonstrate the effect, thus both Rcmp+BL and Mchk+BL achieve several-ten-thousand-fold speedup as expected.

Rcmp+BL vs. Mchk+BL. For ex3 in which minimality tests cannot be omitted, Mchk+BL is about 4 times faster than Rcmp+BL. Although for ex1 (N=5), no difference between Mchk+BL and Rcmp+BL should exist in principle, the former is about 2 times slower than the latter. This is because Mchk+BL has to retain all generated models, thereby causing frequent garbage collection.

8 Related Work

In a tableaux framework, Letz presented *factorization* [6] to prune tableaux trees. Complement splitting (or *folding-down* in [6]) is a restricted way of implementing factorization. It is restricted in the sense that a precedence relation is pre-determined between disjuncts in each disjunction, and that only a disjunct having higher precedence can commit its proof to that of another sibling disjunct with lower precedence, whereas such precedence is not pre-determined in factorization. Although factorization is more powerful than complement splitting, it may also generate nonminimal models without any guide or control.

Lu [7] proposed a minimal model generation procedure which in a sense relaxes the above restriction by adding branching assumptions symmetrically, as in $(a \wedge [\neg b]) \vee (b \wedge [\neg a])$ for a disjunction $a \vee b$. However, his method involves post-determination of the precedence between the disjuncts. This is because mutual proof commitment may occur due to symmetrical branching assumptions, and some possibly open branch are forced to be closed thereby making the proof unsound (and incomplete w.r.t. model finding). If this is the case, some tentatively closed branches have to be re-opened so that the performance would degrade.

Branching lemmas proposed in this paper can still be taken as a restricted implementation of factorization, because it is disabled for a disjunct to generate a branching lemma once a branching assumption of some sibling disjunct is used to prove the disjunct, whether mutual proof commitment actually occurs or not. Nevertheless, our method provides an efficient way of applying factorization to minimal model generation, since it is unnecessary to compute the transitive closure of the factorization relation. The effects of the branching lemma mechanism are summarized as follows: it can (1) suppress the generation of nonminimal models to a great extent, (2) avoid unnecessary minimality tests, and (3) restrict the range of minimality tests on the current model M to models on which committing disjuncts relevant to M appear.

The model checking version of MM-MGTP aims to improve MM-SATCHMO by introducing branching lemmas, and it is also based on complement splitting and constrained searches. Major differences between both systems are the following. MM-SATCHMO stores model constraints as negative clauses and performs minimality tests through conjunctive matching on the negative clauses, thereby being very inefficient in terms of space and time. Our model checking version, on the other hand, is more efficient because model constraints are retained in a model tree in which multiple models can share common paths, and minimality tests are suppressed or restricted by using branching lemmas.

Since the above two systems depend on model constraints which are a kind of memoization, they may consume much memory space, the size of which might increase exponentially in the worst case. This situation is alleviated by Niemelä's method [8]. It can reject every nonminimal model without performing a minimality test against previously found minimal models, by means of the *cut rule* which is essentially equivalent to complement splitting and the *groundedness test* that is an alternative of the constrained search.

The model re-computing version of MM-MGTP takes advantage of Niemelä's method in which it is unnecessary to retain model constraints. However, both systems repeatedly perform groundedness tests rather more expensive than constrained searches. In addition, they necessarily generate each minimal model twice. In the model re-computing version, the latter problem is remedied to some extent by introducing shortened negative clauses. Moreover, due to branching lemmas, it is possible to invoke as few groundedness tests as possible.

9   Conclusion

We have presented an efficient method to construct minimal models by means of branching assumptions and lemmas. Our work was motivated by the two approaches: Bry's method based on complement splitting and constrained searches and Niemelä's method that employs the groundedness test. However both methods may contain redundant computation, which can be suppressed by using branching lemmas in MM-MGTP. The experimental results with MM-MGTP show that orders of magnitude speedup can be achieved for some problems.

Nevertheless, we still need minimality tests when branching lemmas are not applicable. It should be pursued in future work to omit as many minimality tests as possible, for instance, through a static analysis of clauses. It would also be worthwhile to combine our method with other pruning techniques such as folding-up and full factorization, or to apply it to stable model generation.

References

1. Bry, F. and Yahya, A.: Minimal Model Generation with Positive Unit Hyper-Resolution Tableaux. Proc. of 5th Workshop on Theorem Proving with Analytic Tableaux and Related Methods, LNAI 1071, (1996) 143–159
2. Bry, F.: http://www.pms.informatik.uni-muenchen.de/software/MM-SATCHMO/ (1999)
3. Fujita, H. and Hasegawa, R.: Implementing a Model-Generation Based Theorem Prover MGTP in Java. Research Reports on Information Science and Electrical Engineering of Kyushu University, Vol.3, No.1, (1998) 63–68
4. Hasegawa, R. and Fujita, H.: A New Implementation Technique for Model Generation Theorem Provers To Solve Constraint Satisfaction Problems. Research Reports on Information Science and Electrical Engineering of Kyushu University, Vol.4, No.1, (1999) 57–62
5. Inoue, K., Koshimura, M. and Hasegawa, R.: Embedding Negation as Failure into a Model Generation Theorem Prover. Proc. CADE-11, Springer-Verlag, (1992) 400–415
6. Letz, R., Mayer K. and Goller C.: Controlled Integration of the Cut Rule into Connection Tableau Calculi, J. of Automated Reasoning, 13, (1994) 297-337
7. Lu, W.: Minimal Model Generation Based on E-hyper Tableaux. Research reports 20/96, (1996)
8. Niemelä, I.: A Tableaux Calculus for Minimal Model Reasoning. Proc. of 5th Workshop on Theorem Proving with Analytic Tableaux and Related Methods, LNAI 1071, (1996) 278–294
9. Shirai, Y. and Hasegawa, R.: Two Approaches for Finite-Domain Constraint Satisfaction Problem — CP and MGTP —. Proc. of the 12-th Int. Conf. on Logic Programming, MIT Press, (1995) 249–263
10. Stickel, M..: The Path-Indexing Method For Indexing Terms. Technical Note No.473, AI Center, SRI International, (1989)
11. Sutcliffe, G., Suttner, C. and Yemenis, T.: The TPTP Problem Library. Proc. CADE-12, (1994) 252–266
12. Zhou, N.: A Logic Programming Approach to Channel Routing. Proc. 12th Int. Conf. on Logic Programming, MIT Press, (1995) 217–231

FDPLL — A First-Order
Davis-Putnam-Logeman-Loveland Procedure

Peter Baumgartner

Institut für Informatik, Universität Koblenz-Landau
D-56073 Koblenz, Germany
peter@uni-koblenz.de
http://www.uni-koblenz.de/~peter/

Abstract. FDPLL is a directly lifted version of the well-known Davis-Putnam-Logeman-Loveland (DPLL) procedure. While DPLL is based on a splitting rule for case analysis wrt. ground and complementary literals, FDPLL uses a lifted splitting rule, i.e. the case analysis is made wrt. non-ground and complementary literals now.

The motivation for this lifting is to bring together successful first-order techniques like unification and subsumption to the propositionally successful DPLL procedure.

At the heart of the method is a new technique to represent first-order interpretations, where a literal specifies truth values for all its ground instances, unless there is a more specific literal specifying opposite truth values. Based on this idea, the FDPLL calculus is developed and proven as strongly complete.

1 Introduction

The[1] well-known *Davis-Putnam* procedure, as it is usually called, was brought forward in the early 60s by the researchers mentioned in the title [DP60,DLL62], [D63]. Nowadays, the procedure is most successfully applied to decide propositional problems, although it was originally conceived as a method for first-order theorem proving. To this end, successively increased sets of ground instances of first-order clauses are enumerated and fed into the propositional part of the procedure. This latter part is referred to as "propositional DPLL" in the sequel.

With the advent of the resolution calculus, the lifting of inference rules to the first-order level is standard in virtually all calculi and efficient proof procedures for first-order logic — except for Davis-Putnam-Logeman-Loveland methods. Thus, the purpose of this paper is to present a lifted version that fills this gap.

On an abstract level, the advantage of the "lifted" methods compared to the "propositional" methods stems from two sources: first, it is possible with a lifted method to finitely represent infinitely many inferences of the corresponding propositional methods, and, second, much more powerful redundancy elimination techniques are possible, e.g. based on subsumption. The motivation is to bring these advantages to DPLL. The other way round, FDPLL instantiates to propositional DPLL when applied to propositional logic.

---

[1] For a long version of the paper see
http://www.uni-koblenz.de/fb4/publikationen/gelbereihe/.

D. McAllester (Ed.): CADE-17, LNAI 1831, pp. 200–219, 2000.

Brief Description of FDPLL. In order to describe the main idea of FDPLL, it is helpful to refer to the widely-used presentation of propositional DPLL as a calculus with a single *splitting* rule, which carries out a case analysis wrt. a propositional variable A. More exactly, the current clause set S splits into two cases: the one where A is "true", and the other where A is "false", which give rise to simplifications based on the new information. DPLL proceeds by considering different cases until success (some case is a model for S) or failure (each considered case contradicts a clause in S).

The idea in FDPLL is to lift this splitting to the first-order level, i.e. to split with complementary *non-ground* literals like $P(x, y)$ and $\neg P(x, y)$. The difficulty here is that the "usual" way of reading the literals as universally quantified (i.e. $\forall x, y\ P(x, y)$ and $\forall x, y\ \neg P(x, y)$) immediately leads to an *unsound* calculus. Hence, a different reading is adopted: a bit simplified, a literal, say $P(x, y)$, stands *by default* for all its ground instances, say, $P(a, a), P(a, b), P(b, a)$ and $P(b, b)$ (suppose here that only constants a and b are present). However, the presence of a strictly more specific literal (wrt. the instantiation order) than $P(x, y)$ with complementary sign, say $\neg P(x, b)$, gives rise to exceptions of the default reading of $P(x, y)$ by excluding all instances of $P(x, b)$. Symmetrically, $\neg P(x, b)$ stands by default for all its ground instances (with the possibility to have exceptions again). So, the two literals $P(x, y)$ and $\neg P(x, b)$ together stand for $P(a, a), \neg P(a, b), P(b, a)$ and $\neg P(b, b)$, which in turn can be understood as an interpretation $\mathcal{I}$ in the obvious way.

Now, a "case" in FDPLL is just a set of possibly non-ground literals, such that an interpretation can be associated to, as just sketched. Based on this idea, the purpose of the splitting rule of FDPLL can be explained as follows: suppose there is an instance $C\sigma$ of a clause C that is "false" in the interpretation $\mathcal{I}$ associated to the current case (σ is computed by most general unification). Then, a split is attempted with a literal $L \in C\sigma$ in order to "repair" $\mathcal{I}$ towards an interpretation that assigns "true" to L, and hence to $C\sigma$ as well[2]. If this is not possible because of some elementary contradiction between $C\sigma$ and the current case, the current case is refuted ("closed"). Otherwise, two new cases come up, the one extending the current case with L, and the other with $\overline{L}$.

Continuing the example above, suppose the current case is $\{P(x, y), \neg P(x, b)\}$, hence $\mathcal{I} = \{P(a, a),\ \neg P(a, b),\ P(b, a), \neg P(b, b)\}$, and suppose that there is a clause $C = P(x, y) \vee \neg P(x, a)$. The clause instance $C\sigma = P(x, b) \vee \neg P(x, a)$ is "false" in $\mathcal{I}$, where $\sigma = \{y/b\}$ is computed by most general unification of the literals of C and complements of literals of $\mathcal{I}$. Regarding the two literals $P(x, b)$ and $\neg P(x, a)$ in $C\sigma$, only $\neg P(x, a)$ is a candidate for splitting, because $P(x, b)$ is an elementary contradiction to $\neg P(x, b)$ of the current case.

The procedure repeatedly carries out splits in this way and stops if every case is refuted (and reports "unsatisfiable"), or if no clause instance $C\sigma$ of the mentioned kind exists (and reports the current case as a model representation)[3].

---

[2] Actually, this is a bit simplified, but it serves well to illustrate the idea.
[3] There is a second variant of the splitting rule called "Commit" with the purpose to achieve that $\mathcal{I}$ is indeed consistent.

An improvement of this "basic" procedure recovers for certain branch literals the above mentioned universally quantified reading (cf. Section 5). It lifts to the first-order level the well-known propositional DPLL rule for propagating unit clauses. In resolution terminology, the improvement realizes unit-resulting resolution, and the well-known possibility to split clauses on the basis of variable-disjoint subclauses (as in $P(x) \lor Q(y)$).

Properties of FDPLL. Propositional DPLL has certain desirable features: its conceptual simplicity, space efficiency ("one branch at a time"), few inference rules (one is sufficient), efficient and adaptable implementations (the most efficient systematical propositional methods are based on DPLL, e.g. NTAB [CA96] and SATO [Zha97]), existence of non-clausal versions [BBOS98], and, the possibility to immediately extract a model in case that no refutation exists. A goal of this work is to keep these features for the lifted version FDPLL.

FDPLL is in particular space efficient, proof confluent and convergent (i.e. a strong completeness theorem holds). While these properties go without a saying for propositional DPLL, they *are* an issue for certain first-order methods, e.g. tableau and connection calculi (but see [BEF99] for a proof confluent strongly complete connection calculus). Beyond this, FDPLL is known to be a decision procedure for the Bernays-Schönfinkel class, i.e. clause logic without function symbols but constants. This is a non-trivial class, in the sense that most resolution and tableau systems cannot decide it, except in a trivial way by using the finite set of ground clauses.

Structure of the Paper. After stating some preliminaries, the model representation technique is introduced. Based on it, the calculus is developed. Then soundness and and completeness is turned to, followed by a sketch of the mentioned "universal literal" improvement. The subsequent proof procedure proves the existence of a concrete, fair strategy. Finally, some conclusions are drawn, including related work.

2   Preliminaries

The usual notions of first-order logic are applied in a way consistent to [CL73]. A *literal* is an atom or a negated atom. The letters K and L are reserved to denote literals. The *complement* of a literal L is $\overline{L} = A$, if $L = \neg A$ for some atom A, or else $\overline{L} = \neg L$; by $|L|$ the atom of L is denoted, i.e. $|A| = A$ and $|\neg A| = A$ for any atom A. A *clause* is a finite, possibly empty multiset $\{L_1, \ldots, L_n\}$ of literals, usually written as a disjunction $L_1 \lor \cdots \lor L_n$. By a *clause set* always a finite set of clauses is meant. The letters C and D are reserved to denote clauses.

An *interpretation* $\mathcal{I}$ for a given signature Σ is a set of ground Σ-literals such that either $A \in \mathcal{I}$ or $\neg A \in \mathcal{I}$ for every ground Σ-atom A. The signature Σ is always given implicitly by the input clause set under consideration, and the prefix "$\Sigma-$" usually is not written. All the results below hold wrt. such Herbrand interpretations; it only has to be assumed that Σ contains at least one constant symbol (if none is there, some constant a is added artificially).

A ground literal L and a ground clause C is evaluated wrt. an interpretation $\mathcal{I}$ as expected, i.e. $\mathcal{I}(L) = true$ iff $L \in \mathcal{I}$ and $\mathcal{I}(C) = true$ iff $\mathcal{I}(L) = true$ for some $L \in C$. Furthermore, as expected, for a non-ground clause C define $\mathcal{I}(C) = true$ iff $\mathcal{I}(C') = true$ for every ground instance C' of C. As usual, $\mathcal{I} \models X$ means $\mathcal{I}(X) = true$ where X is a literal, clause or clause set (interpreted conjunctively).

A *unifier* for a set Q of terms (or literals) is a substitution δ such that $Q\delta$ is a singleton. The notion of *most general unifier (MGU)* is used in the usual sense [CL73, e.g.], and a respective unification algorithm *unify* is assumed as given. The notation $\sigma = unify(Q)$ means that an MGU σ of Q exists and is computed by *unify* applied to Q.

Quite frequently, a *simultaneous unifier for a set* $\{Q_1, \ldots, Q_n\}$ of unification problems is to be computed, which is a substitution δ that is a unifier for every $Q_1, \ldots, Q_n$. The notion of a *most general* unifier can be defined in the standard way in the simultaneous case as well. Further, a simultaneous most general unifier (simply called MGU as well) can be computed by iterative application of *unify* to $Q_1, \ldots, Q_n$. See [Ede85] for a thorough treatment. Thus, we may suppose as given a simultaneous unification algorithm *s-unify* and write $\sigma = s\text{-}unify(\{Q_1, \ldots, Q_n\})$ in analogy to $\sigma = unify(Q)$ above.

For literals K and L define $K \gtrsim L$, K *is more general than* L, iff there is a substitution σ_K such that $K\sigma_K = L$; K and L are *variants*, written as $K \sim L$, iff $K \gtrsim L$ and $L \gtrsim K$; K is *strictly more general than* L, $K > L$, iff $K \gtrsim L$ and not $K \sim L$. L is also said to be a *strict*, or *proper* instance of K then. If neither $K \gtrsim L$ nor $L \gtrsim K$ then K and L are *incomparable*. Finally, define $L \in^{\sim} N$ iff $L \sim K$ for some $K \in N$, where N is a set of literals.

3   Basic Concepts Related to Literal Sets

As mentioned, interpretations shall be represented by literal sets. This section contains the respective definitions. In the sequel N always denotes a possibly infinite literal set.

Definition 1 (Most Specific Generalization). *A literal* K *is called a* most specific generalization (MSG) *of a literal* L *wrt.* N *iff* $K \gtrsim L$ *and there is no* $K' \in N$ *such that* $K > K' \gtrsim L$.

Notice that nothing is said whether $K, L \in N$ or not.

Example 1. Consider[4] $N_1 = \{P(a, y, u), P(x, b, u)\}$. Then both $P(a, y, u)$ and $P(x, b, u)$ are MSGs of $P(a, b, c)$ wrt. N. This shows that MSGs need not be unique. The literal $P(x, y)$ is not a MSG of $P(y, f(x))$ wrt. $\{P(x, f(y))\}$, because $P(x, y) > P(x, f(y)) \gtrsim P(y, f(x))$.

An MSG $K \in N$ of L wrt. N is a "potential reason" for L to be *true* in the interpretation associated to N, because $K \gtrsim L$ (as said in the introduction).

---

[4] Here and below, the letters $P, Q, R, \ldots$ denote predicate symbols, $a, b, c, \ldots$ denote constants, $f, g, h, \ldots$ denote non-constant function symbols, and $x, y, z, \ldots$ denote variables.

For efficiency reasons in FDPLL it is desirable to have as few such "reasons" as possible. Therefore, *most specific* generalizations are used.

Definition 2 (Productivity). *A literal K produces L wrt. N iff K is a MSG of L wrt. N and there is no $K' \in N$ such that $K > \overline{K'} \gtrsim L$. For a clause C, K produces C wrt. N iff K produces some literal $L \in \tilde{C}$ wrt. N. The set N produces L (resp. C) iff some $K \in N$ produces L (resp. C) wrt. N.*

Referring again to the introduction and above, this definition realizes the possibility to prevent an MSG K of L wrt. N to assign *true* to L, if there is a complementary literal in between (wrt. $\gtrsim$) K and L, as stated. An equivalent, more compact definition of "K produces L wrt. N" is that $K \gtrsim L$ and there is no literal $K' \in N$ such that $|K| > |K'| \gtrsim |L|$.

Example 2. Let $N_2 = \{P(a,y,u),\ \neg P(x,b,u),\ P(a,b,u)\}$. Then, $P(a,b,u)$ produces the literal $P(a,b,f(u))$ wrt. N_2. However, $P(a,y,u)$ does not produce $P(a,b,f(u))$ wrt. N_2 because $P(a,y,u)$ is not an MSG of $P(a,b,f(u))$ wrt. N_2. (since $P(a,b,u) < P(a,y,u)$ is an MSG of $P(a,b,f(u))$ wrt. N_2). The literal $\neg P(x,b,u)$ produces $\neg P(b,b,f(u))$ wrt. N_2 but does not produce $\neg P(a,b,f(u))$ (since $\neg P(x,b,u) > \overline{K'} \gtrsim \neg P(a,b,f(u))$, where $K' = P(a,b,u) \in N_2$).

Definition 3 (Ground Expansion). *Define the* ground expansion *of N as $[\![N]\!] = \{L \mid L$ is a ground literal and N produces $L\}$*

The plan is to identify for a literal set N constructed by FDPLL its ground expansion $[\![N]\!]$ with an interpretation $\mathcal{I}$. Recall from Section 2 that an interpretation is a set of ground literals such that either $A \in \mathcal{I}$ or $\neg A \in \mathcal{I}$ for every ground atom A. However, there is in general no reason for $[\![N]\!]$ to be an interpretation. For instance:

Example 3. The set $N_3 = \{P(a,y,u),\ \neg P(x,b,u)\}$ produces both $P(a,b,c)$ and $\neg P(a,b,c)$. Hence, $[\![N_3]\!]$ is not an interpretation.

Note 1 (Completeness of N). Beyond this *inconsistency* problem, a *completeness* problem arises as well: for instance, $N = \{\}$ does not produce a single literal. The completeness problem can be solved by adding to N an expression $\neg x$, where x is a variable; the "literal" $\neg x \in N$ then acts as a default case to assign *false* to positive literals. Thus, for instance, $[\![\{\neg x, P(a)\}]\!]$ produces every negative literal except $\neg P(a)$ and thus assigns *false* to every positive literal, except $P(a)$[5].

The following definition formalizes these concepts.

Definition 4 (Contradictory, Consistent, Complete). *A literal set N is called* contradictory *iff there are literals $L, K \in N$ such that $L \sim \overline{K}$. The term "non-contradictory" means "not contradictory". N is called* consistent wrt. *a literal L iff N does not produce both L and $\overline{L}$; N is called* consistent *iff N is consistent wrt. every literal L. The term* inconsistent *means "not consistent". N is called* complete *iff for every literal L, N produces L or $\overline{L}$. The term* incomplete *means "not complete".*

---

[5] Of course, instead of "$\neg x$", "x" could be taken as well, which would emphasize the use of negative clauses ("goals") in the calculus.

Example 4. For instance, N_3 from Example 3 is non-contradictory and inconsistent wrt. $P(a, b, u)$ (and hence wrt. $\neg P(a, b, u)$ as well). Adding either $P(a, b, u)$ or $\neg P(a, b, u)$ renders the set consistent (and hence non-contradictory, as is easily seen), and adding both renders the set contradictory and inconsistent wrt. $P(a, y, u)$ again. Each of these sets is incomplete, and adding $\neg x$ achieves completeness.

With these definitions, the intuition so far can be made precise:

Proposition 1 (Interpretation). *If* $\neg x \in N$ *then* N *is complete. If* N *is consistent and complete, then* $[\![N]\!]$ *is an interpretation.*

It can be noted that "productivity" and "modelship" are *not* related on the non-ground level. For instance, take $N = \{\neg x,\ \neg P(a),\ P(b),\ Q(a),\ \neg Q(b)\}$ and $C = P(x) \vee Q(x)$. Then $[\![N]\!] \models C$ but there is no $L \in C$ such that L produces C wrt. N. Conversely, $N = \{\neg x,\ P(a)\}$ produces $\neg P(x)$ but $[\![N]\!] \not\models \neg P(x)$.

As mentioned in the introduction, model candidates shall be given up when being "elementary contradictory" to an (instance of) an input clause. The following definition makes this precise (preliminarily):

Definition 5 (Closed, Open). *A literal set* N *is* closed *by a clause* C *and substitution* δ *iff* $\overline{L} \in^{\sim} N$ *for every* $L \in C\delta$; N *is* closed *by* C *iff* N *is closed by* C *and some substitution* δ. *Finally,* N *is* closed *by a clause set* S *iff* N *is closed by some clause* $C \in S$. *The term "open" means "not closed", and "N is open wrt.* S*" means "N is not closed by S".*

For instance, $N = \{P(x, y),\ Q(a)\}$ is closed by $C = \neg P(x, y) \vee \neg P(y, x) \vee \neg Q(z)$ and $\delta = \{z/a\}$, because for every literal in $C\delta$ there is a complementary variant in N.

Again, as mentioned in the introduction, the substitutions used in FDPLL shall be computed as most general substitutions. This holds in particular for the substitutions δ that allow to close a literal set. This motivates the following definition.

Definition 6 (Branch Unifier). *Let* $C = L_1 \vee \cdots \vee L_n$ *be a clause. A substitution* σ *is called a* branch unifier *of* C *against* N *iff there are pairwise variable disjoint literals* $K_1, \ldots, K_n \in^{\sim} N$, *each variable disjoint from* C, *and such that the following holds:*

(i) $\sigma = s\text{-}unify(\{\{\overline{L_1}, K_1\}, \ldots, \{\overline{L_n}, K_n\}\})$, *and*
(ii) K_i *produces* $\overline{L_i}\sigma$ *wrt.* N, *for* $1 \leq i \leq n$.

If N *is closed by* C *and* σ, *then* σ *is called a* closing branch unifier, *otherwise* σ *is called a* falsifying branch unifier.

Item (i) realizes the mentioned computation at the most general level. Item (ii) guarantees that a clause instance $C\sigma$ is identified, such that (at least one ground instance of) $C\sigma$ is *false* in $[\![N]\!]$ (cf. Lemma 1 below). This is a useful restriction, as a clause instance $C\sigma$ every (ground instance) of which is *true* in $[\![N]\!]$ needs not be considered at all.

Example 5. Let $N_4 = \{P(a, y, u), \neg P(x, b, u), P(a, b, u)\}$ and $C = \neg P(a, c, z) \vee P(z, v, z)$. Take $K_1 = P(a, y, u) \in^{\sim} N_4$ and $K_2 = \neg P(x, b, u') \in^{\sim} N_4$. Observe that $\sigma = \{u/z, u'/z, v/b, x/z, y/c\}$ is a branch unifier of C against N_4, since σ is a simultaneous MGU for $\{\{P(a, c, z), P(a, y, u)\}, \{\neg P(z, v, z), \neg P(x, b, u')\}\}$, and $K_1 = P(a, y, u)$ produces $P(a, c, z)\sigma = P(a, c, z)$ wrt. N_4, and furthermore $K_2 = \neg P(x, b, u')$ produces $\neg P(z, v, z)\sigma = \neg P(z, b, z)$. Further observe that σ is a falsifying branch unifier (i.e. N_4 is not closed by C and σ).

As a negative example, there is no branch unifier of $P(a, b, c)$ against N_4, because although item (i) in Def. 6 is satisfied by taking $K_1 = \neg P(x, b, u)$ and $\sigma = \{x/a, u/c\}$, item (ii) is violated, because $\neg P(x, b, u)$ does not produce $\neg P(a, b, c)$ wrt. N_4.

Branch unifiers are a purely syntactical concept, and existence of branch unifiers for *finite* literal sets N obviously is decidable. The following lemma then, read in the contrapositive direction, guarantees that N is a model for the given clause if no branch unifier exists (provided that N is consistent).

Lemma 1. *Let N be consistent and complete, and C be a clause. If $[\![N]\!] \not\models C$ then there is a branch unifier σ of C against N.*

In order to take advantage of the previous lemma, consistency has to be achieved (there is a respective inference rule in FDPLL). Fortunately, consistency is also a syntactical property, and is decidable in the finite case as well. For the purposes here, the following lemma is sufficient (it can be strengthened):

Lemma 2. *Let N be a non-contradictory literal set. If N is inconsistent then there is a pair of variable disjoint, non-comparable literals $K, L \in^{\sim} N$ with opposite sign (i.e. neither $K \gtrsim \overline{L}$ nor $\overline{L} \gtrsim K$) such that (i) K and $\overline{L}$ are unifiable, i.e. $\sigma = \text{unify}(\{K, \overline{L}\})$ exists, and (ii) neither $K\sigma \in^{\sim} N$ nor $L\sigma \in^{\sim} N$.*

4   The FDPLL Calculus

In this section the inference rules based on branch unifiers are introduced. Recall from the previous section that branch unifiers are either "falsifying" or "closing". However, to close literal sets earlier, hence find shorter refutations, the FDPLL calculus uses a different notion of "closure":

Definition 7 (^{a}-Closed, ^{a}-Open). *Let a be any constant from the signature under consideration (or a "new" constant if none is supplied). By N^a denote the literal set obtained from N by replacing in every literal every occurrence of every variable by a. More formally[6], $L^a = L\gamma$, where $\gamma = \{x/a \mid x \in \text{var}(L)\}$, and $N^a = \{L^a \mid L \in N\}$.*

The literal set N is ^{a}-closed by C and δ iff $\overline{L} \in N^a$, for every $L \in C\delta$. The derived forms, as well as the term "^{a}-open" are defined analogously to "closed" (Def. 5).

So, determining whether N is ^{a}-closed by C is a question of simultaneously matching all literals of C to complementary literals in N^a.

---

[6] The function *var* returns the set of variables occurring in its argument.

Note 2 ("^{a}-Closed" Closes More Branches). It is not too difficult to see that whenever N is closed by C then N is ^{a}-closed by C as well. The converse, however, is not true: for instance, $\{\neg P(x, y)\}$ is ^{a}-closed by $P(x, y) \vee P(x, x)$, but not closed by $P(x, y) \vee P(x, x)$ (because $\neg P(x, x)$ cannot be instantiated to a variant of $\neg P(x, y)$).

Thus, by contraposition, if N is ^{a}-open wrt. C, N is open wrt. C as well.

In the sequel, S always denotes a finite clause set.

Definition 8 (Branch, Branch Set, Selection Functions). *A branch p is a possibly empty, finite set of literals. A branch set $\mathcal{P}$ consists of a finite set of branches. The branch set $\mathcal{P}$ is closed (by C, by S) iff every $p \in \mathcal{P}$ is closed (by C, by S). The term "open" means "not closed". Define $\mathcal{P}$ as ^{a}-closed (^{a}-open) in the expected way by using the ^{a}-versions instead.*

Assume as given a branch selection function sel which maps any ^{a}-open branch set $\mathcal{P}$ wrt. S to one of its ^{a}-open branches. This branch is referred to as the selected branch in $\mathcal{P}$. On ^{a}-closed branch sets, sel may be undefined.

Finally, assume as given a literal selection function litsel(C, p) that maps a clause C and a branch p that is ^{a}-open wrt. C to some literal $L \in C$ such that neither $L \in^\sim p$ nor $\overline{L} \in^\sim p$, provided such a literal exists, and may be undefined otherwise.

Notice that the empty branch set is closed (and hence ^{a}-closed) wrt. every S, and that the branch set $\{\{\}\}$ is ^{a}-open (and hence open) wrt. S, unless S contains the empty clause. The same holds for $\{\{\neg x\}\}$, as no clause contains a "literal" x.

The purpose of the two selection functions will become clear soon. Next, the two inference rules of FDPLL are defined.

Definition 9 (Split Inference Rule). *The following inference rule Split transforms a branch p, a clause C such that p is ^{a}-open wrt. C, and a substitution σ into two new branches:*

$$\text{Split}(C, \sigma) \ \frac{p}{p \cup \{L\} \qquad p \cup \{\overline{L}\}} \quad if \ \begin{cases} (i) & \sigma \text{ is a branch-unifier of } C \text{ against } p, \text{ and} \\ (ii) & \text{for some } L \in C\sigma, \text{ neither } L \in^\sim p \text{ nor} \\ & \overline{L} \in^\sim p, \text{ and} \\ (iii) & L = litsel(C\sigma, p) \end{cases}$$

If for given p, C and σ the conditions (i) and (ii) hold, it is said that the Split inference rule is applicable to p, C and σ, and the result as the set $\{p \cup \{L\}, p \cup \{\overline{L}\}\}$ is denoted by $\text{Split}(p, C, \sigma)$. The literal L in (iii) is called the literal split on.

Note 3 (Purpose of Split). Assume that whenever Split is applied to a branch p, then p is consistent (that this can be achieved is argued for below). By Proposition 1 then, $[\![p]\!]$ is an interpretation. Now, the intuition behind Split is to find a clause C and a branch unifier σ of C against p, such that at least one ground instance of $C\sigma$ is *false* in $[\![p]\!]$. If no such σ exists, by Lemma 1 we can be sure that with $[\![p]\!]$ a model for the clause set has been found. Otherwise, Split is applicable to p, C and some branch unifier σ by the following line of reasoning: since p

is ^{a}-open, hence open (cf. Note 2), σ must be a *falsifying* branch unifier. But then, condition (ii) must be satisfied. For, if (ii) would not be satisfied, for every literal $L \in C\sigma$ it would hold (a) $L \in^\sim p$ or (b) $\overline{L} \in^\sim p$. It is impossible that $L \in^\sim p$ for any $L \in C\sigma$ because then p would produce both L (because $L \in^\sim p$) and $\overline{L}$ (because σ is a branch unifier of C against p, and so p produces $\overline{L}$), and thus p would be inconsistent. Hence case (b) applies for every $L \in C\sigma$, and so σ would be a *closing* branch unifier of C against p, but we know that σ must be a *falsifying* branch unifier. Hence, with this contradiction condition (ii) holds. As a consequence, the literal selection function *litsel* is defined on $C\sigma$ and returns some arbitrarily (i.e. don't-care nondeterministically) selected literal from $C\sigma$ which is used for splitting p into the two new cases as stated.

Since the use of branch unifiers is insisted upon, Split is applicable in a very restricted way only. For instance, referring back to Example 5, Split is *not* applicable to the branch N_4 and the clause $P(a, b, c)$. Non-applicability of Split realizes a search-space reduction.

As said at the beginning of Note 3, it has to be made sure that p is consistent before Split is applied. This is the purpose of the following Commit inference rule (as the branch N_3 in Example 4 shows, consistency does not hold automatically).

Definition 10 (Commit). *The following inference rule* Commit *transforms a branch p, a literal L from p and a substitution σ into two new branches:*

$$\text{Commit}(L, \sigma) \ \frac{p}{p \cup \{L\sigma\} \quad p \cup \{\overline{L\sigma}\}} \quad if \begin{cases} (i) \ \ L \in p, \ and \\ (ii) \ \ \sigma \ = \ \ unify(\{L, \overline{K}\}), \ for \ some \\ \qquad K \in^\sim p, \ variable \ disjoint \ from \ L, \\ \qquad and \\ (iii) \ neither \ L\sigma \in^\sim p \ nor \ \overline{L\sigma} \in^\sim p. \end{cases}$$

If for given p, L and σ the conditions (i) – (iii) hold, it is said that the Commit *inference rule is* applicable *to p, L and σ, and the result as the set $\{p \cup \{L\sigma\}, \{p \cup \{\overline{L\sigma}\}\}$ is denoted by* Commit(p, L, σ). *The literal $L\sigma$ is called the literal* split on.

Note 4 (Purpose of Commit*).* Lemma 2 states (almost directly) that Commit is applicable to p, for some L and σ, whenever p is inconsistent. Thus, by the contrapositive direction, by repeated application of Commit one arrives at a consistent branch eventually.

The converse of Lemma 2 is *not* true: $\{P(x, a, u), \neg P(b, y, a), \neg P(b, a, u)\}$ *is* consistent but Commit is applicable (consider the first two literals). This shows that Commit is possibly applied more often than necessary. As an improvement, condition (iii) in Commit can be replaced by "L produces $L\sigma$ wrt. p and K produces $\overline{L\sigma}$ wrt. p."

Definition 11 (Derivation). *A derivation $\mathcal{D}$ (from a clause set $\mathcal{S}$) is a (possibly infinite) sequence of branch sets $\mathcal{D} = (\mathcal{P}_0 = \{\{\neg x\}\}), \mathcal{P}_1, \ldots, \mathcal{P}_n, \ldots$, such that for $i \geq 0$,*

(i) $\mathcal{P}_{i+1} = (\mathcal{P}_i \setminus \{p_i\}) \cup \text{Split}(p_i, C, \sigma)$ for some clause $C \in \mathcal{S}$ and substitution σ, or

(ii) $\mathcal{P}_{i+1} = (\mathcal{P}_i \setminus \{p_i\}) \cup \text{Commit}(p_i, L, \sigma)$ for some literal L and substitution σ,

where in both cases P_i *is* ^{a}*-open wrt.* S *and* $p_i = sel(P_i)$ *is the selected (hence* ^{a}*-open) branch in* P_i. *A derivation is called a* refutation *(of* S*) iff some* P_i *is* ^{a}*-closed (by* S*). A derivation of* P_n *is a finite derivation that ends in* P_n.

Both Split and Commit are applied to ^{a}-open branches only. Thus, if some P_i is ^{a}-closed, then P_i contains no single ^{a}-open branch and the derivation stops as a refutation.

Example 6 (Derivation). The figure below shows in tree notation a sample derivation from the clause set S consisting of the clauses $C_1 = P(a, y)$ and $C_2 = P(x, b) \vee \neg P(z, y) \vee Q(x, y, z)$.

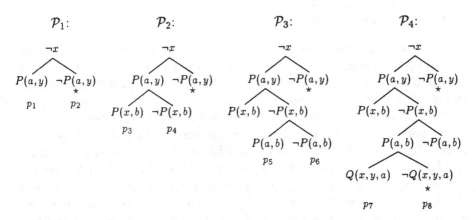

The branch set $P_0 = \{\{\neg x\}\}$ is not depicted; ^{a}-closed branches are marked with a "$\star$". P_1 is obtained from P_0 by a applying Split to $\{\{\neg x\}\}$ and C_1 (and the empty substitution); the branch p_2 is ^{a}-closed (even closed) due to C_1; P_2 is obtained from P_1 by applying Split to p_1 and C_2 (and some σ not made explicit). Neither Split nor Commit is applicable to p_3, so $[\![p_3]\!]$ is an interpretation (cf. Note 4) and $[\![p_3]\!] \models S$ (cf. Note 3). To continue the example suppose that the branch p_4 is selected in P_2. The Commit rule is applicable to p_4, which derives the branches p_5 and p_6. Applying Split to p_5 and C_2 yields p_7 and p_8 (as branch literals in the computation of the branch unifier use variants of $P(a, y)$ and $\neg P(x, b)$).The branch p_8 is ^{a}-closed by C_2, and to p_7, neither Commit nor Split is applicable (in particular, the instance $P(x, b) \vee \neg P(a, b) \vee Q(x, b, a)$ of C_2 which can be obtained by simultaneous unifying away the P-literals of C_2 against $P(a, b)$ and $\neg P(x', b)$ is produced by $Q(x, y, a)$). Thus, $[\![p_7]\!] \models S$. The derivation can continue with P_4 at the branch called p_6 in P_3, which is not shown here.

Note 5 (Regularity). For both inference rules, when applied to a branch p, the literal L split on is "new" to p in the sense that neither L nor $\overline{L}$ is contained in p, not even as a variant. In Split, a respective condition in *litsel*, and in Commit condition (iii) is responsible for this. In other words, a stronger form of "regularity" than the identity-based one used in rigid variable calculi holds; also, it is impossible to derive a contradictory branch (cf. Def 4). Thus, from the completeness of FDPLL (Theorem 2) it follows immediately that FDPLL is a decision procedure for Bernays-Schönfinkel logic (no function symbols except constants), a class that cannot be decided easily by resolution or tableau methods.

This section is concluded with optimizations concerning Split, in particular the literal selection function *litsel* (Def. 8). The basis is the following lemma:

Lemma 3 (Open Path Literal Selection). *Let p be an ^{a}-open branch wrt. S, $C \in S$ be a clause and σ be a branch unifier of C against p. Then, all of the following hold:*

(i) Split *is applicable to p, C and σ.*
(ii) *The set $\mathcal{L} = \{L \in C\sigma \mid \overline{L^a} \notin p^a\}$ is non-empty.*
(iii) $\overline{L} \notin^\sim p$, *for every $L \in \mathcal{L}$.*
(iv) *If p is consistent, then $L \notin^\sim p$, for every $L \in C\sigma$.*

It is clear from the definition of "derivation" that Split is applied to a branch p only if p is ^{a}-open. So, this precondition of the lemma is satisfied whenever Split is attempted. Now assume that C and σ exist as required in the lemma statement, and thus that items (i) – (iv) hold.

Item (i) summarizes what was sketched at the beginning of Note 3.

Suppose that Commit applications are preferred to Split applications (as is realized in the proof procedure in Section 7 below). Then, the branch p is consistent (cf. Note 4), and item (iv) shows that the condition (ii) in the Split inference rule may be equivalently replaced by "for some $L \in C\sigma$, $\overline{L} \notin^\sim p$" (since by item (iv) $L \notin^\sim p$ holds for all literals $L \in C\sigma$). Item (iv) is proven as follows: we are given that σ is a branch-unifier of C against p. This means in particular that each literal $\overline{L}$, where $L \in C\sigma$, is produced by some literal $K \in^\sim p$. Now, if $L \in^\sim p$ would hold, then p would produce L as well. By consistency, however, p cannot produce L.

Next, sensible literal selection by *litsel* is turned to. Concretely, *litsel*$(C\sigma, p)$ should return an element L from the (non-empty) set $\mathcal{L}$ defined in item (ii). Observe that splitting on this literal L is indeed possible, because by item (iii) the modified applicability condition of Split explained in the previous paragraph is satisfied for L. Now, item (ii) expresses that one need not select a literal from $C\sigma$ that is solved in the sense that it contributes to ^{a}-close p. For instance, if $p = \{\neg x, P(x, y)\}$, $C = \neg P(x, x) \vee \neg P(x, b)$ and $\sigma = \epsilon$, then $\mathcal{L} = \{\neg P(x, b)\}$. It is more sensible to select $\neg P(x, b)$ for splitting, because the thus upcoming branch $\{\neg x, P(x, y), P(x, b)\}$ is ^{a}-closed, whereas splitting with $\neg P(x, x)$ leaves the upcoming branch $\{\neg x, P(x, y), P(x, x)\}$ ^{a}-open; with respect to closing branches, $P(x, y)$ and $P(x, x)$ are the same. Selecting literals from $\mathcal{L}$ yields shorter refutations.

5   Universal Literals

Under certain circumstances a literal L occurring in a branch p can be treated, roughly, like a unit clause in resolution, i.e. L then stands for *all* its ground instances — without exception. In the terminology used here, such a literal L that is *universal in a branch p* then produces all instances of L, and any extension of p also containing a literal K with $L > \overline{K}$ can be considered as closed, without any explicit refutation. Thus, many inferences can be saved by closing branches earlier.

The difficulty is to determine criteria under which it is *sound* to consider L as universal in p; since branches can only close earlier, but never later, completeness is preserved trivially, and only soundness is an issue. Possible criteria to derive universal literals correspond to "unit-resulting resolution" and splitting based on variable disjoint subclauses. This is sketched next [7].

The starting point is a stronger condition (than ^{a}-closed) to close branches: assume that a set of universal literals in p, $Univ(p) \subseteq p$, has already been determined, and let C be a clause. Now, p is said to be *closed** by C iff C can be partitioned as $C = C_1 \cup C_2$ such that (i) there is a simultaneous most general unifier σ of every $L \in C_1$ with some literal K s.t. $\overline{K} \in^\sim Univ(p)$ (use a new variant), and (ii) there is a substitution δ such that $\overline{L\delta} \in (p \setminus Univ(p))^a$ for every $L \in C_2\sigma$ (in particular, if $C_2\sigma = \{\}$ then $\delta = \epsilon$ exists trivially). In other words, condition (ii) is the same as saying that $p \setminus Univ(p)$ has to be ^{a}-closed by $C_2\sigma$.

For example, the branch p_4 in Example 6 is closed* (but not ^{a}-closed) by $C = \neg P(y,c) \vee P(x,b)$, using $C_1 = \neg P(y,c)$ and $C_2 = P(x,b)$.

Now, if (i) holds but (ii) does not hold for the considered clause C and branch p, then, as said, $p \setminus Univ(p)$ is not ^{a}-closed by $C_2\sigma$, and there is no *closing* branch unifier of $C_2\sigma$ against $p \setminus Univ(p)$. But it is still possible that there is a *falsifying* branch unifier σ' of $C_2\sigma$ against $p \setminus Univ(p)$. In order to describe how universal literals are derived, assume that this is the case.

It follows that there is at least one literal $L \in C_2\sigma\sigma'$ such that neither $L \in^\sim p$ nor $\overline{L} \in p$ (for this, it has to be assumed that C_1 was chosen maximal, which is a safe assumption). In other words, Split is applicable to p, C_2 (and also to C) and $\sigma\sigma'$ with literal L split on. Now, if L is variable disjoint from the rest, i.e. if $var(L) \cap var(C_2\sigma\sigma' \setminus \{L\}) = \{\}$ holds, then L is determined to be universal in $p \cup \{L\}$, otherwise it is not universal in $p \cup \{L\}$. In the other branch $p \cup \{\overline{L}\}$ coming up in the Split application, the universal literals are just those of p.

The improvements possible by universal literals can now be summarized as follows: expressed abstractly, since less "open*" than ^{a}-open branches exist, their stronger properties can be taken advantage of. For instance, Commit is never applicable when both K and L are universal in p (cf. Def. 10); also, if Split is applicable to p, C and $\sigma\sigma'$, always a literal $L \in C\sigma\sigma'$ exists such that neither L nor $\overline{L}$ is subsumed by a universal literal in p (Formally, Lemma 3 can be strictly strengthened). In other words, instances of universal literals (or their complements) need never be added by a Split. This mirrors "subsumption by a unit clause" in resolution.

Notice as special cases that a ground literal L is trivially universal in every branch containing it, and, more importantly, if $C_2\sigma\sigma' = \{L, \ldots, L\}$ is a singleton when read as a set, then the "other" branch $p \cup \{\overline{L}\}$ is closed* (such as the branch p_2 in Example 6). Hence, no branching is introduced then. It is not too difficult to

---

[7] The long version of the paper contains a full account.

see that as a consequence, when applied to a Horn clause set, FDPLL specializes to positive Hyper-resolution.

It is worth noting that the described technique of reasoning with universal literals can be built-in by only *minor* modifications of the calculus, by attaching boolean labels to the literals in branches to indicate their "universal" status.

6 Soundness and Completeness

Theorem 1 (Soundness of FDPLL). *Let S be a clause set and $\mathcal{D}$ be a refutation of S. Then S is unsatisfiable.*

Proof sketch: consider the last element $\mathcal{P}$ in $\mathcal{D}$, every branch of which is ^{a}-closed. Its ground instantiation $\mathcal{P}^a = \{p^a \mid p \in \mathcal{P}\}$ can be seen as a usual semantic tree $\mathcal{T}$ made up of splits with complementary ground literals (cf. [CL73]). Furthermore, the (finite) set of all those (ground) clauses $C\delta$ that are used for closing the branches in $\mathcal{P}^a$ show that every leaf in $\mathcal{T}$ is a failure node. Now apply the soundness result for usual semantic trees.

Next we turn to completeness. The FDPLL calculus proceeds by further modifying one single branch set, and never "backtracks" to a previously derived branch set. This notion of derivation indicates that to obtain a completeness result, fairness is to be defined as an exhaustive process (up to redundancy) of inference rule applications.

Before going into the details, a general note on this topic can be made: the notion of "derivation" in Def. 11 can be adapted to virtually every confluent rigid variable method. For these methods, the real challenge is to define fairness as just mentioned in such a way that it can be turned into an effective proof procedure. Only few attempts in this direction have been made [BEF99,Bec98]. Coming from the FDPLL side, it seems possible to bring the technique here to e.g. clausal tableaux calculi (by branching on clauses instead of complementary literals).

Definition 12 (Path). *Let $\mathcal{D}$ be a derivation that is not a refutation, written as in Definition 11. Let $\mathcal{P}^\infty = \bigcup_{j \geq 0} \mathcal{P}_j$, the set of all branches ever constructed in $\mathcal{D}$. A path (of $\mathcal{D}$) is a possibly infinite sequence $I = (p_0 = \{\neg x\}) \subset p_1 \subset \cdots \subset p_m \subset \cdots$ of branches in $\mathcal{P}^\infty$ such that for every $i \geq 0$*

(i) $p_i = sel(\mathcal{P}_{s_i})$ is the selected (^{a}-open) branch in some branch set $\mathcal{P}_{s_i}$ in $\mathcal{D}$, and

(ii) $p_{i+1} = p_i \cup \{L_i\}$, for some literal L_i, and

(iii) if in $\mathcal{D}$ the Commit or the Split inference rule is applied to p_i, then I contains a successor element p_{i+1}.

Finally, define the chain limit $\cup I = \bigcup_{i \geq 0} p_i$.

The limit $\cup I$ thus is an "infinitely long" branch, obtained by tracing some branch that is extended infinitely often and remains ^{a}-open. As an important property, $\cup I$ is ^{a}-open, because if $\cup I$ were ^{a}-closed, there is a clause C ^{a}-closing $\cup I$. Since clauses are finite and $\cup I$ is the limit of a chain, some finite $p_j \subset \cup I$ would be ^{a}-closed by C, contradicting item (i) in the definition. It is only noted here without proof that for every derivation $\mathcal{D}$ that is not a refutation a path of $\mathcal{D}$ exists.

Definition 13 (Finishedness, Fairness). *Let $\mathcal{D}$ be a derivation that is not a refutation and I be a path of $\mathcal{D}$, written as in Definition 12. The path I is finished iff for every $i \geq 0$ the following holds:*

(i) *if the* Split *inference rule is applicable to p_i and some clause $C \in \mathcal{S}$ and substitution σ, then $C\sigma$ is produced by p_j (Def. 2) for some $j \geq i$.*

(ii) *if the* Commit *inference rule is applicable to p_i and some literal $L \in p_i$ and substitution σ, then p_j is consistent wrt. $L\sigma$, for some $j \geq i$.*

$\mathcal{D}$ *is* fair *iff (i) $\mathcal{D}$ is a refutation or (ii) some path of $\mathcal{D}$ is finished.*

The purpose of condition (ii) in Definition 13 is to achieve that $\bigcup I$ is consistent wrt. any literal $L\sigma$ identified by a possible Commit application at time point i; similarly, the purpose of condition (i) in Definition 13 is to achieve that $\bigcup I$ produces the clause instance $C\sigma$ identified via a possible Split application at time point i.

That these purposes can be satisfied is a consequence of having a $\bigcup I$ as a *chain* limit (roughly, compactness wrt. the required properties holds then) and that Split and Commit applications achieve (i) and (ii), respectively, whenever violated (cf. Notes 4 and 3 again). However, *actually* carrying out the inferences is only the last resort: it suffices that their *effect* is achieved, namely (in case of Split, e.g.) that the clause $C\sigma$ is produced eventually. Indeed, a Split application might cause a former possible Split application to be impossible. For instance, when a clause $C_3 = \neg P(a,b) \lor R(a)$ is added to the clause set in Example 6, then Split is applicable to p_1 and C_3 (with $\sigma = \{x/R(a),\ y/b\}$), but Split is no longer applicable to C_3 and p_6 (due to $\neg P(a,b) \in p_6$).

These considerations shall serve as a proof sketch for the first main result:

Theorem 2 (Completeness of FDPLL). *Let $\mathcal{S}$ be a clause set and $\mathcal{D}$ be a fair derivation from $\mathcal{S}$. If $\mathcal{D}$ is not a refutation then $\mathcal{S}$ is satisfiable. More specifically, for every finished path I of $\mathcal{D}$, $[\![\bigcup I]\!]$ is an interpretation and $[\![\bigcup I]\!] \models \mathcal{S}$.*

Notice that in the contrapositive direction, the theorem is just a refutational completeness result.

A final remark: the calculus is asymmetric wrt. the rôle of branches. Consider e.g. a branch $p = \{\neg x,\ P(x),\ \neg P(a)\}$. To determine closure, p^a is used, and to extract a model $[\![p]\!]$ is used; inconsistency of p^a (as in the example) is not relevant for model extraction – $P(a)$ is simply *false* in $[\![p]\!]$[8]. For Theorem 2 to hold there is no need to go beyond the Herbrand interpretations as stated in Section 2.

---

[8] The calculus can be slightly improved by considering p as closed if p^a is contradictory (as in the example), although p is ^{a}-open.

7 Proof Procedure

So far, fair derivations are purely abstract mathematical objects, and it still has to be demonstrated that an effective fair strategy exists. In essence, to guarantee fairness, a *maximal term depth bound* is used, which all literals to be split on in Split applications have to obey; starting with a small natural number, the value of this bound is increased only after having exhausted Split within the current value. This inner loop exhaustion always terminates, essentially because variants of literals already present in a branch are never added. Fortunately, Commit need not be subject to such a term depth bound check — it *finitely* exhausts on any (finite) literal set.

These are the essential ingredients of the following concrete proof procedure.

```
 1  funct FDPLL(S) ≡
 3    var I;                                    % for the model representation
 5    funct Satisfiable(p, bound) ≡
 6      % p: the current branch. bound: non-negative integer, the maximal term
 7      % depth admissible in literals for splitting
 8      if ᵃ-Closed(p, S)
 9      then return false
10      else % Try a Commit. First, collect all candidate literals in L:
11            var L := {Lσ | L ∈ p, ∃K ∈ p : σ = unify({L, new(K)}) ≠
12                           undefined ∧ Lσ ∉~ p ∧ L̄σ ∉~ p};
13            if L ≠ {}
14            then % Commit is applicable
15                  var L_c := η L ∈ L : true;        % Select any candidate
16                  if Satisfiable(p ∪ {L_c}, bound)  % Left branch extension
17                  then return true
18                  else return Satisfiable(p ∪ {L̄_c}, bound) % Right branch
19                  fi
20            else % Commit not applicable - try Split. Collect all candidates:
21                  var L := {L | ∃C ∈ S, σ ∈ BranchUnify(C, p) :
22                             L = litsel(Cσ, p)}
23                  if L = {}
24                  then I := p;  % Split is not applicable – got a model in p.
25                       return true
26                  else % L ≠ {}, so Split is applicable
27                       var L_s := η L' ∈ L : ‖L'‖ ≤ bound;
28                           % Select any candidate within depth bound.
29                           % However, it might not exist:
30                       if L_s ≠ undefined
31                       then    % Candidate within depth bound exists
32                             if Satisfiable(p ∪ {L_s}, bound)
33                             then return true
34                             else return Satisfiable(p ∪ {L̄_s}, bound)
35                             fi
36                       else % Hit depth bound - try with higher one:
37                             return Satisfiable(p, bound + 1)
38                       fi fi fi fi.
40    % Body of FDPLL:
```

41   **if** *Satisfiable*$(\{\neg x\}, 0)$
42   **then** <u>return</u> $\mathcal{I}$
43   <u>else</u> <u>return</u> false
44   **fi.**

Some functions remain unspecified: a call of ^{a}-*Closed(p,S)* is supposed to return **true** iff p is ^{a}-closed by $\mathcal{S}$ (cf. Def 5); *new(L)* is supposed to return a "fresh" variant of L, containing no variables used so far. A call of *BranchUnify(C, p)* is supposed to return a possibly empty, finite and complete set of branch unifiers of C against p^9. Finally, $\|L\|$ denotes the depth of $|L|$ as a tree. All these functions can be effectively implemented.

Some comments on the structure of the procedure: *FDPLL* is a wrapper around
Satisfiable. The p parameter of *Satisfiable* is the currently selected branch in an implicitly constructed derivation. The branch selection realized in *Satisfiable* is implicitly a left-to-right depth-first strategy.

The counterpart of ^{a}-closed branches in the definition of "derivation" is a return value of **false** in *Satisfiable*; thus, closed branches are not kept in memory, and so *Satisfiable* realizes a space efficient one-branch-at-a-time approach.

If every incarnation of *Satisfiable* returns *false*, then *FDPLL* returns *false* as well, indicating unsatisfiability of $\mathcal{S}$. *FDPLL* returns a model $\mathcal{I}$ only if some incarnation of *Satisfiable* returns **true** on line 25, because a return value of **true** in *Satisfiable* is immediately propagated to its caller. This happens only if neither Commit nor Split are applicable to p. That $\mathcal{I} \models \mathcal{S}$ holds then follows directly from Notes 3 and 4.

Now some more specific comments on *Satisfiable*: the set comprehension formula on line 11 is just the applicability condition of Commit. An η-expression $\eta L \in \mathcal{L} : \varphi(L)$ (as on lines 15 and 27) returns any $L \in \mathcal{L}$ such that $\varphi(L)$ holds, if such an L exists, and returns <u>undefined</u> else. The set $\mathcal{L}$ on line 21 is assigned a finite set of literals such that whenever Split is applicable to some $C \in \mathcal{S}$ and σ then $litsel(C\sigma, p) \in^\sim \mathcal{L}$. When reaching line 21, p is known to be ^{a}-open (because Line 9 was not reached) and consistent (Note 3 account for this fact). Hence, all of Lemma 3 is applicable, and the improvements for *litsel* discussed there can be taken advantage of.

The *bound* parameter of *Satisfiable* realizes a depth bound, which all literals to be split on in Split applications have to obey. The fairness of the procedure is guaranteed by only increasing *bound* (on line 37) after exhaustion of Split on the currently given value of *bound*. However, Commit need not be subject to such a depth bound check — it *finitely* exhausts on any finite literal set.

In order to save space and concentrate on the most essential issues to be contributed here, the *FDPLL* procedure just described does not use universal literals (cf. Section 5). The long version of the paper contains a procedure with universal literals in full detail, which also realizes the restrictions of Commit and Split mentioned at the end of Section 5. A further built-in improvement is to remove a split rule application from a derivation if in the left derived branch

---

^{9} I.e. it contains modulo renaming every branch unifier of C against p.

$p \cup \{L\}$ the literal L is not needed in the refutation of $p \cup \{L\}$. Hence the right branch $p \cup \{\overline{L}\}$ need not be considered. This well-known improvement[10] realizes (but is more powerful than) the *purity* rule of propositional DPLL. The implementation mentioned in Section 1 refers to the version with these improvements.

Also for space reasons, a correctness proof of *FDPLL* is not possible here. The comments above and in the program may serve as a rough sketch for the following main result. Notice in particular item (iv), stating explicitly the existence of a fair derivation, which was left open in Section 6.

Theorem 3 (Correctness of the FDPLL Procedure). *The FDPLL procedure (with or without universal literals) has the following properties, for any clause set S:*

(i) **Refutational soundness:** *If FDPLL(S) returns false then S is unsatisfiable.*

(ii) **Refutational completeness:** *If S is unsatisfiable then FDPLL(S) returns false.*

(iii) **Model soundness:** *If FDPLL(S) returns a literal set $\mathcal{I}$ then $[\![\mathcal{I}]\!]$ is an interpretation and $[\![\mathcal{I}]\!] \models S$.*

(iv) **Finishedness:** *If FDPLL(S) does not terminate, there is an infinite sequence of incarnations Satisfiable(p_i, b_i) (for $i \geq 0$) such that the sequence $I = (p_0 = \{\neg x\}) \subset p_1 \subset \cdots \subset p_m \subset \cdots$ is a finished path of some fair derivation $\mathcal{D}$ of S which is not a refutation.*

8   Conclusions

A directly lifted, confluent and strongly complete version of the DPLL procedure has been presented. As the theoretical concepts (in particular the model representation) are new, emphasis was put on these rather than on experimental studies.

Related Work. Already in [CL73] a lifted DPLL calculus can be found. It uses the device of "pseudosemantic trees", which, like FDPLL, realizes splits at the non-ground level. Nethertheless, the pseudosemantic tree method is very different: in sharp contrast to FDPLL, a variable is treated rigidly there, i.e. as a placeholder for a (one) not-yet-known term. As a consequence, like in all rigid variable methods, only a very weak regularity condition can be used (cf. Note 5 below). Furthermore, only a weak completeness result is known, which translates into a heavily backtracking oriented proof procedure only.

When FDPLL reports "satisfiable", a model representation has been computed without further processing. This does neither hold for the mentioned method in [CL73], nor for the resolution based methods for model computation in [FL93,FL96]. Typically, the latter attempt to compute a model by enumerating all true *ground* literals, thereby interleaving this enumeration with calls to the resolution procedure again in order to determine the "next" ground literal.

---

[10] Also known as "dependency directed backtracking", "level cut", "condensing" etc.

The probably most advanced first-order tableau system tailored for model computation is RAMCET [Pel99], which is a successor of Peltier's and his co-workers resolution calculus and previous tableaux calculi (see e.g. [CP95]). As a drawback, RAMCET needs additional inference rules for model computation. In particular, the "model explosion" rule seems problematic, as it branches out wrt. the whole signature of the formula set under consideration. FDPLL does not need such a rule. Furthermore, unlike for FDPLL, strong completeness for RAMCET is still unsolved (while proof confluence is trivial), in the sense that no effective fair strategy is known except a trivial one, which needs exponential space — a widespread problem of tableaux calculi.

In the literature several methods are described that are related to FDPLL in the sense that variables are treated in a similar way (cf. "description of FD-PLL" in Section 1). FDPLL was influenced and is intended as a successor of the hyper tableau calculus [Bau98] (which in turn is a successor of the calculus in [BFN96], a calculus in the tradition of Satchmo[MB88]). Among other things, FDPLL improves on this calculus by not needing to store instances of clauses as the derivation proceeds – only the "current interpretation" needs to be kept. Beyond this, FDPLL is conceptually different: like any tableaux calculus, hyper tableau branches on (sub-)formulas, whereas FDPLL branches in a binary way on complementary literals, i.e. uses "cut" as the single inference rule. The latter is more general and "builds in" standard improvements like factorization automatically.

What was said about hyper tableau applies equally to the disconnection method [Bil96]. Also, no model computation result was given for this calculus.

Also related are Plaisted's hyper-linking calculi: the semantic hyper-linking calculus (SHL) [LP92] proceeds by searching in a guided way for (not necessarily ground) instances of input clauses, which are tested for unsatisfiability by a propositional DPLL procedure. Much of what was said about hyper tableau above applies to this calculus as well. In particular, unlike SHL, FDPLL does not interleave two processes "clause instance generation" and "propositional DPLL". The former process occurs in FDPLL only "locally" within the splitting rule, and derived clause instances need not be kept. It seems worth to investigate combinations of SHL and FDPLL, e.g. by replacing propositional DPLL in SHL by FDPLL, or picking up the idea of *guided* instance generation in SHL to improve FDPLL. However, this is future work.

Quite different is the ordered semantic hyper linking (OSHL) calculus [PZ97]. OSHL has many interesting features, for instance "semantical guidance". In the intersection with FDPLL, it can be described as a calculus that applies unit-resulting resolution as long as possible, and then splits with a *ground* literal in order to begin the next round. The main motivation for FDPLL however was to get rid of such ground splits. Therefore, it seems realistic to possibly improve on OSHL by bringing in the non-ground splitting technique of FDPLL.

Future Work. As always, much remains to be done: a part of my FDPLL research plan is an *efficient* implementation. So far, only a slow and *prototypical* implementation in Prolog exists (available from my home page). Although it lacks such crucial features like term indexing, the performance seems promising, in particular for satisfiable or non-Horn problems without equality (there is no built-in equality treatment yet). In the respective subdivisions SAT and NNE of the CASC-16 system competition 1999, FDPLL scored rank 4 of 6 and rank 4 of 10, respectively. From the TPTP library [SSY94], FDPLL can also solve some difficult unsatisfiable problems quite quickly (e.g. ANA002-4, the intermediate value theorem, in 3 seconds). The overall success rate is about 40% (Otter: 52%) for a time limit of 10 minutes.

Other sources for future work are combinations of the techniques described here with hyper-linking calculi, equality treatment and improved termination behavior to name a few. On the theoretical level, the relationship between the model representation capabilities in FDPLL and the *atomic model representations* [GP98] used in the resolution and tableau world should be clarified.

Acknowledgments. I am grateful to the members of our group and to David Plaisted for discussions about FDPLL and comments on the paper. Three referees gave valuable comments and suggestions, in particular concerning improvements of the Commit rule and universal literals.

References

[Bau98]   Peter Baumgartner. Hyper Tableaux — The Next Generation. In Harry de Swaart, editor, *Tableaux-98*, LNAI 1397. Springer, 1998.

[Bec98]   Bernhard Beckert. *Integration und Uniformierung von Methoden des tableaubasierten Theorembeweisens*. Dissertation, University of Karlsruhe, 1998.

[BBOS98] Wolfgang Bibel, Stefan Brüning, Jens Otten, and Thorsten Schaub. Volume I, Chapter 5: Compressions and Extensions. In Wolfgang Bibel and Peter H. Schmitt, editors, *Automated Deduction. A Basis for Applications*, pp. 133–179. Kluwer Academic Publishers, 1998.

[BEF99]   Peter Baumgartner, Norbert Eisinger, and Ulrich Furbach. A confluent connection calculus. In Harald Ganzinger, editor, *CADE-16*, LNAI 1632, pp. 329–343, Trento, Italy, 1999. Springer.

[BFN96]   Peter Baumgartner, Ulrich Furbach, and Ilkka Niemelä. Hyper Tableaux. In *JELIA 96*, LNAI 1126. Springer, 1996.

[Bil96]   Jean-Paul Billon. The Disconnection Method. In P. Miglioli, U. Moscato, D. Mundici, and M. Ornaghi, editors, *Tableaux-96*, LNAI 1071, pp. 110–126. Springer, 1996.

[CA96]   J.M. Crawford and L.D. Auton. Experimental results on the crossover point in random 3sat. *Artificial Intelligence*, 81, 1996.

[CL73]   C. Chang and R. Lee. *Symbolic Logic and Mechanical Theorem Proving*. Academic Press, 1973.

[CP95]   Ricardo Caferra and Nicolas Peltier. Decision Procedures using Model Building techniques. In *Computer Science Logic (CSL '95)*, 1995.

[D63]   Martin Davis. Eliminating the irrelevant from mechanical proofs. In *Proceedings of Symposia in Applied Mathematics – Experimental Arithmetic, High Speed Computing and Mathematics*, volume XV, pp. 15–30. American Mathematical Society, 1963.

[DLL62]   M. Davis, G. Logeman, and D. Loveland. A machine program for theorem proving. *Communications of the ACM*, 5(7), 1962.

[DP60]   M. Davis and H. Putnam. A Computing Procedure for Quantification The-
           ory. *Journal of the ACM*, 7:201–215, 1960.
[Ede85]   Elmar Eder. Properties of Substitutions and Unifications. *Journal of Sym-
           bolic Computation*, 1(1), March 1985.
[FL93]   Christian Fermüller and Alexander Leitsch. Model building by resolution.
           In E. Börger, G. Jäger, H. Kleine-Büning, S. Martini, and M.M. Richter,
           editors, *Computer Science Logic – CSL '92*, LNCS 702, pp. 134–148. Springer,
           1993.
[FL96]   Christian Fermüller and Alexander Leitsch. Hyperresolution and automated
           model building. *Journal of Logic and Computation*, 6(2):173–230, 1996.
[GP98]   Georg Gottlob and Reinhard Pichler. Working with Arms: Complexity Re-
           sults on Atomic Representations of Herbrand Models. In *Proceedings of the
           14th Symposium on Logic in Computer Science*, IEEE, 1998.
[LP92]   S.-J. Lee and D. Plaisted. Eliminating Duplicates with the Hyper-Linking
           Strategy. *Journal of Automated Reasoning*, 9:25–42, 1992.
[MB88]   Rainer Manthey and François Bry. SATCHMO: a theorem prover imple-
           mented in Prolog. In Ewing Lusk and Ross Overbeek, editors, *CADE 9*,
           LNCS 310, pp. 415–434. Springer, 1988.
[Pel99]   N. Peltier. Pruning the search space and extracting more models in tableaux.
           Logic Journal of the IGPL, 7(2):217–251, 1999.
[PZ97]   David A. Plaisted and Yunshan Zhu. Ordered Semantic Hyper Linking. In
           Proceedings AAAI-97, 1997.
[SSY94]   G. Sutcliffe, C. Suttner, and T. Yemenis. The TPTP problem library. In
           Alan Bundy, editor, *CADE 12*, LNAI 814, pp. 192–206, Nancy, France, June
           1994. Springer.
[Zha97]   Hantao Zhang. SATO: An Efficient Propositional Theorem Prover. In
           W. McCune, editor, *CADE 14*, LNAI 1249, pp. 272–275, Springer.

Rigid E-Unification Revisited*

Ashish Tiwari[1], Leo Bachmair[1], and Harald Ruess[2]

[1] Department of Computer Science, SUNY at Stony Brook
Stony Brook, NY 11794, U.S.A
{leo,astiwari}@cs.sunysb.edu
[2] SRI International, Menlo Park, CA 94025, U.S.A
ruess@csl.sri.com

Abstract. This paper presents a sound and complete set of abstract transformation rules for rigid E-unification. Abstract congruence closure, syntactic unification and paramodulation are the three main components of the proposed method. The method obviates the need for using any complicated term orderings and easily incorporates suitable optimization rules. Characterization of substitutions as congruences allows for a comparatively simple proof of completeness using proof transformations. When specialized to syntactic unification, we obtain a set of abstract transition rules that describe a class of efficient syntactic unification algorithms.

1 Introduction

Rigid E-unification arises when tableaux-based theorem proving methods are extended to logic with equality. The general, simultaneous rigid E-unification problem is undecidable [7] and it is not known if a complete set of rigid E-unifiers in the sense of [10] gives a complete proof procedure for first-order logic with equality. Nevertheless complete tableau methods for first-order logic with equality can be designed based on incomplete, but terminating, procedures for rigid E-unification [8]. A simpler version of the problem is known to be decidable and also NP-complete, and several corresponding algorithms have been proposed in the literature (not all of them correct) [9, 10, 5, 8, 11, 6]. In the current paper, we consider this standard, non-simultaneous version of the problem.

Most of the known algorithms for finding a complete set of (standard) rigid unifiers employ techniques familiar from syntactic unification, completion and paramodulation. Practical algorithms also usually rely on congruence closure procedures in one form or another, though the connection between the various techniques has never been clarified. The different methods that figure prominently in known rigid unification procedures—unification, narrowing, superposition, and congruence closure—have all been described in a framework based on transformation rules. We use the recent work on congruence closure as a starting point [12, 4] and formulate a rigid E-unification method in terms of fairly abstract transformation rules.

---

* The research described in this paper was supported in part by the National Science Foundation under grant CCR-9902031.

D. McAllester (Ed.): CADE-17, LNAI 1831, pp. 220–234, 2000.
© Springer-Verlag Berlin Heidelberg 2000

This approach has several advantages. For one thing, we provide a concise and clear explication of the different components of rigid E-unification and the connections between them. A key technical problem has been the integration of congruence closure with unification techniques, the main difficulty being that congruence closure algorithms manipulate term structures over an *extended* signature, whereas unifiers need to be computed over the original signature. We solved this problem by rephrasing unification problems in terms of congruences and then applying proof theoretic methods, that had originally been developed in the context of completion and paramodulation. Some of the new and improved features of the resulting rigid E-unification method in fact depend on the appropriate use of extended signatures.

Almost all the known rigid E-unification algorithms require relatively complicated term orderings. In particular, most approaches go to great length to determine a suitable orientation of equations (between terms to be unified), such as $x \approx fy$, a decision that depends of course on the terms that are substituted (in a "rigid" way) for the variables x and y. But since the identification of a substitution is part of the whole unification problem, decisions about the ordering have to made during the unification process, either by orienting equations non-deterministically, as in [10], or by treating equations as bi-directional constrained rewrite rules (and using unsatisfiable constraints to eliminate wrong orientations) [5]. In contrast, the only orderings we need are simple ones in which the newly introduced constants are smaller than all other non-constant terms. The advantage of such simple orderings is twofold, in that not only the description of the rigid E-unification method itself, but also the corresponding completeness proofs, become simpler.[1] Certain optimizations can be easily incorporated in our method that reduce some of the non-determinism still inherent in the unification procedure. The treatment of substitutions as congruences defined by special kinds of rewrite systems (rather than as functions or morphisms) is a novel feature that allows us to characterize various kinds of unifiers in proof-theoretic terms via congruences.

As an interesting fallout of this work we obtain an abstract description of a class of efficient syntactic unification algorithms based on recursive descent. Other descriptions of these algorithms are typically based on data structures and manipulation of term dags. Since our approach is suitable for abstractly describing sharing, we obtain a pure rule based description.

One motivation for the work presented here has been the generalization of rigid E-unification modulo theories like associativity and commutativity, which we believe are of importance for theorem proving applications. Our approach,

---

[1] A key idea of congruence closure is to employ a concise and simplified term representation via variable abstraction, so that complicated term orderings are no longer necessary or even applicable. There usually is a trade-off between the simplicity of terms thus obtained and the loss of term structure [4]. In the case of rigid unification, we feel that simplicity outweighs the loss of some structure, as the non-determinism inherent in the procedure limits the effective exploitation of a more complicated term structure in any case.

especially because of the use of extensions of signatures and substantially weaker assumptions about term orderings, should more easily facilitate the development of such generalized unification procedures.

We also believe that our way of describing rigid E-unification will facilitate a simpler proof of the fact that the problem is in NP. Previous proofs of membership of this problem in NP "require quite a bit of machinery" [10]. The weaker ordering constraints, a better integration of congruence closure and a rule-based description of the rigid E-unification procedure should result in a simpler proof.

2   Preliminaries

Given a set $\Sigma = \cup_n \Sigma_n$ and a disjoint set V, we define $\mathcal{T}(\Sigma, V)$ as the smallest set containing V and such that $f(t_1, \ldots, t_n) \in \mathcal{T}(\Sigma, V)$ whenever $f \in \Sigma_n$ and $t_1, \ldots, t_n \in \mathcal{T}(\Sigma, V)$. The elements of the sets Σ, V and $\mathcal{T}(\Sigma, V)$ are respectively called *function symbols*, *variables* and *terms* (over Σ and V). The set Σ is called a *signature* and the index n of the set Σ_n to which a function symbol f belongs is called the *arity* of the symbol f. Elements of arity 0 are called *constants*. By $\mathcal{T}(\Sigma)$ we denote the set $\mathcal{T}(\Sigma, \emptyset)$ of all variable-free, or *ground* terms. The symbols $s, t, u, \ldots$ are used to denote terms; $f, g, \ldots$, function symbols; and $x, y, z, \ldots$, variables.

A *substitution* is a mapping from variables to terms such that $x\sigma = x$ for all but finitely many variables x. We use post-fix notation for application of substitutions and use the letters $\sigma, \theta, \ldots$ to denote substitutions. A substitution σ can be extended to the set $\mathcal{T}(\Sigma, V)$ by defining $f(t_1, \ldots, t_n)\sigma = f(t_1\sigma, \ldots, t_n\sigma)$. The *domain* $\mathcal{D}om(\sigma)$ of a substitution σ is defined as the set $\{x \in V : x\sigma \neq x\}$; and the *range* $\mathcal{R}an(\sigma)$ as the set of terms $\{x\sigma : x \in \mathcal{D}om(\sigma)\}$. A substitution σ is *idempotent* if $\sigma\sigma = \sigma$.[2]

We usually represent a substitution σ with domain $\{x_1, \ldots, x_n\}$ as a set of variable "bindings" $\{x_1 \mapsto t_1, \ldots, x_n \mapsto t_n\}$, where $t_i = x_i\sigma$. By a *triangular form* representation of a substitution σ we mean a *sequence* of bindings,

$$[x_1 \mapsto t_1 \; ; \; x_2 \mapsto t_2 \; ; \; \ldots \; ; \; x_n \mapsto t_n],$$

such that σ is the composition $\sigma_1 \sigma_2 \ldots \sigma_n$ of substitutions $\sigma_i = \{x_i \mapsto t_i\}$.

Congruences

An *equation* is a pair of terms, written as $s \approx t$. The *replacement relation* $\rightarrow_{E^g}$ induced by a set of equations E is defined by: $u[l] \rightarrow_{E^g} u[r]$ if, and only if, $l \approx r$ is in E. The *rewrite relation* $\rightarrow_E$ induced by a set of equations E is defined by: $u[l\sigma] \rightarrow_E u[r\sigma]$ if, and only if, $l \approx r$ is in E and σ is some substitution. In other words, the rewrite relation induced by E is the replacement relation induced by $\cup_\sigma E\sigma$, where $E\sigma$ is the set $\{s\sigma \approx t\sigma : s \approx t \in E\}$.

---

[2] We use juxtaposition $\sigma\tau$ to denote function composition, i.e., $x(\sigma\tau) = (x\sigma)\tau$.

If $\to$ is a binary relation, then $\leftarrow$ denotes its inverse, $\leftrightarrow$ its symmetric closure, $\to^+$ its transitive closure and $\to^*$ its reflexive-transitive closure. Thus, $\leftrightarrow^*_{E^g}$ denotes the *congruence relation*[3] induced by E. The *equational theory* of a set E of equations is defined as the relation $\leftrightarrow^*_E$. Equations are often called *rewrite rules*, and a set E a *rewrite system*, if one is interested particularly in the rewrite relation $\to^*_E$ rather than the equational theory $\leftrightarrow^*_E$.

Substitutions as Congruences

It is often useful to reason about a substitution σ by considering the congruence relation $\leftrightarrow^*_{E^g_\sigma}$ induced by the set of equations $E_\sigma = \{x\sigma \approx x : x \in \mathcal{D}om(\sigma)\}$. The following proposition establishes a connection between substitutions and congruences.

Proposition 1. *(a) For all terms $t \in \mathcal{T}(\Sigma, V)$, $t \leftrightarrow^*_{E^g_\sigma} t\sigma$. Therefore, $E\sigma \subseteq \leftrightarrow^*_{(E\cup E_\sigma)^g}$. (b) If the substitution σ is idempotent, then for any two terms $s, t \in \mathcal{T}(\Sigma, V)$, we have $s\sigma = t\sigma$ if, and only if, $s \leftrightarrow^*_{E^g_\sigma} t$.*

Proof. Part (a) is straight-forward and also implies the "only if" direction of part (b). For the "if" direction, note that if $u \leftrightarrow_{E_\sigma} v$, then $u = u[l]$ and $v = v[r]$ for some equation $l \approx r$ or $r \approx l$ in E_σ. Thus, $u\sigma = (u[l])\sigma = u\sigma[l\sigma] \leftrightarrow_{(E_\sigma\sigma)^g} u\sigma[r\sigma] = (u[r])\sigma = v\sigma$. Therefore, if $s \leftrightarrow^*_{E^g_\sigma} t$, then $s\sigma \leftrightarrow^*_{(E_\sigma\sigma)^g} t\sigma$. But if σ is idempotent, then $E_\sigma\sigma$ consists only of trivial equations $t \approx t$, and hence $s\sigma$ and $t\sigma$ are identical.

Theorem 1. *Let σ be an idempotent substitution and $[x_1 \mapsto t_1; \ldots; x_n \mapsto t_n]$ be a triangular form representation of σ. Then the congruences $\leftrightarrow^*_{E^g_\sigma}$ and $\leftrightarrow^*_{(\cup_i E_{\sigma_i})^g}$ are identical, where $\sigma_i = \{x_i \mapsto t_i\}$.*

Proof. It is sufficient to prove that $E_\sigma \subseteq \leftrightarrow^*_{(\cup_i E_{\sigma_i})^g}$ and $E_{\sigma_i} \subseteq \leftrightarrow^*_{E^g_\sigma}$ for all $1 \le i \le n$. If $x_i\sigma \approx x_i$ is an equation in $E\sigma$, then

$$x_i\sigma = x_i\sigma_1 \ldots \sigma_n = x_i\sigma_i \ldots \sigma_n = t_i\sigma_{i+1} \ldots \sigma_n,$$

and therefore, using Proposition 1 part (a),

$$x_i \leftrightarrow^*_{E^g_{\sigma_i}} t_i \leftrightarrow^*_{E^g_{\sigma_{i+1}}} t_i\sigma_{i+1} \leftrightarrow^*_{E^g_{\sigma_{i+2}}} \cdots \leftrightarrow^*_{E^g_{\sigma_n}} t_i\sigma_{i+1} \ldots \sigma_n = x_i\sigma.$$

For the converse, using part of the above proof, we get $t_i \leftrightarrow^*_{(\cup_{k>i} E_{\sigma_i})^g} x_i\sigma \leftrightarrow^*_{E^g_\sigma} x_i$. By induction hypothesis we can assume, $E_{\sigma_k} \subseteq \leftrightarrow^*_{E^g_\sigma}$ for $k > i$, and then the above proof would establish $E_{\sigma_i} \subseteq \leftrightarrow^*_{E^g_\sigma}$.

The theorem indicates that if an idempotent substitution σ can be expressed as a composition $\sigma_1\sigma_2 \ldots \sigma_n$ of finitely many idempotent substitutions σ_i with

---

[3] A congruence relation is a reflexive, symmetric and transitive relation on terms that is also a replacement relation.

disjoint domains, then the congruence induced by E_σ is identical to the congruence induced by $\cup_i E_{\sigma_i}$. We denote by E_σ the set $\cup_i E_{\sigma_i}$.

The *variable dependency ordering* $\succ_V^S$ induced by a set S of equations on the set V of variables is defined by: $x \succ_V^S y$ if there exists an equation $t[x] \approx y$ in S. A finite set S of equations is said to be *substitution-feasible* if (i) the right-hand sides of equations in S are all distinct variables and (ii) the variable dependency ordering $\succ_V^S$ induced by S is well-founded. If S is a substitution-feasible set $\{t_i \approx x_i : 1 \leq i \leq n\}$ such that $x_j \not\succ_V^S x_i$ whenever $i > j$, then the idempotent substitution σ represented by the triangular form $[x_1 \mapsto t_1; \ldots; x_n \mapsto t_n]$ is called the substitution *corresponding to S*. Given an idempotent substitution σ and any triangular form representation $\sigma_1 \sigma_2 \ldots \sigma_n$, the sets E_σ and $\cup_i E_{\sigma_i}$ are substitution-feasible.

Rigid E-Unification

Definition 1. *Let E be a set of equations (over $\Sigma \cup V$) and s and t be terms in $\mathcal{T}(\Sigma, V)$. A substitution σ is called a rigid E-unifier of s and t if $s\sigma \leftrightarrow^*_{(E\sigma)^g} t\sigma$.*

When $E = \emptyset$, rigid E-unification reduces to syntactic unification.

Theorem 2. *An idempotent substitution σ is a rigid E-unifier of s and t if and only if $s \leftrightarrow^*_{(E\cup E_\sigma)^g} t$.*

Proof. Let σ be an idempotent substitution that is a rigid E-unifier of s and t. By definition we have $s\sigma \leftrightarrow^*_{(E\sigma)^g} t\sigma$. Using Proposition 1 part (a), we get

$$s \leftrightarrow^*_{E_\sigma^g} s\sigma \leftrightarrow^*_{(E\cup E_\sigma)^g} t\sigma \leftrightarrow^*_{E_\sigma^g} t.$$

Conversely, suppose σ is an idempotent substitution such that $s \leftrightarrow^*_{(E\cup E_\sigma)^g} t$. Then, $s\sigma \leftrightarrow^*_{(E\sigma \cup E_\sigma \sigma)^g} t\sigma$. But $(E_\sigma)\sigma$ consists of trivial equations of the form $t \approx t$ and hence we have $s\sigma \leftrightarrow^*_{(E\sigma)^g} t\sigma$.

If the substitution σ is not idempotent, the above proof does not go through as the set $(E_\sigma)\sigma$ may contain non-trivial equations. However, we may use Theorem 1 to establish that the congruences induced by $E \cup E_\sigma$ and $E \cup E_\theta$, where $\theta_1 \ldots \theta_n$ is a triangular representation for σ, are identical.

We obtain a characterization of standard E-unification if we replace the congruence induced by $E \cup E_\sigma$ by the congruence induced by $\cup_\sigma E\sigma \cup E_\sigma$ in the above theorem, and a characterization of syntactic unifiers if $E = \emptyset$.

Orderings on Substitutions

Unification procedures are designed to find *most general* unifiers of given terms. A substitution σ is said to be *more general* than another substitution θ with respect to a set of variables V, denoted by $\sigma \preceq^V \theta$, if there exists a substitution σ' such that $x\sigma\sigma' = x\theta$ for all $x \in V$.

A substitution σ is called *more general modulo E^g on V* than θ, denoted by $\sigma \preceq^V_{E^g} \theta$, if there exists a substitution σ' such that $x\sigma\sigma' \leftrightarrow^*_{(E\theta)^g} x\theta$ for all $x \in V$. We also define an auxiliary relation $\sqsubseteq$ between substitutions by $\sigma \sqsubseteq^V_{E^g} \theta$ if $x\sigma \leftrightarrow^*_{(E\theta)^g} x\theta$ for all $x \in V$. Two substitutions σ and θ are said to be *equivalent modulo E^g on V* if $\sigma \preceq^V_{E^g} \theta$ and $\theta \preceq^V_{E^g} \sigma$.

If σ is a rigid E-unifier of s and t, then there exists an idempotent rigid E-unifier of s and t that is more general modulo E^g than σ. Hence, in this paper, we will be concerned only with idempotent unifiers. Comparisons between *idempotent* substitutions can be characterized via congruences.

Theorem 3. *Let σ and θ be idempotent substitutions and V the set of variables in the domain or range of σ. Then, $\sigma \preceq^V_{E^g} \theta$ if and only if $E_\sigma \subseteq \leftrightarrow^*_{E^g \cup E^g_\theta}$.*

Proof. If σ is idempotent then we can prove that $\sigma \preceq^V_{E^g} \theta$ if and only if $\sigma\theta \sqsubseteq^V_{E^g} \theta$. Now assuming $\sigma \preceq^V_{E^g} \theta$, we have $x\sigma\theta \leftrightarrow^*_{(E\theta)^g} x\theta$ for all $x \in V$. But $x\theta \leftrightarrow_{(E_\theta)^g} x$ and $E\theta \subseteq \leftrightarrow^*_{E^g \cup E^g_\theta}$ by Proposition 1. Therefore, it follows that $E_{\sigma\theta} \subseteq \leftrightarrow^*_{E^g \cup E^g_\theta}$. But again by Proposition 1, $x\sigma\theta \leftrightarrow^*_{E^g} x\sigma$ and therefore, $E_\sigma \subseteq \leftrightarrow^*_{E^g \cup E^g_\theta}$.

Conversely, if $E_\sigma \subseteq \leftrightarrow^*_{E^g \cup E^g_\theta}$ then, $E_\sigma\theta \subseteq \leftrightarrow^*_{(E\theta)^g \cup (E_\theta\theta)^g}$, and since the equations in $E_\theta\theta$ are all trivial equations of the form $u \approx u$, it follows that $E_\sigma\theta \subseteq \leftrightarrow^*_{(E\theta)^g}$, which implies $\sigma\theta \sqsubseteq^V_{E^g} \theta$ and hence $\sigma \preceq^V_{E^g} \theta$.

3   Rigid E-Unification

We next present a set of abstract transformation (or transition) rules that can be used to describe a variety of rigid E-unification procedures. By Theorem 2, the problem of finding a rigid E-unifier of two terms s and t amounts to finding a substitution-feasible set S such that $s \leftrightarrow^*_{(E \cup S)^g} t$, and involves (1) constructing a substitution-feasible set S, and (2) verifying that s and t are congruent modulo $E \cup S$. Part (1), as we shall see, can be achieved by using syntactic unification, narrowing and superposition. Efficient techniques for congruence testing via abstract congruence closure can be applied to part (2).

Abstract Congruence Closure

A term t is in *normal form* with respect to a rewrite system R if there is no term t' such that $t \to_R t'$. A rewrite system R is said to be (ground) *confluent* if for all (ground) terms t, u and v with $u \leftarrow^*_R t \to^*_R v$ there exists a term w such that $u \to^*_R w \leftarrow^*_R v$. It is *terminating* if there exists no infinite reduction sequence $t_0 \to_R t_1 \to_R t_2 \cdots$ of terms. Rewrite systems that are (ground) confluent and terminating are called (ground) *convergent*.

Let Γ be a set of function symbols and variables and K be a disjoint set of constants. An *(abstract) congruence closure* (with respect to Γ and K) is a ground convergent rewrite system R over the signature $\Gamma \cup K$[4] such that (i) each

---

[4] We treat variables as constants and in this sense speak of a *ground* convergent system R.

rule in R is either a *D-rule* of the form $f(c_1, \ldots, c_k) \approx c_0$ where f is a k-ary symbol in Γ and $c_0, c_1, \ldots, c_k$ are constants in K, or a *C-rule* of the form $c_0 \approx c_1$ with $c_0, c_1 \in K$, and (ii) for each constant $c \in K$ that is in normal form with respect to R, there exists a term $t \in \mathcal{T}(\Gamma)$ such that $t \to_{R^g}^* c$. Furthermore, if E is a set of equations (over $\Gamma \cup K$) and R is such that (iii) for all terms s and t in $\mathcal{T}(\Gamma)$, $s \leftrightarrow_{E^g}^* t$ if, and only if, $s \to_{R^g}^* \circ \leftarrow_{R^g}^* t$, then R is called an (abstract) congruence closure *for E*.

For example, let $E_0 = \{gfx \approx z, fgy \approx z\}$ and $\Gamma = \{g, f, x, y, z\}$. The set E_1 consisting of the rules $x \approx c_1$, $y \approx c_2$, $z \approx c_3$, $fc_1 \approx c_4$, $gc_4 \approx c_3$, $gc_2 \approx c_5$, $fc_5 \approx c_3$ is an abstract congruence closure (with respect to Γ and $\{c_1, \ldots, c_5\}$) for E_0.

The key idea underlying (abstract) congruence closure is that the constants in K serve as *names* for congruence classes, and equations $f(c_1, \ldots, c_k) \approx c_0$ define relations between congruence classes: a term $f(t_1, \ldots, t_k)$ is in the congruence class c_0 if each t_i is in the congruence class c_i.

The construction of a congruence closure will be an integral part of our rigid E-unification method. We will not list specific transformation rules, but refer the reader to the description in [2] which can be easily adapted to the presentation in the current paper.

For our purposes, transition rules are defined on quintuples $(K, E; V, E?; S)$, where Σ is a given fixed signature, V is a set of variables, K is a set of constants disjoint from $\Sigma \cup V$, and $E, E? and S$ are sets of equations. The first two components of the quintuple represent a partially constructed congruence closure, whereas the third and fourth components are needed to formalize syntactic unification, narrowing and superposition. The substitution-feasible set in the fifth component stores an answer substitution in the form of a set of equations. For a given state $(K, E; V, E?; S)$, we try to find a substitution σ with $\mathcal{D}om(\sigma) \subseteq V$, that is a rigid E-unifier of each equation in the set $E?$. By an *initial* state we mean a tuple $(\emptyset, E_0; V_0, \{s \approx t\}; \emptyset)$ where V_0 is the set of all variables that occur in E_0, s or t. Transition rules specify ways in which one quintuple state can be transformed into another such state. The goal is to successively transform a given initial state to a state in which the fourth component is empty.

C-Closure: $$\frac{(K, E; V, E?; S)}{(K', E'; V, E?; S)}$$

if $K \subset K'$ and E' is an abstract congruence closure (with respect to $\Sigma \cup V$ and K') for E.

Note that we need not choose any term ordering, which is one of the main differences of our approach with most other rigid unification methods.

Syntactic Unification

C-closure can potentially extend the signature by a set of constants. Thus we obtain substitutions (or substitution-feasible sets) and terms over an extended signature that need to be *translated* back to substitutions and terms in the original signature, essentially by replacing these constants by terms from the original signature. For example, consider the abstract congruence closure E_1 for $E_0 = \{gfx \approx z, fgy \approx z\}$ described above, and the substitution-feasible set $S = \{c_3 \approx x, x \approx y\}$. This set can be transformed by replacing the constant c_3 by z to give a substitution-feasible set $\{z \approx x, x \approx y\}$. Unfortunately, this may not be possible always. For example, in the substitution-feasible set $S = \{c_1 \approx x\}$, we can't eliminate the constant c_1 since x is the only term congruent to c_1 modulo E_1, but, the resulting set $\{x \mapsto x\}$ is not substitution-feasible.

We say that a (substitution-feasible) set $S = \{t_i \approx x_i : t_i \in \mathcal{T}(\Sigma \cup K, V), x_i \in V, 1 \leq i \leq n\}$ of rules is E-feasible on V if, there exist a terms $s_i \in \mathcal{T}(\Sigma \cup V)$ with $s_i \leftrightarrow^*_{E^g} t_i$, such that the set $S{\uparrow}_E = \{s_i \approx x_i : 1 \leq i \leq n\}$ is substitution-feasible.

Recall that if σ is a rigid E-unifier of s and t, then there exists a proof $s \leftrightarrow^*_{E^g \cup E^g_\sigma} t$. The transition rules are obtained by analyzing the above hypothetical proof. The rules attempt to deduce equations in E_σ by simplifying the above proof. We first consider the special case when $s \leftrightarrow^*_{E^g_\sigma} t$, and hence s and t are syntactically unifiable. Trivial proofs can be deleted.

Deletion:
$$\frac{(K, E; V, E? \cup \{t \approx t\}; S)}{(K, E; V, E?; S)}$$

If E_σ is a substitution-feasible set and the top function symbols in s and t are identical, then all replacement steps in the proof $s \leftrightarrow^*_{E^g_\sigma} t$ occur inside a non-trivial context, and hence this proof can be broken up into simpler proofs.

Decomposition:
$$\frac{(K, E; V, E? \cup \{f(t_1, \ldots, t_n) \approx f(s_1, \ldots, s_n)\}; S)}{(K, E; V, E? \cup \{t_1 \approx s_1, \ldots, t_n \approx s_n\}; S)}$$

if $f \in \Sigma$ is a function symbol of arity n.

Finally, if the proof $s \leftrightarrow^*_{E^g_\sigma} t$ is a single replacement step (at the root position, and within no contexts), we can eliminate it.

Elimination:
$$\frac{(K, E; V, E? \cup \{x \approx t\}; S)}{(K \cup \{x\}, E \cup E_\theta; V - \{x\}, E?; S \cup E_\theta)}$$

if (i) $\theta = \{x \mapsto t\}$, (ii) the set $E_\theta = \{t \approx x\}$ is E-feasible on V, and (iii) $x \in V$.

Deletion and decomposition are identical to the transformation rules for syntactic unification, c.f. [1]. However, elimination (and narrowing and superposition described below), do *not* apply the substitution represented by E_θ (or $E_\theta{\uparrow}_E$) to the sets $E?$ and S as is done in the corresponding standard rules for syntactic unification. Instead we add the equations E_θ to the second component of the state.

Decomposition, deletion and elimination can be replaced by a single rule that performs full syntactic unification in one step. We chose to spell out the rules above as they provide a method to abstractly describe an efficient quadratic-time syntactic unification algorithm by recursive descent, c.f. [1].

Narrowing and Superposition

The following rule reflects attempts to identify and eliminate steps in a proof $s \leftrightarrow^*_{E^g \cup E^g_\sigma} t$ that use equations in E^g.

$$\text{Narrowing:} \qquad \frac{(K, E; V, E? \cup \{s[l'] \approx t\}; S)}{(K \cup V', E \cup E_\theta; V - V', E? \cup \{s[c] \approx t\}; S \cup E_\theta)}$$

where (i) $l \approx c \in E$, (ii) θ is the most general unifier of l' and l, (iii) the set E_θ is E-feasible on V, (iv) $V' = \mathcal{D}om(\theta) \subset V$, (v) E is an abstract congruence closure with respect to Σ and $K \cup V$, and (vi) either $l' \notin V$ or $l \in V$.

We may also eliminate certain "proof patterns" involving rules in E^g (and E^g_σ) from the proof $s \leftrightarrow^*_{E^g \cup E^g_\sigma} t$ via superposition of rules in E.

$$\text{Superposition:} \qquad \frac{(K, E = E' \cup \{t \approx c, C[t'] \approx d\}; V, E?; S)}{(K \cup V', E' \cup \{t \approx c\} \cup T; V - V', E?; S \cup E_\theta)}$$

if (i) θ is the most general unifier of t and t', (ii) E_θ is E-feasible on V, (iii) $T = E_\theta \cup \{C[c] \approx d\}$, (iv) $V' = \mathcal{D}om(\theta) \subset V$, (v) E is an abstract congruence closure with respect to Σ and $K \cup V$, and (vi) either $t' \notin V$ or $t \in V$.

Narrowing, elimination and superposition add new equations to the second component of the state, which are subsequently processed by C-closure.

We illustrate the transition process by considering the problem of rigidly unifying the two terms fx and gy modulo the set $E_0 = \{gfx \approx z, fgy \approx z\}$. Let E_1 denote an abstract congruence closure $\{x \approx c_1, y \approx c_2, z \approx c_3, fc_1 \approx c_4, gc_4 \approx c_3, gc_2 \approx c_5, fc_5 \approx c_3\}$ for E_0 and K_1 be the set $\{c_1, \ldots, c_5\}$ of constants.

i	K_i	E_i	V_i	$E?_i$	S_i	Rule
0	$\emptyset$	E_0	$\{x, y, z\}$	$\{fx \approx gy\}$	$\emptyset$	C-Closure
1	K_1	E_1	$\{x, y, z\}$	$\{fx \approx gy\}$	$\emptyset$	Narrow
2	$K_1 \cup \{x\}$	$E_1 \cup \{x \approx c_5\}$	$\{y, z\}$	$\{c_3 \approx gy\}$	$\{c_5 \approx x\}$	C-Closure
3	K_2	E_3	$\{y, z\}$	$\{c_3 \approx gy\}$	$\{c_5 \approx x\}$	Narrow
4	$K_2 \cup \{y\}$	$E_3 \cup \{y \approx c_3\}$	$\{z\}$	$\{c_3 \approx c_3\}$	$S_3 \cup \{c_3 \approx y\}]$	Delete
5	K_4	E_4	$\{z\}$	$\emptyset$	S_4	

where $E_3 = \{x \approx c_1, y \approx c_2, z \approx c_3, fc_1 \approx c_3, gc_3 \approx c_3, gc_2 \approx c_1, c_5 \approx c_1, c_4 \approx c_3\}$ is an abstract congruence closure for E_2. Since the set $E?_5$ is empty, the rigid unification process is completed. Any set $S_4\!\uparrow_{E_4}$ is a rigid unifier of fx and gy. For instance, we may choose gy for the constant c_5 and z for c_3 to get the set $S_4\!\uparrow_{E_4} = \{gy \approx x, z \approx y\}$ and the corresponding unifier $[x \mapsto gy; y \mapsto z]$.

Optimizations

A cautious reader might note that the transition rules contain a high degree of non-determinism in the present form. In particular after an initial congruence closure step, every variable x in the third component of a state occurs as a left-hand side of some rule $x \approx c$ in the second component. Consequently, this rule can be used for superposition or narrowing with any rule in the second or fourth component. A partial solution to this problem is to replace all occurrences of c by x in the second component and then delete the rule $x \approx c$. This is correct under certain conditions.

$$\textbf{Compression1:} \quad \frac{(K \cup \{c\}, E \cup \{x \approx c\}; V \cup \{x\}, E?; S)}{(K, E\theta; V \cup \{x\}, E?\theta; S')}$$

if (i) θ is the substitution $\{c \mapsto x\}$, (ii) $E \cup \{x \approx c\}$ is a fully-reduced abstract congruence closure (with respect to Σ and $K \cup V$), (iii) c does not occur on the right-hand side of any rule in E, and (iv) S' is obtained from S by applying substitution θ *only* to the left-hand sides of equations in S.

A related optimization rule is:

$$\textbf{Compression2:} \quad \frac{(K \cup \{c,d\}, E \cup \{c \approx d\}; V, E?; S)}{(K \cup \{d\}, E; V, E?\theta; S')}$$

if (i) θ is the substitution $\{c \mapsto d\}$, (ii) $E \cup \{c \approx d\}$ is a fully-reduced abstract congruence closure (with respect to Σ and $K \cup V$), and (iii) S' is obtained from S by applying substitution θ *only* to the left-hand sides of equations in S.

These two optimization rules can be integrated into the congruence closure phase. More specifically, we assume that application of C-closure rule is always followed by an exhaustive application of the above compression. We refer to this combination as an "Opt-Closure" rule.

To illustrate these new rules, now note that both $x \approx c_1$ and $y \approx c_2$ can be eliminated from the abstract congruence closure E_1 for E_0. We obtain an optimized congruence closure $\{z \approx c_3, fx \approx c_4, gc_4 \approx c_3, gy \approx c_5, fc_5 \approx c_3\}$. Note that we cannot remove $z \approx c_3$ from the above set.

4 Correctness

Let U be an infinite set of constants from which new constants are chosen in opt-closure. If a state $\xi_i = (K_i, E_i; V_i, E?_i; S_i)$ is transformed to a state $\xi_j = (K_j, E_j; V_j, E?_j; S_j)$ by opt-closure, then (i) E_j is an abstract congruence closure (with respect to Σ and $K \cup V$) for E_i and (ii) E_j is contained in a well-founded simplification ordering[5].

We use the symbol $\vdash_{REU}$ to denote the one-step transformation relation induced by opt-closure, deletion, decomposition, elimination, narrowing and superposition. A *derivation* is a sequence of states $\xi_0 \vdash_{REU} \xi_1 \vdash_{REU} \cdots$ with no two consecutive applications of opt-closure.

---

[5] For instance, a simple lexicographic path ordering $\succ$ based on a *partial* precedence on symbols in $\Sigma \cup U \cup V$ for which $f \succ c$, whenever $f \in \Sigma$ and $c \in U \cup V$, and $x \succ c$, whenever $x \in V$ and $c \in U - V$, will suffice.

Theorem 4. *(Termination) All derivations starting from an initial state $(\emptyset, E_0; V_0, \{s \approx t\}; \emptyset)$ are finite.*

Proof. Define a measure associated with a state $(K, E; V, E?; S)$ to be the pair $(|V|, m_{E?})$, where $|V|$ denotes the cardinality of the set V and $m_{E?} = \{\{s, t\} : s \approx t \in E?\}$. These pairs are compared lexicographically using the greater-than relation on the integers in the first component and the two-fold multiset extension of the ordering $\succ$ in the second component. This induces a well-founded ordering on states with respect to which each transition rule is reducing.

Lemma 1. *Let $(K_n, E_n; V_n, E?_n; S_n)$ be the final state of a derivation from $(\emptyset, E_0; V_0, E?_0; \emptyset)$, where $E_0 \cup E?_0$ are equations over $\mathcal{T}(\Sigma, V_0)$. Then*
(a) the set S_n is E_n-feasible on V_0 and
(b) if $E?_n \subseteq \leftrightarrow^*_{(E_n \cup S_n \uparrow_{E_n})^g}$, then $E?_0 \subseteq \leftrightarrow^*_{(E_0 \cup S_n \uparrow_{E_n})^g}$.

Theorem 5 (Soundness). *If $(K_n, E_n; V_n, E?_n; S_n)$ is the final state of a derivation from $(\emptyset, E_0; V_0, E?_0; \emptyset)$, then the set S_n is E_n-feasible and the substitution corresponding to (any) set $S_n \uparrow_{E_n}$ is a rigid E_0-unifier of s and t.*

Proof. The E_n-feasibility of S_n on V_0 follows from Lemma 1. Since $E?_n = \emptyset$, the antecedent of the implication in part (b) of Lemma 1 is vacuously satisfied and hence $E?_0 \subseteq \leftrightarrow^*_{(E_0 \cup S_n \uparrow_{E_n})^g}$.

Note that by Theorem 1, the ground congruence induced by $S_n \uparrow_{E_n}$ is identical to the congruence induced by E_σ, where σ is the idempotent substitution corresponding to the set $S_n \uparrow_{E_n}$. Hence, $s \leftrightarrow^*_{E_0^g \cup E_\sigma^g} t$. Using Theorem 2, we establish that σ is a rigid E_0-unifier of s and t.

Theorem 6 (Completeness). *Let θ be an idempotent rigid E_0-unifier of s and t and V_0 the set of variables in E_0, s and t. Then, there exists a (finite) derivation with initial state $(\emptyset, E_0; V_0, \{s \approx t\}; \emptyset)$ and final state $(K_n, E_n; V_n, \emptyset; S_n)$ where $E_{S_n \uparrow_{E_n}} \subseteq \leftrightarrow^*_{E_0^g \cup E_\theta^g}$.*

Proof. (Sketch) Let $\xi_i = (K_i, E_i; V_i, E?_i; S_i)$ be a state. We say a substitution-feasible set S^i is a *solution* for state ξ_i if S^i is a E_i-feasible on V_i and

$$E?_i \subseteq \leftrightarrow^*_{(E \cup S_i \cup S^i)^g}.$$

Now, given a state ξ_i and a solution S^i for ξ_i, we show how to obtain a new state ξ_j and a solution S^j for ξ_j such that the pair $\langle \xi_j, S^j \rangle$ is *smaller* in a certain well-founded ordering than the pair $\langle \xi_i, S^i \rangle$ and the congruences induced by $E_j \cup S_j \cup S^j$ and $E_i \cup S_i \cup S^i$ are identical. The well-founded ordering will be a lexicographic combination of the ordering on states ξ_i's used in the proof of Theorem 4 and a well-founded ordering on substitution-feasible sets S_i's. If a pair $\langle \xi_i, S^i \rangle$ can not be reduced then we show that $E?_i = \emptyset$. This yields the desired conclusion.

The above reduction of a pair $\langle \xi_i, S^i \rangle$ can be achieved in two ways: (i) by an REU transformation on ξ_i, suitably guided by the given solution S^i, or, (ii) by some simple transformation of the set S^i. The latter transformation rules are defined in the context of the state ξ_i. The initial state is $(S^i, \emptyset)$.

$$R1 : \frac{(D' \cup \{c \approx x\}, C')}{(D', C' \cup \{c \approx x\})} \qquad \text{if } c \in K_i \cup V_i, \ x \not\rightarrow^*_{E_i^g \setminus C'^g} c^6.$$

$$R2 : \frac{(D' \cup \{c \approx x\}, C')}{(D', C')} \qquad \text{if } c \in K_i \cup V_i, \ x \rightarrow^*_{E_i^g \setminus C'^g} c$$

$$R3 : \frac{(D' \cup \{t[l'] \approx x\}, C')}{(D' \cup \{t[c] \approx x\}, C')} \qquad \text{if } l \approx c \in E_i, \ l \leftrightarrow^*_{C'^g} l'$$

$$R4 : \frac{(D' \cup \{t[l'] \approx x\}, C')}{(D' \cup \{t[y] \approx x\}, C')} \qquad \text{if } l \approx y \in D', \ l \leftrightarrow^*_{C'^g} l', \ l \notin K_i \cup V_i$$

These rule construct a generalized congruence closure for the initial set $D' \cup C'$ (modulo the congruence induced by E_i). If (D', C') can be obtained from $(S^i, \emptyset)$ by repeated application of these rules, then the set $D' \cup C'$ is (i) substitution-feasible, (ii) E_i-feasible with respect to V_i, and (iii) equivalent modulo E_i^g on V_i to S^i. The set of rules $R1 - -R4$ is terminating.

5 Specialization to Syntactic Unification

Since rigid unification reduces to syntactic unification when the set E_0 is empty, one pertinent question is what procedure the REU transformation rules yield in this special case? Note that elimination does *not* apply a substitution to the fourth and fifth components of the state, but does perform an *occur check* in condition (ii). This is in the spirit of syntactic unification by recursive descent algorithm which works on term directed acyclic graphs and is a quadratic time-complexity algorithm.

In fact, in the case of syntactic unification, every equation in the second component is of a special form where one side is always a variable. Hence, we can argue that for each $c \in K$, there is *at most* one rule in E of the form $f(\ldots) \rightarrow c$ where $f \in \Sigma$. We may therefore replace superposition by the following rule:

$$\textbf{Det-Decompose:} \quad \frac{(K, E; V, E? \cup \{c \approx t\}; S)}{(K, E; V, E? \cup \{f(c_1, \ldots, c_k) \approx t\}; S)}$$

if there exist exactly one rule $f(c_1, \ldots, c_k) \approx c$ with right-hand side c in E.

In addition, we may restrict narrowing so that the unifier θ is always the identity substitution, that is, narrowing is used to only for simplification of terms in the fourth component $E?$ by equations in the second component E.

We can get various efficient syntactic unification algorithms by using specific strategies over our abstract description. Other descriptions of the quadratic time syntactic unification algorithms are usually based on descriptions of dags and abstract rules that manipulate the dags directly. However, since we can abstractly capture the notion of sharing, we obtain rules for this class of efficient algorithms that work on terms and are very similar to the rules for describing the naive syntactic unification procedures (with a worst case exponential behavior).

6 Summary

We have presented a formulation of rigid E-unification in terms of fairly abstract transformation rules. The main feature is the integration of (abstract) congruence closure with transformation rules for syntactic unification, paramodulation and superposition. The use of an extended signature (inherent in abstract congruence closure) helps to dispense with term orderings over the original signature. An abstract rule-based description facilitates various optimizations. The specialization of the transformation rules to syntactic unification yields a set of abstract transition rules that describe a class of efficient syntactic unification algorithms. Our transformation rules can be derived from proof simplification arguments.

In [10], a congruence closure algorithm is used in a rigid E-unification procedure, but not as a submodule. Congruence closure is used "indirectly" to do ground completion. The work on abstract congruence closure shows that congruence closure actually is ground completion with extension. But for the purpose of rigid E-unification, we don't need to translate the abstract closure to a ground system over the original signature, though we do need to translate the substitutions back to the original signature. Extended signatures also help as we do not need to guess an ordering to orient equations such as $x \approx fa$ when the substitution for x is not yet known. This is a major concern in [10] where the dependence on orderings complicates the unification process.

In [5], the problem of correctly orienting equations is solved by delaying the choice of orientation and maintaining constraints. Constraint satisfiability is required to ensure that orientations are chosen in a consistent manner, and to guarantee the termination of such a procedure.

We would like to point out that the transformation process involves "don't-care" non-determinism (where it does not matter which rule one applies) and "don't-know" non-determinism (where an application of a wrong rule may lead to a failure even if a unifier exists). Whereas opt-closure, deletion and narrowing with identity substitution can be applied "don't-care" non-deterministically, the other rules have to be applied in a "don't-know" non-deterministic manner. The rules for syntactic unification described in Section 5 are "don't-care" non-deterministic.

All algorithms for computing the set of rigid unifiers for a pair of terms can be seen as a combination of top-down and bottom-up method. In a pure bottom-up approach a substitution is guessed non-deterministically: for every variable one tries every subterm that occurs in the given unification problem, see [13] for details. Superposition and narrowing using a rule that contains a variable as its left-hand side captures the bottom-up aspect in our description. A top-down approach is characterized by the use of narrowing to simplify the terms in the goal equations $E?$.

We note that for variables that cannot be eliminated from the left-hand sides of rules using compression1, we need to try a lot of possible substitutions because they can unify with almost all subterms in the second and fourth components. This is the cause of a bottom-up computation for these variables. For other variables, however, we need to try only those substitutions that are produced by some unifier during an application of narrowing or superposition, and hence a top-down approach works for these variables.

We illustrate some of the above observations via an example. Let $E_0 = \{gx \approx x,\ x \approx a\}$, and suppose we wish to find a rigid E_0-unifier of the terms $gfffgfffx$ and $fffx$. The substitution $\{x \mapsto fa\}$ is a rigid E-unifier, but it cannot be obtained unless one unifies the variable x with an appropriate subterm.

We believe that our approach of describing rigid E-unification can be used to obtain an easier proof of the fact that this problem is in NP. We need to show that (i) the length of a maximal derivation from any initial state is bounded by some polynomial in the input size, (ii) each rule can be efficiently applied, and (iii) there are not too many choices between the rules to get the next step in a derivation. It is easy to see that (i) holds. For the second part, a crucial argument involves showing that the test for E-feasibility can be efficiently done. This is indeed the case, but due to space limitations, we don't give a way to do this here.

The notion of an abstract congruence closure is easily extended to handle associative and commutative functions [3]. The use of extended signatures is particularly useful when one incorporates such theories. This leads us to believe that our proposed description of rigid E-unification can be suitably generalized to such applications.

Acknowledgements

We would like to thank David Cyrluk for initial discussions on the problem and the anonymous reviewers for their helpful comments.

References

[1] F. Baader and T. Nipkow. *Term Rewriting and All That*. Cambridge University Press, Cambridge, 1998.

[2] L. Bachmair, C. Ramakrishnan, I.V. Ramakrishnan, and A. Tiwari. Normalization via rewrite closures. In P. Narendran and M. Rusinowitch, editors, *10th Int. Conf. on Rewriting Techniques and Applications*, pages 190–204, 1999. LNCS 1631.

[3] L. Bachmair, I.V. Ramakrishnan, A. Tiwari, and L. Vigneron. Congruence closure modulo associativity and commutativity. In H. Kirchner and C. Ringeissen, editors, *Frontiers of Combining Systems, 3rd Intl Workshop FroCoS 2000*, pages 245–259, 2000. LNAI 1794.

[4] L. Bachmair and A. Tiwari. Abstract congruence closure and specializations. In D. McAllester, editor, *17th Intl Conf on Automated Deduction*, 2000.

[5] G. Becher and U. Petermann. Rigid unification by completion and rigid paramodulation. In B. Nebel and L.D. Fiseher, editors, *KI-94: Advances in Artificial Intelligence, 18th German Annual Conf on AI*, pages 319–330, 1994. LNAI 861.

[6] B. Beckert. A completion-based method for mixed universal and rigid E-unification. In A. Bundy, editor, *12th Intl Conf on Automated Deduction, CADE-12*, pages 678–692, 1994. LNAI 814.

[7] A. Degtyarev and A. Voronkov. The undecidability of simultaneous rigid E-unification. *Theoretical Computer Science*, 166(1–2):291–300, 1996.

[8] A. Degtyarev and A. Voronkov. What you always wanted to know about rigid E-unification. *Journal of Automated Reasoning*, 20(1):47–80, 1998.

[9] J. Gallier, P. Narendran, D. Plaisted, and W. Snyder. Rigid E-unification: Np-completeness and applications to equational matings. *Information and Computation*, 87:129–195, 1990.

[10] J. Gallier, P. Narendran, S. Raatz, and W. Snyder. Theorem proving using equational matings and rigid E-unification. *Journal of the Association for Computing Machinery*, 39(2):377–429, April 1992.

[11] J. Goubault. A rule-based algorithm for rigid E-unification. In G. Gottlob, A. Leitsch, and D. Mundici, editors, *Computational logic and proof theory. Proc. of the third Kurt Godel Colloquium, KGC 93*, pages 202–210, 1993. LNCS 713.

[12] D. Kapur. Shostak's congruence closure as completion. In H. Comon, editor, *8th Intl Conf on Rewriting Techniques and Applications*, pages 23–37, 1997. LNCS 1232.

[13] Eric de Kogel. Rigid E-unification simplified. In P. Baumgartner, R. Hahnle, and J. Posegga, editors, *Theorem Proving with Analytic Tableaux and Related Methods, 4th International Workshop, TABLEAUX '95*, pages 17–30, 1995. LNAI 918.

Connecting Bits with Floating-Point Numbers: Model Checking and Theorem Proving in Practice

Carl-Johan Seger

Intel Strategic CAD Labs
Portland OR
cseger@ichips.intel.com

Abstract. Model checking and theorem proving have largely complementary strengths and weaknesses. Thus, a research goal for many years
has been to find effective and practical ways of combining these approaches. However, this goal has been much harder to reach than originally anticipated, and several false starts have been reported in the literature. In fact, some researchers have gone so far as to question whether
there even exists an application domain in which such a hybrid solution is needed. In this talk I will argue that formal verification of the
floating-point circuits of modern high-performance microprocessors is
such a domain. In particular, when a correctness statement linking the
actual low-level (gate-level) implementation with abstract floating-point
numbers is needed, a combined model checking and theorem proving
based approach is essential. To substantiate the claim, I will draw from
data we have collected during the verification of the floating point units
of several generations of Intel microprocessors. In addition, I will discuss
the in-house formal verification environment we have created that has
enabled this effort with an emphasis on how model checking and theorem
proving have been integrated without sacrificing usability.

D. McAllester (Ed.): CADE-17, LNAI 1831, pp. 235–235, 2000.
© Springer-Verlag Berlin Heidelberg 2000

Reducing Model Checking of the Many to the Few *

E. Allen Emerson and Vineet Kahlon

Department of Computer Sciences
The University of Texas at Austin, Austin TX-78712, USA
{emerson,kahlon}@cs.utexas.edu
http://www.cs.utexas.edu/users/{emerson,kahlon}

Abstract. The *Parameterized Model Checking Problem* (PMCP) is to
determine whether a temporal property is true for every size instance of
a system comprised of many homogenous processes. Unfortunately, it is
undecidable in general. We are able to establish, nonetheless, decidability of the PMCP in quite a broad framework. We consider asynchronous
systems comprised of an arbitrary number of homogeneous copies of a
generic process template. The process template is represented as a synchronization skeleton while correctness properties are expressed using
Indexed CTL*\X. We reduce model checking for systems of arbitrary
size n to model checking for systems of size up to (of) a small *cutoff* size
c. This establishes decidability of PMCP as it is only necessary to model
check a finite number of relatively small systems. Efficient decidability
can be obtained in some cases. The results generalize to systems comprised of multiple heterogeneous classes of processes, where each class is
instantiated by many homogenous copies of the class template (e.g., m
readers and n writers).

1 Introduction

Systems with an arbitrary number of homogeneous processes can be used to
model many important applications. These include classical problems such as
mutual exclusion, readers and writers, as well as protocols for cache coherence
and data communication among others. It is often the case that correctness properties are expected to hold irrespective of the size of the system, as measured
by the number of processes in it. However, time and space constraints permit
us to verify correctness only for instances with a small number of processes.
This makes it impossible to guarantee correctness in general and thus motivates
consideration of automated methods to permit verification for system instances
of arbitrary sizes. The general problem, known in the literature as the *Parameterized Model Checking Problem (PMCP)* is the following: to decide whether a
temporal property is true for every size instance of a given system. This problem

---

* This work was supported in part by NSF grant CCR-980-4737, SRC contract 99-
TJ-685 and TARP project 003658-0650-1999.

D. McAllester (Ed.): CADE-17, LNAI 1831, pp. 236–254, 2000.
© Springer-Verlag Berlin Heidelberg 2000

is known to be undecidable in general [1]. However, by imposing certain stipulations on the organization of the processess we can get a useful framework with a decidable PMCP.

We establish our results in the *synchronization skeleton* framework. Our results apply to systems comprised of multiple heterogeneous *classes* of processes with many homogeneous process *instances* in each class. Thus, given family $(U_1, ..., U_k)$ of k process classes, and tuple $(n_1, ..., n_k)$ of natural numbers, we let $(U_1, ..., U_k)^{(n_1, ..., n_k)}$ denote the concrete system composed of n_1 copies or instances of U_1 through n_k copies or instances of U_k running in parallel asynchronously (i.e., with interleaving semantics). By abuse of notation, we also write $(U_1, ..., U_k)^{(n_1, ..., n_k)}$ for the associated state graph, where each process starts in its designated initial state.

Correctness properties are expressed using a fragment of Indexed CTL*\X. The basic assertions are of the form "for all processes Ah", or "for all processes Eh", where h is an LTL\X formula (built using F "sometimes", G "always", U, "until", but without X "next-time") over propositions indexed just by the processes being quantified over, and A "for all futures" and E "for some future" are the usual path quantifiers. Use of such an indexed, stuttering-insensitive logic is natural for parameterized systems.

We consider correctness properties of the following types:

1. Over all individual processes of single class U_l:
   $\bigwedge_{i_l}$ A$h(i_l)$ and $\bigwedge_{i_l}$ E$h(i_l)$, where i_l ranges over (indices of) individual processes in U_l.

2. Over pairs of different processes of a single class U_l:
   $\bigwedge_{i_l \neq j_l}$ A$h(i_l, j_l)$ and $\bigwedge_{i_l \neq j_l}$ E$h(i_l, j_l)$, where i_l, j_l range over pairs of distinct processes in U_l.

3. Over one process from each of two different classes U_l, U_m:
   $\bigwedge_{i_l, j_m}$ A$h(i_l, j_m)$ and $\bigwedge_{i_l, j_m}$ E$h(i_l, j_m)$, where i_l ranges over U_l and j_m ranges over U_m.

We say that the k-tuple $(c_1, ..., c_k)$ of natural numbers is a *cutoff* of $(U_1, ..., U_k)$ for formula f iff : $\forall (n_1, ..., n_k), (U_1, ..., U_k)^{(n_1, ..., n_k)} \models f$ iff $\forall (m_1, ..., m_k) \preceq (c_1, ..., c_k) : (U_1, ..., U_k)^{(m_1, ..., m_k)} \models f$, where we write $(m_1, ..., m_k) \preceq (c_1, ..., c_k)$ to mean $(m_1, ..., m_k)$ is *component-wise* less than or equal to $(c_1, ..., c_k)$ and $(m_1, ..., m_k) \succeq (c_1, ..., c_k)$ to mean $(c_1, ..., c_k) \preceq (m_1, ..., m_k)$.

In this paper, we show that for systems in the synchronization skeleton framework with transition guards of a particular *disjunctive* or *conjunctive* form, there is a small cutoff. This, in effect, reduces PMCP to ordinary model checking over a relatively few small, finite sized systems. In some cases, depending on the kind of property and guards, we can get an efficient (quadratic in the size of the template processes) solution to PMCP.

Each process class is described by a generic process, a process *template* for the class. A system with k classes is given by templates $(U_1, ..., U_k)$. For such a system, define $c_i = |U_i| + 3$ and $d_i = 2|U_i| + 1$, where $|U_i|$ is the size i.e. the number of local states of template U_i. Then, for both conjunctive and disjunctive guards, cutoffs of $(d_1, ..., d_k)$ and $(c_1, ..., c_k)$ respectively suffice for all three types of formulae described above. These results give decision procedures for PMCP for conjunctive or disjunctive guards. Since these are a broad framework and PMCP is undecidable in general, we view this as quite a positive result. However, the decision procedures are not necessarily efficient ones, although they may certainly be usable on small examples. Because the cutoff is proportional to the sizes of the template processes, the global state graph of the cutoff system is of size exponential in the template sizes, and the decision procedures are also exponential. In the case of disjunctive guards, if we restrict ourselves to the A path quantifiers, but still permit all three type of properties, then the cutoff can be reduced, in quadratic time in the size of the template processes, to something of the form $(1, ..., 2, ..., 1)$ or $(1, ..., 3, ..., 1)$. In fact, depending on the type of property, we can show that it is possible to simplify the guards to ensure that only two or three classes need be retained. On the other hand, for conjunctive guards, if we restrict ourselves to model checking over infinite paths or over finite paths, then sharper cutoffs of the form $(1,...,3,...,1)$, $(1,...,2,...,1)$ or even $(1,...,1)$ can, in some cases, be obtained.

The rest of the paper is organized as follows. Section 2 defines the system model. Section 3 describes how to exploit the symmetry inherent in the model and correctness properties. Sections 4 and 5 prove the results pertaining to disjunctive and conjunctive guards respectively. We show some applications of our results in Section 6. In the concluding Section 7, we discuss related work.

2   The System Model

We focus on systems comprised of multiple heterogeneous *classes* of processes modelled as *synchronization skeletons*(cf. [2]) . Here, an individual concrete process has a transition of the form $l \xrightarrow{g} m$ indicating that the process can transit from local state l to local state m, provided the guard g is true. Each class is specified by giving a generic process *template*. If I is (an) index set $\{1, \ldots, n\}$, then we use U^I, or $(U)^{(n)}$ for short, to denote the concurrent system $U^1 \| \ldots \| U^n$ comprised of the n isomorphic (up to re-indexing) processes U^i running in parallel asynchronously. For a system with k classes associated with the given templates $U_1, U_2, ..., U_k$, we have corresponding (disjoint) index sets $I_1, I_2, \ldots I_k$. Each index set I_j is (a copy of) an interval $\{1, \ldots, c\}$ of natural numbers, denoted $\{1_j, \ldots, n_j\}$ for emphasis[1]. In practice, we assume the k index sets are specified by giving a k-tuple $(n_1, ..., n_k)$ of natural numbers, corresponding to I_1 being (a copy of) interval $\{1, \ldots n_1\}$ through I_k being (a copy of) interval $\{1, \ldots, n_k\}$.

Given family $(U_1, ..., U_k)$ of k template processes, and a k-tuple $(n_1, ..., n_k)$ of natural numbers, we let $(U_1, ..., U_k)^{(n_1, ..., n_k)}$ denote the concrete system composed on n_1 copies of U_1 through n_k copies of U_k running in parallel asynchronously (i.e., with interleaving semantics). A template process $U_l = (S_l, R_l, i_l)$ for class l, is comprised of a finite set S_l of (local) states, a set of transition

---

[1] e.g., if I_1 is a copy of $\{1, 2, 3\}$, the said copy is denoted $\{1_1, 2_1, 3_1\}$. Informally, subscripted index 3_1 means process 3 of class 1; formally, it is the ordered pair $(3, 1)$ as is usual with indexed logics

edges R_l, and an initial(local) state i_l. Each transition R_l is labelled with a guard, a boolean expression over atomic propositions corresponding to local states of other template processes. Then given index i and template process U_l, $U_l^i = (S_l^i, R_l^i, i_l^i)$ is used to denote the ith copy of the template process U_l. Here S_l^i, the state set of U_l^i, R_l^i its transition relation and i_l^i its initial state are obtained from S_l, R_l and i_l respectively by uniformly superscripting the states of U_l with i. Thus, for local states s_l, t_l of S_l, s_l^i, t_l^i denote local states of U_l^i and $(s_l, t_l) \in R_l$ iff $(s_l^i, t_l^i) \in R_l^i$.

Given guards of transitions in the template process, we now describe how to get the corresponding guards for the concrete process U_l^i of $(U_1, ..., U_k)^{(n_1,...,n_k)}$. In this paper, we consider the following two types of guards.

i) Disjunctive guards - of the general form $(a_1 \vee ... \vee b_1) \bigvee ... \bigvee (a_k \vee ... \vee b_k)$, where the various $a_l, ..., b_l$ are (propositions identified with the) local states of template U_l, label each transition $(s_l, t_l) \in R_l$. In concrete process U_l^i of the system $(U_1, ..., U_k)^{(n_1,...,n_k)}$, the corresponding transition $(s_l^i, t_l^i) \in R_l^i$ is then labelled by the guard

$$\bigvee_{r \neq i} (a_l^r \vee ... \vee b_l^r) \vee \bigvee_{j \neq l} (\bigvee_{k \in [1..n_j]} (a_j^k \vee ... \vee b_j^k)),$$

where proposition a_j^k is understood to be true when process k in class U_j i.e. U_j^k is in local state a_j for template process U_j.

ii) Conjunctive guards with initial state - of the general form $(i_1 \vee a_1 \vee ... \vee b_1) \bigwedge ... \bigwedge (i_k \vee a_k \vee ... \vee b_k)$. In concrete process i of class l, U_l^i, in the system $(U_1, ..., U_k)^{(n_1,...,n_k)}$, the corresponding transition is then labelled by the guard

$$\bigwedge_{r \neq i} (i_l^r \vee a_l^r \vee ... \vee b_l^r) \wedge \bigwedge_{j \neq l} (\bigwedge_{k \in [1..n_j]} (i_j^k \vee a_j^k \vee ... \vee b_j^k)).$$

Note that the initial local states of processes must be present in these guards. Thus, the inital state of a process has a "neutral" character so that when process j is in its initial state, it does not prevent progress by another process i. This natural condition permits modelling a broad range of applications (and is helpful technically).

We now formalize the asynchronous concurrent (interleaving) semantics. A process transition with guard g is *enabled* in global state s iff $s \models g$ i.e., g is true over the local states in s. A transition can be fired in global state s iff its guard g is enabled. Let $(U_1, ..., U_k)^{(n_1,...,n_k)} = (S^{(n_1,...,n_k)}, R^{(n_1,...,n_k)}, i^{(n_1,...,n_k)})$ be the global state transition graph of the system instance $(n_1, n_2, ..., n_k)$. A state $s \in S^{(n_1,...,n_k)}$ is written as a $(n_1 + ... + n_k)$-tuple $(u_1^1, ..., u_1^{n_1}, u_2^1, ..., u_k^{n_k})$ where the projection of s onto process i of class l, denoted $s(l, i)$, equals u_l^i, the local state of the ith copy of the template process U_l. The initial state $i^{(n_1,...,n_k)} = (i_1^1, ..., i_k^{n_k})$. A global transition $(s, t) \in R^{(n_1,...,n_k)}$ iff t results from s by firing an enabled transition of some process i.e., there exist i, l such that the guard labelling $(u_l^i, v_l^i) \in R_l^i$ is enabled at s, $s(l, i) = u_l^i$, $t(l, i) = v_l^i$, and for all

$(j,k) \neq (i,l)$, $s(k,j) = t(k,j)$. We write $(U_1, ..., U_k)^{(n_1,...,n_k)} \models f$ to indicate that the global state graph of $(U_1, ..., U_k)^{(n_1,...,n_k)}$ satisfies f at initial state $i^{(n_1,...,n_k)}$.

Finally, for global state s, define $Set(s) = \{t \mid s \text{ contains an indexed local copy of } t\}$. For computation path $x = x_0, x_1, ...$ we define $PathSet(x) = \bigcup_i Set(x_i)$. We say that the sequence of global states $y = y_0, y_1, ...$ is a *stuttering* of computation path x iff there exists a parsing $P_0 P_1 ...$ of y such that for all $j \geq 0$ there is some $r > 0$ with $P_j = (x_j)^r$ (cf. [3]). Also, we extend the definition of projection to include computation sequences as follows: for $i \in [1..n_l]$, the sequence of local states $x_0(l, i), x_1(l, i), ...$ is denoted by $x(l, i)$.

3 Appeal to Symmetry

We can exploit symmetry inherent in the system model and the properties in the spirit of "state symmetry" codified by [8](cf. [16],[12]) to simplify our proof obligation. To establish formulae of types $\bigwedge_{i_l} f(i_l), \bigwedge_{i_l \neq j_l} f(i_l, j_l)$ and $\bigwedge_{i_l, j_m} f(i_l, j_m)$, it suffices to show the results with the formulae replaced by $f(1_l), f(1_l, 2_l)$ and $f(1_l, 1_m)$, respectively. The basic idea is that in a system comprised of fully interchangeable processes 1 through n of a given class, symmetry considerations dictate that process 1 satisfies a property iff each process $i \in [1..n]$ satisfies the property. Proofs are omitted for the sake of brevity.

4 Systems with Disjunctive Guards

In this section, we show how to reduce the PMCP for systems with disjunctive guards, to model checking systems of sizes bounded by a small cutoff, where the size of the cutoff for each process class is essentially the number of local states of individual process template for the class. This yields decidability for this formulation of PMCP, a pleasant result since PMCP is undecidable in full generality. But this result, by itself, does not give us an efficient decision procedure for the PMCP at hand. We go on to show that in the case of universal-path-quantified specification formulae (Ah), efficient decidability can be obtained.

4.1 Properties Ranging over All Processes in a Single Class

We will first establish the

Theorem 4.1.1 (Disjunctive Cutoff Theorem).
Let f be $\bigwedge_{i_l} Ah(i_l)$ or $\bigwedge_{i_l} Eh(i_l)$, where h is an LTL\X *formula and* $l \in [1..k]$. *Then we have the following*

$$\forall (n_1, ..., n_k) \succeq (1, ..., 1) : (U_1, ..., U_k)^{(n_1,...,n_k)} \models f \quad iff$$
$$\forall (d_1, ..., d_k) \preceq (c_1, ..., c_k) : (U_1, ..., U_k)^{(d_1,...,d_k)} \models f,$$

where the cutoff $(c_1, ..., c_k)$ *is given by* $c_l = |U_l| + 2$, *and for* $i \neq l : c_i = |U_i| + 1$.

As a corollary, we will have the

Theorem 4.1.2 (Disjunctive Decidability Theorem). PMCP *for systems with disjunctive guards and single-index assertions as above is decidable in exponential time.*

Proof idea

By the Disjunctive Cutoff Theorem, it is enough to model check each of the exponentially many exponential size state graphs corresponding to systems $(U_1, \ldots, U_k)^{(d_1, \ldots, d_k)}$ for all $(d_1, \ldots, d_k) \preceq (c_1, \ldots, c_k)$. □

For notational brevity, we establish the above results for systems with just two process classes. We begin by proving the following lemmas.

Lemma 4.1.1 (Disjunctive Monotonicity Lemma).

(i) $\forall n \geq 1 : (V_1, V_2)^{(1,n)} \models Eh(1_2)$ *implies* $(V_1, V_2)^{(1,n+1)} \models Eh(1_2)$.

(ii) $\forall n \geq 1 : (V_1, V_2)^{(1,n)} \models Eh(1_1)$ *implies* $(V_1, V_2)^{(1,n+1)} \models Eh(1_1)$.

Proof idea

(i) The idea is that for any computation x of $(V_1, V_2)^{(1,n)}$, there exists an analogous computation y of $(V_1, V_2)^{(1,n+1)}$ wherein the $(n+1)$st copy of template process V_2 stutters in its initial state and the rest of the processes behave as in x.

(ii) This part follows by using a similar argument. □

The following lemma allows reduction in system size, one coordinate at a time.

Lemma 4.1.2 (Disjunctive Bounding Lemma).

(i) $\forall n \geq |V_2| + 2 : (V_1, V_2)^{(1,n)} \models Eh(1_2)$ *iff* $(V_1, V_2)^{(1,c_2)} \models Eh(1_2)$, *where* $c_2 = |V_2| + 2$.

(ii) $\forall n \geq |V_2| + 1 : (V_1, V_2)^{(1,n)} \models Eh(1_1)$ *iff* $(V_1, V_2)^{(1,|V_2|+1)} \models Eh(1_1)$.

Proof

(i) ($\Rightarrow$) Let $x = x_0, x_1, \ldots$ denote a computation sequence of $(V_1, V_2)^{(1,n)}$. Define $Reach = \{s_1, \ldots, s_r\}$ to be the set of all local states of template process V_2 occuring in x. For $s_t \in Reach$, let $t_1, t_2, \ldots, t_m$ be a finite local computation of minimal length in x ending in s_t. Then we use $MinLength(s_t)$ to denote m and $MinComputation(s_t)$ to denote the sequence $t_1, t_2, \ldots, t_{m-1}, (t_m)^\omega$. Let $v = (i_2)^\omega$. If x is an infinite computation sequence and $x(1,1)$ and $x(2,1)$ are finite local computations, then there exists an infinite local computation sequence u, say. In that case, reset $v = u$.

Construct a formal sequence $y = y_0, y_1, \ldots$ of global states of $(V_1, V_2)^{(1,|V_2|+1)}$ from x as follows

1. $y(1,1) = x(1,1)$ and $y(2,1) = x(2,1)$ i.e. the local computation paths in x of process index 1 of classes V_1, V_2 are preserved, and

2. For each state $s_j \in Reach$, we set $y(2, j+1) = MinComputation(s_j)$ i.e. we let the $(j+1)$st copy of V_2 perform a local computation of minimum length in x leading to s_j and then let it stutter in s_j forever. The above condition has the implication that for all $i \geq 1, Set(x_i) \subseteq Set(y_i)$. To see this, let $t \in Set(x_i)$. Then, $MinLength(t) \leq i$. Also, $t \in Set(x_i)$ implies that $t \in Reach$ i.e. $t = s_q$ for some $q \in [1..r]$. Then $y(2, q+1)$ stutters in s_q for all $k \geq MinLength(s_t)$ and therefore for all $k \geq i$, also. Hence $y_i(2, q+1)$ is an indexed copy of s_q, i.e. $t \in Set(y_i)$. Thus for all $i \geq 1, Set(x_i) \subseteq Set(y_i)$.

3. $y(2, c_2) = v$. This ensures that if x is an infinite computation sequence then in y infinitely many local transitions are fired.

However, it might be the case that sequence y violates the interleaving semantics requirement. Clearly, this happens iff the following scenario occurs. Let states $s_p, s_q \in Reach$, be such that $MinComputation(s_p)$ and $MinComputation(s_q)$ are realized by the same local computation of x and suppose that $MinLength(s_p) \leq MinLength(s_q)$. Then, if for $i < MinLength(s_p)$, (t_i, t_{i+1}) is a transition in $MinComputation(s_p)$, $(y_i(2, p+1), y_{i+1}(2, p+1))$ and $(y_i(2, q+1), y_{i+1}(2, q+1))$ are both local transitions driving y_i to y_{i+1}. This violates the interleaving semantics condition requiring that there be atmost one local transition driving each global transition. There are two things to note here. First, for a transition (y_i, y_{i+1}), the violation occurs only for values of $i \leq max_{j \in [1..r]} MinLength(s_j)$ and secondly, for a fixed i, all violations are caused by a unique template transition (s, t) of V_2, namely one which was involved in the transition (x_i, x_{i+1}).

To solve this problem, we construct a sequence of states $w = w_0, w_1, \ldots$ from y by "staggering" copies of the same local transition as described below. Let (y_i, y_{i+1}) be a transition where the interleaving semantics requirement is violated by process indices $in_1, \ldots, in_d$ of V_2 executing indexed copies $(s_2^{in_1}, t_2^{in_1}), \ldots, (s_2^{in_d}, t_2^{in_d})$ respectively of the template transition (s_2, t_2) of V_2. Replace (y_i, y_{i+1}) with a sequence $u_1, u_2, \ldots, u_f$ such that $u_1 = y_i$, $u_f = y_{i+1}$ and for all j, transition (u_j, u_{j+1}) results by executing local transition $(s_2^{in_j}, t_2^{in_j})$. Clearly the interleaving semantics requirement is met as atmost one local transition is executed for each global transition. Also, it is not hard to see that for all j, $Set(y_i) \subseteq Set(u_j)$ and hence for all k, transition (u_k, u_{k+1}) is valid. Finally, note that states with indices other than $in_1, \ldots, in_d$ are made to stutter finitely often in $u_1, \ldots, u_f$ which is allowed since we are considering only formulae without the next-time operator X.

Thus, given a computation path x of $(V_1, V_2)^{(1,n)}$, we have constructed a stuttering computation path w of $(V_1, V_2)^{(1,c_2)}$, such that the local computation sequence $w(2, 1)$ is a stuttering of the local computation sequence $x(2, 1)$. From this path correspondence, we easily have the result.

($\Leftarrow$) The proof follows by repeated application of the Disjunctive Monotonicity Lemma.

(ii) This part follows by using a similar argument.   □

The following lemma allows reduction in system size over multiple coordinates simultaneously (2 coordinates for notational brevity).

Lemma 4.1.3 (Disjunctive Truncation Lemma).

$\forall n_1, n_2 \geq 1 : (U_1, U_2)^{(n_1,n_2)} \models \mathsf{E}h(1_2)$ iff $(U_1, U_2)^{(n'_1,n'_2)} \models \mathsf{E}h(1_2)$, where $n'_2 = min(n_2, |U_2| + 2)$ and $n'_1 = min(n_1, |U_1| + 1)$.

Proof

If $n_2 > |U_2| + 2$, set $V_1 = U_1^{n_1}$ and $V_2 = U_2$. Then, $(U_1, U_2)^{(n_1,n_2)} \models \mathsf{E}h(1_2)$ iff $(V_1, V_2)^{(1,n_2)} \models \mathsf{E}h(1_2)$ iff $(V_1, V_2)^{(1,n'_2)} \models \mathsf{E}h(1_2)$ (by the Disjunctive Bounding Lemma) iff $(U_1, U_2)^{(n_1,n'_2)} \models \mathsf{E}h(1_2)$.

If $n_1 \leq |U_1| + 1$, then $n_1 = n'_1$ and we are done, else set $V_1 = U_2^{n'_2}$ and $V_2 = U_1$. Then, $(U_1, U_2)^{(n_1,n'_2)} \models \mathsf{E}h(1_2)$ iff $(U_2, U_1)^{(n'_2,n_1)} \models \mathsf{E}h(1_1)$ iff $(V_1, V_2)^{(1,n_1)} \models \mathsf{E}h(1_1)$ iff $(V_1, V_2)^{(1,|U_1|+1)} \models \mathsf{E}h(1_1)$ (by the Disjunctive Bounding Lemma) iff $(U_1, U_2)^{(n'_1,n'_2)} \models \mathsf{E}h(1_2)$.   □

An easy but important consequence of the Disjunctive Truncation Lemma is the following

Theorem 4.1.3 (Disjunctive Cutoff Result).

Let f be $\bigwedge_{i_l} \mathsf{A}h(i_l)$ or $\bigwedge_{i_l} \mathsf{E}h(i_l)$, where h is a LTL\X formula and $l \in [1..2]$. Then we have the following

$$\forall (n_1, n_2) \succeq (1, 1) : (U_1, U_2)^{(n_1,n_2)} \models f \quad iff$$
$$\forall (d_1, d_2) \preceq (c_1, c_2) : (U_1, U_2)^{(d_1,d_2)} \models f$$

where the cutoff (c_1, c_2) is given by $c_l = |U_l| + 2$, and for $i \neq l : c_i = |U_i| + 1$.

Proof

By appeal to symmetry and the fact that A and E are duals, it suffices to prove the result for formulae of the type $\mathsf{E}h(1_2)$. The ($\Rightarrow$) direction is trivial. For the ($\Leftarrow$) direction, let $n_1, n_2 \geq 1$. Define $n'_1 = min(n_1, |U_1|+1)$, $n'_2 = min(n_2, |U_2|+2)$. Then, $(U_1, U_2)^{(n_1,n_2)} \models f(1_2)$ iff $(U_1, U_2)^{(n'_1,n'_2)} \models f(1_2)$ by the Disjunctive Truncation Lemma. The latter is true since $(n'_1, n'_2) \preceq (c_1, c_2)$. This proves the cutoff result.   □

The earlier-stated Cutoff *Theorem* re-articulates the above Cutoff *Result* more generally for systems with $k \geq 1$, different classes of processes; since its proof is along similar lines but is notationally more complex, we omit it for the sake of brevity.

4.2 Efficient Decidability for "For All Future" Properties

It can be shown that for "for some future" properties, corresponding to formulae of the type $\bigwedge \mathsf{E}h$, the reduction entailed in the previous result is, in general, the best possible. We omit the proof for the sake of brevity.

However, for universal-path-quantified properties, it is possible to be much more efficient. We will establish the

Theorem 4.2.1 (Reduction Theorem). *Define $V = U_l'$ if for some $l \in [1..k]$, the transition graph for U_l' has a nontrivial strongly connected component else set $V = U_1'$. Then, $(U_1, ..., U_k)^{(c_1, ..., c_k)} \models \bigwedge_{i_l} Ah(i_l)$ iff $(U_l', V)^{(1,1)} \models Ah(1_1)$, where $c_l = |U_l| + 2$, $c_i = |U_i| + 1$ for $i \neq l$ and U_l' is the simplified process that we get from U_l by the reduction technique described below.*

This makes precise our claim that for formulae of the type $\bigwedge_{i_l} Ah(i_l)$, it is possible to give efficient decision procedures for the PMCP at hand, by reducing it to model checking systems consisting of two or three template processes.

To this end, we first prove the following lemma which states that the PMCP problem for the above mentioned properties reduces to model checking just the *single* system instance of size equal to the (small) cutoff (as opposed to all systems of size less than or equal to the cutoff).

Lemma 4.2.1 (Single-Cutoff Lemma).
$\forall n_1, n_2 \geq 1 : (U_1, U_2)^{(n_1, n_2)} \models Ah(1_2)$ iff $(U_1, U_2)^{(c_1, c_2)} \models Ah(1_2)$, where $c_1 = |U_1| + 1$ and $c_2 = |U_2| + 2$.

Proof

($\Rightarrow$) This direction follows easily by instantiating $n_1 = c_1$ and $n_2 = c_2$ on the left hand side.

($\Leftarrow$) Choose arbitrary $k_1, k_2 \geq 1$. Set $k_1' = min(k_1, c_1)$ and $k_2' = min(k_2, c_2)$. Then, $(U_1, U_2)^{(k_1, k_2)} \models Eh(1_2)$ iff $(U_1, U_2)^{(k_1', k_2')} \models Eh(1_2)$ (by the Disjunctive Truncation Lemma) which implies $(U_1, U_2)^{(c_1, c_2)} \models Eh(1_2)$ (by repeated application of the Disjunctive Monotonicity Lemma). Now, by contraposition, $(U_1, U_2)^{(c_1, c_2)} \models Ah(1_2)$ implies $(U_1, U_2)^{(k_1, k_2)} \models Ah(1_2)$. Since k_1, k_2 were arbitrarily chosen, the proof is complete. □

Next, we transform the given template processes and follow that up with lemmas giving the soundness and completeness proofs for the transformation. Given template processes $U_1, ..., U_k$, define $ReachableStates(U_1, ..., U_k) = (S_1', ..., S_k')$, where $S_i' = \{ t \mid t \in S_i,$ such that for some $n_1, n_2, ..., n_k \geq 1,$ there exists a computation path of $(U_1, ..., U_k)^{(n_1, ..., n_k)},$ leading to a global state that contains a local indexed copy of $t \}$. $\forall j \geq 0, \forall l \in [1..k]$, we define P_l^j as follows:

$$P_l^0 = \{i_l\}.$$
$$P_l^{j+1} = P_l^j \bigcup \{p' : \exists p \in P_l^j : \exists p \xrightarrow{g} p' \in R_l \text{ and expression } g \text{ contains a}$$
$$\text{state in } \bigcup_t P_t^j \}.$$

For $l \in [1..k]$, define $P_l = \bigcup_j P_l^j$. Then we have the

Lemma 4.2.2 (Soundness Lemma). *Given* j, *for all* $l \in [1..k]$, *define* $a_l = |P_l^j|$. *Then, there exists a finite computation sequence* $x = x_0, x_1, ..., x_m$ *of* $(U_1, ..., U_k)^{(a_1, ..., a_k)}$, *such that* $\forall l \in [1..k] : \forall s_l \in P_l^j : (\exists p \in [1..a_l] : x_m(l, p) = s_l^p)$.

Proof

The proof is by induction on j. The base case, $j = 0$, is vacuously true. Assume that the result holds for $j \leq u$ and let $y = y_0, y_1, ..., y_t$ be a computation sequence of $(U_1, ..., U_k)^{(r_1, ..., r_k)}$, where $r_l = |P_l^u|$, with the property that $\forall l \in [1..k]: \forall s_l \in P_l^u : (\exists p \in [1..r_l] : x_m(l, p) = s_l^p)$.

Now, assume that $P_l^{u+1} \neq P_l^u$, and let $s_l \in P_l^{u+1} \setminus P_l^u$. Furthermore, let (s_l', s_l) be the transition that led to the inclusion of s_l into P_l^{u+1}. Clearly, $s_l' \in P_l^j$. Then, by the induction hypothesis, $\exists q \in [1..r_l] : y_t(l, q)$ is an indexed copy of s_l'. Consider the sequence $y' = y_0', y_1', ..., y_{2t+1}'$ of states of $(U_1, ..., U_k)^{(r_1, ..., r_l+1, ..., r_k)}$, where for $i \in [1..k] : c \in [1..r_i] : y'(i, c) = y(i, c)(y_t(i, c))^{t+1}$ and $y'(l, r_l + 1) = (i_l^{r_l+1})^t z$, where z is $y(l, q)s_l^{r_l+1}$ with the index q replaced by r_l+1. It can be seen that y' is a valid stuttering computation path of $(U_1, ..., U_k)^{(r_1, ..., r_l+1, ..., r_k)}$, where y_{2t+1}' has the property that $\forall l \in [1..k] : \forall s_l \in P_l^u : \exists p \in [1..r_l] : y_{2t+1}'(l, p) = s_l^p$ and $y_{2t+1}'(l, r_l + 1) = s_l^{r_l+1}$. Repeating the above procedure for all states in $P_l^{u+1} \setminus P_l^u$, we get a computation path with the desired property. This completes the induction step and proves the lemma. □

Lemma 4.2.3 (Completeness Lemma). $(S_1', ..., S_k') = (P_1, ..., P_k)$.

Proof

By the above lemma, $\forall i \in [1..k] : P_i \subseteq S_i'$. If possible, suppose that $(S_1', ..., S_k') \neq (P_1, ..., P_k)$. Then, the set $D = \bigcup_i (S_i' - P_i) \neq \emptyset$. For definiteness, let $s_l \in D \bigcap S_l$. Then by definition of S_l', there exists a finite computation sequence $x = x_0, x_1, ..., x_m$ such that for some i, $x_m(l, i) = s_l^i$. Let $j \in [0..m]$ be the smallest index such that $Set(x_j) \bigcap D \neq \emptyset$. Then, $PathSet(x_0, ..., x_{j-1}) \subseteq \bigcup_i P_i$ which implies that there exists a transition (s_l', s_l) in R_l, with guard g such that $x_{j-1} \models g$. But this implies that for some t, s_l would be included in P_l^t i.e. $s_l \in P_l$, a contradiction to our assumption that $s_l \in D$. Thus $D = \emptyset$ and we are done. □

We now modify the k-tuple of template processes $(U_1, ..., U_k)$ to get the k-tuple $(U_1', ..., U_k')$, where $U_i' = (S_i', R_i', i_i)$, with $(s_i, t_i) \in R_i'$ iff guard g_i labelling (s_i, t_i) in U_i contains an indexed copy of a state in $\bigcup_{i \in [1..k]} S_i'$. Furthermore, any transition in the new system is labelled with g_U, a universal guard that evaluates to true irrespective of the current global state of the system. The motivation behind these definitions is that since for any $n_1, n_2, ..., n_k \geq 1$, no indexed copy of states in $S_i \setminus S_i'$ is reachable in any computation of $(U_1, ..., U_k)^{(n_1, ..., n_k)}$, we can safely delete these states from their respective template process. Also, any guard of a template process involving only states in $S_i \setminus S_i'$, will then always evaluate to false and hence the transition labelled by this guard will never be fired. This justifies deleting such transitions from the transition graph of respective template processes. This brings us to the following Reduction Result, which by appeal to symmetry yields the Reduction Theorem stated before.

Theorem 4.2.2 (Reduction Result). *Define* $V = U_l'$ *if for some* $l \in [1..k]$, *the transition graph for* U_l' *has a nontrivial strongly connected component else* *set* $V = U_1'$. *Then,* $(U_1, ..., U_k)^{(c_1, ..., c_k)} \models Ah(1_p)$ *iff* $(U_p', V)^{(1,1)} \models Ah(1_1)$, *where* $c_p = |U_p| + 2$ *and* $c_i = |U_i| + 1$ *for* $i \neq p$.

Proof

We show that $(U_1, ..., U_k)^{(c_1, ..., c_k)} \models Eh(1_p)$ iff $(U_p', V)^{(1,1)} \models Eh(1_1)$. For definiteness, let $V = U_r'$.

($\Rightarrow$) Define sequence $u = (i_r)^\omega$. If U_r' has a nontrivial strongly connected component, then there exists an infinite path v, say, in its transition graph. In that case, reset $u = v$.

Let $x = x_1, x_2, ...$ be a computation sequence of $(U_1, ..., U_k)^{(c_1, ..., c_k)}$. Define a formal sequence $y = y_1, y_2, ...$ as follows. Set $y(1, 1) = x(p, 1)$ and in case $x(p, 1)$ is a finite computation sequence of length f, say, set $y(2, 1) = (i_r)^f u$ else set $y(2, 1) = (i_r)^\omega$. To prove that y is a valid computation sequence of $(U_p', V)^{(1,1)}$, it suffices to show that all transitions of local path $y(1, 1)$ are valid. This follows from the definition of U_1' and by noting that all states occuring in $x(p, 1)$ are reachable and all transitions in $x(p, 1)$ are labelled by guards whose expressions involve a state in $\bigcup_j S_j$ and hence they occur in R_p'.

($\Leftarrow$) By the Soundness and Completeness lemmas, it follows that there exists a finite computation path $u = u_0, u_1, ..., u_m$ of $(U_1, ..., U_k)^{(|U_1|, ..., |U_k|)}$ starting at $i^{(|U_1|, ..., |U_k|)}$, such that $\forall j \in [1..k] : \forall q_j \in S_j' : \exists t \in [1..|U_j|] : u_m(j, t) = q_j^t$. Let $x = x_0, x_1, ...$ be a computation path of $(U_p', V)^{(1,1)}$. Define a formal sequence $y = y_0, y_1, ...$ of states of $(U_1, ..., U_k)^{(c_1, ..., c_k)}$ as follows. Set $y(p, 1) = ((i_p)^m)x(1, 1)$, $y(r, 1) = ((i_r)^m)x(2, 1)$, $\forall z \in [1..|U_p|] : y(p, z + 1) = u(p, z)$, $\forall z \in [1..|U_r|] : y(r, z + 1) = u(r, z)$, and $\forall j \in [1..k], j \neq p, r : \forall z \in [1..|U_j|] : y(j, z) = u(j, z)(u_m(j, z))^\omega$. Note that, $\forall l \geq m : Set(y_l) = \bigcup_j S_j'$ and hence for all $l \geq m$, all template transitions in $R_1' \cup R_r'$ are enabled in y_l. Thus for all $i \geq m$, all transitions (y_i, y_{i+1}) are valid and hence it follows that y is a stuttering of a valid computation path of $(U_1, ..., U_k)^{(c_1, ..., c_k)}$ with local path $y(1, 1)$ being a stuttering of local path $x(1, 1)$. The path correspondence gives us the result. $\square$

Finally, we get the

Theorem 4.2.3 (Efficient Decidability Theorem). *For systems with disjuctive guards and properties of the type* $\bigwedge_{i_l} Ah(i_l)$, *the PMCP is decidable in time quadratic in the size of the given family* $(U_1, ..., U_k)$, *where size is defined as* $\sum_j (|S_j| + |R_j|)$, *and linear in the size of the Büchi Automaton for* $\neg h(1_l)$.

Proof We first argue that we can construct the simplified system U'_l efficiently. By definition, $\forall j \geq 0 : P^j_l \subseteq P^{j+1}_l$. Let $P^i = \bigcup_l P^i_l$. Then, it is easy to see that, $\forall j \geq 0 : P^j \subseteq P^{j+1}$ and if $P^j = P^{j+1}$, then $\forall i \geq j : P^i = P^j$. Also, $\forall i : P^i \subseteq \bigcup_l \bar{S'_l}$. Thus to evaluate sets P^j_l, for all j, it suffices to evaluate them for values of $j \leq \sum_l |S_l|$. Furthermore, given P^j_l to evaluate P^{j+1}_l, it suffices to make a pass through all transitions leading to states in $S_l \setminus P^j_l$ to check if a guard leading to any of these states contains a state in $\bigcup_l P^j_l$. This can clearly be accomplished in time $\sum_j (|S_j| + |R_j|)$. The above remarks imply that evaluation of sets P^j_l, can be done in time $O((\sum_j (|S_j| + |R_j|))^2)$. Furthermore, given p, whether U'_p has a nontrivial strongly connected component can be decided in time $O(|S'_p| + |R'_p|)$ by constructing all strongly connected components of U'_p. Thus, determining whether such a p exists can be done in time $O(\sum_j (|S_j| + |R_j|))$.

The Reduction Theorem reduces the PMCP problem to model checking for the system $(U'_l, V)^{(1,1)}$, where $V = U'_r$ if for some $r \in [i..k]$, the transition graph for U'_r has a nontrivial strongly connected component else $V = U'_1$. Now, $(U'_l, V)^{(1,1)} \models Ah(1_1)$ iff $(U'_l, V)^{(1,1)} \models \neg E\neg h(1_1)$. Thus it suffices to check whether $(U'_l, V)^{(1,1)} \models E\neg h(1_1)$, for which we use the automata-theoretic approach of [20]. We construct a Büchi Automaton $\mathcal{B}_{\neg h}$ for $\neg h(1_1)$, and check that language of the product Büchi Automaton $\mathcal{P}$, of $(U'_l, V)^{(1,1)}$ and $\mathcal{B}_{\neg h}$ is non-empty(cf [14]). Since the nonemptiness check for $\mathcal{P}$ can be done in time linear in the size of $\mathcal{P}$, and the size of $(U'_l, V)^{(1,1)}$ is $O((\sum_j (|S_j| + |R_j|))^2)$, we are done. $\square$

4.3   Properties Ranging over Pairs of Processes from Two Classes

Using similar kinds of arguments as were used in proving assertions in the sections 4.1 and 4.2, we can prove the following results.

Theorem 4.3.1 (Cutoff Theorem).
Let f be $\bigwedge_{i_l, j_m} Ah(i_l, j_m)$ or $\bigwedge_{i_l, j_m} Eh(i_l, j_m)$, where h is an LTL\X formula and $l, m \in [1..k]$. Then we have the following
$$\forall (n_1, \ldots, n_k) \succeq (1, \ldots, 1) : (U_1, \ldots, U_k)^{(n_1, \ldots, n_k)} \models f \quad iff$$
$$\forall (d_1, \ldots, d_k) \preceq (c_1, \ldots, c_k) : (U_1, \ldots, U_k)^{(d_1, \ldots, d_k)} \models f,$$
where the cutoff $(c_1, \ldots, c_k)$ is given by $c_l = |U_l| + 2$, $c_m = |U_m| + 2$ and for $i \neq l, m : c_i = |U_i| + 1$.

Theorem 4.3.2 (Reduction Theorem).
$(U_1, ..., U_k)^{(c_1, ..., c_k)} \models \bigwedge_{i_l, j_m} Ah(i_l, j_m)$ *iff* $(U'_l, U'_m)^{(1,1)} \models \bigwedge_{i_l, j_m} Ah(i_l, j_m)$,
where $c_l = |U_l| + 2$, $c_m = |U_m| + 2$ and $\forall i \neq l, m : c_i = |U_i| + 1$.

Again, we get the analogous Decidability Theorem and Efficient Decidability Theorem. Moreover, we can specialize these results to apply when $l=m$. This permits reasoning about formulae of the type $\bigwedge_{i_l \neq j_l} Ah(i_l, j_l)$ or $\bigwedge_{i_l \neq j_l} Eh(i_l, j_l)$, for properties ranging over all pairs of processes in a single class l.

5   Systems with Conjunctive Guards

The development of results for conjunctive guards closely resembles that for disjunctive guards. Hence, for the sake of brevity, we only provide a proof sketch for each of the results.

Lemma 5.1 (Conjunctive Monotonicity Lemma).
   (i) $\forall n \geq 1 : (V_1, V_2)^{(1,n)} \models Eh(1_2)$ *implies* $(V_1, V_2)^{(1,n+1)} \models Eh(1_2)$.
   (ii) $\forall n \geq 1 : (V_1, V_2)^{(1,n)} \models Eh(1_1)$ *implies* $(V_1, V_2)^{(1,n+1)} \models Eh(1_1)$.

Proof Sketch The intuition behind this lemma is that for any computation x of $(V_1, V_2)^{(1,n)}$, there exists an analogous computation y of $(V_1, V_2)^{(1,n+1)}$ wherein the $(n+1)$st copy of template process V_2 stutters in its initial state and the rest of the processes behave as in x.   □

Lemma 5.2 (Conjunctive Bounding Lemma).
   (i) $\forall n \geq 2|V_2| + 1 : (V_1, V_2)^{(1,n)} \models Eh(1_2)$ *iff* $(V_1, V_2)^{(1,c_2)} \models Eh(1_2)$, *where*
     $c_2 = 2|V_2| + 1$.
   (ii) $\forall n \geq 2|V_2| : (V_1, V_2)^{(1,n)} \models Eh(1_1)$ *iff* $(V_1, V_2)^{(1,2|V_2|)} \models Eh(1_1)$.

Proof Sketch
Let x be an infinite computation of $(V_1, V_2)^{(1,n)}$. Set $v = (i_2)^\omega$, where i_2 is the initial state of V_2. If none of $x(1,1)$ or $x(2,1)$ is an infinite local computation then there exists $l \neq 1$ such that $x(2,l)$ is an infinite local computation. In that case, reset $v = x(2,l)$. Construct a formal sequence y of $(V_1, V_2)^{(1,c_2)}$ as follows. Set $y(1,1) = x(1,1)$, $y(2,1) = x(2,1)$, $y(2,2) = v$ and $\forall j \in [3..c_2] : y(2,j) = (i_2)^\omega$. Then, it can be proved that y is a stuttering of a valid infinite computation of $(V_1, V_2)^{(1,c_2)}$.

    Now consider the case when $x = x_0 x_1 ... x_d$ is a deadlocked computation sequence of $(V_1, V_2)^{(1,n)}$. Let $S = Set(x_d) \cap S_2$. For each $s \in S$, define an index set I_s as follows. If there exists a unique indexed copy $x_d(2,in)$ of s in x_d set $I_s = \{in\}$ else set $I_s = \{in_1, in_2\}$, where $x_d(2,in_1)$ and $x_d(2,in_2)$ are indexed copies of s and $in_1 \neq in_2$. Let $I = \bigcup_s I_s$. Also, for index j and global state s define $Set(s,j) = \{t | t \in S_1 \cup S_2$ and t has a copy with index other than j in $s\}$.

    Construct a formal sequence $y = y_0, ..., y_d$ of states of $(V_1, V_2)^{(1,2|V_2|+1)}$ by projecting each global state x_i onto process 1 coordinate of V_1 and process $index$ coordinate of V_2, where $index = 1$ or $index \in I$. From our construction, it follows that for all j, $Set(y_j) \subseteq Set(x_j)$. Hence all transitions (y_i, y_{i+1}) are valid. Also for each $i \in [1..(2|V_2| + 1)]$, there exists $j \in [1..n]$ such that $y(2,i)$ is a projection of $x(2,j)$. Then, from our construction, it follows that $Set(x_d, j) = Set(y_d, i)$ and thus process V_2^i is deadlocked in y_d iff V_2^j is deadlocked in x_d. Then from the fact that x_d is deadlocked, we can conclude that y_d is a deadlocked state and hence y is a stuttering of a deadlocked computation of $(V_1, V_2)^{(1,c_2)}$.

    In both cases, when constructing y from x, we preserved the local computation sequence of process V_2^1. This path correspondence gives us the result.   □

Again as before, the following lemma allows reduction in system size over multiple coordinates simultaneously (2 coordinates for notational brevity).

Lemma 5.3 (Conjunctive Truncation Lemma).
$\forall n_1, n_2 \geq 1 : (U_1, U_2)^{(n_1, n_2)} \models Eh(1_2)$ *iff* $(U_1, U_2)^{(n'_1, n'_2)} \models Eh(1_2)$,
where $n'_2 = min(n_2, 2|U_2| + 1)$ *and* $n'_1 = min(n_1, 2|U_1|)$.

Proof Idea
Use the Conjunctive Bounding Lemma and associativity of the $\|$ operator. $\square$

Theorem 5.1 (Conjunctive Cutoff Result).
Let f *be* $\bigwedge_{i_l} Ah(i_l)$ *or* $\bigwedge_{i_l} Eh(i_l)$, *where* h *is a* LTL\X *formula and* $l \in [1..2]$. *Then we have the following*
$$\forall(n_1, n_2) \succeq (1, 1) : (U_1, U_2)^{(n_1, n_2)} \models f \quad \textit{iff}$$
$$\forall(d_1, d_2) \preceq (c_1, c_2) : (U_1, U_2)^{(d_1, d_2)} \models f,$$
where the cutoff (c_1, c_2) *is given by* $c_l = 2|U_l| + 1$, *and for* $i \neq l : c_i = 2|U_i|$.

Proof Sketch Follows easily from the Truncation Lemma. $\square$

More generally, for systems with $k \geq 1$ class of processes we have

Theorem 5.2 (Conjunctive Cutoff Theorem).
Let f *be* $\bigwedge_{i_l} Ah(i_l)$ *or* $\bigwedge_{i_l} Eh(i_l)$, *where* h *is a* LTL\X *formula and* $l \in [1..k]$. *Then we have the following*
$$\forall(n_1, ..., n_k) \succeq (1, ..., 1) : (U_1, ..., U_k)^{(n_1, ..., n_k)} \models f \quad \textit{iff}$$
$$\forall(d_1, ..., d_k) \preceq (c_1, ...c_k) : (U_1, ..., U_k)^{(d_1, ..., d_k)} \models f,$$
where the cutoff $(c_1, ..., c_k)$ *is given by* $c_l = 2|U_l| + 1$, *and for* $i \neq l : c_i = 2|U_i|$.

Although the above results yield decidability for PMCP in the Conjunctive guards case, the decision procedures are not efficient.

We now show that if we limit path quantification to range over *infinite* paths only (i.e. ignore deadlocked paths); or *finite* paths only; then we can give an efficient decision procedure for this version of the PMCP. We use A_{inf} for "for all infinite paths", E_{inf} for "for some infinite path", A_{fin} for "for all finite paths", and E_{fin} for "for some finite path".

Theorem 5.3 (Infinite Conjunctive Reduction Theorem).
For any LTL\X *formula* h *and* $l \in [1..k]$, *we have*
(i) $\forall(n_1, ..., n_k) \succeq (1, ..., 1) : (U_1, ..., U_k)^{(n_1, ..., n_k)} \models \bigwedge_{i_l} E_{inf}h(i_l)$, *iff*
     $(U_1, ..., U_k)^{(c_1, ..., c_k)} \models E_{inf}h(1_l)$;
(ii) $\forall(n_1, ..., n_k) \succeq (1, ..., 1) : (U_1, ..., U_k)^{(n_1, ..., n_k)} \models \bigwedge_{i_l} A_{inf}h(i_l)$, *iff*
     $(U_1, ..., U_k)^{(c_1, ..., c_k)} \models A_{inf}h(1_l)$,
where $(c_1, ..., c_k) = (1, ..., \underbrace{2}_{l}, ..., 1)$.

Proof Sketch

To obtain (a), by appeal to symmetry, it suffices to establish that for each $(n_1, \ldots, n_k) \succeq (1, \ldots, 1) : (U_1, \ldots, U_k)^{(n_1, \ldots, n_k)} \models \mathsf{E}_{\mathsf{inf}} h(1_l)$ iff $(U_1, \ldots, U_k)^{(c_1, \ldots, c_k)}$ $\models \mathsf{E}_{\mathsf{inf}} h(1_l)$. Using the duality between $\mathsf{A}_{\mathsf{inf}}$ and $\mathsf{E}_{\mathsf{inf}}$ on both sides of the latter equivalence, we can also appeal to symmetry to obtain (b). We establish the latter equivalence as follows.

($\Rightarrow$) Let $x = x_0 \xrightarrow{b_0, g_0} x_1 \xrightarrow{b_1, g_1} \ldots$ denote an infinite computation of $(U_1, \ldots, U_k)^{(n_1, \ldots, n_k)}$, where b_i indicates which process fired the transition driving the system from global states x_i to x_{i+1} and g_i is the guard enabling the transition. Since x is infinite, it follows that there exists some process such that the result of projecting x onto that process results in a stuttering of an infinite local computation of the process. By appeal to symmetry, we can without loss of generality, assume that for each process class U_p, if a copy of U_p in $(U_1, \ldots, U_k)^{(n_1, \ldots, n_k)}$ has the above property then that copy is in fact the concrete process U_p^1 in case $p \neq l$ and the concrete process U_p^2 in case $p = l$ and local computation $x(l, 1)$ is finite.

Define a (formal) sequence $y = y_0 \xrightarrow{b'_0, g'_0} y_1 \xrightarrow{b'_1, g'_1} \ldots$ by projecting each global state x_i onto process 1 coordinate for each class U_p for $p \neq l$ and onto process coordinates 1 and 2 for process class U_l to get a state y_i. We let $b'_i = 1_l$ if b_i $= 1_l$, $b'_i = 2_l$ if $b_i = 2_l$, else set $b'_i = \epsilon$, while g'_i is the syntactic guard resulting from g_i by deleting all conjuncts corresponding to indices not preserved in the projection. Then, by our construction and the fact that x was an infinite computation, we have that y denotes a stuttering of a genuine infinite computation of $(U_1, \ldots, U_k)^{(c_1, \ldots, c_k)}$. To see this, note that for any i such that $y_i \neq y_{i+1}$, the associated (formal) transitions have their guard g'_i true, since for conjunctive guards g_i and their projections g'_i we have $x_i \models g_i$ implies $y_i \models g'_i$, and can thus fire in $(U_1, \ldots, U_k)^{(c_1, \ldots, c_k)}$. For any stuttering i where $y_i = y_{i+1}$, the (formal) transition is labelled by $b'_i = \epsilon$.

Thus, given infinite computation path of $(U_1, \ldots, U_k)^{(n_1, \ldots, n_k)}$, there exists a stuttering of an infinite computation path of $(U_1, \ldots, U_k)^{(c_1, \ldots, c_k)}$, such that the local computation path of U_l^1 is the same in both. This path correspondence proves the result.

($\Leftarrow$) Let $y = y_0, y_1, \ldots$ be an infinite computation path of $(U_1, \ldots, U_k)^{(c_1, \ldots, c_k)}$. Then, consider the sequence of states $= x_0, x_1, \ldots,$, where $x(l, 1) = y(l, 1)$, $x(l, 2) = y(l, 2)$ and $\forall (k, j) \neq (l, 1), (l, 2) : x(k, j) = (i_k^j)^\omega$. Let g_i be the guard labelling the transition $s_l^1 \to t_l^1$ in state y_i. Then all the other processes are in their initial states in x_i, and since the guards do allow initial states of all template processes as "nonblocking" states in that their being present in the global state does not falsify any guards, we have $x_i \models g_i$.

Thus, given infinite computation path y of $(U_1, \ldots, U_k)^{(c_1, \ldots, c_k)}$, there exists an infinite computation path x of $(U_1, \ldots, U_k)^{(n_1, \ldots, n_k)}$, such that the local computation path of U_l^1 is the same in both. This path correspondence easily gives us the desired result.   □

In a similar fashion, we may prove the following result.

Theorem 5.4 (Finite Conjunctive Reduction Theorem).
For any LTL\X formula h, and $l \in [1..k]$ we have
  (i) $\forall(n_1, \ldots, n_k) \succeq (1, \ldots, 1) : (U_1, \ldots, U_k)^{(n_1, \ldots, n_k)} \models \bigwedge_{i_l} \mathsf{E}_{\mathsf{fin}} h(i_l)$, *iff*
    $(U_1, \ldots, U_k)^{(1, \ldots, 1)} \models \mathsf{E}_{\mathsf{fin}} h(1_l)$;
  (ii) $\forall(n_1, \ldots, n_k) \succeq (1, \ldots, 1) : (U_1, \ldots, U_k)^{(n_1, \ldots, n_k)} \models \bigwedge_{i_l} \mathsf{A}_{\mathsf{fin}} h(i_l)$, *iff*
    $(U_1, \ldots, U_k)^{(1, \ldots, 1)} \models \mathsf{A}_{\mathsf{fin}} h(1_l)$.

Note that the above theorem permits us to verify safety properties efficiently. Informally, this is because if there is a finite path leading to a "bad" state in the system $(U_1, \ldots, U_k)^{(n_1, \ldots, n_k)}$, then there exists a finite path leading to a bad state in $(U_1, \ldots, U_k)^{(1, \ldots, 1)}$. Thus, checking that there is no finite path leading to bad state in $(U_1, \ldots, U_k)^{(n_1, \ldots, n_k)}$ reduces to checking it for $(U_1, \ldots, U_k)^{(1, \ldots, 1)}$.

We can use this to obtain an Efficient Conjunctive Decidability Theorem. Moreover, the results can be readily extended to formulae with multiple indices as in the disjunctive guards case.

6 Applications

Here, we consider a solution to the *mutual exclusion* problem. The template process is given below. Initially, every process is in local state N, the non-critical

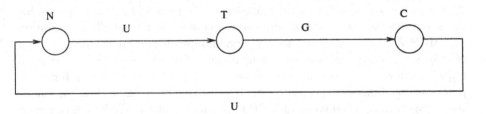

region. $U = T \vee N \vee C$ denotes the universal guard, which is always true independent of the local states of other processes. If a process wants to enter the critical section C, it goes into the trying region T which it can always do since U is always true. Guard $G = N \vee T$, instantiated for process i of n processes, takes the conjunctive form $\bigwedge_{j \neq i}(N_j \vee T_j)$. When G is true, no other process is in the critical section, and the transition from T to C can be taken. Note that all guards are conjunctive with neutral(i.e., non-blocking) initial state N. Thus, by the Finite Conjunctive Reduction Theorem for multi-indexed properties, PMCP for all sizes n with the mutual exclusion property $\bigwedge_{i,j,i \neq j} \mathsf{A}_{\mathsf{fin}} \mathsf{G}\neg(C_i \wedge C_j)$ can be reduced to checking a 2-process instance. Using the Conjunctive Cutoff Theorem, the starvation-freedom property $\bigwedge_i \mathsf{A}(\mathsf{G}(T_i \Rightarrow \mathsf{F}C_i))$ can be checked by a 7-process instance. In this simple example, mutual exclusion is maintained but starvation-freedom fails.

7   Concluding Remarks

PMCP is, in general, undecidable[1]. However, under certain restrictions, a variety of positive results have been obtained. Early work includes [15] which uses an abstract graph of exponential size "downstairs" to capture the behaviour of arbitrary sized parameterized asynchronous programs "upstairs" over Fetch-and-Add primitives; however, while it caters for partial automation, the completeness of the method is not established, and it is not clear that it can be made fully automatic. A semi-automated method requiring construction of a *closure process* which represents computations of an arbitrary number of processes is described in [4]; it is shown that, if for some k, $C||U^k$ is appropriately bisimilar to $C||U^{k+1}$, then it suffices to check instances of size at most k to solve the PMCP. But it is not shown that such a cutoff k exists, and the method is not guaranteed to be complete. Kurshan and McMillan [13] introduce the related notion of a *process invariant* (cf. [22]). Ip and Dill [12] describe another approach to dealing with many processes using an abstract graph; it is sound but not guaranteed to be complete; [18] proposes a similar construction for verification of safety properties of cache coherence protocols, which is also sound but not complete. A theme is that most these methods suffer, first, from the drawback of being only partially automated and hence requiring human ingenuity, and, second, from being sound but not guaranteed complete (i.e., a path "upstairs" maps to a path "downstairs", but paths downstairs do not necessarily lift). Other methods can be fully automated but do not appear to have a clearly defined class of protocols on which they are guaranteed to terminate successfully (cf. [5], [21], [19]).

For systems comprised of CCS processes, German and Sistla [10] combine the automata-theoretic method with process closures to permit efficient solution to PMCP for single index properties, modulo deadlock. But efficient solution is only yielded for processes in a single class. Even for systems of the form $C||U^n$ a doubly exponential decision procedure results, which likely limits its practical use. Emerson and Namjoshi [7] show that in a single class (or client-server) synchronous framework PMCP is decidable but with PSPACE-complete complexity. Moreover, this framework is undecidable in the asynchronous case. A different type of parameterized reasoning about time bounds is considered in [9].

In some sense, the closest results might be those of Emerson and Namjoshi [6] who for the token ring model, reduce reasoning, for multi-indexed temporal logic formulae, for rings of arbitrary size to rings up to a small cutoff size. These results are significant in that, like ours, correctness over all sizes holds iff correctness of (or up to) the small cutoff size holds. But these results were formulated only for a single process class and, for a restricted version of the token ring model, namely one where the token cannot be used to pass values. Also, related are the results of Attie and Emerson [2]. In the context of program synthesis, rather than program verification, it is shown how certain 2-process solutions to synchronization problems could be inflated to n-process solutions. However, the correspondence is not an "iff", but is established in only one direction for conjunctive-type guards. Disjunctive guards are not considered, nor are multiple process classes.

We believe that our positive results on PMCP are significant for several reasons. Because PMCP solves (a major aspect of) the state explosion problem and the scalability problem in one fell swoop, many researchers have attempted to make it more tractable, despite its undecidability in general. Of course, PMCP seems to be prone to undecidability in practice as well, as is evidenced by the wide range of solution methods proposed that are only partially automated or incomplete or lack a well-defined domain of applicability. Our methods are fully automated returning a yes/no answer, they are sound and complete as they rely on establishing exact (up to stuttering) correspondences (yes upstairs iff yes downstairs). In many cases, our methods are efficient, making the problem genuinely tractable. An additional advantage, is that downstairs we have a small system of cutoff size that, but for its size, looks like a system of size n. This contrasts with methods that construct an abstract graph downstairs which may have a complex and non-obvious organization.

References

1. K. Apt and D. Kozen. Limits for automatic verification of finite-state concurrent systems. *Information Processing Letters*, 15, pages 307-309, 1986.
2. P.C. Attie and E.A. Emerson. Synthesis of Concurrent Systems with Many Similar Processes. *ACM Transactions on Programming Languages and Systems*, Vol. 20, No. 1, January 1998, pages 51-115.
3. M.C. Browne, E.M. Clarke and O. Grumberg. Reasoning about Networks with Many Identical Finite State Processes. *Information and Control*, 81(1), pages 13-31, April 1989.
4. E.M. Clarke and O. Grumberg. Avoiding the State Explosion Problem in Temporal Logic Model Checking Algorithms. In *Proceedings of the Sixth Annual ACM Symposium on Principles of Distributed Computing*, pages 294-303, 1987.
5. E.M. Clarke, O. Grumberg and S. Jha. Verifying Parameterized Networks using Abstracion and Regular Languages. In *CONCUR '95: Concurrency Theory, Proceedings of the 6th International Conference*, LNCS 962, pages 395-407, Springer-Verlag, 1995.
6. E.A. Emerson and K.S. Namjoshi. Reasoning about Rings. In *Conference Record of POPL '95: 22nd ACM SIGPLAN-SIGACT Symposium on Principles of Programming Languages*, pages 85-94, 1995.
7. E.A. Emerson and K.S. Namjoshi. Automatic Verification of Parameterized Synchronous Systems. In *Computer Aided Verification, Proceedings of the 8th International Conference*. LNCS , Springer-Verlag, 1996.
8. E.A. Emerson and A.P. Sistla. Symmetry and Model Checking. In *Computer Aided Verification, Proceedings of the 5th International Conference*. LNCS 697, Springer-Verlag, 1993.
9. E. Emerson and R. Trefler, Parametric Quantitative Temporal Reasoning. LICS 1999, 336-343.
10. S.M. German and A.P. Sistla. Reasoning about Systems with Many Processes. *J. ACM*, 39(3), July 1992.

11. C. Ip and D. Dill. Better verification through symmetry. In *Proceedings of the 11th International Symposium on Computer Hardware Description Languages and their Applications.*1993.
12. C. Ip and D. Dill, Verifying Systems with Replicated Components in Murphi, pp. 147-158 CAV 1996.
13. R.P. Kurshan and L. McMillan. A Structural Induction Theorem for Processes. In *Proceedings of the Eight Annual ACM Symposium on Principles of Distributed Computing*, pages 239-247, 1989.
14. O. Lichtenstein and A. Pnueli. Checking that finite state concurrent programs satisfy their linear specifications. In *Conference Record of POPL '85: 12nd ACM SIGPLAN-SIGACT Symposium on Principles of Programming Languages*, pages 97-107, 1985.
15. B. Lubachevsky. An Approach to Automating the Verification of Compact Parallel Coordination Programs I.*Acta Informatica 21*, 1984.
16. K. McMillan, Verification of Infinite State Systems by Compositional Model Checking, CHARME'99.
17. A. Pnueli. The Temporal Logic of Programs. In *Proceedings of the eighteenth Symposium on Foundations of Computer Science.* 1977.
18. F. Pong and M. Dubois. A New Approach for the Verification of Cache Coherence Protocols. *IEEE Transactions on Parallel and Distributed Systems*, August 1995.
19. A. P. Sistla, Parameterized Verification of Linear Networks Using Automata as Invariants, CAV, 1997, 412-423.
20. M. Vardi and P. Wolper. An Automata-theoretic Approach to Automatic Program Verification. In *Proceedings, Symposium on Logic in Computer Science*, pages 332-344, 1986.
21. I. Vernier. Specification and Verification of Parameterized Parallel Programs. In *Proceedings of the 8th International Symposium on Computer and Information Sciences*, Istanbul, Turkey, pages 622-625,1993.
22. P. Wolper and V. Lovinfosse. Verifying Properties of Large Sets of Processes with Network Invariants. In J. Sifakis(ed) *Automatic Verification Metods for Finite State Systems*, Springer-Verlag, LNCS 407, 1989.

Simulation Based Minimization

Doron Bustan and Orna Grumberg

Computer Science Dept.
Technion, Haifa 32000, Israel
orna@cs.technion.ac.il

Abstract. This[1] work presents a minimization algorithm. The algorithm receives a Kripke structure M and returns the smallest structure that is simulation equivalent to M. The *simulation equivalence* relation is weaker than bisimulation but stronger than the simulation preorder. It strongly preserves ACTL and LTL (as sub-logics of ACTL*).
We show that every structure M has a unique up to isomorphism *reduced* structure that is simulation equivalent to M and smallest in size.
We give a Minimizing Algorithm that constructs the reduced structure. It first constructs the quotient structure for M, then eliminates transitions to little brothers and finally deletes unreachable states.
The first step has maximal space requirements since it is based on the simulation preorder over M. To reduce these requirements we suggest the Partitioning Algorithm which constructs the quotient structure for M without ever building the simulation preorder. The Partitioning Algorithm has a better space complexity but might have worse time complexity.

1 Introduction

Temporal logic model checking is a method for verifying finite-state systems with respect to propositional temporal logic specifications. The method is fully automatic and quite efficient in time, but is limited by its high space requirements. Many approaches to beat the *state explosion problem* of model checking have been suggested, including abstraction, partial order reduction, modular methods, and symmetry ([8]). All are aimed at reducing the size of the model (or Kripke structure) to which model checking is applied, thus, extending its applicability to larger systems.

Abstraction methods, for instance, hide some of the irrelevant details of a system and then construct a reduced structure. The abstraction is required to be *weakly preserving*, meaning that if a property is true for the abstract structure then it is also true for the original one. Sometimes we require the abstraction to be *strongly preserving* so that, in addition, a property that is false for the abstract structure, is also false for the original one.

In a similar manner, for modular model checking we construct a reduced abstract environment for a part of the system that we wish to verify. In this case

---

[1] The full version of this paper including proofs of correctness can be found in [6].

D. McAllester (Ed.): CADE-17, LNAI 1831, pp. 255–270, 2000.

as well, properties that are true (false) of the abstract environment should be true (false) of the real environment.

It is common to define equivalence relations or preorders on structures in order to reflect strong or weak preservation of various logics. For example, language equivalence (containment) strongly (weakly) preserves the linear-time temporal logic LTL. Other relations that are widely used are the *bisimulation equivalence* [15] and the *simulation preorder* [14]. The former guarantees strong preservation of branching-time temporal logics such as CTL and CTL* [7]. The latter guarantees weak preservation of the universal fragment of these logics (ACTL and ACTL* [10]).

Bisimulation has the advantage of preserving more expressive logics. However, this is also a disadvantage since it requires the abstract structure to be too similar to the original one, thus allowing less powerful reductions. The simulation preorder, on the other hand, allows more powerful reductions, but it provides only weak preservation. Language equivalence provides strong preservation and large reduction, however, its complexity is exponential while the complexity to compute bisimulation and simulation is polynomial.

In this paper we investigate the *simulation equivalence* relation that is weaker than bisimulation but stronger than the simulation preorder and language equivalence. Simulation equivalence strongly preserves ACTL*, and also strongly preserves LTL and ACTL as sublogics of ACTL*. Both ACTL and LTL are widely used for model checking in practice.

As an equivalence relation that is weaker than bisimulation, it can derive smaller minimized structure. For example, the structure in part 2 of Figure 1 is minimized with respect to simulation equivalence. In comparison, the minimized structure with respect to bisimulation is the structure in part 1 of Figure 1 and the minimized structure with respect to language equivalence is the structure in part 3 of Figure 1 .

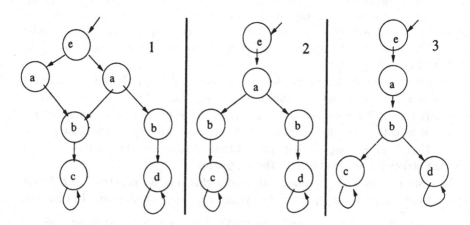

Fig. 1. Different minimized structures with respect to different equivalence relations

Given a Kripke structure M, we would like to find a structure M' that is simulation equivalent to M and is the smallest in size (number of states and transitions).

For bisimulation this can be done by constructing the *quotient structure* in which the states are the equivalence classes with respect to bisimulation. Bisimulation has the property that if one state in a class has a successor in another class then all states in the class have a successor in the other class. Thus, in the quotient structure there will be a transition between two classes if every (some) state in one class has a successor in the other. The resulting structure is the smallest in size that is bisimulation equivalent to the given structure M.

The quotient structure for simulation equivalence can be constructed in a similar manner. There are two main difficulties, however. First, it is not true that all states in an equivalence class have successors in the same classes. As a result, if we define a transition between classes whenever *all* states of one have a successor in the other, then we get the $\forall$−quotient structure. If, on the other hand, we have a transition between classes if there *exists* a state of one with a successor in the other, then we get the $\exists$−quotient structure. Both structures are simulation equivalent to M, but the $\forall$−quotient structure has fewer transitions and therefore is preferable.

The other difficulty is that the quotient model for simulation equivalence is *not* the smallest in size. Actually, it is not even clear that there is a unique smallest structure that is simulation equivalent to M.

The first result in this paper is showing that every structure has a *unique up to isomorphism* smallest structure that is simulation equivalent to it. This structure is *reduced*, meaning that it contains no simulation equivalent states, no little brothers (states that are smaller by the simulation preorder than one of their brothers), and no unreachable states.

Our next result is presenting the Minimizing Algorithm that given a structure M constructs the reduced structure for M. Based on the maximal simulation relation over M, the algorithm first builds the $\forall$−quotient structure with respect to simulation equivalence. Then it eliminates transitions to little brothers. Finally, it removes unreachable states. The time complexity of the algorithm is $O(|S|^3)$. Its space complexity is $O(|S|^2)$ which is due to the need to hold the simulation preorder in memory.

Since our main concern is space requirements, we suggest the Partitioning Algorithm which computes the quotient structure without ever computing the simulation preorder. Similarly to [13], the algorithm starts with a partition Σ_0 of the state space to classes whose states are equally labeled. It also initializes a preorder H_0 over the classes in Σ_0. At iteration $i + 1$, Σ_{i+1} is constructed by splitting classes in Σ_i. The relation H_{i+1} is updated based on Σ_i, Σ_{i+1} and H_i.

When the algorithm terminates (after k iterations) Σ_k is the set of equivalence classes with respect to simulation equivalence. These classes form the states of the quotient structure. The final H_k is the maximal simulation preorder over the states of the quotient structure. Thus, the Partitioning Algorithm replaces the first step of the Minimizing Algorithm . Since every step in the Minimizing

Algorithm further reduces the size of the initial structure, the first step handles the largest structure. Therefore, improving its complexity influences most the overall complexity of the algorithm.

The space complexity of the Partitioning Algorithm is $O(|\Sigma_k|^2 + |S| \cdot log(|\Sigma_k|))$. We assume that in most cases $|\Sigma_k| << |S|$, thus this complexity is significantly smaller than that of the Minimizing Algorithm . Unfortunately, time complexity will probably become worse (depending on the size of Σ_k). It is bounded by $O(|S|^2 \cdot |\Sigma_k|^2 \cdot (|\Sigma_k|^2 + |R|))$. However, since our main concern is the reduction in memory requirements, the Partitioning Algorithm is valuable.

Other works also suggest minimization algorithms. In [13], the quotient structure with respect to bisimulation is constructed without first building the bisimulation relation. We follow a similar approach. However, in our case states may remain in the same class even when they do not have successors in the same classes. Thus, our analysis is more complicated and requires both Σ_i and H_i. Symbolic bisimulation minimization is suggested in [5]. In [4] a minimized structure with respect to bisimulation is generated directly out of the text. In [9] a bisimulation minimization is applied to the intersection of the system automaton and the specification automaton. The algorithm from [13] is used. [12] shows that eliminating little brothers results in a simulation equivalent structure. However, the paper does not consider the minimization problem.

Several works minimize a structure in a compositional way, preserving language containment [2] or a given CTL formula [1]. Minimizing with respect to a given formula may result in a more power reduction, however it requires to determine the checked formula in advance.

The rest of the paper is organized as follows. Section 2 gives our basic definitions. Section 3 defines reduced structures and shows that every structure has a unique simulation equivalent reduced structure. Section 4 presents the Minimizing Algorithm . Finally, Section 5 describes the Partitioning Algorithm and discusses its space and time complexity.

2   Preliminaries

Let AP be a set of atomic propositions. A *Kripke structure* M over AP is a four tuple $M = (S, s_0, R, L)$ where S is a finite set of states; $s_0 \in S$ is the initial state; $R \subseteq S \times S$ is the transition relation that must be *total*, i.e., for every state $s \in S$ there is a state $s' \in S$ such that $R(s, s')$; and $L : S \to 2^{AP}$ is a function that labels each state with the set of atomic propositions true in that state.

The *size* $|M|$ of a Kripke structure M is the pair $(|S|, |R|)$. We say that $|M| \leq |M'|$ if $|S| \leq |S'|$ or $|S| = |S'|$ and $|R| \leq |R'|$.

Given two structures M and M' over AP, a relation $H \subseteq S \times S'$ is a *simulation relation* [14] over $M \times M'$ iff the following conditions hold:

1. $(s_0, s_0') \in H$.
2. For all $(s, s') \in H$, $L(s) = L'(s')$ and

$$\forall t[(s, t) \in R \to \exists t'[(s', t') \in R' \wedge (t, t') \in H]].$$

We say that M' *simulates* M (denoted by $M \preceq M'$) if there exists a simulation relation H over $M \times M'$.

The logic ACTL* [10] is the universal fragment of the powerful branching-time logic CTL*. ACTL* consists of the temporal operators **X** (next-time), **U** (until) and **R** (release) and the universal path quantifier **A** (for all paths). For lack of space the formal definition is omitted. It can be found in [8].

The following lemma and theorem have been proven in [10].

Lemma 1. $\preceq$ *is a preorder on the set of structures.*

Theorem 2. *Suppose* $M \preceq M'$. *Then for every ACTL* formula f, $M' \models f$ implies $M \models f$.*

Given two Kripke structures M, M', we say that M is *simulation equivalent* to M' iff $M \preceq M'$ and $M' \preceq M$. It is easy to see that this is an equivalence relation. By Theorem 2 , if M and M' are simulation equivalent then they are equivalent with respect to $ACTL^*$. However, they are not equivalent with respect to CTL^*.

A simulation relation H over $M \times M'$ is *maximal* iff for all simulation relations H' over $M \times M'$, $H' \subseteq H$. In [10] it has been shown that if there is a simulation relation over $M \times M'$ then there is a *unique* maximal simulation over $M \times M'$.

3   The Reduced Structure

Given a Kripke structure M, we would like to find a *reduced* structure that will be simulation equivalent to M and smallest in size. In this section we show that a reduced structure always exists. Furthermore, we show that all reduced structures of M are *isomorphic* to each other.

Let M be a Kripke structure. The *maximal simulation relation* over $M \times M$ always exists and is denoted by H_M. We need the following two definitions in order to characterize reduced structures.

Two states $s_1, s_2 \in M$ are *simulation equivalent* iff $(s_1, s_2) \in H_M$ and $(s_2, s_1) \in H_M$.

A state s_1 is a *little brother* of a state s_2 iff there exists a state s_3 such that:

- $(s_3, s_2) \in R$ and $(s_3, s_1) \in R$.
- $(s_1, s_2) \in H_M$ and $(s_2, s_1) \notin H_M$.

Definition 3. *A Kripke structure M is* reduced *if:*

1. *There are no simulation equivalent states in M.*
2. *There are no states s_1, s_2 such that s_1 is a little brother of s_2.*
3. *All states in M are reachable from s_0.*

Theorem 4. *: Let M M' be two reduced Kripke structures. Then the following two statements are equivalent:*

1. *M and M' are simulation equivalent.*
2. *M and M' are isomorphic.*

The proof that 2 implies 1 is straight forward. In the rest of this section we assume that M and M' are reduced Kripke structures. We will show that if $M \preceq M'$ and $M' \preceq M$ then M and M' are isomorphic.

We use $H_{MM'}$ and $H_{M'M}$ to denote the maximal simulation relations over $M \times M'$ and $M' \times M$ respectively. The *composed* relation $H_{MM'M} \subseteq S \times S$ is defined by $H_{MM'M} = \{(s_1, s_2) | \exists s' \in S'. (s_1, s') \in H_{MM'} \wedge (s', s_2) \in H_{M'M}\}$.

Lemma 5. *The composed relation $H_{MM'M}$ is a simulation relation.*

For the reduced Kripke structures M and M', we define the *matching relation* $f \subseteq S' \times S$ as follows:

$$(s', s) \in f \text{ iff } (s', s) \in H_{M'M} \text{ and } (s, s') \in H_{MM'}.$$

We show that f is an isomorphism between M' and M, i.e., f is an one to one and onto total function that preserves the state labeling and the transition relation.

Lemma 6. *Let $f \subseteq S' \times S$ be the matching relation. Then f is an one to one, onto, and total function from S' to S.*

Proof Sketch : First we need to prove that f is a function from S' to S. We assume to the contrary that there are different states $s_1, s_2 \in S$ and $s' \in S'$ such that $(s', s_1) \in f$ and $(s', s_2) \in f$. We show that $(s_1, s_2) \in H_{MM'M}$ and $(s_2, s_1) \in H_{MM'M}$. Since $H_{MM'M}$ is included in H_M, this contradicts the assumption that M is reduced. The proof that f^{-1} is a function from S to S' is similar. Thus, we conclude that f is one to one.

Next, we prove that f is onto, i.e. for every state s in S there exists a state s' in S' such that $(s', s) \in f$. The proof is by induction on the distance of $s \in S$ from the initial state. (since all states are reachable, the distance is bounded by $|S|$). Again we use the composed relation $H_{MM'M}$ to show that if f is not onto then M' is not reduced.

Similarly, we can show that f^{-1} is onto and therefore f is total. $\square$

Lemma 7. *For all $s' \in S'$, $L'(s') = L(f(s'))$. Furthermore, for all $s_1', s_2' \in S'$, $(s_1', s_2') \in R'$ iff $(f(s_1'), f(s_2')) \in R$.*

Thus, we conclude Theorem 4 .

Theorem 8. *Let M be a non-reduced Kripke structure, then there exists a reduced Kripke structure M' such that M, M' are simulation equivalent and $|M'| < |M|$.*

In order to prove Theorem 8 , we present in the next sections an algorithm that receives a Kripke structure M and computes a reduce Kripke structure M', which is simulation equivalent to $|M|$, such that $|M'| \leq |M|$. Moreover, if M is not reduced then $|M'| < |M|$.

Lemma 9. *Let M' be a reduced Kripke structure. For every M that is simulation equivalent to $|M'|$, if M and M' are not isomorphic then $|M'| < |M|$.*

4   The Minimizing Algorithm

In this section we present the Minimizing Algorithm that gets a Kripke structure M and computes a reduced Kripke structure M' which is simulation equivalent to M and $|M'| \leq |M|$. If M is not reduced then $|M'| < |M|$.

The algorithm consists of three steps. First, a quotient structure is constructed in order to eliminate equivalent states. The resulting quotient model is simulation equivalent to M but may not be reduced. The next step disconnects little brothers and the last one removes all unreachable states.

In each step of the algorithm, if the resulting structure differs from the original one then the resulting one is strictly smaller than the original structure.

4.1   The ∀−Quotient Structure

In order to compute a simulation equivalent structure that contains no equivalent states, we compute the ∀−$quotient\ structure$ with respect to the simulation equivalence relation. We fix M to be the original Kripke structure. We denote by $[s]$ the equivalence class which includes s.

Definition 10. The ∀−$quotient\ structure$ $M_q\ =<\ S_q, R_q, s_{0_q}, L_q\ >$ of M is defined as follow:

- S_q is the set of the equivalence classes of the simulation equivalence. (We will use Greek letters to represent equivalence classes).
- $R_q = \{(\alpha_1, \alpha_2) | \forall s_1 \in \alpha_1 \ \exists s_2 \in \alpha_2. \ (s_1, s_2) \in R\}$
- $s_{0_q} = [s_0]$.
- $L_q([s]) = L(s)$.

The transitions in M_q are ∀-transitions, in which there is a transition between two equivalence classes iff $every$ state of the one has a successor in the other. We could also define ∃-transitions, in which there is a transition between classes if there exists a state in one with a successor in the other. Both definitions result in a simulation equivalent structure. However, the former has smaller transition relation and therefore it is preferable.

Note that, $|S_q| \leq |S|$ and $|R_q| \leq |R|$. If $|S_q| = |S|$, then every equivalence class contains a single state. In this case, R_q is identical to R and M_q is isomorphic to M. Thus, when M and M_q are not isomorphic, $|S_q| < |S|$.
Next, we show that M and M_q are simulation equivalent.

Definition 11. Let $G \subseteq S$ be a set of states. A state $s_m \in G$ is maximal in G iff there is no state $s \in \bar{G}$ such that $(s_m, s) \in H_M$ and $(s, s_m) \notin H_M$.

Definition 12. Let α be a state of M_q, and t_1 a successor of some state in α. The set $G(\alpha, t_1)$ is defined as follow:

$$G(\alpha, t_1) = \{t_2 \in S | \exists s_2 \in \alpha \wedge (s_2, t_2) \in R \wedge (t_1, t_2) \in H_M\}.$$

Intuitively, $G(\alpha, t_1)$ is the set of states that are greater than t_1 and are successors of states in α. Notice that since all state in α are simulation equivalent, every state in α has at least one successor in $G(\alpha, t_1)$.

Lemma 13. *Let α, t_1 be as defined in Definition 12 . Then for every maximal state t_m in $G(\alpha, t_1)$, $[t_m]$ is a successor of α.*

Proof : Let t_m be a maximal state in $G(\alpha, t_1)$, and let $s_m \in \alpha$ be a state such that t_m is a successor of s_m. We prove that for every state $s \in \alpha$, there exists a successor $t \in [t_m]$, which implies that $[t_m]$ is a successor of α.

$s, s_m \in \alpha$ implies $(s_m, s) \in H_M$. This implies that there exists a successor t of s such that $(t_m, t) \in H_M$. By transitivity of the simulation relation, $(t_1, t) \in H_M$. Thus $t \in G(\alpha, t_1)$. Since t_m is maximal in $G(\alpha, t_1)$, $(t, t_m) \in H_M$. Thus, t and t_m are simulation equivalent and $t \in [t_m]$.   □

Theorem 14. *The structures M and M_q are simulation equivalent.*

Proof Sketch : It is straight forward to show that $H' = \{(\alpha, s) | s \in \alpha\}$ is a simulation relation over $M_q \times M$. Thus, $M_q \preceq M$.

In order to prove that $M \preceq M_q$ we choose $H' = \{(s_1, \alpha)|$ there exists a state $s_2 \in \alpha$ such that $(s_1, s_2) \in H_M\}$. Clearly, $(s_0, s_{0_q}) \in H'$ and for all $(s, \alpha) \in H'$, $L(s) = L_q(\alpha)$.

Assume $(s_1, \alpha_1) \in H'$ and let t_1 be a successor of s_1. We prove that there exists a successor α_2 of α_1 such that $(t_1, \alpha_2) \in H'$. We distinguish between two cases:

1. $s_1 \in \alpha_1$. Let t_m be a maximal state in $G(\alpha_1, t_1)$, then Lemma 13 implies that $(\alpha_1, [t_m]) \in R_q$. Since t_m is maximal in $G(\alpha_1, t_1)$, $(t_1, t_m) \in H_M$ which implies $(t_1, [t_m]) \in H'$.
2. $s_1 \notin \alpha_1$. Let $s_2 \in \alpha_1$ be a state such that $(s_1, s_2) \in H_M$. Since $(s_1, s_2) \in H_M$ there is a successor t_2 of s_2 such that $(t_1, t_2) \in H_M$. The first case implies that there exists an equivalence class α_2 such that $(\alpha_1, \alpha_2) \in R_q$ and $(t_2, \alpha_2) \in H'$. By $(t_2, \alpha_2) \in H'$ we have that there exists a state $t_3 \in \alpha_2$ such that $(t_2, t_3) \in H_M$. By transitivity of simulation $(t_1, t_3) \in H_M$. Thus, $(t_1, \alpha_2) \in H'$.   □

4.2 Disconnecting Little Brothers

Our next step is to disconnect the little brothers from their fathers. As a result of applying this step to a Kripke structure M with no equivalent states, we get a Kripke structure M' satisfying:

1. M are M' are simulation equivalent.
2. There are no equivalent states in M'.
3. There are no little brothers in M'.
4. $|M'| \leq |M|$, and if M and M' are not identical, then $|M'| < |M|$.

```
change := true
while (change = true) do
    Compute the maximal simulation relation H_M
    change := false
    If there are s_1, s_2, s_3 ∈ S such that s_1 is a little brother of s_2
            and s_3 is the father of both s_1 and s_2 then
        change := true
        R = R \ {(s_3, s_1)}
    end
end
```

Fig. 2. The Disconnecting Algorithm.

In Figure 2 we present an iterative algorithm which disconnects little brothers and results in M'.

Since in each iteration of the algorithm one edge is removed, the algorithm will terminate after at most $|R|$ iterations. We will show that the resulting structure is simulation equivalent to the original one.

Lemma 15. Let $M' = < S', R', s'_0, L' >$ be the result of the Disconnecting Algorithm on M. Then M and M' are simulation equivalent.

Proof Sketch : We prove the lemma by induction on the number of iterations.
Base: at the beginning M and M are simulation equivalent.
Induction step: Let M'' be the result of the first i iterations and H'' be the maximal simulation over $M'' \times M''$. Let M' be the result of the $(i+1)$th iteration where $R' = R'' \setminus \{(s''_1, s''_2)\}$. Assume that M and M'' are simulation equivalent. It is straight forward to see that $H' = \{(s'_1, s''_2) | (s''_1, s''_2) \in H''\}$ is a simulation relation over $M' \times M''$. Thus, $M' \preceq M''$.

To show that $M'' \preceq M'$ we prove that $H' = \{(s''_1, s'_2) | (s''_1, s''_2) \in H''\}$ is a simulation relation. Clearly, $(s''_0, s'_0) \in H'$ and for all $(s''_1, s'_2) \in H''$, $L''(s''_1) = L'(s'_2)$.

Suppose $(s''_1, s'_2) \in H'$ and t''_1 is a successor of s''_1. Since H'' is a simulation relation, there exists a successor t''_2 of s''_2 such that $(t''_1, t''_2) \in H''$. This implies that $(t''_1, t'_2) \in H'$. If $(s'_2, t'_2) \in R'$ then we are done. Otherwise, (s''_2, t''_2) is removed from R'' because t''_2 is a little brother of some successor t''_3 of s''_2. Since (s''_2, t''_2) is the only edge removed at the $(i+1)$th iteration, $(s'_2, t'_3) \in R'$. Because t''_2 is a little brother of t''_3 then $(t''_2, t''_3) \in H''$. By transitivity of the simulation relation, $(t''_1, t''_3) \in H''$, thus $(t''_1, t'_3) \in H'$.   □

We proved that the result M' of the Disconnecting Algorithm is simulation equivalent to the original structure M. Note that M' has the same set of states as M. We now show that the maximal simulation relation over M is identical to the maximal simulation relations for all intermediate structures M'' (including M'), computed by the Disconnecting Algorithm. Since there are no simulation equivalent states in M, there are no such states in M' as well.

Lemma 16. *Let $M' = < S, R', s_0, L >$ be the result of the Disconnecting Algorithm on M and let $H' \subseteq S' \times S'$ be the maximal simulation over $M' \times M'$. Then, $H_M = H'$.*

The lemma is proved by induction on the number of iterations.

As a result of the last lemma, the Disconnecting Algorithm can be simplified significantly. The maximal simulation relation is computed once on the original structure M and is used in all iterations. If the algorithm is executed symbolically (with BDDs) then this operation can be performed efficiently in one step:

$$R' = R - \{(s_1, s_2) | \exists s_3 : (s_1, s_3) \in R \wedge (s_2, s_3) \in H_M \wedge (s_3, s_2) \notin H_M\}.$$

4.3   The Algorithm

We now present our algorithm for constructing the reduced structure for a given one.

1. Compute the $\forall$-quotient structure M_q of M and
        the maximal simulation relation H_M over $M_q \times M_q$.
2. $R' = R_q - \{(s_1, s_2) | \exists s_3 : (s_1, s_3) \in R_q \wedge (s_2, s_3) \in H_M\}$
3. Remove all unreachable states.

Fig. 3. The Minimizing Algorithm

Note that, in the second step we eliminate the check $(s_3, s_2) \notin H_M$. This is based on the fact that M_q does not contain simulation equivalent states. Removing unreachable states does not change the properties of simulation with respect to the initial states. The size of the resulting structure is equal to or smaller than the original one. Similarly to the first two steps of the algorithm, if the resulting structure is not identical then it is strictly smaller in size.

We have proved that the result of the Minimizing Algorithm M' is simulation equivalent to the original structure M. Thus we can conclude that Theorem 8 is correct.

Figure 4 presents an example of the three steps of the Minimizing Algorithm applied to a Kripke structure.

1. Part 1 contains the original structure, where the maximal simulation relation is (not including the trivial pairs):
   $\{(2,3), (3,2), (11,2), (11,3), (4,5), (6,5), (7,8), (8,7), (9,10), (10,9)\}$.
   The equivalence classes are : $\{\{1\}, \{2,3\}, \{11\}, \{4\}, \{5\}, \{6\}, \{7,8\}, \{9,10\}\}$.
2. Part 2 presents the $\forall$-structure M_q. The maximal simulation relation H_M is (not including the trivial pairs):
   $H_M = \{(\{11\}, \{2,3\}), (\{4\}, \{5\}), (\{6\}, \{5\})\}$.
3. $\{11\}$ is a little brother of $\{2,3\}$ and $\{1\}$ is their father. Part 3 presents the structure after the removal of the edge $(\{1\}, \{11\})$.
4. Finally, part 4 contains the reduced structure, obtained by removing the unreachable states.

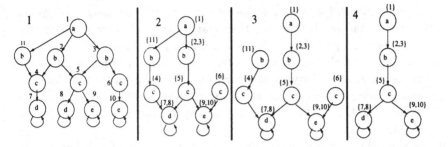

Fig. 4. An example of the Minimizing Algorithm

4.4 Complexity

The complexity of each step of the algorithm depends on the size of the Kripke structure resulting from the previous step. In the worst case the Kripke structure does not change, thus all three steps depend on the original Kripke structure. Let M be the given structure. We analyze each step separately (a naive analysis):

1. First, the algorithm constructs equivalence classes. To do that it needs to compute the maximal simulation relation. [3, 11] showed that this can be done in time $O(|S| \cdot |R|)$. Once the algorithm has the simulation relation, the equivalence classes can be constructed in time $O(|S|^2)$. Next, the algorithm constructs the transition relation. This can be done in time $O(|S| + |R|)$. As a whole, building the quotient structure can be done in time $O(|S| \cdot |R|)$.
2. Disconnecting little brothers can be done in $O(|S|^3)$.
3. Removing unreachable states can be done in $O(|R|)$.

As a whole the algorithm works in time $O(|S|^3)$

The space bottle neck of the algorithm is the computation of the maximal simulation relation which is bounded by $|S|^2$.

5 Partition Classes

In the previous section, we presented the Minimizing Algorithm . The algorithm consists of three steps, each of which results in a structure that is smaller in size. Since the first step handles the largest structure, improving its complexity will influence most the overall complexity of the algorithm.

In this section we suggest an alternative algorithm for computing the set of equivalence class. The algorithm avoids the construction of the simulation relation over the original structure. As a result, it has a better space complexity, but its time complexity is worse. Since the purpose of the Minimizing Algorithm is to reduce space requirements, it is more important to reduce its own space requirement.

5.1 The Partitioning Algorithm

Given a structure M, we would like to build the equivalence classes of the simulation equivalence relation, without first calculating H_M. Our algorithm, called the *Partitioning Algorithm* , starts with a *partition* Σ_0 of S to classes. The classes in Σ_0 differ from one another only by their state labeling. In each iteration, the algorithm refines the partition and forms a new set of classes. We use Σ_i to denote the set of the classes obtained after i iterations. In order to refine the partitions we build an *ordering* relation H_i over $\Sigma_i \times \Sigma_i$ which is updated in every iteration according to the previous and current partitions (Σ_{i-1} and Σ_i) and the previous ordering relation (H_{i-1}). Initially, H_0 includes only the identity pairs (of classes).

In the algorithm, we use *succ(s)* for the set of successors of s. Whenever Σ_i is clear from the context, $[s]$ is used for the equivalence class of s. We also use a function Π that associates with each class $\alpha \in \Sigma_i$ the set of classes $\alpha' \in \Sigma_{i-1}$ that contain a successor of some state in α.

$$\Pi(\alpha) = \{[t]^{i-1} | \exists s \in \alpha. \ (s,t) \in R\}$$

We use English letters to denote states, capital English letters to denote sets of states, Greek letters to denote equivalence classes, and capital Greek letters to denote sets of equivalence classes. The Partitioning Algorithm is presented in Figure 5 .

Definition 17. *The partial order $\leq_i$ on S is defined by: $s_1 \leq_i s_2$ implies, $L(s_1) = L(s_2)$ and if $i > 0$, $\forall t_1[(s_1,t_1) \in R \rightarrow \exists t_2[(s_2,t_2) \in R \wedge ([t_1],[t_2]) \in H_{i-1}]]$. In case $i = 0$, $s_1 \leq_0 s_2$ iff $L(s_1) = L(s_2)$.*
Two states s_1, s_2 are i–equivalent iff $s_1 \leq_i s_2$ and $s_2 \leq_i s_1$.

In the rest of this section we explain how the algorithm works. There are three invariants which are preserved during the execution of the algorithm.

Invariant 1: For all states $s_1, s_2 \in S$, s_1 and s_2 are in the same class $\alpha \in \Sigma_i$ iff s_1 and s_2 are i–equivalent.
Invariant 2: For all states $s_1, s_2 \in S$, $s_1 \leq_i s_2$ iff $([s_1], [s_2]) \in H_i$.
Invariant 3: H_i is transitive.

Σ_i is a set of equivalence classes with respect to the i–equivalence relation. In the ith iteration we split the equivalence classes of Σ_{i-1} so that only states that are i-equivalent remain in the same class.

A class $\alpha \in \Sigma_{i-1}$ is repeatedly split by choosing an arbitrary state $s_p \in \alpha$ (called the *splitter*) and identifying the states in α that are i–equivalent to s_p. These states form an i–equivalence class α' that is inserted to Σ_i.

α' is constructed in two steps. First we calculate the set of states $GT \subseteq \alpha$ that contains all states s_g such that $s_p \leq_i s_g$. Next we calculate the set of states $LT \subseteq \alpha$ that contains all states s_l such that $s_l \leq_i s_p$. The states in the intersection of GT and LT are the states in α that are i–equivalent to s_p.

H_i captures the partial order $\leq_i$, i.e., $s_1 \leq_i s_2$ iff $([s_1], [s_2]) \in H_i$. Note that the sequence $\leq_0, \leq_1, \ldots$ satisfies $\leq_0 \supseteq \leq_1 \supseteq \leq_2 \supseteq \ldots$. Therefore, if $s_1 \leq_i s_2$ then $s_1 \leq_{i-1} s_2$. Thus, $([s_1], [s_2]) \in H_i$ implies $([s_1], [s_2]) \in H_{i-1}$. Based on that, when constructing H_i it is sufficient to check $(\alpha'_1, \alpha'_2) \in H_i$ only in case $\alpha_2 \supseteq \alpha'_2$, $\alpha_1 \supseteq \alpha'_1$, and $(\alpha_1, \alpha_2) \in H_{i-1}$.

Initialize the algorithm:
    change := *true*
    for each label $a \in 2^{AP}$ construct $\alpha_a \in \Sigma_0$ such that $s \in \alpha_a \Leftrightarrow L(s) = a$.
$H_0 = \{(\alpha, \alpha) | \alpha \in \Sigma_0\}$
while *change* $= true$ do begin
    change := *false*
    refine Σ:
        $\Sigma_{i+1} := \emptyset$
            for each $\alpha \in \Sigma_i$ do begin
            while $\alpha \neq \emptyset$ do begin
                choose s_p such that $s_p \in \alpha$
                $GT := \{s_g | s_g \in \alpha \wedge \forall t_p \in succ(s_p)\ \exists t_g \in succ(s_g).\ ([t_p], [t_g]) \in H_i\}$
                $LT := \{s_l | s_l \in \alpha \wedge \forall t_l \in succ(s_l)\ \exists t_p \in succ(s_p).\ ([t_l], [t_p]) \in H_i\}$
                $\alpha' := GT \cap LT$
                if $\alpha \neq \alpha'$ then *change* := *true*
                $\alpha := \alpha \setminus \alpha'$
                Add α' as a new class to Σ_{i+1}.
        end
      end
    update H:
        $H_{i+1} = \emptyset$
        for every $(\alpha_1, \alpha_2) \in H_i$ do begin
            for each $\alpha_2', \alpha_1' \in \Sigma_{i+1}$ such that $\alpha_2 \supseteq \alpha_2'$, $\alpha_1 \supseteq \alpha_1'$ do begin
                $\Phi = \{\phi | \exists \xi \in \Pi(\alpha_2')\ (\phi, \xi) \in H_i\}$
                if $\Phi \supseteq \Pi(\alpha_1')$ then
                    insert (α_1', α_2') to H_{i+1}
                else
                    change := true
            end
      end
end

Fig. 5. The Partitioning Algorithm

For suitable α_1' and α_2', we first construct the set Φ of classes that are "smaller" than the classes in $\Pi(\alpha_2')$. By checking if $\Phi \supseteq \Pi(\alpha_1')$ we determine whether every class in $\Pi(\alpha_1')$ is "smaller" than some class in $\Pi(\alpha_2')$, in which case (α_1', α_2') is inserted to H_i.

When the algorithm terminates, $\leq_i$ is the maximal simulation relation and the i-equivalence is the simulation equivalence relation over $M \times M$. Moreover, H_i is the maximal simulation relation over the corresponding quotient structure M_q.

The algorithm runs until there is no change both in the partition Σ_i and in the relation H_i. A change in Σ_i is the result of a partitioning of some class $\alpha \in \Sigma_i$. The number of changes in Σ_i is bounded by the number of possible partitions, which is bounded by $|S|$.

A change in H_i results in the relation $\leq_{i+1}$ which is contained in $\leq_i$ and smaller in size, i.e., $| \leq_i | > | \leq_{i+1} |$. The number of changes in H_i is therefore bounded by $| \leq_0 |$, which is bounded by $|S|^2$. Thus, the algorithm terminates after at most $|S|^2 + |S|$ iterations. Note that, it is possible that in some iteration i, Σ_i will not change but H_i will, and in a later iteration $j > i$, Σ_j will change again.

Example: In this example we show how the Partitioning Algorithm is applied to the Kripke structure presented in Figure 6 .

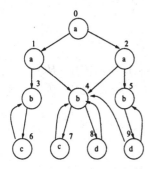

Fig. 6. An example structure

- We initialize the algorithm as follows:
  $\Sigma_0 = \{\alpha_0, \beta_0, \gamma_0, \delta_0\}$, $H_0 = \{(\alpha_0, \alpha_0), (\beta_0, \beta_0), (\gamma_0, \gamma_0), (\delta_0, \delta_0)\}$,
  where $\alpha_0 = \{0, 1, 2\}, \beta_0 = \{3, 4, 5\}, \gamma_0 = \{6, 7\}, \delta_0 = \{8, 9\}$.
- The first iteration results in the relations:
  $\Sigma_1 = \{\alpha_1, \alpha_2, \beta_1, \beta_2, \beta_3, \gamma_0, \delta_0\}$,
  $H_1 = \{(\alpha_1, \alpha_1), (\alpha_2, \alpha_2), (\beta_1, \beta_1), (\beta_2, \beta_2), (\beta_3, \beta_3), (\beta_1, \beta_2), (\beta_3, \beta_2), (\gamma_0, \gamma_0),$
  $(\delta_0, \delta_0)\}$, where $\alpha_1 = \{0\}, \alpha_2 = \{1, 2\}, \beta_1 = \{3\}, \beta_2 = \{4\}, \beta_3 = \{5\}, \gamma_0 = \{6, 7\}, \delta_0 = \{8, 9\}$.
- The second iteration results in the relations:
  $\Sigma_2 = \{\alpha_1, \alpha_2, \beta_1, \beta_2, \beta_3, \gamma_1, \gamma_2, \delta_0\}$,
  $H_2 = \{(\alpha_1, \alpha_1), (\alpha_2, \alpha_2), (\beta_1, \beta_1), (\beta_2, \beta_2), (\beta_3, \beta_3),$
  $\qquad (\beta_1, \beta_2), (\beta_3, \beta_2), (\gamma_1, \gamma_1), (\gamma_2, \gamma_2), (\gamma_1, \gamma_2), (\delta_0, \delta_0)\}$,
  where $\alpha_1 = \{0\}, \alpha_2 = \{1, 2\}, \beta_1 = \{3\}, \beta_2 = \{4\}, \beta_3 = \{5\}, \gamma_1 = \{6\}, \gamma_2 = \{7\}, \delta_0 = \{8, 9\}$.
- The third iteration results in the relations:
  $\Sigma_3 = \Sigma_2, H_3 = H_2$ - $change = false$.
  The equivalence classes are:
  $\alpha_1 = \{0\}, \alpha_2 = \{1, 2\}, \beta_1 = \{3\}, \beta_2 = \{4\}, \beta_3 = \{5\}, \gamma_1 = \{6\}, \gamma_2 = \{7\}, \delta_0 = \{8, 9\}$

Since the third iteration results in no change to the computed partition or ordering relation, the algorithm terminates. Σ_2 is the final set of equivalence classes which constitutes the set S_q of states of M_q. H_2 is the maximal simulation relation over $M_q \times M_q$. The proof of correctness of the algorithm can be fount in the full version.

5.2   Space and Time Complexity

The space complexity of the Partitioning Algorithm depends on the size of Σ_i. We assume that the algorithm applied to Kripke structures with some redundancy, thus $|\Sigma_i| << |S|$.

We measure the space complexity with respect to the size of the three following relations:

1. The relation R.
2. The relations H_i whose size depends on Σ_i. We can bound the size of H_i by $|\Sigma_i|^2$.
3. A relation that relates each state to its equivalence class. Since every state belongs to a single class, the size of this relation is $O(|S| \cdot log(|\Sigma_i|))$.

In the ith iteration we do not need to keep all $H_0, H_1, \ldots$ and $\Sigma_0, \Sigma_1, \ldots,$ since we only refer to H_i, H_{i+1} and Σ_i, Σ_{i+1}. By the above we conclude that the total space complexity is $O(|R| + |\Sigma_k|^2 + |S| \cdot log(|\Sigma_k|))$

In practice, we often do not hold the transition relation R in the memory. Rather we use it to provide, whenever needed, the set of successors of a given state. Thus, the space complexity is $O(|\Sigma_k|^2 + |S| \cdot log(|\Sigma_k|))$. Recall that the space complexity of the naive algorithm for computing the equivalence classes of the simulation equivalence relation is bounded by $|S|^2$, which is the size of the simulation relation over $M \times M$. In case $|\Sigma_k| << |S|$, the Partitioning Algorithm achieve a much better space complexity.

As we already mentioned, the algorithm runs at most $|S|^2$ iterations. In every iteration it performs one **refine** and one **update**. **refine** can be done in $O(|\Sigma_k|^3 + |\Sigma_k| \cdot |R|)$ and **update** can be done in $O(|\Sigma_k|^2 \cdot (|\Sigma_k|^2 + |R|))$. Thus the total time complexity is $O(|S|^2 \cdot |\Sigma_k|^2 \cdot (|\Sigma_k|^2 + |R|))$.

References

1. A. Aziz, T.R. Shiple, V. Singhal, and A.L. Sangiovanni-Vincetelly. Formula-dependent equivalence for compositional CTL model checking. In D. Dill, editor, *proceedings of the Sixth Conference on Computer Aided Verification (CAV'94)*, volume 818 of *LNCS*, pages 324–337, 1994.
2. A. Aziz, V. Singhal, G.M. Swamy, and R.K. Brayton. Minimizing interacting finite state machines: A compositional approach to language containment. In *Proceedings of the International Conference on Computer Design*, pages 255–261, 1994.
3. B. Bloom and R. Paige. Transformational design and implementation of new efficient solution to the ready simulation problem. In *Science of Computer Programming*, volume 24, pages 189–220, 1996.
4. A. Bouajjani, J.-C. Fernandez, and N. Halbwachs. Minimal model generation. In E.M Clarke and R.P. Kurshan, editors, *Computer-Aided Verification*, pages 197–203, New York, June 1990. Springer-Verlag.
5. A. Bouali and R. de Simone. Symbolic bisimulation minimisation. In G. V. Bochmann and D. K. Probst, editors, *Proceedings of the 4th Conference on Computer-Aided Verification*, volume 663 of *LNCS*, pages 96–108. Springer Verlag, July 1992.

6. D. Bustan and O. Grumberg. Simulation based minimization. Technical Report TR #CS-2000-04, Computer Science Department, Technion, Haifa, April 2000.
7. E. M. Clarke and E. A. Emerson. Synthesis of synchronization skeletons for branching time temporal logic. In *Logic of Programs: Workshop, Yorktown Heights, NY, May 1981*, volume 131 of *LNCS*. SPRINGER VERLAG, 1981.
8. E.M. Clarke, O. Grumberg, and D.A. Peled. *Model Checking*. MIT press, December 1999.
9. K. Fisler and M. Vardi. Bisimulation minimization in an automata-theoretic verification framework. In *FMCAD*, pages 115–132, 1998.
10. O. Grumberg and D.E. Long. Model checking and modular verification. *ACM Trans. on Programming Languages and Systems*, 16(3):843–871, 1994.
11. M.R. Henzinger, T.A. Henzinger, and P.W. Kopke. Computing simulation on finite and infinite graphs. In *Proc. Symp. Foundations of Computer Science*, pages 453–462, 1995.
12. A. Kucera and R. Mayr. Simulation preorder on simple process algebras. In *International Colloquium on Automata, Languages and Programing*, volume LNCS 1644, 1999.
13. D. Lee and M. Yannakakis. Online minimization of transition systems. In *Proceedings of the 24th ACM Symp. on Theory of Computing*, 1992.
14. R. Milner. An algebraic definition of simulation between programs. In *Proc. of the 2nd IJCAI*, pages 481–489, London, UK, 1971.
15. D. Park. Concurrency and automata on infinite sequences. In *5th GI-Conference on Theoretical Computer Science*, pages 167–183. Springer-Verlag, 1981. LNCS 104.

Rewriting for
Cryptographic Protocol Verification

Thomas Genet[1] and Francis Klay[2]

[1] IRISA / Université de Rennes
Campus de Beaulieu, F-35042 Rennes Cedex
genet@irisa.fr
[2] France Telecom
avenue Pierre Marzin, F-22307 Lannion Cedex
Francis.Klay@cnet.francetelecom.fr

Abstract. On a case study, we present a new approach for verifying cryptographic protocols, based on rewriting and on tree automata techniques. Protocols are operationally described using Term Rewriting Systems and the initial set of communication requests is described by a tree automaton. Starting from these two representations, we automatically compute an over-approximation of the set of exchanged messages (also recognized by a tree automaton). Then, proving classical properties like confidentiality or authentication can be done by automatically showing that the intersection between the approximation and a set of prohibited behaviors is the empty set. Furthermore, this method enjoys a simple and powerful way to describe intruder work, the ability to consider an unbounded number of parties, an unbounded number of interleaved sessions, and a theoretical property ensuring safeness of the approximation.

Introduction

In this paper, we present a new way of verifying cryptographic protocols. We do not aim here at discovering attacks on the protocol but our goal is to prove that there is not any, which is a more difficult problem. In practice, positive proofs of security properties on cryptographic protocols are highly desirable results since they give a better guarantee on the reliability of the protocol than any amount of passed tests. In [9], a decidable approximation of the set of descendants (reachable terms) was presented. In this paper, we propose to apply those theoretical results to the verification of cryptographic protocols. Our case study is the Needham-Schroeder Public Key protocol [19] (NSPK for short). We chose this particular example for two reasons. First of all, this protocol is real but can be easily understood. The second reason is that, in spite of its apparent simplicity and robustness, and in spite of several verification attempts, this protocol designed in 1978 was proved insecure only in 1995 by G. Lowe [13] and in 1996 by C. Meadows [17]. In particular, G. Lowe found a smart attack invalidating the main security properties of the protocol. In this paper, we will use the corrected version of the NSPK protocol also proposed by G. Lowe in [14].

D. McAllester (Ed.): CADE-17, LNAI 1831, pp. 271–290, 2000.
© Springer-Verlag Berlin Heidelberg 2000

Starting from a TRS representing the protocol and a tree automaton recognizing the initial set of communication requests, we automatically compute a superset of the set of exchanged messages by over-approximating the set of reachable terms. This model – also a tree automaton – takes into account an unbounded number of parties, an unbounded number of interleaved sessions as well as a powerful intruder activity description. For building this model, we needed to extend the approximation technique of [9], initially designed to approximate functional programs encoded by left-linear TRSs, to the more general class of TRSs (possibly non left-linear) with associative and commutative symbols.

In section 1, we recall basic definitions of terms, term rewriting systems, and tree automata. In section 2, we recall the technique for approximating the set of descendants for left-linear term rewriting systems and regular set of terms [9]. In section 3, we shortly present the Needham-Schroeder Public Key protocol, comment on its expected properties and propose an encoding into a term rewriting system in section 4. However, the term rewriting system describing the NSPK is not left-linear, has Associative and Commutative (AC for short) symbols and, consequently, is out of the scope of the basic approximation technique of [9]. Thus, in section 5, we show how to extend our technique to the case of non left-linear and AC TRSs. We also describe the application of approximation to NSPK and show how to prove confidentiality and authentication properties. Finally, in section 6, we conclude, compare with other approaches and present ongoing developments.

1   Preliminaries

We now introduce some notations and basic definitions. Comprehensive surveys can be found in [7] for term rewriting systems, in [3] for tree automata and tree language theory, and in [11] for connections between regular tree languages and term rewriting systems.

Terms, Substitutions, Rewriting Systems

Let $\mathcal{F}$ be a finite set of symbols associated with an arity function, $\mathcal{X}$ be a countable set of variables, $\mathcal{T}(\mathcal{F}, \mathcal{X})$ the set of terms, and $\mathcal{T}(\mathcal{F})$ the set of ground terms (terms without variables). Positions in a term are represented as sequences of integers. The set of positions in a term t, denoted by $\mathcal{P}os(t)$, is ordered by lexicographic ordering $\prec$. The empty sequence ϵ denotes the top-most position. If $p \in \mathcal{P}os(t)$, then $t|_p$ denotes the subterm of t at position p and $t[s]_p$ denotes the term obtained by replacement of the subterm $t|_p$ at position p by the term s. For any term $s \in \mathcal{T}(\mathcal{F}, \mathcal{X})$, we denote by $\mathcal{P}os_{\mathcal{F}}(s)$ the set of functional positions in s, i.e. $\{p \in \mathcal{P}os(s) \mid p \neq \epsilon \text{ and } \mathcal{R}oot(s|_p) \in \mathcal{F}\}$ where $\mathcal{R}oot(t)$ denotes the symbol at position ϵ in t. A *ground context* is a term of $\mathcal{T}(\mathcal{F} \cup \{\Box\})$ with exactly one occurrence of $\Box$, where $\Box$ is a special constant not occurring in $\mathcal{F}$. For any term $t \in \mathcal{T}(\mathcal{F})$, $C[t]$ denotes the term obtained after replacement of $\Box$ by t in the ground context $C[\ ]$. The set of variables of a term t is denoted by $\mathcal{V}ar(t)$. A term is linear if any variable of $Var(t)$ has exactly one occurrence in t. A substitution is a mapping σ from $\mathcal{X}$ into $\mathcal{T}(\mathcal{F}, \mathcal{X})$, which can uniquely be extended to an endomorphism of $\mathcal{T}(\mathcal{F}, \mathcal{X})$. Its domain $\mathcal{D}om(\sigma)$ is $\{x \in X \mid x\sigma \neq x\}$.

A term rewriting system $\mathcal{R}$ is a set of *rewrite rules* $l \to r$, where $l, r \in \mathcal{T}(\mathcal{F}, \mathcal{X})$, $l \notin \mathcal{X}$, and $Var(l) \supseteq Var(r)$. A rewrite rule $l \to r$ is *left-linear* (resp. *right-linear*) if the left-hand side (resp. right-hand side) of the rule is linear. A rule is linear if it is both left and right-linear. A TRS $\mathcal{R}$ is linear (resp. left-linear, right-linear) if every rewrite rule $l \to r$ of $\mathcal{R}$ is linear (resp. left-linear, right-linear).

The relation $\to_{\mathcal{R}}$ induced by $\mathcal{R}$ is defined as follows: for any $s, t \in \mathcal{T}(\mathcal{F}, \mathcal{X})$, $s \to_{\mathcal{R}} t$ if there exist a rule $l \to r$ in $\mathcal{R}$, a position $p \in \mathcal{P}os(s)$ and a substitution σ such that $l\sigma = s|_p$ and $t = s[r\sigma]_p$. The reflexive transitive closure of $\to_{\mathcal{R}}$ is denoted by $\to_{\mathcal{R}}^*$. The set of $\mathcal{R}$-descendants of a set of ground terms E is denoted by $\mathcal{R}^*(E)$ and $\mathcal{R}^*(E) = \{t \in \mathcal{T}(\mathcal{F}) \mid \exists s \in E \text{ s.t. } s \to_{\mathcal{R}}^* t\}$.

Automata, Regular Tree Languages

Let $\mathcal{Q}$ be a finite set of symbols, with arity 0, called *states*. $\mathcal{T}(\mathcal{F} \cup \mathcal{Q})$ is called the set of *configurations*. A *transition* is a rewrite rule $c \to q$, where $c \in \mathcal{T}(\mathcal{F} \cup \mathcal{Q})$ and $q \in \mathcal{Q}$. A *normalized transition* is a transition $c \to q$ where $c = q' \in \mathcal{Q}$ or $c = f(q_1, \ldots, q_n)$, $f \in \mathcal{F}$, $ar(f) = n$, and $q_1, \ldots, q_n \in \mathcal{Q}$. A bottom-up non-deterministic finite tree automaton (tree automaton for short) is a quadruple $\mathcal{A} = \langle \mathcal{F}, \mathcal{Q}, \mathcal{Q}_f, \Delta \rangle$, where $\mathcal{Q}_f \subseteq \mathcal{Q}$ and Δ is a set of normalized transitions. A tree automaton is *deterministic* if there are no two rules with the same right hand side. The rewriting relation induced by Δ is denoted either by $\to_\Delta$ or by $\to_{\mathcal{A}}$. The tree language recognized by $\mathcal{A}$ is $\mathcal{L}(\mathcal{A}) = \{t \in \mathcal{T}(\mathcal{F}) \mid \exists q \in \mathcal{Q}_f \text{ s.t. } t \to_{\mathcal{A}}^* q\}$. For a given $q \in \mathcal{Q}$, the tree language recognized by $\mathcal{A}$ and q is $\mathcal{L}(\mathcal{A}, q) = \{t \in \mathcal{T}(\mathcal{F}) \mid t \to_{\mathcal{A}}^* q\}$. A tree language (or a set of terms) E is *regular* if there exists a bottom-up tree automaton $\mathcal{A}$ such that $\mathcal{L}(\mathcal{A}) = E$. The class of regular tree languages is closed under boolean operations $\cup, \cap, \backslash$, and inclusion is decidable. A $\mathcal{Q}$-substitution is a substitution $\sigma : \mathcal{X} \mapsto \mathcal{Q}$. Let $\Sigma(\mathcal{Q}, \mathcal{X})$ be the set of $\mathcal{Q}$-substitutions. For every transition, there exists an equivalent set of normalized transitions. Normalization consists in decomposing a transition $s \to q$, into a set $Norm(s \to q)$ of normalized transitions. The method consists in abstracting subterms s' of s s.t. $s' \notin \mathcal{Q}$ by states of $\mathcal{Q}$. We first define the abstraction function as follows:

Definition 1. *Let $\mathcal{F}$ be a set of symbols, and $\mathcal{Q}$ a set of states. For a given configuration $s \in \mathcal{T}(\mathcal{F} \cup \mathcal{Q}) \setminus \mathcal{Q}$, an abstraction of s is a mapping α:*

$$\alpha : \{s|_p \mid p \in \mathcal{P}os_{\mathcal{F}}(s)\} \mapsto \mathcal{Q}$$

The mapping α is extended on $\mathcal{T}(\mathcal{F} \cup \mathcal{Q})$ by defining α as identity on $\mathcal{Q}$, i.e. $\forall q \in \mathcal{Q} : \alpha(q) = q$.

Definition 2. *Let $\mathcal{F}$ be a set of symbols, $\mathcal{Q}$ a set of states, $s \to q$ a transition s.t. $s \in \mathcal{T}(\mathcal{F} \cup \mathcal{Q})$ and $q \in \mathcal{Q}$, and α an abstraction of s. The set $Norm_\alpha(s \to q)$ of normalized transitions is inductively defined by:*

1. *if $s = q$, then $Norm_\alpha(s \to q) = \emptyset$, and*
2. *if $s \in Q$ and $s \neq q$, then $Norm_\alpha(s \to q) = \{s \to q\}$, and*
3. *if $s = f(t_1, \ldots, t_n)$, then $Norm_\alpha(s \to q) =$*
   $\{f(\alpha(t_1), \ldots, \alpha(t_n)) \to q\} \cup \bigcup_{i=1}^n Norm_\alpha(t_i \to \alpha(t_i))$.

Example 1. Let $\mathcal{F} = \{f, g, a\}$ and $\mathcal{A} = \langle \mathcal{F}, Q, Q_f, \Delta \rangle$, where $Q = \{q_0, q_1, q_2, q_3, q_4\}$, $Q_f = \{q_0\}$, and $\Delta = \{f(q_1) \to q_0, g(q_1, q_1) \to q_1, a \to q_1\}$.

• The languages recognized by q_1 and q_0 are the following: $\mathcal{L}(\mathcal{A}, q_1)$ is the set of terms built on $\{g, a\}$, i.e. $\mathcal{L}(\mathcal{A}, q_1) = \mathcal{T}(\{g, a\})$, and $\mathcal{L}(\mathcal{A}, q_0) = \mathcal{L}(\mathcal{A}) = \{f(x) \mid x \in \mathcal{L}(\mathcal{A}, q_1)\}$.

• Let $s = f(g(q_1, f(a)))$, and α_1 be an abstraction of s, mapping $g(q_1, f(a))$ to q_2, $f(a)$ to q_3 and a to q_4. The normalization of transition $f(g(q_1, f(a))) \to q_0$ with abstraction α_1 is the following: $Norm_{\alpha_1}(f(g(q_1, f(a))) \to q_0) = \{f(q_2) \to q_0, g(q_1, q_3) \to q_2, f(q_4) \to q_3, a \to q_4\}$.

2   Approximation Technique

For a regular set of terms $E \subseteq \mathcal{T}(\mathcal{F})$, although there exists some restricted classes of TRSs $\mathcal{R}$ such that $\mathcal{R}^*(E)$ is regular (see [5, 21, 4, 12]), this is not the case in general [11, 12]. In [9], for any tree automaton $\mathcal{A}$ (s.t. $\mathcal{L}(\mathcal{A}) \supseteq E$) and for any left-linear TRS $\mathcal{R}$, it is proposed to build an approximation automaton $\mathcal{T}_\mathcal{R}{\uparrow}(\mathcal{A})$ such that $\mathcal{L}(\mathcal{T}_\mathcal{R}{\uparrow}(\mathcal{A})) \supseteq \mathcal{R}^*(E)$. The quality of the approximation highly depends on an approximation function called γ which define some *folding positions*: subterms who can be approximated. We now briefly recall the construction of $\mathcal{T}_\mathcal{R}{\uparrow}(\mathcal{A})$ [9]:

Let $\mathcal{R}$ be a left-linear term rewriting system and $\mathcal{A} = \langle \mathcal{F}, Q, Q_f, \Delta \rangle$ a tree automaton such that $E = \mathcal{L}(\mathcal{A})$ (or even $E \subseteq \mathcal{L}(A)$). First, we infinitely extend the set of states Q of $\mathcal{A}$ with an infinite number of new states, initially not occurring in Q. Note that since we do not modify Δ nor Q_f (in particular, they remain finite), the language recognized by $\mathcal{A}$ is the same. On the other hand, it is always possible to come back to a finite set of states for $\mathcal{A}$ by restricting Q to the set of *accessible states*, i.e. states q such that $\mathcal{L}(\mathcal{A}, q) \neq \emptyset$.

Starting from $\mathcal{A}_0 = \mathcal{A}$, we incrementally build a finite number of tree automata $\mathcal{A}_i = \langle \mathcal{F}, Q, Q_f, \Delta_i \rangle$ with $i \geq 0$ such that $\forall i \geq 0 : \mathcal{L}(\mathcal{A}_i) \subset \mathcal{L}(\mathcal{A}_{i+1})$ until we get an automaton $\mathcal{A}_k$ with $k \in \mathbb{N}$ such that $\mathcal{L}(\mathcal{A}_k) \supseteq \mathcal{R}^*(\mathcal{L}(\mathcal{A}_0))$, i.e. $\mathcal{L}(\mathcal{A}_k) \supseteq \mathcal{R}^*(E)$. We denote by $\mathcal{T}_\mathcal{R}{\uparrow}(\mathcal{A})$ this automaton $\mathcal{A}_k$. To construct $\mathcal{A}_{i+1}$ from $\mathcal{A}_i$, the technique consists in finding a term s in $\mathcal{L}(\mathcal{A}_i)$ such that $s \to_\mathcal{R} t$ and $t \notin \mathcal{L}(\mathcal{A}_i)$, and then in building Δ_{i+1} such that $\mathcal{L}(\mathcal{A}_i) \subset \mathcal{L}(\mathcal{A}_{i+1})$ and $t \in \mathcal{L}(\mathcal{A}_{i+1})$.

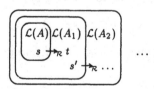

Since $\mathcal{A}_i$ and $\mathcal{A}_{i+1}$ only differs by their respective transitions sets, to ensure $\mathcal{L}(\mathcal{A}_i) \subset \mathcal{L}(\mathcal{A}_{i+1})$ it is enough to construct Δ_{i+1} such that it strictly contains

Δ_i. In order to have also $t \in \mathcal{L}(\mathcal{A}_{i+1})$ it is necessary to add some transitions to Δ_i to obtain Δ_{i+1}. This can be viewed as a *completion step* between the two term rewriting systems: the set of transitions Δ_i of $\mathcal{A}_i$ and $\mathcal{R}$. If there exists a term s in $\mathcal{L}(\mathcal{A}_i)$ such that $s \twoheadrightarrow_{\mathcal{R}} t$, by definition of $\twoheadrightarrow_{\mathcal{R}}$, there exists a rule $l \to r$, a ground context $C[]$ and a substitution (a match) σ such that $s = C[l\sigma] \twoheadrightarrow_{\mathcal{R}} C[r\sigma] = t$. On the other hand, by construction of tree automata, $s = C[l\sigma] \in \mathcal{L}(\mathcal{A}_i)$ means that (1) there exists a state $q \in \mathcal{Q}$ such that $l\sigma \to^*_{\mathcal{A}_i} q$ and (2) $C[q] \to^*_{\mathcal{A}_i} q'$ such that $q' \in \mathcal{Q}_f$. Hence, from (1) we know that we have following *critical pair* between transitions of $\mathcal{A}_i$ and rules of $\mathcal{R}$:

$$
\begin{array}{ccc}
l\sigma & \xrightarrow{\quad\mathcal{R}\quad} & r\sigma \\[2pt]
\mathcal{A}_i \downarrow {\scriptstyle *} & & \\[6pt]
q & &
\end{array}
$$

Since every transition of $\mathcal{A}_i$ is in $\mathcal{A}_{i+1}$ (i.e. $\Delta_i \subseteq \Delta_{i+1}$), for the term t to be recognized by $\mathcal{A}_{i+1}$, it is enough to ensure that (3) $r\sigma \to^*_{\mathcal{A}_{i+1}} q$. This is sufficient since we can then rewrite $t = C[r\sigma]$ into $C[q]$ and from (2) we get that $C[q] \to^*_{\mathcal{A}_{i+1}} q'$, since $\Delta_i \subseteq \Delta_{i+1}$. Finally, since $q' \in \mathcal{Q}_f$, $t \in \mathcal{L}(\mathcal{A}_{i+1})$.

To ensure (3), we need to add some transitions to Δ_{i+1}, i.e. join the critical pair:

$$
\begin{array}{ccc}
l\sigma & \xrightarrow{\quad\mathcal{R}\quad} & r\sigma \\[2pt]
A_i \downarrow {\scriptstyle *} & & \vdots {\scriptstyle *} \\[6pt]
q & \xleftarrow{\quad\quad} & A_{i+1}
\end{array}
$$

A direct solution to have $r\sigma \to^*_{\mathcal{A}_{i+1}} q$ is to have a transition of the form $r\sigma \to q$ in $\mathcal{A}_{i+1}$. However, this is not compatible with the standard normalized form of the tree automata we use here[1]. Thus, before adding $r\sigma \to q$ to transitions of $\mathcal{A}_i$, we normalize it first thanks to the $Norm_\alpha$ function (see definition 2). Hence, $\Delta_{i+1} = \Delta_i \cup Norm_\alpha(r\sigma \to q)$. We give here an example of completion process on a simple TRS

Example 2. Let $\mathcal{F} = \{f, g, a\}$ and $\mathcal{R}$ the one rule TRS $\mathcal{R} = \{f(g(x)) \to g(f(x))\}$. Let $\mathcal{A}_0 = \langle \mathcal{F}, \mathcal{Q}, \mathcal{Q}_f, \Delta_0 \rangle$ such that $\mathcal{Q}_f = \{q_f\}$ and $\Delta_0 = \{f(q_f) \to q_f, g(q_a) \to q_f, a \to q_a\}$. We have $\mathcal{L}(\mathcal{A}_0) = f^*(g(a))$. Between $\mathcal{R}$ and transitions of $\mathcal{A}_0$ there exists a critical pair:

$$
\begin{array}{ccc}
f(g(q_a)) & \xrightarrow{\quad\mathcal{R}\quad} & g(f(q_a)) \\[2pt]
\mathcal{A}_0 \downarrow {\scriptstyle *} & & \\[6pt]
q_f & &
\end{array}
$$

---

[1] keeping tree automata in standard normalized form allows, in particular, to apply usual algorithms: intersection, union, etc.

The Q-substitution used here is $\sigma = \{x \mapsto q_a\}$. As defined before, we have $\Delta_1 = \Delta_0 \cup Norm_\alpha(g(f(q_a))) \to q_f)$. Let α be the abstraction function such that $\alpha(f(q_a)) = q_{new}$ where q_{new} is a state not occurring in transitions of $\mathcal{A}_0$. Then, we have $\Delta_1 = \Delta_0 \cup \{g(q_{new}) \to q_f, f(q_a) \to q_{new}\}$.

Except in some simple decidable case, this completion procedure is not guaranteed to converge but, instead, may infinitely add new transitions and thus generate an infinite number of tree automata $\mathcal{A}_1, \mathcal{A}_2$, etc. However, choosing particular values for α may force the completion process to converge by approximating infinitely many transitions by finite sets of more general transitions. Those particular abstraction functions are associated with *approximation functions* denoted by γ, defining some folding positions: positions in the right hand side of rules where subterms are approximated by regular languages: for each completion step from $\mathcal{A}_i$ to $\mathcal{A}_{i+1}$ involving a rewrite step $l\sigma \to_{\mathcal{R}} r\sigma$, a folding position p is a position in r which is assigned a state q' such that we only ensure $\mathcal{L}(\mathcal{A}_{i+1}, q') \supseteq \{r\sigma|_p\}$ instead of strict equality: $\mathcal{L}(\mathcal{A}_{i+1}, q') = \{r\sigma|_p\}$. This comes from the fact that the same state q' can be used for recognizing different terms obtained by different positions, rules or substitutions. The role of the approximation function is to relate $r\sigma|_p$ and the state q'. Folding positions depend on the applied rule $l \to r$ and on the substitution σ. Furthermore, since in our setting a rewriting step $s = C[l\sigma] \to_{\mathcal{R}} C[r\sigma] = t$ is modeled by a completion step on the critical pair $l\sigma \to_{\mathcal{R}} r\sigma$ and $l\sigma \to_{\mathcal{A}_i} q$, q is also a parameter of the approximation function. Finally, the approximation function γ maps every triple $(l \to r, q, \sigma)$ to a sequence of states (one for each position in $\mathcal{P}os_{\mathcal{F}}(r)$) used for the normalization of the transition $r\sigma \to q$.

Definition 3. *Let Q be a set of states and Q^* the set of sequences $q_1 \cdots q_k$ of states in Q. An approximation function is a mapping $\gamma : \mathcal{R} \times Q \times \Sigma(Q, \mathcal{X}) \mapsto Q^*$, such that $\gamma(l \to r, q, \sigma) = q_1 \cdots q_k$, where $k = Card(\mathcal{P}os_{\mathcal{F}}(r))$.*

From every $\gamma(l \to r, q, \sigma) = q_1 \cdots q_k$, we can associate $q_1, \ldots, q_k$ to positions $p_1, \ldots, p_k$ in $\mathcal{P}os_{\mathcal{F}}(r)$. This can be done by defining the corresponding abstraction function α on the restricted domain $\{r\sigma|_p \mid \forall l \to r \in \mathcal{R}, \forall p \in \mathcal{P}os_{\mathcal{F}}(r), \forall \sigma \in \Sigma(Q, \mathcal{X})\}$:

$$\alpha(r\sigma|_{p_i}) = q_i$$

for all $p_i \in \mathcal{P}os_{\mathcal{F}}(r) = \{p_1, \ldots, p_k\}$, s.t. $p_i \prec p_{i+1}$ for $i = 1 \ldots k - 1$ (where $\prec$ is the lexicographic ordering). In the following, we will note $Norm_\gamma$ the normalization function whose α value is defined according to γ as above.

Starting from a left-linear TRS $\mathcal{R}$, a tree automaton $\mathcal{A}$ and an approximation function γ, the *algorithm* for building the approximation automaton $\mathcal{T}_{\mathcal{R}}\uparrow(\mathcal{A})$ is the following. First, set $\mathcal{A}_0$ to $\mathcal{A}$. Then, to construct $\mathcal{A}_{i+1}$ from $\mathcal{A}_i$:

1. search for a critical pair, i.e. a state $q \in Q$, a rewrite rule $l \to r$ and a substitution $\sigma \in \Sigma(Q, \mathcal{X})$ such that $l\sigma \to_{\mathcal{A}_i}^* q$ and $r\sigma \not\to_{\mathcal{A}_i}^* q$.
2. $\mathcal{A}_{i+1} = \mathcal{A}_i \cup Norm_\gamma(r\sigma \to q)$.

This process is iterated until it stops on a tree automaton A_k such that $\forall q \in Q$, $\forall l \to r \in \mathcal{R}$ and $\forall \sigma \in \Sigma(Q, \mathcal{X})$ if $l\sigma \to^*_{A_k} q$ then $r\sigma \to^*_{A_k} q$. Then, $\mathcal{T}_\mathcal{R}\uparrow(A) = A_k$. The fact that Q and $\Sigma(Q, \mathcal{X})$ may be infinite is not a problem in practice since, for finding a critical pair, we can restrict Q to the finite set of accessible states in A_i, without changing $\mathcal{L}(A_i)$ nor $\mathcal{L}(A_{i+1})$. We now recall a theorem of [9].

Theorem 1. *(Completeness) Given a tree automaton A and a left-linear TRS $\mathcal{R}$, for any approximation function γ,*

$$\mathcal{L}(\mathcal{T}_\mathcal{R}\uparrow(A)) \supseteq \mathcal{R}^*(\mathcal{L}(A))$$

The γ function fix the quality of the approximation. For example, one of the roughest approximation is obtained with a constant γ function mapping every triple $(l \to r, \sigma, q)$ to sequences of q' a unique state of Q: $\forall l \to r \in \mathcal{R}, \forall \sigma \in \Sigma(Q, \mathcal{X}), \forall q \in Q : \gamma(l \to r, \sigma, q) = q' \cdots q'$. On the opposite, the best approximation consists in mapping every triple $(l \to r, \sigma, q)$ to sequences of distinct states. However, although any rough approximation built with the first γ is guaranteed to terminate, this is not necessarily the case for the second one.

On a practical point of view, the fact that completeness of the approximation construction does not depend on the chosen γ (Theorem 1) is a very interesting property. Indeed, it guarantees that for any approximation function, $\mathcal{T}_\mathcal{R}\uparrow(A)$ is a safe model of $\mathcal{R}^*(E)$, in the sense of abstract interpretation.

Example 3. Back to the example 2, adding to A_0 transitions $\{g(q_{new}) \to q_f, f(q_a) \to q_{new}\}$ to obtain A_1 brings another critical pair:

$$f(g(q_{new})) \xrightarrow{\mathcal{R}} g(f(q_{new}))$$

$$A_1 \Big\downarrow {\scriptstyle *}$$

$$q_f$$

Like in the previous example, we build Δ_2 by adding $Norm_\alpha(f(g(q_{new})) \to q_f)$ to Δ_1. However, if α maps $g(q_{new})$ to another state q'_{new} not occurring in Δ_1, we add some new transitions and get another critical pair, and the process may go on for ever. Instead, we can here define an approximation function γ in a simple and static way, for example: $\forall \sigma \in \Sigma(Q, \mathcal{X}), \forall q \in Q : \gamma(f(g(x)) \to g(f(x)), q, \sigma) = q_{new}$. Since $\mathcal{P}os_\mathcal{F}(g(f(x))) = \{1\}$ is a singleton, note that the γ function maps triple of the form $(f(g(x)) \to g(f(x)), q, \sigma)$ to sequences of states of length one. This γ function defines a very rough approximation since the same state q_{new} is used for every normalization, whatever values q and σ may be. Thanks to this approximation function γ, the completion terminates. The value of Δ_1 remain the same but, for the next completion step, we have $Norm_\gamma(f(g(q_{new})) \to q_f) = \{g(q_{new}) \to q_f, f(q_{new}) \to q_{new}\}$. Thus, $\Delta_2 = \Delta_1 \cup \{f(q_{new}) \to q_{new}\}$, there is no new critical pair between Δ_2 and rule $f(g(x)) \to g(f(x))$, and we have $\mathcal{L}(A_2) = f^*(g(f^*(a)))$.

Once $\mathcal{T}_\mathcal{R}\uparrow(A)$ is obtained, it is easy to verify some reachability properties on $\mathcal{R}$ and E. It can be shown for example that a regular set of terms F cannot be reached from terms of E by $\rightarrow_{\mathcal{R}}{}^*$. This can be done by showing that $\mathcal{L}(\mathcal{T}_{\mathcal{R}}\uparrow$ $(\mathcal{A})) \cap F = \emptyset$. We will apply this to the verification of the Needham-Schroeder Public Key Protocol in section 5.

3   Needham-Schroeder Public Key Protocol

In this section, we present our case study on the Needham-Schroeder Public Key protocol (NSPK). More precisely, we here use the fixed version of the protocol [14] without key server. Key servers have been discarded here for the sake of simplicity. Note that attacks from [14] have been found on the NSPK without key servers. Moreover, the approximation technique have also been successfully applied to the protocol with key servers.

The NSPK protocol aim at mutual authentication of two agents, an initiator A and a responder B, separated by an insecure network. Mutual authentication means that, when a protocol session is completed between two agents, they should be assured of each other's identity. In general, the main property expected for this kind of protocol is to prevent an intruder from impersonating one of the two agents. This protocol is based on an exchange of *nonces* (usually fresh random numbers or time stamps) and on *asymmetric* encryption of messages: every agent has a *public key* (for encryption) and a *private key* (for decryption). Every public key is supposed to be known by any agent whereas, the private key of agent X is supposed to be only known by X. Thus, in this setting, we suppose that messages encrypted with the public key of X can only be decrypted and read by X. Here is a description of the three steps of the fixed version of protocol, borrowed from [14]:

1. $A \hookrightarrow B : \{N_A, A\}_{K_B}$
2. $B \hookrightarrow A : \{N_A, N_B, B\}_{K_A}$
3. $A \hookrightarrow B : \{N_B\}_{K_B}$

In the first step, A tries to initiate a communication with B: A creates a nonce N_A and sends to B a message, containing N_A as well as his identity, encrypted with the public key of B: K_B. Then, in the second step, B sends back to A a message encrypted with the public key of A, containing the nonce N_A that B received, a new nonce N_B, and B's identity. Finally, in the last step, A returns the nonce N_B he received from B. If the protocol is completed, mutual authentication of the two agents is ensured:

- as soon as A receives the message containing the nonce N_A, sent back by B at step 2., A *believes* that this message was really built and sent by B. Indeed, N_A was encrypted with the public key of B and, thus, B is the only agent that is able to send back N_A,
- similarly, when B receives the message containing the nonce N_B, sent back by A at step 3., B *believes* that this message was really built and sent by A.

Another property that may be expected for this kind of protocol is *confidentiality* of nonces. In particular, if nonces remain confidential, they can be used later as keys for symmetric encryption of communications between A and B. Thus, confidentiality of nonces may also be of interest.

A cryptographic protocol is supposed to resist to any attack of an intruder. In particular for NSPK, we intend to show that, for agents respecting the protocol, and whatever the intruder may do,

- nonces and private keys remain confidential (confidentiality),
- if an agent X believes that a message was built by another agent Y, then the message was effectively built by Y (authentication).

4   Encoding the Protocol and the Intruder

In this section, we show how to model NSPK by a TRS. First, we present the signature $\mathcal{F}$ and the terms of $\mathcal{T}(\mathcal{F})$ used for representing agents, messages, keys, etc. Each agent is labeled by a unique identifier, let L_{agt} be the set of agent labels (terms representing agent labels will be given later). For any agent label $l \in L_{agt}$, the term $agt(l)$ will denote the agent whose label is l. The term $mesg(x, y, c)$ will represent a message whose header refers agent x as emitter, agent y as receiver and whose contents is c. The term $pubkey(a)$ denotes the public key of agent a and $encr(k, a, c)$ denotes the result of encryption of content c by key k. In this last term, a is a flag recording who has performed the encryption. This field is not used by the protocol rules but will be used for verification. The term $N(x, y)$ represents a nonce generated by agent x for identifying a communication with y. We also use an AC binary symbol $\sqcup$ in order to represent sets. For example the term $x \sqcup (y \sqcup z)$ (equivalent modulo AC to $(x \sqcup y) \sqcup z$) will represent the set $\{x, y, z\}$.

Starting from a set of initial requests, our aim is to compute a tree automaton recognizing an over-approximation of all sent messages. The approximation also contains some terms signaling either communication requests or established communications. For example, a term of the form $goal(x, y)$ means that x expect to open a communication with y. A term of the form $c\_init(x, y, z)$ means that x believes to have initiated a communication with y, but, in reality x communicates with z. Conversely, a term $c\_resp(y, x, z)$ means that y believes to have responded to a communication request coming from x but z is the real author of the request.

Then, encoding of the protocol into AC rewrite rules[2] is straightforward: each step of the protocol is described thanks to a rewrite rule whose left-hand

---

[2] We describe here our encoding in a general way. However, for the particular case of NSPK, encoding could have be done without the AC-symbol $\sqcup$, since $\sqcup$ is only needed when the sending of a message depends on the reception of two (or more) distinct messages, i.e. rules of the form: $m_1 \sqcup m_2 \rightarrow m_3$. In general, those rules are necessary to modelize protocols, but it is not the case for this simple version of NSPK.

side is a precondition on the current state (set of received messages and communication requests), and the right-hand side represents the message to be sent (and sometimes established communication) if the precondition is met. The sent message is added to the current state. As a result, every rewrite rule we use is a 'cumulative rule', i.e. of the form $l \rightarrow l \sqcup X$. Thus, for commodity, we choose to use the short-hand LHS for the term l occurring in the right-hand side. For instance, the rule $mesg(x, y, c) \rightarrow LHS \sqcup c\_init(x, y, y)$ will represent the rule: $mesg(x, y, c) \rightarrow mesg(x, y, c) \sqcup c\_init(x, y, y)$. Now for each step of the protocol, we give the corresponding rewrite rule. The encoding into TRS is longer than the initial protocol specification of section 3 because it is more complete. For instance, whereas the initial specification only informally define how to check the content of messages and how to deal with communication requests, these points are formally defined in our specification with rewrite rules. Furthermore, the initial specification can be viewed as a trace of a correct execution of the NSPK protocol for two specific agents A and B. Thus, this specification cannot be directly used in a more general context where some other agents also use the protocol. Hence, another difference between our specification and the initial specification of section 3 is that agents' identities of initial specification (A and B) have been abstracted by term with variables of the form $agt(x)$, $agt(y)$. In the following, x, y, z, u, v, $x2$, $x3$ and $z2$ are supposed to be variables since we consider an unbounded number of agents and transactions.

1. $A \hookrightarrow B : \{N_A, A\}_{K_B}$. The emission of the first message is encoded by the rule:

   $goal(x, y) \rightarrow LHS \sqcup mesg(x, y, encr(pubkey(y), x, [N(x, y), x]))$

   The meaning of this rule is the following: if an agent x wants to establish a communication with y then x sends a message to y whose contents is encrypted with public key of y. The contents is here represented by a list (build with classical operators *cons* and *null*) containing a nonce $N(x, y)$ produced by x for y as well as x's identity. For commodity, lists will be represented in the usual way, for example a list of the form $cons(u, cons(v, null))$ will be denoted by $[u, v]$.

2. $B \hookrightarrow A : \{N_A, N_B, B\}_{K_A}$.

   $mesg(x, agt(u), encr(pubkey(agt(u)), z, [v, agt(x2)])) \rightarrow$
   $LHS \sqcup mesg(agt(u), agt(x2), encr(pubkey(agt(x2)), agt(u),$
   $\qquad\qquad [v, N(agt(u), agt(x2)), agt(u)]))$

   The second message is sent by an agent $agt(u)$ when he receives the first message from an agent $agt(x2)$ whose identity is enclosed in the message[3]. Note that in those rules, we achieve some kind of type checking on the content of the message. For instance, in the left-hand side of this rule, by expecting

---

[3] In this protocol, agent's identity contained in the header of the message (x in our example) is never used, since it may have been corrupted by an intruder. However, this information is sometimes used, for example in the extended version of NSPK where a key server is also involved.

the message content pattern $[v, agt(x2)]$ instead of a more general pattern like $[v, x3]$, we check that this element of the message is an agent's identity. The role of this kind of type checking is important since it permits to avoid some attacks based on type confusion like those described in [17].

3. $A \hookrightarrow B : \{N_B\}_{K_B}$. This step is encoded by the rule:
$mesg(x, agt(y), encr(pubkey(agt(y)), z2, [N(agt(y), agt(z)), u, agt(z)])) \rightarrow$
$LHS \sqcup mesg(agt(y), agt(z), encr(pubkey(agt(z)), [u]))$
$\qquad \sqcup c\_init(agt(y), agt(z), z2)$

When agent $agt(y)$ receives from $agt(z)$ the nonce $N(agt(y), agt(z))$ he has built for $agt(z)$ then he performs two actions. The first action is to send the last protocol message to $agt(z)$. The second action consists in reporting the communication $agt(y)$ thinks to have established with $agt(z)$. However, the reality may be different and the identity of the real author of the message, $z2$, is used for filling the third field of the c_init term.

4. In the last step of the protocol, no message is sent but when an agent receives the last message of the protocol sent at step 3., he reports a communication where he has the responder role.
$mesg(x, agt(y), encr(pubkey(agt(y)), z2, [N(agt(y), z)])) \rightarrow$
$LHS \sqcup c\_resp(agt(y), z, z2)$

To prove the authentication property on the protocol, we need to prove that any couple of agents can securely establish a communication through the network, whatever the behavior of other agents and the behavior of an intruder may be. Thus, we assume that there is an unbounded number of agent labels in L_{agt} but we will observe more precisely two agents, namely agents labeled by A and B. For the unbounded number of other agent labels we will use integers built on usual operators 0 and s (successor). Hence, $L_{agt} = \{A, B\} \cup \mathbb{N}$ and the initial set of terms E is the set of terms of the form $goal(agt(x), agt(y))$ where $x, y \in L_{agt}$. In other words, E is the set of all communication requests

- from A or B towards any other agent $agt(i)$ with $i \in \mathbb{N}$, and
- from $agt(i)$ with $i \in \mathbb{N}$ towards A or B, and
- from any agent $agt(i)$ to any agent $agt(j)$, $i, j \in \mathbb{N}$, and
- from A to B, B to A, A to A and B to B.

Note that we work in a very general setting where we also take into account the case where an agent use the protocol to authenticate himself. It is clear that self-authentication of an agent may be not of practical interest, but, if it happens we want to verify that the intruder cannot take advantage of it to build an attack. The set E is recognized by the following tree automaton $\mathcal{A}_0$. The final state of $\mathcal{A}_0$ is q_{net} and here is the set of transitions:

$0 \rightarrow q_{int}$	$agt(q_B) \rightarrow q_{agtB}$	$goal(q_{agtA}, q_{agtI}) \rightarrow q_{net}$
$s(q_{int}) \rightarrow q_{int}$	$q_{net} \sqcup q_{net} \rightarrow q_{net}$	$goal(q_{agtI}, q_{agtA}) \rightarrow q_{net}$
$A \rightarrow q_A$	$goal(q_{agtA}, q_{agtB}) \rightarrow q_{net}$	$goal(q_{agtB}, q_{agtI}) \rightarrow q_{net}$
$B \rightarrow q_B$	$goal(q_{agtB}, q_{agtA}) \rightarrow q_{net}$	$goal(q_{agtI}, q_{agtB}) \rightarrow q_{net}$
$agt(q_{int}) \rightarrow q_{agtI}$	$goal(q_{agtA}, q_{agtA}) \rightarrow q_{net}$	$goal(q_{agtI}, q_{agtI}) \rightarrow q_{net}$
$agt(q_A) \rightarrow q_{agtA}$	$goal(q_{agtB}, q_{agtB}) \rightarrow q_{net}$	

Description of the Intruder

In this last automaton, the state q_{net} is a special state representing both the network and the fact base containing communication requests and communication reports. As in many other verification approach of cryptographic protocols, the intruder is supposed to have a total control on the network. In particular, the intruder is assumed to know every message sent on the network. In our approach this assumption is a bit stronger: the intruder *is* the network. A direct consequence of this choice is that the knowledge of the intruder and every message that the intruder can build is supposed to always remain on the network. Furthermore, we suppose that agents $agt(i)$ with $i \in \mathbb{N}$ (i.e. every agent that is not A or B) may be dishonest and deliberately give to the intruder their private key as well as the content of any message they send or receive. The intruder can also disassemble messages or build new ones from his knowledge. Rewrite rules are the simplest way to describe how an intruder can decrypt or disassemble components of a message. Since the agents $agt(i)$ with $i \in \mathbb{N}$ are fool enough to give their private keys to the intruder, he can decrypt the messages encrypted with their public keys. On the opposite, we assume that the intruder has no means of guessing the private key of A or B. Here are the corresponding rules which can be applied on the AC-term representing the network, i.e. the intruder knowledge:

$$
\begin{aligned}
&cons(x,y) \sqcup z \to LHS \sqcup x && / * \text{Disassembling} * / \\
&cons(x,y) \sqcup z \to LHS \sqcup y \\
&mesg(x,y,z) \sqcup u \to LHS \sqcup z \\
&encr(pubkey(agt(0)),y,z) \sqcup u \to LHS \sqcup z && / * \text{Decrypting} * / \\
&encr(pubkey(agt(s(x))),y,z) \sqcup u \to LHS \sqcup z
\end{aligned}
$$

On the other hand, intruder's ability to build new messages from its knowledge is shortly defined thanks to some tree automaton transitions. Since q_{net} is the state of $\mathcal{A}_0$ recognizing all the messages on the network, and since in our setting the knowledge of the intruder *is* the network, q_{net} is also the state recognizing the knowledge of the intruder. First, we assume that the intruder knows the identity of every agent of the network, as well as their public keys.

$$
\begin{aligned}
agt(q_{int}) &\to q_{net} & agt(q_A) &\to q_{net} & agt(q_B) &\to q_{net} \\
pubkey(q_{agtI}) &\to q_{net} & pubkey(q_{agtA}) &\to q_{net} & pubkey(q_{agtB}) &\to q_{net}
\end{aligned}
$$

Agents $agt(i)$ with $i \in \mathbb{N}$ give the intruder the nonces they produce for other agents:

$$
N(q_{agtI}, q_{agtA}) \to q_{net} \qquad N(q_{agtI}, q_{agtB}) \to q_{net} \qquad N(q_{agtI}, q_{agtI}) \to q_{net}
$$

Finally, starting from components he already knows or will obtain later (i.e. terms in q_{net}), the intruder can combine them into lists with the *cons* operator, encrypt them with anything (including keys) he knows with operator *encr*, build messages with operator *mesg*, etc. in order to enrich his knowledge (the language recognized by q_{net}). Note, however, that the second field of the operator *encr* (which is a flag) cannot be corrupted by the intruder and always refer to q_{agtI} the real author of the encryption, i.e. the intruder.

$$cons(q_{net}, q_{net}) \rightarrow q_{net} \qquad null \rightarrow q_{net} \qquad encr(q_{net}, q_{agtI}, q_{net}) \rightarrow q_{net}$$
$$mesg(q_{net}, q_{net}, q_{net}) \rightarrow q_{net}$$

There are several things to notice here. First, the initial description of $\mathcal{L}(\mathcal{A}_0, q_{net})$ is as wide and loose as possible: roughly, it authorizes the intruder to build nearly every term of $\mathcal{T}(\mathcal{F})$ except terms containing nonces built by A or B, i.e. terms containing subterms of the form $N(agt(A), agt(x))$ or $N(agt(B), agt(y))$. This can be automatically obtained by a complement operation. This kind of specification is quite natural with regards to intruder description since it is much more simpler and more convincing to specify what cannot be built by the intruder than to precisely and totally define what he can do. Consequently, the language recognized by state q_{net} is loose and it may also contain strangely formed messages whose effect on the protocol can hardly be predicted, for example:

$$mesg(agt(A), agt(B), encr(pubkey(agt(B)), agt(0),$$
$$[encr(pubkey(agt(A)), agt(0),$$
$$[N(agt(0), agt(A))]), N(agt(0), agt(B))]))$$

i.e. a message of the form $agt(A) \hookrightarrow agt(B) : \{\{N_{agt(0)}\}_{K_{agt(A)}}, N_{agt(0)}\}_{K_{agt(B)}}$. The language recognized by q_{net} contains also, for instance, terms representing repeated encryption (an unbound number) which are important to consider for cryptographic protocols verification:

$$encr(pubkey(agt(A)), agt(s(0)), encr(pubkey(agt(B)), agt(0), encr(\dots$$

The last thing to remark here is that during approximation construction, new messages or messages components m obtained by rewriting are added to the language recognized by automaton $\mathcal{A}_i$ as new transitions into $\mathcal{A}_{i+1}$ s.t. $m \rightarrow^*_{\mathcal{A}_{i+1}}$ q_{net} and thus can be used 'dynamically' as new base components for intruder's message constructions.

To sum up, we have here described a model where we consider an unbounded number of agents executing an unbounded number of protocol sessions in parallel. In particular, note that if there exists an attack based on parallel protocol sessions between, say four agents A, B, C and D, this attack will appear in the model: C and D can be represented by two 'dishonest' agents, say $agt(i)$ and $agt(j)$ with $i, j \in \mathbb{N}$ and $i \neq j$ since all 'dishonest' agents are able to respect the protocol.

5   Approximation and Verification

Extensions of Approximations to AC non Left-Linear TRSs

In this section, we show how to extend the approximation construction to this larger class of TRSs. Roughly, the problem with non left-linear rules is the following: let $f(x, x) \rightarrow g(x)$ be a rule of $\mathcal{R}$ and let $\mathcal{A}$ be a tree automaton whose set of transitions contains $f(q_1, q_1) \to q_0$ and $f(q_2, q_3) \to q_0$. Although we can construct a valid substitution $\sigma = \{x \mapsto q_1\}$ for matching the rewrite rule on the first transition, it is not the case for the second one. The semantics of a completion between rule $f(x, x) \to g(x)$ and transition $f(q_2, q_3) \to q_0$ would be to find the common language of terms recognized both by q_2 and q_3. This can be obtained by computing a new tree automaton $\mathcal{A}'$ with a set of states $\mathcal{Q}'$ such that $\mathcal{Q}'$ is disjoint from states of $\mathcal{A}$ and $\exists q \in \mathcal{Q}' : \mathcal{L}(\mathcal{A}', q) = \mathcal{L}(\mathcal{A}, q_2) \cap \mathcal{L}(\mathcal{A}, q_3)$. Then, to end the completion step it would be enough to add transitions of $\mathcal{A}'$ to $\mathcal{A}$ with the new transition $g(q) \to q_0$. However, adding transitions of $\mathcal{A}'$ to $\mathcal{A}$ also adds $\mathcal{Q}'$ to states of $\mathcal{A}$. Thus, we add new states to $\mathcal{A}$ and, in some cases, this may lead to non-termination of the approximation construction.

On the other hand, one can remark that the non-linearity problem would disappear with deterministic automata since for any deterministic automaton $\mathcal{A}_{det}$ and for all states q, q' of $\mathcal{A}_{det}$ we trivially have $\mathcal{L}(\mathcal{A}, q) \cap \mathcal{L}(\mathcal{A}, q') = \emptyset$. However, determinization of a tree automaton may result in an exponential blow-up of the number of states [3]. Thus, we chose here to use *locally deterministic* tree automata: non-deterministic tree automata with some *deterministic states*, i.e. states q such that there is no two rules $t \to q$ and $t \to q'$ with $q \neq q'$. Hence, for all deterministic state q, we have $\forall q' \neq q : \mathcal{L}(\mathcal{A}, q) \cap \mathcal{L}(\mathcal{A}, q') = \emptyset$. During the approximation construction, if all states, matched by a non-linear variable of the left-hand side of a rule, are deterministic then it is enough to build critical pairs where non linear variables of the left-hand side are mapped to the same state. For instance, in the last example, it is enough to build the first critical pair, add the transition $g(q_1) \to q_0$, and keep q_2, q_3 deterministic, i.e. such that $\mathcal{L}(\mathcal{T}_\mathcal{R}\uparrow(\mathcal{A}), q_2) \cap \mathcal{L}(\mathcal{T}_\mathcal{R}\uparrow(\mathcal{A}), q_3) = \emptyset$. We now show the completeness of this algorithm on locally deterministic tree automata.

For all term t non linear, let us denote by t_{lin} the term t linearized, i.e. where all occurrences of non linear variables are replaced by disjoint variables. For example, if $t = f(x, y, g(x, x))$, then $t_{lin} = f(x', y, g(x'', x'''))$.

Definition 4. *(States Matching) Let $\mathcal{A}$ be a tree automaton, $\mathcal{Q}$ its set of states, $t \in \mathcal{T}(\mathcal{F}, \mathcal{X})$ a non linear term, and $\{p_1, \ldots, p_n\} \subseteq \mathcal{Pos}(t)$ the set of positions of a non linear variable x in t. We say that states $q_1, \ldots, q_n \in \mathcal{Q}$ are matched by x iff $\exists \sigma \in \Sigma(\mathcal{Q}, \mathcal{X})$ s.t. $t_{lin}\sigma \to_\mathcal{A}^* q \in \mathcal{Q}$, and $t_{lin}\sigma|_{p_1} = q_1, \ldots, t_{lin}\sigma|_{p_n} = q_n$*

Theorem 2. *(Completeness Extended to non Left-Linear TRS) Let $\mathcal{A}$ be a tree automaton, $\mathcal{R}$ a TRS, $\mathcal{T}_\mathcal{R}\uparrow(\mathcal{A})$ the corresponding approximation automaton and $\mathcal{Q}$ its set of states. For all non left-linear rule $l \to r \in \mathcal{R}$, for all non linear variable x of l, for all states $q_1, \ldots, q_n \in \mathcal{Q}$ matched by x, if either $q_1 = \ldots = q_n$ or $\mathcal{L}(\mathcal{T}_\mathcal{R}\uparrow(\mathcal{A}), q_1) \cap \ldots \cap \mathcal{L}(\mathcal{T}_\mathcal{R}\uparrow(\mathcal{A}), q_n) = \emptyset$ then*

$$\mathcal{L}(\mathcal{T}_\mathcal{R}\uparrow(\mathcal{A})) \supseteq \mathcal{R}^*(\mathcal{L}(\mathcal{A}))$$

Proof. (Sketch)(See [10] for a detailed proof) Assume that there exists a term t such that $t \in \mathcal{R}^*(\mathcal{L}(\mathcal{A}))$ and $t \notin \mathcal{L}(\mathcal{T}_\mathcal{R}\uparrow(\mathcal{A}))$. Let $s \in \mathcal{L}(\mathcal{A})$ such that $s \to_\mathcal{R}^* t$. On the rewrite chain from s to t, let t_1, t_2 be the first two terms such that $t_1 \in \mathcal{L}(\mathcal{T}_\mathcal{R}\uparrow(\mathcal{A}))$, $t_1 \to_\mathcal{R} t_2$ and $t_2 \notin \mathcal{L}(\mathcal{T}_\mathcal{R}\uparrow(\mathcal{A}))$. We then show that the rule

$l \to r \in \mathcal{R}$ applied for rewriting t_1 into t_2 is necessarily a non left-linear rule (otherwise t_2 would be in $\mathcal{L}(\mathcal{T}_{\mathcal{R}}{\uparrow}\,(\mathcal{A}))$). Then, we obtain that there exists a subterm u of t_1 matched by all occurrences of a non linear variable x in l and there exists at least two distinct states q, q' of $\mathcal{T}_{\mathcal{R}}{\uparrow}\,(\mathcal{A})$ such that $u \to^*_{\mathcal{T}_{\mathcal{R}}{\uparrow}(\mathcal{A})} q$ and $u \to^*_{\mathcal{T}_{\mathcal{R}}{\uparrow}(\mathcal{A})} q'$. This contradicts the hypothesis of the theorem since q and q' are matched by x in l and $\mathcal{L}(\mathcal{T}_{\mathcal{R}}{\uparrow}\,(\mathcal{A}), q) \cap \mathcal{L}(\mathcal{T}_{\mathcal{R}}{\uparrow}\,(\mathcal{A}), q') \supseteq \{u\} \neq \emptyset$. $\square$

In our framework, states matched by non-linear variables are easily kept deterministic. For example, in the NSPK specification, non-linear variables always match terms A, B, $i \in \mathbb{N}$ (representing agent labels) which are initially recognized by q_A, q_B, q_{int}, respectively. Those states are initially deterministic and this property is trivially preserved during completion since agent labels do not occur in right-hand side of rules and thus agent labels do not occur in new transitions to be added. However, when necessary, we can also automatically check this property on $\mathcal{T}_{\mathcal{R}}{\uparrow}\,(\mathcal{A})$ by proving that $\mathcal{L}(\mathcal{T}_{\mathcal{R}}{\uparrow}\,(\mathcal{A}), q_1) \cap \ldots \cap \mathcal{L}(\mathcal{T}_{\mathcal{R}}{\uparrow}\,(\mathcal{A}), q_n) = \emptyset$, for each non linear variable x of a rule matching distinct states $q_1, \ldots, q_n$.

For dealing with the AC symbols, the extension is straightforward. Since approximation can deal with non terminating TRS, we can explicitly define the AC-behavior of a symbol. Thus, we replace in $\mathcal{F}$ the (implicit) AC-symbol $\sqcup$ by a non-AC symbol U and add to $\mathcal{R}$ the following left-linear rules defining explicitly the AC behavior of U:

$$x \text{ U } y \to y \text{ U } x \qquad (x \text{ U } y) \text{ U } z \to x \text{ U } (y \text{ U } z) \qquad x \text{ U } (y \text{ U } z) \to (x \text{ U } y) \text{ U } z$$

Approximation Function

Let $\mathcal{R}$ and $\mathcal{A}_0$ be respectively the set of all rewrite rules and the tree automaton given above. Our aim is now to compute a tree automaton $\mathcal{T}_{\mathcal{R}}{\uparrow}\,(\mathcal{A}_0)$ recognizing a superset of $\mathcal{R}^*(\mathcal{L}(\mathcal{A}_0))$ and thus, to over-approximate the network, i.e. the set of all possible sent messages (as well as the set of communication reports). We now give the approximation function γ, defining the folding positions for $\mathcal{R}$ and $\mathcal{A}_0$. For approximation, the first choice we have made is to confuse dishonest agents $(agt(i)$ with $i \in \mathbb{N})$ together. In other words, in our approximation, no difference is made between agents $agt(i)$ and $agt(j)$ for any $i, j \in \mathbb{N}$. However, we still distinguish between $agt(A)$, $agt(B)$ and any agent $agt(i)$ with $i \in \mathbb{N}$. In a similar manner, we collapse together all the messages sent and received by dishonest agents but we still do not confuse messages involving $agt(A)$ or $agt(B)$. For example, the approximation function used for the rule ①, i.e.

$$goal(x, y) \to LHS \text{ U } mesg(x, y, encr(pubkey(y), x, [N(x, y), x]))$$

is such that there are only seven distinct values for γ (The detail of sequences of new states used for each value can be found in [10] with the complete specification.):

i	$\gamma(①, q_{net}, \{x \mapsto q_{agtA}, y \mapsto q_{agtB}\})$	ii	$\gamma(①, q_{net}, \{x \mapsto q_{agtB}, y \mapsto q_{agtA}\})$
iii	$\gamma(①, q_{net}, \{x \mapsto q_{agtA}, y \mapsto q_{agtA}\})$	iv	$\gamma(①, q_{net}, \{x \mapsto q_{agtB}, y \mapsto q_{agtB}\})$
v	$\gamma(①, q_{net}, \{x \mapsto q_{agtI}, y \mapsto q_{agtA}\})$	vi	$\gamma(①, q_{net}, \{x \mapsto q_{agtI}, y \mapsto q_{agtB}\})$
vii	$\gamma(①, q_{net}, \{y \mapsto q_{agtI}\})$		

According to case (i) all messages generated thanks to rule 1, where x is the agent labeled by A and y is the agent labeled by B, are decomposed using the same states defined by the sequence $\gamma(\textcircled{1}, q_{net}, \{x \mapsto q_{agtA}, y \mapsto q_{agtB}\})$. Similarly, the case (vii) means that all messages generated thanks to rule 1, where y is an agent labeled by $i \in \mathbb{N}$ and x is any agent, are decomposed using states of the same sequence $\gamma(\textcircled{1}, q_{net}, \{y \mapsto q_{agtI}\})$. Thus, no difference is made, for example, between messages sent by $agt(A)$ to $agt(i)$, messages sent by $agt(B)$ to $agt(j)$, and messages sent for by $agt(i)$ to $agt(j)$ for any $i, j \in \mathbb{N}$. This is in fact natural since all messages sent to a dishonest agent are captured and factorized by the same intruder.

Verification

We use a prototype, based on a tree automata library [9, 8] developed in ELAN [2], which permits to automatically compute approximations for a given $\mathcal{R}$, $\mathcal{A}_0$ and an approximation function γ. Thanks to the approximation function given above, we obtain a finite tree automaton $\mathcal{T}_{\mathcal{R}}{\uparrow}(\mathcal{A}_0)$, with about 130 states and 340 transitions, recognizing a regular superset of $\mathcal{R}^*(\mathcal{L}(\mathcal{A}_0))$. See [10] for the complete specification and for a complete listing of the automaton $\mathcal{T}_{\mathcal{R}}{\uparrow}(\mathcal{A}_0)$.

Thanks to this automaton, we can directly verify that NSPK has the confidentiality and authentication property. For confidentiality, it is enough to verify that the intruder cannot capture a nonce of the form $N(agt(x), agt(y))$ where $x, y \in \{A, B\}$. Since in our model the intruder emits all his knowledge on the network (as explained in section 4), this can be done by checking that the intruder cannot emit a nonce of the form $N(agt(A), agt(B))$, $N(agt(B), agt(A))$, ... i.e. that the intersection between $\mathcal{T}_{\mathcal{R}}{\uparrow}(\mathcal{A}_0)$ and the automaton $\mathcal{A}_{conf}$ is empty. The final state of $\mathcal{A}_{conf}$ is q_{net} and its transitions are:

$$
\begin{array}{lll}
A \rightarrow q_A & agt(q_B) \rightarrow q_{agtB} & N(q_{agtA}, q_{agtA}) \rightarrow q_{net} \\
B \rightarrow q_B & N(q_{agtA}, q_{agtB}) \rightarrow q_{net} & N(q_{agtB}, q_{agtB}) \rightarrow q_{net} \\
agt(q_A) \rightarrow q_{agtA} & N(q_{agtB}, q_{agtA}) \rightarrow q_{net} & q_{net} \cup q_{net} \rightarrow q_{net}
\end{array}
$$

The intersection can be automatically computed and we obtain a tree automaton whose set of states is empty, i.e. the recognized language is empty. Hence, there is no term of $\mathcal{L}(\mathcal{A}_{conf})$ in $\mathcal{L}(\mathcal{T}_{\mathcal{R}}{\uparrow}(\mathcal{A}_0))$ nor in $\mathcal{R}^*(\mathcal{L}(\mathcal{A}_0))$. Similarly, the cases where authentication is corrupted can be described by the following automaton $\mathcal{A}_{aut}$ whose final state is q_{net} and transitions are:

$$
\begin{array}{lll}
0 \rightarrow q_{int} & c\_init(q_{agtA}, q_{agtB}, q_{agtI}) \rightarrow q_{net} & c\_init(q_{agtA}, q_{agtA}, q_{agtI}) \rightarrow q_{net} \\
s(q_{int}) \rightarrow q_{int} & c\_init(q_{agtA}, q_{agtB}, q_{agtA}) \rightarrow q_{net} & c\_resp(q_{agtA}, q_{agtA}, q_{agtI}) \rightarrow q_{net} \\
A \rightarrow q_A & c\_resp(q_{agtB}, q_{agtA}, q_{agtI}) \rightarrow q_{net} & c\_init(q_{agtA}, q_{agtA}, q_{agtB}) \rightarrow q_{net} \\
B \rightarrow q_B & c\_resp(q_{agtB}, q_{agtA}, q_{agtB}) \rightarrow q_{net} & c\_resp(q_{agtA}, q_{agtA}, q_{agtB}) \rightarrow q_{net} \\
agt(q_{int}) \rightarrow q_{agtI} & c\_init(q_{agtB}, q_{agtA}, q_{agtI}) \rightarrow q_{net} & c\_init(q_{agtB}, q_{agtB}, q_{agtI}) \rightarrow q_{net} \\
agt(q_A) \rightarrow q_{agtA} & c\_init(q_{agtB}, q_{agtA}, q_{agtB}) \rightarrow q_{net} & c\_resp(q_{agtB}, q_{agtB}, q_{agtI}) \rightarrow q_{net} \\
agt(q_B) \rightarrow q_{agtB} & c\_resp(q_{agtA}, q_{agtB}, q_{agtI}) \rightarrow q_{net} & c\_init(q_{agtB}, q_{agtB}, q_{agtA}) \rightarrow q_{net} \\
q_{net} \cup q_{net} \rightarrow q_{net} & c\_resp(q_{agtA}, q_{agtB}, q_{agtA}) \rightarrow q_{net} & c\_resp(q_{agtB}, q_{agtB}, q_{agtA}) \rightarrow q_{net}
\end{array}
$$

encoding all the cases where there is a distortion in communication reports between the belief of the parties and the reality, for example terms of the form $c\_init(agt(A), agt(B), agt(k))$ for $k \in \mathbb{N} \cup \{A\}$ meaning that $agt(A)$ think to have established a communication with B but, in reality, he has been fooled and he communicates with some $agt(i)$ with $i \in \mathbb{N}$ or with himself. The intersection between $\mathcal{T_R}{\uparrow}(\mathcal{A}_0)$ and the automaton $\mathcal{A}_{aut}$ is also empty (see [10] for traces of execution).

6   Conclusion

In this paper, we have shown an application of descendant approximation to cryptographic protocols verification. We have obtained a positive proof of authentication and confidentiality of NSPK. Moreover, applying the same approximation mechanism on the flawed NSPK specification of [19] has led to some non-empty intersections with $\mathcal{A}_{conf}$ and $\mathcal{A}_{aut}$, signaling violation of confidentiality and authentication properties.

An interesting aspect of this method is that it takes advantage of theorem proving and a form of abstract interpretation called approximation. The basic deduction mechanism, coming from the domain of theorem proving, provide some simple and efficient tools – tree automata – to manipulate infinite objects. On the other hand, approximation simplifies the proof in such a way that it can be automatically computed afterwards.

Compared to other rewriting based verification techniques like proofs by consistency or proofs by induction, properties that can be proved with the approximation technique are clearly more restricted: they could be qualified as 'regular properties'. However, by restricting attention to 'regular properties', we obtain a verification technique that enjoys many interesting practical properties: termination of the TRS is not needed, TRS may include AC symbols, proofs are obtained by intersections with $\mathcal{T_R}{\uparrow}(\mathcal{A}_0)$ (automatically and quickly computed), construction of $\mathcal{T_R}{\uparrow}(\mathcal{A}_0)$ is automatic, incremental and can be guaranteed to terminate by a good choice of the γ approximation function (like in the NSPK case above or in a fully automatic way like in [9]). Constructing an approximation function does not require any particular skill in formal proof since it only consists in pointing out some sets of objects (represented here by states recognizing regular sets of terms) to be merged together in order to build an approximated model. In the NSPK case, the γ approximation has been entirely given by hand but it is systematic: for each distinct value of the co-domain of γ the user has to give a sequence of fresh states used for normalizing new transitions. For historical reasons, this step is manual in our prototype but will be automated in the new implementation of this tool which is in progress.

We can also compare this technique with other verification techniques used for verifying cryptographic protocols. The first main difference to be pointed out is that our technique is not designed for discovering attacks. From approximation $\mathcal{T_R}{\uparrow}(\mathcal{A}_0)$, we can derive some information on the context of those attacks but it is approximate and should be studied with a theorem prover or a model-checker to re-construct an exact trace of the attack. Model-checking is, in fact, particularly well suited for attack discovery as showed by the many flaws discovered by G. Lowe [15]. Furthermore, when attacks are no longer found, model-checking can also be used to verify cryptographic protocols by lifting the properties proved on a finite domain to an unbounded. However, the lifting has to be done by hand like G. Lowe did in [14] or, in a more automatic way, by abstract interpretation like it is done by D. Bolignano in [1]. Although we started with a different formalism and used a different technique, our approach is very close to D. Bolignano's one. In particular, approximation functions can be seen as particular abstract interpretations. Nevertheless, approximations enjoys a property that abstract interpretations have not in general: safety of our abstract model (approximations) is implicit and guaranteed by Theorem 1 for every approximation function γ.

Automated theorem proving has also been widely used for cryptographic protocols verification. The NRL Protocol Analyzer, developed by C. Meadows [17], uses narrowing. L. Paulson applied induction proof and the theorem prover Isabelle/HOL to the verification of cryptographic protocols [20]. Those two theorem proving approaches achieve a very detailed verification of protocols. In particular, they provide one of the most convincing answer to the problem of freshness. In counterpart, the proofs may diverge and the main difficulty remain to inject the right lemma at the right moment in order to make the proof converge. Thus, automation of this kind of method remains partial. Furthermore, proofs are long, complex and they require a user with a strong practical experience of the prover. A more recent work is due to C. Weidenbach [22] who gave a positive proof for the Neuman-Stubblebine protocol thanks to the theorem prover SPASS. His technique is based on saturation of sets of horn clauses, which is related to the descendant computation we here use. For a restricted class of clauses called semi-linear, saturation can be computed exactly. However, when the protocol specification cannot be encoded into semi-linear clauses the saturation process may diverge. Thus, specifications must be modified in order to ensure termination of the process. In our framework, no restriction is set on the TRSs we use but, instead, we defined an over-approximation technique in order to tackle the divergence problem.

In [6], G. Denker, J. Meseguer and C. Talcott proposed to encode the NSPK into object-oriented TRSs. This encoding is executable and is used for detecting attacks in the initial version of the protocol by testing. Using objects is clearly a great advantage for a better clarity and readability of the encoding. Nevertheless, since rewriting remains the operational model of object oriented rewriting, it should be possible to extend approximations to objects and thus benefit of the clarity of object oriented specifications.

In [18], D. Monniaux also use tree automata and a completion mechanism for verifying cryptographic protocols. With regards to our work, an important difference is that his method can only deal with a bounded number of agents and a bounded number of protocol sessions. On a more technical point of view, unlike our approach, rewriting is only used for estimating intruder knowledge and not for encoding the protocol itself. Moreover, his completion mechanism is limited to the decidable and well known case of collapsing rules[4] covered by the decidable and more general case of right-linear and monadic rules [21]. However, this approach is interesting since it shows a possible way for combining tree automata and state-transition models for abstract interpretation of protocols: tree automata and completion for abstracting structures and state-transition models for representing the notion of time in the abstract model.

In the approximation model we consider for NSPK, time is totally collapsed, i.e. every message is considered to be permanently sent and received at every moment. Collapsing time let us easily consider an infinite number of protocol sessions in a finite model. Although this does not raise problems for proving confidentiality or authentication properties on NSPK, this is not the case in general. For instance, in electronic commerce protocols like SET [16], there is little hope to prove any security property on an abstract model with no time since freshness plays a central role. A direct solution is to consider several states for the network (i.e. of intruder knowledge) for different steps of the protocol instead of collapsing all states in one.

Our main goal is to be able to handle protocol as complex as SET. To achieve this goal, the first thing to consider is to formally define the concepts used in cryptographic protocols (keys, nonces, agents, ...) in order to get a natural protocol language description and an automatic translator to the encoding presented in this paper. The second point would be, on the one hand, to extend the present work with conditional rules in order to get a more powerful behavior description language and, on the other hand, to handle other tree grammars to get finer approximations.

Finally, we think that approximations could be used for the verification of systems different from cryptographic protocols. Rewriting based approximations seems to be a way to combine, in the same formalism, automated theorem proving techniques and abstract interpretation: theorem proving for proving properties needing high level proof techniques – like induction – and approximations for proving the remaining parts of the proof where abstract interpretation and model-checking are enough.

Acknowledgments

We would like to thank Pascal Brisset for discussion about cryptographic protocols and Pierre-Etienne Moreau for technical help with ELAN.

References

1. D. Bolignano. Towards a Mechanization of Cryptographic Protocol Verification. In *Proc. 9th CAV Conf., Haifa (Israel)*, volume 1254 of *LNCS*. Springer-Verlag, 1997.

---

[4] right-hand side of a collapsing rule is reduced to a variable occurring in its left-hand side.

2. P. Borovanský, C. Kirchner, H. Kirchner, P.-E. Moreau, and M. Vittek. ELAN: A logical framework based on computational systems. In *Proc. 1st WRLA*, volume 4 of *ENTCS*, Asilomar (California), 1996.
3. H. Comon, M. Dauchet, R. Gilleron, F. Jacquemard, D. Lugiez, S. Tison, and M. Tommasi. Tree automata techniques and applications. http://l3ux02.univ-lille3.fr/tata/, 1997.
4. J. Coquidé, M. Dauchet, R. Gilleron, and S. Vágvölgyi. Bottom-up tree pushdown automata and rewrite systems. In R. V. Book, editor, *Proc. 4th RTA Conf., Como (Italy)*, volume 488 of *LNCS*, pages 287–298. Springer-Verlag, 1991.
5. M. Dauchet and S. Tison. The theory of ground rewrite systems is decidable. In *Proc. 5th LICS Symp., Philadelphia (Pa., USA)*, pages 242–248, June 1990.
6. G. Denker, J. Meseguer, and C. Talcott. Protocol Specification and Analysis in Maude. In *Proc. 2nd WRLA Workshop, Pont à Mousson (France)*, 1998.
7. N. Dershowitz and J.-P. Jouannaud. *Handbook of Theoretical Computer Science*, volume B, chapter 6: Rewrite Systems, pages 244–320. Elsevier Science Publishers B. V. (North-Holland), 1990. Also as: Research report 478, LRI.
8. T. Genet. Tree Automata Library. http://www.loria.fr/ELAN/.
9. T. Genet. Decidable approximations of sets of descendants and sets of normal forms. In *Proc. 9th RTA Conf., Tsukuba (Japan)*, volume 1379 of *LNCS*, pages 151–165. Springer-Verlag, 1998.
10. T. Genet and F. Klay. Rewriting for cryptographic protocols verification (extended version). Technical report, INRIA, 2000. http://www.irisa.fr/lande/genet/publications.html.
11. R. Gilleron and S. Tison. Regular tree languages and rewrite systems. *Fundamenta Informaticae*, 24:157–175, 1995.
12. F. Jacquemard. Decidable approximations of term rewriting systems. In H. Ganzinger, editor, *Proc. 7th RTA Conf., New Brunswick (New Jersey, USA)*, pages 362–376. Springer-Verlag, 1996.
13. G. Lowe. An Attack on the Needham-Schroder Public-Key Protocol. *IPL*, 56:131–133, 1995.
14. G. Lowe. Breaking and fixing the Needham-Schroeder public-key protocol using CSP and FDR. In *Proc. 2nd TACAS Conf., Passau (Germany)*, volume 1055 of *LNCS*, pages 147–166. Springer-Verlag, 1996.
15. G. Lowe. Some New Attacks upon Security Protocols. In *9th Computer Security Foundations Workshop*. IEEE Computer Society Press, 1996.
16. Mastercard & Visa. Secure Electronic Transactions. http://www.visa.com/set/, 1996.
17. C. A. Meadows. Analyzing the Needham-Schroeder Public Key Protocol: A comparison of two approaches. In *Proc. 4th ESORICS Symp., Rome (Italy)*, volume 1146 of *LNCS*, pages 351–364. Springer-Verlag, 1996.
18. D. Monniaux. Abstracting Cryptographic Protocols with Tree Automata. In *Proc. 6th SAS, Venezia (Italy)*, 1999.
19. R. M. Needham and M. D. Schroeder. Using Encryption for Authentication in Large Networks of Computers. *CACM*, 21(12):993–999, 1978.
20. L. Paulson. Proving Properties of Security Protocols by Induction. In *10th Computer Security Foundations Workshop*. IEEE Computer Society Press, 1997.
21. K. Salomaa. Deterministic Tree Pushdown Automata and Monadic Tree Rewriting Systems. *J. of Computer and System Sciences*, 37:367–394, 1988.
22. C. Weidenbach. Towards an Automatic Analysis of Security Protocols. In *Proc. 16th CADE Conf., Trento, (Italy)*, volume 1632 of *LNAI*, pages 378–382. Springer-Verlag, 1999.

System Description: *SAT
A Platform for the Development of Modal Decision Procedures*

Enrico Giunchiglia and Armando Tacchella

DIST, Università di Genova
Viale Causa 13 – 16145 Genova, Italy
{enrico,tac}@dist.unige.it

Abstract. *SAT is a platform for the development of modal decision procedures. Currently, *SAT features decision procedures for the normal modal logic K(m) and for the classical modal logic E(m). *SAT embodies a state of the art SAT solver, and includes techniques for optimizing automated deduction in modal and temporal logics. Owing to its modular design and to the extensive reuse of software components, *SAT provides an open, easy to maintain, yet efficient implementation framework.

1 Introduction

In this paper we present *SAT, a platform for the development of SAT-based decision procedures. By SAT-based we mean built on top of a SAT solver in the spirit of [1]. Currently, *SAT features SAT-based decision procedures for the normal modal logic K(m), and for the classical modal logic E(m) [9, 2].

The *SAT propositional engine is an embedded version of SATO 3.2, one of the most efficient SAT checkers publicly available [3]. We chose SATO because it is a fast propositional reasoner and it features many optimizations that we exploited in *SAT. We also implemented other optimizations that speed up modal reasoning, like:

- early investigation of modal successors [1],
- internal optimized clause form conversions [2], and
- caching structures and retrieval algorithms [4].

*SAT has been designed to be modular and to allow for an easy integration of new decision procedures and optimizations. The system is implemented in C and extensively reuses software components from state-of-the-art systems, i.e., SATO, and the GLU library of data types from the VIS model checking system [5]. The GLU library provides *SAT with efficient implementations of, e.g., lists, hash-tables, sparse-matrices. Taking SATO and GLU off-the-shelf, we inherit and exploit

* We wish to thank Fausto Giunchiglia, Peter Patel-Schneider and Roberto Sebastiani for useful discussions related to *SAT. This work is supported by MURST.

D. McAllester (Ed.): CADE-17, LNAI 1831, pp. 291–296, 2000.
© Springer-Verlag Berlin Heidelberg 2000

in *SAT several years of experience in building highly optimized data structures and algorithms for automated deduction in propositional and temporal logics.

*SAT source code, documentation and experimental results are available on the WWW at:

http://www.mrg.dist.unige.it/~tac/StarSAT.html

2   Algorithms

For sake of clarity, we introduce some preliminary notions. The set of formulas is constructed starting from a given set of propositional letters and applying the 0-ary operators $\top$ and $\bot$ (representing truth and falsity respectively); the unary operators $\neg$ and $\Box$; and the binary operators $\wedge$, $\vee$, $\supset$ and $\equiv$.[1] A *modal logic* is a set of formulas (called *theorems*) closed under tautological consequence. A formula φ is *consistent* in a modal logic L (or *L-consistent*) if $\neg\varphi$ is not a theorem of L, i.e., if $\neg\varphi \notin$ L. By *atom* we mean a propositional letter or a formula of the form $\Box\varphi$. A *literal* is either an atom or the negation of an atom. An *assignment* is any conjunction μ of literals such that for any pair ψ, ψ' of conjuncts in μ, it is not the case that $\psi = \neg\psi'$. An assignment μ *satisfies* a formula φ if μ entails φ by propositional reasoning.

Consider a formula φ. Let L be a modal logic. Whether φ is L-consistent can be determined by implementing two mutually recursive procedures:

- LSAT(φ) for the generation of assignments satisfying φ, and
- LCONSIST(μ) for testing the L-consistency of each generated assignment μ.

The procedure LSAT is independent of the particular modal logic L considered, and can be based on any propositional decision procedure (see [2]). Indeed, the logic specific reasoning is delegated to LCONSIST. Currently, *SAT features the procedures ECONSIST and KCONSIST playing the role of LCONSIST for the logics E(m) and K(m) respectively. For lack of space, we present only the LSAT algorithm here. For ECONSIST and KCONSIST, see [2] and also [4].

The LSAT procedure implemented in *SAT is based on SATO 3.2 [3], an efficient implementation of the Davis-Putnam-Longemann-Loveland (DP) procedure [6]. Figure 1 shows an high-level description of LSAT. In the Figure:

- *cnf*(φ) is a set of clauses obtained from φ by applying a conversion to conjunctive normal form (CNF) based on renaming (see, e.g., [7]).
- *choose-literal*(Φ, μ) returns a literal occurring in Φ and chosen according to some heuristic criterion.
- if l is a literal, $\bar{l}$ stands for A if $l = \neg A$, and for A if $l = \neg A$;
- for any literal l and set Φ of clauses, *assign*(l, Φ) is the set of clauses obtained from Φ by (i) deleting the clauses in which l occurs as a disjunct, and (ii) eliminating $\bar{l}$ from the others.

---

[1] For simplicity, we consider the case with only one modality.

function LSAT(φ)   return LSAT$_{DP}$($cnf(\varphi)$, T).

function LSAT$_{DP}$(Φ, μ)
    if $\Phi = \emptyset$ then return LCONSIST(μ);   /* base */
    if $\emptyset \in \Phi$ then return *False*;   /* backtrack */
    if { a unit clause $\{l\}$ is in Φ }   /* unit */
        then return LSAT$_{DP}$($assign(l,\Phi),\mu \wedge l$);
    if not LCONSIST(μ) then return *False*;   /* early pruning */
    $l := choose\text{-}literal(\Phi,\mu)$;
    return LSAT$_{DP}$($assign(l,\Phi),\mu \wedge l$) or   /* split */
        LSAT$_{DP}$($assign(\bar{l},\Phi),\mu \wedge \bar{l}$).

Fig. 1. LSAT and LSAT$_{DP}$

As can be observed, the procedure LSAT$_{DP}$ in Figure 1 is the DP-procedure modulo (i) the call to LCONSIST(μ) when it finds an assignment μ satisfying the input formula ($\Phi = \emptyset$), and (ii) the *early pruning* step, i.e., a call to LCONSIST(μ) that forces backtracking after each unit propagation when incomplete assignments are not L-consistent. Early pruning prevents *SAT from *thrashing*, i.e., from repeatedly generating different assignments that contain a same inconsistent kernel [1,8].

3   Implementation and Features

The *SAT modular architecture is depicted in Figure 2. The thickest external box represents the whole system and, inside it, each solid box represents a different module. By module, we mean a set of routines dedicated to a specific task.[2] The dashed horizontal lines single out the four main parts of *SAT:

INTERFACE: The modules KRIS, KSATC, LWB, and TPTP are parsers for different input syntaxes. The module TREES stores the input formula as a tree, at the same time performing some simple preprocessing (e.g. pushing negations down to atoms).

DATA: The module DAGS (for *Directed Acyclic Graphs*) implements the main data structures of *SAT. The input formula is preprocessed and stored as a DAG. A Look Up Table (LUT), mapping each atom $\Box\psi$ into a newly introduced propositional letter C_ψ is built. Then, each modal atom is replaced by the corresponding propositional letter. The initial preprocessing allows to map trivially equivalent[3] modal atoms into a single propositional letter, thus fostering the detection of (un)satisfiable subformulae [1,10].

ENGINE: This part includes the module SAT, the propositional core of *SAT. Since SAT implements a DP algorithm, techniques like semantic branching,

---

[2] As a matter of fact, each module corresponds to a file in *SAT distribution package.
[3] Technically, the preprocessing maps a formula φ into a formula φ' which is logically equivalent to φ in any classical modal logic (see [9] for the definition of classical modal logic).

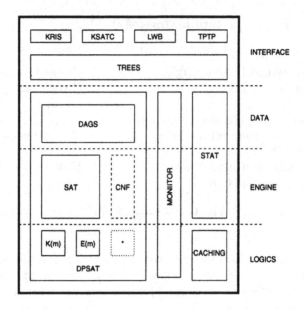

Fig. 2. *SAT modular architecture

boolean constraint propagation (BCP) and heuristic guided search are inherited for free by *SAT. The dashed box (labeled CNF) stands for a set of DPSAT routines implementing CNF conversions based on renaming. CNF routines allow *SAT to handle any formula even if the SAT decider accepts CNF formulae only.

LOGICS: Currently, *SAT features the two modules K(m) and E(m), implementing KCONSIST and ECONSIST respectively. The dotted box "*" is a placeholder for other L-consistency modules that will be implemented in the near future.
CACHING implements data structures and retrieval algorithms that are used to optimize the L-consistency checking routines contained in the logic dependent modules (see [2] for caching in E(m) and [4] for caching in K(m)).

The modules DPSAT, MONITOR and STAT span across different parts of *SAT. DPSAT interfaces the inner modules between them. The result is that these modules are loosely coupled and can be modified/replaced (almost) independently from each other. MONITOR records information about *SAT performance, e.g, cpu time, memory consumption, number of L-consistency checks. STAT explores the preprocessed input formula and provides information like number of occurrences of a variable, number of nested boxes. This information is used by different modules, e.g., for dimensioning the internal data structures.

To understand the behavior of *SAT, let φ be the formula $\neg(\neg\Box(\Box C_2 \wedge \Box C_1) \wedge \Box C_2)$. *SAT first stores φ as an intermediate representation (provided by TREES) where it undergoes some preliminary transformations. In our case, φ becomes $(\Box(\Box C_2 \wedge \Box C_1) \vee \neg\Box C_2)$. Then, the building of the internal representation (provided by DAGS) causes lexical normalization $((\Box C_2 \wedge \Box C_1)$ would be

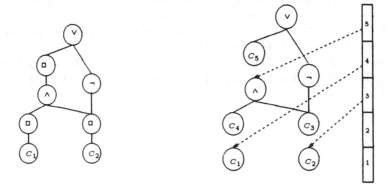

Fig. 3. Internal representation of concepts in *SAT

rewritten into $(\Box C_1 \wedge \Box C_2))$ and propositional simplification (e.g., $(C \vee C)$ would be rewritten into C) to be performed on φ. The resulting formula is represented by the data structure depicted in Figure 3 (left). Next, *SAT creates the LUT and replaces each modal atom with the corresponding propositional letter. The result is depicted in Figure 3 (right), where the numbers appearing in the LUT have the obvious meaning. Notice that the top-level formula $\varphi_0 = C_5 \vee \neg C_3$ is now purely propositional. If SAT accepts only CNF formulae then (i) for every LUT entry C_ψ, both ψ and $\neg\psi$ are converted to CNF and (ii) the top level formula φ_0 is replaced by its CNF conversion. Finally, the core decision process starts. SAT is properly initialized and called with φ_0 as input. Once a satisfying truth assignment is found, a logic dependent module (e.g. K(m)) is called to check its L-consistency. The recursive tests are built in constant time using the LUT to reference the subformulae. The process continues until no more truth assignments are possible or a model is found ([2] details this process for K(m), E(m) and several other classical modal logics).

References

1. F. Giunchiglia and R. Sebastiani. Building decision procedures for modal logics from propositional decision procedures - the case study of modal K. In *Proc. CADE-96*, LNAI, 1996.
2. E. Giunchiglia, F. Giunchiglia, and A. Tacchella. SAT-Based Decision Procedures for Classical Modal Logics. To appear in *Journal of Automated Reasoning*.
3. H. Zhang. SATO: An efficient propositional prover. In *Proc. CADE-97*, volume 1249 of *LNAI*, 1997.
4. E. Giunchiglia and A. Tacchella. Subset-matching Size-bounded Caching for Satisfiability in Modal Logics, 2000. Submitted.
5. VIS-group. VIS: A system for verification and synthesis. In *Proc. CAV-96*, LNAI, 1996.
6. M. Davis, G. Longemann, and D. Loveland. A machine program for theorem proving. *Journal of the ACM*, 5(7), 1962.

7. D.A. Plaisted and S. Greenbaum. A Structure-preserving Clause Form Translation. *Journal of Symbolic Computation*, 2:293–304, 1986.
8. I. Horrocks. Optimizing Tableaux Decision Procedures for Description Logics. PhD thesis, University of Manchester, 1997.
9. B. F. Chellas. *Modal Logic – an Introduction*. Cambridge University Press, 1980.
10. U. Hustadt and R.A. Schmidt. On evaluating decision procedures for modal logic. In *Proc. IJCAI-97*, 1997.

System Description: DLP

Peter Patel-Schneider

Bell Labs Research, Murray Hill, NJ, U.S.A.
pfps@research.bell-labs.com

DLP (Description Logic Prover) is an experimental description logic knowledge representation system. DLP implements an expressive description logic that includes propositional dynamic logic as a subset. DLP provides a simple interface allowing users to build knowledge bases of descriptions in this description logic, but, as an experimental system, DLP does not have a full user interface.

Because of the correspondence between description logics and propositional modal logics, DLP can serve as a reasoner for several propositional modal logics. As well as propositional dynamic logic, the logic underlying DLP contains fragments that are in direct correspondence to the propositional modal logics $K_{(m)}$ and $K4_{(m)}$. DLP provides an interface that allows direct satisfiability checking of formulae in $K_{(m)}$ and $K4_{(m)}$. Using a standard encoding, the interface also allows satisfiability checking of formulae in $KT_{(m)}$ and $S4_{(m)}$.

DLP is available via the WWW at http://www.bell-labs.com/user/pfps. DLP is implemented in SML/NJ. The current version of DLP, version 4.1, includes a number of new optimisations and options not included in previous versions.

One of the purposes in building DLP was to investigate various optimisations for description logic systems. A number of these optimisations have appeared in various description logic systems [1, 3, 7]. As there is still need to investigate optimisations further and to develop new optimisation techniques, DLP has a number of compile-time options to select various description logic optimisations.

DLP implements the description logic in Figure 1. In the syntax chart A is an atomic concept; C and D are arbitrary concepts; P is an atomic role; R and S are arbitrary roles; and n is an integer. There is an obvious correspondence between most of the constructs in this description logic and propositional dynamic logic, which is given in the chart.

Implementation

DLP uses the now-standard method for subsumption testing in description logics, namely translating subsumption tests into satisfiability tests and checking for satisfiability using an optimised tableaux method. DLP was designed from the beginning to be an experimental system. As a result, much more attention has been paid to making the internal algorithms correct and efficient in the worst-case than to reducing constant factors. Similarly, the internal data structures have been chosen for their flexibility rather than having the absolute best modification and access speeds. Some care has been taken to make the internal

D. McAllester (Ed.): CADE-17, LNAI 1831, pp. 297–301, 2000.
© Springer-Verlag Berlin Heidelberg 2000

	DL Syntax	PDL Syntax	Semantics
Concepts	A	A	$A^\mathcal{I} \subseteq \Delta^\mathcal{I}$
(Formulae)	$\top$	T	$\Delta^\mathcal{I}$
	$\bot$	F	$\emptyset$
	$\neg C$	$\sim C$	$\Delta^\mathcal{I} - C^\mathcal{I}$
	$C \sqcap D$	$C \wedge D$	$C^\mathcal{I} \cap D^\mathcal{I}$
	$C \sqcup D$	$C \vee D$	$C^\mathcal{I} \cup D^\mathcal{I}$
	$\exists R.C$	$\langle R \rangle C$	$\{d \in \Delta^\mathcal{I} : R^\mathcal{I}(d) \cap C^\mathcal{I} \neq \emptyset\}$
	$\forall R.C$	$[R] C$	$\{d \in \Delta^\mathcal{I} : R^\mathcal{I}(d) \subseteq C^\mathcal{I}\}$
	$\geq nP$		$\{d \in \Delta^\mathcal{I} : \mid R^\mathcal{I}(d) \mid \geq n\}$
	$\leq nP$		$\{d \in \Delta^\mathcal{I} : \mid R^\mathcal{I}(d) \mid \leq n\}$
	$P{:}n$		$\{d \in \Delta^\mathcal{I} : R^\mathcal{I}(d) \ni n\}$
Roles	P	P	$P^\mathcal{I} \subseteq \Delta^\mathcal{I} \times \Delta^\mathcal{I}$
(Modalities)	$R \sqcup S$	$R \cup S$	$R^\mathcal{I} \cup S^\mathcal{I}$
(Actions)	$R \circ S$	$R ; S$	$R^\mathcal{I} \circ S^\mathcal{I}$
	R/C	$R ; C?$	$R^\mathcal{I} \cap (\Delta^\mathcal{I} \times C^\mathcal{I})$
	R^+	$R ; R^*$	$\bigcup_{n>1} R^{\mathcal{I}^n}$

Fig. 1. Simplified Syntax for DLP

data structures reasonably fast, however—there is considerable use of binary maps and hash tables instead of lists to store sets, for example.

DLP is implemented in SML/NJ instead of a language like C so that it can be more-easily changed. There is some price to be paid for this, as SML/NJ does not allow some of the low-level optimisations possible in languages like C. Further, DLP is implemented in a mostly-functional fashion. The only non-functional portions of the satisfiability checker in DLP have to do with unique storage of formulae, and caching of several kinds of information. All this caching is monotone, i.e., it does not have be undone during a proof, or even between proofs. Nonetheless, DLP is quite fast on several problem sets, including the Tableaux'98 propositional modal logic comparison benchmark [9] and several collections of hard random formulae in **K** [10, 8, 11].

Optimisation Techniques

Many of the optimisation techniques in DLP have already appeared in various description logic systems. The most complete description of these optimisations can be found in Ian Horrocks' thesis [7]. The basic algorithm in DLP is a simple tableau algorithm that searches for a model that demonstrates the satisfiability of a description logic description or, equivalently, a propositional modal logic formula. The algorithm process modal constructs by building successor nodes with attached formulae that represent related possible worlds. The algorithm incorporates the usual control mechanism to guarantee termination, including a check for equality of the formulae at nodes to guarantee termination for transitive roles (modalities).

Before the model search algorithm in DLP starts, incoming formulae are converted into a normal form, and common sub-formulae are uniquely stored. This conversion detects analytically satisfiable sub-formulae. This unique storage of formulae also allows values to be efficiently given to any sub-formula in the formula, not just propositional variables. This can result in search failures being detected much earlier than would otherwise be the case.

DLP performs semantic branching search. When DLP decides to branch, it picks a formula and assigns that formula to true and false in turn instead of picking a disjunction and assigning each of its disjuncts to true in turn. Semantic branching is guaranteed to explore each section of the search space at most once, as opposed to syntactic branching, and this is important in propositional modal logics as the generation and analysis of successors can result in large overlap in the search space when using syntactic branching.

DLP looks for formulae whose value is determined by the current set of assignments, and immediately gives these formulae the appropriate value. This technique can result in dramatic reductions in the search space, particularly in the presence of semantic branching.

For every sub-formula DLP keeps track of which choice points lead to the deduction of that sub-formula. When backtracking to a choice point, DLP checks to see if the current search failure depends on that choice; if it does not, the alternative branch need not be considered, as it would just lead to the same failure. This technique, often called backjumping [2], can dramatically reduce the search space, but does have some overhead.

During a satisfiability check successor nodes with the same set of formulae as a previously-encountered node are often generated. As all that matters is whether the node is satisfiable or not, DLP caches and reuses their status. Care has to be taken to ensure that caching does not interfere with the rest of the algorithm, particularly the determination of dependencies and loop analysis. Caching does require that information about each node generated be retained for a longer period of time than required for a basic depth-first implementation of the satisfiability checker. However, caching can produce dramatic gains in speed.

There are many heuristic techniques that can be used to determine which sub-formula to branch on first. However, these techniques require considerable information to be computed for each sub-formula of the unexpanded disjunctions. Further, the heuristic techniques available have mostly been devised for non-modal logics and are not necessarily suitable for modal logics. Nonetheless, DLP includes some simple heuristics to guide its search, mostly heuristics for more-effective backjumping.

New Techniques

Version 4.1 of DLP includes quite a number of new techniques to improve its performance.

In previous versions of DLP, the cache did not include dependency information, which meant that a conservative approximation to this information had to be made, possibly resulting in less-than-optimal backjumping. The formula cache has now been expanded to incorporate the dependency information needed in backjumping, so that caching does not interfere with backjumping. Of course, this does increase the size of cache entries.

The low-level computations in DLP used to be quite expensive for very large formulae. If the formula was also difficult to solve, this cost would be masked by the search time, but if the formula was easy to solve, the low-level computation cost would dominate the solution time. Version 4.1 of DLP dramatically reduces the time taken for low-level computations both by reducing the amount of heuristic information generated when there are many clauses active and also by caching some of this information so that it does not have to be repeatedly computed. Of course, DLP is still much slower on large-but-easy formulae than provers that use imperative techniques, but such provers are much harder to build and debug than DLP.

DLP used to completely generate assignments for the current node before investigating any modal successors. The current version of DLP has an option to investigate modal successors whenever a choice point is encountered, a technique taken from KSATC [5]. This option can be beneficial but often increases solution times.

DLP can now retain not only the status of nodes, but the model found if the node is satisfiable. This model can be used to restart the search when reinvestigating modal successors, reducing the time overhead for early investigation of modal successors—at the cost of considerably increasing the space required for the cache. DLP can now also return a model for satisfiable formulae.

DLP now incorporates a variant of dynamic backtracking [4]. When jumping over a choice point, a determination is made as to whether any invalidated branch(es) from that choice point depends on the choice being changed. If it does not, then the search ignores the invalidated branch(es) when the choice point is again encountered.

Summary

DLP has not been used in any actual applications, and as an experimental system, it is unlikely to receive any such use. DLP has been used to classify a version of the Galen medical knowledge base [12]. DLP performed capably on this knowledge base, creating the subsumption partial order in 210 seconds on a Sparc Ultra 1-class machine. DLP has also been tested on several sets of benchmarks, including the Tableaux'98 comparison benchmarks [6] and several collections of hard random modal formulae [10, 8, 11]. DLP is the fastest modal decision procedure for many of these tests.

As it is an experimental system, I did not expect DLP to be particularly fast on hard problems. It was gratifying to me that it is competitive with existing propositional modal reasoners including FaCT and KSATC. My current plan for DLP is to incorporate inverse roles (converse modalities), a change that requires considerable modification to the implementation of the system.

References

1. F. Baader, E. Franconi, B. Hollunder, B. Nebel, and H.-J. Profitlich. An empirical analysis of optimization techniques for terminological representation systems or: Making KRIS get a move on. In Bernhard Nebel, Charles Rich, and William Swartout, editors, *Principles of Knowledge Representation and Reasoning: Proceedings of the Third International Conference (KR'92)*, pages 270–281. Morgan Kaufmann Publishers, San Francisco, California, October 1992. Also available as DFKI RR-93-03.

2. A. B. Baker. *Intelligent Backtracking on Constraint Satisfaction Problems: Experimental and Theoretical Results*. PhD thesis, University of Oregon, 1995.

3. P. Bresciani, E. Franconi, and S. Tessaris. Implementing and testing expressive description logics: a preliminary report. In Gerard Ellis, Robert A. Levinson, Andrew Fall, and Veronica Dahl, editors, *Knowledge Retrieval, Use and Storage for Efficiency: Proceedings of the First International KRUSE Symposium*, pages 28–39, 1995.

4. M. L. Ginsberg. Dynamic backtracking. *Journal of Artificial Intelligence Research*, 1:25–46, 1993.

5. E. Giunchiglia, F. Giunchiglia, R. Sebastiani, and A. Tacchella. More evaluation of decision procedures for modal logics. In Anthony G. Cohn, Lenhart Schubert, and Stuart C. Shapiro, editors, *Principles of Knowledge Representation and Reasoning: Proceedings of the Sixth International Conference (KR'98)*, pages 626–635. Morgan Kaufmann Publishers, San Francisco, California, June 1998.

6. A. Heuerding and S. Schwendimann. A benchmark method for the propositional modal logics K, KT, and S4. Technical report IAM-96-015, University of Bern, Switzerland, October 1996.

7. I. Horrocks. *Optimising Tableaux Decision Procedures for Description Logics*. PhD thesis, University of Manchester, 1997.

8. I. Horrocks, P. Patel-Schneider, and R. Sebastiani. An Analysis of Empirical Testing for Modal Decision Procedures. *Logic Journal of the IGPL*, 2000.

9. I. Horrocks and P. Patel-Schneider. FaCT and DLP. In Harrie de Swart, editor, *Automated Reasoning with Analytic Tableaux and Related Methods: International Conference Tableaux'98*, number 1397 in Lecture Notes in Artificial Intelligence, pages 27–30, Berlin, May 1998. Springer-Verlag.

10. I. Horrocks and P. Patel-Schneider. Performance of DLP on random modal formulae. In *Proceedings of the 1999 Description Logic Workshop*, pages 120–124, July 1999.

11. P. Patel-Schneider and I. Horrocks. DLP and FaCT. In Neil V. Murray, editor, *Automated Reasoning with Analytic Tableaux and Related Methods: International Conference Tableaux'99*, number 1617 in Lecture Notes in Artificial Intelligence, pages 19–23, Berlin, June 1999. Springer-Verlag.

12. A. Rector, S. Bechhofer, C. A. Goble, I. Horrocks, W. A. Nowlan, and W. D. Solomon. The GRAIL concept modelling language for medical terminology. *Artificial Intelligence in Medicine*, 9:139–171, 1997.

Two Techniques to Improve Finite Model Search

Gilles Audemard, Belaid Benhamou, and Laurent Henocque

Laboratoire d'Informatique de Marseille
Centre de Mathematiques et d'Informatique
39, Rue Joliot Curie - 13453 Marseille cedex 13 - France
Tel: 04 91 11 36 25 - Fax: 04 91 11 36 02
{audemard,benhamou,henocque}@lim.univ-mrs.fr

Abstract. This article introduces two techniques to improve the propagation efficiency of CSP based finite model generation methods. One approach consists in statically rewriting some selected clauses so as to trigger added constraint propagations. The other approach uses a dynamic lookahead strategy to both filter out inconsistent domain values and select the most appropriate branching variable according to a first fail heuristic.

1 Introduction

Many methods have been implemented to deal with many-sorted or uni-sorted theories: FINDER [7], FMSET [3], SATO [8], SEM [11], FMC [5] are known systems which solved some open problems.

The method SEM (System for Enumerating Models) introduced by J. Zhang and H. Zhang in [11] is one of the most powerful known methods for solving problems expressed as many-sorted theories.

The goal of this article is to explore ways to improve SEM by increasing the propagations it performs (i.e. the number of inferred negative ground literals) so as to reduce the search space and overall computation time. A first possible improvement is a static preprocessing which automatically rewrites clauses having a specific structure.

A second improvement consists in a dynamic domain filtering achieved by using a lookahead at some nodes of the search tree. This lookahead procedure uses unit propagation and detects incompatible assignments (e.g. trying $f(0) = 0$, then $f(0) = 1$ ...). This filtering is augmented by the introduction of a new heuristic, in the spirit of the SATZ propositional solver (see [4]).

This article is organized as follows: Section 2 defines the first-order logic theories accepted as input language and the background of the SEM algorithm. In section 3 we study two techniques which improve SEM efficiency. In section 4, we compare our work with other methods on mathematicals problems. Section 5 concludes.

D. McAllester (Ed.): CADE-17, LNAI 1831, pp. 302–308, 2000.

2   Background and SEM Description

The theories accepted as input by the model generator SEM are many sorted first order theories, with equality, without existential quantifiers, in clause normal form (CNF). Since we are interested in finite models only, all sorts are finite. Because all the variables are universally quantified, the quantifiers are usually omitted.

We call the *degree* of a literal the number of its functional symbol occurrences. We call a *cell* the *ground term* $f(e_1, ...e_k)$ where all e_i are sort elements. An interpretation of a theory maps each cell to a value from the appropriate sort. A model of a theory is an interpretation which satisfies all its clauses.

As an initial preprocessing stage, SEM expands the original theory axioms to the set of their terminal instances (i.e. ground clauses), by substituting for each logical variable all the members of the appropriate sort. SEM's finite model search algorithm is described in figure 1. It uses the following parameters: A the set of assignments, B the set of unassigned cells and their possible values and C the set of ground clauses. The function *Propa* of the search algorithm propagates the assignment from A to C. This simplifies C and may force some cells in B to become assigned. It modifies (A, B, C) until a fixed point is reached or an inconsistency is detected, and returns the modified triple (A, B, C) upon success. For a full description of SEM and the propagation algorithm one can refer to [9] and [11].

Function Search(A, B, C): Return Boolean
        If $B = \emptyset$ Then **Return** TRUE
        Choose and delete (ce_i, D_i) from B
        If $D_i = \emptyset$ Then **Return** FALSE
        for All $e \in D_i$ Do
            $(A', B', C') = Propa(A \cup (ce_i, e), B, C)$
            If $C' \neq False$ Then Search(A', B', C')

Algorithm 1: SEM Search Algorithm

3   Two Domain Filtering Techniques

SEM's propagation algorithm allows propagation of negative assignments only when literals with the form $ce! = e$ exist in the set of clauses (ce is a cell and e an element). Otherwise, only positive facts (value assignments) are propagated. This leads to an increase in the number of decision points necessary for the search, and potentially increases run times. Because SEM performs some amount of value elimination using negative facts, one approach consists in favoring it by rewriting some clauses to give them the appropriate structure. This static technique is performed in a preprocessing phase.

The second approach is dynamic, involving computations performed at each decision node. It consists in using a lookahead technique, called unit propagation, to eliminate selected values from the domains of selected cells.

3.1 Clauses Transformation

SEM performs value elimination when clauses contain negative literals of degree two. We can thus rewrite some selected clauses, using a flattening technique as in FMSET [3] to rewrite clauses to logically equivalent clauses containing negative literals of degree two. Because such a transformation introduces auxiliary variables, and thus increases the number of ground clause instances, such a rewriting is a tradeoff. Candidate clauses are thus carefully selected: we restrict the rewriting process to clauses of degree 3. This transformation allows to drastically reduce on some problems the number of decision points and the execution times as shown in section 4, the results obtained with this rewriting technique are listed under the name CTSEM (Clause Transformation in SEM).

Definition 1. *A reducible clause is a clause which contains a literal with the following pattern:* $f(x_1, ...x_m, g(x_{k+1}, ..., x_l), x_{m+1}...x_k) = x_0$ *where* f, g *are two functional symbols and* $x_{i \in \{1..l\}}$ *are variables. Such a literal is called reducible.*

By using the clause transformation algorithm described in [3], we can rewrite each reducible literal to the form: $f(x_1, ..., x_m, v, x_{m+1}...x_k) = x_0 \lor v \neq g(x_{k+1}, ..., x_l)$. This preserves the semantics of the clause, and introduces the negative literal $v \neq g(x_{k+1}, ..., x_l)$. It requires the introduction of an auxiliary variable v.

Example 1. The literal $h(h(x, y), x) = y$ is reducible and can be transformed to its logical equivalent $h(v, x) = y \lor v \neq h(x, y)$. Now, the ground clause $h(0, 1) \neq 0 \lor h(0, 0) = 1$ exists in the set of ground clauses. When we assign $h(0, 0)$ to 0 the second literal of the previous clause becomes false. So SEM can propagate the fact $h(0, 1) \neq 0$. This eliminates 0 from the domain of the cell $h(0, 1)$.

3.2 Value Elimination

At a given node of the search tree, let B equal the set of yet unassigned cells and C_A the set of axioms simplified by assignments of cells in A. In other words $C_A = C'$ such that $(A', B', C') = Propa(A, B, C)$. Let $(ce, D_{ce}) \in B$ and let $e \in D_{ce}$. If $Propa(A \cup \{(ce, e)\}, B, C)$ leads to inconsistency, then we can remove the value e from D_{ce}.

However, such unit propagations are time consuming. We must restrict the number of calls to $Propa$ in order to obtain an efficient method. We use here a property similar to the one introduced for SAT problems in [2]. After the call to $Propa(A \cup \{(ce, e)\}, B, C)$, there are two possibilities:

- $C_{A \cup \{(ce,e)\}} = False$ and then $D_{ce} = D_{ce} - \{e\}$: Value elimination.
- $C_{A \cup \{(ce,e)\}} \neq False$: the value assignments (ci, e_i) propagated during the process are compatible with the current assignment and would not lead to value elimination if tried later.

This drastically reduces the number of possible candidates for propagation, and minimizes the number of calls to *Propa*. Formally, we have the following propositions :

Proposition 1. *Let* $(ce, e) \in B$, *if* $C_{A \cup \{(ce,e)\}} \models \bot$ *then* C_A *is equivalent to* $C_A \wedge (ce \neq e)$.

This property is used to eliminate values from the cell domains.

Proposition 2. *Let* $(ce, e) \in B$, *if* $C_{A \cup \{(ce,e)\}} \models (ce_1, e_1), ...(ce_n, e_n)$ *and if* $C_{A \cup \{(ce,e)\}} \not\models \bot$ *then* $\forall i \in \{1...n\} | ce_i \in B$, $C_{A \cup \{(ce_i,e_i)\}} \not\models \bot$

This property avoids to propagate useless facts. This allows to perform fewer calls to the *Propa* procedure.

An additional possibility to reduce the number of unit propagations is to select which cells must be tried. We note $T \subseteq B$ the set of cells which are candidates for unit propagation. Because of symmetries (LNH), only the cells with indices less or equal than mdn need to be considered. We call those cells *mdn cells*. The results obtained using this dynamic filtering technique are listed in section 4 under the name VESEM.

3.3 A First Fail Heuristic

In its original version, SEM chooses as the next cell the one with the smallest domain, and tries to not increment mdn. We note H these previous conditions. Then the heuristic chooses as the next variable to instantiate, the one that both satisfies conditions H and that maximizes the count of the number of propagations done on each cell for all their possible values. This approach is similar to the one described in [4] for propositional logic. The algorithm of this heuristic and value elimination process is shown in the algorithm 2.

Remark 1. In algorithm 2, Mark[ce,e]=True means that the value e of the cell ce can be suppressed. The number Nb equals the number of propagations.

4 Experimentations

We compare SEM, VESEM (SEM + Value Elimination), CTSEM (SEM + Clause Transformation preprocessing) and CTVESEM (SEM + Clause Transformation + Value Elimination) on a set of well known problems. Run times are in seconds. All experiments were carried out under Linux on a K6II 400 PC with 128 MB of RAM. A '+' indicates that a program fails to solve a problem in less than two hours.

Function Up_Heuristic(A, B, C): Return Next cell to choose
For All $(ce, D) \in T$ Do
    For All $e \in D$ Do Mark[ce, e]=True
For All $(ce, D) \in T$ Do
    $Nb = 0$
    For All $e \in D$ such that Mark[ce, e]=True Do
        (A', B', C')=Propa($A \cup (ce, e), B, C$)
        For All (ce', e') propagated Do Mark[ce', e']=False
                $Nb = Nb + 1$
    If $C' = False$ Then
        $D = D - \{e\}$
        If $|D| = 1$ Then return ce
    Else
        $w(ce) = w(ce) + Nb$
Return ce with the smallest domain and maximising w

Algorithm 2: Value Elimination and Heuristic

4.1 Quasigroup Problems

A quasigroup is a binary operator '.' such that the equations $a.x = b$ and $x.a = b$ have an unique solution for all a, b. We deal here with idempotent quasigroups, statisfying the additional axiom $(x.x = x)$. Adding different extra axioms leads to several problem instances, fully described in [6]. None of these axioms are reducible. The results obtained with quasigroups are listed in table 1.

The results show that VESEM always explores fewer nodes than SEM. The amount of memory required to solve these problems is the same with both algorithms. Because of the cost of computing the heuristic, computation times are not significantly improved in general except on one example (QG6). Only two examples (QG7 and QG1) exhibit results slightly worse with VESEM than with SEM. Although the quasigroup problems do not clearly prove a superiority of VESEM, they show that the value elimination and lookahead strategy generally results in a favorable tradeoff and should be used.

4.2 Group and Ring Problems

We compare VESEM and CTSEM and CTVESEM to SEM on a list of group and ring problems described by J. Zhang in [10]. The results are listed in table 2. Our algorithms explore fewer nodes than SEM. The lookahead strategy implemented in VESEM generally leads to improved computation times. The execution time ratio is sometimes very important: about 60 for NG and GRP.

CTSEM and VESEM not only solve problems faster, but solve problems of much higher orders (NG, GRP, RU). To the best of our knowledge, it is the first time that a program ever computes a finite model for NG34 and RU24 or proves the inconsistency of GRP38.

The program CTVESEM combining both techniques (Clause Transformation and Value Elimination) visits fewer search tree nodes. But, almost all the values suppressed (leading to skipped nodes) are due to the clause rewriting technique.

Table 1. Quasigroup Problems - Comparison.

Problem	Nb Model	SEM		VESEM	
		Time	Nodes	Time	Nodes
QG1 7	4	22	411	24	194
QG2 6	0	1.4	17	1.4	9
QG2 7	3	63	871	59	401
QG3 9	0	7.4	48 278	6.3	40 015
QG3 10	0	1416	7 948 372	1335	3 558 564
QG4 9	74	6.3	38 407	5.3	17 116
QG4 10	0	1263	6 946 603	1 099	2 941 094
QG5 14	0	83	320 728	53	106 703
QG5 15	0	2031	7 518 920	1306	2 251 311
QG6 11	0	40	840 542	2.3	13 690
QG6 12	0	2519	50 290 872	142	929 781
QG7 13	2	14.5	69 053	16	37 132
QG7 14	0	443	2 015 778	528	1 107 404

Thus, adding value elimination to clause transformation is redundant and results in increased computation times. All results obtained and a fully detailed description of the different algorithms described in this paper are available in [1].

Table 2. Ring and Group Problems - Comparison

Problem	Nb Models	SEM		VESEM		CTSEM		CTVESEM	
		Time	Nodes	Time	Nodes	Time	Nodes	Time	Nodes
AG 28	162	328	642 103	321	76 663	336	57 941	394	41 859
AG 32	2 295	940	2 037 525	956	624 304	968	101 356	1 272	76 393
NG 28	51	6 934	8 359 103	806	108 120	432	100 036	1 105	88 832
NG 29	0	+		752	94 417	489	108 922	1 191	82 519
NG 34	3	+		5 450	504 182	3 469	478 337	+	
GRP 31	0	3 831	2 751 805	272	21 821	97	14 711	378	24 691
GRP 32	2 712	+		1 620	740 797	529	35 546	2 204	93 420
GRP 38	0	+		6 690	584 374	3 480	442 039	+	
RU 19	1	4 591	2 720 769	1729	94 326	848	197 953	1 666	15 741
RU 20	21	+		2 904	370 652	3 678	609 320	2 957	336 612
RU 24	445	+		5 029	434 006	+		5 019	366 597
RNA 14	0	592	646 421	+		354	131 355	426	45 162
RNA 15	0	592	646 421	1 021	150 538	513	144 613	682	56 623
RNA16	?	+		+		+		+	
RNB 17	0	15	13 148	20	6 389	15	2 287	21	1 309
RNB 18	0	16	13 238	36	2 171	16	2 377	34	852

5   Conclusion

We introduce two techniques that can be used to improve CSP approaches to
finite model generation of first order theories. Their efficiency stems from the
introduction of negative facts in the clause transformation technique case (CT-
SEM), and from the elimination of domain values at some node of the search
tree in the dynamic filtering case (VESEM).

The behaviour of the algorithms on the AG and RNA problems suggests to
search for improvements in the heuristic strategy associated with the lookahead
procedure in VESEM, and also to eliminate more isomorphic subspaces than is
actually done with the LNH heuristic used in those programs. VESEM seems
to provide the basis for a general algorithm for finite model search of first order
theories.

References

[1] G. Audemard, B. Benhamou, and L. Henocque. Two techniques to improve finite
    model search. Technical report, Laboratoire d'Informatique de Marseille, 1999. ac-
    cessible electronicaly at http://www.cmi.univ-mrs.fr/~audemard/publi.html.

[2] G. Audemard, B. Benhamou, and P. Siegel. La mÅéthode d'avalanche aval: une
    mÅéthode ÅénumÅérative pour sat. In JNPC, pages 17–25, 1999.

[3] B. Benhamou and L. Henocque. A hybrid method for finite model search in
    equational theories. Fundamenta Informaticae, 39(1-2):21–38, June 1999.

[4] Chu Min Li and Anbulagan. Heuristics based on unit propagation for satisfiability
    problems. In Proceedings of the 15th International Joint Conference on Artificial
    Intelligence (IJCAI-97), pages 366–371, August 23–29 1997.

[5] Nicolas Peltier. A new method for automated finite model building exploiting
    failures and symmetries. Journal of Logic and Computation, 8(4):511–543, 1998.

[6] J. Slaney, M. Fujita, and M. Stickel. Automated reasoning and exhaustive search:
    Quasigroup existence problems. Computers and Mathematics with Applications,
    29(2):115–132, 1993.

[7] J. Slanley. Finder: Finite domain enumerator. version 3 notes and guides. Tech-
    nical report, Austrian National University, 1993.

[8] H. Zhang and Mark E. Stickel. Implementing the davis-putnam algorithm by tries.
    Technical report, Department of Computer Science, University of Iowa,1994.

[9] J. Zhang and H. Zhang. Constraint propagation in model generation. In proceed-
    ings of CP95, Marseille 1995.

[10] Jian Zhang. Constructing finite algebras with FALCON. Journal of Automated
     Reasoning, 17(1):1–22, August 1996.

[11] Jian Zhang and Hantao Zhang. SEM: a system for enumerating models. In Chris S.
     Mellish, editor, Proceedings of the Fourteenth IJCAI, pages 298–303, 1995.

Eliminating Dummy Elimination

Jürgen Giesl[1] and Aart Middeldorp[2]

[1] Computer Science Department
University of New Mexico, Albuquerque, NM 87131, USA
giesl@cs.unm.edu
[2] Institute of Information Sciences and Electronics
University of Tsukuba, Tsukuba 305-8573, Japan
ami@is.tsukuba.ac.jp

Abstract. This paper is concerned with methods that automatically prove termination of term rewrite systems. The aim of dummy elimination, a method to prove termination introduced by Ferreira and Zantema, is to transform a given rewrite system into a rewrite system whose termination is easier to prove. We show that dummy elimination is subsumed by the more recent dependency pair method of Arts and Giesl. More precisely, if dummy elimination succeeds in transforming a rewrite system into a so-called simply terminating rewrite system then termination of the given rewrite system can be directly proved by the dependency pair technique. Even stronger, using dummy elimination as a preprocessing step to the dependency pair technique does not have any advantages either. We show that to a large extent these results also hold for the argument filtering transformation of Kusakari et al.

1 Introduction

Traditional methods to prove termination of term rewrite systems are based on simplification orders, like polynomial interpretations [6, 12, 17], the recursive path order [7, 14], and the Knuth-Bendix order [9, 15]. However, the restriction to simplification orders represents a significant limitation on the class of rewrite systems that can be proved terminating. Indeed, there are numerous important and interesting rewrite systems which are not *simply terminating*, i.e., their termination cannot be proved by simplification orders. Transformation methods (e.g. [5, 10, 11, 16, 18, 20–22]) aim to prove termination by transforming a given term rewrite system into a term rewrite system whose termination is easier to prove. The success of such methods has been measured by how well they transform non-simply terminating rewrite systems into simply terminating systems, since simply terminating systems were the only ones where termination could be established automatically.

In recent years, the dependency pair technique of Arts and Giesl [1, 2] emerged as the most powerful automatic method for proving termination of rewrite systems. For any given rewrite system, this technique generates a set of constraints which may then be solved by standard simplification orders. In this way, the

D. McAllester (Ed.): CADE-17, LNAI 1831, pp. 309–323, 2000.

power of traditional termination proving methods has been increased significantly, i.e., the class of systems where termination is provable mechanically by the dependency pair technique is much larger than the class of simply terminating systems. In light of this development, it is no longer sufficient to base the claim that a particular transformation method is successful on the fact that it may transform non-simply terminating rewrite systems into simply terminating ones. In this paper we compare two transformation methods, dummy elimination [11] and the argument filtering transformation [16], with the dependency pair technique. With respect to dummy elimination we obtain the following results:

1. If dummy elimination transforms a given rewrite system $\mathcal{R}$ into a simply terminating rewrite system $\mathcal{R}'$, then the termination of $\mathcal{R}$ can also be proved by the most basic version of the dependency pair technique.
2. If dummy elimination transforms a given rewrite system $\mathcal{R}$ into a *DP simply terminating* rewrite system $\mathcal{R}'$, i.e., the termination of $\mathcal{R}'$ can be proved by a simplification order in combination with the dependency pair technique, then $\mathcal{R}$ is also DP simply terminating.

These results are constructive in the sense that the constructions in the proofs are solely based on the termination proof of $\mathcal{R}'$. This shows that proving termination of $\mathcal{R}$ directly by dependency pairs is never more difficult than proving termination of $\mathcal{R}'$. The second result states that dummy elimination is useless as a preprocessing step to the dependency pair technique. Not surprisingly, the reverse statements do not hold. In other words, as far as automatic termination proofs are concerned, dummy elimination is no longer needed.

The recent argument filtering transformation of Kusakari, Nakamura, and Toyama [16] can be viewed as an improvement of dummy elimination by incorporating ideas of the dependency pair technique. We show that the first result above also holds for the argument filtering transformation. The second result does not extend in its full generality, but we show that under a suitable restriction on the argument filtering applied in the transformation of $\mathcal{R}$ to $\mathcal{R}'$, DP simple termination of $\mathcal{R}'$ also implies DP simple termination of $\mathcal{R}$.

The remainder of the paper is organized as follows. In the next section we briefly recall some definitions and results pertaining to termination of rewrite systems and in particular, the dependency pair technique. In Section 3 we relate the dependency pair technique to dummy elimination. Section 4 is devoted to the comparison of the dependency pair technique and the argument filtering transformation. We conclude in Section 5.

2   Preliminaries

An introduction to term rewrite systems (TRSs) can be found in [4], for example. We first introduce the dependency pair technique. Our presentation combines features of [2, 13, 16]. Apart from the presentation, all results stated below are due to Arts and Giesl. We refer to [2, 3] for motivations and proofs. Let $\mathcal{R}$ be a (finite) TRS over a signature $\mathcal{F}$. As usual, all root symbols of left-hand sides of rewrite rules are called *defined*, whereas all other function symbols are *constructors*. Let $\mathcal{F}^\sharp$ denote the union of $\mathcal{F}$ and $\{f^\sharp \mid f \text{ is a defined symbol of } \mathcal{R}\}$ where $f^\sharp$ has the same arity as f. Given a term $t = f(t_1, \ldots, t_n) \in \mathcal{T}(\mathcal{F}, \mathcal{V})$ with f defined, we write $t^\sharp$ for the term $f^\sharp(t_1, \ldots, t_n)$. If $l \to r \in \mathcal{R}$ and t is a subterm of r with defined root symbol then the rewrite rule $l^\sharp \to t^\sharp$ is called a *dependency pair* of $\mathcal{R}$. The set of all dependency pairs of $\mathcal{R}$ is denoted by $\mathrm{DP}(\mathcal{R})$. In examples we often write F for $f^\sharp$.

For instance, consider the following well-known one-rule TRS $\mathcal{R}$ from [8]:

$$f(f(x)) \to f(e(f(x))) \tag{1}$$

Here f is defined, e is a constructor, and $\mathrm{DP}(\mathcal{R})$ consists of the two dependency pairs

$$F(f(x)) \to F(e(f(x))) \qquad F(f(x)) \to F(x)$$

An *argument filtering* [2] for a signature $\mathcal{F}$ is a mapping π that associates with every n-ary function symbol an argument position $i \in \{1, \ldots, n\}$ or a (possibly empty) list $[i_1, \ldots, i_m]$ of argument positions with $1 \leqslant i_1 < \cdots < i_m \leqslant n$. The signature $\mathcal{F}_\pi$ consists of all function symbols f such that $\pi(f)$ is some list $[i_1, \ldots, i_m]$, where in $\mathcal{F}_\pi$ the arity of f is m. Every argument filtering π induces a mapping from $\mathcal{T}(\mathcal{F}, \mathcal{V})$ to $\mathcal{T}(\mathcal{F}_\pi, \mathcal{V})$, also denoted by π:

$$\pi(t) = \begin{cases} t & \text{if } t \text{ is a variable,} \\ \pi(t_i) & \text{if } t = f(t_1, \ldots, t_n) \text{ and } \pi(f) = i, \\ f(\pi(t_{i_1}), \ldots, \pi(t_{i_m})) & \text{if } t = f(t_1, \ldots, t_n) \text{ and } \pi(f) = [i_1, \ldots, i_m]. \end{cases}$$

Thus, an argument filtering is used to replace function symbols by one of their arguments or to eliminate certain arguments of function symbols. For example, if $\pi(f) = \pi(F) = [1]$ and $\pi(e) = 1$, then we have $\pi(F(e(f(f(x))))) = F(f(x))$. However, if we change $\pi(e)$ to $[]$, then we obtain $\pi(F(e(f(f(x))))) = F(e)$.

A *preorder* (or *quasi-order*) is a transitive and reflexive relation. A *rewrite preorder* is a preorder $\succsim$ on terms that is closed under contexts and substitutions. A *reduction pair* [16] consists of a rewrite preorder $\succsim$ and a compatible well-founded order $>$ which is closed under substitutions. Here compatibility means that the inclusion $\succsim \cdot > \subseteq >$ or the inclusion $> \cdot \succsim \subseteq >$ holds. In practice, $>$ is often chosen to be the strict part $\succ$ of $\succsim$ (or the order where $s > t$ iff $s\sigma \succ t\sigma$ for all ground substitutions σ). The following theorem presents the (basic) dependency pair approach of Arts and Giesl.

Theorem 1. *A TRS $\mathcal{R}$ over a signature $\mathcal{F}$ is terminating if and only if there exists an argument filtering π for $\mathcal{F}^\sharp$ and a reduction pair $(\succsim, >)$ such that $\pi(\mathcal{R}) \subseteq \succsim$ and $\pi(\mathrm{DP}(\mathcal{R})) \subseteq >$.*

Because rewrite rules are just pairs of terms, $\pi(\mathcal{R}) \subseteq \succsim$ is a shorthand for $\pi(l) \succsim \pi(r)$ for every rewrite rule $l \to r \in \mathcal{R}$. In our example, when using $\pi(e) = []$, the inequalities $f(f(x)) \succsim f(e)$, $F(f(x)) > F(e)$, and $F(f(x)) > F(x)$

resulting from the dependency pair technique are satisfied by the recursive path order, for instance. Hence, termination of this TRS is proved.

Rather than considering all dependency pairs at the same time, like in the above theorem, it is advantageous to treat groups of dependency pairs separately. These groups correspond to *clusters* in the *dependency graph* of $\mathcal{R}$. The nodes of the dependency graph are the dependency pairs of $\mathcal{R}$ and there is an arrow from node $l_1^\sharp \to t_1^\sharp$ to $l_2^\sharp \to t_2^\sharp$ if there exist substitutions σ_1 and σ_2 such that $t_1^\sharp \sigma_1 \to_{\mathcal{R}}^* l_2^\sharp \sigma_2$. (By renaming variables in different occurrences of dependency pairs we may assume that $\sigma_1 = \sigma_2$.) The dependency graph of $\mathcal{R}$ is denoted by $\mathrm{DG}(\mathcal{R})$. We call a non-empty subset $\mathcal{C}$ of dependency pairs of $\mathrm{DP}(\mathcal{R})$ a *cluster* if for every two (not necessarily distinct) pairs $l_1^\sharp \to t_1^\sharp$ and $l_2^\sharp \to t_2^\sharp$ in $\mathcal{C}$ there exists a non-empty path in $\mathcal{C}$ from $l_1^\sharp \to t_1^\sharp$ to $l_2^\sharp \to t_2^\sharp$.

Theorem 2. *A TRS $\mathcal{R}$ is terminating if and only if for every cluster $\mathcal{C}$ in $\mathrm{DG}(\mathcal{R})$ there exists an argument filtering π and a reduction pair $(\succsim, >)$ such that $\pi(\mathcal{R}) \subseteq \succsim$, $\pi(\mathcal{C}) \subseteq \succsim \cup >$, and $\pi(\mathcal{C}) \cap > \neq \varnothing$.*

Note that $\pi(\mathcal{C}) \cap > \neq \varnothing$ denotes the situation that $\pi(l^\sharp) > \pi(t^\sharp)$ for at least one dependency pair $l^\sharp \to t^\sharp \in \mathcal{C}$.

In the above example, the dependency graph only contains an arrow from $\mathsf{F}(f(x)) \to \mathsf{F}(x)$ to itself and thus $\{\mathsf{F}(f(x)) \to \mathsf{F}(x)\}$ is the only cluster. Hence, with the refinement of Theorem 2 the inequality $\mathsf{F}(f(x)) > \mathsf{F}(\mathsf{e})$ is no longer necessary. See [3] for further examples which illustrate the advantages of regarding clusters separately.

Note that while in general the dependency graph cannot be computed automatically (since it is undecidable whether $t_1^\sharp \sigma \to_{\mathcal{R}}^* l_2^\sharp \sigma$ holds for some σ), one can nevertheless approximate this graph automatically, cf. [1–3, "estimated dependency graph"]. In this way, the criterion of Theorem 2 can be mechanized.

Most classical methods for automated termination proofs are restricted to simplification (pre)orders, i.e., to (pre)orders satisfying the subterm property $f(\ldots t \ldots) \succ t$ or $f(\ldots t \ldots) \succsim t$, respectively. Hence, these methods cannot prove termination of TRSs like (1), as the left-hand side of its rule is embedded in the right-hand side (so the TRS is not simply terminating). However, with the development of the dependency pair technique now the TRSs where an automated termination proof is potentially possible are those systems where the inequalities generated by the dependency pair technique are satisfied by simplification (pre)orders.

A straightforward way to generate a simplification preorder $\succeq$ from a simplification order $\succ$ is to define $s \succeq t$ if $s \succ t$ or $s = t$, where $=$ denotes syntactic equality. Such relations $\succeq$ are particularly relevant, since many existing techniques generate simplification *orders* rather than *preorders*. By restricting ourselves to this class of simplification preorders, we obtain the notion of DP simple termination.

Definition 1. *A TRS $\mathcal{R}$ is called DP simply terminating if for every cluster $\mathcal{C}$ in $\mathrm{DG}(\mathcal{R})$ there exists an argument filtering π and a simplification order $\succ$ such that $\pi(\mathcal{R} \cup \mathcal{C}) \subseteq \succeq$ and $\pi(\mathcal{C}) \cap \succ \neq \varnothing$.*

Simple termination implies DP simple termination, but not vice versa. For example, the TRS (1) is DP simply terminating, but not simply terminating. The above definition coincides with the one in [13] except that we use the real dependency graph instead of the *estimated* dependency graph of [1-3]. The reason for this is that we do not want to restrict ourselves to a particular computable approximation of the dependency graph, for the same reason that we do not insist on a particular simplification order to make the conditions effective.

3 Dummy Elimination

In [11], Ferreira and Zantema defined an automatic transformation technique which transforms a TRS $\mathcal{R}$ into a new TRS dummy($\mathcal{R}$) such that termination of dummy($\mathcal{R}$) implies termination of $\mathcal{R}$. The advantage of this transformation is that non-simply terminating systems like (1) may be transformed into simply terminating ones. Thus, after the transformation, standard techniques may be used to prove termination.

Below we define Ferreira and Zantema's dummy elimination transformation. While our formulation of dummy($\mathcal{R}$) is different from the one in [11], it is easily seen to be equivalent.

Definition 2. *Let $\mathcal{R}$ be a TRS over a signature $\mathcal{F}$. Let e be a distinguished function symbol in $\mathcal{F}$ of arity $m \geqslant 1$ and let $\diamond$ be a fresh constant. We write $\mathcal{F}_\diamond$ for $(\mathcal{F} \setminus \{e\}) \cup \{\diamond\}$. The mapping* cap: $\mathcal{T}(\mathcal{F}, \mathcal{V}) \to \mathcal{T}(\mathcal{F}_\diamond, \mathcal{V})$ *is inductively defined as follows:*

$$\mathrm{cap}(t) = \begin{cases} t & \textit{if } t \in \mathcal{V}, \\ \diamond & \textit{if } t = e(t_1, \ldots, t_m), \\ f(\mathrm{cap}(t_1), \ldots, \mathrm{cap}(t_n)) & \textit{if } t = f(t_1, \ldots, t_n) \textit{ with } f \neq e. \end{cases}$$

The mapping dummy *assigns to every term in $\mathcal{T}(\mathcal{F}, \mathcal{V})$ a subset of $\mathcal{T}(\mathcal{F}_\diamond, \mathcal{V})$, as follows:*

dummy$(t) = \{\mathrm{cap}(t)\} \cup \{\mathrm{cap}(s) \mid s$ *is an argument of an e symbol in $t\}$.*

Finally, we define

dummy$(\mathcal{R}) = \{\mathrm{cap}(l) \to r' \mid l \to r \in \mathcal{R} \text{ and } r' \in \text{dummy}(r)\}.$

The mappings cap and dummy are illustrated in Figure 1, where we assume that the numbered contexts do not contain any occurrences of e. Ferreira and Zantema [11] showed that dummy elimination is sound.

Theorem 3. *Let $\mathcal{R}$ be a TRS. If* dummy($\mathcal{R}$) *is terminating then $\mathcal{R}$ is terminating.*

For the one-rule TRS (1), dummy elimination yields the TRS consisting of the two rewrite rules

$$\mathsf{f}(\mathsf{f}(x)) \to \mathsf{f}(\diamond) \qquad \mathsf{f}(\mathsf{f}(x)) \to \mathsf{f}(x)$$

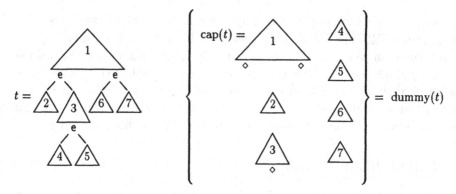

Fig. 1. The mappings cap and dummy.

In contrast to the original system, the new TRS is simply terminating and its termination is easily shown automatically by standard techniques like the recursive path order. Hence, dummy elimination can transform non-simply terminating TRSs into simply terminating ones. However, as indicated in the introduction, nowadays the right question to ask is whether it can transform non-DP simply terminating TRSs into DP simply terminating ones. Before answering this question we show that if dummy elimination succeeds in transforming a TRS into a simply terminating TRS then the original TRS is DP simply terminating. Even stronger, whenever termination of dummy($\mathcal{R}$) can be proved by a simplification order, then the *same* simplification order satisfies the constraints of the dependency pair approach. Thus, the termination proof using dependency pairs is not more difficult or more complex than the one with dummy elimination.

Theorem 4. *Let $\mathcal{R}$ be a TRS. If* dummy($\mathcal{R}$) *is simply terminating then $\mathcal{R}$ is DP simply terminating.*

Proof. Let $\mathcal{F}$ be the signature of $\mathcal{R}$. We show that $\mathcal{R}$ is DP simply terminating even without considering the dependency graph refinement. So we define an argument filtering π for $\mathcal{F}^{\sharp}$ and a simplification order $\succ$ on $\mathcal{T}(\mathcal{F}^{\sharp}_{\pi}, \mathcal{V})$ such that $\pi(\mathcal{R}) \subseteq \succeq$ and $\pi(\text{DP}(\mathcal{R})) \subseteq \succ$. The argument filtering π is defined as follows: $\pi(e) = []$ and $\pi(f) = [1, \ldots, n]$ for every n-ary symbol $f \in (\mathcal{F} \setminus \{e\})^{\sharp}$. Moreover, if e is a defined symbol, we define $\pi(e^{\sharp}) = []$. Let $\sqsupset$ be any simplification order that shows the simple termination of dummy($\mathcal{R}$). We define the simplification order $\succ$ on $\mathcal{T}(\mathcal{F}^{\sharp}_{\pi}, \mathcal{V})$ as follows: $s \succ t$ if and only if $s' \sqsupset t'$ where $(\cdot)'$ denotes the mapping from $\mathcal{T}(\mathcal{F}^{\sharp}_{\pi}, \mathcal{V})$ to $\mathcal{T}(\mathcal{F}_{\diamond}, \mathcal{V})$ that first replaces every marked symbol F by f and afterwards replaces every occurrence of the constant e by $\diamond$. Note that $\succ$ and $\sqsupset$ are essentially the same. It is very easy to show that $\pi(t)' = \pi(t^{\sharp})' = \text{cap}(t)$ for every term $t \in \mathcal{T}(\mathcal{F}, \mathcal{V})$. Let $l \to r \in \mathcal{R}$. Because cap(l) $\to$ cap(r) is a rewrite rule in dummy($\mathcal{R}$), we get $\pi(l)' = \text{cap}(l) \sqsupset \text{cap}(r) = \pi(r)'$ and thus $\pi(l) \succ \pi(r)$. Hence $\pi(\mathcal{R}) \subseteq \succ$ and thus certainly $\pi(\mathcal{R}) \subseteq \succeq$. Now let $l^{\sharp} \to t^{\sharp}$ be a dependency pair of $\mathcal{R}$, originating from the rewrite rule $l \to r \in \mathcal{R}$. From $t \trianglelefteq r$ ($\trianglelefteq$ denotes the subterm relation) we easily infer the existence of a term $u \in$ dummy(r) such that cap(t) $\trianglelefteq u$. Since cap(l) $\to u$ is a rewrite rule in dummy($\mathcal{R}$), we have

$\pi(l^{\sharp})' = \mathrm{cap}(l) \sqsupset u$. The subterm property of $\sqsupset$ yields $u \sqsupseteq \mathrm{cap}(t) = \pi(t^{\sharp})'$. Hence $\pi(l^{\sharp})' \sqsupset \pi(t^{\sharp})'$ and thus $\pi(l^{\sharp}) \succ \pi(t^{\sharp})$. We conclude that $\pi(\mathrm{DP}(\mathcal{R})) \subseteq \succ$. $\square$

The previous result states that dummy elimination offers no advantage compared to the dependency pair technique. On the other hand, dependency pairs succeed for many systems where dummy elimination fails [1, 2] (an example is given in the next section). One could imagine that dummy elimination may nevertheless be helpful in combination with dependency pairs. Then to show termination of a TRS one would first apply dummy elimination and afterwards prove termination of the transformed TRS with the dependency pair technique. In the remainder of this section we show that such a scenario cannot handle TRSs which cannot already be handled by the dependency pair technique directly. In short, dummy elimination is useless for automated termination proofs. We proceed in a stepwise manner. First we relate the dependency pairs of $\mathcal{R}$ to those of $\mathrm{dummy}(\mathcal{R})$.

Lemma 1. *If $l^{\sharp} \to t^{\sharp} \in \mathrm{DP}(\mathcal{R})$ then $\mathrm{cap}(l)^{\sharp} \to \mathrm{cap}(t)^{\sharp} \in \mathrm{DP}(\mathrm{dummy}(\mathcal{R}))$.*

Proof. In the proof of Theorem 4 we observed that there exists a rewrite rule $\mathrm{cap}(l) \to u$ in $\mathrm{dummy}(\mathcal{R})$ with $\mathrm{cap}(t) \trianglelefteq u$. Since $\mathrm{root}(\mathrm{cap}(t))$ is a defined symbol in $\mathrm{dummy}(\mathcal{R})$, $\mathrm{cap}(l)^{\sharp} \to \mathrm{cap}(t)^{\sharp}$ is a dependency pair of $\mathrm{dummy}(\mathcal{R})$. $\square$

Now we prove that reducibility in $\mathcal{R}$ implies reducibility in $\mathrm{dummy}(\mathcal{R})$.

Definition 3. *Given a substitution σ, the substitution σ_{cap} is defined as $\mathrm{cap} \circ \sigma$ (i.e., the composition of cap and σ where σ is applied first).*

Lemma 2. *For all terms t and substitutions σ, we have $\mathrm{cap}(t\sigma) = \mathrm{cap}(t)\sigma_{\mathrm{cap}}$.*

Proof. Easy induction on the structure of t. $\square$

Lemma 3. *If $s \to^{*}_{\mathcal{R}} t$ then $\mathrm{cap}(s) \to^{*}_{\mathrm{dummy}(\mathcal{R})} \mathrm{cap}(t)$.*

Proof. It is sufficient to show that $s \to_{\mathcal{R}} t$ implies $\mathrm{cap}(s) \to^{*}_{\mathrm{dummy}(\mathcal{R})} \mathrm{cap}(t)$. There must be a rule $l \to r \in \mathcal{R}$ and a position p such that $s|_{\pi} = l\sigma$ and $t = s[r\sigma]_p$. If p is below the position of an occurrence of e, then we have $\mathrm{cap}(s) = \mathrm{cap}(t)$. Otherwise, $\mathrm{cap}(s)|_p = \mathrm{cap}(l\sigma) = \mathrm{cap}(l)\sigma_{\mathrm{cap}}$ by Lemma 2. Thus, $\mathrm{cap}(s) \to_{\mathrm{dummy}(\mathcal{R})} \mathrm{cap}(s)[\mathrm{cap}(r)\sigma_{\mathrm{cap}}]_p = \mathrm{cap}(s)[\mathrm{cap}(r\sigma)]_p = \mathrm{cap}(t)$. $\square$

Next we show that if there is an arrow between two dependency pairs in the dependency graph of $\mathcal{R}$ then there is an arrow between the corresponding dependency pairs in the dependency graph of $\mathrm{dummy}(\mathcal{R})$.

Lemma 4. *Let s, t be terms with defined root symbols. If $s^{\sharp}\sigma \to^{*}_{\mathcal{R}} t^{\sharp}\sigma$ for some substitution σ, then $\mathrm{cap}(s)^{\sharp}\sigma_{\mathrm{cap}} \to^{*}_{\mathrm{dummy}(\mathcal{R})} \mathrm{cap}(t)^{\sharp}\sigma_{\mathrm{cap}}$.*

Proof. Let $s = f(s_1, \ldots, s_n)$. We have $s^{\sharp}\sigma = f^{\sharp}(s_1\sigma, \ldots, s_n\sigma)$. Since $f^{\sharp}$ is a constructor, no step in the sequence $s^{\sharp}\sigma \to^{*}_{\mathcal{R}} t^{\sharp}\sigma$ takes place at the root position and thus $t^{\sharp} = f^{\sharp}(t_1, \ldots, t_n)$ with $s_i\sigma \to^{*}_{\mathcal{R}} t_i\sigma$ for all $1 \leqslant i \leqslant n$. We obtain $\mathrm{cap}(s_i)\sigma_{\mathrm{cap}} = \mathrm{cap}(s_i\sigma) \to^{*}_{\mathrm{dummy}(\mathcal{R})} \mathrm{cap}(t_i\sigma) = \mathrm{cap}(t_i)\sigma_{\mathrm{cap}}$ for all $1 \leqslant i \leqslant n$ by Lemmata 2 and 3. Hence $\mathrm{cap}(s)^{\sharp}\sigma_{\mathrm{cap}} \to^{*}_{\mathrm{dummy}(\mathcal{R})} \mathrm{cap}(t)^{\sharp}\sigma_{\mathrm{cap}}$. $\square$

Finally we are ready for the main theorem of this section.

Theorem 5. *Let* $\mathcal{R}$ *be a TRS. If* dummy($\mathcal{R}$) *is DP simply terminating then* $\mathcal{R}$ *is DP simply terminating.*

Proof. Let $\mathcal{C}$ be a cluster in the dependency graph of $\mathcal{R}$. From Lemmata 1 and 4 we infer the existence of a corresponding cluster, denoted by dummy($\mathcal{C}$), in the dependency graph of dummy($\mathcal{R}$). By assumption, there exists an argument filtering π' and a simplification order $\sqsupseteq$ such that $\pi'(\text{dummy}(\mathcal{R}) \cup \text{dummy}(\mathcal{C})) \subseteq \sqsupseteq$ and $\pi'(\text{dummy}(\mathcal{C})) \cap \sqsupseteq \neq \varnothing$. Let $\mathcal{F}$ be the signature of $\mathcal{R}$. We define an argument filtering π for $\mathcal{F}^{\sharp}$ as follows: $\pi(f) = \pi'(f)$ for every $f \in (\mathcal{F} \setminus \{e\})^{\sharp}$, $\pi(e) = []$ and, if e is a defined symbol of $\mathcal{R}$, $\pi(e^{\sharp}) = []$. Slightly different from the proof of Theorem 4, let $(\cdot)'$ denote the mapping that just replaces every occurrence of the constant e by $\diamond$ and every occurrence of $e^{\sharp}$ by $\diamond^{\sharp}$. It is easy to show that $\pi(t)' = \pi'(\text{cap}(t))$ for every term $t \in \mathcal{T}(\mathcal{F}, \mathcal{V})$ and $\pi(t^{\sharp})' = \pi'(\text{cap}(t)^{\sharp})$ for every term $t \in \mathcal{T}(\mathcal{F}, \mathcal{V})$ with a defined root symbol. Similar to Theorem 4, we define the simplification order $\succ$ on $\mathcal{F}_{\pi}$ as $s \succ t$ if and only if $s' \sqsupseteq t'$. We claim that π and $\succ$ satisfy the constraints for $\mathcal{C}$, i.e., $\pi(\mathcal{R} \cup \mathcal{C}) \subseteq \succeq$ and $\pi(\text{dummy}(\mathcal{C})) \cap \succ \neq \varnothing$. If $l \to r \in \mathcal{R}$, then $\text{cap}(l) \to \text{cap}(r) \in \text{dummy}(\mathcal{R})$ and thus $\pi(l)' = \pi'(\text{cap}(l)) \sqsupseteq \pi'(\text{cap}(r)) = \pi(r)'$. Hence $\pi(l) \succeq \pi(r)$. If $l^{\sharp} \to t^{\sharp} \in \mathcal{C}$, then $\text{cap}(l)^{\sharp} \to \text{cap}(t)^{\sharp} \in \text{dummy}(\mathcal{C})$ by Lemma 1 and thus $\pi(l^{\sharp})' = \pi'(\text{cap}(l)^{\sharp}) \sqsupseteq \pi'(\text{cap}(t)^{\sharp}) = \pi(t^{\sharp})'$. Hence $\pi(l^{\sharp}) \succeq \pi(t^{\sharp})$ and if $\pi'(\text{cap}(l)^{\sharp}) \sqsupset \pi'(\text{cap}(t)^{\sharp})$, then $\pi(l^{\sharp}) \succ \pi(t^{\sharp})$. □

We stress that the proof is constructive in the sense that a DP simple termination proof of dummy($\mathcal{R}$) can be automatically transformed into a DP simple termination proof of $\mathcal{R}$ (i.e., the orders and argument filterings required for the DP simple termination proofs of dummy($\mathcal{R}$) and $\mathcal{R}$ are essentially the same). Thus, the termination proof of dummy($\mathcal{R}$) is not simpler than a direct proof for $\mathcal{R}$.

Theorem 5 also holds if one uses the *estimated* dependency graph of [1–3] instead of the real dependency graph. As mentioned in Section 2, such a computable approximation of the dependency graph must be used in implementations, since constructing the real dependency graph is undecidable in general. The proof is similar to the one of Theorem 5, since again for every cluster in the estimated dependency graph of $\mathcal{R}$ there is a corresponding one in the estimated dependency graph of dummy($\mathcal{R}$).

4   Argument Filtering Transformation

By incorporating argument filterings, a key ingredient of the dependency pair technique, into dummy elimination, Kusakari, Nakamura, and Toyama [16] recently developed the argument filtering transformation. In their paper they proved the soundness of their transformation and they showed that it improves upon dummy elimination. In this section we compare their transformation to the dependency pair technique. We proceed as in the previous section. First we recall the definition of the argument filtering transformation.

Definition 4. *Let π be an argument filtering, f a function symbol, and $1 \leqslant i \leqslant$ arity(f). We write $f \perp_\pi i$ if neither $i \in \pi(f)$ nor $i = \pi(f)$. Given two terms s and t, we say that s is a preserved subterm of t with respect to π and we write $s \trianglelefteq_\pi t$, if $s \trianglelefteq t$ and either $s = t$ or $t = f(t_1, \ldots, t_n)$, s is a preserved subterm of t_i, and $f \not\perp_\pi i$.*

Definition 5. *Given an argument filtering π, the argument filtering $\bar{\pi}$ is defined as follows:*

$$\bar{\pi}(f) = \begin{cases} \pi(f) & \text{if } \pi(f) = [i_1, \ldots, i_m], \\ [\pi(f)] & \text{if } \pi(f) = i. \end{cases}$$

The mapping AFT_π assigns to every term in $\mathcal{T}(\mathcal{F}, \mathcal{V})$ a subset of $\mathcal{T}(\mathcal{F}_\pi, \mathcal{V})$, as follows:

$$\mathrm{AFT}_\pi(t) = \{\pi(t) \mid \bar{\pi}(t) \text{ contains a defined symbol}\} \cup \bigcup_{s \in S} \mathrm{AFT}_\pi(s)$$

with S denoting the set of outermost non-preserved subterms of t. Finally, we define

$$\mathrm{AFT}_\pi(\mathcal{R}) = \{\pi(l) \to r' \mid l \to r \in \mathcal{R} \text{ and } r' \in \mathrm{AFT}_\pi(r) \cup \{\pi(r)\}\}.$$

Consider the term t of Figure 1. Figure 2 shows $\mathrm{AFT}_\pi(t)$ for the two argument filterings with $\pi(\mathbf{e}) = [1]$ and $\pi(\mathbf{e}) = 2$, respectively, and $\pi(f) = [1, \ldots, n]$ for every other n-ary function symbol f. Here we assume that all numbered contexts contain defined symbols, but no occurrence of $\mathbf{e}$.

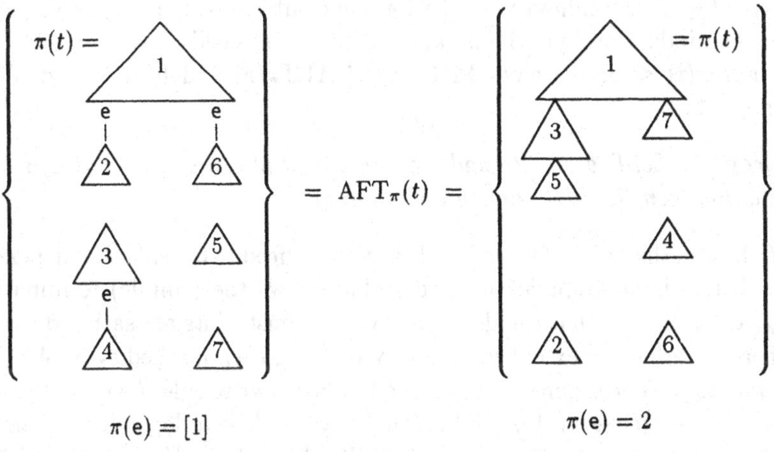

Fig. 2. The mappings π and AFT_π.

So essentially, $\mathrm{AFT}_\pi(t)$ contains $\pi(s)$ for $s = t$ and for all (maximal) subterms s of t which are eliminated if the argument filtering π is applied to t.

However, one only needs terms $\pi(s)$ in $\mathrm{AFT}_\pi(t)$ where s contained a defined symbol outside eliminated arguments (otherwise the original subterm s cannot have been responsible for a potential non-termination). Kusakari et al. [11] proved the soundness of the argument filtering transformation.

Theorem 6. *If* $\mathrm{AFT}_\pi(\mathcal{R})$ *is terminating then* $\mathcal{R}$ *is terminating.*

We show that if $\mathrm{AFT}_\pi(\mathcal{R})$ is simply terminating then $\mathcal{R}$ is DP simply terminating and again, a termination proof by dependency pairs works with the *same* argument filtering π and the simplification order used to orient $\mathrm{AFT}_\pi(\mathcal{R})$. Thus, the argument filtering transformation has no advantage compared to dependency pairs. We start with two easy lemmata.[1]

Lemma 5. *Let s and t be terms. If $s \trianglelefteq_\pi t$ then $\pi(s) \trianglelefteq \pi(t)$.*

Proof. By induction on the definition of $\trianglelefteq_\pi$. If $s = t$ then the result is trivial. Suppose $t = f(t_1, \ldots, t_n)$, $s \trianglelefteq_\pi t_i$, and $f \not{\perp}_\pi i$. The induction hypothesis yields $\pi(s) \trianglelefteq \pi(t_i)$. Because $f \not{\perp}_\pi i$, $\pi(t_i)$ is a subterm of $\pi(t)$ and thus $\pi(s) \trianglelefteq \pi(t)$ as desired.   □

Lemma 6. *Let r be a term. For every subterm t of r with a defined root symbol there exists a term $u \in \mathrm{AFT}_\pi(r)$ such that $\pi(t) \trianglelefteq u$.*

Proof. We use induction on the structure of r. In the base case we must have $t = r$ and we take $u = \pi(r)$. Note that $\pi(r) \in \mathrm{AFT}_\pi(r)$ because $\mathrm{root}(\bar\pi(r)) = \mathrm{root}(r)$ is defined. In the induction step we distinguish two cases. If $t \trianglelefteq_\pi r$ then we also have $t \trianglelefteq_{\bar\pi} r$ and hence $\bar\pi(t) \trianglelefteq \bar\pi(r)$ by Lemma 5. As $\mathrm{root}(\bar\pi(t)) = \mathrm{root}(t)$ is defined, the term $\bar\pi(r)$ contains a defined symbol. Hence $\pi(r) \in \mathrm{AFT}_\pi(r)$ by definition and thus we can take $u = \pi(r)$. In the other case t is not a preserved subterm of r. This implies that $t \trianglelefteq s$ for some outermost non-preserved subterm s of r. The induction hypothesis, applied to $t \trianglelefteq s$, yields a term $u \in \mathrm{AFT}_\pi(s)$ such that $\pi(t) \trianglelefteq u$. We have $\mathrm{AFT}_\pi(s) \subseteq \mathrm{AFT}_\pi(r)$ and hence u satisfies the requirements.   □

Theorem 7. *Let $\mathcal{R}$ be a TRS and π an argument filtering. If $\mathrm{AFT}_\pi(\mathcal{R})$ is simply terminating then $\mathcal{R}$ is DP simply terminating.*

Proof. Like in the proof of Theorem 4 there is no need to consider the dependency graph. Let $\succ$ be a simplification order that shows the (simple) termination of $\mathrm{AFT}_\pi(\mathcal{R})$. We claim that the dependency pair constraints are satisfied by π and $\succ$, where π and $\succ$ are extended to $\mathcal{F}^\sharp$ by treating each marked symbol F in the same way as the corresponding unmarked f. For rewrite rules $l \to r \in \mathcal{R}$ we have $\pi(l) \succ \pi(r)$ as $\pi(l) \to \pi(r) \in \mathrm{AFT}_\pi(\mathcal{R})$. Let $l^\sharp \to t^\sharp$ be a dependency pair of $\mathcal{R}$, originating from the rewrite rule $l \to r$. We show that $\pi(l) \succ \pi(t)$ and hence,

---

[1] Argumentations similar to the proofs of Lemma 6 and Theorem 7 can also be found in [16, Lemma 4.3 and Theorem 4.4]. However, [16] contains neither Theorem 7 nor our main Theorem 8, since the authors do not compare the argument filtering transformation with the dependency pair approach.

$\pi(l^\sharp) \succ \pi(t^\sharp)$ as well. We have $t \trianglelefteq r$. Since root(t) is a defined function symbol by the definition of dependency pairs, we can apply Lemma 6. This yields a term $u \in \mathrm{AFT}_\pi(r)$ such that $\pi(t) \trianglelefteq u$. The subterm property of $\succ$ yields $u \succeq \pi(t)$. By definition, $\pi(l) \to u \in \mathrm{AFT}_\pi(\mathcal{R})$ and thus $\pi(l) \succ u$ by compatibility of $\succ$ with $\mathrm{AFT}_\pi(\mathcal{R})$. Hence $\pi(l) \succ \pi(t)$ as desired. $\square$

Note that in the above proof we did not make use of the possibility to treat marked symbols differently from unmarked ones. This clearly shows why the dependency pair technique is much more powerful than the argument filtering transformation; there are numerous DP simply terminating TRSs which are no longer DP simply terminating if we are forced to interpret a defined function symbol and its marked version in the same way. As a simple example, consider

$$\mathcal{R}_1 = \left\{ \begin{array}{ll} x - 0 & \to x \\ x - \mathsf{s}(y) \to \mathsf{p}(x - y) & \quad 0 \div \mathsf{s}(y) \to 0 \\ \mathsf{p}(\mathsf{s}(x)) \to x & \quad \mathsf{s}(x) \div \mathsf{s}(y) \to \mathsf{s}((x - y) \div \mathsf{s}(y)) \end{array} \right\}.$$

Note that $\mathcal{R}_1$ is not simply terminating as the rewrite step $\mathsf{s}(x) \div \mathsf{s}(\mathsf{s}(x)) \to \mathsf{s}((x-\mathsf{s}(x))\div\mathsf{s}(\mathsf{s}(x)))$ is self-embedding. To obtain a terminating TRS $\mathrm{AFT}_\pi(\mathcal{R}_1)$, the rule $\mathsf{p}(\mathsf{s}(x)) \to x$ enforces $\mathsf{p} \not\perp_\pi 1$ and $\mathsf{s} \not\perp_\pi 1$. From $\mathsf{p} \not\perp_\pi 1$ and the rules for $-$ we infer that $\pi(-) = [1,2]$. But then, for all choices of $\pi(\div)$, the rule $\mathsf{s}(x) \div \mathsf{s}(y) \to \mathsf{s}((x-y) \div \mathsf{s}(y))$ is transformed into one that is incompatible with a simplification order. So $\mathrm{AFT}_\pi(\mathcal{R}_1)$ is not simply terminating for any π. (Similarly, dummy elimination cannot transform this TRS into a simply terminating one either.) On the other hand, DP simple termination of $\mathcal{R}_1$ is easily shown by the argument filtering $\pi(\mathsf{p}) = 1$, $\pi(-) = 1$, $\pi(-^\sharp) = [1,2]$, and $\pi(f) = [1,\ldots,\mathrm{arity}(f)]$ for every other function symbol f in combination with the recursive path order. This example illustrates that treating defined symbols and their marked versions differently is often required in order to benefit from the fact that the dependency pair approach only requires *weak* decreasingness for the rules of $\mathcal{R}_1$.

The next question we address is whether the argument filtering transformation can be useful as a preprocessing step for the dependency pair technique. Surprisingly, the answer to this question is yes. Consider the TRS

$$\mathcal{R}_2 = \left\{ \begin{array}{lll} \mathsf{f}(\mathsf{a}) \to \mathsf{f}(\mathsf{c}(\mathsf{a})) & \mathsf{f}(\mathsf{a}) \to \mathsf{f}(\mathsf{d}(\mathsf{a})) & \mathsf{e}(\mathsf{g}(x)) \to \mathsf{e}(x) \\ \mathsf{f}(\mathsf{c}(x)) \to x & \mathsf{f}(\mathsf{d}(x)) \to x & \\ \mathsf{f}(\mathsf{c}(\mathsf{a})) \to \mathsf{f}(\mathsf{d}(\mathsf{b})) & \mathsf{f}(\mathsf{c}(\mathsf{b})) \to \mathsf{f}(\mathsf{d}(\mathsf{a})) & \end{array} \right\}.$$

This TRS is not DP simply terminating which can be seen as follows. The dependency pair $\mathsf{E}(\mathsf{g}(x)) \to \mathsf{E}(x)$ constitutes a cluster in the dependency graph of $\mathcal{R}_2$. Hence, if $\mathcal{R}_2$ were DP simply terminating, there would be an argument filtering π and a simplification order $\succ$ such that (amongst others)

$$\begin{array}{ll} \pi(\mathsf{f}(\mathsf{a})) \succeq \pi(\mathsf{f}(\mathsf{c}(\mathsf{a}))) & \pi(\mathsf{f}(\mathsf{a})) \succeq \pi(\mathsf{f}(\mathsf{d}(\mathsf{a}))) \\ \pi(\mathsf{f}(\mathsf{c}(x))) \succeq x & \pi(\mathsf{f}(\mathsf{d}(x))) \succeq x \\ \pi(\mathsf{f}(\mathsf{c}(\mathsf{a}))) \succeq \pi(\mathsf{f}(\mathsf{d}(\mathsf{b}))) & \pi(\mathsf{f}(\mathsf{c}(\mathsf{b}))) \succeq \pi(\mathsf{f}(\mathsf{d}(\mathsf{a}))) \end{array}$$

From $\pi(\mathsf{f}(\mathsf{c}(x))) \succeq x$ and $\pi(\mathsf{f}(\mathsf{d}(x))) \succeq x$ we infer that $\mathsf{f} \not\perp_\pi 1$, $\mathsf{c} \not\perp_\pi 1$, and $\mathsf{d} \not\perp_\pi 1$. Hence $\pi(\mathsf{f}(\mathsf{a})) \succeq \pi(\mathsf{f}(\mathsf{c}(\mathsf{a})))$ and $\pi(\mathsf{f}(\mathsf{a})) \succeq \pi(\mathsf{f}(\mathsf{d}(\mathsf{a})))$ can only be satisfied if $\pi(c) = \pi(d) = 1$. But then $\pi(f(c(a))) \succeq \pi(f(d(b)))$ and $\pi(f(c(b))) \succeq \pi(f(d(a)))$ amount to either $f(a) \succeq f(b)$ and $f(b) \succeq f(a)$ (if $\pi(f) = [1]$) or $a \succeq b$ and $b \succeq a$ (if $\pi(f) = 1$). Since $f(a) \neq f(b)$ and $a \neq b$ the required simplification order does not exist.

On the other hand, if $\pi(e) = 1$ then $\mathrm{AFT}_\pi(\mathcal{R}_2)$ consists of the first six rewrite rules of $\mathcal{R}$ together with $g(x) \to x$. One easily verifies that there are no clusters in $\mathrm{DG}(\mathrm{AFT}_\pi(\mathcal{R}_2))$ and hence $\mathrm{AFT}_\pi(\mathcal{R}_2)$ is trivially DP simply terminating.

Definition 6. *An argument filtering π is called* collapsing *if $\pi(f) = i$ for some defined function symbol f.*

The argument filtering in the previous example is collapsing. In the remainder of this section we show that for non-collapsing argument filterings the implication "$\mathrm{AFT}_\pi(\mathcal{R})$ is DP simply terminating $\Rightarrow \mathcal{R}$ is DP simply terminating" is valid. Thus, using the argument filtering transformation with a non-collapsing π as a preprocessing step to the dependency pair technique has no advantages.

First we prove a lemma to relate the dependency pairs of $\mathcal{R}$ and $\mathrm{AFT}_\pi(\mathcal{R})$.

Lemma 7. *Let π be a non-collapsing argument filtering. If $l^\sharp \to t^\sharp \in \mathrm{DP}(\mathcal{R})$ then $\pi(l)^\sharp \to \pi(t)^\sharp \in \mathrm{DP}(\mathrm{AFT}_\pi(\mathcal{R}))$.*

Proof. By definition there is a rewrite rule $l \to r \in \mathcal{R}$ and a subterm $t \trianglelefteq r$ with defined root symbol. According to Lemma 6 there exists a term $u \in \widehat{\mathrm{AFT}}_\pi(r)$ such that $\pi(t) \trianglelefteq u$. Thus, $\pi(l) \to u \in \mathrm{AFT}_\pi(\mathcal{R})$. Since π is non-collapsing, $\mathrm{root}(\pi(t)) = \mathrm{root}(t)$. Hence, as $\mathrm{root}(t)$ is defined, $\pi(l)^\sharp \to \pi(t)^\sharp$ is a dependency pair of $\mathrm{AFT}_\pi(\mathcal{R})$. □

Example $\mathcal{R}_2$ shows that the above lemma is not true for arbitrary argument filterings. The reason is that $e(g(x))^\sharp \to e(x)^\sharp$ is a dependency pair of $\mathcal{R}$, but with $\pi(e) = 1$ there is no corresponding dependency pair in $\mathrm{AFT}_\pi(\mathcal{R})$.

The next three lemmata will be used to show that clusters in $\mathrm{DG}(\mathcal{R})$ correspond to clusters in $\mathrm{DG}(\mathrm{AFT}_\pi(\mathcal{R}))$.

Definition 7. *Given an argument filtering π and a substitution σ, the substitution σ_π is defined as $\pi \circ \sigma$ (i.e., σ is applied first).*

Lemma 8. *For all terms t, argument filterings π, and substitutions σ, $\pi(t\sigma) = \pi(t)\sigma_\pi$.*

Proof. Easy induction on the structure of t. □

Lemma 9. *Let $\mathcal{R}$ be a TRS and π a non-collapsing argument filtering. If $s \to_\mathcal{R}^* t$ then $\pi(s) \to_{\mathrm{AFT}_\pi(\mathcal{R})}^* \pi(t)$.*

Proof. It suffices to show that $\pi(s) \to_{\mathrm{AFT}_\pi(\mathcal{R})}^* \pi(t)$ whenever $s \to_\mathcal{R}^* t$ consists of a single rewrite step. Let $s = C[l\sigma]$ and $t = C[r\sigma]$ for some context C, rewrite rule $l \to r \in \mathcal{R}$, and substitution σ. We use induction on C. If C is the empty context, then $\pi(s) = \pi(l\sigma) = \pi(l)\sigma_\pi$ and $\pi(t) = \pi(r\sigma) = \pi(r)\sigma_\pi$ according to

Lemma 8. As $\pi(l) \to \pi(r) \in \mathrm{AFT}_\pi(\mathcal{R})$, we have $\pi(s) \to_{\mathrm{AFT}_\pi(\mathcal{R})} \pi(t)$. Suppose $C = f(s_1, \ldots, C', \ldots, s_n)$ where C' is the i-th argument of C. If $f \perp_\pi i$ then $\pi(s) = \pi(t)$. If $\pi(f) = i$ (which is possible for constructors f) then $\pi(s) = \pi(C'[l\sigma])$ and $\pi(t) = \pi(C'[r\sigma])$, and thus we obtain $\pi(s) \to^*_{\mathrm{AFT}_\pi(\mathcal{R})} \pi(t)$ from the induction hypothesis. In the remaining case we have $\pi(f) = [i_1, \ldots, i_m]$ with $i_j = i$ for some j and hence $\pi(s) = f(\pi(s_{i_1}), \ldots, \pi(C'[l\sigma]), \ldots, \pi(s_{i_m}))$ and $\pi(t) = f(\pi(s_{i_1}), \ldots, \pi(C'[r\sigma]), \ldots, \pi(s_{i_m}))$. In this case we obtain $\pi(s) \to^*_{\mathrm{AFT}_\pi(\mathcal{R})} \pi(t)$ from the induction hypothesis as well. $\qquad\square$

The following lemma states that if two dependency pairs are connected in $\mathcal{R}$'s dependency graph, then the corresponding pairs are connected in the dependency graph of $\mathrm{AFT}_\pi(\mathcal{R})$ as well.

Lemma 10. *Let $\mathcal{R}$ be a TRS, π a non-collapsing argument filtering, and s, t be terms with defined root symbols. If $s^\sharp \sigma \to^*_\mathcal{R} t^\sharp \sigma$ for some substitution σ then $\pi(s)^\sharp \sigma_\pi \to^*_{\mathrm{AFT}_\pi(\mathcal{R})} \pi(t)^\sharp \sigma_\pi$.*

Proof. We have $s = f(s_1, \ldots, s_n)$ and $t = f(t_1, \ldots, t_n)$ for some n-ary defined function symbol f with $s_i \sigma \to^*_\mathcal{R} t_i \sigma$ for all $1 \leqslant i \leqslant n$. Let $\pi(f) = [i_1, \ldots, i_m]$. This implies $\pi(s\sigma)^\sharp = f^\sharp(\pi(s_{i_1}\sigma), \ldots, \pi(s_{i_m}\sigma))$ and $\pi(t\sigma)^\sharp = f^\sharp(\pi(t_{i_1}\sigma), \ldots, \pi(t_{i_m}\sigma))$. From the preceding lemma we know that $\pi(s_{i_j}\sigma) \to^*_{\mathrm{AFT}_\pi(\mathcal{R})} \pi(t_{i_j}\sigma)$ for all $1 \leqslant j \leqslant m$. Hence, using Lemma 8, $\pi(s)^\sharp \sigma_\pi = \pi(s\sigma)^\sharp \to^*_{\mathrm{AFT}_\pi(\mathcal{R})} \pi(t\sigma)^\sharp = \pi(t)^\sharp \sigma_\pi$. $\qquad\square$

Now we can finally prove the main theorem of this section.

Theorem 8. *Let $\mathcal{R}$ be a TRS and π a non-collapsing argument filtering. If $\mathrm{AFT}_\pi(\mathcal{R})$ is DP simply terminating then $\mathcal{R}$ is DP simply terminating.*

Proof. Let $\mathcal{C}$ be a cluster in $\mathrm{DG}(\mathcal{R})$. According to Lemmata 7 and 10, there is a corresponding cluster in $\mathrm{DG}(\mathrm{AFT}_\pi(\mathcal{R}))$, which we denote by $\pi(\mathcal{C})$. By assumption, there exist an argument filtering π' and a simplification order $\succ$ such that $\pi'(\mathrm{AFT}_\pi(\mathcal{R}) \cup \pi(\mathcal{C})) \subseteq \succeq$ and $\pi'(\pi(\mathcal{C})) \cap \succ \neq \varnothing$. We define an argument filtering π'' for $\mathcal{R}$ as the composition of π and π'. For a precise definition, let $\flat$ denote the unmarking operation, i.e., $f^\flat = f$ and $(f^\sharp)^\flat = f$ for all $f \in \mathcal{F}$. Then for all $f \in \mathcal{F}^\sharp$ we define

$$\pi''(f) = \begin{cases} [i_{j_1}, \ldots, i_{j_k}] & \text{if } \pi(f^\flat) = [i_1, \ldots, i_m] \text{ and } \pi'(f) = [j_1, \ldots, j_k], \\ i_j & \text{if } \pi(f^\flat) = [i_1, \ldots, i_m] \text{ and } \pi'(f) = j, \\ i & \text{if } \pi(f) = i. \end{cases}$$

It is not difficult to show that $\pi''(t) = \pi'(\pi(t))$ and $\pi''(t^\sharp) = \pi'(\pi(t)^\sharp)$ for all terms t without marked symbols. We claim that π'' and $\succ$ satisfy the constraints for $\mathcal{C}$, i.e., $\pi''(\mathcal{R} \cup \mathcal{C}) \subseteq \succeq$ and $\pi''(\mathcal{C}) \cap \succ \neq \varnothing$. These two properties follow from the two assumptions $\pi'(\mathrm{AFT}_\pi(\mathcal{R}) \cup \pi(\mathcal{C})) \subseteq \succeq$ and $\pi'(\pi(\mathcal{C})) \cap \succ \neq \varnothing$ in conjunction with the obvious inclusion $\pi(\mathcal{R}) \subseteq \mathrm{AFT}_\pi(\mathcal{R})$. $\qquad\square$

Theorem 8 also holds for the *estimated* dependency graph instead of the real dependency graph.

5   Conclusion

In this paper, we have compared two transformational techniques for termination proofs, viz. dummy elimination [11] and the argument filtering transformation [16], with the dependency pair technique of Arts and Giesl [1–3]. Essentially, all these techniques transform a given TRS into new inequalities or rewrite systems which then have to be oriented by suitable well-founded orders. Virtually all well-founded orders which can be generated automatically are simplification orders. As our focus was on *automated* termination proofs, we therefore investigated the strengths of these three techniques when combined with simplification orders.

To that end, we showed that whenever an automated termination proof is possible using dummy elimination or the argument filtering transformation, then a corresponding termination proof can also be obtained by dependency pairs. Thus, the dependency pair technique is more powerful than dummy elimination or the argument filtering transformation on their own.

Moreover, we examined whether dummy elimination or the argument filtering transformation would at least be helpful as a preprocessing step to the dependency pair technique. We proved that for dummy elimination and for an argument filtering transformation with a non-collapsing argument filtering, this is not the case. In fact, whenever there is a (pre)order satisfying the dependency pair constraints for the rewrite system resulting from dummy elimination or a non-collapsing argument filtering transformation, then the *same* (pre)order also satisfies the dependency pair constraints for the original TRS.

As can be seen from the proofs of our main theorems, this latter result even holds for arbitrary (i.e., non-simplification) (pre)orders. Thus, in particular, Theorems 5 and 8 also hold for *DP quasi-simple termination* [13]. This notion captures those TRSs where the dependency pair constraints are satisfied by an arbitrary simplification preorder $\underset{\sim}{\succ}$ (instead of just a preorder $\succeq$ where the equivalence relation is syntactic equality as in *DP simple termination*).

Future work will include a further investigation on the usefulness of collapsing argument filtering transformations as a preprocessing step to dependency pairs. Note that our counterexample $\mathcal{R}_2$ is DP quasi-simply terminating (but not DP simply terminating). In other words, at present it is not clear whether the argument filtering transformation is useful as a preprocessing step to the dependency pair technique if one admits arbitrary simplification *pre*orders to solve the generated constraints. However, an extension of Theorem 8 to DP quasi-simple termination and to *collapsing* argument filterings π is not straightforward, since clusters of dependency pairs in $\mathcal{R}$ may disappear in $\mathrm{AFT}_\pi(\mathcal{R})$ (i.e., Lemma 7 does not hold for collapsing argument filterings). We also intend to examine the relationship between dependency pairs and other transformation techniques such as "freezing" [20].

Acknowledgements. Jürgen Giesl is supported by the DFG under grant GI 274/4-1. Aart Middeldorp is partially supported by the Grant-in-Aid for Scientific Research C(2) 11680338 of the Ministry of Education, Science, Sports and Culture of Japan.

References

1. T. Arts and J. Giesl, *Automatically Proving Termination where Simplification Orderings Fail*, Proc. 7th TAPSOFT, Lille, France, LNCS 1214, pp. 261–273, 1997.
2. T. Arts and J. Giesl, *Termination of Term Rewriting Using Dependency Pairs*, Theoretical Computer Science 236, pp. 133–178, 2000. Long version available at www.inferenzsysteme.informatik.tu-darmstadt.de/~reports/ibn-97-46.ps.
3. T. Arts and J. Giesl, *Modularity of Termination Using Dependency Pairs*, Proc. 9th RTA, Tsukuba, Japan, LNCS 1379, pp. 226–240, 1998.
4. F. Baader and T. Nipkow, *Term Rewriting and All That*, Cambridge University Press, 1998.
5. F. Bellegarde and P. Lescanne, *Termination by Completion*, Applicable Algebra in Engineering, Communication and Computing 1, pp. 79–96, 1990.
6. A. Ben Cherifa and P. Lescanne, *Termination of Rewriting Systems by Polynomial Interpretations and its Implementation*, Science of Computer Programming 9, pp. 137–159, 1987.
7. N. Dershowitz, *Orderings for Term-Rewriting Systems*, Theoretical Computer Science 17, pp. 279–301, 1982.
8. N. Dershowitz, *Termination of Rewriting*, Journal of Symbolic Computation 3, pp. 69–116, 1987.
9. J. Dick, J. Kalmus, and U. Martin, *Automating the Knuth Bendix Ordering*, Acta Informatica 28, pp. 95–119, 1990.
10. M.C.F. Ferreira, *Termination of Term Rewriting: Well-foundedness, Totality and Transformations*, Ph.D. thesis, Utrecht University, The Netherlands, 1995.
11. M.C.F. Ferreira and H. Zantema, *Dummy Elimination: Making Termination Easier*, Proc. 10th FCT, Dresden, Germany, LNCS 965, pp. 243–252, 1995.
12. J. Giesl, *Generating Polynomial Orderings for Termination Proofs*, Proc. 6th RTA, Kaiserslautern, Germany, LNCS 914, pp. 426–431, 1995.
13. J. Giesl and E. Ohlebusch, *Pushing the Frontiers of Combining Rewrite Systems Farther Outwards*, Proc. 2nd FROCOS, 1998, Amsterdam, The Netherlands, Studies in Logic and Computation 7, Research Studies Press, Wiley, pp. 141–160, 2000.
14. S. Kamin and J.J. Lévy, *Two Generalizations of the Recursive Path Ordering*, unpublished manuscript, University of Illinois, USA, 1980.
15. D.E. Knuth and P. Bendix, *Simple Word Problems in Universal Algebras*, in: Computational Problems in Abstract Algebra (ed. J. Leech), Pergamon Press, pp. 263–297, 1970.
16. K. Kusakari, M. Nakamura, and Y. Toyama, *Argument Filtering Transformation*, Proc. 1st PPDP, Paris, France, LNCS 1702, pp. 48–62, 1999.
17. D. Lankford, *On Proving Term Rewriting Systems are Noetherian*, Report MTP-3, Louisiana Technical University, Ruston, USA, 1979.
18. A. Middeldorp, H. Ohsaki, and H. Zantema, *Transforming Termination by Self-Labelling*, Proc. 13th CADE, New Brunswick (New Jersey), USA, LNAI 1104, pp. 373–387, 1996.
19. J. Steinbach, *Simplification Orderings: History of Results*, Fundamenta Informaticae 24, pp. 47–87, 1995.
20. H. Xi, *Towards Automated Termination Proofs Through "Freezing"*, Proc. 9th RTA, Tsukuba, Japan, LNCS 1379, pp. 271–285, 1998.
21. H. Zantema, *Termination of Term Rewriting: Interpretation and Type Elimination*, Journal of Symbolic Computation 17, pp. 23–50, 1994.
22. H. Zantema, *Termination of Term Rewriting by Semantic Labelling*, Fundamenta Informaticae 24, pp. 89–105, 1995.

Extending Decision Procedures with Induction Schemes

Deepak Kapur[1*] and Mahadavan Subramaniam[2]

[1] Department of Computer Science
University of New Mexico, Albuquerque, NM
kapur@cs.unm.edu
[2] HAL Computer Systems, Fujitsu Inc, Campbell, CA
subu@hal.com

Abstract. Families of function definitions and conjectures based in quantifier-free decidable theories are identified for which inductive validity of conjectures can be decided by the *cover set* method, a heuristic implemented in a rewrite-based induction theorem prover *Rewrite Rule Laboratory (RRL)* for mechanizing induction. Conditions characterizing definitions and conjectures are syntactic, and can be easily checked, thus making it possible to determine a priori whether a given conjecture can be decided. The concept of a $\mathcal{T}$-based function definition is introduced that consists of a finite set of terminating complete rewrite rules of the form $f(s_1, \cdots, s_m) \to r$, where $s_1, \cdots, s_m$ are interpreted terms from a decidable theory $\mathcal{T}$, and r is either an interpreted term or has non-nested recursive calls to f with all other function symbols from $\mathcal{T}$. Two kinds of conjectures are considered. *Simple* conjectures are of the form $f(x_1, \cdots x_m) = t$, where f is $\mathcal{T}$-based, x_i's are distinct variables, and t is interpreted in $\mathcal{T}$. *Complex* conjectures differ from simple conjectures in their left sides which may contain many function symbols whose definitions are $\mathcal{T}$-based and the nested order in which these function symbols appear in the left sides have the *compatibility property* with their definitions.

The main objective is to ensure that for each induction subgoal generated from a conjecture after selecting an induction scheme, the resulting formula can be simplified so that induction hypothesis(es), whenever needed, is applicable, and the result of this application is a formula in $\mathcal{T}$. Decidable theories considered are the quantifier-free theory of Presburger arithmetic, congruence closure on ground terms (with or without associative-commutative operators), propositional calculus, and the quantifier-free theory of constructors (mostly, free constructors as in the case of finite lists and finite sequences). A byproduct of the approach is that it can predict the structure of intermediate lemmas needed for automatically deciding this subclass of conjectures. Several examples over lists, numbers and of properties involved in establishing the number-theoretic correctness of arithmetic circuits are given.

* Partially supported by the National Science Foundation Grant nos. CCR-9712396, CCR-9712366, CCR-9996150, and CDA-9503064.

D. McAllester (Ed.): CADE-17, LNAI 1831, pp. 324–345, 2000.

1   Introduction

Inductive reasoning is ubiquitous in verifying properties of computations realized in hardware and software. Automation of inductive reasoning is hampered by the fact that proofs by induction need an appropriate selection of variables for performing induction and a suitable induction scheme, as well as intermediate lemmas. It is well-known that inductive reasoning often needs considerable user guidance, because of which its automation has been a major challenge. A lot of effort has been spent on mechanizing induction in theorem provers (e.g., *Nqthm*, *ACL2*, *RRL*, *INKA*, *Oyster-Clam*), and induction heuristics in these provers have been successfully used to establish several nontrivial properties. However, the use of induction as a mechanized rule of inference is seriously undermined due to the lack of automation in using this rule. Many reasoning tools including model checkers (in conjunction with decision procedures), invariant generators, deductive synthesis tools preclude induction for lack of automation. This severely limits the reasoning capability of these tools. In many cases inductive properties are established outside these tools manually.

For hardware circuit descriptions, need for inductive reasoning arises when reasoning is attempted about a circuit description parameterized by data width and/or generic components. In protocol verification, induction is often needed when a protocol has to be analyzed for a large set of processors (or network channels). Inductive reasoning in many such cases is not as challenging as in software specifications as well as recursive and loop programs.

This paper is an attempt to address this limitation of these automated tools while preserving their automation, and without having the full generality of a theorem prover. It is shown how decision procedures for simple theories about certain data structures, e.g., numbers, booleans, finite lists, finite sequences, can be enhanced to include induction techniques with the objective that proofs employing such techniques can be done automatically. The result is an extended decision procedure with a built-in induction scheme, implying that an inductive theorem prover can be run in push-button mode as well. We believe the proposed approach can substantially enhance the reasoning power of tools built using decision procedures and model-checkers without losing the advantage of automation.

This cannot be done, however, in general. Conditions are identified on function definitions and conjectures which guarantee such automation. It becomes possible to determine *a priori* whether a given conjecture can be decided automatically, thus predicting the success or failure of using a theorem proving strategy. That is the main contribution of the paper. A byproduct of the proposed approach is that in case of a failure of the theorem proving strategy, it can predict for a subclass of conjectures, the structure of lemmas needed for proof attempts to succeed.

The proposed approach is based on two main ideas: First, terminating recursive definitions of function symbols as rewrite rules oriented using a well-founded ordering, can be used to generate induction schemes providing useful induction hypotheses for proofs by induction; this idea is the basis for the *cover set* method proposed in [16] and implemented in a rewrite-rule based induction theorem prover *Rewrite Rule Laboratory (RRL)* [14]. Second, for inductive proofs of conjectures satisfying certain conditions, induction schemes generated from $\mathcal{T}$-*based* recursive function definitions (a concept characterized precisely below), lead to subgoals in $\mathcal{T}$ (after the application of induction hypotheses), where $\mathcal{T}$ is a decidable theory. These conditions are based on the structure of the function definitions and the conjecture.

The concept of a $\mathcal{T}$-*based* function definition is introduced to achieve the above objective. It is shown that conjectures of the form $f(x_1, \cdots, x_m) = r$, where f has a $\mathcal{T}$-based function definition, x_i's are distinct variables, and r is an *interpreted* term in $\mathcal{T}$ (i.e., r includes only function symbols from $\mathcal{T}$), can be decided using the cover set method[1]. The reason for focusing on such simple conjectures is that there is only one induction scheme to be considered by the cover set method. It might be possible to relax this restriction and consider more complicated conjectures insofar as they suggest one induction scheme and the induction hypothesis(es) is applicable to the subgoals after using the function definitions for simplification.

Decidable theories considered are the quantifier-free theory of Presburger arithmetic, congruence closure on ground terms (with or without associative-commutative operators), propositional calculus as well as the quantifier-free theory of constructors (mostly free constructors as in the case of finite lists and finite sequences). For each such theory, decision procedures exist, and *RRL*, for instance, has an implementation of them integrated with rewriting [7, 8].

Below, we review two examples providing an overview of the proposed approach, the subclass of conjectures and definitions which can be considered.

1.1   A Simple Conjecture

Consider the following very simple but illustrative example.

        (C1): double(m) = m + m,

where double is recursively defined using the rewrite rules:

        1. double(0)   --> 0,
        2. double(s(x)) --> s(s(double(x))).

A proof by induction with m as the induction variable and using the standard induction scheme (i.e., Peano's principle of mathematical induction over numbers), leads to one basis goal and one induction step goal. For the basis subgoal, the substitution m <- 0 gives double(0) = 0 + 0 which simplifies using the definition of double to a valid formula in Presburger arithmetic.

In the step subgoal, the conclusion generated using substitution m <- s(x) is double(s(x)) = s(x) + s(x), with the induction hypothesis got by the substitution m <- x, being double(x) = x + x.

---

[1] As will be shown later, it is not necessary to require that each argument to f be a distinct variable; instead, non-induction arguments can be interpreted terms that do not include variables in inductive positions of f.

By the second rule in the definition of **double**, the formula simplifies, the induction hypothesis applies, resulting again in a valid formula $s(s(x + x)) = s(x) + s(x)$ in Presburger arithmetic. Hence, (C1) is valid using Presburger arithmetic and the induction scheme of **double**.

Similarly, a conjecture **double(m) = m** can be decided to be false: the basis case will go through, but the formula resulting from the induction step and the application of the induction hypothesis is not valid.

The main features of the above conjectures and the definition of **double** are:

1. unambiguous induction scheme using which the formula can be decided,
2. induction hypotheses are strong enough to be applicable to induction subgoals, and finally
3. the formulas resulting from subgoals after applying the definition and the induction hypotheses are decidable.

Properties of $\mathcal{T}$-based definitions and simple conjectures ensure the above.

1.2   A Complex Conjecture

For considering complex conjectures including many function symbols with $\mathcal{T}$-based definitions, it becomes necessary to consider the interaction among their definitions based on their nesting order in conjectures. This aspect is captured by the *compatibility* property of function definitions (which is precisely characterized in a later section). The key insight is similar to the one observed of a simple conjecture. Compatible function definitions can be viewed as composing into a single $\mathcal{T}$-based function definition so that a complex conjecture can be viewed as being a simple conjecture in terms of the composed function as illustrated below.

$$(C2): \log(\exp2(m)) = m.$$

The definitions of the functions **log**, **exp2** (logarithm and exponentiation to the base 2 respectively) are as follows. Following the mathematical convention, **log** is defined on positive numbers only[2].

```
1. log(s(0))      --> 0,
2. log(x + x)     --> s(log(x)),
3. log(s(x + x)) --> s(log(x)).

4. exp2(0)        --> s(0),
5. exp2(s(x))     --> exp2(x) + exp2(x).
```

Unlike a simple conjecture, the left side of (C2) is a nested term. Again, there is only one induction variable **m**, and the induction scheme used for attempting

---

[2] This implies that the induction schemes generated using the definition of **log** can be used to decide the validity of conjectures over positive numbers only. For a detailed discussion of the use of cover sets and induction schemes derived from definitions such as **log**, please refer to [9].

a proof by induction is the principle of mathematical induction for numbers (suggested by the cover set of exp2, as explained later in section 2.1).

There is one basis goal and one induction step subgoal. The basis subgoal is log(exp2(0)) = 0. The left side rewrites using the definitions of exp2 and then log, resulting in a valid formula in Presburger arithmetic.

In the induction step, the conclusion is log(exp2(s(x))) = s(x) with the hypothesis being log(exp2(x)) = x. By the definition of exp2, exp2(s(x)) simplifies to exp2(x) + exp2(x). This subgoal will simplify to a formula in Presburger arithmetic if log(exp2(x) + exp2(x)) rewrites to s(log(exp2(x))) either as a part of the definition of log or as an intermediate lemma, and then, the induction hypothesis can apply. Such interaction between the definitions of log and exp2 is captured by *compatibility*. Since the definition of log includes such a rule, the validity of the induction step case and hence the validity of (C2) can be decided.

The validity of a closely related conjecture,

(C2'): exp2(log(m)) = m,

can be similarly decided since exp2 is compatible with log. An induction proof can be attempted using m as the induction variable as before. However, m can take only positive values since the function log is defined only for these. The induction scheme used is different from the principle of mathematical induction. Instead, it is based on the definition of log. There is a basis case corresponding to the number s(0), and two step cases corresponding to m being a positive even or a positive odd number respectively (this scheme is derived from the cover set of log as explained later in section 2.1).

In one of the induction step subgoals, the left side of the conclusion, exp2(log(s(x + x))) = s(x + x), rewrites by the definition of log to exp2(s(log(x))) which then rewrites to exp2(log(x)) + exp2(log(x)) to which the hypothesis exp2(log(x)) = x applies to produce the inconsistent Presburger arithmetic formula x + x = s(x + x).

As stated above, if log and exp2 are combined to develop the definition of the composed function log(exp2(x)) from their definitions, then (C2) is a simple conjecture about the composed function. Further, the definition of the composed function can be proved to be $\mathcal{T}$-based as well. So the decidability of the conjecture follows.

The notion of compatibility among the definitions of function symbols can be generalized to a *compatible sequence* of function symbols $f_1, \cdots, f_d$ where each f_i is compatible with f_{i+1} at j_i-th argument, $1 \leq i \leq d-1$. A conjecture $l = r$ can then be decided if the sequence of function symbols from the root of l to the innermost function symbol forms a compatible sequence, and r is an interpreted term in $\mathcal{T}$.

The proposed approach is discussed below in the framework of our theorem prover *Rewrite Rule Laboratory* (*RRL*), but the results should apply to other induction provers that rely on decision procedures and support heuristics for selecting induction schemes, e.g., Boyer and Moore's theorem prover *Nqthm*,

ACL2, and *INKA*. And, the proposed approach can be integrated in tools based on decision procedures and model checking.

The main motivation for this work comes from our work on verifying properties of generic, parameterized arithmetic circuits, including adders, multipliers, dividers and square root [12, 10, 11, 13]. The approach is illustrated on several examples including properties arising in proofs of arithmetic circuits, as well as commonly used properties of numbers and lists involving defined function symbols. A byproduct of this approach is that if a conjecture with the above mentioned restriction cannot be decided, structure of intermediate lemmas needed for deciding it can be predicted. This can aid in automatic lemma speculation.

1.3   Related Work

Boyer and Moore while describing the integration of linear arithmetic into *Nqthm* [3] discussed the importance of reasoning about formulas involving defined function symbols and interpreted terms. Many examples of such conjectures were discussed there. They illustrated how these examples can be done using the interaction of the theorem prover and the decision procedure. In this paper we have focussed on automatically deciding the validity of such conjectures. Most of the examples described there can be automatically decided using the proposed approach.

Fribourg [4] showed that properties of certain recursive predicates over lists expressed as logic programs along with numbers, can be decided. Most of the properties established there can be formulated as equational definitions and decided using the proposed approach. The procedure in [4] used bottom-up evaluation of logic programs which need not terminate if successor operation over numbers is included. The proposed approach does not appear to have this limitation.

Gupta's dissertation [1] was an attempt to integrate (a limited form of) inductive reasoning with a decision procedure for propositional formulas, e.g., ordered BDDs. She showed how properties about a certain subclass of circuits of arbitrary data width can be verified automatically. Properties automatically verified using her approach constitute a very limited subset, however.

2   Cover Set Induction

The cover set method is used to mechanize well-founded induction in *RRL*, and has been used to successfully perform proofs by induction in a variety of nontrivial application domains [12, 10, 11]. For attempting a proof by induction of a conjecture containing a subterm $t = f(x_1, \cdots, x_m)$, where each x_i is a distinct variable, an induction scheme from a complete definition of f given as a set of terminating rewrite rules, is generated as follows. There is one induction subgoal corresponding to each terminating rule in the definition of f. The induction conclusion is generated using the substitution from the left side of a rule, and an induction hypothesis is generated using the substitution from each recursive function call in the right side of the rule. Rules without recursive calls in their right sides lead to subgoals without any induction hypotheses (basis steps).

The recursive definitions of function symbols appearing in a conjecture can thus be used to come up with an induction scheme. Heuristics have been developed and implemented in RRL, which in conjunction with failure analysis of induction schemes and backtracking in case of failure, have been found appropriate for prioritizing induction schemes, automatically selecting the "most appropriate" induction scheme (thus selecting induction variables), and generating the proofs of many conjectures.

2.1 Definitions and Notation

Let $T(F, X)$ denote a set of terms where F is a finite set of function symbols and X is a set of variables. A term is either a variable $x \in X$, or a function symbol $f \in F$ followed by a finite sequence of terms, called arguments of f. Let $Vars(t)$ denote the variables appearing in a term t. The *subterms* of a term are the term itself and the subterms of its arguments. A *position* is a finite sequence of positive integers separated by ".".'s, which is used to identify a subterm in a term. The subterm of t at the position denoted by the empty sequence ϵ is t itself. If $f(t_1, \cdots, t_m)$ is a subterm at a position p then t_j is the subterm at the position $p.j$. Let $depth(t)$ denote the depth of t; $depth(t)$ is 0 if t is a variable or a constant (denoted by a function symbol with arity 0). $depth(f(t_1, \cdots, t_m)) = maximum(depth(t_i)) + 1$ for $1 \leq i \leq m$.

A term $f(t_1, \cdots, t_m)$ is called *basic* if each t_i is a distinct variable.

A substitution θ is a mapping from a finite set of variables to terms, denoted as $\{x_1 \leftarrow t_1, \cdots, x_m \leftarrow t_m\}$, $m \geq 0$, and $x_i's$ are distinct. θ applied on $s = f(s_1, \cdots, s_m)$ is $f(\theta(s_1), \cdots, \theta(s_m))$. Term s *matches* t under θ if $\theta(s) = t$. Terms s and t *unify* under θ if $\theta(s) = \theta(t)$.

A rewrite rule $s \rightarrow t$ is an ordered pair of terms (s, t) with $Vars(t) \subseteq Vars(s)$. A rule $s \rightarrow t$ is *applicable* to a term u iff for some substitution θ and position p in u, $\theta(s) = u|_p$. The application of the rule *rewrites* u to $u[p \leftarrow \theta(t)]$, the term obtained after replacing the subterm at position p in u by $\theta(t)$. A rewrite system R is a finite set of rewrite rules. R induces a relation among terms denoted $\rightarrow_R$. $s \rightarrow_R t$ denotes rewriting of s to t by a single application of a rule in R. $\rightarrow_R^+$ and $\rightarrow_R^*$ denote the transitive and the reflexive, transitive closure of $\rightarrow_R$.

The set F is partitioned into *defined* and *interpreted* function symbols. An interpreted function symbol comes from a decidable theory $\mathcal{T}$. A defined function symbol is defined by a finite set of terminating rewrite rules, and its definition is assumed to be complete. Term t is *interpreted* if all the function symbols in it are from $\mathcal{T}$. Underlying decidable theories are quantifier-free theories.

A equation $s = t$ is *inductively valid* (valid, henceforth) iff for each variable in it, whenever any ground term of the appropriate type is substituted into $s = t$, the instantiated equation is in the equational theory (modulo the decidable theory $\mathcal{T}$) of the definitions of function symbols in s, t. This is equivalent to $s = t$ holding in the initial model of the equations corresponding to the definitions and the decidable theory $\mathcal{T}$.

Given a complete function definition as a finite set of terminating rewrite rules $\{l_i \rightarrow r_i \mid l_i = f(s_1, \cdots, s_m), \ 1 \leq i \leq k\}$, the main steps of cover set method are

1. *Generating a Cover Set from a Function Definition:* A cover set associated with a function f is a finite set of triples. For a rule $l \rightarrow r$, where $l = f(s_1, \cdots, s_m)$ and $f(t_1^i, \cdots, t_m^i)$ is the ith recursive call to f in the right side r, the corresponding triple is $\langle \langle s_1, \cdots, s_m \rangle, \{\cdots, \langle t_1^i, \cdots, t_m^i \rangle, \cdots\}, \{\} \rangle$[3]. The second component of a triple is the empty set if there is no recursive call to f in r.

The cover sets of double, exp2 and log obtained from their definitions in section 1 are given below.

```
Cover(double):  {<<0>, {}, {}>, <<s(x)>, {<x>}, {}>,
Cover(exp2):    {<<0>, {}, {}>, <<s(x)>,{<x>}, {}>,
Cover(log):     {<<s(0)>, {}, {}>,
                <<x + x>, {<x>}, {}>, <<s(x + x)>, {<x>}, {}>}.
```

2. *Generating Induction Schemes using Cover Sets:* Given a conjecture C, a basic term $t = f(x_1, \cdots, x_m)$ appearing in C can be chosen for generating an induction scheme from the cover set of f. The variables in argument positions in t over which the definition of f recurses are called *induction variables* and the corresponding positions in t are called the *inductive* (or *changeable*) positions; other positions are called the *unchangeable* positions [2].

An induction scheme is a finite set of induction cases, each of the form $\langle \sigma_c, \{\theta_i\} \rangle$ generated from a cover set triple $\langle \langle s_1, \cdots, s_m \rangle, \{\cdots, \langle t_1^i, \cdots, t_m^i \rangle, \cdots\}, \{\} \rangle$ as follows [4]: $\sigma_c = \{x_1 \leftarrow s_1, \cdots, x_m \leftarrow s_m\}$, and , $\theta_i = \{x_1 \leftarrow t_1^i, \cdots, x_m \leftarrow t_m^i\}$.[5]

The induction scheme generated from the cover sets of double, exp2 is the principle of mathematical induction. The scheme generated from the cover set of log is different since the function log is defined over positive numbers only. There is one basis step—$\langle \{x \leftarrow s(0)\}, \{\} \rangle$, and $\langle \{x \leftarrow s(0)\}, \{\} \rangle$. There are two induction steps —$\langle \{x \leftarrow m + m\}, \{\{x \leftarrow m\}\} \rangle$, and $\langle \{x \leftarrow s(m + m)\}, x \leftarrow m \rangle$. The variable m is a positive number.

---

[3] The third component in a triple is a condition under which the conditional rewrite rule is applicable; for simplicity, we are considering only unconditional rewrite rules, so the third component is empty to mean that the rule is applicable whenever its left side matches. The proposed approach extends to conditional rewrite rules as well. See [16, 15, 13].

[4] The variables in a cover set triple are suitably renamed if necessary.

[5] To generate an induction scheme, it suffices to unify the subterm t in a conjecture C with the left side of each rule in the definition of f as well as with the recursive calls to f in the right side of the rule. This is always possible in case t is a basic term; but it even works if only variables in the induction positions of t are distinct.

3. *Generating Induction Subgoals using an Induction Scheme:* Each induction case generates an induction subgoal: σ_c is applied to the conjecture to generate the induction conclusion, whereas each substitution θ_i applied to the conjecture generates an induction hypothesis. Basis subgoals come from induction cases whose second component is empty.

The reader can consult examples (C1) and (C2) discussed above, and see how induction subgoals are generated using the induction schemes generated from the cover sets of double and exp2.

3 $\mathcal{T}$-Based Definitions and Simple Conjectures

Definition 1. *A definition of a function symbol f is $\mathcal{T}$-based in a decidable theory $\mathcal{T}$ iff for each rule $f(t_1, \cdots, t_m) \to r$ in the definition, each t_i, $1 \le i \le m$, is an interpreted term in $\mathcal{T}$, any recursive calls to f in r only have interpreted terms as arguments, and the abstraction of r defined as replacing recursive calls to f in r by variables is an interpreted term in $\mathcal{T}^6$.*

For examples, the definitions of double, log and exp2 given in Section 1 are $\mathcal{T}$-based over Presburger arithmetic. So is the definition of * given using rules,

```
1.  x * 0      --> 0,
2.  x * s(y) --> x + (x * y).
```

We abuse the notation slightly and call the functions themselves as being $\mathcal{T}$-based whenever their definitions are $\mathcal{T}$-based.

In order to use $\mathcal{T}$-based definitions for generating induction schemes, they should be complete as well as terminating over $\mathcal{T}$. For a brief discussion of how to perform such checks, see [9, 6]. It should be easy to see that terms in the cover set generated from a $\mathcal{T}$-based function definition are interpreted in $\mathcal{T}$.

3.1   Simple Conjectures

Definition 2. *A term is $\mathcal{T}$-based if it contains variables, interpreted function symbols from $\mathcal{T}$ and function symbols with $\mathcal{T}$-based definitions.*

Definition 3. *A conjecture $f(x_1, \cdots, x_m) = r$, where f has a $\mathcal{T}$-based definition, $x_i's$ are distinct variables and r is interpreted in $\mathcal{T}$, is called simple.*

Note that both sides of a simple conjecture are $\mathcal{T}$-based.

For example, the conjecture (C1): double(m) = m + m about double is simple over Presburger arithmetic, whereas the conjecture (C2): log(exp2(m)) = m about log is not simple over Presburger arithmetic.

For a simple conjecture, the cover set method proposes only one induction scheme, which is generated from the cover set derived from the definition of f.

---

[6] If r includes occurrences of cond, a special built-in operator in *RRL* for doing simulated conditional rewriting and automatic case analysis, then the first argument to cond is assumed to be an interpreted boolean term in $\mathcal{T}$.

Theorem 4. *A simple conjecture C over a decidable theory $\mathcal{T}$ can be decided using the cover set method.*

Proof. Given $f(x_1, \cdots, x_m) = r$, where r is interpreted in $\mathcal{T}$, from the cover set associated with the definition of f, an induction scheme can be generated and a proof can be attempted.

Since $\sigma_c(f(x_1, \cdots, x_m)) = l_i$ for some rule $l_i \rightarrow r_i$ in the definition of f, the left side of a basis subgoal

$$\sigma_c(f(x_1, \cdots, x_m)) = \sigma_c(r),$$

rewrites using the rule to r_i, an interpreted term in $\mathcal{T}$. The result is a decidable formula in $\mathcal{T}$. This part of the proof exploits the fact that the right side of a simple conjecture, r, is an interpreted term in $\mathcal{T}$.

For each induction step subgoal derived from a rule $l_j \rightarrow r_j$ in the definition of f where $r_j = h(\cdots, f(\cdots), \cdots)$, with recursive calls to f, the conclusion is $\sigma_c(f(x_1, \cdots, x_m)) = \sigma_c(r)$; $\sigma_c(f(x_1, \cdots, x_m)) = l_j$ with $\theta_i(f(x_1, \cdots, x_m)) = \theta_i(r)$ being an induction hypothesis corresponding to each recursive call to f in r_j. The left side of the conclusion simplifies by the corresponding rule to r_j which includes an occurrence of $\theta_i(f(x_1, \cdots, x_m))$ as a subterm at a position p_i in r_j. The application of these hypotheses generates the formula $r_j[p_1 \leftarrow \theta_1(r), \cdots, p_k \leftarrow \theta_k(r)] = \sigma_c(r)$ of $\mathcal{T}$, since the abstraction of r_j after recursive calls to f have been replaced by variables, is an interpreted term in $\mathcal{T}$.

Since every basis and induction step subgoal generated by the cover set method can be decided in $\mathcal{T}$, the conjecture C can be decided by the cover set method.□

As the above proof suggests, a slightly more general class of simple conjectures can be decided. Not all the arguments to f need be distinct variables. It suffices if the inductive positions in f are distinct variables, and the other positions are interpreted and do not contain variables appearing in the inductive positions. The above proof would still work.

For example, the following conjecture

```
(C3): append(n, nil) = n,
```

is not simple. The validity of the conjecture (C3) can be decided over the theory of free constructors `nil` and `cons` for lists. The definition of **append** is

```
1. append(nil,x)      --> x,
2. append(cons(x,y),z) --> cons(x,append(y,z)).
```

The requirement that unchangeable positions in a conjecture do not refer to the induction variables, seems essential for the above proof to work, as otherwise the application of the induction hypotheses may get blocked.

For example, the cover set method fails to disprove a conjecture such as

```
append(m, m) = m.
```

from the definition of the function **append**. An inductive proof attempt based on the cover set of the function **append** results in an induction step subgoal with the conclusion

append(cons(x, y), cons(x, y)) = cons(x, y),

and the hypothesis **append(y, y) = y**. The conclusion rewrites to cons(x, append(y, cons(x, y))) = cons(x, y) to which the hypothesis cannot be applied. Therefore, the cover set method fails since the induction step subgoal cannot be established.

4   Complex $\mathcal{T}$-Based Conjectures

To decide more complex conjectures by inductive methods, the choice of induction schemes have to be limited as well as the interaction among the function definitions have to be analyzed. In [15], such an analysis is undertaken to predict the failure of proof attempts a priori without actually attempting the proof. The notion of **compatibility of function definitions**, an idea illustrated in Section 1, is introduced for characterizing this interaction and for identifying intermediate steps in a proof which get blocked in the absence of additional lemmas.

In this section, we use related concepts to identify conditions under which conjectures such as (C2), more complex than the generalized simple conjectures discussed in section 3, can be decided. We first consider the interaction among two function symbols. This is subsequently extended to consider the interaction among a sequence of function symbols.

Definition 5. *A $\mathcal{T}$-based term t is composed if*

1. *t is a basic term $f(x_1, \cdots, x_m)$, where f is $\mathcal{T}$-based and $x_i's$ are distinct variables or*
2. *(a) $t = f(s_1, \cdots, t', \cdots, s_m)$, where t' is composed and is in an inductive position of a $\mathcal{T}$-based function f, and each s_i is an interpreted term, and*
   (b) variables x_i's appearing in the inductive positions of the basic subterm (in the innermost position) of t do not appear elsewhere in t. Other variables in unchangeable positions of the basic subterm can appear elsewhere in t.

For example, the left side of the conjecture (C2), log(exp2(m)), is a composed term of depth 2.

The first requirement in the above definition can be relaxed as in the case of simple conjectures. Only the variables in the inductive positions of a basic subterm in t have to be distinct; the terms interpreted in $\mathcal{T}$ can appear in the unchangeable positions of the basic subterm.

Given a conjecture of the form $l = r$, where l is composed and r is interpreted, it is easy to see that there is only one basic subterm in it whose outermost symbol is $\mathcal{T}$-based. The cover set method thus suggests only one induction scheme. We will first consider conjectures such as (C2) in which the left side l is of depth 2; later, conjectures in which l is of higher depth, are considered.

For a conjecture $f(t_1, \cdots, g(x_1, \cdots, x_k), \cdots, t_m) = r$, the interaction between the right sides of rules defining g and the left side of rules defining f must be considered, as seen in the proof of the conjecture (C2). The interaction is formalized below in the property of *compatibility*.

Definition 6. *A definition of f is compatible with a definition of g in its i-th argument in $\mathcal{T}$ iff for each right side r_g of a rule defining g, the following conditions hold*

1. *whenever r_g is interpreted, then $f(x_1, \cdots, r_g, \cdots, x_m)$ rewrites to an interpreted term in $\mathcal{T}$, and*
2. *whenever $r_g = h(s_1, \cdots, g(t_1, \cdots, t_k), \cdots, s_n)$, having a single recursive call to g, the definition of f rewrites $f(x_1, \cdots, h(s_1, \cdots, y, \cdots, s_n), \cdots, x_m)$ to $h'(u_1, \cdots, f(x_1, \cdots, y, \cdots, x_m), \cdots, u_n)$, where x_i's are distinct variables, h, h' are interpreted symbols in $\mathcal{T}$, and s_i, u_j's are interpreted terms of $\mathcal{T}$.[7] In case r_g has many recursive calls to g, say $h(s_1, \cdots, g(t_1, \cdots, t_k), \cdots, g(v_1, \cdots, v_k), \cdots, s_n)$, then the definition of f rewrites $f(x_1, \cdots, h(s_1, \cdots, y, \cdots, z, \cdots, s_n), \cdots, x_m)$ to $h'(u_1, \cdots, f(x_1, \cdots, y \cdots, x_m), \cdots, f(x_1, \cdots, z, \cdots, x_m), \cdots, u_n)$.*

The definition of a function f is compatible with a definition of g iff it is compatible with g in every argument position.

As will be shown later, the above requirements on compatibility lead to the function symbol f to be distributed over the interpreted terms to have g as an argument so that the induction hypothesis(es) can be applied.[8]

The above definition is also applicable for capturing the interaction between an interpreted function symbol and a $\mathcal{T}$-based function symbol. For example, the interpreted symbol + in Presburger arithmetic is compatible with * (in both arguments) because of the associativity and commutativity properties of +, which are valid formulas in $\mathcal{T}$.

As stated and illustrated in the introduction, the compatibility property can be viewed as requiring that the composition of f with g has a $\mathcal{T}$-based definition. Space limitations do not allow us to elaborate on this interpretation of compatibility property.

For ensuring the compatibility property, any lemmas already proved about f can be used along with the definition of f. The requirements for showing compatibility can be used to speculate bridge lemmas as well.

The conjecture (C2) is of depth 2. The above insight can be generalized to complex conjectures in which the left side is of arbitrary depth. A conjecture in which a composed term of depth d is equated to an interpreted term, can

---

[7] The requirement on the definition of f can be relaxed by including bridge lemmas along with the defining rules of f.

[8] In [5], we have given a more abstract treatment of these conditions. The above requirement is one way of ensuring conditions in [5].

be decided if all the function symbols from the root to the position p of the basic subterm in its left side can be pushed in so that the induction hypothesis is applicable. The notion of compatibility of a function definition with another function definition is extended to a *compatible sequence* of definitions of function symbols. In a compatible sequence of function symbols $\langle f_1, \cdots, f_d \rangle$, each f_i is compatible with f_{i+1} at j_i-th argument, $1 \leq i \leq d - 1$.

For example, consider the following conjecture

$$(C4): \; \texttt{bton(pad0(ntob(m)))} = \texttt{m}.$$

Functions **bton** and **ntob** convert binary representations to decimal representations and vice versa, respectively. The function **pad0** adds a leading binary zero to a bit vector. These functions are used to reason about number-theoretic properties of parameterized arithmetic circuits [12,10]. Padding of output bit vectors of one stage with leading zeros before using them as input to the next stage is common in multiplier circuits realized using a tree of carry-save adders. An important property that is used while establishing the correctness of such circuits is that the padding does not affect the number output by the circuit. The underlying decidable theory is the combination of the quantifier-free theories of bit vectors with free constructors **nil, cons** and **b0, b1**, to stand for binary 0 and 1, and Presburger arithmetic.

In the definitions below, bits increase in significance in the list with the first element of the list being the least significant. Definitions of **bton, ntob,** and **pad0** are $\mathcal{T}$-based.

```
1.  bton(nil)             --> 0,
2.  bton(cons(b0, y1)) --> bton(y1) + bton(y1),
3.  bton(cons(b1, y1)) --> s(bton(y1) + bton(y1)),

4.  ntob(0)                    --> cons(b0, nil),
5.  ntob(s(0))                 --> cons(b1, nil),
6.  ntob(s(s(x2+x2)))          --> cons(b0,ntob(s(x2))),
7.  ntob(s(s(s((x2+x2))))) --> cons(b1,ntob(s(x2))),

8.  pad0(nil)          --> cons(b0, nil),
9.  pad0(cons(b0, y)) --> cons(b0, pad0(y)),
10. pad0(cons(b1, y)) --> cons(b1, pad0(y)).
```

The function **pad0** is compatible with **ntob**; **bton** is compatible with **pad0** as well as **ntob**. However, **ntob** is not compatible with **bton** since **ntob(s(bton(y1) + bton(y1)))** cannot be rewritten using the definition of **ntob**. However, bridge lemmas,

$$\texttt{ntob(bton(y1) + bton(y1))} = \texttt{cons(b0, ntob(bton(y1)))}$$
$$\texttt{ntob(s(bton(y1) + bton(y1)))} = \texttt{cons(b1, ntob(bton(y1)))}$$

can be identified such that along with these lemmas, **ntob** is compatible with **bton**.

A proof attempt of (C4) leads to two basis and two step subgoals based on the cover set of **ntob**. The first basis subgoal where **m <- 0**, is

$$\texttt{bton(pad0(ntob(0)))} = 0.$$

The subterm $\texttt{ntob(0)}$ rewrites using the definition of $\texttt{ntob}$ to $\texttt{cons(b0, nil)}$, then $\texttt{pad0(cons(b0, nil)))}$ rewrites to $\texttt{cons(b0, cons(b0, nil)))}$, and finally, $\texttt{bton(pad0(ntob(0)))}$ rewrites to $\texttt{0 + 0 + 0 + 0}$, simplifying the above equation to a valid formula in Presburger arithmetic. The second basis subgoal is similar.

Consider the first induction step subgoal. The conclusion is

$$\texttt{bton(pad0(ntob(s(s(x2 + x2)))))} = \texttt{s(s(x2 + x2))}$$

with the hypothesis being

$$\texttt{bton(pad0(ntob(s(x2))))} = \texttt{s(x2)}.$$

The subterm $\texttt{ntob(s(s(x2 + x2)))}$ in the the left side of the conclusion rewrites to $\texttt{cons(b0, ntob(s(x2)))}$ by the definition of $\texttt{ntob}$; the subterm $\texttt{pad0(cons(b0, ntob(s(x2))))}$ then rewrites to $\texttt{cons(b0, pad0(ntob(s(x2))))}$. Term $\texttt{bton(pad0(ntob(s(s(x2 + x2)))))}$ thus rewrites to $\texttt{bton(pad0(ntob(s(x2))))}$ $\texttt{+ bton(pad0(ntob(s(x2))))}$, on which the hypothesis is applicable. The result is a valid formula $\texttt{s(x2) + s(x2)} = \texttt{s(s(x2 + x2))}$ in Presburger arithmetic. It can be shown that the second step subgoal also simplifies to a valid formula in Presburger arithmetic.

Every induction subgoal can be decided, and hence (**C4**) can be decided.

The reader would have noticed that the compatibility requirement ensures that all the function symbols are pushed over interpreted symbols for the induction hypothesis to be applicable.

Note: For understanding the proof below, it would be helpful to concurrently consult the proofs of examples (**C2**) in Section 1 as well as of (**C4**) above.

Theorem 7. *The validity of a conjecture $l = r$, where l is a composed term and r is interpreted in $\mathcal{T}$, can be decided by the cover set method if the sequence of function symbols $\langle f_d, f_{d-1}, \cdots, f_2, f_1 \rangle$ from the outermost function symbol f_d of l to the basic subterm $f_1(x_1, \cdots, x_m)$ is compatible.*

Proof. By induction on the depth d of l.
Basis case ($d = 2$): Consider a conjecture $l = r$ where $l = f_2(t_1, \cdots, f_1(x_1, \cdots, x_m), \cdots, t_k)$ and $\langle f_2, f_1 \rangle$ form a compatible sequence (i.e., f_2 is compatible with f_1 in its argument position), each t_j, $1 \leq j \leq k$, and r are interpreted terms in $\mathcal{T}$. Recall that any induction variable x_i appearing in an inductive position of f_1 does not occur in any t_j.

The cover set method uses the induction scheme generated from the cover set associated with the definition of f_1.

Consider a basis subgoal $\sigma_c(l) = \sigma_c(r)$ generated from a rule $l_1 \to r_1$ in the definition of f_1, where r_1 does not have any recursive calls to f_1. Since $\sigma_c(f_1(x_1, \cdots, x_m)) = l_1$, $\sigma_c(l)$ rewrites to $f_2(\sigma_c(t_1), \cdots, r_1, \cdots, \sigma_c(t_k))$. Since t_i does not include any induction variable, $\sigma_c(t_i) = t_i$, implying $f_2(\sigma_c(t_1), \cdots, r_1, \cdots, \sigma_c(t_k)) = f_2(t_1, \cdots, r_1, \cdots, t_k)$. Because of compatibility of f_2 with f_1,

$f_2(t_1, \cdots, r_1, \cdots, t_k)$ rewrites to an interpreted term. The basis subgoal therefore simplifies to a formula in $\mathcal{T}$.

Consider an induction step subgoal generated from a rule $l_2 \to r_2$ where $r_2 = h_1(s_1, \cdots, f_1(v_1, \cdots, v_m), \cdots, s_n)$ with a single recursive call to f_1 (for simplicity). Let the conclusion be $\sigma_c(f_2(t_1, \cdots, f_1(x_1, \cdots, x_m) \cdots, t_k)) = \sigma_c(r)$ and the induction hypothesis be $\theta_i(f_2(t_1, \cdots, , f_1(x_1, \cdots, x_m), \cdots, t_k)) = \theta_i(r)$, where $\sigma_c(f_1(x_1, \cdots, x_m)) = l_2$ and $\theta_i(f_1(x_1, \cdots, x_m)) = f_1(v_1, \cdots, v_m)$. The left side of the conclusion rewrites to $f_2(t_1, \cdots, r_2, \cdots, t_k)$ (just as in the basis case). As per definition of compatibility of f_2 with f_1, $f_2(y_1, \cdots, h_1(s_1, \cdots, y, \cdots, s_n), \cdots, y_k)$ rewrites to $h_2(s'_1, \cdots, f_2(y_1, \cdots, y, \cdots, y_k), \cdots, s'_n)$ where y_i and y are distinct variables, and h_2, s_j's and s'_j's are in $\mathcal{T}$. This means that the left side of the simplified conclusion $f_2(t_1, \cdots, r_2, \cdots, t_k)$ rewrites by the same sequence of rules to $h_2(\delta(s'_1), \cdots, \delta(f_2(y_1, \cdots, y, \cdots, y_k)), \cdots, \delta(s'_n))$, where $\delta(y_i) = t_i, 1 \le i \le k$, and $\delta(y) = f_1(v_1, \cdots, v_m)\}$. The hypothesis applies since $\theta_i(t_j) = t_j$ for all t_j (recall that there are no x_i's in t_j's), and $\theta_i(f_1(x_1, \cdots, x_m)) = f_1(v_1, \cdots, v_m)$, which simplifies the conclusion to $h_2(\delta(s'_1), \cdots, \theta_i(r), \cdots, \delta(s'_n)) = \sigma_c(r)$, a formula in $\mathcal{T}$.

The above proof step assumed a single recursive call in r_2 and the application of a single induction hypothesis. The proof generalizes when there are multiple recursive calls in r_2 and many possibly different hypotheses have to be applied.

Induction Step case: Assume that the statement of the theorem for all conjectures $l' = r'$, where l' is a composed term of depth $d' < d$, and r' is interpreted.

The main idea in this proof is to use the fact that a conjecture $l = r$ in which the composed term $l = f_d(t_1, \cdots, l_{d-1}, \cdots, t_k)$ is of depth d, uses the same induction scheme as a related conjecture $l_{d-1} = c$, where l_{d-1} is a composed term of depth $d-1$, and c is an interpreted term. By the induction hypothesis, $l_{d-1} = c$ can be decided since all subgoals, including basis and induction steps, can be decided. Because of the compatibility of f_d with f_{d-1}, the outermost symbol of l_{d-1}, it can be shown that each subgoal of $l = r$ using the same induction scheme can also be decided. In the basis step, the instantiated conjecture rewrites to a formula in $\mathcal{T}$, and in the induction step, f_d can be pushed over the interpreted symbols to surround f_{d-1} so that the hypothesis is again applicable, resulting in a formula in $\mathcal{T}$. More details follow.

The same basic term $f_1(x_1, \cdots, x_k)$ in l_{d-1} used for generating an induction scheme for $l_{d-1} = c$ is also used for generating an induction scheme for $l = r$. By the induction hypothesis, each of the subgoals generated from $l_{d-1} = c$ using this induction scheme can be decided in $\mathcal{T}$. For $l = r$ as well, a subgoal $\sigma_c(f_d(t_1, \cdots, l_{d-1}, \cdots, t_k)) = \sigma_c(r)$ using the same substitution can be decided.

Consider a basis subgoal $\sigma_c(l_{d-1}) = \sigma_c(c)$ where $\sigma_c(l_{d-1})$ simplifies to the interpreted term u through a sequence of rewrite steps using the definitions of the $\mathcal{T}$-based function symbols in l_{d-1}. (The $\mathcal{T}$-based function symbols in l_{d-1} are successively eliminated in a bottom up fashion starting with f_1 until finally f_{d-1} rewrites to u by a rule of the form $f_{d-1}(\cdots, r_g, \cdots) \to u$ in the definition of f_{d-1}.) By the compatibility of f_d with f_{d-1}, $f_d(t_1, \cdots, u, \cdots, t_k)$ rewrites to an interpreted term, say u', implying that the basis subgoal $\sigma_c(l) = \sigma_c(r)$ simplifies to $u' = \sigma_c(r)$, a formula in $\mathcal{T}$.

Consider an induction step subgoal with the conclusion $\sigma_c(l_{d-1}) = \sigma_c(c)$ and the hypothesis $\theta_i(l_{d-1}) = \theta_i(c)$, generated from a rule in the definition of f_1 whose right side has a single recursive call to f_1 (for simplicity). The left side of the conclusion simplifies through a sequence of rewrite steps using the definitions of the $\mathcal{T}$-based function symbols in l_{d-1} to a term of the form $h_{d-1}(s_1', \cdots, \theta_i(l_{d-1}), \cdots s_n')$ where $s_j's$ and h_{d-1} are interpreted. The $\mathcal{T}$-based functions in l_{d-1} are successively pushed over interpreted function symbols in a bottom up fashion until finally f_{d-1} is pushed using a rule of the form $f_{d-1}(y_1, h_{d-2}(s_1, \cdots, y, \cdots, s_n), \cdots, y_k) \rightarrow h_{d-1}(s_1', \cdots, f_{d-1}(y_1, \cdots, x, \cdots, y_k), \cdots, s_n')$, to get the left side of the hypothesis. By compatibility of f_d with f_{d-1}, $f_d(z_1, \cdots, h_{d-1}(s_1', \cdots, z, \cdots, s_n') \cdots, z_k)$ rewrites to $h_d(s_1'', \cdots, f_d(z_1, \cdots, z, \cdots, z_k), \cdots, s_n'')$. where h_d and $s''_j's$ are interpreted. This implies that in the corresponding induction step subgoal, the left side of the conclusion $\sigma_c(f_d(t_1, \cdots, l_{d-1}, \cdots, t_k))$ will simplify to $h_d(s_1'' \cdots, \theta_i(f_d(t_1, \cdots l_{d-1}, \cdots, t_k)), \cdots, s_n'')$, a term containing the left side of the hypothesis. The application of the hypothesis simplifies the conclusion to $h_d(s_1'' \cdots, \theta_i(r), \cdots, s_n'') = \sigma_c(r)$, a formula in $\mathcal{T}$. $\square$

5   Relaxing Linearity Requirement: Nonlinear Conjectures

To cover a larger class of formulas, we discuss conditions for deciding a conjecture with multiple occurrences of induction variables in its left side.

Definition 8. *A conjecture* $f(s_1, \cdots, s_m) = r$, *where* $f(s_1, \cdots, s_m)$ *is a* $\mathcal{T}$-based *term, r is interpreted in* $\mathcal{T}$, *and for* $1 \leq i \leq m$, *either s_i is interpreted in* $\mathcal{T}$, *or* $s_i = g_i(x_1, \cdots, x_n)$ *is a basic term, is called basic nonlinear if some variable has multiple occurrences in* l.

In a basic nonlinear conjecture, induction variables (as well as noninduction variables) appearing as arguments in basic terms can be shared. For example, the conjecture below is basic nonlinear,

```
(C5): append(blast(m), last(m)) = m,
```

where **last** returns the singleton list containing the last element of a list, and **blast** returns the input list without the last element.

```
1. last(cons(x, nil))         --> cons(x, nil).
2. last(cons(x, cons(y, z))) --> last(cons(y, z)),
3. blast(cons(x,nil))         --> nil,
4. blast(cons(x,cons(y,z))) --> cons(x,blast(cons(y,z))).
```

To decide such a conjecture, additional conditions become necessary. First, since there can be many induction schemes possible, one each generated from the cover set of a basic term, it is required that they can be merged into a single induction scheme [2,9] (the case when each cover set generates the same induction scheme trivially satisfies this requirement). The second requirement is similar to that of compatibility: f above must be *simultaneously compatible* with each g_i. In the definition below, we assume, for simplicity, that there is at most one recursive call in the function definitions.

Definition 9. *The definition of f is simultaneously compatible in $\mathcal{T}$ with the definitions of g and h in its ith and j^{th} arguments, where $i \neq j$ if for each right side r_g and r_h of the rules in the definitions of g and h, respectively:*

1. *whenever r_g and r_h are interpreted in $\mathcal{T}$, $f(x_1, \cdots, r_g, \cdots, r_h, \cdots, x_m)$ rewrites to an interpreted term in $\mathcal{T}$, and*
2. *whenever $r_g = h_1(\cdots, g(\cdots), \cdots)$ and $r_h = h_2(\cdots, h(\cdots), \cdots)$, the definition of f rewrites $f(x_1, \cdots, h_1(\cdots, x, \cdots), \cdots, h_2(\cdots, y, \cdots), \cdots, x_m)$ to $h_3(\cdots, f(x_1, \cdots, x, \cdots, y, \cdots, x_m), \cdots)$.*

For example, **append** is simultaneously compatible with **blast** in its first argument and **last** in its second argument.

Theorem 10. *A basic nonlinear conjecture $f(\cdots, g(x_1, \cdots, x_n), \cdots, h(x_1, \cdots, x_n), \cdots) = r$, such that $x_1, \cdots, x_n$ do not appear elsewhere in the left side of the conjecture and the remaining arguments of f are interpreted terms, can be decided by the cover set method if f is simultaneously compatible with g and h at i^{th} and j^{th} arguments, respectively, and the induction schemes suggested by $g(x_1, \cdots, x_n)$ and $h(x_1, \cdots, x_n)$ can be merged.*

The proof is omitted due to lack of space; it is similar to the proof of the basis case of Theorem 7. The main steps are illustrated using a proof of (C5).

The induction schemes suggested by the basic terms `blast(m)`, `last(m)` in (C5) are identical. There is one basis subgoal and one induction step subgoal. The basis subgoal obtained by `m <- cons(x, nil)`,

append(blast(cons(x, nil)), last(cons(x, nil))) = cons(x, nil),

simplifies to a valid formula by the definitions of `blast` and `last`, and then by the definition of `append`.

In the step subgoal, the induction conclusion is

append(blast(cons(x,cons(y,z))),last(cons(x,cons(y,z)))) = cons(x,cons(y,z)),

with the hypothesis being,

append(blast(cons(y, z)), last(cons(y, z))) = cons(y, z).

The left side of the conclusion rewrites by the definitions of `last`, `blast` to append(cons(x, blast(cons(y, z))),
last(cons(y, z)))) which rewrites using the definition of `append` to cons(x, append(blast(cons(y, z)), last(cons(y, z)))) to which the hypothesis applies, leading to the valid formula cons(x, cons(y, z)) = cons(x, cons(y, z)).

The notion of simultaneous compatibility and the above theorem generalize to complex nonlinear conjectures, similar to the complex conjecture (C4) discussed in Section 4, in which a conjecture includes a sequence of simultaneously compatible function symbols. Because of space limitations, we cannot discuss this in detail here. The example below illustrates the idea to some extent. The underlying theory is that of free constructors with 0, s. The function symbol + is assumed to have the usual recursive definition: 0 + y --> y, s(x) + y --> s(x + y). The equation is:

$$\text{(C6):} \quad \text{mod2}(x) + (\text{half}(x) + \text{half}(x)) = x,$$

is a complex nonlinear conjecture with the following definitions of half and mod2.

1. half(0)   --> 0,
2. half(s(0))   --> 0,
3. half(s(s(x))) --> s(half(x)).

4. mod2(0)   --> 0,
5. mod2(s(0))   --> s(0),
6. mod2(s(s(x))) --> mod2(x).

For + to be compatible with half in both its arguments, an intermediate lemma (either the commutativity of + or x + s(y) = s(x + y)) is needed as well.[9]

It can be a priori determined that (C6) can be decided by the cover set method since the basic terms half(x), mod2(x) suggest the same induction scheme, and the function symbol + is simultaneously compatible with mod2, + as well as half) in the presence of the above lemma about +.

6   Bootstrapping

As discussed above, simple and complex conjectures with $\mathcal{T}$-based function symbols can be decided using the cover set method, giving an extended decision procedure and an extended decidable theory. In this section, we outline preliminary ideas for bootstrapping this extended decidable theory with the definitions of $\mathcal{T}$-based function symbols and the associated induction schemes, to define and decide a larger class of conjectures.

Definition 11. *A definition of a function symbol f is extended $\mathcal{T}$-based for a decidable theory $\mathcal{T}$ if for each rule, $f(t_1, \cdots t_m) \to r$ in the definition, where $t_i's$ are interpreted over $\mathcal{T}$, the only recursive call to f in r, if any, has only $\mathcal{T}$-based terms as arguments, and the abstraction of r after replacing the recursive call to f by a variable, is either an interpreted term over $\mathcal{T}$, or a basic term $g(\cdots)$ where g has an (extended) $\mathcal{T}$-based definition.*

---

[9] If + is defined by recursing on the second argument, even then commutativity of + or s(x) + y = s(x + y) is needed.

For example, exp denoting exponentiation, defined below, is extended $\mathcal{T}$-based over Presburger arithmetic. For rules defining *, please refer to the beginning of Section 3.

```
1. exp(x, 0)    --> s(0),
2. exp(x, s(y)) --> x * exp(x, y).
```

Unlike simple conjectures, an inductive proof attempt of an extended $\mathcal{T}$-based conjecture may involve multiple applications of the cover set method. Induction may be required to decide the validity of the induction subgoals. In order to determine a priori this, the number of recursive calls in any rule in an extended $\mathcal{T}$-based definition, is restricted to be at most one. The abstracted right side r could be an interpreted term in $\mathcal{T}$, or a basic term with an extended $\mathcal{T}$-based function.

Theorem 12. *A simple extended $\mathcal{T}$-based conjecture $f(x_1, \cdots, x_m) = r$, where f is an extended $\mathcal{T}$-based function, and r is interpreted over $\mathcal{T}$, can be decided by the cover set method.*

The key ideas are suggested in the disproof of an illustrative conjecture about exp:

$$(C7): \quad exp(s(0), m)   = s(m).$$

In the proof attempt of (C7), with induction variable m, there is one basis and one step subgoal. The basis subgoal,

$$exp(s(0), 0) = s(0)$$

rewrites by definition of exp to the valid formula s(0) = s(0). In the step subgoal, the conclusion

$$exp(s(0), s(y)) = s(s(y)),$$

rewrites by definition of exp to s(0) * exp(s(0), y) = s(s(y)), to which the hypothesis, exp(s(0), y) = s(y), applies to give s(0) * s(y) = s(s(y)), which then rewrites by definition of * to s(0) * y = s(y), a simple conjecture which can be decided to be false by the cover set method.

Complex extended $\mathcal{T}$-based conjectures can be similarly defined, and conditions for deciding their validity can be developed. This is currently being explored.

7   Conclusion

This paper describes how inductive proof techniques implemented in existing theorem provers, such as *RRL*, can be used to decide a subclass of equational conjectures. Sufficient conditions for such automation are identified based on the structure of the conjectures and the definitions of the function symbols appearing in the conjectures as well as interaction among the function definitions.

The basic idea is that if the conditions are met, the induction subgoals automatically generated a conjecture by the cover set method simplify to formulas in a decidable theory. This is first shown for simple conjectures with a single function symbol recursively defined using interpreted terms in a decidable theory. Subsequently, this is extended to complex conjectures with nested function symbols by defining the notion of compatibility among their definitions. The compatibility property ensures that in induction subgoals, function symbols can be pushed inside the instantiated conjectures using definitions and bridge lemmas, so as to enable the application of the induction hypotheses, leading to decidable subgoals.

It is shown that certain nonlinear conjectures with multiple occurrences of induction variables can also be decided by extending the notion of compatibility to that of simultaneous compatibility of a function symbol to many function symbols. Some preliminary ideas on bootstrapping the proposed approach are discussed by considering conjectures with function symbols that are defined in terms of other recursively defined function symbols.

Our preliminary experience regarding the effectiveness of the proposed conditions is encouraging. Several examples about properties of lists and numbers as well as properties used to establish the number-theoretic correctness of arithmetic circuits have been successfully tried.

Some representative conjectures, both valid and nonvalid formulas, decided by the proposed approach are given below. With each conjecture, the annotations indicate whether it is simple or complex, as discussed above, its validity and the underlying decidable subtheories. Conjectures are annotated as being nonlinear if they contain multiple basic terms with the same induction variables. For example, the conjectures 12-16 below are nonlinear since they have multiple basic terms with the induction variable x. However, conjectures are 7-9 are not nonlinear since they do not contain multiple basic terms even though they contain multiple occurrences of the variable x. In conjectures 18-20, the underlying theory is Presburger arithmetic extended with the function symbol $*$.

Conjectures 16 and 17 establish the correctness of a restricted form of ripple-carry and carry-save adders respectively. The arguments to the two adders are restricted to be the same in these conjectures. This restriction can be relaxed, and the number-theoretic correctness of parameterized ripple-carry and carry-save adders[12, 10] can be done using the proposed approach. In addition, several intermediate lemmas involved in the proof of multiplier circuits and the SRT divider circuit [10, 11] can be handled.

```
1. half(double(x))       = x,       [Complex, valid, Presburger]
2. mod2(double(x))       = 0,       [Complex, valid, Presburger]
3. half(mod2(x))         = 0,       [Complex, valid, Presburger]
4. log(mod2(x))          = 0,       [Complex, valid, Presburger]
5. exp2(log(x))          = x,       [Complex, inval, Presburger]
6. log(exp2(x))          = x,       [Complex, valid, Presburger]
7. x * log(mod2(x))      = 0,       [Complex, valid, Presburger]
8. x * mod2(double(x))   = 0,       [Complex, valid, Presburger]
9. memb(x, delete(x, y)) = false,   [Complex, valid, lists]
```

```
10. bton(pad0(ntob(x)))    = x,         [Complex, valid, lists]
11. last(ntob(double(x))) = 0,          [Complex, valid, lists]
12. length(append(x, y)) - (length(x) + length(y)) = 0,
                           [Complex, valid, nonlinear, Presburger, Lists]
13. rotate(length(x), x)   = x,
                           [Complex,valid,nonlinear,Presburger,Lists]
14. length(nth(x, y))      <= length(x),
                           [Complex,valid,nonlinear,Presburger,Lists]
15. length(delete(x, y)) <= length(y),
                           [Complex,valid,nonlinear,Presburger,Lists]
16. bton(carry-saveadder(ntob(x), ntob(x), ntob(x)))   = x + x + x,
                           [Complex, valid, nonlinear, Presburger, Bitvectors]
17. bton(ripple-carryadder(ntob(x), ntob(x), ntob(x))  = x + x + x.
                           [Complex, valid, nonlinear, Presburger, Bitvectors]
18. exp(1, x)              = x,
                           [Simple, valid, Presburger extend by *]
19. exp(1, x)              = s(x),
                           [Simple, inval, Presburger extend by *]
20. exp(x, mod2(double(y))) = s(0)
                           [Complex,valid, Presburger extend by *]
```

Inductive reasoning plays a central role in several nontrivial applications, but induction techniques are hardly supported in many reasoning tools, primarily due to the intense manual intervention required to perform inductive proofs in general. The proposed approach can be used to integrate induction proof methods in other reasoning tools and selectively invoke these methods to significantly enhance the reasoning capabilities of these tools without compromising automation. For instance, procedures implementing the cover set induction method can be integrated as a component decision procedure in a cooperating decision procedures framework; it can be invoked to check the validity of inductive subgoals.

To make the proposed approach more effective, it should be generalized to decide more general quantifier-free formulas as well as mechanically generate subsidiary conditions under which a given quantifier-free formula is valid. Such an investigation has been initiated and preliminary results are discussed in [5]. It is also necessary to consider decidability of formulas that require nested induction. Another promising direction for extending this work is to use the proposed approach to guide generation of intermediate lemmas.

Acknowledgements: Thanks to Jürgen Giesl and the referees for useful comments on an earlier draft of the paper.

References

1. A. Gupta, *Inductive Boolean Function Manipulation: A Hardware Verification Methodology for Automatic Induction*. Ph.D. Thesis, Department of Computer Science, Carnegie Mellon University, Pittsburgh, 1994.

2. R.S. Boyer and J S. Moore, *A Computational Logic*. ACM Monographs in Computer Science, 1979.

3. R.S. Boyer and J S. Moore, "Integrating decision procedures into heuristic theorem provers: A case study of linear arithmetic", *Machine Intelligence* 11, J.E. Hayes, D. Mitchie and J. Richards (eds), 1988.

4. L. Fribourg, "Mixing list recursion and arithmetic", Proc. *Seventh Symp. on Logic in Computer Science*, 1992.

5. J. Giesl and D. Kapur, *Decidable Classes of Inductive Theorems*. Technical Report, Department of Computer Science, University of New Mexico, Feb. 2000.

6. D. Kapur, "Automated tools for analyzing completeness of specifications," Proc. *1994 Intl. Symp. on Software Testing and Analysis (ISSTA)*, Seattle, WA, August 1994, 28-43.

7. D. Kapur, "Rewriting, decision procedures and lemma speculation for automated hardware verification," Proc. *10th Intl. Conf. Theorem Proving in Higher Order Logics*, LNCS 1275, 1997.

8. D. Kapur and X. Nie, "Reasoning about numbers in Tecton," Proc. *8th Intl. Symp. Methodologies for Intelligent Systems, (ISMIS'94)*, North Carolina, October 1994.

9. D. Kapur and M. Subramaniam, "New uses of linear arithmetic in inductive theorem proving," *J. Automated Reasoning*, 16 (1-2), 1996, 39-78.

10. D. Kapur and M. Subramaniam, "Mechanically verifying a family of multiplier circuits," Proc. *Computer Aided Verification (CAV'96)*, New Jersey, Springer LNCS 1102 (eds. Alur & Henzinger), 1996, 135-146.

11. D. Kapur and M. Subramaniam "Mechanizing reasoning about arithmetic circuits: SRT division," Proc. *17th FSTTCS*, LNCS (eds. Sivakumar & Ramesh), 1997.

12. D. Kapur and M. Subramaniam, "Mechanical verification of adder circuits using powerlists," *J. of Formal Methods in System Design*, Nov. 1998.

13. D. Kapur and M. Subramaniam, "Using an induction prover for verifying arithmetic circuits," to appear in *J. of Software Tools for Technology Transfer*, Springer Verlag, March 1999.

14. D. Kapur, and H. Zhang, "An overview of Rewrite Rule Laboratory (RRL)," *J. of Computer and Mathematics with Applications*, 29, 2, 1995, 91-114.

15. M. Subramaniam, *Failure Analyses in Inductive Theorem Provers*. Ph.D. Thesis, Department of Computer Science, University at Albany, State University of New York, 1997.

16. H. Zhang, D. Kapur, and M.S. Krishnamoorthy, "A mechanizable induction principle for equational specifications," Proc. *9th Intl. Conf. Automated Deduction (CADE)*, Springer LNCS 310, (eds. Lusk and Overbeek), Chicago, 1988, 250-265.

Complete Monotonic Semantic Path Orderings

Cristina Borralleras[1], Maria Ferreira[2], and Albert Rubio[3]*

[1] Univ. de Vic, Dept. L.S.I.
C/ de Miramarges, s/n 08500 Vic, Spain
cristina.borralleras@uvic.es
[2] Univ. Nova de Lisboa, Dept. Informática
2825-114 Caparica, Portugal
cf@di.fct.unl.pt
[3] Universitat Politècnica de Catalunya, Dept. LS,
Jordi Girona 1, 08034 Barcelona, Spain
rubio@lsi.upc.es

Abstract. Although theoretically it is very powerful, the *semantic path ordering* (SPO) is not so useful in practice, since its monotonicity has to be proved *by hand* for each concrete *term rewrite system* (TRS).

In this paper we present a *monotonic* variation of SPO, called MSPO. It characterizes termination, i.e., a TRS is terminating if and only if its rules are included in some MSPO. Hence MSPO is a *complete* termination method.

On the practical side, it can be easily automated using as ingredients standard interpretations and general-purpose orderings like RPO. This is shown to be a sufficiently powerful way to handle several non-trivial examples and to obtain methods like *dummy elimination* or *dependency pairs* (without the dependency graph refinement) as particular cases. Finally, we obtain some positive modularity results for termination based on MSPO.

1 Introduction

Rewrite systems are sets of rules (directed equations) used to compute by repeatedly replacing parts of a given formula with equal ones until the simplest possible form is obtained. Depending on the kind of objects that are rewritten there are different kinds of rewrite systems, like *string rewrite systems* (Thue or semi-Thue systems) or *term rewrite systems* (TRS; see [DJ90,Klo92,BN98] for detailed surveys).

Termination is a fundamental property for most applications of rewrite systems. Termination of a TRS is, in general, an undecidable property, even for one rule TRSs. Termination of TRSs can be proved by showing that the induced rewrite relation is included in a well-founded ordering on terms. If the ordering $\succ$ is also monotonic and stable under substitutions, i.e., a *reduction ordering*, then it suffices to check that $l \succ r$ for every rule $l \to r$ in the system.

---

* partially supported by the CICYT project HEMOSS ref. TIC98-0949-C02-01.

D. McAllester (Ed.): CADE-17, LNAI 1831, pp. 346–364, 2000.
© Springer-Verlag Berlin Heidelberg 2000

Monotonic orderings including the subterm relation are called *simplification orderings*, and their well-foundedness follows from Kruskal's theorem [Kru60]. Inside this class, path orderings, and in particular the *recursive path ordering* (RPO) [Der82], have received a special attention (see [Der87,DJ90,BN98]). Unfortunately, although these orderings are simple and easy to use, they turn out, in many cases, to be a weak termination proving tool, as there are many TRSs that are terminating but are not contained in any simplification ordering, i.e. they are not *simply terminating*.

To avoid this problem, many different transformation methods have been developed, e.g. [BD86,BL90,Zan94,FZ95,Ste95,Xi98,KNT99]. By transforming the TRS into a set of ordering constraints, the *dependency pair* method [AG97,AG00] has become a successful general technique for proving termination of (non-simply terminating) TRSs.

As an alternative to transformation methods, more powerful term orderings can be used. Due to its simplicity, the Semantic Path Ordering (SPO) ([KL80]) becomes a potential well-known candidate: in SPO the scheme of RPO is generalized by replacing the precedence on function symbols by any (well-founded) underlying (quasi-)ordering involving the whole term and not only its head symbol. Although the simplicity of the presentation is kept, this makes the ordering much more powerful. In fact, for every terminating TRS there is some SPO that includes its rewrite relation.

Unfortunately, SPO is not so useful in practice. Due to the generalization, the monotonicity property is lost, even if the underlying ordering is monotonic. Hence, it is not sufficient to check that the rules are included in the ordering to ensure termination, since this does not imply that each rewrite step is inside the ordering. In order to ensure termination of a TRS R, the user is responsible for proving (by hand) monotonicity restricted to all terms s and t such that s rewrites to t with R in one step.

In this paper we present a monotonic version of the SPO. On the one hand, this *monotonic semantic path ordering* (MSPO) is still very powerful for proving termination of TRSs: on the theoretical side, it also characterizes termination, i.e. a TRS is terminating if and only if the rules are included in some MSPO; and on the practical side, it generalizes most of the (automatable) termination proof methods. On the other hand, termination can be automatically checked once the ingredients, i.e. the underlying (base) quasi-orderings, of the MSPO are provided.

The first and only other, as far as we know, monotonic version of SPO is due to Geser [Ges92]. On the one hand, this proposal is not as general as ours (in fact, as we will show, it does not characterize termination), and on the other hand, it is less suitable for practical implementations. In section 5.3 a detailed comparison with this work is provided.

We are not only interested in using MSPO for checking termination, but also for proving termination automatically. Since we cannot expect to automatically generate an adequate MSPO for a given TRS whenever it exists, we have studied particular classes of underlying quasi-orderings, which can be automatically generated. As a hint of the power of the resulting family of MSPO's, it is shown that some known methods like *dummy elimination* [FZ95] or *dependency pairs* [AG97,AG00] (without the dependency graph refinement) are particular instances.

Using these classes of underlying quasi-orderings the termination proofs of several non-simply terminating, but terminating, term rewriting systems can be fully automated. However, some heuristics for choosing the underlying quasi-ordering have to be developed in order to make the method more effective in practice. On the other hand, due to its generality and its additional flexibility, MSPO provides, in some cases, simpler termination proofs than the dependency pair method (see example 6). A first system based on MSPO has been developed by which many examples, including the ones in the paper, have been checked. The software and examples are available at www.lsi.upc.es/~albert.

Additionally, applying known abstract sufficient conditions ensuring modularity of termination [Ohl94,Gra94], modularity results for termination based on MSPO are obtained. In particular, for TRSs proved terminating using MSPO with the aforementioned classes of underlying quasi-orderings, termination is proved to be modular for disjoint systems and for finite constructor-sharing systems. Note that these modularity properties are crucial for many practical applications of automatic termination proof systems.

Formal definitions and basic tools are introduced in section 2. In section 3 we present and study the monotonic semantic path ordering. Section 4 is devoted to examples. Other termination methods are analyzed in section 5. Section 6 presents an ordering constraint solving approach for proving termination using MSPO. In section 7 modularity results are presented. Some conclusions are given in section 8.

2   Preliminaries

In the following we consider that $\mathcal{F}$ is a set of function symbols, $\mathcal{X}$ a set of variables and $\mathcal{T}(\mathcal{F}, \mathcal{X})$ is the set of terms built from $\mathcal{F}$ and $\mathcal{X}$. Let s and t be arbitrary terms in $\mathcal{T}(\mathcal{F}, \mathcal{X})$, let f be a function symbol in $\mathcal{F}$ and let σ be a substitution. A (strict partial) ordering $\succ$ is a transitive irreflexive relation. It is *monotonic* if $s \succ t$ implies $f(\ldots s \ldots) \succ f(\ldots t \ldots)$, and *stable under substitution* if $s \succ t$ implies $s\sigma \succ t\sigma$. Monotonic orderings that are stable under substitutions are called *rewrite orderings*. A *reduction ordering* is a rewrite ordering that is *well-founded*: there are no infinite sequences $t_1 \succ t_2 \succ \ldots$

The reflexive-transitive closure of a binary relation $\rightarrow$ is denoted by $\xrightarrow{*}$ and the transitive closure by $\xrightarrow{+}$.

A term rewrite system (TRS) is a (possibly infinite) set of rules $l \rightarrow r$ where l and r are terms. Given a TRS R, s rewrites to t with R, denoted by $s \rightarrow_R t$, if there is some rule $l \rightarrow r$ in R, $s|_p = l\sigma$ for some position p and substitution σ and $t = s[r\sigma]_p$. A TRS R is terminating if there exists no infinite sequence $t_1 \rightarrow_R t_2 \rightarrow_R \ldots$ Thus, the transitive closure $\xrightarrow{+}_R$ of any terminating TRS is a reduction ordering. Furthermore, reduction orderings characterize termination of TRSs.

Theorem 1. *A rewrite system R is terminating if and only if all rules are contained in a reduction ordering $\succ$, i.e., $l \succ r$ for every $l \to r \in R$.*

Another interesting property of reduction orderings is that they can be combined with the subterm relation $\rhd$ without loosing well-foundedness. However, note that in general, monotonicity will be lost, since we only add the subterm relation, and not its monotonic closure.

Proposition 1. *If $\succ$ is a reduction ordering then $\succ \cup \rhd$ is well-founded.*

Given a relation $\succ$, the multiset extension of $\succ$ on finite multisets, denoted by $\twoheadrightarrow$, is defined as the smallest transitive relation containing

$$X \cup \{s\} \twoheadrightarrow X \cup \{t_1, \ldots, t_n\} \qquad \text{if } s \succ t_i \text{ for all } i \in \{1 \ldots n\}$$

If $\succ$ is a well-founded ordering on terms then $\twoheadrightarrow$ is a well-founded ordering on finite multisets of terms.

A *quasi-ordering* $\succeq$ is a transitive and reflexive binary relation. Its inverse is denoted by $\preceq$. Its *strict part* $\succ$ is the strict ordering $\succeq \setminus \preceq$ (i.e, $s \succ t$ iff $s \succeq t$ and $s \not\preceq t$). Its *equivalence* $\sim$ is $\succeq \cap \preceq$. Note that $\succeq$ is the disjoint union of $\succ$ and $\sim$, and that if $=$ denotes syntactic equality then $\succ \cup =$ is a quasi-ordering whose strict part is $\succ$.

Notation: In the remainder of this paper, $\succeq$ (possibly with subscripts) will always denote a quasi-ordering.

The following definitions for $\succeq$ will be used:

1. $\succeq$ is *monotonic* if $\succ$ is.
2. $\succeq$ is *well-founded* if $\succ$ is.
3. $\succeq$ is *stable under substitutions* if $\succ$ is and $s\sigma \succeq t\sigma$ whenever $s \succeq t$.
4. $\succeq$ is *quasi-monotonic* if $f(\ldots, s, \ldots) \succeq f(\ldots, t, \ldots)$ whenever $s \succeq t$.
5. $\succeq$ is a *quasi-reduction quasi-ordering* if it fulfills the above properties 2, 3 and 4.

Note that if $\succeq$ is quasi-monotonic then $\succ$ is not necessarily monotonic.

The *lexicographic combination* of $\succeq_1, \ldots, \succeq_n$, denoted by $(\succeq_1, \ldots, \succeq_n)_{lex}$, is defined as usual as, $s(\succeq_1, \ldots, \succeq_n)_{lex} t$ iff either $s \succ_i t$ for some i and $s \succeq_j t$ for all $j < i$, or $s \succeq_i t$ for all i.

If all $\succeq_i$ are well-founded (stable under substitutions) then its lexicographic combination also is.

A *precedence* $\succeq_{\mathcal{F}}$ is a well-founded quasi-ordering on $\mathcal{F}$.

Given a precedence $\succeq_{\mathcal{F}}$, the *recursive path ordering* (RPO), denoted by $\succ_{rpo}$, is defined recursively as follows:

$s = f(s_1, \ldots, s_m) \succ_{rpo} t$ iff

1. $s_i \succeq_{rpo} t$,   for some $i = 1, \ldots, m$,   or
2. $t = g(t_1, \ldots, t_n)$ with $f \succ_{\mathcal{F}} g$ and $s \succ_{rpo} t_i$ for all $i = 1 \ldots n$ or
3. $t = g(t_1, \ldots, t_n)$ with $f \succeq_{\mathcal{F}} g$ and $\{s_1, \ldots, s_m\} \twoheadrightarrow_{rpo} \{t_1, \ldots, t_n\}$.

where $\succeq_{rpo}$ is defined as $\succ_{rpo} \cup =$.

3   A Monotonic Semantic Path Ordering

Let us now recall the definition of the semantic path ordering. Then we will analyze an example of non-monotonicity, which will provide the intuition behind the monotonic version we will propose.

Definition 1. *Given $\succeq_Q$, called the underlying (or base) quasi-ordering, the semantic path ordering (SPO) [KL80], denoted as $\succ_{spo}^Q$, is defined as*

$$s = f(s_1, \ldots, s_m) \succ_{spo}^Q t \quad iff$$

1. $s_i \succeq_{spo}^Q t$, for some $i = 1, \ldots, m$, or
2. $s \succ_Q t = g(t_1, \ldots, t_n) \quad and \quad s \succ_{spo}^Q t_i$ for all $i = 1, \ldots, n$, or
3. $s \succeq_Q t = g(t_1, \ldots, t_n) \quad and \quad \{s_1, \ldots, s_m\} \succ\!\!\succ_{spo}^Q \{t_1, \ldots, t_n\},$

where $\succeq_{spo}^Q$ is defined as $\succ_{spo}^Q \cup =$.

The semantic path ordering is well-defined, which can be easily proved by induction on sum of the sizes of s and t, and fulfills the following property.

Lemma 1 ([KL80]). *If $\succeq_Q$ is well-founded and stable under substitutions then $\succ_{spo}^Q$ is a well-founded ordering stable under substitutions.*

But, as said before, the semantic path ordering is, in general, non-monotonic, even when $\succeq_Q$ is quasi-monotonic (in fact, the same problem appears if $\succeq_Q$ is monotonic). This is shown in the following example.

Example 1. Consider the following quasi-ordering $\succeq_Q$ defined for all terms s and t as: (i) $f(s) \succeq_Q g(t)$; (ii) $f(s) \succeq_Q f(t)$ iff $s \succeq_Q t$; and (iii) $g(s) \succeq_Q g(t)$, and reflexive in variables and constants.

This quasi-ordering is well-founded, since its strict part is $u[f(s)] \succ_Q u[g(t)]$ for all non-empty contexts u containing only the symbol f and for all terms s and t (note that the length of any decreasing sequence is at most the number of f's above the first g symbol of the initial term in the sequence); it is stable under substitutions and quasi-monotonic. However, the induced SPO is not monotonic: by case 1 of SPO we have $g(f(a)) \succ_{spo}^Q f(a)$, but by adding the context $f([])$ onto both terms we have $f(g(f(a))) \not\succ_{spo}^Q f(f(a))$ (in fact, in this case we can even prove $f(f(a)) \succ_{spo}^Q f(g(f(a)))$).

Analyzing this example we can observe that, even if $\succeq_Q$ is quasi-monotonic (or monotonic), since in case 1 of SPO we do not require $s \succeq_Q t$, it may happen that by adding some context u, if we do not have $u[s] \succeq_Q u[t]$, we cannot apply any of the cases. Then, in order to ensure monotonicity of $\succ_{spo}^Q$, we need to be sure that $u[s] \succeq_Q u[t]$ for any context. A way to obtain this is to require always $s \succeq_Q t$, provided that $\succeq_Q$ is quasi-monotonic, that is defining a new ordering $\succ_M$ as $s \succ_M t$ iff and only if $s \succeq_Q t$ and $s \succ_{spo}^Q t$.

However, requiring always $s \succeq_Q t$ can be a bit too strong and, in fact, from the example above we only need to have $u[s] \succeq_Q u[t]$ for every (non-empty) context u. Hence instead of requiring $s \succeq_Q t$, we will ask for something weaker that ensures $u[s] \succeq_Q u[t]$.

Definition 2. *We say that* $\succeq_I$ *is* quasi-monotonic *on* $\succeq_Q$ *(or* $\succeq_Q$ *is quasi-monotonic wrt.* $\succeq_I$*) if*

$$s \succeq_I t \quad implies \quad f(\ldots s \ldots) \succeq_Q f(\ldots t \ldots)$$

for all terms s and t and function symbols f.

Definition 3. *A pair* $\langle \succeq_I, \succeq_Q \rangle$ *is called a* quasi-reduction pair *if* $\succeq_I$ *is quasi-monotonic,* $\succeq_Q$ *is well-founded, both are stable under substitutions and* $\succeq_I$ *is quasi-monotonic on* $\succeq_Q$.

Now we define the *monotonic semantic path ordering* (MSPO):

Definition 4. *Let* $\langle \succeq_I, \succeq_Q \rangle$ *be a quasi-reduction pair. The corresponding monotonic semantic path ordering, denoted by* $\succ_{mspo}$*, is defined as:*

$$s \succ_{mspo} t \quad if \ and \ only \ if \quad s \succeq_I t \quad and \quad s \succ_{spo}^{Q} t$$

for all terms s and t.

Theorem 2. $\succ_{mspo}$ *is a reduction ordering.*

Proof. Well-foundedness follows from the fact that $\succ_{mspo} \subseteq \succ_{spo}^{Q}$ and $\succ_{spo}^{Q}$ is well-founded. Transitivity and stability under substitutions follow respectively from the transitivity and the stability of $\succeq_I$ and $\succ_{spo}^{Q}$.

For monotonicity we have to show that $s \succ_{mspo} t$ implies $f(\ldots, s, \ldots) \succ_{mspo} f(\ldots, t, \ldots)$, that is $f(\ldots, s, \ldots) \succeq_I f(\ldots, t, \ldots)$ and also $f(\ldots, s, \ldots) \succ_{spo}^{Q} f(\ldots, t, \ldots)$, for all terms s and t and function symbols f.

By definition of $\succ_{mspo}$, $s \succ_{mspo} t$ implies $s \succeq_I t$. Hence, by the quasi-monotonicity of $\succeq_I$, we have $f(\ldots, s, \ldots) \succeq_I f(\ldots, t, \ldots)$. On the other hand, by quasi-monotonicity of $\succeq_I$ on $\succeq_Q$, we have $f(\ldots, s, \ldots) \succeq_Q f(\ldots, t, \ldots)$. Then, by definition of $\succ_{mspo}$, $s \succ_{mspo} t$ implies $s \succ_{spo}^{Q} t$ and, therefore we have $\{\ldots s \ldots\} \succ\succ_{spo}^{Q} \{\ldots t \ldots\}$, which implies $f(\ldots, s, \ldots) \succ_{spo}^{Q} f(\ldots, t, \ldots)$ by case (3). $\square$

The previous theorem shows that $\succ_{mspo}$ provides a correct method for proving termination of term rewriting systems. The following result shows that it is also a complete method, i.e., for any terminating term rewriting there is a monotonic semantic path ordering that includes its rules. Therefore $\succ_{mspo}$ characterizes termination. The proof is very similar to the completeness proof of SPO and also to the completeness proof of other methods like semantic labeling in [MHH96].

Theorem 3. *A rewrite system R is terminating if and only if there exists some quasi-reduction pair* $\langle \succeq_I, \succeq_Q \rangle$*, s.t.* $l \succ_{mspo} r$ *for every rule* $l \to r$ *in R.*

Proof. The right to left implication follows from Theorem 2. For the left to right, we will build an appropriate pair $\langle \succeq_I, \succeq_Q \rangle$ for any terminating TRS R.

Let $\succeq_I$ be $\to_R^*$ and let $\succeq_Q$ be $(\to_R \cup \rhd)^*$. By definition of rewriting $\succeq_I$ is quasi-monotonic and both $\succeq_I$ and $\succeq_Q$ are stable under substitutions. By termination of R and Property 1, the strict part of $\succeq_Q$, i.e. $(\to_R \cup \rhd)^+$, is well-founded. Finally, since $\succeq_I \subseteq \succeq_Q$, and $\succeq_I$ is quasi-monotonic, we have that $\succeq_I$ is quasi-monotonic on $\succeq_Q$.

Therefore, since $\langle \succeq_I, \succeq_Q \rangle$ is a quasi-reduction pair, we only have to prove that $l \succ_{mspo} r$ for every rule $l \to r$ in R. By definition, we have $l \succeq_I r$ and $l \succ_Q r$, and, moreover, for every subterm r' of r we have $l \succ_Q r'$, since $l \to_R r \rhd r'$. Therefore, we have $l \succeq_I r$ and $l \succ_{spo}^Q r$, by repeatedly applying case 2, which implies $l \succ_{mspo} r$.   □

The quasi-reduction pair condition is quite tight: as shown in the following example, if $\succeq_Q$ is required to be a quasi-reduction quasi-ordering then even SPO does not characterize termination, i.e. there are terminating sets of rules that are not included in any SPO with an underlying quasi-reduction quasi-ordering.

Example 2. There is no quasi-reduction quasi-ordering $\succeq_Q$ such that the following TRS is included in the generated SPO.

$$
\begin{aligned}
a &\to b \\
f(b) &\to g(a) \\
h(a) &\to h(f(b))
\end{aligned}
$$

First, note that the TRS is terminating (it will be shown using MSPO in example 3). Now we show that these rules cannot be included in SPO with a quasi-reduction quasi-ordering $\succeq_Q$. To this end it is important to remark that in SPO to have $c \succeq_{spo}^Q t$ for some constant c and term t we need $c \succ_Q t'$ for all t' subterm of t.

By definition we need $a \succ_Q b$ to be able to include the first rule. Then to include the second one we need $f(b) \succ_Q g(a)$ and $f(b) \succeq_Q a$, in order to conclude by case 2 first and then by case 2 or 3 for the recursive call with $f(b) \succ_{spo}^Q a$, since any other possibility requires $b \succeq_Q a$, which contradicts the first assumption. Now we have (i) $a \succ_Q b$, (ii) $f(b) \succ_Q g(a)$ and (iii) $f(b) \succeq_Q a$. We proceed with the third rule. If $a \succeq_{spo}^Q h(f(b))$ we need $a \succ_Q f(b)$, which contradicts assumption (iii). Otherwise we need $h(a) \succeq_Q h(f(b))$. By assumption (iii) and quasi-monotonicity of $\succeq_Q$, we have $h(f(b)) \succeq_Q h(a)$, which implies $h(a) \not\succ_Q h(f(b))$. Therefore we have to apply case 3 of SPO, and hence we need $a \succ_{spo}^Q f(b)$, which requires $a \succ_Q f(b)$ contradicting assumption (iii).

Theorem 3 shows the theoretical power of the ordering as a termination proof method, but in order to make it more useful in practice in the following two sections we will show general methods to obtain quasi-reduction pairs $\langle \succeq_I, \succeq_Q \rangle$.

3.1   Building $\succeq_I$

We consider $\succeq_I$ obtained by combining an interpretation I on terms with some quasi-reduction quasi-ordering $\succeq_B$, called the *basic* quasi-ordering, which can be obtained by well-known practical general-purpose methods like the path orderings or polynomial interpretations. For the interpretation I, as a general property, we require the preservation of the quasi-monotonicity and stability under substitutions of the basic quasi-ordering. Below some particular such interpretations which are suitable for practical applications are provided. These interpretations are not original; they have been used in many different transformation based termination methods, and in particular a slightly restricted version of them, called *argument filtering systems* (AFS), has been used in a similar way for the dependency pair method [AG00].

We will consider interpretations as mappings from terms to terms $I : \mathcal{T}(\mathcal{F}, \mathcal{X})$ $\to \mathcal{T}(\mathcal{F}', \mathcal{X}')$, although, of course, we can also consider interpretation from terms to multisets of terms or any other domain provided that the required properties are fulfilled.

¿From now on we consider that $\succeq_I$ is defined as follows

$$s \succeq_I t \quad \text{if and only if} \quad I(s) \succeq_B I(t)$$

Note that this does not imply any loss of generality since I can be the identity mapping. Additionally, it follows that $s \succ_I t$ if and only if $I(s) \succ_B I(t)$.

For any I, if $\succeq_B$ is a quasi-ordering then $\succeq_I$ is a quasi-ordering which is well-founded if $\succeq_B$ also is. For quasi-monotonicity and stability under substitutions this is not always the case. Let us give some examples of interpretations that do preserve these properties.

Each symbol f can be interpreted either by a projection on a single argument, denoted by the pair $(f(x_1, \ldots, x_n), x_i)$, or else by a function symbol f_I applied to an arbitrary sequence obtained from the arguments of f, denoted by the pair $(f(x_1, \ldots, x_n), f_I(x_{i_1}, \ldots, x_{i_k}))$, for some $k \geq 0$ and $i_1, \ldots, i_k \in \{1, \ldots, n\}$. Additionally we consider I to be the identity for variables (although it can be any bijection).

Note that in the AFS the second kind of interpretations are restricted to have $f_I \notin \mathcal{F}$ and all x_{i_j} to be different variables in $\{x_1, \ldots x_n\}$, which is not the case here.

We assume that there is only one pair for each symbol. Usually the identity pairs will be omitted. Thus the interpretation I is recursively defined from these pairs as, $I(x) = x$ and $I(f(t_1, \ldots, t_n))$ is

- $I(t_i)$ if we have the pair $(f(x_1, \ldots, x_n), x_i)$, or
- $f_I(I(t_{i_1}), \ldots, I(t_{i_k}))$ if we have the pair $(f(x_1, \ldots, x_n), f_I(x_{i_1}, \ldots, x_{i_k}))$.

It is easy to show that these interpretations preserve quasi-monotonicity and stability under substitutions (recall that the stability of quasi-orderings requires also the stability of their strict parts).

Proposition 2. *Let I be an interpretation as defined above.*

1. *If $\succeq_B$ is quasi-monotonic then $\succeq_I$ is quasi-monotonic.*
2. *If $\succeq_B$ is stable under substitutions then $\succeq_I$ is stable under substitutions.*

3.2 Building $\langle \succeq_I, \succeq_Q \rangle$

In this section we show how quasi-reduction pairs $\langle \succeq_I, \succeq_Q \rangle$ can be obtained. First, as basic cases, we present two possible quasi-orderings $\succeq_Q$ fulfilling the quasi-monotonicity requirement for a given $\succeq_I$.

Proposition 3. *Let $\succeq_I$ be a quasi-reduction quasi-ordering. Then*

1. $\langle \succeq_I, \succeq_I \rangle$ *is a quasi-reduction pair.*
2. *Let $\succeq_{\mathcal{F}}$ be a precedence on $\mathcal{F}$ and let $\succeq_T^{\mathcal{F}}$ be defined as $s \succeq_T^{\mathcal{F}} t$ iff $top(s) \succeq_{\mathcal{F}}$ $top(t)$. Then $\langle \succeq_I, \succeq_T^{\mathcal{F}} \rangle$ is a quasi-reduction pair.*

Now we show how to obtain new quasi-reduction pairs from one or several given quasi-reduction pairs. Hence, we can start by pairs as in the proposition above and then (repeatedly) apply the following properties to obtain more suitable quasi-reduction pairs.

First we define what a renaming quasi-ordering is.

Definition 5. *Let N be a mapping from $\mathcal{F}$ to $\mathcal{F}$, called a renaming in $\mathcal{F}$, and let f_N denote $N(f)$. We extend N to terms, obtaining a head renaming map, in the following way $N(f(t_1, \ldots, t_m)) = f_N(t_1, \ldots, t_m)$ for every symbol f in $\mathcal{F}$.*

Given $\succeq$ and a renaming map in $\mathcal{F}$, the renaming quasi-ordering $\succeq^N$ is defined as $s \succeq^N t$ if and only if either $s = t \in \mathcal{X}$ or $N(s) \succeq N(t)$.

Note that the renaming map is only applied to the head symbol of the term, and not to the arguments, and that, to preserve stability under substitutions, it is not defined for variables. This notion of renaming already appears in the dependency pair method, where the head symbol of both terms in each dependency pair is always renamed (see section 5.2 for details).

Proposition 4. $\langle \succeq_I, \succeq_Q \rangle$ *is a quasi-reduction pair if*

1. $\langle \succeq_I, \succeq_{Q_0} \rangle$ *is a quasi-reduction pair and $\succeq_Q$ is well-founded and stable under substitutions and $\succeq_{Q_0} \subseteq \succeq_Q$; or*
2. $\langle \succeq_I, \succeq_{Q_0} \rangle$ *is a quasi-reduction pair and $\succeq_Q$ is $\succeq_{Q_0}^N$ for some renaming map N in $\mathcal{F}$; or*
3. $\langle \succeq_I, \succeq_{Q_i} \rangle$ *is a quasi-reduction pair for all $i \in \{1 \ldots n\}$ and $\succeq_Q$ is $(\succeq_{Q_1}, \ldots, \succeq_{Q_n})_{lex}$.*

Let us show how propositions 3 and 4 can be used to build quasi-reduction pairs. Assume that $\succeq_I$ is a quasi-reduction quasi-ordering, $\succeq_{\mathcal{F}}$ is a precedence on $\mathcal{F}$ and N is a renaming map in $\mathcal{F}$. Then, by proposition 3.1, $\langle \succeq_I, \succeq_I \rangle$ is a quasi-reduction pair and, by proposition 4.2, $\langle \succeq_I, \succeq_I^N \rangle$ is a quasi-reduction pair as well. Now since, by proposition 3.2, $\langle \succeq_I, \succeq_T^{\mathcal{F}} \rangle$ is a quasi-reduction pair, we can conclude by proposition 4.3, that $\langle \succeq_I, (\succeq_I^N, \succeq_T^{\mathcal{F}})_{lex} \rangle$ is a quasi-reduction pair.

4 Examples

In the examples we will always provide the quasi-reduction pair $\langle \succeq_I, \succeq_Q \rangle$ and in some cases the details of the checking of $\succ_{mspo}$ will be included. First we give the definition of $\succeq_Q$ using $\succeq_I$ and then the definition of $\succeq_I$. In all cases we use the methods described in sections 3.1 and 3.2. Since for the basic quasi-ordering $\succeq_B$ of $\succeq_I$ we always use RPO, to avoid confusion, its precedence will be denoted by $\succeq_P$ (note that we can use other precedences to build $\succeq_Q$), and for simplicity we will directly give its strict part $\succ_P$.

Example 3. The following TRS comes from example 2

$$a \to b$$
$$f(b) \to g(a)$$
$$h(a) \to h(f(b))$$

In this case for $\succeq_Q$ we use $\succeq_I^N$ with the renaming map N which is the identity except $N(f) = F$. For $\succeq_I$ we use the interpretation I generated by the pairs: $(f(x), b)$ and $(g(x), b)$; and RPO with the precedence $h \succ_P F$, $F \succ_P g$, $F \succ_P a$, $a \succ_P b$. Note that we have added to the signature the symbol F.

For the first rule we have $a \succeq_I b$ since $I(a) = a \succ_{rpo} b = I(b)$ and since $N(a) = a$ and $N(b) = b$, we have $a \succ_Q b$, which implies $a \succ_{spo}^Q b$ by case 2 of SPO, and hence $a \succ_{mspo} b$.

For the second rule $I(f(b)) = b = I(g(a))$, and hence $f(b) \succeq_I g(a)$. To prove $f(b) \succ_{spo}^Q g(a)$, since $I(N(f(b))) = F(b) \succ_{rpo} b = I(N(g(a)))$, which implies $f(b) \succ_Q g(a)$, by case 2 of SPO we only need to check $f(b) \succ_{spo}^Q a$, which follows again by case 2.

For the third rule $I(h(a)) = h(a) \succ_{rpo} h(b) = I(h(f(b)))$ and hence $h(a) \succeq_I h(f(b))$. To prove $h(a) \succ_{spo}^Q h(f(b))$, since $I(N(h(a))) = h(a) \succ_{rpo} h(b) = I(N(h(f(b))))$, which implies $h(a) \succ_Q h(f(b))$, by case 2 of SPO we only need to check $h(a) \succ_{spo}^Q f(b)$. This follows again by case 2, since $I(N(h(a))) = h(a) \succ_{rpo} F(b) = I(N(f(b)))$, and hence $h(a) \succ_Q f(b)$, and $h(a) \succ_{spo}^Q b$ follows as well by case 2.

Example 4. In the following example of nested recursion (from [FZ95]) we use a precedence $\succeq_F$ as first component in $\succeq_Q$.

$$f(g(x)) \to g(f(f(x)))$$
$$f(h(x)) \to h(g(x))$$

We can take as $\succeq_Q$ the lexicographic combination $(\succeq_T^F, \succeq_I^N)_{lex}$ with the precedence $f \succ_F g$ and $f \succ_F h$, and the renaming map N which is the identity except $N(f) = F$. For $\succeq_I$, the interpretation is given by $(f(x), x)$ and $(h(x), a)$; and the basic quasi-ordering is RPO generated by the empty precedence. Note that we have added to the signature the function symbol F and the constant a.

We only show here that $f(g(x)) \succ_{mspo} g(f(f(x)))$. First, we have $f(g(x)) \succeq_I g(f(f(x)))$, since $I(f(g(x))) = g(x) = I(g(f(f(x))))$. To prove $f(g(x)) \succ_{spo}^Q$

$g(f(f(x)))$ we apply case 2, since $f \succ_{\mathcal{F}} g$ and hence $f(g(x)) \succ_Q g(f(f(x)))$. For the recursive call $f(g(x)) \succ^Q_{spo} f(f(x))$ we apply case 2 as well, since $I(N(f(g(x)))) = F(g(x)) \succ_{rpo} F(x) = I(N(f(f(x))))$ and hence $f(g(x)) \succ_Q g(f(f(x)))$. For the recursive call $f(g(x)) \succ^Q_{spo} f(x)$ we apply again case 2, since $I(N(f(g(x)))) = F(g(x)) \succ_{rpo} F(x) = I(N(f(x)))$, and for the recursive call $f(g(x)) \succ^Q_{spo} x$ we apply twice case 1.

Example 5. In the following non-simply terminating example (from [AG97]) we use a precedence $\succeq_{\mathcal{F}}$ as second component in $\succeq_Q$.

$$
\begin{aligned}
le(0, y) &\to true \\
le(s(x), 0) &\to false \\
le(s(x), s(y)) &\to le(x, y) \\
minus(0, y) &\to 0 \\
minus(s(x), y) &\to if(le(s(x), y), s(x), y) \\
if(true, s(x), y) &\to 0 \\
if(false, s(x), y) &\to s(minus(x, y)) \\
quot(0, s(y)) &\to 0 \\
quot(s(x), s(y)) &\to s(quot(minus(x, y), s(y)))
\end{aligned}
$$

We can take as $\succeq_Q$ the lexicographic combination $(\succeq_I, \succeq^{\mathcal{F}}_T)_{lex}$ with the precedence $le \succ_{\mathcal{F}} true$, $le \succ_{\mathcal{F}} false$, $minus \succ_{\mathcal{F}} if$ and $if \succ_{\mathcal{F}} s$. For $\succeq_I$, the interpretation is given by $(le(x, y), b)$, $(true, b)$, $(false, b)$, $(if(x, y, z), y)$ and $(minus(x, y), x)$; and the basic quasi-ordering is RPO generated by the precedence $quot \succ_P s$, $s \succ_P b$ and $s \succ_P 0$.

We only show here that $minus(s(x), y) \succ_{mspo} if(le(s(x), y), s(x), y)$. First $minus(s(x), y) \succeq_I if(le(s(x), y), s(x), y)$ holds since $I(minus(s(x), y)) = s(x) = I(if(le(s(x), y), s(x), y))$. To prove $minus(s(x), y) \succ^Q_{spo} if(le(s(x), y), s(x), y)$, we have $minus(s(x), y) \succ_Q if(le(s(x), y), s(x), y)$, since $minus \succ_{\mathcal{F}} if$. Then by case 2 of SPO we only need to check the recursive call $minus(s(x), y) \succ^Q_{spo} le(s(x), y)$, $minus(s(x), y) \succ^Q_{spo} s(x)$ and $minus(s(x), y) \succ^Q_{spo} y$. The last two follow from case 1 of SPO. For the first one we have $minus(s(x), y) \succ_Q le(s(x), y)$, since $I(minus(s(x), y) = s(x) \succ_{rpo} b = I(le(s(x), y))$, and as before we can conclude by case 2.

Example 6. The following system is an automatic translation of a prolog program that computes the Ackermann function. This example comes from Claus Claves Master's thesis, in the context of the development of TALP [OCM00], an automated termination proof tool for logic programs based on rewriting techniques [1]. This TRS can be proved terminating by the dependency pair method (see section 5.2), but using a further refinement of the dependency graph introduced in [AG98] (see also [AG00]).

---

[1] This example was posed to us by Claude Marche.

As shown below, this example can be easily handled by MSPO with a very simple quasi-reduction pair.

$$
\begin{aligned}
ack\_in(0,n) &\rightarrow ack\_out(s(n)) \\
ack\_in(s(m),0) &\rightarrow u11(ack\_in(m,s(0))) \\
u11(ack\_out(n)) &\rightarrow ack\_out(n) \\
ack\_in(s(m),s(n)) &\rightarrow u21(ack\_in(s(m),n),m) \\
u21(ack\_out(n),m) &\rightarrow u22(ack\_in(m,n)) \\
u22(ack\_out(n)) &\rightarrow ack\_out(n)
\end{aligned}
$$

We can take as $\succeq_Q$ the lexicographic combination $(\succeq_T^{\mathcal{F}}, \succeq_I^N)_{lex}$ with the precedence $ack\_in \succ_{\mathcal{F}} ack\_out$, $ack\_in \succ_{\mathcal{F}} u11$, $ack\_in \succ_{\mathcal{F}} s$, $ack\_in \succ_{\mathcal{F}} s$, $ack\_in \sim_{\mathcal{F}} u21$ and $u21 \succ_{\mathcal{F}} u22$; and with the renaming map N which is the identity except $N(ack\_in) = Ack\_in$ and $N(u21) = U21$. For $\succeq_I$, the interpretation is given by $(ack\_out(x),a)$, $(ack\_in(x,y),a)$, $(Ack\_in(x,y),acku(x))$, $(u21(x,y),a)$, $(U21(X,Y),acku(Y))$,$(u11(X),a)$, and $(u22(X),a)$ and the basic quasi-ordering is RPO generated by the precedence $s \succ_P a$ and $0 \succ_P a$.

5 Generalizing Other Termination Proof Methods

As an application of the provided methods to generate suitable quasi-orderings $\succeq_Q$, we will show how two known termination proof methods, namely *dummy elimination* [FZ95] and the *dependency pairs* [AG97], can be seen as a particular instance of the monotonic semantic path ordering. Note that as a side effect, this provides a new simple proof of their correctness. Finally we study Geser's proposal and show that it is strictly weaker than ours.

5.1 Dummy Elimination

Dummy elimination consists of a transformation which eliminates function symbols from a signature replacing them by a constant ($\diamond$ in our notation); terms and rewrite rules are transformed accordingly. The soundness result states that a TRS R, defined over $\mathcal{T}(\mathcal{F} \cup \mathcal{F}_a, \mathcal{X})$, where $\mathcal{F}_a$ contains symbols of arity ≥ 1 which are to be eliminated, is terminating if the transformed TRS $E(R)$, defined over $\mathcal{T}(\mathcal{F} \cup \{\diamond\}, \mathcal{X})$, is terminating.

Let I_a be the interpretation defined by the pairs $(g(x_1,\ldots,x_n),\diamond_g)$ for every symbol g to be eliminated (and the identity pair for all other symbols). The system $E(R)$ is given by

$$
E(R) = \{\mathtt{cap}(l) \rightarrow u \mid u \in \{\mathtt{cap}(r)\} \cup \mathtt{dec}(r),\ l \rightarrow r \in R\}
$$

where $\mathtt{cap}(s) = I_a(s)$ and $\mathtt{dec}(s)$ contains $I_a(s')$ for all s' subterm of s just below a function symbol to be eliminated (for details see [FZ95]). For example, $R = \{f(f(x)) \rightarrow f(g(f(x)))\}$ is transformed, via the elimination of g, in the system $E(R) = \{f(f(x)) \rightarrow f(\diamond);\ f(f(x)) \rightarrow f(x)\}$, where $\diamond$ is the constant replacing g.

In the following we show that whenever termination of R can be shown using dummy elimination, we can find a simple quasi-reduction pair $\langle \succeq_I, \succeq_Q \rangle$ s.t. R is contained in MSPO.

Let $\succ$ be the reduction ordering containing $E(R)$, and $\succeq$ be $\succ \cup =$. Then we define $\succeq_I$ by means of the interpretation I_a defined above and $\succeq$ as basic quasi-ordering, i.e. $s \succeq_I t$ iff $I_a(s) \succeq I_a(t)$. On the other hand, we define $\succeq_Q$ as $s \succeq_Q t$ iff $I_a(s)(\succeq \cup \rhd)^* I_a(t)$. Note that if $\succ$ is a simplification ordering like RPO, the subterm relation is already included.

Now we show that $\langle \succeq_I, \succeq_Q \rangle$ is a quasi-reduction pair. Since $\succ$ is a reduction ordering, $\succeq$ is a quasi-reduction quasi-ordering, and, by proposition 2, $\succeq_I$ also is. As already said, since $\succ$ is a reduction ordering, we have that $(\succeq \cup \rhd)^*$ is a well-founded quasi-ordering stable under substitutions, and since $\succeq_I \subseteq \succeq_Q$, by propositions 3.1 and 4.1, we conclude that $\langle \succeq_I, \succeq_Q \rangle$ is a quasi-reduction pair.

Finally we show that if $E(R)$ is contained in $\succ$ then R is contained in $\succ_{mspo}$. Since $\mathsf{cap}(l) \to \mathsf{cap}(r) \in E(R)$ we have $I_a(l) = \mathsf{cap}(l) \succ \mathsf{cap}(r) = I_a(r)$ and hence $l \succeq_I r$. To show that $l \succ_{spo}^Q r$ we prove that $l \succ_{spo}^Q r'$ for all r' subterm of r, by induction on $|r'|$. We have that $u \rhd \mathsf{cap}(r')$ for some $u \in \{\mathsf{cap}(r)\} \cup \mathsf{dec}(r)$. By definition, $\mathsf{cap}(l) \to u \in E(R)$, and hence $\mathsf{cap}(l) \succ u$. Consequently, since the strict part of $(\succeq \cup \rhd)^*$ includes $(\succ \cup \rhd)^+$, we have $I_a(l) = \mathsf{cap}(l)(\succ \cup \rhd)^+ \mathsf{cap}(r') = I_a(r')$ and thus, $l \succ_Q r'$. By induction hypothesis $l \succ_{spo}^Q r'_i$ for all arguments r'_i of r', and therefore, by case 2, we have $l \succ_{spo}^Q r'$.

Recently, the *argument filtering* transformation method [KNT99] has been proposed. The basic idea of this method is the same as in dummy elimination but using more general term transformations, like the AFS in the dependency pair method, or the ones presented here. In fact, they coincide with the interpretations we have given but requiring that f and f_I are the same symbol. Therefore, we can prove, in the same way as for dummy elimination, that this method is a particular instance of MSPO, since we can take $\succeq_I$ and $\succeq_Q$ as before, but with this new interpretations.

Finally, let us remark that, by using interpretation from terms to multisets of terms, we can also show that the *distribution elimination* method [Zan94] is a particular case of MSPO. In our case the restrictions imposed on this method to be correct are necessary to assure that $\succeq_I$ and $\succeq_Q$ are stable under substitutions.

5.2 Dependency Pairs

We consider here the plain dependency pair method, i.e. the method without using what is called the *dependency graph* refinement (see [AG97,AG00] for details). In section 6 some ideas about how this refinement can be incorporated to our method will be given.

In this method for a given TRS R the signature $\mathcal{F}$ is split into two sets: the constructor symbols set C and the defined symbols set D. Defined symbols are those heading the left hand side of a rule in R and constructor symbols are all others.

Let N be a renaming map in $\mathcal{F}$ defined as $N(f) = F$ for all symbols $f \in D$ and the identity for the others. The dependency pairs of a rule $l \to r$ in R

is the set of pairs $\langle N(l), N(r') \rangle$ for every subterm r' of r headed by a symbol in D. For example, the rule $f(f(x)) \rightarrow f(g(f(x)))$ has the dependency pairs $\langle F(f(x)), F(g(f(x))) \rangle$ and $\langle F(f(x)), F(x) \rangle$.

Then R is terminating if there is a quasi-reduction quasi-ordering $\succeq$ such that for every rule $l \rightarrow r$ we have $l \succeq r$ and for every dependency pair $\langle s, t \rangle$ of $l \rightarrow r$ we have $s \succ t$. Note that in practice these conditions are expressed as an ordering constraint, called in the rest of the paper the *dependency pair constraint*.

In the following we show that whenever termination of R can be shown using dependency pairs, we can a find a simple quasi-reduction pair $\langle \succeq_I, \succeq_Q \rangle$ obtained from the ordering used by the dependency method, s.t. R is contained in MSPO.

Let $\succeq_{\mathcal{F}}$ be a precedence on $\mathcal{F}$ where $f \succeq_{\mathcal{F}} g$ iff $f \in D$ (hence $f \succ_{\mathcal{F}} g$ iff $f \in D$ and $g \in C$). Then we take $\succeq_I$ as $\succeq$ and $\succeq_Q$ as $(\succeq_T^{\mathcal{F}}, \succeq_I^N)_{lex}$. By propositions 3 and 4 (2 and 3), we have that $\langle \succeq_I, \succeq_Q \rangle$ is a quasi-reduction pair.

Finally we show that R is contained in $\succ_{mspo}$. We have directly that $l \succeq_I r$ since $l \succeq r$ in the dependency pairs proof. To show that $l \succ_{spo}^Q r$ we prove that $l \succ_{spo}^Q r'$ for all r' subterm of r, by induction on $|r'|$. If $top(r') \in C$ then $top(l) \succ_{\mathcal{F}} top(r')$ and hence $l \succ_Q r'$. Otherwise $top(l) \succeq_{\mathcal{F}} top(r')$ and there is a dependency pair $\langle N(l), N(r') \rangle$, s.t. $N(l) \succ N(r')$, which implies $l \succ_I^N r'$, and hence $l \succ_Q r'$. Since, by induction hypothesis $l \succ_{spo}^Q r'_i$ for all arguments r'_i of r', by case 2 of SPO, we have $l \succ_{spo}^Q r'$.

Note that we have only used case 2 of SPO for proving that R is contained in $\succ_{mspo}$. Moreover, note that the precedence we have used to build $\succeq_Q$ is quite weak in its strict part. We believe that by using better precedences as a first component of $\succeq_Q$, e.g. adding strict comparisons between the defined symbols, we can capture easily part of the power of the dependency graph (see section 6).

5.3 Geser's Monotonic Semantic Path Ordering

Now we analyze Geser's proposal [Ges92] for a monotonic SPO. We give here the strict part of his definition.

Definition 6. *Let $\succeq_Q$ be a quasi-reduction quasi-ordering.*

$$s \succ_G t \quad \text{iff} \quad s \succ_{spo}^Q t \text{ and } f(\ldots, s, \ldots) \succeq_Q f(\ldots, t, \ldots) \text{ for all } f \in \mathcal{F}$$

Although this version is an important step in the right direction, it has two main weaknesses. First of all, the requirement of $\succeq_Q$ to be a quasi-reduction quasi-ordering is too strong. As shown in example 2, this makes SPO loose completeness, and hence Geser's proposal cannot be complete. Furthermore, in a more practical view, with such a $\succeq_Q$ neither the dummy elimination technique nor the dependency pair technique can be included (note that, for instance, the renaming mapping does not preserve quasi-monotonicity). On the other hand, with respect to efficiency of an implementation of the method, a termination proof requires a huge number of comparisons with $\succeq_Q$, since for every rule $l \rightarrow r$,

every symbol f and every argument position of this symbol $f(\ldots,l,\ldots) \succeq_Q$
$f(\ldots,r,\ldots)$ has to be checked. Note that instead we only have to check $l \succeq_I r$.
Now we show that $\succ_G$ is included in $\succ_{mspo}$.

Lemma 2. *Let $\succeq_Q$ be a quasi-reduction quasi-ordering. There exists some $\succeq_I$,
s.t. $\langle \succeq_I, \succeq_Q \rangle$ is a quasi-reduction pair and $\succ_G \subseteq \succ_{mspo}$.*

Proof. We take $\succeq_I$ as $s \succeq_I t$ iff $f(\ldots,s,\ldots) \succeq_Q f(\ldots,t,\ldots)$ for all $f \in \mathcal{F}$. It is
obvious that in this case $\succ_G \subseteq \succ_{mspo}$. Then, we have to prove that $\langle \succeq_I, \succeq_Q \rangle$
is a quasi-reduction pair, that is, $\succeq_I$ is quasi-monotonic on $\succeq_Q$, which follows
by definition of $\succeq_I$, and $\succeq_I$ is a quasi-monotonic quasi-ordering stable under
substitutions, which follows directly from the properties of $\succeq_Q$.   □

6   Constraints

Using MSPO, we can translate our termination problem into an *ordering constraint solving problem* (similar to the ones given in [Com90] for lexicographic
path ordering, except that here variables are universally quantified), which is,
in general, more suitable for automation. This translation is simply based on
applying the definition of MSPO, and SPO, to the rules of the TRS.

Let R be a set of rules $\{l_i \to r_i \mid 1 \le i \le n\}$. We consider the following initial
MSPO-constraint:

$$l_1 \succ_{mspo} r_1 \wedge \ldots \wedge l_n \succ_{mspo} r_n$$

Which is transformed by applying the definition of MSPO into the conjunction
of two constraints I_C and SPO_C

$$I_C : \quad l_1 \succeq_I r_1 \wedge \ldots \wedge l_n \succeq_I r_n$$
$$SPO_C : l_1 \succ^Q_{spo} r_1 \wedge \ldots \wedge l_n \succ^Q_{spo} r_n$$

Now the definition of SPO is applied to the second part of the constraint. This
is formalized by means of constraint transformation rules:

$$s \succeq^Q_{spo} t \Longrightarrow \top \qquad\qquad\qquad\quad \text{if } s \equiv t$$
$$s \succeq^Q_{spo} t \Longrightarrow s \succ^Q_{spo} t \qquad\qquad\quad \text{if } s \not\equiv t$$
$$x \succ^Q_{spo} t \Longrightarrow \bot$$
$$s \succ^Q_{spo} x \Longrightarrow \top \qquad\qquad\qquad \text{if } s \not\equiv x \in Vars(s)$$
$$s = f(s_1,\ldots,s_m) \succ^Q_{spo} g(t_1,\ldots,t_n) = t \Longrightarrow$$
$$s_1 \succeq^Q_{spo} t \vee \ldots \vee s_m \succeq^Q_{spo} t \vee$$
$$(s \succ_Q t \wedge s \succ^Q_{spo} t_1 \wedge \ldots \wedge s \succ^Q_{spo} t_n) \vee$$
$$(s \succeq_Q t \wedge \{s_1,\ldots,s_m\} \gg^Q_{spo} \{t_1,\ldots,t_n\})$$

Where $\{s_1,\ldots,s_m\} \gg^Q_{spo} \{t_1,\ldots,t_n\}$ is translated into a constraint over $\succ^Q_{spo}$
and $\succeq^Q_{spo}$.

It is easy to see that these transformation rules are terminating and confluent.
Moreover, the resulting normal form is an ordering constraint over $\succ_Q$ and $\succeq_Q$.
Then after computing the disjunctive normal form, the initial constraint SPO_C

has been translated into a disjunction of constraints over $\succ_Q$ and $\succeq_Q$ each one of the form

$$Q_C : \quad s_1 \succ_Q t_1 \wedge \ldots \wedge s_p \succ_Q t_p \wedge s_1' \succeq_Q t_1' \wedge \ldots \wedge s_q' \succeq_Q t_q'$$

where none of the terms are variables.

Now we have to find a reduction pair $\langle \succeq_I, \succeq_Q \rangle$ satisfying I_C and one of these constraints Q_C. Note that, from what we have seen in section 5.2, the dependency pair constraint can be obtained by applying always case 2 of SPO, which means that some Q_C obtained above represents this path. Since there are several possible Q_C, it may happen, as in example 6, that the path chosen by the dependency pair method is not the easiest one.

To solve the constraints I_C and Q_C some simplification techniques are necessary. A first simple example of such a simplification, is obtained by considering that $\succeq_Q$ is a lexicographic combination $\langle \succeq_T^{\mathcal{F}}, \succeq_M \rangle$ with $\langle \succeq_I, \succeq_M \rangle$ being a quasi-reduction pair. Then we take the constraint Q_C, and define $top(s) \succeq_{\mathcal{F}} top(t)$ iff either $s \succeq_Q t$ or $s \succ_Q t$ is in Q_C. Now, if $\succ_{\mathcal{F}}$ is the strict part of $\succeq_{\mathcal{F}}$ and $\sim_{\mathcal{F}}$ its equivalence, we can simplify the constraint Q_C into M_C by the following rules:

$$
\begin{aligned}
s \succ_Q t &\Longrightarrow \top & \text{if } top(s) \succ_{\mathcal{F}} top(t) \\
s \succeq_Q t &\Longrightarrow \top & \text{if } top(s) \succ_{\mathcal{F}} top(t) \\
s \succ_Q t &\Longrightarrow s \succ_M t & \text{if } top(s) \sim_{\mathcal{F}} top(t) \\
s \succeq_Q t &\Longrightarrow s \succeq_M t & \text{if } top(s) \sim_{\mathcal{F}} top(t)
\end{aligned}
$$

At this point, in general, it is interesting to define $\succeq_M$ by means of a renaming N of all symbols heading s or t in some $s \succ_M t$ or $s \succeq_M t$ in M_C, which leads to a final renamed constraint called N_C.

Solving the constraints I_C and N_C, i.e., finding an adequate quasi-reduction pair, is very similar to solve the dependency pair constraints. The simplest solution is to consider the quasi-reduction pair $\langle \succeq_I, \succeq_I \rangle$, but we believe that, if necessary, we will be able to use refinements like the dependency graph in our constraint solver.

7   Modularity

In this section we present some modularity results for MSPO which are obtained applying known abstract sufficient conditions ensuring modularity of termination. We consider disjoint unions of TRSs, i.e., systems that do not share any symbol, and constructor sharing unions of TRSs, i.e., systems which share only constructors.

A TRS R is called terminating under non-deterministic collapses, denoted $\mathcal{C}_\varepsilon$-terminating, if $R \cup \{G(x,y) \to x, G(x,y) \to y\}$ terminates for some new symbol G. For $\mathcal{C}_\varepsilon$-termination we have the following results:

- [Ohl94] $\mathcal{C}_\varepsilon$-termination is a modular property for disjoint unions of TRSs.
- [Gra94] $\mathcal{C}_\varepsilon$-termination is a modular property for constructor-sharing unions of finite TRSs.

Lemma 3. *If* $\trianglerighteq \subseteq \succeq_B$ *and* R *is included in* $\succ_{mspo}$ *then* R *is* C_ε*-terminating.*

Note that since $\succeq_B$ includes the subterm relation and G is a new symbol, we have $I(G(x,y)) = G(x,y)$ and hence $G(x,y) \succeq_I x$ and $G(x,y) \succeq_I y$. Therefore, since by case 1 both rules are included in SPO, we can conclude that $G(x,y) \succ_{mspo} x$ and $G(x,y) \succ_{mspo} y$.

Corollary 1. *Let* $\succ^1_{mspo}$ *and* $\succ^2_{mspo}$ *be two MSPO whose basic orderings include the subterm relation and let* R_1 *and* R_2 *be TRSs that are included in* $\succ^1_{mspo}$ *and* $\succ^2_{mspo}$ *respectively.*

- *If* R_1 *and* R_2 *are disjoint then* $R_1 \cup R_2$ *is* C_ε*-terminating, and thus terminating.*
- *If* R_1 *and* R_2 *share only constructors and are finite then* $R_1 \cup R_2$ *is* C_ε*-terminating, and thus terminating.*

Similarly in [GO98] modularity results for disjoint and constructor sharing unions of TRSs proved terminating using the dependency pair method are presented. For disjoint systems the results are the same, but for constructor sharing unions the restriction is not imposed on the finiteness of the systems but on the treatment of the shared symbols in the termination proof.

8   Conclusion

In this paper we have described a new ordering-based general method for proving termination of TRSs. MSPO is based on the well-known SPO, but unlike SPO it is monotonic, which makes it useful in practice. It is a complete method, i.e., it characterizes termination. The method generalizes, in a simple way, many known methods based on transformations. In the case of the dependency pairs method, which is by now one of the most successful general methods applied in practice, we have only shown that we generalize it without the "dependency graph" refinement. These kind of "operational" refinements do not fit so well in our framework, but, by considering the termination of TRSs as an ordering constraint solving problem over MSPO, we believe that, if necessary, we will be able to use these refinements in our constraint solver. On the other hand, due to its additional flexibility, we have seen that MSPO provides, in some cases, simpler termination proofs than the dependency pair method. Thus, in order to study the behavior of MSPO, we are developing a termination system based on this method. Currently this system can only check termination once the ingredients for the MSPO are provided, but our aim is to fully automate the termination proofs, by using some heuristics for selecting ingredients to be tried.

Besides its application to first-order term rewriting, the fact that the method is defined by means of orderings, for which the properties to be fulfilled are well-known, opens the door to other important classes like AC-rewriting (i.e., rewriting modulo associativity and commutativity axioms) or higher-order rewriting, for which the lack of general methods is more important. The idea is to combine with MSPO some recent results obtained for RPO in the AC-case [Rub99] and in the HO-case [JR99], since all them share the same structure and are based on orderings.

Finally, apart from the presented modularity results, we are studying other kinds of combinations. In particular, we are interested in the so called *hierarchical combinations* (see [AG98] related results for the dependency pair method), since reusing termination proofs or proving the termination of the TRS by splitting it in different parts may be crucial in practice.

Acknowledgments: We want to thank Pilar Nivela, Roberto Nieuwenhuis, Enno Ohlebusch and Claude Marche for their help in the development of this work and to Thomas Arts, Jürguen Giesl and Aart Middeldorp for their comments on an earlier version of this work.

References

[AG97] T. Arts and J. Giesl. Automatically proving termination where simplification orderings fail. In *TAPSOFT: 7th International Joint Conference on Theory and Practice of Software Development*, LNCS 1214, pages 261–272. Springer-Verlag, 1997.

[AG98] T. Arts and J. Giesl. Modularity of termination using dependency pairs. In T. Nopkow, editor, *9th International Conference on Rewriting Techniques and Applications (RTA)*, LNCS 1379, pages 226–240, Tsukuba, Japan, 1998. Springer-Verlag.

[AG00] T. Arts and J. Giesl. Termination of term rewriting using dependency pairs. *Theoretical Computer Science*, 236:133–178, 2000.

[BD86] Leo Bachmair and Nachum Dershowitz. Commutation, transformation, and termination. In Jörg H. Siekmann, editor, *8th International Conference on Automated Deduction (CADE)*, LNCS 230, pages 5–20, Oxford, England, 1986. Springer-Verlag.

[BL90] Françoise Bellegarde and Pierre Lescanne. Termination by completion. *Applicable Algebra in Engineering, Communication and Computing*, 1:79–96, 1990.

[BN98] F. Baader and T. Nipkow. *Term Riwriting and all that.* Cambridge University Press, 1998.

[Com90] Hubert Comon. Solving symbolic ordering constraints. *International Journal of Foundations of Computer Science*, 1(4):387–411, 1990.

[Der82] Nachum Dershowitz. Orderings for term-rewriting systems. *Theoretical Computer Science*, 17(3):279–301, 1982.

[Der87] Nachum Dershowitz. Termination of rewriting. *Journal of Symbolic Computation*, 3:69–116, 1987.

[DJ90] Nachum Dershowitz and Jean-Pierre Jouannaud. Rewrite systems. In Jan van Leeuwen, editor, *Handbook of Theoretical Computer Science*, volume B: Formal Models and Semantics, chapter 6, pages 244–320. Elsevier Science Publishers B.V., 1990.

[FZ95] Maria Ferreira and Hans Zantema. Dummy elimination: making termination easier. In H. Reichel, editor, *10th Int. Conf. on Fundamentals of Computation Theory*, LNCS 965, pages 243–252. Springer-Verlag, 1995.

[Ges92]   Alfons Geser. On a monotonic semantic path ordering. Technical Report
          92-13, Ulmer Informatik-Berichte, Universität Ulm, Ulm, Germany, 1992.
[GO98]   J. Giesl and E. Ohlebusch. Pushing the frontiers of combining rewrite systems
          farther outwards. In *Proceedings of the Second International Workshop on
          Frontiers of Combining Systems (FroCoS '98)*, Logic and Computation Series,
          pages 141–160, Amsterdam, The Netherlands, 1998. Research Studies Press,
          John Wiley & Sons.
[Gra94]   B. Gramlich. Generalized sufficient conditions for modular termination of
          rewriting. *Applicable Algebra in Engineering, Communication and Comput-
          ing*, 5:131–158, 1994.
[JR99]   Jean-Pierre Jouannaud and Albert Rubio. The higher-order recursive path
          ordering. In *14th IEEE Symposium on Logic in Computer Science (LICS)*,
          pages 402–411, Trento, Italy, 1999.
[KL80]   S. Kamin and J.-J. Levy. Two generalizations of the recursive path ordering.
          Unpublished note, Dept. of Computer Science, Univ. of Illinois, Urbana, IL,
          1980.
[Klo92]   J.W. Klop. Term rewriting systems. In S. Abramsky, D.M. Gabbay, and
          T. S. E. Maibaum, editors, *Handbook of Logic in Computer Science*, volume 2,
          pages 1–116. Oxford University Press, 1992.
[KNT99]   K. Kusakari, M. Nakamura, and Y. Toyama. Argument filtering transfor-
          mation. In *Proceedings of the International Conference on Principles and
          Practice of Declarative Programming (PPDP'99)*, LNCS 1702, pages 47–61.
          Springer-Verlag, 1999.
[Kru60]   Joseph B. Kruskal. Well-quasi-ordering, the Tree Theorem, and Vazsonyi's
          conjecture. *Transactions of the American Mathematical Society*, 95:210–225,
          May 1960.
[MHH96]   A. Middeldorp, Ohsaki H., and Zantema H. Transforming termination by
          self-labelling. In *Proceedings of the 13th International Conference on Au-
          tomated Deduction*, volume 1104 of *LNAI*, pages 373–386, New Brunswick,
          1996. Springer-Verlag.
[OCM00]   E. Ohlebusch, C. Claves, and C. Marché. Talp: a tool for the termination
          analysis of logic programs. In *Proceedings of the 11th conference on Rewriting
          Techniques and Applications*, LNCS. Springer-Verlag, 2000.
[Ohl94]   E. Ohlebusch. On modularity of termination of term rewriting systems. *The-
          oretical Computer Science*, 136(2):333–360, 1994.
[Rub99]   A. Rubio. A fully syntactic AC-RPO. In P. Narendran and M. Rusinowitch,
          editors, *Tenth International Conference on Rewriting Techniques and Ap-
          plications (RTA)*, LNCS 1631, pages 133–147, Trento, Italy, 1999. Springer-
          Verlag.
[Ste95]   J. Steinbach. Automatic termination proofs with transformation orderings.
          In J. Hsiang, editor, *6th International Conference on Rewriting Techniques
          and Applications (RTA)*, LNCS 914, pages 11–25, Kaiserslautern, Germany,
          1995.
[Xi98]   H. Xi. Towards automated termination proofs through freezing. In *9th Int.
          Conf. on Rewriting Techniques and Applications (RTA)*, LNCS 1379, pages
          271–285, Tsukuba, Japan, 1998.
[Zan94]   H. Zantema. Termination of term rewriting: interpretation and type elimi-
          nation. *Journal of Symbolic Computation*, 17:23–50, 1994.

Stratified Resolution

Anatoli Degtyarev[1] and Andrei Voronkov[2]

[1] Manchester Metropolitan University
a.degtiarev@doc.mmu.ac.uk
[2] University of Manchester
voronkov@cs.man.ac.uk

Abstract. We introduce a calculus of *stratified resolution*, in which special attention is paid to clauses that "define" relations. If such clauses are discovered in the initial set of clauses, they are treated using the rule of *definition unfolding*, i.e. the rule that replaces defined relations by their definitions. Stratified resolution comes with a new, previously not studied, notion of redundancy: a clause to which definition unfolding has been applied can be removed from the search space. To prove completeness of stratified resolution with redundancies we use a novel technique of *traces*.

1 Introduction

In this article we introduce two versions of *stratified resolution*, — a resolution calculus with special rules for handling hierarchical definitions of relations. Stratified resolution generalizes SLD-resolution for Horn clauses to a more general case, where clauses may be non-Horn but "Horn with respect to a set of relations".

EXAMPLE 1 Suppose we try to establish inconsistency of a set of clauses S containing a recursive definition of a relation *split* that splits a list of conferences into two sublists: of deduction-related conferences, and of all other conferences.

$split([x \,|\, y], [x \,|\, z], u) :- deduction(x), split(y, z, u).$
$split([x \,|\, y], z, [x \,|\, u]) :- \neg deduction(x), split(y, z, u).$
$split([], [], []).$

Suppose that S also contains other clauses, for example

$\neg split(x, y, z) \lor conference\_list(x).$

If we use ordered resolution with negative selection (as most state-of-the-art systems would do), we face several choices in selecting the order and negative literals in clauses. For example, if we choose the order in which every literal with the relation *deduction* is greater than any literal with the relation *split*, then we must select either $\neg deduction(x)$ or $\neg split(y, z, u)$ in the first clause. It seems

D. McAllester (Ed.): CADE-17, LNAI 1831, pp. 365–384, 2000.

much more natural to select $split(\llbracket x \mid y \rrbracket, \llbracket x \mid z \rrbracket, u)$ instead, then we can use the first clause in the same way it would be used in logic programming. Likewise, if we always try to select a negative literal in a clause, the literal $\neg split(y, z, u)$ will be selected in the second clause, which is most likely a wrong choice, since then any resolvent with the second clause will give us a larger clause.

Let us now choose an ordering in which the literals $split(\llbracket x \mid y \rrbracket, \llbracket x \mid z \rrbracket, u)$ and $split(\llbracket x \mid y \rrbracket, z, \llbracket x \mid u \rrbracket)$ are maximal in their clauses, and select these literals. Consider the fourth clause. If we select $\neg split(x, y, z)$ in it, we can resolve this literal with all three clauses defining $split$. It would be desirable to select $conference\_list(x)$ in it (if our ordering allows us to do so), since a resolvent upon $conference\_list(x)$ is likely to instantiate x to a nonvariable term t, and then the literal $\neg split(t, y, z)$ can be resolved with only two, one or no clauses at all, depending on the form of t.

In all cases, it seems reasonable to choose an ordering and selection function in such a way that the first three clauses will be used as a *definition* of $split$ so that we *unfold* this definition, i.e. replace the heads of these clauses with their bodies. Such ordering would give us the best results if we have a right strategy of negative selection which says: select $\neg split(t, r, s)$ only if t is instantiated enough, or if we have no other choice.

In order to implement this idea we have to be able to formalize the right notion of a "definition" in a set of clauses. Such a formalization is undertaken in our paper, in the form of a calculus of *stratified resolution*. Stratified resolution is based on the following ideas that can be tracked down to earlier ideas developed in logic programming.

1. Logic programming is based on the idea of using definite clauses as definitions of relations. Similar to the notion of definite clause, we introduce a more general notion of a set of clauses *definite w.r.t. a set of relations*. These relations are regarded as defined by this set of clauses.
2. In logic programming, relations are often defined in terms of other relations. The notion of *stratification* [5, 1, 8] allows one to formalize the notion "*P* is defined in terms of *Q*". We use a similar idea of stratification, but in our case stratification must be related to a reduction ordering $\succ$ on literals.

Consider another example.

EXAMPLE 2 The difficult problem is to find automatically the right ordering that makes the atom in the head of a "definition" greater than atoms in the body of this definition. Consider, for example, clauses defining reachability in a directed graph, where the graph is formalized by the binary relation *edge*:

$reachable(x, y) :- edge(x, y).$
$reachable(x, z) :- edge(x, y), reachable(y, z).$

There is no well-founded ordering stable under substitutions that makes the atom $reachable(x, z)$ greater than $reachable(y, z)$. So the standard ordered resolution with negative selection cannot help us in selecting the "right" ordering. The theory developed in this paper allows one to select only the literal $reachable(x, z)$ in this clause despite that this literal is not the greatest.

In addition to intelligent selection of literals in clauses, stratified resolution has a new kind of redundancy, not exploited so far in automated deduction. To explain this kind of redundancy, let us come back to Example 1. Suppose we have a clause

$$\neg split([cade, www, lpar], y, z). \tag{1}$$

Stratified resolution can resolve this clause with the first two clauses in the definition of $split$, obtaining two new clauses

$$deduction(cade), split([www, lpar], y, z);$$
$$\neg deduction(cade), split([www, lpar], y, z).$$

In resolution-based theorem proving these two clauses would be *added* to the search space. We prove that they can *replace* clause (1) thus making the search space smaller.

When the initial set of clauses contains no definitions or cannot be stratified, stratified resolution becomes the ordinary ordered resolution with negative selection. However, sets of clauses which contain definitions and can be stratified in our sense are often met in practice, since they correspond to definitions (maybe recursive) of relations of a special form. For example, a majority of TPTP problems can be stratified.

This paper is organized as follows. In Section 2 we define the ground version of stratified resolution and prove its soundness and completeness. Then in Section 3 we define stratified resolution with redundancies, a calculus in which a clause can be removed from the search space after a definition unfolding has been applied to it. Then in Section 4 we define a nonground version of stratified resolution with redundancies.

In this paper we deal with first-order logic without equality, we will only briefly discuss equality in Section 6.

Related Work

There are not so many papers in the automated deduction literature relevant to our paper. Our formal system resembles SLD-resolution [6]. When the initial set of clauses is Horn, our stratified resolution with redundancies becomes SLD-resolution. The possibility for arbitrary selection for Horn clauses, and even in the case of equational logic was proved in [7]. For one of our proofs we used a renaming technique introduced in [2,3].

2 Stratified Resolution: The Ground Case

As usual, we begin with a propositional case, lifting to the general case will be standard.

Throughout this paper, we denote by $\mathcal{L}$ a set of propositional atoms. The literal complementary to a literal L is denoted by $\overline{L}$. For every set of atoms $\mathcal{P}$, we call a $\mathcal{P}$-*atom* any atom in $\mathcal{P}$, and a $\mathcal{P}$-*literal* any literal whose atom is in $\mathcal{P}$. A *clause* is a finite multiset of literals. We denote literals by L, M and clauses by C, D, maybe with indices. The empty clause is denoted by $\Box$. We use the standard notation for multisets, for example write C_1, C_2 for the multiset union of two clauses C_1 and C_2, and write $L \in C$ if the literal L is a member of the clause C. For two ground clauses C_1 and C_2, we say that C_1 *subsumes* C_2, if C_1 is a submultiset of C_2. This means, for example, that A, A does *not* subsume A.

Let $\succ$ be a total well-founded ordering on $\mathcal{L}$. We extend this ordering on literals in $\mathcal{L}$ in the standard way, such that for every atom A we have $\neg A \succ A$ and there is no literal L such that $\neg A \succ L \succ A$. If L is a literal and C is a clause, we write $L \succ C$ if for every literal $M \in C$ we have $L \succ M$. Usually, we will write a clause as a disjunction of its literals. As usual, we write $L \succeq L'$ if $L \succ L'$ or $L = L'$ and write $L \succeq C$ if for every literal $M \in C$ we have $L \succeq M$. In this paper we assume that $\succ$ always denotes a fixed well-founded ordering.

We will now define a notion of selection function which makes a slight deviation from the standard notions (see e.g. [3]. However, the resulting inference system will be standard, and several following definitions will become simpler. We call a *selection function* any function σ on the set of clauses such that (i) $\sigma(C)$ is a set of literals in C, (ii) $\sigma(C)$ is nonempty whenever C is nonempty, and (iii) if $A \in \sigma(C)$ and A is a positive literal, then $A \succeq C$. If $L \in \sigma(C)$, we say that L is *selected by* σ in C. When we use a selection function, we underline selected literals, so when we write $\underline{A} \vee C$, this means that A (and maybe some other literals) are selected in $A \vee C$.

In our proofs we will use the result on completeness of the inference system of ordered binary resolution with negative selection (but our main inference system will be different). Let us define this inference system.

DEFINITION 3 Let σ be a selection function. The *inference system* $R_\sigma^\succ$ consists of the following inference rules:

1. *Positive ordered factoring*:

$$\frac{C \vee A \vee \underline{A}}{C \vee A},$$

where A is a positive literal.

2. *Binary ordered resolution with selection:*

$$\frac{C \vee \underline{A} \quad D \vee \underline{\neg A}}{C \vee D},$$

where $A \succ C$.

The following theorem is well-known (for example, a stronger statement can be found in [3]).

THEOREM 4 *Let S be a set of clauses. Then S is unsatisfiable if and only $\square$ is derivable from S in $R_\sigma^\succ$.*

The following two definitions are central to our paper.

DEFINITION 5 Let $\mathcal{P} \subseteq \mathcal{L}$ be a set of atoms. A set S of clauses is *Horn with respect to* $\mathcal{P}$, if every clause of S contains at most one positive $\mathcal{P}$-literal. A clause D is called *definite with respect to* $\mathcal{P}$ if $\mathcal{D}$ contains exactly one positive $\mathcal{P}$-literal. Let $P \vee L_1 \vee \ldots \vee L_n$ be a clause definite w.r.t. $\mathcal{P}$ such that $P \in \mathcal{P}$. We will sometimes denote this clause by $P :- \overline{L}_1, \ldots, \overline{L}_n$.

DEFINITION 6 (Stratification) We call a $\succ$-*stratification of* $\mathcal{L}$ any finite sequence $\mathcal{L}_n \succ \ldots \succ \mathcal{L}_0$ of subsets of $\mathcal{L}$ such that

1. $\mathcal{L} = \mathcal{L}_0 \cup \ldots \cup \mathcal{L}_n$;
2. If $m > n$, $A \in \mathcal{L}_m$, and $B \in \mathcal{L}_n$, then $A \succ B$.

We will denote $\succ$-stratifications by $\mathcal{L}_n \succ \ldots \succ \mathcal{L}_0$ and call them simply stratifications.

From now on we assume a fixed stratification of $\mathcal{L}$ of the form

$$\mathcal{Q}_n \succ \mathcal{P}_n \succ \mathcal{Q}_{n-1} \succ \mathcal{P}_{n-1} \succ \ldots \succ \mathcal{Q}_1 \succ \mathcal{P}_1 \succ \mathcal{Q}_0. \qquad (2)$$

We denote $\mathcal{P} = \mathcal{P}_n \cup \ldots \cup \mathcal{P}_1$ and $\mathcal{Q} = \mathcal{Q}_n \cup \ldots \cup \mathcal{Q}_0$ and use this notation throughout the paper. Atoms in $\mathcal{P}$ will be denoted by P (maybe with indices).

Let C be a clause definite w.r.t. $\mathcal{P}$. Then C contains a positive literal $P \in \mathcal{P}_i$, for some i. We say that C *admits stratification (2)*, if all atoms occurring in C belong to $\mathcal{P}_i \cup \mathcal{Q}_{i-1} \cup \ldots \cup \mathcal{P}_1 \cup \mathcal{Q}_0$. Note that every such clause has the form

$$P :- \underbrace{P_1, \ldots, P_k}_{\substack{\text{atoms in} \\ \mathcal{P}_i \cup \ldots \cup \mathcal{P}_1}}, \underbrace{L_1, \ldots, L_l}_{\substack{\mathcal{Q}_{i-1} \cup \ldots \cup \mathcal{Q}_0\text{-} \\ \text{literals}}} .$$

EXAMPLE 7 Consider the set consisting of four clauses: $A \vee B$, $A \vee \neg B$, $\neg A \vee B$, and $\neg A \vee \neg B$. This set is Horn with respect to $\{A\}$ and also with respect to $\{B\}$, but not with respect to $\{A, B\}$. This set of clauses admits stratification $\emptyset \succ \{A\} \succ \{B\}$, in which A is considered as a relation defined in terms of B, but also admits $\emptyset \succ \{B\} \succ \{A\}$, in which B is considered as defined in terms of A. This example shows that it is hard to expect to find the "greatest" (in any sense) stratification.

Let us fix a well-founded order $\succ$, a selection function σ, and a $\succ$-stratification (2).

DEFINITION 8 (Stratified Resolution) The inference system of *stratified resolution*, denoted $\mathcal{SR}$, consists of the following inference rules.

1. *Positive ordered factoring*:

$$\frac{C \vee A \vee \underline{A}}{C \vee A},$$

where A is a positive literal, and C contains no positive $\mathcal{P}$-literals.

2. *Binary ordered resolution with selection:*

$$\frac{C \vee \underline{A} \quad D \vee \neg \underline{A}}{C \vee D},$$

where $A \succ C$ and C, D, A contain no positive $\mathcal{P}$-literals.

3. *Definition unfolding:*

$$\frac{C \vee P \quad D \vee \underline{\neg P}}{C \vee D},$$

where $P \in \mathcal{P}$ and D contains no positive $\mathcal{P}$-literals. Note that in this rule we do not require that P be selected in $C \vee P$.

THEOREM 9 (Soundness and Completeness of Stratified Resolution)
Let S be a set of clauses Horn w.r.t. $\mathcal{P}$. Let, in addition, every clause in S definite w.r.t. $\mathcal{P}$ admits stratification (2). Then S is unsatisfiable if and only if $\square$ is derivable from S in the system of stratified resolution.

PROOF. Soundness is obvious. To prove completeness, we use a technique of [3], (see also [2]) for proving completeness of resolution with free selection for Horn sets. Let us explain the idea of this proof. The inference rules of stratified resolution reminds us of the rules of ordered resolution with negative selection, but with a nonstandard selection function: in any clause $P \vee C$ definite w.r.t. $\mathcal{P}$ the literal P is selected. We could use Theorem 4 on completeness of ordered resolution with negative selection if P was maximal in $P \vee C$, since we could then select P in $P \vee C$ using a standard selection function. However, P is not necessarily maximal: there can be other $\mathcal{P}$-literals in C greater than P. What we do is to "rename" the clause $P \vee C$ so that P becomes greater than C.

Formally, let $\mathcal{P}'$ be a new set of atoms of the form P^n, where $P \in \mathcal{P}$, and n is a natural number. Denote by $\mathcal{L}'$ the set of atoms $\mathcal{P}' \cup \mathcal{Q}$. We will refer to the set $\mathcal{L}'$ as the *new language* as opposite to the *original language* $\mathcal{L}$. Define a mapping $\rho : \mathcal{L}' \longrightarrow \mathcal{L}$ by (i) $\rho(P^n) = P$ for any natural number n, (ii) $\rho(A) = A$ if $A \in \mathcal{Q}$. Extend the mapping to literals and clauses in the new language in a natural way: $\rho(\neg A) = \neg \rho(A)$ and $\rho(L_1 \vee \ldots \vee L_n) = \rho(L_1) \vee \ldots \vee \rho(L_n)$.

For every clause C Horn w.r.t. $\mathcal{P}$ of the original language we define a set of clauses C^ρ in the new language as follows.

1. If C is definite w.r.t. $\mathcal{P}$, then C has the form $P \vee \neg P_1, \ldots, \vee \neg P_k \vee D$, where $k \geq 0$ and D has no $\mathcal{P}$-literals. We define C^ρ as the set of clauses

$$\{ P^{1+n_1+\ldots+n_k} \vee \neg P_1^{n_1} \vee \ldots \vee \neg P_k^{n_k} \vee D \mid n_1, \ldots, n_k \text{ are natural numbers } \}.$$

2. Otherwise, C has the form $\neg P_1 \vee \ldots \vee \neg P_k \vee D$, where $k \geq 0$ and D has no $\mathcal{P}$-literals. We define C^ρ as the set of clauses

$$\{\neg P_1^{n_1} \vee \ldots \vee \neg P_k^{n_k} \vee D \mid n_1, \ldots, n_k \text{ are natural numbers }\}.$$

For any set of clauses S Horn w.r.t. $\mathcal{P}$ we define S^ρ as the set of clauses $\bigcup_{C \in S} C^\rho$. We prove the following:

(3) S is satisfiable if and only if so is S^ρ.

Suppose S is satisfiable, then some valuation $\tau : \mathcal{L} \longrightarrow \{true, false\}$ satisfies S. Define a valuation $\tau' : \mathcal{L}' \longrightarrow \{true, false\}$ such that $\tau'(A) = \tau(A)$, if $A \in \mathcal{Q}$, and $\tau'(P^n) = \tau(P)$. It is not hard to argue that τ' satisfies S^ρ.

   Now we suppose that S is unsatisfiable and show that S^ρ is unsatisfiable, too. We apply induction on the number k of $\mathcal{Q}$-literals occurring in S.

1. *Case $k = 0$.* Then S is a set of Horn clauses. This case has been considered in [3]. (The idea is that P^n is interpreted as P has derivation by SLD-resolution in n steps).
2. *Case $k > 0$.* For any set of clauses T and literal L, denote by T/L the set of clauses obtained from T be removing all clauses containing L and removing from the rest of clauses all occurrences of literals $\overline{L}$. Note that if L is a $\mathcal{Q}$-literal occurring in T, then T/L contains less $\mathcal{Q}$-literals than T. It is not hard to argue that T is unsatisfiable if and only so are both T/L and $T/\overline{L}$. Take any $\mathcal{Q}$-literal L occurring in S. Since S is unsatisfiable, so are both S/L and $S/\overline{L}$. By the induction hypothesis, $(S/L)^\rho$ and $(S/\overline{L})^\rho$ are unsatisfiable, too. It is easy to see that $(S/L)^\rho = S^\rho/L$ and $(S/\overline{L})^\rho = S^\rho/\overline{L}$, so both S^ρ/L and $S^\rho/\overline{L}$ are unsatisfiable. But then S^ρ is unsatisfiable too.

The proof of (3) is completed. Let us continue the proof of Theorem 9. Define an order $\succ'$ on $\mathcal{L}'$ as follows:

1. $\succ'$ and $\succ$ coincide on $\mathcal{Q}$-literals.
2. If L is a $\mathcal{Q}$-atom and P is a $\mathcal{P}$-atom, then for every n we let $L \succ' P^n$ if and only if $L \succ P$.
3. $P_1^{n_1} \succ' P_2^{n_2}$ if $P_1 \in \mathcal{P}_i$, $P_2 \in \mathcal{P}_j$ and $i > j$.
4. $P_1^{n_1} \succ' P_2^{n_2}$ if P_1 and P_2 belong to the same set $\mathcal{P}_i$ and $n_1 > n_2$.
5. $P_1^n \succ' P_2^n$ if P_1 and P_2 belong to the same set $\mathcal{P}_i$ and $P_1 \succ P_2$.

Using the selection function σ on clauses of the original language we will now define a selection function σ' on clauses of the new language. We will only be interested in the behavior of σ' on clauses Horn w.r.t. $\mathcal{P}'$, so we do not define it for other clauses.

1. If a clause C is definite w.r.t. $\mathcal{P}'$, then σ' selects in C the maximal literal.
2. If C contains no positive $\mathcal{P}$-literals, then σ' selects a literal L in C if and only if σ selects $\rho(L)$ in $\rho(C)$.

Note that our ordering $\succ$ is defined in such a way that for any clause $(P^n \vee C) \in S^\rho$ definite w.r.t. $\mathcal{P}'$ we have $P^n \succ C$, and hence P^n is always selected in this clause.

It is not hard to argue that σ' is a selection function. To complete the proof we use Theorem 4 on completeness of the system $R_{\sigma'}^{\succ}$ applied to S^ρ.

Consider a derivation of a clause D' from S^ρ with respect to $R_{\sigma'}^{\succ}$. We show simultaneously that:

(4)   there exists a derivation of the clause $D = \rho(D')$ from S in the system $\mathcal{SR}$;

(5)   if $D' \notin S^\rho$, then D' contains no positive $\mathcal{P}'$-literal.

We apply induction on the length of the derivation of D'. If $D' \in S^\rho$, then $\rho(D') \in S$, so the induction base is straightforward. Assume now that the derivation of D' is obtained by an inference from $D_1', \ldots, D_n'$. We will show that D can be obtained by an inference in $\mathcal{SR}$ from $\rho(D_1'), \ldots, \rho(D_n')$, this will imply (4), since $\rho(D_1'), \ldots, \rho(D_n')$ can be derived in $\mathcal{SR}$ from S by the induction hypothesis. Consider the following cases.

1. *Case: $D' = C' \vee A$ is derived from $C' \vee A \vee \underline{A}$ by positive ordered factoring.* Then $A \succeq' C'$. Note that A cannot be a $\mathcal{P}'$-atom: $C' \vee P^n \vee P^n \notin S^\rho$ because all clauses in S^ρ are Horn w.r.t. $\mathcal{P}'$, and $C' \vee P^n \vee P^n$ cannot be derived by the induction hypothesis. Therefore, A is a $\mathcal{Q}$-atom, and hence $\rho(C' \vee A \vee A) = \rho(C') \vee A \vee A$.

   Moreover, C' contains no positive literal P^n because in this case the clause $(\rho(C') \vee A \vee A) \in S$ is definite w.r.t. $\mathcal{P}$, so it admits the stratification, so $P \succ A$, which contradicts to $A \succ' P^n$.

   Therefore, $C' \vee A$ contains no positive $\mathcal{P}'$-literal. Then by our definition of σ' and $\succ'$, A is maximal and selected in $\rho(C') \vee A \vee A$, so we can apply positive ordered factoring and derive $\rho(C') \vee A$ in $\mathcal{SR}$. It is easy to see that $\rho(C') \vee A = \rho(C' \vee A)$ and that $\rho(C' \vee A)$ contains no positive $\mathcal{P}'$-literal.

2. *Case: $D = C_1' \vee C_2'$ is obtained from $C_1' \vee \underline{A}$ and $C_2' \vee \neg A$ by binary ordered resolution with selection.* Denote $C_1 = \rho(C_1')$ and $C_2 = \rho(C_2')$. Consider two cases.

   (a) *Case: $A \in \mathcal{Q}$.* Then $\rho(A) = A$. In this case we show that

   $$\frac{C_1 \vee \underline{A} \quad C_2 \vee \underline{\neg A}}{C_1 \vee C_2}$$

   is an inference in $\mathcal{SR}$ by binary ordered resolution with selection. Since $A \succ' C_1'$, then $A \succ C_1$. By our definition of σ', A ($\neg A$) is selected by σ in $C_1 \vee A$ (in $C_2 \vee \neg A$) because A ($\neg A$) is selected by σ' in $C_1 \vee A$ ($C_2 \vee \neg A$). It remains to check that C_1, C_2 contain no positive $\mathcal{P}$-literals. This is done exactly as in case of positive ordered factoring.

   (b) *Case: $A = P^n$.* We prove that

   $$\frac{C_1 \vee P \quad C_2 \vee \underline{\neg P}}{C_1 \vee C_2}$$

is an inference in $\mathcal{SR}$ by definition unfolding. Indeed, since $\neg P^n$ is selected by σ' in $C_2' \vee \neg P^n$, then $\neg P$ is selected by σ in $C_2 \vee \neg P$.

As in the previous cases, we can prove that C_1', C_2' contains no positive literal of the form P^m.

Now we can conclude the proof of Theorem 9. Suppose that S in unsatisfiable. By (3), S^ρ is unsatisfiable. Then by Theorem 4 the empty clause $\square$ is derivable from S^ρ in $R_{\sigma'}^{\succ'}$. Hence, by (4) $\rho(\square)$ is derivable from S in $\mathcal{SR}$. But $\rho(\square) = \square$, so $\square$ is derivable from S in $\mathcal{SR}$.

We will now show that the condition on clauses to admit stratification is essential.

EXAMPLE 10 This example is taken from [7]. Consider the following set of clauses:

$$\begin{array}{lll} \neg Q \vee \underline{R} & \neg P \vee \underline{Q} & \neg R \vee \underline{\neg Q} \\ \neg Q \vee \underline{\neg P} & \neg P \vee \underline{\neg R} & \\ \neg R \vee P & R \vee Q \vee P & \end{array}$$

This clause is unsatisfiable and definite w.r.t. P. Consider the ordering $R \succ Q \succ P$. However, the empty clause cannot be derived from it, even if tautologies are allowed. Indeed, the conclusion of any inference by stratified resolution is subsumed by one of the clauses in this set. The problem is that the clause $R \vee P \vee Q$ admits no $\succ$-stratification.

3   Redundancies in Stratified Resolution

In this section we make stratified resolution into an inference system on sets of clauses and add two kinds of inferences that remove clauses from sets. Then we prove completeness of the resulting system using a new technique of *traces*.

The derivable objects in the inference system of stratified resolution with redundancies are sets of clauses. Each inference has exactly one premise S and conclusion S', we will denote such an inference by $S \triangleright S'$. We assume an ordering $\succ$, stratification and selection function be defined as in the case of $\mathcal{SR}$.

In this section we always assume that S_0 is an initial set of clauses Horn w.r.t. $\mathcal{P}$ and having the following property: for every $P \in \mathcal{P}$ S_0 contains only a finite number of clauses in which P is a positive literal.

DEFINITION 11 (Stratified Resolution with Redundancies) The inference system of *stratified resolution with redundancies*, denoted $\mathcal{SRR}$ consists of the following inference rules.

1. Suppose that $(C \vee A \vee \underline{A}) \in S$ and A is a positive literal. Then

$$S \triangleright S \cup \{C \vee A\},$$

is an inference by *positive ordered factoring*.

2. Suppose that $\{C \vee \underline{A}, D \vee \neg \underline{A}\} \subseteq S$, where $A \succ C$ and C, D, A contain no positive $\mathcal{P}$-literals. Then

$$S \rhd S \cup \{C \vee D\}$$

is an inference by *binary ordered resolution*.

3. Suppose $(C \vee \underline{\neg P}) \in S$, where $P \in \mathcal{P}$. Furthermore, suppose that $P \vee D_1, \ldots, P \vee D_k$ are all clauses in S containing P positively. Then

$$S \rhd (S - \{C \vee \underline{\neg P}\}) \cup \{C \vee D_1, \ldots, C \vee D_n\}$$

is an inference by *definition rewriting*. Note that this inference *deletes* the clause $C \vee \underline{\neg P}$. Moreover, if S contains no clauses containing the positive literal P, then $C \vee \underline{\neg P}$ is deleted from the search space and not replaced by any other clause.

4. Suppose $\{C, D\} \in S$, $C \neq D$ and C subsumes D. Then

$$S \rhd S - \{D\}$$

is an inference by *subsumption*.

For the new calculus, we have to change the notion of derivation into a new one. We call a *derivation* from a set of clauses S_0 any sequence of inferences:

$$S_0 \rhd S_1 \rhd S_2 \rhd \ldots \, ,$$

possibly infinite. The derivation is said to *succeed* if some S_i contains $\square$, and *fail* otherwise.

Consider a derivation $\mathbf{S} = S_0 \rhd S_1 \rhd \ldots$. The set $\bigcup_i \bigcap_{j \geq i} S_j$ is called the *limit* of this derivation and denoted by $\lim_{\mathbf{S}}$. The derivation is called *fair* if the following three conditions are fulfilled:

1. If $(C \vee A \vee \underline{A})$ belongs to the limit of $\mathbf{S}$ and $A \succ C$, then some inference in $\mathbf{S}$ is the inference by positive ordered factoring

$$S_i \rhd S_i \cup \{C \vee A\}.$$

2. If $C \vee \underline{A}, D \vee \underline{\neg A}$ belong to the limit of $\mathbf{S}$, $A \succ C$ and C, D, A contain no positive $\mathcal{P}$-literals, then some inference in $\mathbf{S}$ is the inference by binary ordered resolution

$$S_i \rhd S_i \cup \{C \vee D\}.$$

3. Let a clause $C \vee \underline{\neg P}$ belongs to the limit of $\mathbf{S}$ and $P \in \mathcal{P}$. Then some inference in $\mathbf{S}$ is the inference by definition rewriting

$$S_i \rhd (S_i - \{C \vee \underline{\neg P}\}) \cup \{C \vee D_1, \ldots, C \vee D_n\}.$$

We call a selection function σ *subsumption-stable* if it has the following property. Suppose C subsumes D, L is a literal in D selected by σ and $L \in C$. Then L is also selected in C by σ. It is evident that subsumption-stable selection functions exist.

THEOREM 12 (Soundness and Completeness of $\mathcal{SRR}$) *Let S_0 be a set of clauses Horn w.r.t. $\mathcal{P}$ and every clause definite w.r.t. $\mathcal{P}$ in S_0 admits stratification (2). Let σ be a subsumption-stable selection function. Consider derivations in $\mathcal{SRR}$. (i) If any derivation from S_0 succeeds, then S_0 is unsatisfiable. (ii) If S_0 is unsatisfiable, then every fair derivation from S succeeds.*

PROOF. Soundness is easy to prove. Let a derivation $S_0 \triangleright \ldots \triangleright S_n$ succeed. It is easy to see that every clause in every S_i is a logical consequence of S_0. Then $\square$ is a logical consequence of S_0, and hence S_0 is unsatisfiable.

To prove completeness is more difficult. We will introduce several notions, prove some intermediate lemmas, and then come back to the proof of this theorem.

The main obstacle in the proof of completeness is that some clauses can be deleted from the search space, when we apply definition rewriting or subsumption. If we only had subsumption, the proof could be done by standard methods based on clause orderings. However, clause rewriting can rewrite clauses into larger ones. For example, consider a set S of clauses that contains two clauses definite w.r.t. $\mathcal{P}$: $P_1 :- P_2$ and $P_2 :- P_1$. Then the following is a valid derivation:

$$S \cup \{\neg P_1\} \triangleright S \cup \{\neg P_2\} \triangleright S \cup \{\neg P_1\} \triangleright \ldots,$$

so independently of a choice of ordering $\succ$ one of these inferences replaces a smaller clause by a bigger one.

To prove completeness we will use a novel technique of *traces*. This technique was originally introduced in [10], where it was used to prove completeness of a system of modal logic with subsumption. Intuitively, a trace is whatever remains of a clause when the clause is deleted. Unlike [10], in this paper a trace can be an infinite set of clauses.

Suppose that S_0 is the set of initial clauses, Horn w.r.t. $\mathcal{P}$. Consider any clause D that contains no positive $\mathcal{P}$-literals. Consider the following one-player game that builds a tree of clauses. Initially, the tree contains one root node D. We call a node C *closed* if either $C = \square$ or the literal selected in C by σ is a Q-literal. An *open node* is any leaf that is not closed. At every step of the game the player has two choices of moves:

1. *(Subsumption move).* Select any leaf C (either closed or open) and add to C as the child node any clause C' such that C' subsumes C and $C' \neq C$.
2. *(Expansion move).* Select an open leaf C. If the literal selected in C by σ is a negative $\mathcal{P}$-literal $\neg P$, then C has the form $\neg P \vee C'$. (As we shall see later, no clause in the tree can contain a positive $\mathcal{P}$-literal, so in this case a negative $\mathcal{P}$-literal is always selected). Let all clauses in S_0 that contain the positive literal P be $P \vee D_1, \ldots, P \vee D_n$. Then add to this node as children all nodes $C' \vee D_1, \ldots, C' \vee D_n$.

Using the property that S_0 is Horn w.r.t. $\mathcal{P}$ it is not hard to argue that no clause in the tree can contain a positive $\mathcal{P}$-literal, so an open leaf always contains a selected negative $\mathcal{P}$-literal.

A game is *fair* if every non-closed node is selected by the player. Let us call a *tree for D* any tree obtained as a limit of a fair game. We call a *cover for D* the set of all closed leaves in any tree T for D. We will say that the cover is *obtained from* the tree T and denote the cover by cl_T.

EXAMPLE 13 Suppose that the set of clauses S_0 is

$$\{P_i \vee \neg Q_i \mid i = 1, 2, \ldots\} \cup \{P_i \vee \neg P_{i+1} \mid i = 1, 2, \ldots\},$$

where all P_i's are $\mathcal{P}$-literals and all Q_i's are $\mathcal{Q}$-literals. Suppose that the selection function σ always selects a $\mathcal{P}$-literal in a clause that contains at least one $\mathcal{P}$-literal. Then one possible tree for $\underline{\neg P_1} \vee Q_1$ is as follows:

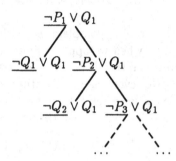

In this tree no subsumption move was applied. The cover for $\underline{\neg P_1} \vee Q_1$ obtained from this tree is the set of clauses

$$\{\underline{\neg Q_i} \vee Q_1 \mid i = 1, 2, \ldots\}.$$

Let us prove some useful properties of covers.

LEMMA 14 *Every cover for a clause C is also a cover for any clause $C \vee D$.*

PROOF. Let S be a cover for C and T be the tree from which this cover is obtained. Apply a subsumption move to $C \vee D$

and extend this tree by the tree T below C. Evidently, we obtain a tree for $C \vee D$ and the cover obtained from this tree is also S.

The following straightforward lemma asserts that it is enough to consider trees with no repeated subsumption moves.

LEMMA 15 *For every cover S for a clause C there exists a tree T for C such that $S = cl_T$ and no subsumption move in T was applied to a node obtained by a subsumption move.*

PROOF. Let T' be a tree for C such that $S = cl_{T'}$. Suppose that T' contains a node obtained by a subsumption move so that a subsumption move is applied to this node as well:

$$D_1 \vee D_2 \vee D_3$$
$$|$$
$$D_1 \vee D_2$$
$$|$$
$$D_1$$

Evidently, these two subsumption moves can be replaced by one subsumption move:

$$D_1 \vee D_2 \vee D_3$$
$$|$$
$$D_1$$

In this way we can eliminate all sequences of subsumption moves and obtain a required tree T.

Let C be a clause and $\mathbf{S}$ be a derivation. We call a *trace* of C in $\mathbf{S}$ any cover S for C such that $S \subseteq \lim_{\mathbf{S}}$. When we speak about a trace of C in $\mathbf{S}$ we do not assume that C occurs in $\mathbf{S}$.

LEMMA 16 *Let $\mathbf{S} = S_0 \triangleright S_1 \triangleright \dots$ be a fair derivation. Then every clause D occurring in any S_i has a trace in $\mathbf{S}$.*

PROOF. We will play a game which builds a tree T such that cl_T is a trace of D in $\mathbf{S}$. We will define a sequence of trees $T_1, T_2, \dots$ such that each T_{j+1} is obtained from T_j by zero or more moves, and T is the limit of this sequence. The initial tree T_1 consists of the root node D. Each tree T_{j+1} is obtained from T_j by inspecting the inference $S_j \triangleright S_{j+1}$ as follows.

1. Suppose this inference is a subsumption inference

$$S_j \triangleright S_j - \{C \vee D\},$$

where $C \in S$ and $C \vee D$ is a leaf node in T_j. Then make a subsumption move for each such leaf, adding C to it as the child.

2. Suppose this inference is a definition rewriting inference

$$S_j \triangleright S_j - \{C \vee \underline{\neg P}\} \cup \{C \vee D_1, \ldots, C \vee D_n\},$$

such that $C \vee \underline{\neg P}$ is a leaf in T_j. Then make an expansion move for each such leaf, adding to it the clauses $C \vee D_1, \ldots, C \vee D_n$ as children.
3. In all other cases make no move at all.

Let us make several observations about the tree T and the game.

1. *Every leaf in any tree T_j is a member of some S_k.* This is proved by induction on j. For T_1 this holds by the definition of T_1, since it contains one leaf $D \in S_i$. By our construction, every leaf of any T_{j+1} which is not a leaf of T_j is a member of S_{j+1}.
   It remains to prove that every node which is a leaf in both T_j and T_{j+1} belongs to S_{j+1}. Take any such node C, by the induction hypothesis it belongs to S_j. Suppose, by contradiction, $C \notin S_{j+1}$, then C has been removed by either subsumption or definition rewriting inference. But by our construction, in both cases C will become a nonleaf in T_{j+1}, so we obtain a contradiction.
2. *The game is fair.* Suppose, by contradiction, that an open leaf has never been selected during the game. Then this leaf contains a clause $C \vee \neg P$. Let T_j be any tree containing this open leaf. Then the leaf belongs to all trees T_k for $k \geq j$. By the previous item, $C \vee \underline{\neg P} \in S_k$ for all $k \geq j$. Therefore, the clause $C \vee \neg P$ belongs to the limit of $\mathbf{S}$, but definition rewriting has not been applied to it. This contradicts to the fairness of $\mathbf{S}$.
3. *Every closed leaf C in T belongs to the limit of $\mathbf{S}$.* Let this leaf first appeared in a tree T_j. Then it belongs to all trees T_k for $k \geq j$. By the first item, $C \in S_k$ for all $k \geq j$, but then $C \in \lim_{\mathbf{S}}$.

Now consider cl_T. Since T was built using a fair game, cl_T is a cover for D. But we proved that every element of cl_T belongs to the limit of $\mathbf{S}$, hence cl_T is also a trace of D in $\mathbf{S}$.

LEMMA 17 *Let $\mathbf{S} = S_0 \triangleright S_1 \triangleright \ldots$ be a fair derivation. Further, let E be a clause derivable from S_0 in $\mathcal{SR}$. Then D has a trace in $\mathbf{S}$.*

PROOF. The proof is by induction on the derivation of E in $\mathcal{SR}$. When the derivation consists of 0 inferences, $E \in S_0$. By Lemma 16 every clause of every S_i has a trace in $\mathbf{S}$, so E has a trace.

When the derivation has at least one inference, we consider the last inference of the derivation. By the induction hypothesis, all premises of this inference have traces in $\mathbf{S}$. We consider three cases corresponding to the inference rules of stratified resolution. In all cases, using Lemma 15 we can assume that no tree contains a subsumption move applied to a node which itself is obtained by a subsumption move.

1. *The last inference is by positive ordered factoring.*

$$\frac{C \vee A \vee \underline{A}}{C \vee A},$$

Consider a trace S of $C \vee A \vee \underline{A}$ in **S** and a tree T from which this trace was obtained. Since A is selected by σ in C, there are only two possibilities for T:

(a) *Case: T consists of one node $C \vee A \vee A$.* Then $\{C \vee A \vee A\}$ is a trace, and hence $(C \vee A \vee A) \in \text{lim}_\mathbf{S}$. Since **S** is fair, some S_i contains the clause $C \vee A$ obtained from $C \vee A \vee A$ by positive ordered factoring. Then by Lemma 16, $C \vee A$ has a trace in **S**.

(b) *Case: the first move of T is a subsumption move.* This subsumption move puts a clause C' which subsumes $C \vee A \vee A$ as the child to $C \vee A \vee A$. Consider two cases.

    i. *Case: C' has a form $C'' \vee A \vee A$, where C'' subsumes C.* Since the selection function is subsumption-stable, A is selected in $C'' \vee A \vee A$ by σ, so the tree T contains no node below $C'' \vee A \vee A$. Then $(C'' \vee A \vee A) \in \text{lim}_\mathbf{S}$. Since **S** is fair, some S_i contains the clause $C'' \vee A$ obtained from $C'' \vee A \vee A$ by positive ordered factoring. By Lemma 16, $C'' \vee A$ has a trace in **S**. But $C'' \vee A$ subsumes $C \vee A$, so by Lemma 14 this trace is also a trace of $C \vee A$.

    ii. *Case: C' subsumes $C \vee A$.* Note that cl_T is also a trace of C', so by Lemma 14 cl_T is also a trace of $C \vee A$.

2. *The last inference is by binary ordered resolution with selection:*

$$\frac{C \vee \underline{A} \quad D \vee \neg \underline{A}}{C \vee D},$$

where $A \succ C$ and C, D, A contain no positive $\mathcal{P}$-literals. Consider traces S_1, S_2 of $C \vee A$ and $D \vee \neg A$, respectively and trees T_1, T_2 from which these traces were obtained. We consider two simple cases and then the remaining case

(a) If the first move of T_1 is a subsumption move that replaces $C \vee A$ by a clause C' that subsumes C, then C' has a trace in **S**, but C subsumes $C \vee D$, so by Lemma 14 this trace is also a trace of $C \vee D$.

(b) Likewise, if the first move of T_2 is a subsumption move that replaces $D \vee \neg A$ by a clause D' that subsumes D, then D' has a trace in **S**, so by Lemma 14 this trace is also a trace of $C \vee D$.

(c) If neither of the previous cases takes place, then either T_1 consists of one node or the top move in T_1 is a subsumption move:

such that C' subsumes C. Note that in the latter case, since the selection function is subsumption-stable, A is selected in $C' \vee A$. In both cases the limit of $\mathbf{S}$ contains a clause $C' \vee \underline{A}$ such that C' subsumes C (in the former case we can take C as C').

Likewise, we can prove that the limit of $\mathbf{S}$ contains a clause $D' \vee \neg\underline{A}$ such that D' subsumes D. Since $\mathbf{S}$ is fair, some S_i contains the clause $C' \vee D'$ obtained from $C' \vee \underline{A}$ and $D' \vee \neg\underline{A}$ by binary ordered resolution with selection. By Lemma 16, $C' \vee D'$ has a trace in $\mathbf{S}$. But $C' \vee D'$ subsumes $C \vee D$, so by Lemma 14 $C \vee D$ has a trace in $\mathbf{S}$.

3. *The last inference is by definition unfolding:*

$$\frac{C \vee P \quad D \vee \neg\underline{P}}{C \vee D} \, ,$$

where $P \in \mathcal{P}$. Consider a S of $D \vee \neg\underline{P}$, and a tree T from which this trace was obtained.

(a) If the first move of T is a subsumption move that replaces $D \vee \neg P$ by a clause D' that subsumes D, then D' has a trace in $\mathbf{S}$, so by Lemma 14 this trace is also a trace of $C \vee D$.

(b) If the previous case does not take place, there are two possibilities for the top move of T: it is either an expansion move or a subsumption move followed by an expansion move. We consider the former case, the latter case is similar. The top of the tree T has the form

Denote the subtree of T rooted at $C \vee D$ by T'. Note that $cl_{T'}$ is a cover of $C \vee D$. But since T' is a subtree of T, we have $cl_{T'} \subseteq cl_T$. Since cl_T is a trace, we also have $cl_T \subseteq \lim_{\mathbf{S}}$, then $cl_{T'} \subseteq \lim_{\mathbf{S}}$, and hence $cl_{T'}$ is a trace. So $C \vee D$ has a trace in $\mathbf{S}$.

The proof is completed.

We can now easily complete the proof of completeness of stratified resolution with redundancies.

PROOF (of Theorem 12, Continued). Suppose S_0 is unsatisfiable. Take any fair derivation $\mathbf{S} = S_0 \rhd S_1 \rhd \dots$. By Theorem 9 there exists a derivation of $\Box$ from clauses in S_0 by stratified resolution. By Lemma 17, $\Box$ has a trace in $\mathbf{S}$, i.e. a cover whose members belong to the limit of $\mathbf{S}$. By the definition of a cover, $\Box$ has only one cover $\{\Box\}$. Then $\Box$ belongs to the limit of $\mathbf{S}$, and hence $\mathbf{S}$ is successful.

The proof can be easily modified for a system with redundancies in which clauses containing positive $\mathcal{P}$-literals can also be subsumed and in which tautologies are deleted.

4   Nonground Case

In this section we introduce the general, nonground, version of stratified resolution with redundancies.

Most of the definitions given below are obtained by a simple modification of the definition for the ground case. In this section we treat $\mathcal{L}$, $\mathcal{P}$, and $\mathcal{Q}$ as sets of *relations* but not atoms. To define stratification, we use a *precedence relation* $\succ_{\mathcal{L}}$, i.e. a total ordering on $\mathcal{L}$. We call a $\mathcal{P}$-literal any literal whose relation belongs to $\mathcal{P}$, and similar for other sets of relations instead of $\mathcal{P}$. The notions of *Horn set w.r.t. $\mathcal{P}$*, *definite clause w.r.t. $\mathcal{P}$*, *stratification*, and *clause admitting a stratification* are the same as in the ground case.

Now we have to define an ordering $\succ$ corresponding to our notion of stratification. We require $\succ$ to be well-founded ordering on atoms, stable under substitutions (i.e. $A \succ B$ implies $A\theta \succ B\theta$) and total on ground atoms. In addition, we require that $\succ$ is compatible with the precedence relation in the following sense: if $A_1 \succ_{\mathcal{L}} A_2$, then $A_1(t_1, \ldots, t_n) \succ A_2(s_1, \ldots, s_m)$ for all relations A_1, A_2 and terms $t_1, \ldots, t_n$ and $s_1, \ldots, s_m$. It is obvious how to modify recursive path orderings or Knuth-Bendix orderings so that they satisfy this requirement.

In the definition of selection function for nonground case we revert in fact to one of earlier notions given in [4]. We define a (nonground) *selection function* to be a function σ on the set of clauses such that (i) $\sigma(C)$ is a set of literals in C, (ii) $\sigma(C)$ is nonempty whenever C is nonempty, and (iii) at least one negative literal or otherwise all maximal literals must be selected. Similar to the ground case we assume that selection functions are subsumption-stable in the following sense: if a literal L is selected in $C \vee L$ and D is a submultiset of C, then L is selected in $D \vee L$.

Let us fix an order $\succ$, a selection function σ, and stratification (2).

DEFINITION 18 (Stratified Resolution with Redundancies) The inference system of *stratified resolution with redundancies*, denoted $\mathcal{SRR}$ consists of the following inference rules.

1. *Positive ordered factoring* is the following inference rule:

$$S \triangleright S \cup \{(C \vee A_1)\theta\}$$

   such that (i) S contains a clause $C \vee \underline{A_1} \vee \underline{A_2}$, (ii) A_1, A_2 are positive literals, and (iii) θ is a most general unifier of A_1 and A_2.
2. *Binary ordered resolution with selection* is the following inference rule:

$$S \triangleright S \cup \{(C \vee D)\theta\}$$

   such that (i) S contains clauses $C \vee \underline{A_1}$ and $D \vee \underline{\neg A_2}$, (ii) $A_1\theta \not\subseteq C\theta$, (iii) C, D, A_1 contain no positive $\mathcal{P}$-literals, and (iv) θ is a most general unifier of A_1 and A_2.

3. *Definition rewriting.* Suppose $(C \vee \underline{\neg P}) \in S$, where $P \in \mathcal{P}$. Furthermore, suppose that $P_1 \vee D_1, \ldots, P_k \vee D_k$ are all clauses in S such that P_i is unifiable with P. Then

$$S \triangleright (S - \{C \vee \neg P\}) \cup \{(C \vee D_1)\theta_1, \ldots, (C \vee D_n)\theta_n\},$$

where each θ_i is a most general unifier of P and P_i, is an inference by *definition rewriting.*

THEOREM 19 *Let S be a set of clauses Horn w.r.t. $\mathcal{P}$. Let, in addition, every clause in S definite w.r.t. $\mathcal{P}$ admit stratification (2). Then S is unsatisfiable if and only if every fair derivation from S in the system $\mathcal{SRR}$ contains the empty clause.*

The proof can be obtained by standard lifting with a special care of selection function.

5   How to Select a Stratification

Example 7 shows that a set of clauses may admit several different stratifications. How can choose a "good" stratification?

Suppose that a set of clauses S is Horn w.r.t. $\mathcal{P}$. Then we can always use the stratification

$$\emptyset \succ \mathcal{P} \succ \mathcal{Q},$$

in which any $\mathcal{P}$-atom is greater than any $\mathcal{Q}$-atom. Unfortunately, this stratification may be not good enough, since it gives us too little choice for selecting positive $\mathcal{Q}$-literals. Let us illustrate this for clauses of Example 1. Assume that $\mathcal{P}$ is $\{split\}$. We obtain the stratification

$$\emptyset \succ \{split\} \succ \{deduction, conference\_list\}.$$

This stratification does not allow us to select the literal $conference\_list(x)$ in

$$\neg split(x, y, z) \vee conference\_list(x),$$

while intuitively it should be the right selection.

This observation shows that it can be better to split the set of atoms into as many strata as possible. Then we will have more options for selecting positive $\mathcal{Q}$-atoms in clauses. In Example 1, a more flexible stratification would be

$$\{conference\_list\} \succ \{split\} \succ \{deduction\}.$$

We are planning experiments with the choice of stratification using our theorem prover Vampire [9].

6   Conclusion

We will mention several interesting open problems associated with stratified resolution.

In our proofs, we could not get rid of an annoying condition on selection function to be subsumption-stable.

PROBLEM 1 *Is stratified resolution with redundancies complete with selection functions that are not subsumption-stable?*

It is quite possible this can be done, but require a different proof technique.

PROBLEM 2 *Find new techniques for proving completeness of stratified resolution with redundancies.*

PROBLEM 3 *Find a powerful generalization of definite resolution for logic with equality.*

There is quite a simple generalization based on the following transformation of clauses definite w.r.t. $\mathcal{P}$: we can replace any clause $P(t_1,\ldots,t_n)$:- C by $P(x_1,\ldots,x_n)$:- $x_1 = t_1,\ldots,x_n = t_n, C$. Then we can define stratified resolution and prove its completeness in essentially the same way as before. However, in this case *any* clause containing $\neg P(s_1,\ldots,s_n)$ will be unifiable with the head of *any* clause defining P, so the gain of using definition rewriting is not obvious any more.

The standard semantics of stratified logic programs are based on nonmonotonic reasoning. Stratified resolution makes one think of a logic that combines nonmonotonic reasoning with the monotonic resolution-based reasoning. Such a logic, its semantics and ways of automated reasoning in it, should be investigated. So we have

PROBLEM 4 *Find a combination of stratified resolution with nonmonotonic logics.*

Stratified resolution is different from ordered resolution with negative selection in that it allows one to select heads of clauses, even when they are not maximal in their clauses. It is interesting to see if this can give us new decision procedures for decidable fragments of predicate calculus. So, we have

PROBLEM 5 *Find new decision procedures based on stratified resolution.*

7   Recent Developments

Recently, Robert Nieuwenhuis proposed a new interesting method for proving completeness of stratified resolution. His method gives a positive answer to Problem 1: the restriction on selection functions to be subsumption-stable is not

384 Anatoli Degtyarev and Andrei Voronkov

necessary. In a way, it also answers to Problem 2, since his technique is rather different from ours. He also proposed a more concise definition of stratification and clauses admitting stratification in terms of well quasi orderings.

Acknowledgments

The idea of this paper has appeared due to discussions of the second author during CADE-16 with Harald Ganzinger and Tanel Tammet. Harald Ganzinger explained that the SATURATE system, tries to use an ordering under which the heads of "definition-like" clauses are the greatest in their clauses, and hence can be selected. This allows one to use definitions in the way they are often used in mathematics. Tanel Tammet explained that he tried to implement in his prover GANDALF "rewriting on the clause level", again for clauses that look like definitions, though the implementation was not ready by CASC-16.

References

1. K.R. Apt and H.A. Blair. Towards a theory of declarative knowledge. In J. Minker, editor, *Foundations of Deductive Databases and Logic Programming*, pages 89–148. Morgan Kaufmann, 1988.
2. L. Bachmair and H. Ganzinger. A theory of resolution. Research report MPI-I-97-2-005, Max-Planck Institut für Informatik, Saarbrücken, Germany, 1997. To appear in Handbook of Automated Reasoning, edited by A. Robinson and A. Voronkov.
3. L. Bachmair and H. Ganzinger. A theory of resolution. In A. Robinson and A. Voronkov, editors, *Handbook of Automated Reasoning*. Elsevier Science and MIT Press, 2000. To appear.
4. L. Bachmair, H. Ganzinger, C. Lynch, and W. Snyder. Basic paramodulation. *Information and Computation*, 121:172–192, 1995.
5. A. Van Gelder. Negation as failure using tight derivations for general logic programs. In J. Minker, editor, *Deductive Databases and Logic Programming*, pages 149–177. Morgan Kaufmann, 1988.
6. R.A. Kowalski and D. Kuehner. Linear resolution with selection function. *Artificial Intelligence*, 2:227–260, 1971.
7. C. Lynch. Oriented equational logic is complete. *Journal of Symbolic Computations*, 23(1):23–45, 1997.
8. H. Przymusinski. On the declarative semantics of deductive databases and logic programs. In J. Minker, editor, *Foundations of Deductive Databases and Logic Programming*, pages 193–216. Morgan Kaufmann, 1988.
9. A. Riazanov and A. Voronkov. Vampire. In H. Ganzinger, editor, *Automated Deduction—CADE-16. 16th International Conference on Automated Deduction*, volume 1632 of *Lecture Notes in Artificial Intelligence*, pages 292–296, Trento, Italy, July 1999.
10. A. Voronkov. How to optimize proof-search in modal logics: a new way of proving redundancy criteria for sequent calculi. Preprint 1, Department of Computer Science, University of Manchester, February 2000. To appear in LICS'2000.

Support Ordered Resolution

Bruce Spencer and Joseph D. Horton

Faculty of Computer Science, University of New Brunswick
P.O. Box 4400, Fredericton, New Brunswick, Canada E3B 5A3
bspencer@unb.ca, jdh@unb.ca
http://www.cs.unb.ca

Abstract. In a binary tree representation of a binary resolution proof, rotating some tree edge reorders two adjacent resolution steps. When rotation is not permitted to disturb factoring, and thus does not change the size of the tree, it is invertible and defines an equivalence relation on proof trees. When one resolution step is performed later than another after every sequence of such rotations, we say that resolution supports the other.

For a given ordering on atoms, or on atom occurrences, a *support ordered* proof orders its resolution steps so that the atoms are resolved consistently with the given order without violating the support relation between nodes. Any proof, including the smallest proof tree, can be converted to a support ordered proof by rotations. For a total order, the support ordered proof is unique. The support ordered proof is also a rank/activity proof where atom occurrences are ranked in the given order.

Procedures intermediate between literal ordered resolution and support ordered resolution are considered. One of these, 1-weak support ordered resolution, allows to resolve on a non-maximal literal only if it is immediately followed by both a factoring and a resolution on some greater literal. In a constrained experiment where literal ordered resolution solves only six of 408 TPTP problems with difficultly between 0.11 and 0.56, 1-weak support ordered resolution solves 75.

1 Introduction

Automated theorem provers, in their search for a proof, must balance the deductive power of a calculus, telling what can be derived from a given point in the search, with restriction strategies, telling which deductions are to be avoided. Clearly the restriction strategy must not remove all of the choices that eventually lead to a proof, at least not without the user's being aware of its incompleteness. But even so, the restriction strategy may remove all shortest proofs, leading to another undesirable effect: the theorem prover takes longer to find a longer proof. An ideal restriction strategy would reduce the space to one richly populated with only short proofs, be simple to implement and quick to check. This is an unrealistic ideal. In this paper we give a reduction strategy is that is quick to check, simple to implement, admits smallest proofs trees, and is almost as

D. McAllester (Ed.): CADE-17, LNAI 1831, pp. 385–400, 2000.

restrictive as literal ordered resolution, a widely used and successful restriction. Literal ordered resolution was used in the CADE-16 System Competition [9] by at least Bliksem, Gandalf, HOTTER, SPASS and Vampire.

In our setting we represent proofs as binary trees, labeled by clauses according to Robinson's resolution method[6]. A node is labeled both by a clause, referring to the conclusion drawn at this point by the resolution, and if it is not a leaf, by the atom that was resolved upon to give this conclusion. Our measure of size is the number of nodes in this tree. Often theorem provers build sequences of formulae where each deductively follows from previous ones. This sequence represents a traversal of a directed acyclic graph (dag) that underlies the tree. The size of an underlying dag is a more natural measure of proof size than the size of the tree. But often if one dag is smaller than another then the tree expanded from the first dag is smaller than the tree expanded from the other.

Support ordered resolution depends on the notion of support between two resolution steps, first defined in [4] and defined for proof trees in [7]. When compared with the literal ordered restriction, used by many theorem provers and explored in [3], support ordered resolution is less restrictive, in that it admits a very specific additional resolution step. On the other hand the support ordered restriction does not increase the size of the smallest tree, unlike literal ordered resolution which may restrict all smallest trees, and in some cases admit only exponentially larger trees and dags, as shown in Example 5 below.

Rotating some tree edge reorders two adjacent resolution steps. When rotation is not permitted to disturb factoring, and thus does not change the size of the tree, it is invertible and defines an equivalence relation on proof trees. For a given total order on atoms, there is a single support ordered tree in each rotation equivalence class. Since the equivalence classes typically contain an exponential number of trees, support ordered resolution substantially reduces the search space.

It is interesting to find a restriction of a resolution calculus that admits a smallest deduction (tree) while substantially reducing the search space, as support ordered resolution does. It is also interesting in that it brings together two apparently different restrictions, the rank/activity restriction[5] and literal ordered resolution. A given ordering on atoms can be used to set the ranks of literals in each clause, and then the rank/activity proof is the support ordered proof.

Viewed as a generalization of other restrictions, we can identify a number of other special cases of support ordered resolution. These suggest themselves as candidates for experiments. One set of these experiments has been done for the special case called 1-weak support ordered resolution, or 1-wso. This proof format depends on a restriction that can be quickly checked on partially closed binary resolution trees. Recall that the literal ordered restriction allows a resolution only on a maximal atom in each clause; 1-wso allows maximal resolutions and also allows a resolution on some non-maximal atom but then requires an immediate merge on a greater atom from different parents followed by a resolution step on that merged atom. Our experiments indicate that 1-wso provides deductive / reductive tradeoff that is worth taking.

In the following sections we provide necessary background including the recent notion of support between nodes in a binary resolution tree[7]. This is followed by the introduction of support ordered resolution, the restriction of resolution closely related to literal ordered resolution but weakened in those situations where it conflicts with the support of one node for another. We also describe support ordered resolution as a generalization of both rank/activity and literal ordered resolution. This suggests a space of possible strategies. We describe one proof procedure in this space, 1-weak support ordered resolution which is a slight addition to a typical literal ordered resolution theorem prover. We also give the results of our experiments.

2 Background

A binary tree is a set of nodes and edges, where each edge joins a *parent* node to a *child* node, and where each node has one child or has zero and is then called the *root*, and each node has two parents or has zero and is then called a *leaf*. The descendant (ancestor) relation is the reflexive, transitive closure of child (parent).

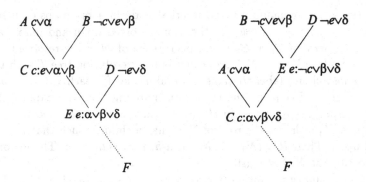

Fig. 1. A binary tree rotation

Given the binary tree fragment T on the left of Figure 1, a rotation is the reassignment of edges so that the tree T' on the right of Figure 1 is produced. The parent C of E becomes the child of E and the parent B of C becomes the parent of E. In other words, the edges (B, C) and (C, E) are replaced by (B, E) and (E, C). If E has a child F in T, then C takes that child in T', or equivalently the edge (E, F) is replaced by (C, F).

We use standard definitions [2] for atom, literal, substitution, unifier and most general unifier. A *clause* is a multiset of literals. The clause C *subsumes* the clause D if there exists a substitution θ such that $C\theta \subseteq D$ (as sets, not as multisets). A *variable renaming substitution* is one in which every replacement of a variable maps to another variable, and no two variables map to the same variable. Two clauses C and D are *equal up to variable renaming* if there exists a variable renaming substitution θ such that $C\theta = D$ (as multisets). Two clauses are *standardized apart* if no variable occurs in both. Given two *parent* clauses $C_1 \vee a_1 \vee \ldots \vee a_m$ and $C_2 \vee \neg b_1 \vee \ldots \vee \neg b_n$ which are standardized apart (a variable renaming substitution may be required) a *resolvent* is the clause $(C_1 \vee C_2)\theta$ where θ is a most general unifier of $\{a_1, \ldots, a_m, b_1, \ldots, b_n\}$. The *atom resolved upon* is $a_1\theta$, and the set of *resolved literals* is $\{a_1, \ldots, a_m, \neg b_1, \ldots, \neg b_n\}$.

Definition 1. *A* binary resolution tree, *or* brt *on a set S of input clauses is a binary tree where each node N in the tree is labeled by a* clause label, *denoted $cl(N)$. The clause label of a leaf node is an instance of a clause in S, and the clause label of a non-leaf is the resolvent of the clause label of its parents. A non-leaf node is also labeled by an* atom label, $al(N)$, *equal to the atom resolved upon. The clause label of the root is called the* result *of the tree, $result(T)$. A binary resolution tree is* closed *if its result is the empty clause, $\square$.*

Our resolution is based on Robinson's original resolution, which we use to define resolution mapping and history path. The resolution mapping tells what happens to each literal in a given resolution step, and the history path tells what happens to it from the leaf where it is introduced to the node where it is resolved away.

The *resolution mapping* ρ at an internal node in a brt maps each resolved literal, $a_1, \ldots, a_m, \neg b_1, \ldots, \neg b_n$, to the atom resolved upon and maps each unresolved member c of C_1 or C_2 to the occurrence of $c\theta$ in the resolvent.

Let the nodes $(N_0, \ldots, N_n)$ occur in a binary resolution tree T such that N_0 is a leaf whose clause label contains a literal a, and for each $i = 1, \ldots, n$, N_{i-1} is a parent of N_i. Let ρ_i be the resolution mapping from the parents of N_i to N_i. Also let $\rho_i \ldots \rho_2 \rho_1 a$ occur in $cl(N_i)$, so that a is not resolved away at any N_i. Suppose N_n either is the root of T, or has a child N such that $\rho_n \ldots \rho_1 a$ is resolved upon. Then $P = (N_0, \ldots, N_n)$ is a *history path* for a. The history path is said to *close* at N if N exists.

Let T be a binary resolution tree as in Figure 1 with an edge (C, E) between internal nodes such that E has a parent C and C has two parents A and B. Further, assume that no history path through A closes at E. Then the result of a *rotation* on this edge is the binary resolution tree T' defined by resolving $cl(B)$ and $cl(D)$ on $al(E)$ giving $cl(E)$ in T' and then resolving $cl(E)$ with $cl(A)$ on $al(C)$ giving $cl(C)$ in T'. Any history path closed at C in T is closed at C in T'; similarly any history path closed at E in T is closed at E in T'. Also, the child of E in T, if it exists, is the child of C in T'.

Two trees T_1 and T_2 are *rotation equivalent* if T_1 is the result of a rotation of an edge in T_2, or if T_1 and T_2 are both rotation equivalent to another tree.

The condition on a rotation that no history path through A closes at E is important, because otherwise the rotation would disturb factoring and thus more resolutions might be needed. Consider the left tree in Figure 2. There are history paths for e through both A and B that close at E. The tree after the edge rotation, shown on the right side of Figure 2 has an unresolved occurrence e in the result.

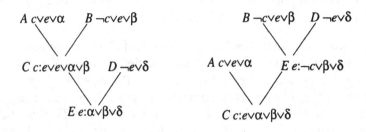

Fig. 2. Not a brt rotation

The calculus of binary resolution trees consists of the following:

1 A node labeled by (an instance of) an input clause is a brt.

2 If T is a brt and θ is a variable substitution then $T\theta$ is a brt formed by replacing each label l in T by $l\theta$.

3 Suppose T_1 and T_2 are brts and no variable appears in both, R_1 and R_2 are the clause labels of the roots of T_1 and T_2 respectively, and R is the clause formed by resolving R_1 and R_2 on atom A with substitution θ. Then T is a brt formed by creating a new node N with atom label A, clause label R and the roots of $T_1\theta$ and $T_2\theta$ are R's parents.

4 If T is a brt and (C, E) is an edge then T' formed by rotating the edge (C, E) is a brt (shown in Figure 1).

Because not all rotations are allowed, sometimes a node N in a brt T remains below another node M, under all sequences of rotations. When this occurs, we say that N *supports* M. Support is a transitive relation, and support cannot be circular. Although it is not exploited in this paper, support can be determined from the history paths of T [7].

3   Support Ordered Resolution

To simplify the discussion we assume that the literal ordering is independent of sign, and thus is an atom ordering. The same results can be developed for literal ordering. Because we use atom ordering, the internal nodes of a brt are ordered in atom ordering on the atom labels of the nodes, *i.e.* the atom resolved upon at that node.

Definition 2. *Given an ordering $\preceq$ of atoms and a binary resolution tree T, we say that a node N is support ordered if no descendant of N has higher order than N unless it supports N. T is support ordered if all its nodes are.*

In effect, support ordering is the lexical composition of the support relation and atom ordering.

Theorem 1. *For a given partial ordering $\preceq$ on atoms and a given brt T, there is a support ordered proof tree T^* that is rotation equivalent to T. If $\preceq$ is total, T^* is unique.*

Proof. We proceed by induction on the size of T. If T has one or three nodes, T is trivially support ordered. Suppose T has k nodes. Consider N a node ordered highest in $\preceq$ not supporting any other node in T. Rotate edges above N so that both parents of N are leaves. These rotations are possible because N supports no nodes. Let C_N be the clause label of N in the resulting tree T_0. From this tree, remove the parents L_1 and L_2 of N, so that N is a leaf, and call the resulting smaller tree T_1. By induction there exists T_1^*, a support ordered binary resolution tree of $k-2$ nodes that is rotation equivalent to T_1. N is a leaf of T_1^*. Construct T^* by replacing the parents L_1 and L_2 of N in T^* so that the resolution done at N is the same as in T_0. Because T and T^* are rotation equivalent, the support relations in T^* and T are the same. All nodes in T^* that are also in T_1^* are support ordered, with the possible exception of N. But since any descendants of N that are ordered higher than N are supports of N, N is support ordered.

To argue uniqueness, consider for each leaf L the highest ordered descendant D_L that supports no other descendants of L. Such a node must exist because circular support relations are not possible. Since $\preceq$ is total, D_L is unique for L. But D_L must be the child of L in T^*; otherwise T^* is not support ordered. Thus each leaf of T^* has a unique child. For any node N in T^*, other than the root, this argument can be combined with an induction on the height of the tree above N to show that N has a unique child. Thus T^* is unique. $\square$

To be useful, a restriction must be applicable to a partial proof. Also the check should require only a simple computation, preferably one with low complexity (constant or linear time) and require information that is local to the proof step. Therefore we define weak support ordered resolution, which can be checked quickly, locally, and on a partial proof. Unfortunately the check cannot be made with just the partial proof we have so far – more steps must be done. We will revisit this problem.

Definition 3. *Given an ordering $\preceq$ and a brt T, an edge between a parent N_1 and its child N_2 in a brt is weak support ordered if the N_1 is ordered higher than N_2 or if the edge is not rotatable. T is weak support ordered if every edge of T is.*

Support ordered resolution is strictly more restrictive than weak support ordered, in that it admits strictly fewer proofs. Consider the following brt: The nodes on a branch are (N_d, N_a, N_c, N_b) where each is a parent of the next. Let a be the atom resolved at N_a, and similarly for b, c, d, and let the atom ordering dictate that a should be resolved before b, *etc.* Suppose N_a supports N_d and N_b supports N_c, so a must be resolved after d and b after c. Thus according to the support relation, we have only two choices to resolve first, d or c, and according to the atom order, we must resolve c first. The resulting support ordered tree is illustrated in the left of Figure 3. All edges in this tree are also weak support ordered; the only rotatable edge is (N_a, N_c) but it conforms to the atom ordering without being rotated. But weak support order applys the atom ordering only locally. A rotation equivalent tree, shown on the right of Figure 3 has the branch (N_c, N_b, N_d, N_a) which is also weak support ordered, since the only rotatable edge (N_b, N_d) and it also conforms. Thus wso is a weaker restriction than support order.

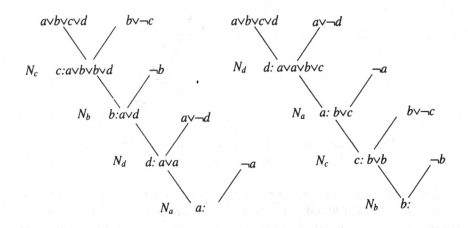

Fig. 3. A support ordered tree and a weak support ordered tree

4   Relation to the Rank/Activity Restrictions

The rank/activity restriction [5], or r/a, can be stated in terms of history paths and where they close.

Given a rank function r that orders atoms in each clause, a brt T is defined to be r-compliant if for each leaf L of T the literals of L are resolved away either in r-order, or in the opposite order only if there is another history path that closes with the higher ordered literal's history path, but does not intersect lower ordered one. That is, for each pair l_1, l_2 of literals in L that close at descendants d_1 and d_2 respectively, if $r(l_2) < r(l_1)$ then either (maximal case) d_2 is a descendant of d_1 or (non-maximal case) d_1 is a descendant of d_2 and some history path that does not intersect the path for l_2 also closes at d_1. This is illustrated in Figure 4. The ovals represent nodes in the brt's and the lines with ground symbols represent history paths and where they close. The tree on the left illustrates a maximal resolution, where $l_2 \preceq l_1$ so that l_1 is maximal and should be resolved first. The tree on the right is a non-maximal resolution, where l_1 is resolved later, only after it shares nodes with another history path for l_1. Note that the shared node may be d_2, and that d_1 may be the child of d_2, as it will be in the case of 1-wso resolution, below.

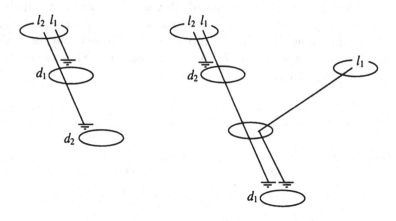

Fig. 4. A maximally and a non-maximally ordered resolution

Alternately one can state the r/a restriction in terms of an activity level associated with each literal. Initially all literals are active, *i.e.* are available for resolution. When a literal of a given rank is resolved, all literals of higher rank are turned off, and are turned back on only if merged at this or some later resolution step. In the definition above, if l_1 is resolved before l_2, l_2 is turned off, and then the other history path closing at d_2 is the one that re-activates l_2.

Note that in [5] it is the lower ranked literal that gets turned off. This arbitrary decision was reversed for this paper to be more consistent with the usual description of literal ordered resolution.

Thus the literal ordered resolution restriction is a special case of the r/a restriction, in which the ranks of literals in a clause are specified according to an ordering on the atoms, and non-maximal resolutions are not used. Since there is no reactivation of literals, there is no chance that a resolution step on a non-maximal literal will end in a refutation proof – it can never be resolved away. Thus a procedure to compute literal ordered resolution proofs needs to consider only maximal literals and has much less choice at each step; this leads to a much reduced search space, and accounts for the speed and success of such procedures. R/A procedures have considerable fan out since non-maximal literals are often active as well.

Yet in terms of proof size, the reduced choice of literal ordered resolution can lead to the elimination of all proofs rotation equivalent to smallest proof tree. In Example 5 the smallest proof, found when reverse alphabetical ordering of the atoms is used, has 15 resolutions, but the proof has 32,767 resolutions when the order is alphabetical. This example, for $n=4$, generalizes to 2^n clauses with n

$\{\ a,\quad b1,\quad c11,\quad d111\}$
$\{\ a,\quad b1,\quad c11, \neg d111\}$
$\{\ a,\quad b1, \neg c11,\quad d112\}$
$\{\ a,\quad b1, \neg c11, \neg d112\}$
$\{\ a, \neg b1,\quad c12,\quad d121\}$
$\{\ a, \neg b1,\quad c12, \neg d121\}$
$\{\ a, \neg b1, \neg c12,\quad d122\}$
$\{\ a, \neg b1, \neg c12, \neg d122\}$
$\{\neg a,\quad b2,\quad c21,\quad d211\}$
$\{\neg a,\quad b2,\quad c21, \neg d211\}$
$\{\neg a,\quad b2, \neg c21,\quad d212\}$
$\{\neg a,\quad b2, \neg c21, \neg d212\}$
$\{\neg a, \neg b2,\quad c22,\quad d221\}$
$\{\neg a, \neg b2,\quad c22, \neg d221\}$
$\{\neg a, \neg b2, \neg c22,\quad d222\}$
$\{\neg a, \neg b2, \neg c22, \neg d222\}$

Fig. 5. 32767 alphabetical a-ordered resolutions are required

literals each, which can take from $2^n - 1$ to $2^{2^n-1} - 1$ resolutions depending on the ordering.

In terms of deletion strategies, both rank/activity and literal ordered resolution retain completeness when used with tautology deletion. In fact, rank/activity can also be used with the regular and the surgery-minimal restrictions [7] which are strictly more restrictive. Literal ordered resolution with the surgery-minimal restriction is not complete.

Subsumption deletion works well with literal ordered resolution, but only a weakened form works with rank/activity – if a clause with some inactive literals is used as the subsuming clause, it must have all of its literals reactivated. This can be seen just by observing that otherwise the subsuming clause would not be able to draw a non-strictly stronger conclusion in cases where the subsumed clause could.

With respect to the ordering on literals, the r/a restriction does not require that the ordering (or rank) of literals in a clause is consistent with an overall literal ordering. In fact the ordering in r/a is applied on the literal occurrences and identical literals occuring in different clauses need not be ranked the same. Thus the ordering can be total without being liftable. An ordering is said to be liftable if $a \preceq b$ iff $a\theta \preceq b\theta$ for all substitutions θ. In literal ordered resolution, the orderings must be liftable to maintain completeness. Since liftable orderings are often are not total (but see [3]), in these cases the restriction cannot choose a unique maximal literal, leading to fan out in the search space.

Lock resolution [1], incidentally, is closely related. In lock resolution, as in rank/activity, the ranks are assigned to literal occurrences, chosen in any order, not according to an overall literal ordering. Like literal ordered resolution, lock resolution uses only the maximal case of the resolutions in Figure 4. Unfortunately lock resolution does not retain completeness with either tautology deletion or subsumption.

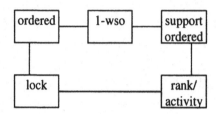

Fig. 6. A space of restrictions. The horizontal axis ranges from allowing only maximal resolutions to allowing non-maximal supported resolutions. The vertical axis indicates that ordering is defined on literal occurrences vs. defined on the literal set

From this discussion we imagine a space of strategies, shown in Figure 6, where the systems on the top all depend on orderings that can not in general be liftable, thus are not total, and those on the bottom depend on arbitrary ranks within a clause, which can be total. Those on the left depend only on the maximal case of the resolutions in Figure 4, and those on the right, support ordered and rank/activity, use both cases. The system labelled 1-wso is one system in this space, described in the next section. With 1-wso, a very restrictive form of the non-maximal case is permitted: a non-maximal atom may be resolved on, but then a merge and a resolution on a greater atom must follow immediately.

5   Building 1-Weak Support Ordered Proofs

One should not construct support ordered proofs directly from the definition, since the support part of the restriction cannot be determined until the tree is completed. The rank/activity calculus with the ranks set according to the literal or atom ordering should be used instead. It then computes support ordered proofs. Even so, the number of allowable deductions at each stage may be too high, especially in the early stages when all literals are active.

On the other hand the advantages of r/a may be useful: preservation of the smallest proof and compatibility with the surgery-minimal restriction.

The 1-weak support ordered resolution keeps some of the advantages of both. It uses the maximal case and a restricted form of the non-maximal case, from Figure 4. Because the maximal case is sufficient to ensure completeness of the procedure, we are free to further restrict the non-maximal case. In the 1-wso non-maximal case, a non-maximal atom may be resolved but only if some greater atoms occur in each parent and those greater atoms can be merged or factored. Then the factoring is performed, even if a substitution is required. Moreover, the resulting clause is forced to resolve in a conventional way on this factored literal, whether or not it is maximal in the clause and whether or not a non-maximal resolution would otherwise be permitted.

We defined wso resolution to make a slightly modified support condition which was easier to compute. Although weak support ordered resolution looked promising, we noted the restriction could not be applied to a partial proof without knowing more information. Specifically when a resolution on a non-maximal literal is made, it is not immediately knowable that this new node will eventually be supported by the resolution on a greater literal that has been deferred. This guarantee is now made by 1-wso by forcing such a supporting resolution to be the next step. One can see that this new node supports the non-maximal one because the rotation between these two nodes would disturb the merge, as in Figure 2. We need to refer the reader to [7] for the proof that no other sequence of rotations could somehow invert these nodes.

Because the distance from the resolution producing the merge or factor is one node away from the resolution on the merged literal, we call the resulting system 1-weak support ordered resolution.

There is a surprising distinction betweem wso and 1wso. While they are closely related, neither is a restriction of the other. In Figure 7, the tree on the left is wso but not 1-wso. It is not 1-wso because the non-maximal resolution on b does not end up merging a greater literal. It is wso because of the two possible rotations, the upper one would unorder the nodes and the lower one would disturb a merge on a. The tree on the right of Figure 7 is 1-wso but not wso. It is 1-wso because after the non-maximal resolution on c, the resolution on the greater, merged atom b follows immediately. But it is not wso, because the lower resolutions should be inverted, according to the atom ordering. In Figure 7 wso and support order correspond, so the tree on the left is support ordered while the tree on the right is not.

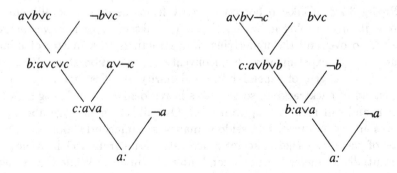

Fig. 7. A wso and a 1-wso tree

One of the design decisions of 1-wso was to prevent a great increase in the number of choices beyond literal ordered resolution. The rather complex condition for a non-maximal resolution, and the identification of the forced literal for the next resolution help to prevent such an increase. So 1-wso resembles ordered resolution in that only one literal in each clause is available for maximal resolution. Contrast this with rank/activity which exactly computes support ordered resolution but has all literals available for resolution initially, so it must rely on other heuristics to limit the search.

Rank/activity guarantees that some shortest proof tree is in its search space. 1-wso does not, but it does guarantee that some tree in its search space is non-strictly smaller than the smallest one in the search space for support ordered resolution. For the problem in Figure 5, a 1-wso tree of size 1511 nodes is found by the theorem prover described below. While this is much greater than the minimal 15, it is better that 32767. There may be a smaller 1-wso tree for this example.

The 1-wso system inherits full subsumption from literal ordered resolution between clauses without forced literals. We may include in this any clause with a forced literal that happens to be maximal, which is the most common case. If a clause has a forced literal, and its other literals are subsumed by some clause C with no forced literals, then the subsumption should also be allowed; any clause generated when the forced literal is resolved upon will also be subsumed by C. A clause with a forced literal can also be subsumed by a clause with a forced literal if the forced literals are not different, *i.e.* if they correspond in the subsumption's subset test. The subsumption is not allowed if a clause with a forced non-maximal literal tries to subsume a clause with a different literal either forced or maximal, because the candidate subsuming clause is not adequate to be used in place of the candidate subsumed clause.

6   Experiments

The experiments were conducted on a 400MHz Pentium II with 64 Mb RAM running RedHat Linux 6.0 with a theorem prover called pBliksem written by the first author in about 5000 lines of Prolog. The experiments depended on SWI-Prolog 3.2.8. pBliksem borrows heavily from the design of Bliksem, but as it is written in Prolog it has no claims to Bliksem's speed. Nevertheless it is possible to overhaul the underlying data structures in a matter of minutes or hours, and to experiment with unconventional resolution steps. Written in Prolog it has a degree of dependability and clarity. It relies mainly on forward calls, instead of backtracking, so it makes heavy demands on Prolog's garbage collection and tail recursive optimization. Overall it can manage about 5000 inferences in ten minutes, but seldom manages significantly more inferences, because of memory usage. The inference rate varies from 150 inferences per second initially, to under ten per second after ten minutes. While this is clearly not a competitive system, there is nothing to suggest that the results we have obtained with this experimental system would not also be obtained by a better theorem prover. In particular the inferences it chooses are almost identical to those chosen by Bliksem. Occasionally the choices to be made are not determined by the settings of the flags and parameters, and in these tied cases the systems may chose differently.

The experiments were performed on about 80% of the problems in TPTP-v2.3.0 [8] with difficulties in the range strictly above zero and strictly below 0.6. We selected 408 of the 528 such problems. It was configured to use Bliksem's liftable literal order, in which lexicographical ordering is used, but literals that are lexically identical up to a variable are not comparable. The clause ordering depends first on the complexity of the clause, which is the sum of the number of function symbols, predicate symbols and variables. If the complexity of clauses is identical then the sizes of the underlying brts are compared. Two configurations of the theorem prover were tried: literal ordered resolution and 1-wso resolution. Ten minutes of computing time was given to each problem.

Literal ordered resolution solved only six problems, where as 1-wso solved 75, including the six. 1-wso used about 10% more time than literal ordered resolution to solve those six. The numbers of non-maximal resolutions used are given in Table 1. This table shows that non-maximal resolutions were used in most proofs, except the six ones solved by the literal ordered resolution system. Most did not require many non-maximal resolutions but a few did: BOO035-1 required 20 and SYN074-1 required 19.

Number of non-maximal resolutions in the tree	Number of solved	Percentage
0	6	8%
1	32	43%
2	14	19%
3	11	15%
>3	12	16%

Table 1. Counting non-maximal resolutions used by 1-wso

The 1-wso program did best on the problems with difficulty between 0.2 and 0.3 and worst on the problems between 0.3 and 0.4, as shown by Table 2. Both above and below these ranges almost 20% percent of the problems were solved, indicating that the correlation between the difficulty measure and the success of the algorithm is rather small. Perhaps this is because the difficulty of a theorem is a very hard property to measure.

The times taken to solve the problems, reported in Table 3, were scattered somewhat logarithmically with more toward the upper end of the time allocated. Each factor of two allowed about 10 more problems to be solved.

Table 4 shows the number of inferences generated by the prover, which would not be affected by the inefficiency of our implementation. The effect of allowing more inferences is similar to allowing more time, but there is a slightly larger

Difficulty d	Number of Problems	Solved	%
$0.1 \leq d < 0.2$	66	12	18%
$0.2 \leq d < 0.3$	84	32	38%
$0.3 \leq d < 0.4$	170	15	9%
$0.4 \leq d < 0.5$	48	8	17%
$0.5 \leq d < 0.6$	40	8	20%
	Total 408		

Table 2. Difficulty of problems and number solved by 1-wso

Time range (seconds)	Number of problems
$0 \leq t \leq 1$	13
$1 \leq t < 2$	4
$2 \leq t < 4$	2
$4 \leq t < 8$	1
$8 \leq t < 16$	11
$16 \leq t < 32$	2
$32 \leq t < 64$	6
$64 \leq t < 128$	13
$128 \leq t < 256$	9
$256 \leq t < 512$	12
$512 \leq t < 600$	2

Table 3. Times taken to solve problems

shift to the upper end of the range. Each factor of two allowed about 12 more problems to be solved.

Inferences Generated g	Number of problems
$0 < g < 10$	1
$10 \leq g < 20$	0
$20 \leq g < 40$	2
$40 \leq g < 80$	9
$80 \leq g < 160$	7
$160 \leq g < 320$	6
$320 \leq g < 640$	11
$640 \leq g < 1280$	11
$1280 \leq g < 2560$	16
$2560 \leq g < 5120$	11
$5120 \leq g \leq 5357$	1

Table 4. Number of inferences required to solve problems

Table 5 shows the number of problems of each category solved. The 1-wso procedure seems to be much better at some categories than others. Note that there is no special treatment of the equality literal in our implementation.

Category	Number of problems	Solved	Percent
BOO	5	1	20%
CAT	11	0	0%
COL	15	2	13%
FLD	26	0	0%
GEO	57	19	33%
GRP	20	7	35%
HEN	5	5	100%
LCL	33	23	70%
LDA	6	1	17%
NUM	3	1	33%
PLA	15	0	0%
PUZ	1	0	0%
RNG	6	0	0%
ROB	1	0	0%
SET	113	11	10%
SYN	91	15	5%
	408	75	18%

Table 5. Problems solved in each category

7   Conclusions

A good restriction strategy in an automated theorem prover depends balancing the economy of choice with the economy of cuts. Literal ordered resolution strongly limits choices but cuts away many proofs, sometimes leaving only very big proofs. Support ordered resolution, or rank/activity, allows more choices, and cuts away all but one of a set of rotation equivalent proofs, and leaves a smallest proof tree. 1-weak support ordered resolution is a step away from literal ordered resolution toward support ordered resolution, allowing very specific extra choices. It does not eliminate all the redundancy of rotation equivalence, nor does it preserve the smallest proof, but it is guaranteed to find non-strictly smaller proofs than literal ordered resolution, occasionally finding much smaller ones. Experiments indicate that the balance here is in favor of increasing choice beyond literal ordered resolution to increase the number of possible proofs. This leaves open the question whether and how to move toward support ordered resolution, allowing modestly more choices while leaving more and smaller proofs in the search space.

Acknowledgements

The authors are grateful to NSERC of Canada for funding, and to the anonymous reviewers for their constructive comments. The first author did some initial work while visiting the Max-Planck-Institut für Informatik, Saarbrücken.

References

1. R. S. Boyer. *Locking: A Restriction of Resolution*. PhD thesis, University of Texas at Austin, 1971.
2. Chin-Liang Chang and Richard Char-Tung Lee. *Symbolic Logic and Mechanical Theorem Proving*. Academic Press, New York and London, 1973.
3. Hans de Nivelle. Resolution games and non-liftable resolution orderings. *Collegium Logicum, Annals of the Kurt Gödel Society*, 2:1–20, 1996.
4. J. D. Horton and B. Spencer. Clause trees: a tool for understanding and implementing resolution in automated reasoning. *Artificial Intelligence*, 92:25–89, 1997.
5. J. D. Horton and B. Spencer. Rank/activity: a canonical form for binary resolution. In C. Kirchner and H. Kirchner, editors, *Automated Deduction – CADE-15*, number 1421 in Lecture Notes in Artificial Intelligence, pages 412–426. Springer-Verlag, Berlin, July 1998.
6. J. A. Robinson. A machine-oriented logic based on the resolution principle. *J. ACM*, 12:23–41, 1965.
7. B. Spencer and J. D. Horton. Efficient procedures for detecting and restoring minimality, an extension of the regular restriction of resolution. *Journal of Automated Reasoning*, to appear.
8. G. Sutcliffe, C. Suttner, and T. Yemenis. The TPTP problem library. In D. Kapur, editor, *Automated Deduction CADE-12*, number 814 in Lecture Notes in Artificial Intelligence, pages 252–266. Springer-Verlag, Berlin, 1994.
9. Geoff Sutcliffe. The CADE-16 system competition. *Journal of Automated Reasoning*, 24:371–396, 2000.

System Description: IVY*

William McCune[1] and Olga Shumsky[2]

[1] Mathematics and Computer Science Division
Argonne National Laboratory, U.S.A.
mccune@mcs.anl.gov
http://www.mcs.anl.gov/~mccune
[2] Department of Electrical and Computer Engineering
Northwestern University, U.S.A.
shumsky@ece.nwu.edu

Abstract. IVY is a verified theorem prover for first-order logic with equality. It is coded in ACL2, and it makes calls to the theorem prover Otter to search for proofs and to the program MACE to search for countermodels. Verifications of Otter and MACE are not practical because they are coded in C. Instead, Otter and MACE give detailed proofs and models that are checked by verified ACL2 programs. In addition, the initial conversion to clause form is done by verified ACL2 code. The verification is done with respect to finite interpretations.

1 Introduction

Our theorem provers Otter [6,7,10] and EQP [4,8] and our model searcher MACE [3,5] are being used for practical work in several areas. Therefore, we wish to have very high confidence that the proofs and models they produce are correct. However, these are high-performance programs, coded in C, with many tricks, hacks, and optimizations, so formal verification of the programs is not practical.

Instead, our approach is to have the C programs give their results explicitly as detailed proof objects or models, and to have separate *checker programs* check the results. The checker algorithms are relatively simple and straightforward, so it is practical to apply program verification techniques to them. In particular, we use the ACL2 program verification system to prove that if the checker program accepts a proof, then the proof is correct.

Otter can convert first order formulas into clauses (by normal form translation and Skolemization), but it is not able to include these preprocessing steps as part of the proof objects. Therefore, we have recoded the clause form translator in ACL2 and proved its soundness directly. The result is a hybrid system, named IVY, that (1) is driven by ACL2 code, (2) calls ACL2 functions for the preprocessing, (3) calls an external program to search for a proof or a model,

---

* This work was supported by the Mathematical, Information, and Computational Sciences Division subprogram of the Office of Advanced Scientific Computing Research, U.S. Department of Energy, under Contract W-31-109-Eng-38.

D. McAllester (Ed.): CADE-17, LNAI 1831, pp. 401–405, 2000.

and (4) calls ACL2 checker functions to check the results. The top-level soundness theorems have the form: *If IVY claims a proof, then the input formula is a theorem.* A weakness of the verification method is that the soundness proofs are with respect to finite interpretations. In Section 6 we discuss an approach for all interpretations.

ACL2 (A Computational Logic for Applicative Common Lisp) [2, 1], is the successor to the Boyer-Moore theorem prover. ACL2 is a specification/programming language, based on Common Lisp, together with an environment for proving theorems about the programs. Its strength is automated support for proving inductive theorems about recursively defined programs.

2 Specification of the Logic

We use ACL2 to define a first-order logic, and this becomes the specification for our verification. The definitions of *well-formed term* and *well-formed formula* are straightforward. We next define the semantics of our logic by defining *interpretation* of a first-order language. This part is nonstandard, because we restrict ourselves to finite interpretations; see Section 6. Finally, we define *evaluation* of a formula in an interpretation. The evaluation function is a pair of mutually recursive functions, in which one recurses through the structure of formulas, and the other (called for quantified formulas) recurses through the elements of the domain of the interpretation. In particular, the function (FEVAL F I) evaluates formula F in interpretation I.

3 The Proof Procedure

The proof search procedure is standard for first-order resolution/paramodulation theorem provers. Starting with the negation of a conjecture, we (1) convert to negation-normal form, (2) rename bound variables, (3) Skolemize, (4) move universal quantifiers to the top, (5) convert to conjunctive normal form, (6) search for a refutation (or a model), and (7) check the refutation (or model).

Steps 1, 2, 4, and 5 produce an equivalent formula, and Skolemization produces an equiconsistent formula.

In IVY, the preprocessing steps (1–5) are coded in ACL2, the search step (6) is accomplished by calling Otter or MACE, and the checker step (7) is coded in ACL2.

4 Soundness Theorems

The function to convert formulas to negation-normal form is (NNF F), and the soundness theorem states that NNF produces an equivalent formula:

```
(EQUAL (FEVAL (NNF F) I)
       (FEVAL F I)).
```

The soundness theorems for steps 2, 4, and 5 of the proof procedure are similar. The soundness theorem for Skolemization is more complicated, because we have to extend the interpretation with the new Skolem symbols:

```
(EQUAL (FEVAL (SKOLEMIZE F) (SKOLEMIZE-EXTEND F I))
       (FEVAL F I)).
```

Steps 6 and 7 of the proof procedure are combined in an ACL2 function `(REFUTE-N-CHECK F)` which calls Otter (see Sec. 5) and the checker function. If Otter finds a refutation, and if the checker accepts the refutation, `REFUTE-N-CHECK` returns `FALSE` (the contradictory formula of our logic); otherwise `REFUTE-N-CHECK` returns the input formula `F`. Hence, it always produces an equivalent formula, and the soundness theorem is

```
(EQUAL (FEVAL (REFUTE-N-CHECK F) I)
       (FEVAL F I)).
```

All of the preprocessing functions, `REFUTE-N-CHECK`, and a few other functions are composed into a top-level function `(PROVED F)`, which takes the positive form of a conjecture, checks that it is well-formed and closed, negates it, and applies the proof procedure. The top-level soundness theorem is

```
(IMPLIES (PROVED F)
         (AND (WFF F)
              (NOT (FREE-VARS F))
              (FEVAL F I))).
```

In other words, if IVY claims a proof of a conjecture F, then F is a closed well-formed formula that is true in all (finite) interpretations. Of course, to accept this theorem, a user must accept our ACL2 definition of first-order logic and the soundness of the ACL2 system. But the point is that the user doesn't have to trust Otter, which does the hard part of the work.

The other side of the problem, searching for countermodels, is easier because checking a claimed model produced by the C program MACE is done by simply evaluating the negation of the conjecture in the claimed model. The top-level function `(COUNTERMODEL F)` is analogous to `(PROVED F)`: it checks that the conjecture F is closed and well formed, negates it, preprocesses it, calls MACE to search for a finite model, and checks that the negation of F is true in any model found by MACE. The soundness theorem for `(COUNTERMODEL F)` is nearly trivial, because the evaluation property we need to prove is checked by `COUNTERMODEL`:

```
(IMPLIES (COUNTERMODEL F)
         (AND (WFF F)
              (NOT (FREE-VARS F))
              (NOT (FEVAL F (COUNTERMODEL F))))).
```

In other words, if IVY claims a countermodel to a conjecture F, then F is a closed well-formed formula that is false in some interpretation.

5   Interface to the C Code

The function REFUTE-N-CHECK takes the universal closure of a conjunction of clauses and returns an equivalent formula. First it transforms the input formula into an initial proof object. Next it calls the function EXTERNAL-PROVER which augments the initial proof object with additional steps that represent some derivation (a derivation of the empty clause if we are lucky). Then it checks that each step of the proof object follows from preceding steps.

In the ACL2 environment, EXTERNAL-PROVER is a *defstub*, that is, we tell ACL2 that it exists but that we don't know any other properties of it. We use ACL2 to prove properties of REFUTE-N-CHECK (e.g., soundness), but these properties are necessarily independent of EXTERNAL-PROVER.

At run time, a Common Lisp function EXTERNAL-PROVER is loaded along with the ACL2 code, and the Common Lisp version of EXTERNAL-PROVER overrides the ACL2 defstub.[1] The Common Lisp version of EXTERNAL-PROVER contains operating system calls to build an input file for Otter, run Otter, and read and process Otter's output. If the Common Lisp version of EXTERNAL-PROVER returns a proof object that is not well formed or is unsound, the check fails, and REFUTE-N-CHECK returns its input.

A similar situation holds when searching for a countermodel with MACE. A defstub EXTERNAL-MODELER is used in the ACL2 environment when defining functions and proving properties, and a Common Lisp version of EXTERNAL-MODELER, which calls MACE, is loaded at run time.

It is possible to use the preprocessing and proof checking functions of IVY with other first-order resolution/paramodulation provers and model searchers, provided they produce appropriate proof objects or models. (The format for proof object can be found in [9].) This can be accomplished by simply rewriting the Common Lisp version of the EXTERNAL-PROVER or EXTERNAL-MODELER to call the desired program.

6   The Finite Domain Assumption

Our approach of proving soundness with respect to finite interpretations is certainly questionable. Consider, for example, the sentence

```
(IMP (ALL X (ALL Y (IMP (= (F X) (F Y))
                        (= X Y))))
     (ALL X (EXISTS Y (= (F Y) X)))),
```

that is, one-to-one functions are onto. It is *not* valid, but it is true for finite domains. Could IVY claim to have a proof of such a nontheorem?

We conjecture that it could not—that the weakness is in the metaproof method rather than the first-order proof procedure. Nonetheless, we are pursuing the following general approach that covers infinite interpretations.

---

[1] According to the ACL2 designers, having an ACL2 function call a Common Lisp function in this way is not officially endorsed, but it is acceptable in this situation.

ACL2 has an encapsulation feature that allows it to reason safely about incompletely specified functions. We believe we can use encapsulation to abstract the finiteness.[2] In our current specification, the important way in which finiteness enters the picture is by the definition of FEVAL-D, which recurses through the domain. This function, in effect, expands universally quantified formulas into conjunctions and existentially quantified formulas into disjunctions. Instead of FEVAL-D, we can consider a constrained function that chooses an element of the domain, if possible, that makes a formula true. When evaluating an existentially quantified formula, we substitute the chosen element for the existentially quantified variable and continue evaluating. (Evaluation of universally quantified variables requires some fiddling with negation.) However, proving the soundness of Skolemization may present complications in this approach.

7 Performance and Availability

Aside from the overhead of starting up ACL2, the performance of IVY is essentially the same as the performance of Otter's autonomous mode or MACE with its default settings. IVY cannot yet accept parameters to be passed to Otter or MACE. The latest version of IVY is available from http://www.mcs.anl.gov/~mccune/ivy. A more complete paper on IVY can be found in [9].

References

1. M. Kaufmann, P. Manolios, and J Moore, editors. *Computer-Aided Reasoning: ACL2 Case Studies*. Kluwer Academic, 2000. To appear.
2. M. Kaufmann and J Moore. An industrial strength theorem prover for a logic based on Common Lisp. *IEEE Transactions on Software Engineering*, 23(4):203–213, 1997.
3. W. McCune. A Davis-Putnam program and its application to finite first-order model search: Quasigroup existence problems. Tech. Report ANL/MCS-TM-194, Argonne National Laboratory, Argonne, IL, May 1994.
4. W. McCune. EQP. http://www.mcs.anl.gov/~AR/eqp/, 1994.
5. W. McCune. MACE: Models and Counterexamples. http://www.mcs.anl.gov/AR/mace/, 1994.
6. W. McCune. Otter. http://www.mcs.anl.gov/AR/otter/, 1994.
7. W. McCune. Otter 3.0 Reference Manual and Guide. Tech. Report ANL-94/6, Argonne National Laboratory, Argonne, IL, 1994.
8. W. McCune. 33 basic test problems: A practical evaluation of some paramodulation strategies. In Robert Veroff, editor, *Automated Reasoning and its Applications: Essays in Honor of Larry Wos*, chapter 5, pages 71–114. MIT Press, 1997.
9. W. McCune and O. Shumsky. IVY: A preprocessor and proof checker for first-order logic. Preprint ANL/MCS-P775-0899, Argonne National Laboratory, Argonne, IL, 1999. To appear in [1].
10. W. McCune and L. Wos. Otter: The CADE-13 competition incarnations. *J. Automated Reasoning*, 18(2):211–220, 1997.

---

[2] This approach was suggested by Matt Kaufmann.

System Description: SystemOnTPTP

Geoff Sutcliffe

School of Information Technology, James Cook University, Townsville, Australia
geoff@cs.jcu.edu.au

Abstract. SystemOnTPTP is a WWW interface that allows an ATP problem to be easily and quickly submitted in various ways to a range of ATP systems. The interface uses a suite of currently available ATP systems. The interface allows the problem to be selected from the TPTP library or for a problem written in TPTP syntax to be provided by the user. The problem may be submitted to one or more of the ATP systems in sequence, or may be submitted via the SSCPA interface to multiple systems in parallel. SystemOnTPTP also can provide system recommendations for a problem.

SystemOnTPTP is a WWW interface that allows an ATP problem to be easily and quickly submitted in various ways to a range of ATP systems. The interface uses a suite of currently available ATP systems, which are maintained in a database structure. The interface is generated directly from the database, and thus is as current as the database. The interface is a single WWW page, in three parts: the problem specification part, the mode specification part, and the system selection part. A user

Fig. 1. Problem specification

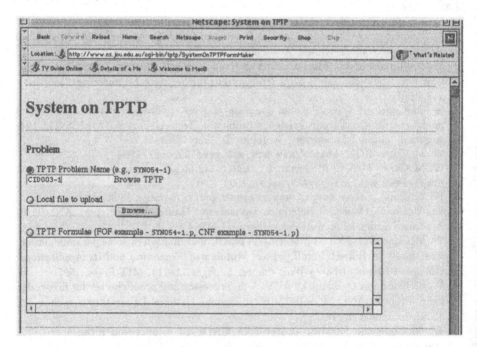

specifies the problem first, optionally selects systems for use, and then specifies the mode of use.

Figure 1 shows the problem specification part of the WWW page. There are three ways that the problem can be specified. First, a problem in the TPTP library [SS98] can be specified by name. The interface provides a tool for browsing the TPTP problems if desired. Second, a file, containing a problem written in TPTP syntax, on the same computer as the WWW browser can be specified for uploading. The interface provides an option for browsing the local disk and selecting the file. Third, the problem formulae can be provided directly in a text window. Links to example TPTP files are provided to remind the user of the TPTP syntax if required.

Figure 2 shows the start of the system selection part of the interface. There is one line for each system in the suite, indicating the system name and version, a default CPU time limit, the default tptp2X transformations for the system, the default tptp2X format for the system, and the default command line for the system. To select a system the user selects the corresponding tickbox. The default values in the text fields can be modified, if required.

Fig. 2. System selection

ATP System	Time Limit	Transform	Format	Command
Bliksem 1.10A	100 seconds	rm_equality:rstfp	bliksem	bliksem %s
CoDe 3.00	100 seconds	none	code	code3 -FE -o- -d- -e-
E 0.51	100 seconds	rm_equality:rstfp	tptp	eprover --tptp-forma
FDP 0.9	100 seconds	none	protein	fdp-caso %s %d
Fiesta 2	100 seconds	rm_equality:rstfp	dedam	fiesta-wrapper %s
GLiDeS 0.0	100 seconds	none,rm_equality:stf	glides	GLiDeS %s

Figure 3 shows the mode specification part of the interface. The lefthand side contains information and the RecommendSystems button for obtaining system recommendations for the specified problem. System recommendations are generated

Fig. 3. Mode specification

System Recommendations	Solve the Problem at JCU		
System recommendations are free. The recommendations are based on the systems' results for the TPTP. There is no guarantee that they are the best systems for this particular problem.	Solution attempts are limited to users with passwords plus a few users without passwords. If you submit without a valid password your job may be rejected if the machine is too busy. Password:		
Note: Submitting your ATP system's latest performance data is the best way to get it recommended.	**Output mode** ○ Quiet ● Interesting ○ Verbose	**Parallel mode** ○ Selected ○ Naive ... ○ SSCPA ... ● Eager SSCPA ...	300 seconds All selected systems 3 systems
Recommend Systems	Run Selected Systems	Run Parallel	

Our server does not output results until all tasks are completed. Be patient while the provers do their thing.

as follows: ATP problems have been classified into 14 disjoint specialist problem classes (SPCs) according to problem characteristics such as effective order, use of equality, and syntactic form. In a once off analysis phase for each SPC, performance data for ATP systems (not necessarily all in the suite), for some carefully selected TPTP problems in the SPC, is analyzed. Systems that solve a subset of the problems solved by another system are discarded. The remaining systems are recorded as recommended, in order of the number of problems solved in the SPC. Later at run time, when system recommendations for a specified problem (not necessary one of those used in the analysis phase, or not even from the TPTP) are requested, the problem is classified into its SPC and the corresponding system recommendations are returned.

The righthand side of the mode specification part of the interface gives information, options, and the submit buttons for using the ATP systems sequentially or in parallel. When a problem is submitted using either of these submit buttons, the ATP systems are executed on a server at James Cook University. Due to resource restrictions, only one public user may submit at a time. A password field is provided that allows privileged users to submit at any time. The Output mode options specify how much output is returned during processing. In Quiet mode only the final result is returned, giving the type of result and the time taken. The result is either that a proof was found, that it was established that no proof can be found, or that the systems gave up trying for an unknown reason. In Interesting mode information about the progress of the submission is returned; see below for an example. In Verbose mode a full trace of the submission is returned, including the problem in TPTP format, the problem after transformation and formatting for the systems, and all standard output produced by the systems. The RunSelectedSystems button sequentially gives the specified problem to each of the systems selected in the system selection part of the interface. For each selected system, the problem is transformed and formatted using the tptp2X utility as specified in the system selection. The transformed and formatted problem is given to the system using the specifed command line, with a CPU time limit as specified for the system.

The Parallel mode options specify the type of parallelism to use when a problem is submitted using the RunParallel button. All of the parallel modes perform competition parallelism [SS94], i.e., multiple systems are run in parallel on the machine (using UNIX multitasking if there are less available CPUs than systems) and when any one gets a deciding result all of the systems are killed. The differences between the modes are which systems are used and the individual time limits imposed on each system's execution. A limit on the total CPU time that can be taken by the executing systems is specified in the seconds field of the interface. In Naive selected mode all of the systems selected in the system selection part of the interface are run in parallel with equal CPU time limits (the appropriate fraction of the total time limit). In Naive mode the specified number of systems, taken in alphabetical order from the selection list, are run in parallel with equal time limits. In SSCPA mode the system recommendation component is used to get system recommendations for the specified problem. The suite of systems is then checked for versions of the recommended systems, in order of recommendation, until the specified number of systems have been found, or the recommendations are exhausted. The systems are then run in parallel with equal time limits. In Eager SSCPA mode the systems used are the same as for SSCPA mode, but the individual system time limits are calculated by repeatedly dividing the total time limit by two, and allocating the values to the systems in order. In this manner the highest recommended system gets half the total time limit, the next system gets a quarter, and so on, with the last two systems getting equal time limits. The motivations and effects of these parallel modes are discussed in [SS99]. The effectiveness of SSCPA was demonstrated in the CADE-16 ATP System Competition [Sut00].

Figure 4 shows the interesting output from submission of the TPTP problem CID003-1 in the Eager SSCPA mode. The execution is first transferred from the WWW server onto a SUN workstation where the ATP systems are installed. The system recommendation component is then invoked. The problem is identified as being real 1st order, having some equality, in CNF, and Horn. Five systems are recommended for the SPC: E 0.32, E-SETHEO 99csp, Vampire 0.0, OtterMACE 437, and Gandalf

Fig. 4. Interesting output for CID003-1 in Eager SSCPA mode

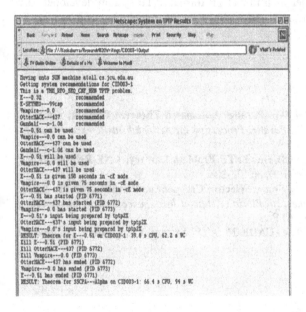

c-1.0d. The suite of systems is then checked for versions of these systems, and it is found that versions of four of them, E 0.51, Vampire 0.0, OtterMACE 437, and Gandalf c-1.0d, are in the suite. The submission required three systems, so E 0.51, Vampire 0.0, and OtterMACE 437 are used. The individual system time limits out of the total specified limit of 300 seconds are then computed, 150 seconds for E 0.51 and 75 seconds each for Vampire 0.0 and OtterMACE 437. The problem is then transformed and formatted for each of the systems, and the systems are run in parallel. E 0.51 finds a proof after 39.8 seconds CPU time, 62.2 seconds wall clock time, at which stage all of the systems are killed.

SystemOnTPTP is implemented by `perl` scripts that generate the WWW interface from the system database, accept the submission from the browser, extract the system recommendations, invoke the tptp2X utility to transform and reformat the problem, and control the execution of the systems. The interface is available at:
http://www.cs.jcu.edu.au/cgi-bin/tptp/SystemOnTPTPFormMaker

SystemOnTPTP makes it easy for users to easily and quickly submit a problem in TPTP syntax to an appropriate ATP system. The user is absolved of the responsibilities and chores of selecting systems to use, installing the systems, transforming and formatting the problem for the systems, and controlling their execution. This user friendly environment is particularly appropriate for ATP system users who want to focus on the problem content rather than the mechanisms of ATP. The interface is not designed for, and is therefore not suitable for, users who wish to submit a batch of problems to a particular ATP system. Such users should obviously install that ATP system on their own computer, which would also allow use of the system's own input format rather than the TPTP format. ATP system developers are invited to submit their systems and performance data for inclusion and use in the interface.

References

SS94 Suttner C.B., Schumann J. (1994), **Parallel Automated Theorem Proving**, Kanal L., Kumar V., Kitano H., Suttner C., *Parallel Processing for Artificial Intelligence 1*, pp.209-257, Elsevier Science.

SS98 Sutcliffe G., Suttner C.B. (1998),**The TPTP Problem Library: CNF Release v1.2.1**, *Journal of Automated Reasoning* 21(2), pp.177-203.

SS99 Sutcliffe G., Seyfang D. (1999), **Smart Selective Competition Parallelism ATP**, Kumar A., Russell I., *Proceedings of the 12th Florida Artificial Intelligence Research Symposium* (Orlando, USA), pp.341-345, AAAI Press.

Sut00 Sutcliffe G. (To appear),**The CADE-16 ATP System Competition**, *Journal of Automated Reasoning*.

System Description: PTTP+GLiDeS
Semantically Guided PTTP

Marianne Brown and Geoff Sutcliffe

School of Information Technology
James Cook University
{marianne,geoff}@cs.jcu.edu.au
http://www.cs.jcu.edu.au/~marianne/GLiDeS.html

Introduction

PTTP+GLiDeS is a semantically guided linear deduction theorem prover, built from PTTP [9] and MACE [7]. It takes problems in clause normal form (CNF), generates semantic information about the clauses, and then uses the semantic information to guide its search for a proof.

In the last decade there has been some work done in the area of semantic guidance, in a variety of first order theorem proving paradigms: SCOTT [8] is based on OTTER and is a forward chaining resolution system; CLIN-S [3] uses hyperlinking; RAMCS [2] uses constrained clauses to allow it to search for proofs and models simultaneously; and SGLD [11] is a chain format linear deduction system based on Graph Construction. Of these, CLIN-S and SGLD need to be supplied with semantics by the user. SCOTT uses FINDER [8] to generate models, and RAMCS generates its own models.

The Semantic Guidance

PTTP+GLiDeS uses a semantic pruning strategy that is based upon the strategy that can be applied to linear-input deductions. In a completed linear-input refutation, all centre clauses are FALSE in all models of the side clauses. This leads to a semantic pruning strategy that, at every stage of a linear-input deduction, requires all centre clauses in the deduction so far to be FALSE in a model of the side clauses. To implement this strategy it is necessary to know which are the potential side clauses, so that a model can be built. A simple possibility is to choose a negative top clause from a set of Horn clauses, in which case the mixed clauses are the potential side clauses. More sensitive analysis is also possible [4, 10]. Linear-input deduction and this pruning strategy are complete only for Horn clauses. Unfortunately, the extension of this pruning strategy to linear deduction, which is also complete for non-Horn clauses, is not direct. The possibility of ancestor resolutions means that centre clauses may be TRUE in a model of the side clauses.

In PTTP+GLiDeS, rather than placing a constraint on entire centre clauses, a semantic constraint is placed on certain literals of the centre clauses: The input clauses other than the chosen top clause of a linear deduction are named

D. McAllester (Ed.): CADE-17, LNAI 1831, pp. 411–416, 2000.
© Springer-Verlag Berlin Heidelberg 2000

the *model clauses*. In a completed linear refutation, all centre clause literals that have resolved against input clause literals are required to be FALSE in a model of the model clauses. TRUE centre clause literals must be resolved against ancestor clause literals.

PTTP+GLiDeS implements linear deduction using the Model Elimination [6] (ME) paradigm. ME uses a chain format, where a chain is an ordered list of A- and B-literals. The disjunction of the B-literals is the clause represented by the chain. Input chains are generated from the input clauses and are composed entirely of B-literals. The chains that form the linear sequence are called the centre chains. A-literals are used in centre chains to record information about ancestor clause in the deduction. The input chains that are resolved against centre chains are called side chains.

PTTP+GLiDeS maintains a list of all the A-literals created throughout the entire deduction. This list is called the A-list. The pruning strategy requires that at every stage of the deduction, there must exist at least one ground instance of the A-list that is FALSE in a model of the model clauses. The result is that only FALSE B-literals are extended upon, and TRUE B-literals must reduce. Figure 1 shows an example of a PTTP+GLiDeS refutation. The problem clauses are $\{\sim money \vee tickets(buy), \sim tickets(sell) \vee money, money \vee tickets(X), \sim money \vee \sim tickets(X)\}$. The clause $\sim money \vee \sim tickets(X)$ is chosen to form the top chain, so that the other three clauses are the model clauses. The model M is $\{money, tickets(buy), \sim tickets(sell)\}$. The A-list is shown in braces under the centre chains.

Since the work described in [1], PTTP+GLiDeS has been enhanced to order the use of side chains, using the model of the model clauses. The model is used to give a score to each clause as follows: If there are N ground domain instances of a clause C with k literals, then for each literal L, n_L is the number of TRUE instances of L within the N ground instances. L is given the score $\frac{n_L}{N}$. The score for the clause C is $\frac{1}{N*k} \sum_{L=1}^{k} n_L$. The clause set is then ordered in descending order of scores. This gives preference to clauses that have more TRUE literal instances in the model. The use of these clauses leads to early pruning and forces the deduction into areas more likely to lead to a proof.

Implementation

PTTP+GLiDeS consists of a modified version of PTTP version 2e and MACE v1.3.3, combined with a csh script. It requires the problem to be presented in both PTTP and OTTER formats. The OTTER format file is processed so that it contains only the model clauses, and is used by MACE.

Initially the domain size for the model to be generated by MACE is set to equal the number of constants in the problem. If a model of this size cannot be found, the domain size is reset to 2 and MACE is allowed to determine the domain size. If no model is found PTTP+GLiDeS exits. If a model is found, the modified PTTP is started. The modified PTTP uses the model to reorder the clause set, then transforms the reordered clauses into Prolog procedures

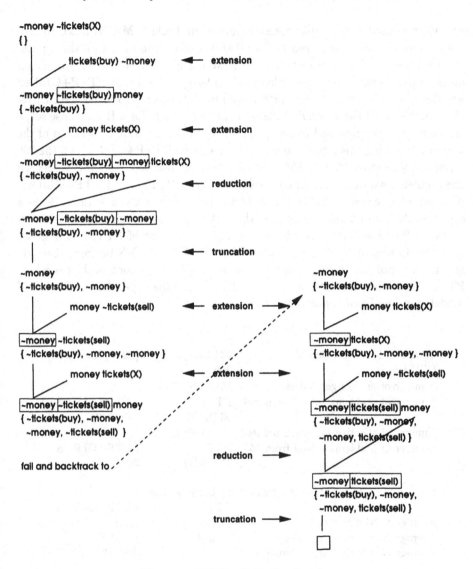

Fig. 1. A PTTP+GLiDeS refutation

that implement the ME deduction and maintain the A-list. A semantic check is performed on the A-list after each extension and reduction operation. If the A-list does not have an instance in which every literal evaluates to FALSE in the model provided by MACE, then the extension or reduction is rejected.

Performance

Testing was carried out on 541 "difficult" problems from the TPTP problem library [12] v2.1.0. Both PTTP and PTTP+GLiDeS were tested on the same problems under the same conditions. Experiments where carried out on a SunSPARC 20 server using ECLiPSe v3.7.1 as the Prolog engine. A CPU time limit of 300 seconds was used. The results are summarized in Table 1. MACE failed to generate a model in 272 cases, and so PTTP+GLiDeS couldn't attempt those problems. Of those 269 problem where models were generated, worst performance is on Horn problems: all of the problems solved by PTTP and not PTTP+GLiDeS are Horn. MACE tends to generate trivial models (positive literals TRUE) for Horn problems. If the top centre clause is negative then, for a Horn clause set, a trivial model does not lead to any pruning. With the additional overhead of the semantic checking this leads to poor performance by PTTP+GLiDeS. Of the 269 models produced by MACE, 155 were effective, i.e., resulted in some pruning. Of the problems with effective models solved by both PTTP and PTTP+GLiDeS, in 13 out of 18 cases PTTP+GLiDeS had a lower inference count; in some cases significantly lower. This is shown by the fact that the average number of inferences for PTTP+GLiDeS is 2.5 times smaller than that of PTTP. This shows that the pruning is having a positive effect. PTTP+GLiDeS performs best on non-Horn problems. Table 2 shows some results for non-Horn problems where PTTP+GLiDeS performed better than PTTP. For these problems even trivial models can be of assistance.

Total number of problems:	541 (311/230)	(Horn/Non-Horn)
CPU time limit:	300 s	
Number of models generated:	269 (227/42)	
Number of problems solved from 269:	PTTP	PTTP+GLiDeS
	66 (60/6)	59 (51/8)
Number of effective models generated:	155 (120/35)	
Number of problems solved from 155:	PTTP	PTTP+GLiDeS
	21 (16/5)	20 (13/7)

For the 18 problems (from 155) solved by both systems:		
	PTTP	PTTP+GLiDeS
Average CPU Time:	34.24	69.18
Average Number of Inferences:	119634.28	47812.22
Average Number of Rej. Inferences:		3896.78

Table 1. Summary of experimental data.

With respect to ordering of the clause set, experiments have been carried out using both ascending and descending with respect to the truth score. Initially it was thought that ordering the clause set in ascending order of truth score (from 'less TRUE' to 'more TRUE') would lead the search away from pruning and therefore towards the proof. This turns out not to be the case. While the results are not statistically significantly different in terms of rejected inferences and inferences, descending ordering solved 4 more problems overall, of which 3 had effective models. As solving problems is the most significant measure of a theorem prover's ability this shows that pruning early is more effective.

| Problem | Model | PTTP+GLiDeS | | | PTTP | |
		CPU Inferences Time		Rejected Inferences	CPU Inferences Time	
CAT003-3	non-Trivial	68.5	64232	10451	TIMEOUT	
CAT012-3	Trivial	32.0	49220	4174	54.6	175367
GRP008-1	Trivial	248.3	404198	3937	TIMEOUT	
SYN071-1	non-Trivial	70.1	84908	27653	262.8	832600

Table 2. Results for some non-Horn problems where PTTP+GLiDeS out-performs PTTP.

Conclusion

In those cases where a strongly effective model has been obtained, results are good. This leads to the question, "what makes a model effective?" At present the first model generated by MACE is used. If the characteristics of a strongly effective model can be quantified then it should be possible to generate many models and select the one most likely to give good performance.

PTTP is not a high performance implementation of ME, and thus the performance of PTTP and PTTP+GLiDeS is somewhat worse than that of current state-of-the-art ATP systems. This work has used PTTP to establish the viability of the semantic pruning strategy. It is planned to implement the pruning strategy in the high performance ME implementation SETHEO [5], in the near future.

On the completeness issue, this prover prunes away proofs which contain complementary A-literals on different branches of the tableau. In the few cases examined to date, another proof that conforms to this extended admissibility rule has always been found. Whether there is always another such proof is not known.

References

1. M. Brown and G. Sutcliffe. PTTP+GLiDeS: Guiding Linear Deductions with Semantics. In N. Foo, editor, *Advanced Topics in Artifical Intelligence: 12th Australian Joint Conference on Artificial Intelligence, AI'99*, number 1747 in LNAI, pages 244–254. Springer-Verlag, 1999.
2. R. Caferra and N. Peltier. Extending Semantic Resolution via Automated Model Building: Applications. In C.S. Mellish, editor, *Proceedings of the 14th International Joint Conference on Artificial Intelligence*, pages 328–334. Morgan Kaufmann, 1995.
3. H. Chu and D. Plaisted. Semantically Guided First-order Theorem Proving using Hyper-linking. In A. Bundy, editor, *Proceedings of the 12th International Conference on Automated Deduction*, number 814 in LNAI, pages 192–206. Springer-Verlag, 1994.

4. D.A. de Waal and J.P. Gallagher. The Applicability of Logic Programming Analysis and Transformation to Theorem Proving. In A. Bundy, editor, *Proceedings of the 12th International Conference on Automated Deduction*, number 814 in LNAI, pages 207–221. Springer-Verlag, 1994.

5. R. Letz, J. Schumann, S. Bayerl, and W. Bibel. SETHEO: A High-Performance Theorem Prover. *Journal of Automated Reasoning*, 8(2):183–212, 1992.

6. D.W. Loveland. A Simplified Format for the Model Elimination Theorem-Proving Procedure. *Journal of the ACM*, 16(3):349–363, 1969.

7. W.W. McCune. A Davis-Putnam Program and its Application to Finite First-Order Model Search: Quasigroup Existence Problems. Technical Report Technical Report ANL/MCS-TM-194, Argonne National Laboratory, Argonne, USA, 1994.

8. J.K. Slaney. SCOTT: A Model-Guided Theorem Prover. In R. Bajcsy, editor, *Proceedings of the 13th International Conference on Artificial Intelligence*, pages 109–114. Morgan-Kaufman, 1993.

9. M.E. Stickel. A Prolog Technology Theorem Prover: A New Exposition and Implementation in Prolog. Technical Report Technical Note 464, SRI International, Menlo Park, USA, 1989.

10. G. Sutcliffe. Linear-Input Subset Analysis. In D. Kapur, editor, *Proceedings of the 11th International Conference on Automated Deduction*, number 607 in LNAI, pages 268–280, Saratoga Springs, NY, USA, June 1992. Springer-Verlag.

11. G. Sutcliffe. The Semantically Guided Linear Deduction System. In D. Kapur, editor, *Proceedings of the 11th International Conference on Automated Deduction*, number 607 in LNAI, pages 677–680, Saratoga Springs, NY, USA, June 1992. Springer-Verlag.

12. G. Sutcliffe and C.B. Suttner. The TPTP Problem Library: CNF Release v1.2.1. *Journal of Automated Reasoning*, 21(2):177–203, 1998.

A Formalization of a Concurrent Object Calculus up to α-Conversion

Guillaume Gillard

INRIA Sophia Antipolis

Abstract. We experiment a method for representing a concurrent object calculus in the Calculus of Inductive Constructions. Terms are first defined in de Bruijn style, then names are re-introduced in binders. The terms of the calculus are formalized in the mechanized logic by suitable subsets of the de Bruijn terms; namely those whose de Bruijn indices are relayed beyond the scene. The α-equivalence relation is the Leibnitz equality and the substitution functions can be defined as sets of partial rewriting rules on these terms. We prove induction schemes for both the terms and some properties of the calculus which internalize the re-naming of bound variables. We show that despite the fact the terms which formalize the calculus are not generated by a last fixed point relation, we can prove the desired inversion lemmas. We formalize the computational part of the semantic and a simple type system of the calculus. Finally, we prove a subject reduction theorem and see that the specifications and proofs have the nice feature of not mixing de Bruijn technical manipulations with *real* proofs.

1 Introduction

Providing a satisfactory method to encode the binding operators of a programming language when we want to formalize it in a Logical Framework is still a challenge. Although many different methods have been proposed so far, none seems completely satisfactory. From de Bruijn codes to higher order encoding each method has is advantages and disadvantages, its supporters and its detractors. A major problem raised by all these methods is that theorems and theirs proofs become highly linked with the chosen encoding. In other words, if it is sometimes possible to have specifications and theorems close to the unmechanized version (which is not the case for de Bruijn encoding), the proof structures are themselves very different from the informal ones. In this paper, we show that the method proposed in [Gor94] for representing binders in mechanized logic can successfully be extended to a large calculus with different kinds of binders. Beside, we show that, once some work has been done with de Bruijn indices, *real* proofs do not manipulate them moreover, they are similar to the unmechanized ones.

We have chosen to formalize the **concς-calculus** [GH98], a concurrent object calculus consisting of M. Abadi and L. Cardelli's imperative object calculus **impς** [AC96] extended with primitives from the π-calculus [MPW92]. This calculus

D. McAllester (Ed.): CADE-17, LNAI 1831, pp. 417–432, 2000.
© Springer-Verlag Berlin Heidelberg 2000

was introduced as a possible formalism for modeling computations based on concurrent processes and objects. We think that formal proofs of properties of protocols can be realized within proof-assistant if we have good methods for encoding such calculus. Our choice of this calculus is motivated by its size, its different kinds of binders and its good expressiveness, thus giving an idea of the real problems that arise when we encode such formalisms in computational logics.

The COQ system we use for our implementation is a proof-assistant based on the calculus of inductive constructions [Wer94], a higher order logic with dependent types and inductive definitions. All the proofs have been done with the user-interface CtCoq [BBC+97]. This paper has been written so as to be understood by people not familiar with the COQ system. We use mathematical notations instead of the COQ syntax and we only show significant parts of large COQ encodings. Please refer to [Gil00] for a full presentation of our technical results.

Organization of the Paper: In section 2 we formalize the concς-calculus in COQ. In section 3 we prove a more powerful induction theorem for the terms of the calculus which internalizes the renaming of bound names. In section 4 we formalize the semantics of the concς-calculus in COQ and we produce an efficient induction principle for the semantic relation. In section 5 we give a simple type system for the calculus and discuss the problems raised by the inversion lemmas generated by the COQ system on this example. Section 6 presents the statement and the proof of the subject reduction theorem. Section 7 is a short discussion about the formalizing technique we used. Finally, section 8 draws some conclusions.

2   The concς-Calculus

This calculus was first introduced by A. Gordon and P. Hankin. It is the imperative object calculus **imp**ς of M. Abadi and L. Cardelli in which objects are located at addresses, extended with a parallel composition and the name restriction operator from the π-calculus. The reader interested in a more detailed presentation of this calculus should refer to [GH98].

2.1   Informal Syntax of the Calculus

We assume that there are infinite disjoint sets of *references, variables* and *labels*. We distinguish references, representing addresses of stored object (channels in the π-calculus) from variables, representing intermediate values (variables in the λ-calculus). Let us call both notions *names*. The expressions of the language are defined as follows:

In the method $\varsigma(x)b$, and in the expression *let* $x = a$ *in* b the variable x is bound in b. In a restriction, $(\nu p).a$ the reference p is bound in a. The notation $a = b$ means that the terms a and b are equal up to bound names renaming and reordering of the labeled components of objects.

Table 1. Syntax of the informal concς-calculus

l	labels
u,v::=	result
x,y,z	variables
p,q	references
a,b,c::=	terms
u,v	result
$p \mapsto d$	denomination
$p.l$	method select
$p.l \Leftarrow \varsigma(x).b$	method update
$clone(p)$	cloning
$let\ x = a\ in\ b$	let
$a \upharpoonright b$	parallel composition
$\nu p.a$	restriction
d::=	denotations
$[l_i = \varsigma(x_i)a_i{}^{i \in 1..n}]$	object

2.2 De Bruijn Specification

We define a de Bruijn syntax (table 2) in which free names n, m are encoded by *named* names [dB94]. Variables **x,y,z** are either *free variables* x,y,z or de Bruijn variables *(dvar i)*, *(dvar j)*, labels l,l_i are named and references **p,q** are either *free references* p,q or de Bruijn references *(dref i),(dref j)*. We assume there are infinite disjoint sets of *references, variables* and *labels* and that equality is decidable on each of these sets.

Table 2. de Bruijn formalism

$$DB \stackrel{def}{=} \mathbf{x} \mid \mathbf{p} \mid \mathbf{p} \mapsto ODB \mid DB.l \mid$$

$$DB.l \Leftarrow_{db} DB \mid clone(DB) \mid let_{db}\ DB\ in\ DB \mid DB \upharpoonright DB \mid \nu_{db}.DB$$

$$ODB \stackrel{def}{=} [] \mid [l : DB :: ODB]_{db}$$

Except if otherwise stated, in the sequel of the paper we no longer refer to *references, variables, labels, results* of the informal calculus described in 2.1. De Bruijn terms (DB) a,b,c are called *dbterm*, lists of *dbterms* (ODB) are called *denotation* and for readability reasons, *var* and *ref* will be use for free variables and free references respectively in all our formal definitions.

The binding constructors here are $\Leftarrow_{db}$, let_{db} (in their second argument) and $[\ ]_{db}$ (the self variable of each method is bound by the *object-constructor*) for de Bruijn variables. The ν_{db} operator is the only binding constructor for de Bruijn references. Objects are represented by lists. Although sets seem closer to the idea of objects (a collection of attributes and methods), we cannot define object as sets because sets in COQ are specified as predicates, and predicates cannot be used in the type of a constructor of an inductive set. Moreover, COQ provides an efficient tool for generating induction scheme for mutual inductive definitions when some type is a list of another [1].

Our syntax is a bit more general than the one proposed by A. Gordon for his concς-calculus in the sense that we allow cloning, method calling, and method updating not only for results but for all terms of our syntax. This choice was motivated because *dbterm* are less nested than they would have been with a de Bruijn result type. Since the concς-calculus terms will eventually be identified by an inductively defined subset of *dbterm* this will have no consequences on its formalization in COQ.

Thanks to de Bruijn indices we do not need an *alpha-equivalence* notion and $a = b$ means that the *dbterm* a and b are equal in the sense of the Leibnitz equality. De Bruijn formalization for binders takes off the syntax its intuitive meaning. We shall show how to recover it later on (see 2.5).

As we have de Bruijn indices for both references and variables, we define two degree functions (computing the usual notion of degree for a term with de Bruijn indices [dB94]), one for each kind. We omit their formal definitions here so that readers do not get confused with too many technical details (see [Gil00] for more details). We generalize the notion of *closed* de Bruijn terms to our calculus such that a *dbterm* is said to be *closed* when both of its degrees are zero.

2.3 Function as Binders

Abstraction and Instantiation Functions. We define a variable abstraction function $Abst_v$. For a given *dbterm* a, $Abst_v(a\ i\ x)$ is computed by substituting in a all the occurrences of the variable x by the de Bruijn variable $(dvar\ i)$. The substitution is defined recursively on the *dbterms* so as the de Bruijn indices substituted is increased by one each time a binder is met. In a dual way, we define an instantiation function $Inst_v$. $Inst_v(a\ i\ x)$ is computed by substituting all the occurrences of $(dvar\ i)$ in a by the variable x. Similarly, we define $Abst_r$ and $Inst_r$ on *references*.

Functions as Constructors. We define new functions on *dbterms let*, *res*, and *sigma* behaving like the de Bruijn binding operators except that they use names in their arguments (see table 3).

In the following, we shall write *let* $x := a$ in b for $(let\ x\ a\ b)$, $\nu p.a$ for $(res\ p\ a)$ and $\varsigma(x).a$ for $(sigma\ x\ a)$. We will also drop the $_{db}$ mark in a constructor when it can be guessed from the context.

2.4 Substitution and α-Equivalence

To relegate the de Bruijn indices of the underlying terms behind the scene we also need to define two new substitution functions $Subst_v$ and $Subst_r$. Intuitively

---

[1] Using the Scheme tactic

Table 3. Functions as constructor

$$(res\ p\ a) \stackrel{def}{=} \nu_{db}.Abst_r(a\ 0\ p) \qquad (sigma\ x\ a) \stackrel{def}{=} Abst_v(a\ 0\ x)$$
$$(let\ x\ a\ b) \stackrel{def}{=} let_{db}\ a\ in\ Abst_v(a\ 0\ x)$$

these functions are defined to rename free *names* in *dbterms*. Their definitions (see definition table 4) use de Bruijn indices in their bodies but, with some work, we shall manipulate them without referring to de Bruijn indices (see 3.2). We write $a[x/y]_v$ and $a[p/q]_r$ for $Subst_v(a\ y\ x)$ and $Subst_r(a\ q\ p)$ respectively.

Table 4. Substitution functions

$$Subst_v(a\ x\ y) \stackrel{def}{=} Inst_v(Abst_v(a\ 0\ x)\ 0\ y)\quad Subst_r(a\ q\ p) \stackrel{def}{=} Inst_r(Abst_r(a\ 0\ q)\ 0\ p)$$

If we think of the ν, *let* and ς function as constructors and $Subst_v$ and $Subst_r$ as renaming functions, we prove that α-equivalent *dbterms* are encoded in our formalism by a unique *dbterm*.

2.5   Formalization of the Syntax

The result type $(u,v,)$ is defined as the disjoint union of our free names:

$$result \stackrel{def}{=} var\ |\ ref$$

The inductive predicate Term (table 5) defines the subset of *dbterm* which formalizes the concς-calculus. The proof of correctness of this encoding is straightforward (omitted) if one thinks of *let*, *res* and *sigma* as constructors.

¿From now on, we shall call *Term*, a *dbterm* having the *Term* property and write $\forall a : Term.(P\ a)$ as a short hand for $\forall a : dbterm.(Term\ a) \Rightarrow (P\ a)$ in the translation of our COQ notations. *Terms* are based on a de Bruijn formalism but de Bruijn indices are hidden in the syntax by the *let*, *res* and *sigma* functions.

3   An Induction Principle for the concς-Calculus

In the sequel, for the sake of clarity, we only show one (significant) case of the theorems. A more detailed presentation is available in [Gil00].

Table 5. formalization of the concς-calculus in COQ

$Term : dbterm \rightarrow Prop :=$

Resu: $\forall r : result.(Term\ r)$
— Deno: $\forall p : ref.\forall obj : denotation.\ (OTerm\ obj) \Rightarrow (Term\ p \mapsto obj)$
— Msel: $\forall l : labels.\forall u : result.\ (Term\ u.l)$
— Mupd: $\forall u : result.\forall a : dbterm.\forall l : labels.\forall x : var.$
         $(Term\ a) \Rightarrow (Term\ u.l \Leftarrow \varsigma(x).a)$
— Clone: $\forall u : result.(Term\ (clone\ u))$
— Let: $\forall a, b : dbterm.\forall x : var.(Term\ a) \Rightarrow (Term\ b) \Rightarrow (Term\ let\ x := a\ in\ b)$
— Par: $\forall a, b : dbterm.(Term\ a) \Rightarrow (Term\ b) \Rightarrow (Term\ a \upharpoonright b)$
— Res: $\forall a : dbterm.\forall p : ref.\ (Term\ a) \Rightarrow (Term\ \nu p.a)$

with
$OTerm : denotation \rightarrow Prop :=$

Mnul: $(OTerm\ [])$
— Mocons: $\forall l : label.\forall a : dbterm.\forall obj : denotation.\forall x : var.$
           $(Term\ a) \Rightarrow (OTerm\ obj) \Rightarrow (OTerm\ (l : \varsigma(x).a :: obj))$

3.1   The Induction Scheme Generated by the COQ System

It appears that the induction scheme generated by COQ (table 6) for the predicate Term is not powerful enough for our purpose[2].

Table 6. Induction scheme generated by COQ

$Term\_ind :=$
$\forall P : dbterm \rightarrow Prop.$
   $\cdots$
   $\langle \forall x : var.\forall a, b : Term.(P\ a) \Rightarrow (P\ b) \Rightarrow (P\ let\ x := a\ in\ b) \rangle \Rightarrow$
   $\cdots$
$\forall a, b : Term.(P\ a\ b).$

For example, it is not clear how one can derive the fact that $Terms$ are closed under the substitution $Subst_v$ and $Subst_r$ with it. We shall not be able to deduce $(Term\ (let\ x := a\ in\ b)[z/y]_v)$ from $(Term\ a[z/y]_v)$ and $(Term\ b[z/y]_v)$ because we have not the necessary informations on x, y and z to compute $(let\ x := a\ in\ b)[z/y]_v$. In this COQ formalization of the concς-calculus α-equivalence terms are equal. We want to integrate within the induction scheme of Term the fact that bound $names$, in $Terms$, can always be renamed.

---

[2] Actually we need to use the Scheme tactic to generate an efficient principle

3.2   An Intermediate Induction Scheme for *Terms*

We define a new function *length* on *dbterms* which computes the numbers of constructors appearing in a term. The order $\leq_{length}$ induced on *dbterms* by this function is well founded and we show that renaming *names* in a *dbterm* does not change its length. Using the general induction theorem for well founded relations with respect to $\leq_{length}$ we prove a more powerful induction scheme than *Term_ind* for the Term relation (table 7).

Table 7. Intermediate induction scheme

Term_length_ind:=
$\forall P : dbterm \to Prop.$
$\dots$

$$\forall x : var.\forall a, b : Term.(P\ a) \Rightarrow \langle \forall b' : Term.\ length(b) = length(b') \Rightarrow (P\ b') \rangle \Rightarrow$$
$$(P\ let\ x := a\ in\ b) \rangle \Rightarrow$$

$\dots$

$\forall a, b : Term.(P\ a\ b).$

With this theorem, as the length of *dbterms* is invariant for the renaming functions, we prove that substitution can be propagated inside binders for *Terms* if the *side conditions* are satisfied (see table 8).

Table 8. substitution rewriting rules for the *let* binder

let_rw1: $\forall x, y, z : var.\forall a, b : Term.x \neq y \Rightarrow x \neq z \Rightarrow$
$\qquad (let\ x := a\ in\ b)[z/y] = let\ x := a[z/y]\ in\ b[z/y].$
let_rw2: $\forall x, y : var.\forall a, b : Term.(let\ x := a\ in\ b)[y/x] = let\ x := a[y/x]\ in\ b.$

3.3   The Full Induction Scheme for *Terms*

The use of the *length* function inside *Term_length_ind* is not satisfactory because this is not natural. We prove a final induction scheme on *Term* (table 9) using the *Term_length_ind* theorem and the properties of the substitutions functions we have deduced from it.

This induction scheme internalizes the property that bound *names* can always be chosen outside any set of *names* in the context.

Example: *Given a property P on Term, we prove that it holds for all Terms using the term-induction theorem. In order to prove that $(P\ let\ x := a\ in\ b)$ holds, we select a finite set X and try to solve our goal under the assumptions $(P\ a)$, $(P\ b)$ and $x \notin X$. Giving the set X amounts to specify that x is a fresh variable.*

Table 9. Induction scheme for the *Term* predicate

Term_induction:=
$\forall P : dbterm \to Prop.$
...

$(\forall a : Term.(P\ a) \Rightarrow \langle \exists X : set.(Finite\ X) \wedge$
$\quad \forall x : var.\forall b : Term.\ x \notin X \Rightarrow (P\ b) \Rightarrow (P\ \ let\ x := a\ in\ b\rangle)) \Rightarrow$
...

$\forall a, b : Term.(P\ a\ b).$

4   Semantics of the concς-Calculus

The semantics of the calculus is given by a reduction relation and a structural congruence. The formalization of the reduction rules in COQ is natural and we can prove an induction scheme which internalizes α-renaming.

4.1   Rules for the Semantics

Informal Semantics. Terms of the calculus are interpreted either as *processes* or as *expressions*. *Expressions* and *processes* are concurrent computations but an expression is expected to return a result while a process is not. As opposed to many concurrent calculi the parallel composition ($\upharpoonright$) is not commutative. The term $a \upharpoonright b$ is an expression in which a and b run in parallel. Its result is the result returned by b; any result returned by a is discarded. The structural congruence ($\equiv$), except from the unusual behavior for ($\upharpoonright$) is standard. The reduction relation ($\to_{red}$) (table 10) is analogous to the β-reduction for the λ-calculus. The structural congruence relation allows the rearrangement inside a term so that reduction may be applied. Please refer to [GH98] for motivations and more details on this semantics.

Table 10. Reduction relation: $a \to b$

For the first three rules, let $d = [l_i = \varsigma(x_i)b_i^{i \in 1..n}]$.

$$(p \mapsto d) \upharpoonright p.l_j \to (p \mapsto d) \upharpoonright b_j[p/x_j] \qquad \text{if } j \in 1..n$$
$$(p \mapsto d) \upharpoonright (p.l_j \Leftarrow \varsigma(x).b) \to (p \mapsto d') \upharpoonright p \qquad \text{if } j \in 1..n$$
$$\qquad d' = [l_j := \varsigma(x).b, l_i = \varsigma(x_i)b_i\ ^{i \in (1..n)-j}]$$
$$(p \mapsto d) \upharpoonright (clone\ p) \to (p \mapsto d) \upharpoonright \nu q.((q \mapsto d) \upharpoonright q)\ \text{if } q \notin fn(d)$$
$$let\ x := p\ in\ a \to a[p/x]$$
$$\nu p.a \to \nu p.a' \qquad \text{if } a \to a'$$
$$(a \upharpoonright b) \to (a' \upharpoonright b) \qquad \text{if } a \to a'$$
$$(b \upharpoonright a) \to (b \upharpoonright a') \qquad \text{if } a \to a'$$
$$let\ x := a\ in\ b \to let\ x := a'\ in\ b \qquad \text{if } a \to a'$$
$$a \to b \qquad \text{if } a \equiv a', b \equiv b', a' \to b'$$

The notations $fn(a)$ and $fv(a)$ denote respectively the sets of free names and free variables in the expression a. The expression $a[p/x]$ is the notation for the substitution of the reference p for each free occurrence of the variable x in the expression a.

Formalization in COQ. We use two inductive definitions to formalize the above relation in COQ. The first one (table 11) is a restriction to *effective* reductions in terms. The second one (table 12) is the complete formalization in the COQ system of the semantics (proof omitted here). We use this trick to prevent looping in the proofs. In the sequel of this paper we shall focus on the first definition.

The COQ formalizations of both relations are the natural translations of the rules in table 10 into inductive definitions ($\rightarrow_{red}$ and $\rightarrow_{eval}$). The $\rightarrow_{red}$ relation is defined for *Terms* and not *dbterms* so, for every *dbterm* a appearing in the COQ definition of $\rightarrow_{red}$ (*Term a*) must hold.

Table 11. Formalization of the $\rightarrow_{red}$ relation in COQ

$\rightarrow_{red}$: $dbterms \rightarrow dbterms \rightarrow Prop :=$
...
Let_red1: $\forall a : Term.\forall p : ref.(let\ x := p\ in\ a) \rightarrow_{red} a[p/x]_*$
...

For any given term a, $Subst_*(a\ x\ p)$, written $a[p/x]_*$, is computed by substituting all the occurrences of the variable x by the reference p in a. In the COQ system, $Subst_*$ must be defined on *dbterms* and de Bruijn indices are used in the body of this function. With the help of technical lemmas, we show that $Subst_*$ restricted to *Terms* can be manipulated without dealing with de Bruijn indices.

Table 12. Formalization of the reduction relation $\rightarrow_{eval}$ in COQ

$\rightarrow_{eval}$: $dbterm \rightarrow dbterm \rightarrow Prop :=$
Eval: $\forall a, a', b, b, : Term.(a \equiv a') \Rightarrow (b \equiv b') \Rightarrow (a' \rightarrow_{red} b') \Rightarrow (a \rightarrow_{eval} b)$

4.2   Induction Scheme for the Semantics

As before, for the sake of clarity, we only show one (significant) part of the theorems. Courageous readers could refer to [Gil00] for a more detailed presentation.

We can extend the induction scheme for the $\rightarrow_{red}$ relation as we did for the *Term* predicate. In the induction scheme generated by the COQ system (table 13) we do not have any informations for bound *names*.

Table 13. $\rightarrow_{red}$ induction scheme generated by COQ

Red_ind:=
$\forall P : dbterm \rightarrow dbterm \rightarrow Prop.$
$\ldots$

$\quad (\forall p : ref.\forall a, b : Term.(a \rightarrow_{red} b) \Rightarrow (P\ a\ b) \Rightarrow (P\ \nu p.a\ \nu p.b)) \Rightarrow$
$\ldots$

$\forall a, b : dbterm.(a \rightarrow_{red} b) \Rightarrow (P\ a\ b).$

Following the idea of the section 3 we can produce an extended induction scheme which internalizes the α-renaming of bound *names* in proofs. By using the general well founded induction theorem for a suitable order on pairs of *dbterm* we prove an intermediate theorem (table 14) in which the length of *dbterm* is introduced.

Table 14. $\rightarrow_{red}$ induction scheme with the *length* function

Red_length_ind:=
$\forall P : dbterm \rightarrow dbterm \rightarrow Prop.$
$\ldots$

$(\forall p : ref.\forall a, b : Term.$
$\langle \forall a', b' : Term.$
$length(a) = length(a') \Rightarrow length(b) = length(b') \Rightarrow (a' \rightarrow_{red} b') \Rightarrow (P\ a'\ b')\rangle$
$\quad \Rightarrow (a \rightarrow_{red} b) \Rightarrow (P\ \nu p.a\ \nu p.b)) \Rightarrow$
$\ldots$

$\forall a, b : dbterm.(a \rightarrow_{red} b) \Rightarrow (P\ a\ b).$

Finally , we prove the induction scheme (table 15) in which bound *names* can be chosen *outside* the set of *names* in the context. To prove the theorem, we first need an intermediary lemma stating that $\rightarrow_{red}$ is closed for *names* renaming. More precisely, we show that $\rightarrow_{red}$ is closed for *names* renaming provided that *names* are renamed in new *names*.

Table 15. $\rightarrow_{red}$ induction scheme

Red_induction:=
$\forall P : dbterm \rightarrow dbterm \rightarrow Prop.$
$\ldots$

$\langle \exists Q : set.(Finite\ Q)\ \wedge$
$\quad (\forall p : ref.\forall a, b : Term.p \notin Q \Rightarrow (a \rightarrow_{red} b) \Rightarrow (P\ a\ b) \Rightarrow (P\ \nu p.a\ \nu p.b))\rangle \Rightarrow$
$\ldots$

$\forall a, b : dbterm.(a \rightarrow_{red} b) \Rightarrow (P\ a\ b).$

5 Well-Formed Terms

The concς-calculus can be typed to distinguish expressions from processes. This very basic types system has only two types *Exp* and *Proc* standing for expressions and processes respectively. Basically, this typing system only ensures that proper processes cannot appear in a context expecting an expression and that references are correctly handled in a term. A term a is defined as an *expression* or a *process* if $a : Exp$ and $a : Proc$, respectively.

5.1 Definition

The typing rules are defined in table 16. T stands for either *Exp* or *Proc*. The *domain* of a term a, $dom(a)$ is the set of the free references representing the addresses of an object. Please refer to [GH98] for a general overview.

Table 16. The *well − formed* relation

(Well Result)   (Well Clone)   (Well Res)

$$\frac{}{u : Exp} \qquad \frac{}{clone(u) : Exp} \qquad \frac{a : T \quad p \in dom(a)}{\nu p.a : T}$$

(Well Select)   (Well Concur)   (Well Update)

$$\frac{}{u.l : Exp} \qquad \frac{a : Exp}{a : Proc} \qquad \frac{b : Exp \quad dom(b) = \emptyset}{u.l \Leftarrow \varsigma(x)b : Exp}$$

(Well Let)   (Well Par)

$$\frac{a : Exp \quad b : Exp \quad dom(b) = \emptyset}{let\ x := a\ in\ b : Exp} \qquad \frac{a : Proc \quad b : T \quad dom(a) \cap dom(b) = \emptyset}{a \upharpoonright b : T}$$

(Well Object)

$$\frac{b_i : Exp \quad dom(b_i) = \emptyset \quad \forall i \in 1..n}{p \mapsto [l_i = \varsigma(x_i)b_i^{i \in 1..n}] : Proc}$$

The COQ formalization of the *well-formed* relation is its natural translation as an inductive definition *well_formed*, given table 16. *well_formed* is defined for *Term* and not *dbterms*. We must insure than for every term a, $(Term\ a)$ holds in the COQ definition.

5.2 Inversion

In the activity of proofs, inversion theorems are as important as induction schemes. In the usual cases, inversion theorems automatically generated by the proof assistants are those expected because the syntax of the calculi are defined in terms of a least fixed point. In our formalism, binders are functions on top

Table 17. Formalization of *well_formed* in COQ

$$\text{flag} \stackrel{def}{=} Exp \mid Proc$$

well_formed : *dbterms* → *flag* → *Prop* :=
...
Well_Res: $\forall a : Term.\forall p : ref.\forall T : flag.(well\_formed\ a\ T) \Rightarrow$
$p \in dom(a) \Rightarrow (well\_formed\ \nu p.a\ T).$
...

of a de Bruijn syntax thus from the equality $\nu p.a = \nu q.b$ it is not possible to deduce than $p = q$ and $a = b$. If the inversion theorems generated by COQ are used roughly they introduced news terms not directly related to anything in the proof. In table 18 we present the inversion theorem for the *Well_Res* constructor of the *well_formed* property generated by COQ [3].

Table 18. Inversion lemma for *well_formed* generated by COQ

$\forall P : dbterm \rightarrow flag \rightarrow Prop.$
$\forall a : Term.\forall p : ref.\forall T : flag.$
$\langle \forall q : names.\forall b : Term.\ \nu p.a = \nu q.b \Rightarrow q \in dom(b) \Rightarrow (well\_formed\ b\ T) \Rightarrow (P\ a\ T) \rangle$
$\Rightarrow (well\_formed\ \nu p.a\ T) \Rightarrow (P\ a\ T).$

In a proof in which $(well\_formed\ \nu p.a\ T)$ is amongst the assumptions, using this theorem will not add $(well\_formed\ a\ T)$ in the hypothesis as expected. Similarly to induction schemes, the *right* inversion lemmas must be proved. Fortunately, it is sufficient to derive a specialized lemma for each constructors of the inductive definition. Then the COQ system provides tactics to use them properly[4]. Because we can produce one lemma for each constructor, their formulation remains simple (see table 19 as an example).

Table 19. Inversion lemma for the *Well_Res* constructor

Lemma *well_res_inv*: $\forall P : dbterm \rightarrow flag \rightarrow Prop.$
$\forall a : Term.\forall p : ref.\forall T : flag$
$(p \in dom(a) \Rightarrow (well\_formed\ a\ T) \Rightarrow (P\ a\ T)) \Rightarrow (well\_formed\ \nu p.a\ T) \Rightarrow (P\ a\ T).$

To prove *well_res_inv* we use a property stating than if $\nu p.a = \nu p.b$ holds then $a = b$ holds (such properties has to be proved for each of our binders). To complete the proof, we must show that if $(well\_formed\ a\ T)$ holds for a term a then

---

[3] The COQ system generates a general inversion theorem for *well_formed*; this is a specialized version

[4] The *Inversion ... using ...*, tactic

(*well_formed* $a[p/q]$ T) holds for any q and any p such that $p \notin Q$ for a finite set Q. In other words, we need to prove than *well_formed* is closed under reference (name in general) renaming. Again, this is not surprising. This property of the relation *well_formed* should also be checked when we are reasoning *up to* *α-conversion* during informal proofs, through this is most often omitted.

6 Subject Reduction Theorem

We show that *well_formed* terms are closed for the $\rightarrow_{red}$ relation. The formulation of this theorem (see table 20) is exactly the same as its unmechanized version appearing in [GH98]. The proof is done by induction on $\rightarrow_{red}$ using the extended induction theorem (see table 15). For each induction case, there is an hypothesis of the form (*well_formed* a T). We use our inversion lemmas to extract informations on sub-terms of a from it.

Table 20. Subject reduction theorem

Theorem srt:
$\forall a, b : dbterm. \forall T : flag.$ (*well_formed* a T) $\Rightarrow$ ($a \rightarrow_{red} b$) $\Rightarrow$
(*well_formed* b T) $\wedge$ $dom(a) = dom(b)$.

The proof of the theorem is very closed to the informal proof with implicit renamings of bound names. We do not manipulate de Bruijn indices neither are we doing α-renaming. All the lemmas used during the proof have a semantic contents.

7 Discussion and Related Work

The size of the different parts of the COQ code is summarize in the table below. In the column of Term we consider all the formalizations and proofs necessary for using the *Terms*. It includes the properties for the α-conversion and the renaming, the proofs of the extended induction principles the Term property most of the lemmas we have proved with the theorem *Term_induction*. We classify in the column of well_formed and $\rightarrow_{red}$ all the COQ codes which deal with the corresponding property (induction schemes, inversion lemmas, behavior of *Subst*⋆ and *dom*). The srt column stand for the subject reduction theorem COQ codes part and the total column include all the previous ones plus some general lemmas (mainly set theory theorems) which do not use de Bruijn indices.

	Term	well_formed	$\rightarrow_{red}$	srt	total
lines of COQ code	7 700	2 400	2 500	1000	14 600
% of de Bruijn code	65%	10%	25%	-	40%

The percentage of the de Bruijn code in proofs is high during the setting of this technique. In fact, large de Bruijn codes mainly concern the *Abst* and *Inst* functions. But, once we have completely mastered the behaviors of *Terms* we do not use de Bruijn indices.

Given a property $P : Term \to \dots \to Term \to Prop$, we must show that there exists a finite set X such that $m \notin X \Rightarrow (P\ a_1 \dots a_n) \Rightarrow (P\ a_1[m/n] \dots a_n[m/n])$ to get an induction principle which internalizes name renaming and the expected inversion lemmas. Checking that P is closed by renaming of names can be laborious in COQ whereas this is assumed for *on paper* proofs. Moreover, as we have experimented during this development, it clears the way for further proofs on the P property.

Related work. Among all the works formalizing the variable-binding operators in calculi none, as far as we know, uses the technique we have used here. Daniel Hirschkoff has encoded a polyadic π-calculus with de Bruijn numbers and proved many bisimulation results [Hir97]. Bruno Barras [Bar95] formalizes COQ in COQ with de Bruijn indices. In both approaches de Bruijn indices appear in almost all theorems and specifications. We think this is not natural. L. Henry-Gréard [Hen98] uses R. Pollack and J. McKinna technique [MP93] to formalize the π-calculus and prove a subject reduction theorem for it. In this technique, closer to the *on paper* formalism, there are two kinds of names, one for free ones and another for bound one. We think this is not completely natural. J. Despeyroux has investigated a higher-order approach in which the lambda abstraction of the logic is used for binding free variables of the calculus [Des]. See [DH94], for a general approach of this technique in COQ. F. Honsel, M. Miculan and I. Scagnetto [FI98] have encoded the π-calculus in COQ following a higher order approach. They use *Co inductive* types in their encoding of bisimulation. Although second order techniques are very efficient, we think that proofs using these techniques are very different from proofs on paper.

8   Conclusion and Future Work

We have formalized a concurrent object calculus in the COQ system with names in binders using a technique proposed by A. Gordon [Gor94]. We have shown that defining properties on *Terms*, namely those who formalize the concς-calculus in the COQ system, is very natural and easy because we just need to rewrite them using the COQ syntax. Under the assumption that a given property P is invariant under the renaming of names, the induction theorem generated by COQ for P can be strengthened to internalize α-renaming of bound variables. In spite of our syntax is not generated by a last fixed point we have inversion lemmas for P but they must be proved. The proofs of these theorems as the proof of the subject reduction theorem, are de Bruijn indices free. Moreover, the proofs dealing with *real* property of the calculus follow the general guideline of their *on paper* matching piece.

The main drawback of this approach is that each time we have to define functions on *Terms* we have to define them on *dbterms* first, then prove that they behave as expected on *Terms*. We believe that with a good understanding of the behavior of a function on *Terms*, it is not hard to give its definition on *dbterms*. We claim that this weakness does not overcome the advantages of the method. In fact, new functions on our syntax will probably use functions we have already defined, allowing re-use of our COQ proofs (as it is done for the function *Subst*_* which appears in $\rightarrow_{red}$).

For property P, the strengthened induction theorems could be a large term. It is interesting to develop tools for generating it automatically because this extended induction scheme is mechanically derivable (not provable) from P. Another reasonable development could be to include tactics for automating, on *Terms*, the computation steps of functions. We have done some preliminary work in this direction.

Acknowledgments

I specially thank Joëlle Despeyroux for enlightening discussions about this work.

References

[AC96]   Martin Abadi and Luca Cardelli. *A Theory of Objects*. Monographs in Computer Science. Springer-Verlag, 1996.

[Bar95]   B. Barras. Coq en Coq. Mémoire du DEA informatique, mathématiques et applications, École Polytechnique, 1995. INRIA research report RR-3026, October 1996.

[BBC⁺97] Janet Bertot, Yves Bertot, Yann Coscoy, Healfdene Goguen, and Francis Montagnac. *User guide to the CtCoq proof environment*, October 1997. Inria technical report, RT-0210.

[dB94]   N.G. de Bruijn. Lambda calculus notation with nameless dummies, a tool for automatic formula manipulation, with application to the church-rosser theorem. In J.H. Geuvers R.P. Nederpelt and R.C. de Vrijer, editors, *Selected Papers on Automath*, volume 133, pages 375–388. Studies in Logic, 1994. Reprinted.

[Des]   Joëlle Despeyroux. A higher-order specification of the pi-calculus. Presented at the Modelisation and Verification seminar, Marseille, Dec 98. Submitted for publication, March.

[DH94]   Joëlle Despeyroux and André Hirschowitz. Higher-order syntax and induction in Coq. In F. Pfenning, editor, *Proceedings of the fifth Int. Conf. on Logic Programming and Automated Reasoning (LPAR 94)*, volume 822, pages 159–173. Springer-Verlag LNAI, July 1994. Also appears as INRIA Research Report RR-2292 (June 1994).

[FI98]   M.Miculan F.Honsell and I.Scagnetto. Pi calculus in (co)inductive type theories. Technical report, Universita' di Udine, September 1998.

[GH98]   A. Gordon and P. Hankin. A concurrent object calculus: reduction and typing. In *Proceedings of the HLCL'98 Conference*. Elsevier ENTCS, 1998.

[Gil00]   Guillaume Gillard. A full formalization of a concurrent object calculus up to alpha-conversion. draft, January 2000. Available at ftp://ftp-sop.inria.fr/certilab/ps/conc_calculus.ps.

[Gor94]   A. Gordon.   A mechanisation of name-carrying syntax up to alpha-conversion. In *Proceedings of the 6th int. workshop on Higher Order Logic Theorem Proving and its Applications, Vancouver,* Springer-Verlag LNCS 780, pages 413–425, 1994.

[Hen98]   Loïc Henry Greard. A proof of type preservation for the pi-calculus in Coq. Research Report RR-3698, Inria, December 1998. Also available in the Coq Contrib library.

[Hir97]   Daniel Hirschkoff. A full formalization of pi-calculus theory in the Calculus of Constructions. In Amy Felty and Elsa Gunter, editors, *Proceedings of the International Conference on Theorem Proving in Higher Order Logics,* Murray Hill, New Jersey, August 1997.

[MP93]   James McKinna and Robert Pollack. Pure Type Sytems formalized. In M. Bezem and J. F. Groote, editors, *Proceedings of the International Conference on Typed Lambda Calculi and Applications,* pages 289–305. Springer-Verlag LNCS 664, March 1993.

[MPW92] R. Milner, R. Parrow, and J. Walker. A calculus of mobile processes, (part I and II). *Information and Computation,* 100:1–77, 1992.

[Wer94]   B. Werner. *Une Théorie des Constructions Inductives.* PhD thesis, Université Paris VII, Mai. 1994.

A Resolution Decision Procedure for Fluted Logic

Renate A. Schmidt[1] and Ullrich Hustadt[2]

[1] Department of Computer Science, University of Manchester
Manchester M13 9PL, United Kingdom
schmidt@cs.man.ac.uk

[2] Centre for Agent Research and Development, Manchester Metropolitan University
Manchester M1 5GD, United Kingdom
U.Hustadt@doc.mmu.ac.uk

Abstract. Fluted logic is a fragment of first-order logic without function symbols in which the arguments of atomic subformulae form ordered sequences. A consequence of this restriction is that, whereas first-order logic is only semi-decidable, fluted logic is decidable. In this paper we present a sound, complete and terminating inference procedure for fluted logic. Our characterisation of fluted logic is in terms of a new class of so-called fluted clauses. We show that this class is decidable by an ordering refinement of first-order resolution and a new form of dynamic renaming, called separation.

1 Introduction

Fluted logic is of interest for a number of reasons. One of our main motivations for studying fluted logic is the continuation of the programme of characterising first-order decidability by resolution methods. There are various ways of defining decidable fragments of first-order logic. Fragments considered until the sixties usually involve some form of restriction on quantification. In prefix classes such as the Bernays-Schönfinkel class, the initially extended Ackermann class, the initially extended Gödel class the quantifier prefixes are restricted, to $\exists^*\forall^*$, $\exists^*\forall\exists^*$ and $\exists^*\forall\forall\exists^*$. In the guarded and loosely guarded fragments, which were introduced more recently, quantifiers are restricted to conditional quantifiers of the form $\exists \overline{y} G(\overline{x}, \overline{y}) \wedge \varphi$ or $\forall \overline{y} G(\overline{x}, \overline{y}) \to \varphi$, where $G(\overline{x}, \overline{y})$ is a guard formula satisfying certain restrictions ($G(\overline{x}, \overline{y})$ is an atom in the case of the guarded fragment). In Maslov's class K (more precisely, in the dual class $\overline{K}$) there is a restriction on universal quantification. Other decidable classes such as the monadic class and FO^2 are defined over predicate symbols with bounded arity. By contrast, the restriction of first-order logic which ensures decidability for fluted logic is an ordering on variables and arguments. With the exception of fluted logic, the mentioned logics have been studied in the context of resolution and superposition, see for example Joyner [18], Fermüller, Leitsch, et al. [5,6], Bachmair, Ganzinger and Waldmann [3], de Nivelle [4], Ganzinger and de Nivelle [7], Hustadt and Schmidt [15].

D. McAllester (Ed.): CADE-17, LNAI 1831, pp. 433–448, 2000.
© Springer-Verlag Berlin Heidelberg 2000

Another reason for our interest in fluted logic is the relationship to non-classical logics. Extended modal logics and expressive description logics play an increasingly important role in various areas of computer science. Fluted logic may be viewed as a generalisation of modal logic, just as the guarded fragments can. The properties fluted logic is known to share with modal logics are decidability and the finite model property [21, 22, 24]. From a modal perspective an advantage of fluted logic over the guarded fragment is that relational atoms may be negated. This means that logics such as Boolean modal logic [8] and other enriched modal logics [9, 12, 13], as well as expressive description logics like $\mathcal{ALB}$ (without converse) [16], which cannot be embedded in the guarded fragment, can be embedded in fluted logic. Interestingly, translations of propositional modal formulae by both the relational translation and a variation of the functional translation (described and used in [11, 14]) are fluted formulae. This raises the question whether the results of Ohlbach and Schmidt [19, 26] can be generalised to fluted logic. The answer to this question is negative, though. Already the use of the quantifier exchange operator, which swaps existential and universal quantifiers in a non-standard fashion [19], leads to loss of soundness. A counter example is the relational translation of the second formula in the class of branching K-formulae, defined in [10, Prop. 6.5].

Historically, fluted logic arose as a byproduct of the predicate functor logic introduced by Quine [25]. Adding various combinatory operators to fluted logic defines a lattice of fluted logics, in which fluted logic is the weakest logic and first-order logic with equality is the most expressive logic. (The combinatory operators are equality, binary converse, permutation of arguments, addition of vacuous arguments, fusions of arguments, and composition of binary atoms.) In a series of papers [21, 22, 24] Purdy studies the decision problem of fluted logics in this lattice, and establishes the limit of decidability to be the boundary of the ideal generated the fluted logic with binary converse and equality [24]. This logic is the most expressive decidable logic in the lattice of fluted logics. In [23] Purdy describes an application in computational linguistics of fluted logics for modelling ordinary English.

In this paper we characterise fluted logic by a new class of clauses, called the class of *fluted clauses*. We present a decision procedure for this class which is based on an ordering refinement of resolution and an additional *separation rule*. This is a new inference rule which does dynamic renaming. It replaces a clause $C \vee D$ by two clauses $\neg A \vee C$ and $A \vee D$, where A is an atom with a newly introduced predicate symbol. The rule is sound, in general, and resolution extended by this rule remains complete, if for any set N of clauses the number of applications of separation in any derivation from N is finitely bounded. Separation is essential for our decision procedure, since it allows us to transform certain problematic fluted clauses into so-called *strongly fluted clauses*. A strongly fluted clause is a fluted clause that contains a literals which includes all the variables of the clause. When inference is restricted to such literals (i) the number of variables in any derivable clause is finitely bounded, in particular, the number of variables does not exceed the number of variables in the original clause set. To show termination, it is usually sufficient [18] to show in addition that (ii) there is a bound on the depth of terms occurring in derived clauses. Because separation introduces new predicate names during the derivation, in our case we also need to show that (iii) there is a bound on the number of applications of the separation rule. Exhibiting (ii) and (iii), along with verifying the deductive closure of the class of (strongly) fluted clauses are the most difficult parts of the termination proof. The difficulty can be attributed to the fact that the depth of terms can grow during the derivation, as is the case for some other solvable clausal classes, for example, those associated with Maslov's class $\overline{K}$ [5, 14].

The paper is organised as follows. Fluted logic is defined in Section 2. Section 3 gives a brief description of the general ordered resolution calculus. The class of fluted clauses is defined in Section 4. In Section 5 we specify how fluted formulae can be translated into sets of fluted clauses. The new separation rule is defined in Section 6. In Section 7 we define an ordering refinement and prove termination. We conclude with some remarks about the complexity of the class. Because of the space limitations all proofs had to be omitted, but can be found in [17].

Throughout, our notational convention is the following: x, y, z are the letters reserved for first-order variables, s, t, u, v for terms, a, b for constants, f, g, h for function symbols, and p, q, r, P, Q, R for predicate symbols. The Greek letters φ, ψ, ϕ are reserved for formulae. A is the letter reserved for atoms, L for literals, and C, D for clauses. For sets of clauses we use the letter N.

2   Fluted Logic

Let $\mathcal{P}$ be a finite set of predicate symbols and let $X_m = \{x_1, \ldots, x_m\}$ be an *ordered* set of variables. An *atomic fluted formula* of $\mathcal{P}$ over X_i is an n-ary atom $P(x_l, \ldots, x_i)$, with $l = i - n + 1$ and $n \leq i$. *Fluted formulae* are defined inductively as follows:

1. Any atomic fluted formula over X_i is a fluted formula over X_i.
2. $\exists x_{i+1}\varphi$ and $\forall x_{i+1}\varphi$ are fluted formulae over X_i, if φ is a fluted formula over X_{i+1}.
3. Any Boolean combination of fluted formulae over X_i is a fluted formula over X_i. That is, $\varphi \to \psi$, $\neg\varphi$, $\varphi \wedge \psi$, etcetera, are fluted formulae over X_i, if both φ and ψ are.

By definition, for any formula φ, if there is a variable renaming h such that $h(\varphi)$ is a fluted formula according to the above definition then φ is a fluted formula. In this paper the assumption is that all fluted formulae are closed.

The semantics of fluted logic is defined like the semantics of first-order logic.

Three examples of fluted formulae from a linguistic or knowledge representation setting are the following. (mwmc is short for 'married couple all of whose children are married'.)

$$\forall x_1(\text{cheese-eater}(x_1) \leftrightarrow \exists x_2(\text{cheese}(x_2) \wedge \text{eats}(x_1, x_2)))$$

$$\forall x_1(\text{cheese-lover}(x_1) \leftrightarrow \forall x_2(\text{cheese}(x_2) \rightarrow (\text{eats}(x_1, x_2) \wedge \text{likes}(x_1, x_2))))$$

$$\forall x_1 x_2(\text{mwmc}(x_1, x_2) \leftrightarrow (\text{married}(x_1, x_2) \wedge$$
$$\forall x_3(\text{have-child}(x_1, x_2, x_3) \rightarrow \exists x_4 \text{married}(x_3, x_4))))$$

The first formula can be expressed by a multi-modal formula, while the second can only be expressed in a modal logic with an enriched language like Boolean modal logic or description logics with role negation. Because guards may only have a certain polarity the second formula does not belong to the guarded or loosely guarded fragments. The third formula is also not guarded and does not belong to Maslov's class $\overline{K}$ either. On the other hand, the formulae

$$\forall x_1 x_2(\text{married}(x_1, x_2) \wedge \forall x_3(\text{is-child}(x_3, x_1, x_2) \rightarrow \text{doctor}(x_3))))$$

$$\forall x_1 x_2(\text{married}(x_1, x_2) \wedge \exists x_3(\text{have-child}(x_1, x_2, x_3) \rightarrow \exists x_4 \text{married}(x_4, x_3)))$$

$$\forall x_1 x_2 x_3(\text{ancestor}(x_1, x_2) \wedge \text{ancestor}(x_2, x_3)) \rightarrow \text{ancestor}(x_1, x_3)$$

are not fluted formulae, because in all instances the ordering of the arguments is violated in some atom.

3   Resolution

The usual definition of clausal logic is assumed. A *literal* is an atom or the negation of an atom. The former is said to be a *positive literal* and the latter a *negative literal*. In this paper *clauses* are assumed to be multisets of literals, and will be denoted by $P(x) \vee P(x) \vee \neg R(x, y)$, for example. The components in the variable partition of a clause are called *variable-disjoint* or *split components*, that is, split components do not share variables. A clause which cannot be split further will be called a *maximally split clause*. The *condensation* $\text{cond}(C)$ of a clause C is a minimal subclause of C which is a factor of C. We take equality of clauses (or formulae) to be equality modulo variable renaming. Two clauses (or formulae) that are equal modulo variable renaming are said to be *variants* of each other.

We say an expression is *functional* if it contains a constant or a non-nullary function symbol. Otherwise it is called *non-functional*. An expression is *shallow* if it does not contain a non-constant functional term. The set of variables of an expression E will be denoted by $\text{var}(E)$.

Next, we briefly recall the definition of ordered resolution from Bachmair and Ganzinger [1, 2]. Derivations are controlled through an admissible ordering $\succ$. In the full calculus a second parameter, a selection function, may be used, but for the results of this paper it is not essential.

By definition, an ordering $\succ$ is *admissible*, if (i) it is a total well-founded ordering on the set of ground literals, (ii) for any atoms A and B, it satisfies: $\neg A \succ A$, and $B \succ A$ implies $B \succ \neg A$, and (iii) it is stable under the application

Deduce: $\dfrac{N}{N \cup \{\mathrm{cond}(C)\}}$ if C is a factor or resolvent of premises in N.

Delete: $\dfrac{N \cup \{C\}}{N}$ if C is redundant.

Split: $\dfrac{N \cup \{C \vee D\}}{N \cup \{C\} \mid N \cup \{D\}}$ if C and D are variable-disjoint.

Resolvents and factors are computed with:

Ordered resolution: $\dfrac{C \vee A_1 \quad \neg A_2 \vee D}{(C \vee D)\sigma}$

provided (i) σ is the most general unifier of A_1 and A_2, (ii) $A_1\sigma$ is strictly maximal with respect to $C\sigma$, and (iii) $\neg A_2\sigma$ is maximal with respect to $D\sigma$.

Ordered factoring: $\dfrac{C \vee A_1 \vee A_2}{(C \vee A_1)\sigma}$

provided (i) σ is the most general unifier of A_1 and A_2, and (ii) $A_1\sigma$ is maximal with respect to $C\sigma$.

Fig. 1. The calculus R

of substitutions. (An ordering is said to be *liftable* if it satisfies (iii).) The multiset extension of $\succ$ provides an admissible ordering on clauses. A literal L is said to be *(strictly) maximal* with respect to a clause C if for any literal L' in C, $L' \not\succ L$ ($L' \not\succeq L$.) A literal in a clause C is said to be *eligible* if it is maximal with respect to C. An ordering is *compatible with a given complexity measure* c_L on ground literals, if $c_L \succ c_{L'}$ implies $L \succ L'$ for any two ground literals L and L'.

Let R be the resolution calculus defined by the rules of Figure 1. The completeness proof sanctions a global notion of *redundancy*, with which additional don't-care non-deterministic simplification and deletion rules can be supported. Essentially, a ground clause is redundant in a set N with respect to the ordering $\succ$ if it follows from smaller instances of clauses in N, and a non-ground clause is redundant in N if all its ground instances are redundant in N. For example, any tautologous clause is redundant.

A (*theorem proving*) *derivation* from a set N of clauses is a finitely branching tree with root N constructed by applications of the expansion rules. A derivation T is a *refutation* if for every path $N(= N_0), N_1, \ldots$, the clause set $\bigcup_j N_j$ contains the empty clause. A derivation T from N is called *fair* if for any path $N(= N_0), N_1, \ldots$ in the tree T, with *limit* $N_\infty = \bigcup_j \bigcap_{k \geq j} N_k$, it is the case that each clause C that can be deduced from non-redundant premises in N_∞ is contained in some set N_j.

Theorem 1 ([3]). *Let N be a set of clauses and let T be a fair R-derivation from N (up to redundancy). Then, N is unsatisfiable iff for every path $N(= N_0), N_1, \ldots$, the clause set $\bigcup_j N_j$ contains the empty clause.*

It should be noted that inferences with ineligible literals are not forbidden, but are provably redundant. In other words, only inferences with eligible literals need to be performed for soundness and completeness.

Strictly, the "Split" rule is inessential for the results of this paper, though it may have some computational advantages (we comment on this in the final section). However, the inclusion of splitting allows for a more concise presentation of fluted clauses.

4   Fluted Clauses

This section introduces the class of fluted clauses into which fluted formulae can be translated. Without loss of generality we consider only maximally split clauses.

In fluted clauses the arguments of literals have a characteristic form which will be described with the help of a sequence notation. $(\overline{u}_i)$ will denote a finite, possibly empty, sequence $(u_i, u_{i+1}, \ldots, u_m)$ of terms. In this paper unless specified otherwise each non-empty sequence $(\overline{u}_i)$ is assumed to end with u_m. Thus, the sequences $(\overline{u}_1), (\overline{u}_2), \ldots, (\overline{u}_m)$ are linearly ordered by (the converse of) the 'is a proper suffix of' relationship. Note that $(\overline{u}_m) = u_m$. Given that $(\overline{u}_i) = (u_i, \ldots, u_m)$,

$(\overline{u}_i, t)$ will denote the sequence $(u_i, \ldots, u_m, t)$,

$f(\overline{u}_i)$ will denote the term $f(u_i, \ldots, u_m)$,

$P(\overline{u}_i)$ will denote the atom $P(u_i, \ldots, u_m)$,

$C(\overline{u}_i)$ will denote a (possibly empty) clause of literals
            of the form $(\neg)P(\overline{u}_i)$.

If $(\overline{u}_i)$ is the empty sequence then $f(\overline{u}_i)$, $P(\overline{u}_i)$ and $C(\overline{u}_i)$ respectively denote a constant, a propositional literal and a (possibly empty) propositional clause. $(\overline{u}_i)$ is said to be the *argument sequence* of $f(\overline{u}_i)$, $P(\overline{u}_i)$ and $C(\overline{u}_i)$. A sequence with n elements will be called an n-sequence.

Assume m is a non-negative integer, and $X_m = \{x_1, \ldots, x_m\}$ is a set of m ordered variables. We refer to a sequence of terms $\overline{u} = (u_1, \ldots, u_n)$ as a *fluted sequence* over X_m, if the following conditions are all satisfied: (i) $n > m$, (ii) $u_1 = x_1, \ldots, u_m = x_m$, (iii) the number of variables occurring in $(u_{m+1}, \ldots, u_n)$ is m, and (iv) for every k with $m < k \leq n$, there is an i with $1 \leq i < k$ such that $u_k = f(u_i, \ldots, u_{k-1})$ for some function symbol f. The sequence $(x_1, \ldots, x_m)$ will be called the *variable prefix* of $\overline{u}$. Examples of fluted sequences are:

(a),   a fluted sequence over $X_0 = \emptyset$,

$(x_1, x_2, x_3, f(x_1, x_2, x_3))$,

$(x_1, x_2, x_3, f(x_2, x_3), g(x_1, x_2, x_3, f(x_2, x_3)))$,

$(x_1, x_2, f(x_1, x_2), g(f(x_1, x_2)), h(x_2, f(x_1, x_2), g(f(x_1, x_2))))$.

However, $(x_1, x_2, x_3, f(x_2, x_3))$ is not a fluted sequence, as condition (iii) is violated.

By definition, a clause C is a *fluted clause* over X_m if one of the following holds.

(FL0) C is a (possibly empty) propositional clause.

(FL1) C is not empty, $\mathrm{var}(C) = X_m$, and for any literal L in C, there is some i where $1 \leq i \leq m$ such that the argument sequence of L is $(x_i, x_{i+1}, \ldots, x_m)$.

(FL2) C is functional and not empty, $\mathrm{var}(C) = X_m$, and for any literal L in C the argument sequence of L is either $(x_i, x_{i+1}, \ldots, x_m)$ or $(u_j, u_{j+1}, \ldots, u_n)$, where $1 \leq i \leq m$ and $(u_j, u_{j+1}, \ldots, u_n)$ is a suffix of some fluted sequence $\bar{u} = (u_1, \ldots, u_n)$ over $\{x_k, \ldots, x_m\}$, for some k with $1 \leq k \leq m$. $\bar{u}$ will be referred to as the *fluted sequence associated* with L. (By 4. of Lemma 1 below there can be just one fluted sequence associated with a given literal.)

(FL3) C is not empty, $\mathrm{var}(C) = X_{m+1}$, and for any literal L in C, the argument sequence of L is either $(x_1, x_2, \ldots, x_m)$ or $(x_i, \ldots, x_m, x_{m+1})$, where $1 \leq i \leq m$.

A fluted clause will be called a *strongly fluted clause* if it is either ground or has a literal which contains all the variables of the clause.

It may be helpful to consider some examples. The clause

$$P(x_1, x_2, x_3, x_4, x_5) \vee Q(x_1, x_2, x_3, x_4, x_5) \vee \neg R(x_4, x_5) \vee S(x_5).$$

satisfies the scheme (FL1), and is defined over five variables. Examples of fluted clauses of type (FL3) which are defined over two (!) variables are:

$$Q(x_1, x_2) \vee \neg P(x_1, x_2, x_3) \vee \neg R(x_2, x_3) \vee S(x_3)$$
$$Q(x_1, x_2) \vee \neg R(x_2, x_3) \vee S(x_3)$$

The following are fluted clauses of type (FL2), where $(\bar{x}_1) = (x_1, \ldots, x_4)$.

$$R(x_2, f(x_1, x_2)) \vee S(f(x_1, x_2))$$
$$Q(x_1, x_2) \vee R(x_2) \vee P(x_1, x_2, f(x_1, x_2)) \vee R(x_2, f(x_1, x_2)) \vee S(f(x_1, x_2))$$
$$Q(x_3, x_4) \vee R(x_4) \vee P(x_4, g(\bar{x}_1), f(x_4, g(\bar{x}_1))) \vee R(f(x_4, g(\bar{x}_1)))$$
$$Q(\bar{x}_1) \vee P(f(\bar{x}_1, h(\bar{x}_1)), g(\bar{x}_1, h(\bar{x}_1)), f(\bar{x}_1, h(\bar{x}_1)))$$
$$\vee R(x_4, h(\bar{x}_1), g'(x_4, h(\bar{x}_1)))$$

A few remarks are in order. First, the non-functional subclause of a (FL2)-clause will be denoted by ∇. Note that ∇ satisfies (FL1), in other words, clauses of the form (FL1) are building blocks of (FL2)-clauses. Second, clauses of the form (FL3) are defined to be fluted clauses over m variables, even though they contain $m+1$ variables. This may seem a bit strange, but this definition ensures a direct association of fluted formulae over m variables to fluted clauses over m variables. Third, no fluted clause can simultaneously satisfy any two of (FL0), (FL1), (FL2) and (FL3). Fourth, using the previously introduced notation a schematic description of the non-propositional shallow clauses (FL1) and (FL3) is:

$$C(\overline{x}_1) \vee C(\overline{x}_2) \vee \ldots \vee C(x_m) \quad (= \nabla)$$
$$C(\overline{x}_1) \vee C(\overline{x}_1, x_{m+1}) \vee C(\overline{x}_2, x_{m+1}) \vee \ldots \vee C(x_m, x_{m+1}) \vee C(x_{m+1})$$

Fifth, strongly fluted clauses have special significance in connection with termination of resolution, particularly with respect to the existence of a bound on the number of variables in any clause. Under the refinement we will use the eligible literals are literals which contain all the variables of the clause. So, the number of variables in resolvents of strongly fluted premises will always be less than or equal to the number of variables in any of the parent clauses.

The next results give some properties of fluted sequences and strongly fluted clauses.

Lemma 1. *Let $\overline{u}$ be a fluted sequence over X_m. Then:*

1. *There is an element u_k of $\overline{u}$ such that $u_k = f(u_1, \ldots, u_{k-1})$, for some f.*
2. *If u_n is last element of $\overline{u}$ then $\mathrm{var}(u_n) = X_m$.*
3. *$\overline{u}$ is uniquely determined by its last element.*
4. *If $(u_j, u_{j+1}, \ldots, u_n)$ is a suffix of $\overline{u}$ then $(u_1, \ldots, u_{j-1})$ is uniquely determined by $(u_j, u_{j+1}, \ldots, u_n)$.*

By the definition of (FL2)-clauses:

Lemma 2. *Let L be any literal of a (FL2)-clause defined over X_m. Then, all occurrences of variable sequences in L are suffixes of $(x_1, \ldots, x_m)$.*

Lemma 3. *Let C be a fluted clause over m variables. C is strongly fluted iff 1. C satisfies exactly one of the conditions (FL0), (FL1), (FL2), or 2. C satisfies condition (FL3), and it contains a literal with $m + 1$ variables.*

In other words, with the exception of certain (FL3)-clauses all fluted clauses include at least one literal which contains all the variables of the clause.

5 From Fluted Formulae to Fluted Clauses

Our transformation of fluted formulae into clausal form employs a standard renaming technique, known as structural transformation or renaming, see for example [20]. For any first-order formula φ, the definitional form obtained by introducing new names for subformulae at positions in Λ will be denoted by $\mathrm{Def}_\Lambda(\varphi)$.

Theorem 2. *Let φ be a first-order formula. For any subset Λ of the set of positions of φ, 1. φ is satisfiable iff $\mathrm{Def}_\Lambda(\varphi)$ is satisfiable, and 2. $\mathrm{Def}_\Lambda(\varphi)$ can be computed in polynomial time.*

In this paper we assume the clausal form of a first-order formula φ, written $\mathrm{Cls}(\varphi)$, is computed by transformation into conjunctive normal form, outer Skolemisation, and clausifying the Skolemised formula.

By introducing new literals for each non-literal subformula position, any given fluted formula can be transformed into a set of strongly fluted clauses.

Lemma 4. *Let φ be any fluted formula. If Λ contains all non-literal subformula positions of φ then $\mathrm{ClsDef}_\Lambda(\varphi)$ is a set of strongly fluted clauses (provided the newly introduced literals have the form $(\neg)Q_\lambda(\overline{x}_i)$).*

Transforming any given fluted formula into a set of *fluted* clauses requires the introduction of new symbols for all quantified subformulae.[1]

Lemma 5. *Let φ be any fluted formula over m ordered variables. If Λ contains at least the positions of any subformulae $\exists x_{i+1}\psi$, $\forall x_{i+1}\psi$, then $\mathrm{ClsDef}_\Lambda(\varphi)$ is a set of fluted clauses (again, provided the new literals have the form $(\neg)Q_\lambda(\overline{x}_i)$).*

6 Separation

The motivation for introducing separation is that the class of fluted clauses is not closed under resolution. In particular, resolvents of non-strongly fluted (FL3)-clauses are not always fluted and can cause (potentially) unbounded variable chaining across literals. This is illustrated by considering resolution between $P_1(x_1, x_2) \vee Q_1(x_2, x_3) \vee R(x_2, x_3)$ and $\neg R(x_1, x_2) \vee P_2(x_1, x_2) \vee Q_2(x_2, x_3)$, which produces the resolvent $P_1(x_1, x_2) \vee Q_1(x_2, x_3) \vee P_2(x_2, x_3) \vee Q_2(x_3, x_4)$. We note that it contains four variables, whereas the premises each contain only three variables. The class of strongly fluted clauses is also not closed under resolution. Fortunately, however, inferences with two strongly fluted clauses always produce fluted clauses, and non-strongly fluted clauses are what we call separable and can be restored to strongly fluted clauses.

Consider the resolvent $C = P(x_1, x_2) \vee P(x_2, x_3)$ of the strongly fluted clauses $P(x_1, x_2) \vee R(x_1, x_2, x_3)$ and $\neg R(x_1, x_2, x_3) \vee P(x_2, x_3)$. C is a fluted clause of type (FL3), but it is not strongly fluted, as none of its literals contains all the variables of the clause. Consequently, the literals are incomparable under an admissible ordering (in particular, a liftable ordering), because the literals have a common instance, for example $C\{x_1 \mapsto a, x_2 \mapsto a, x_3 \mapsto a\} = P(a, a) \vee P(a, a)$. The 'culprits' are the variables x_1 and x_3. Because they do not occur together in any literal, C can be separated and replaced by the following two clauses, where q is a new predicate symbol.

$$\neg q(x_2) \vee P(x_1, x_2)$$
$$q(x_2) \vee P(x_2, x_3)$$

---

[1] More generally, it requires at least the introduction of new symbols for all positive occurrences of universally quantified subformulae, all negative occurrences of existentially quantified subformulae, and all quantified subformulae with zero polarity. But then inner Skolemisation needs to be used, first Skolemising the deepest existential formulae.

The first clause is of type (FL1) (and thus strongly fluted) and the second is a strongly fluted clause of type (FL3).

In the remainder of this section we will formally define separation and consider under which circumstances soundness and completeness hold. In the next section we will show how separation can be used to stay within the class of fluted clauses.

Let C be an arbitrary (not necessarily fluted) clause. C is *separable* if it can be partitioned into two non-empty subclauses D_1 and D_2 such that $\mathrm{var}(D_1) \not\subseteq \mathrm{var}(D_2)$ and $\mathrm{var}(D_2) \not\subseteq \mathrm{var}(D_1)$. For example, the clauses $P(x_1, x_2) \vee Q(x_2, x_3)$ and $P(x_1) \vee Q(x_2)$ are separable, but $P(x_1, x_2) \vee Q$ and $P(x_1, x_2) \vee Q(x_2, x_3) \vee R(x_1, x_3)$ are not. (The last clause is not fluted.)

Theorem 3. *Let $C \vee D$ be a separable clause such that $\mathrm{var}(C) \not\subseteq \mathrm{var}(D)$, $\mathrm{var}(D) \not\subseteq \mathrm{var}(C)$, and $\mathrm{var}(C) \cap \mathrm{var}(D) = \{x_1, \ldots, x_n\}$ for $n \geq 0$. Let q be a fresh predicate symbol with arity n (q does not occur in N). Then, $N \cup \{C \vee D\}$ is satisfiable iff $N \cup \{\neg q(x_1, \ldots, x_n) \vee C, q(x_1, \ldots, x_n) \vee D\}$ is satisfiable.*

On the basis of this theorem we can define the following replacement rule:

Separate:
$$\frac{N \cup \{C \vee D\}}{N \cup \{\neg q(x_1, \ldots, x_n) \vee C, q(x_1, \ldots, x_n) \vee D\}}$$

provided (i) $C \vee D$ is separable such that $\mathrm{var}(C) \not\subseteq \mathrm{var}(D)$ and $\mathrm{var}(D) \not\subseteq \mathrm{var}(C)$, (ii) $\mathrm{var}(C) \cap \mathrm{var}(D) = \{x_1, \ldots, x_n\}$ for $n \geq 0$, and (iii) q does not occur in N, C or D.

C and D will be referred to as the *separation components* of $C \vee D$.

Lemma 6. *The replacements of a separable clause C each contain less variables than C.*

Even though it is possible to define an ordering under which the replacement clauses are strictly smaller than the original clause, and consequently, $C \vee D$ is redundant in $N \cup \{\neg q(x_1, \ldots, x_n) \vee C, q(x_1, \ldots, x_n) \vee D\}$, in general, "Separate" is not a simplification rule in the sense of Bachmair-Ganzinger. Nevertheless, we can prove the following.

Theorem 4. *Let R^{sep} denote the extension of R with the separation inference rule. Let N be a set of clauses and let T be a fair R^{sep}-derivation from N such that separation is applied only finitely often in any path of T. Then N is unsatisfiable iff for every path $N(= N_0), N_1, \ldots$, the clause set $\bigcup_j N_j$ contains the empty clause.*

More generally, this theorem holds also if R^{sep} is based on ordered resolution (or superposition) with selection.

By Lemma 3 separable fluted clauses have the form

$$C(\overline{x}_1) \vee C(\overline{x}_i, x_{m+1}) \vee \ldots \vee C(x_m, x_{m+1}) \vee C(x_{m+1}), \tag{1}$$

where $(\overline{x}_1)$ is a non-empty m-sequence, $\mathcal{C}(\overline{x}_1)$ is not empty, and i is the smallest integer $1 < i \leq m$ such that $\mathcal{C}(\overline{x}_i, x_{m+1})$ is not empty.

Let sep be a mapping from separable fluted clauses of the form (1) to sets of clauses defined by

$$\text{sep}(C) = \{\neg q(\overline{x}_i) \vee \mathcal{C}(\overline{x}_1),$$
$$q(\overline{x}_i) \vee \mathcal{C}(\overline{x}_i, x_{m+1}) \vee \ldots \vee \mathcal{C}(x_m, x_{m+1}) \vee \mathcal{C}(x_{m+1})\}$$

where q is a fresh predicate symbol uniquely associated with C and all its variants. Further, let $\text{sep}(N) = \bigcup\{\text{sep}(C) \mid C \in N\}$. For example:

$$\text{sep}(P(x_1, x_2) \vee Q(x_2, x_3)) = \{\neg q(x_2) \vee P(x_1, x_2), \; q(x_2) \vee Q(x_2, x_3)\}.$$

Lemma 7. *The separation of a separable fluted clause (1) is a set of strongly fluted clauses.*

Lemma 8. *For fluted clauses a separation inference step can be performed in linear time.*

7   Termination

In this section we define a minimal resolution calculus R^{sep} and prove that it provides a decision procedure for fluted logic.

The ordering $\succ$ of R^{sep} is required to be any admissible ordering compatible with the following complexity measure. Let $\succ_s$ denote the proper superterm ordering. Define the complexity measure of any literal L by $c_L = (\text{ar}(L), \max(L), \text{sign}(L))$, where $\text{ar}(L)$ is the arity (of the predicate symbol) of L, $\max(L)$ is a $\succ_s$-maximal term occurring in L, and $\text{sign}(L) = 1$, if L is negative, and $\text{sign}(L) = 0$, if L is positive. The ordering on the complexity measures is given by the lexicographic combination of $>$, $\succ_s$, and $>$ (where $>$ is the usual ordering on the non-negative integers).

Let R^{sep} be any calculus in which (i) derivations are generated by strategies applying "Delete", "Split", "Separate", namely, $N \cup \{C\}/N \cup \text{sep}(C)$, and "Deduce" in this order, (ii) no application of "Deduce" with identical premises and identical consequence may occur twice on the same path in derivations, and (iii) the ordering is based on $\succ$, defined above.

Now we address the question as to whether the class of fluted clauses is closed under R^{sep}-inferences.

Lemma 9. *A factor of a strongly fluted clause C is again a strongly fluted clause of the same type as C.*

In fact, any (unordered) factor of a strongly fluted clause C is again a strongly fluted clause of the same type.

The next lemma is the most important technical result.

Lemma 10. *Let $C = C' \vee A_1$ and $D = \neg A_2 \vee D'$ be (FL2)-clauses. Suppose A_1 and $\neg A_2$ are eligible literals in C and D, respectively, and suppose σ is the most general unifier of A_1 and A_2. Then:*

1. *$C\sigma$, $D\sigma$ and $C\sigma \vee D\sigma$ are (FL2)-clauses.*
2. *For any functional literal $L\sigma$ in $C\sigma \vee D\sigma$, the fluted sequence associated with $L\sigma$ is the σ-instance of a fluted sequence $\bar{v}$ associated with some literal L' in $C \vee D$.*

Lemma 11. *Let $C = C' \vee A_1$ and $D = \neg A_2 \vee D'$ be strongly fluted clauses. Suppose A_1 and $\neg A_2$ are eligible literals in C and D, respectively, and suppose σ is the most general unifier of A_1 and A_2. Then $C\sigma \vee D\sigma$ is a strongly fluted clause.*

Lemma 12. *Let C, D and σ be as in Lemma 11. Then, $|\mathrm{var}(C\sigma \vee D\sigma)| \leq \max\{|\mathrm{var}(C)|, |\mathrm{var}(D)|\}$.*

Lemma 13. *Removing any subclause from a fluted clause produces a fluted clause.*

This cannot be said for strongly fluted clauses, in particular, not for clauses of the form (FL3). For all other forms the statement is also true for strongly fluted clauses, namely, removing any subclause from strongly fluted clauses produces strongly fluted clauses. Consequently:

Lemma 14. *The condensation of any (strongly) fluted clause is a (strongly) fluted clause.*

Lemma 15. *The resolvent of any two strongly fluted clauses is a strongly fluted clause, or, it is only a fluted clause, if one of the premises is a (FL3)-clause.*

Lemma 16. *Any maximally split, condensed and separated factor or resolvent of strongly fluted clauses is strongly fluted.*

This proves that the class of (strongly) fluted clauses is closed under $\mathsf{R}^{\mathrm{sep}}$-resolution with eager application of condensing, splitting and separation.

In the next three lemmas, N is assumed to be a finite set of fluted clauses (which will be transformed into a set of strongly fluted clauses during the derivation, see Lemma 7). Our goal is to exhibit the existence of a term depth bound of all inferred clauses, as well as the existence of a bound on the number of variables occurring in any inferred clause. The latter follows immediately from Lemmas 6 and 12.

Lemma 17. *All clauses occurring in an $\mathsf{R}^{\mathrm{sep}}$-derivation from N contain at most $m + 1$ variables, where m is the maximal arity of any predicate symbol in N.*

The definition of fluted clauses places no restriction on the level of nesting of functional terms. But:

Lemma 18. *A bound on the maximal term depth of clauses derived by* $\mathsf{R}^{\mathsf{sep}}$ *from* N *is* m, *where* m *is the maximal arity of any predicate symbol in* N.

Because the signature is extended dynamically during the derivation, it remains to show that separation cannot be performed infinitely often.

Lemma 19. *The number of applications of the "Separate"-rule in an* $\mathsf{R}^{\mathsf{sep}}$- *derivation from* N *is bounded.*

Now, we can state the main theorem of this paper.

Theorem 5. *Let* φ *be any fluted formula and* $N = \mathrm{ClsDef}_\Lambda(\varphi)$, *where* Def_Λ *satisfies the restrictions of Lemma 5. Then:*

1. *Any* $\mathsf{R}^{\mathsf{sep}}$-*derivation from* N *(up to redundancy) terminates.*
2. φ *is unsatisfiable iff the* $\mathsf{R}^{\mathsf{sep}}$-*saturation (up to redundancy) of* N *contains the empty clause.*

The final theorem gives a rough estimation of an upper bound for the space requirements.

Theorem 6. *The number of maximally split, condensed strongly fluted clauses in any* $\mathsf{R}^{\mathsf{sep}}$-*derivation from* N *is an* $O(m)$-*story exponential, where* m *is the maximal arity of any predicate symbol in* N.

8 Concluding Remarks

Developing a resolution decision procedure for fluted logic turned out to be more complicated than expected. Even though to begin with, clauses are simple in the sense that no nesting of non-nullary function symbols occurs (Lemma 4), the class of fluted clauses is rather complex. It is thus natural to ask whether there is a less complex clausal class which corresponds to fluted logic. The complexity of the class is a result of the ordering we have proposed. This ordering is unusual as it first considers the arity of a literal, while more conventional ordering refinements first consider the depth of a literal. A conventional ordering has the advantage that term depth growth can be avoided completely. This would induce a class of clauses which can be described by these schemes:

$$\text{propositional clauses} \tag{2}$$

$$C(\overline{x}_1) \vee C(\overline{x}_2) \vee \ldots \vee C(x_m) \quad (= \nabla) \tag{3}$$

$$\nabla \vee C(\overline{x}_1, f(\overline{x}_1)) \vee C(\overline{x}_2, f(\overline{x}_1)) \vee \ldots \vee C(x_m, f(\overline{x}_1)) \vee C(f(\overline{x}_1)) \tag{4}$$

$$C(\overline{x}_1) \vee C(\overline{x}_1, x_{m+1}) \vee C(\overline{x}_2, x_{m+1}) \vee \ldots \vee C(x_m, x_{m+1}) \vee C(x_{m+1}) \tag{5}$$

The difference between this class and the class of fluted clauses defined in Section 4 is scheme (4). Clauses satisfying scheme (4) are (FL2)-clauses, but not every (FL2)-clause has the form (4). With separation (on (5)-clauses without an embracing literal) it is possible to stay within the confines of this class. However, the danger with the separation rule is that it could be applied infinitely often. It is open whether there is a clever way of applying the separation rule so that only finitely many new predicate symbols are introduced. For fluted logic with binary converse an example giving rise to an unbounded derivation is the following.

$$P_2(x_1, x_2, x_3) \vee P_1(f(x_1, x_2, x_3), x_3)$$

$$\neg P_2(x_1, x_2, x_3) \vee \neg P_1(x_1, x_2) \vee P_0(x_2, x_3)$$

$$\neg Q_1(x_2) \vee \neg P_1(x_1, x_2)$$

We do not know whether an alternative form of the separation rule could help.

Noteworthy about fluted logic and the proposed method is that, in order to establish an upper bound on the number of variables in derived clauses, a truly dynamic renaming rule is needed (namely separation). Though renaming is a standard technique for transforming formulae into well-behaved clausal classes, it is usually applied in advance, see for example [7, 15]. From a theoretical point of view whenever it is possible to do the renaming transformations as part of preprocessing, it is sensible to do so. The above example illustrates what could go wrong otherwise. It should be added though that there are instances where renaming on the fly is useful [27]. For fluted logic it is open whether there is a resolution decision procedure which does not require dynamic renaming. Going by the experience with other solvable classes, for example, Maslov's class $\overline{K}$ [5, 14], where renaming is only necessary when liftable ordering refinements are used, one possibility for avoiding dynamic renaming may be by using a refinement which is based on a non-liftable ordering. However, it would seem that the problems described in Section 6 are the same with non-liftable orderings. Even if it turns out that there is a resolution decision procedure which does not use separation, one could imagine that the separation rule can have a favourable impact on the performance of a theorem prover, for, with separation the size of clauses can be kept small, which is generally desirable, and for fluted logic separation is a cheap operation (Lemma 8).

As noted earlier, the splitting rule is not essential for the results of this paper. The separation rule already facilitates some form of 'weak splitting', because, if C and D are variable disjoint and non-ground subclauses of $C \vee D$ then separation will replace it by $q \vee C$ and $\neg q \vee D$, where q is a new propositional symbol. A closer resemblance to the splitting rule can be achieved by making q minimal in $q \vee C$ and selecting $\neg q$ in $\neg q \vee D$. Nevertheless, splitting has the advantage that more redundancy elimination operations are possible, for example forward subsumption.

The realisation of a practical decision procedure for fluted logic would require a modest extension of one of the many available first-order theorem provers which are based on ordered resolution with an implementation of the separation rule. Modern theorem provers such as Spass [28] are equipped with a wide range of simplification rules so that reasonable efficiency could be expected.

Acknowledgements

We wish to thank Bill Purdy and the referees for valuable comments.

References

1. L. Bachmair and H. Ganzinger. Rewrite-based equational theorem proving with selection and simplification. *J. Logic Computat.*, 4(3):217–247, 1994.
2. L. Bachmair and H. Ganzinger. Resolution theorem proving. In J. A. Robinson and A. Voronkov, eds., *Handbook of Automated Reasoning*. Elsevier, 2000. To appear.
3. L. Bachmair, H. Ganzinger, and U. Waldmann. Superposition with simplification as a decision procedure for the monadic class with equality. In *Proc. Third Kurt Gödel Colloquium (KGC'93)*, vol. 713 of *LNCS*, pp. 83–96. Springer, 1993.
4. H. de Nivelle. A resolution decision procedure for the guarded fragment. In *Automated Deduction—CADE-15*, vol. 1421 of *LNAI*, pp. 191–204. Springer, 1998.
5. C. Fermüller, A. Leitsch, T. Tammet, and N. Zamov. *Resolution Method for the Decision Problem*, vol. 679 of *LNCS*. Springer, 1993.
6. C. G. Fermüller, A. Leitsch, U. Hustadt, and T. Tammet. Resolution theorem proving. In J. A. Robinson and A. Voronkov, eds., *Handbook of Automated Reasoning*. Elsevier, 2000. To appear.
7. H. Ganzinger and H. de Nivelle. A superposition decision procedure for the guarded fragment with equality. In *Fourteenth Annual IEEE Symposium on Logic in Computer Science*, pp. 295–303. IEEE Computer Society Press, 1999.
8. G. Gargov and S. Passy. A note on Boolean modal logic. In P. P. Petkov, ed., *Mathematical Logic: Proceedings of the 1988 Heyting Summerschool*, pp. 299–309. Plenum Press, 1990.
9. V. Goranko and S. Passy. Using the universal modality: Gains and questions. *J. Logic Computat.*, 2(1):5–30, 1992.
10. J. Y. Halpern and Y. Moses. A guide to completeness and complexity for modal logics of knowledge and belief. *Artificial Intelligence*, 54:319–379, 1992.
11. A. Herzig. A new decidable fragment of first order logic, 1990. In Abstracts of the Third Logical Biennial, Summer School & Conference in Honour of S. C. Kleene, Varna, Bulgaria.
12. I. L. Humberstone. Inaccessible worlds. *Notre Dame J. Formal Logic*, 24(3):346–352, 1983.
13. I. L. Humberstone. The modal logic of 'all and only'. *Notre Dame J. Formal Logic*, 28(2):177–188, 1987.
14. U. Hustadt and R. A. Schmidt. An empirical analysis of modal theorem provers. *J. Appl. Non-Classical Logics*, 9(4), 1999.
15. U. Hustadt and R. A. Schmidt. Maslov's class K revisited. In *Automated Deduction—CADE-16*, vol. 1632 of *LNAI*, pp. 172–186. Springer, 1999.
16. U. Hustadt and R. A. Schmidt. Issues of decidability for description logics in the framework of resolution. In *Automated Deduction in Classical and Non-Classical Logics*, vol. 1761 of *LNAI*, pp. 192–206. Springer, 2000.
17. U. Hustadt and R. A. Schmidt. A resolution decision procedure for fluted logic. Technical Report UMCS-00-3-1, University of Manchester, UK, 2000.
18. W. H. Joyner Jr. Resolution strategies as decision procedures. *J. ACM*, 23(3):398–417, 1976.
19. H. J. Ohlbach and R. A. Schmidt. Functional translation and second-order frame properties of modal logics. *J. Logic Computat.*, 7(5):581–603, 1997.
20. D. A. Plaisted and S. Greenbaum. A structure-preserving clause form translation. *J. Symbolic Computat.*, 2:293–304, 1986.
21. W. C. Purdy. Decidability of fluted logic with identity. *Notre Dame J. Formal Logic*, 37(1):84–104, 1996.

22. W. C. Purdy. Fluted formulas and the limits of decidability. *J. Symbolic Logic*, 61(2):608–620, 1996.
23. W. C. Purdy. Surrogate variables in natural language. To appear in M. Böttner, ed., *Proc. of the Workshop on Variable-Free Semantics*, 1996.
24. W. C. Purdy. Quine's 'limits of decision'. *J. Symbolic Logic*, 64:1439–1466, 1999.
25. W. V. Quine. Variables explained away. In *Proc. American Philosophy Society*, vol. 104, pp. 343–347, 1960.
26. R. A. Schmidt. Decidability by resolution for propositional modal logics. *J. Automated Reasoning*, 22(4):379–396, 1999.
27. G. S. Tseitin. On the complexity of derivations in propositional calculus. In A. O. Slisenko, ed., *Studies in Constructive Mathematics and Mathematical Logic, Part II*, pp. 115–125. Consultants Bureau, New York, 1970.
28. C. Weidenbach. SPASS, 1999. http://spass.mpi-sb.mpg.de.

ZRES: The Old Davis–Putnam Procedure Meets ZBDD

Philippe Chatalic and Laurent Simon

Laboratoire de Recherche en Informatique
U.M.R. CNRS 8623
Université Paris-Sud, 91405 Orsay Cedex, France
{chatalic,simon}@lri.fr

Abstract. ZRES is a propositional prover based on the original procedure of Davis and Putnam, as opposed to its modified version of Davis,
Logeman and Loveland, on which most of the current efficient SAT
provers are based. On some highly structured SAT instances, such as
the well known Pigeon Hole and Urquhart problems, both proved hard
for resolution, ZRES performs very well and surpasses all classical SAT
provers by an order of magnitude.

1 The DP and DLL Algorithms

Stimulated by hardware progress, many more and more efficient SAT solvers have
been designed during the last decade. It is striking that most of the complete
solvers are based on the procedure of Davis, Logeman and Loveland (DLL for
short) presented in 1962 [11]. The DLL procedure may roughly be described
as a backtrack procedure that searches for a model. Each step amounts to the
extension of a partial interpretation by choosing an assignment for a selected
variable. The success of this procedure is mainly due to its space complexity,
since making choices only results in simplifications. However, the number of
potential extensions remains exponential. Therefore, if the search space cannot
be pruned by clever heuristics, this approach becomes intractable in practice.

The picture is very different with DP, the original Davis–Putnam algorithm
[3]. DP is able to determine if a propositional formula f, expressed under conjunctive normal form (CNF), is satisfiable or not. Assuming the reader is familiar
with propositional logic, DP may be roughly described as follows [9]:

I.   Choose a propositional variable x of f.
II. Replace all the clauses which contain the literal x (or $\neg x$) by all
     binary resolvents (on x) of these clauses (cut elimination of x),
     and remove all subsumed clauses.
III. a. If the new set of clauses is reduced to the empty clause, then the
        original set is unsatisfiable.
     b. If it is empty, then the original formula is satisfiable.
     c. Otherwise, repeat steps I-III for this new set of clauses.

D. McAllester (Ed.): CADE-17, LNAI 1831, pp. 449–454, 2000.

As opposed to DLL, DP avoids making choices by considering the two possible instantiations of a variable simultaneously. It amounts to a sequence of cut eliminations. Since the number of clauses generated at each step may grow exponentially, it is widely acknowledged in the literature as inefficient, although no real experimental study has been conducted to confirm this point. Dechter and Rish [5] have first pointed out some instances for which DLL is not appropriate and where DP obtains better results. A more comprehensive experimentation has been conducted in [2], to evaluate DP on a variety of instances. If this study confirms the superiority of DLL on random instances, it also shows that for structured instances, DP may outperform some of the best DLL procedures. Substantial progress for DLL are due to better heuristics. For DP, significant improvements are possible thanks to efficient data structures for representing very large sets of clauses.

Several authors have pointed out that resolution-based provers (like DP and DLL) are intrinsically limited, since they have found instances that require an exponential number of resolution steps to be solved (e.g. [9, 10, 14]). This is the case for the Pigeon Hole [10] and for the Urquhart problem [14]. They suggest that more powerful proof systems have to be used practically to solve such problems efficiently. However, all these results are based on the implicit hypothesis that successive resolutions in DP and DLL are performed one by one.

This paper presents the ZRES system, which is an implementation of the DP algorithm that is able to perform several resolutions in a single step. As a result, ZRES is able to solve instances of such hard problems much more efficiently than the best current DLL provers.

2   Reviving DP

The crucial point of DP is step II, which tends to generate a very large number of clauses. Eliminating subsumed clauses at this step induces a significant overhead but still pays off.

Efficient Data Structures. In [2], Trie structures are used to represent sets of clauses. Until now, they seem to remain the state-of-the-art data structures for subsumption checking [15, 4]. Tries allow the factorization of clauses beginning in the same way, according to a given order on literals. In ZRES we further generalize this principle, to allow the factorization of the end as well as the beginning of clauses simultaneously. Sets of clauses are thus represented by means of directed acyclic graphs (DAG) instead of trees (by Tries).

Using DAG to represent a boolean formula has been intensively investigated in many works on binary decision diagrams (BDD) [1, 7]. Many variants of BDD have been proposed but all attempt to compute the BDD encoding of the formula, expressed in Shannon normal form. From the SAT point of view and since the resulting BDD characterizes the formula validity and satisfiability, this construction is *de facto* more difficult than testing for satisfiability.

The approach followed by ZRES is quite different since, instead of computing the Shannon normal form, we use here BDD-like structures only to represent sets of clauses. Practically, the set of clauses is represented by a ZBDD (a variant of BDD [12]), proved useful for manipulations of large sets. Sets are represented by their characteristic (boolean) functions and basic set operations can thus be performed as boolean operations on ZBDD. Moreover, it has been shown that the size of such a ZBDD is not directly related to the size of the corresponding set. Since the cost of basic set operations only depend on the size of the considered ZBDD, this hints to performing the cut elimination step of the DP algorithm directly at the set level.

Another way to represent f in the cut elimination step of DP is to factorize x and $\neg x$ among its clauses. The formula f can then be rewritten as $(x \vee f_x^+) \wedge (\neg x \vee f_x^-) \wedge f_{x'}$, where f_x^+ (resp. f_x^-) is the CNF obtained from the set of clauses containing x (resp. $\neg x$), after factorization, and where $f_{x'}$ denotes the set of clauses containing neither x nor $\neg x$. The second step of the algorithm then amounts to put the formula $(f_x^+ \vee f_x^-) \wedge f_{x'}$ into CNF. This can be done in 3 stages. First, distribute the set of clauses f_x^+ over f_x^-. Second, eliminate tautologies and subsumed clauses from the resulting clauses. Third, compute the union of the remaining clauses with those of $f_{x'}$, while deleting subsumed clauses. The two first stages could be performed successively, using standard operations on ZBDD. However, the ZBDD used in ZRES have a special semantics and thus, a more efficient algorithm, called *clause-distribution*, can be designed. This operation guarantees that, during the bottom-up construction of the result, each intermediate ZBDD is free of tautologies and subsumed clauses. Tautologies are eliminated on the fly and subsumed clauses are deleted by a set difference operation, at each level. Similarly, in the third stage of the cut elimination, subsumed clauses may be deleted while computing the union of the two sets of clauses. This new algorithm takes full advantage of the data structure used to represent sets of clauses.

3 Experimental Results

ZRES[1] is written in C, using the Cudd package [13] which provides us with basic ZBDD operations as well as useful dynamic reordering functions. We have tested ZRES on two classes of hard problems for resolution: Hole and Urquhart. Those tests have been performed on a Linux Pentium-II 400MHz[2] with 256MB. Our results are compared, when possible, with those of two DLL implementations: ASAT [8], which is a good-but-simple DLL implementation, and SATO 3.0 [15], which includes many optimizations and recent improvements, such as backjumping and conflict memorization. Cpu times are given in seconds. We assume that an instance that cannot be solved in less than 10000 seconds counts for 10000.

---

[1] ZRES is available at http://www.lri.fr/~simon/research/zres.

[2] On the DIMACS [6] machine scale benchmark, our tests have granted this machine a user time saving of 305%, in comparison with the Sparc10.41, given as a reference.

The Hole Problem. The following table describes the obtained results on different instances of the Hole problem by ASAT, SATO and ZRES. For ZRES, the *std* and *cd* columns describe respectively the time obtained without (resp. with) the *clause-distribution* operation.

Instances	Var. Nb.	Cl. Nb.	ASAT	SATO	ZRES *std*	ZRES *cd*
Hole-09	90	415	11.94	8.90	1.87	1.01
Hole-10	110	561	141.96	80.94	3.26	1.61
Hole-11	132	738	1960.77	7373.65	5.76	2.65
Hole-12	156	949	10000	10000	10.18	4.06
Hole-20	420	4221	–	–	654.8	69
Hole-30	930	13981	–	–	10000	1102
Hole-40	1640	32841	–	–	–	9421

ZRES clearly surpasses both ASAT and SATO. While DLL algorithms can't solve instances of Hole-n for $n > 11$, ZRES manages to solve much larger instances[3]. As we can see, the speedup induced by the *clause-distribution* operation is significant on such instances.

Our experiments have shown that this problem is very sensitive to the heuristic function used to choose the cut variable. Surprisingly, the best results were obtained using a heuristic function that tends to maximize the number of clauses produced by the cut. Other heuristics, such as in [2], did not allow to solve those Hole instances.

The Urquhart Problem. Urquhart has described a class of problems based on properties of expander graphs [14]. Actually, each Urq-n is a class of problems where the number of clauses and variables is not fixed but only bounded to a specific interval. In the last table, MnV (resp. MnC) denotes the mean number of variables (resp. clauses) for a set of instances of a given class. Contrary to Hole, Urquhart problem does not seem sensitive to the heuristics used. The results of ASAT and SATO on 100 Urq-3 instances attest the hardness of these problems:

System	Total cpu time	#resolved	Mean cpu time (resolved)
ASAT	404 287	69	1366
SATO	776 364	26	1398
ZRES *cd*	69.2	100	0.69

Solving instances for greater values of n seems out of the scope of these systems. On the other hand, ZRES performs quite well on such instances. The following table gives the mean time on 1000 instances for greater values of n. Note that we do not give the *std* time because the speedup due to the *clause-distribution* operation is not relevant for this problem.

---

[3] We even solved the Hole-55 instances, with 3080 variables and 84756 clauses, in less than 2 days.

Instances	MnV	MnC	Mean cpu-time
Urq-3	42	364	0.57
Urq-4	77	705	1.72
Urq-5	123	1143	4.25
Urq-6	178	1665	8.88
Urq-7	242	2299	16.5
Urq-8	317	3004	29.6
Urq-9	403	3837	48.8

About the Compression Power of ZBBD. The previous examples illustrate quite well that DP, associated with the ZBDD encoding of sets of clauses, may in some cases be more effective than DLL. The experiment on the Hole-40 shows that for some of the cut eliminations, the number of clauses corresponding to f_x^+ and f_x^- may exceed 10^{60}. Clearly, the result of such a cut could not be computed efficiently without using such an extremely compact data structure. The ability of ZBDD to capture redundancies in sets of clauses suits the DP algorithm particularly well. Indeed, additional redundancies are produced during each cut elimination, when each clause of f_x^+ is merged with each clause of f_x^-. Moreover, unlike in random instances, we think that such redundancies may also be found in structured instances, corresponding to real-world problems.

In order to appreciate the compression power of ZBDD structures, it is interesting to consider *level of compression* which may be characterized by the ratio *nb of literals/nb of nodes*. We have recorded its successive values on 1000 random 3-SAT instances of 42 variables and 180 clauses, on wich DP is known to be a poor candidate [2]. For such instances, the initial value of the ratio is about 2, then it increases up to 6.15 and eventually decreases together with the number of clauses. In contrast, on Hole-40, this ratio varies from 10 to more that 10^{60}. Similarly, on Urq-10 it may exceed 10^{64}. Pigeon and Urquart classes however correspond to extreme cases. We have also tested ZRES on some other instances of the SAT DIMACS base [6]. Results are particularly interesting. For some instances the compression level is much more important than for random instances (more than 10^8), while on others, like *ssa* or $flat - 50$ it is very close to that of random instances. It is however striking that the latter instances, even if they correspond to concrete problems, have been generated in a random manner. This seems to confirm that our hypothesis (structured problems generally have regularities) is well founded.

4   Conclusion

Dechter and Rish [5] were the first to revive the interest in the original Davis–Putnam procedure for SAT. But DP also proves useful for knowledge compilation or validation techniques, which was our initial motivation [2]. The introduction of ZBDD brings significant improvements allowing ZRES to deal with huge sets of clauses. It leads us to completely reconsider the performances of the cut elimination, which can be performed independently of the number of handled clauses.

It is thus able to solve two hard problems out of the scope of other resolution-based provers. Although such examples might be considered somewhat artificial, their importance in the study of the complexity of resolution procedures must not be forgotten. On other examples, such like DIMACS ones, results are not so good, but important compression level, due to ZBDD, can be observed on real-world instances. The strength of ZRES definitely comes from its ability to capture regularities in sets of clauses. Although it has a no chance to compete on random instances, which lack such regularities, it might be a better candidate for solving real-world problems.

Further improvements are possible. The Hole example pointed out that standard heuristics for DP are not always appropriate for ZRES. We are studying new heuristics, based on the structure of ZBDD rather than on what they represent. One may also investigate more adapted reordering algorithms, which take advantage of the particular semantics of the ZBDD used in ZRES. Eventually, DP and DLL may be considered as complementary approaches. An interesting idea is to design an hybrid algorithm integrating both DP and DLL in ZRES.

References

1. R.E. Bryant. Graph - based algorithms for boolean function manipulation. *IEEE Trans. on Comp.*, 35(8):677–691, 1986.
2. Ph. Chatalic and L. Simon. Davis and putnam 40 years later: a first experimentation. Technical Report 1237, LRI, Orsay, France, 2000. *Submitted to the Journal of Automated Reasoning.*
3. M. Davis and H. Putnam. A computing procedure for quantification theory. *Journal of the ACM*, pages 201–215, 1960.
4. Johan de Kleer. An improved incremental algorithm for generating prime implicates. In *AAAI'92*, pages 780–785, 1992.
5. R. Dechter and I. Rish. Directional resolution: The Davis–Putnam procedure, revisited. In *Proceedings of KR-94*, pages 134–145, 1994.
6. The DIMACS challenge benchmarks.
   ftp://ftp.rutgers.dimacs.edu/challenges/sat.
7. R. Drechsler and B. Becker. *Binary Decision Diagram: Theory and Implementation.* Kluwer Academic Publisher, 1998.
8. Olivier Dubois. Can a very simple algorithm be efficient for SAT?
   ftp://ftp.dimacs.rutgers.edu/pub/challenges/sat/contributed/dubois.
9. Zvi Galil. On the complexity of regular resolution and the Davis–Putnam procedure. *Theorical Computer Science*, 4:23–46, 1977.
10. A. Haken. The intractability of resolution. *Theorical Computer Science*, 39:297–308, 1985.
11. G. Logeman M. Davis and D. Loveland. A machine program for theorem-proving. *Communications of the ACM*, pages 394–397, 1962.
12. S. Minato. Zero-suppressed bdds for set manipulation in combinatorial problems. In *30th ACM/IEEE Design Automation Conference*, 1993.
13. F. Somenzy. Cudd release 2.3.0. http://bessie.colorado.edu/~fabio.
14. A. Urquhart. Hard examples for resolution. *Journal of the ACM*, 34:209–219, 1987.
15. Hantao Zhang. SATO: An efficient propositional prover. In *CADE-14*, LNCS 1249, pages 272–275, 1997.

System Description:
MBASE, an Open Mathematical Knowledge Base

Andreas Franke and Michael Kohlhase

FB Informatik, Universität des Saarlandes
{afranke,kohlhase}@ags.uni-sb.de

Abstract. In this paper we describe the MBASE system, a web-based, distributed mathematical knowledge base. This system is a mathematical service in MATHWEB that offers a universal repository of formalized mathematics where the formal representation allows semantics-based retrieval of distributed mathematical facts.

1 Introduction

Around 1994, an anonymous (but well-known) group of authors put forward the "QED Manifesto" [QED95], which advocates building up a mathematical knowledge base (and supporting software systems) as a kind of "human genome project" for the deduction community. Unfortunately, the vision has failed to catch on in spite of a wave of initial interest. In our view this is largely due to the lack of supporting software, as well as to the ensuing debate on the "right" logical formalism.

In this paper we describe the MBASE system, a web-based mathematical knowledge base (see http://www.mathweb.org/mbase). It offers a the infrastructure for a universal, distributed repository of formalized mathematics. Since it is independent of a particular deduction system and particular logic[1], the MBASE system can be seen as an attempt to revive the QED initiative from an infrastructure viewpoint. The system is realized as a mathematical service in the MATHWEB system [FK99], an agent-based implementation of a mathematical software bus for distributed theorem proving.

We will start with a description of the system from the implementation point of view in the next section (we have described the data model and logical issues in [KF00]). In section 3, we will take a brief look at the interface protocols based on the OPENMATH and KQML standards (see [FHJ+99,Koh00]). This reliance of Internet standards for communication makes MBASE an open system, and the implementation presented in this paper just one of its possible instances.

2 Architecture

The MBASE system is realized as a distributed set of MBASE servers (see figure 1). Each MBASE server consists of a Relational Data Base Management

---

[1] See [KF00] for the logical issues related to supporting multiple logical languages while keeping a consistent overall semantics.

D. McAllester (Ed.): CADE-17, LNAI 1831, pp. 455–459, 2000.

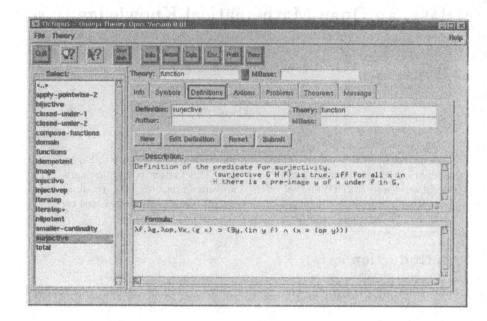

Fig. 1. System Architecture

System (RDBMS) e.g. ORACLE connected to a MOZART process (yielding a MATHWEB service) via a standard data base interface (in our case JDBC). For browsing the MBASE content, any MBASE server provides an http server (see http://mbase.mathweb.org:8000 for an example) that dynamically generates presentations based on HTML or XML forms.

This architecture combines the storage facilities of the RDBMS with the flexibility of the concurrent, logic-based programming language Oz [Smo95], of which MOZART is a distributed implementation (see http://www.mozart-oz.org). Most importantly for MBASE, MOZART offers a mechanism called **pickling**, which allows for a limited form of persistence: MOZART objects can be efficiently transformed into a so-called pickled form, which is a binary representation of the (possibly cyclic) data structure. This can be stored in a byte-string and efficiently read by the MOZART application effectively restoring the object. This feature makes it possible to represent complex objects (e.g. logical formulae) as Oz data structures, manipulate them in the MOZART engine, but at the same time store them as strings in the RDBMS. Moreover, the availability of "Ozlets" (MOZART functors) gives MBASE great flexibility, since the functionality of MBASE can be enhanced at run-time by loading remote functors. For instance complex data base queries can be compiled by a specialized MBASE client, sent (via the Internet) to the MBASE server and applied to the local data e.g. for specialized searching (see [Duc98] for a related system and the origin of this idea).

MBASE supports transparent distribution of data among several MBASE servers (see [KF00] for details). In particular, an object O residing on an MBASE server S can refer to (or depend on) an object O' residing on a server S'; a query to O that needs information about O' will be delegated to a suitable query to the server S'. We distinguish two kinds of MBASE servers depending on the data they contain: *archive servers* contain data that is referred to by other MBASEs, and *scratch-pad* MBASEs that are not referred to. To facilitate caching protocols, MBASE forces archive servers to be *conservative*, i.e. only such changes to the data are allowed, that the induced change on the corresponding logical theory is a conservative extension. This requirement is not a grave restriction: in this model errors are corrected by creating new theories (with similar presentations) shadowing the erroneous ones. Note that this restriction does not apply to the non-logical data, such as presentation or description information, or to scratchpad MBASEs making them ideal repositories for private development of mathematical theories, which can be submitted and moved to archive MBASEs once they have stabilized.

3   Interfaces

The primary interface language of MBASE is the XML-based markup language OMDOC [Koh00], a document-centered extension of the emerging OPENMATH standard [CC98] for mathematical objects. For instance the definition of a double function would be of the following form.

```
<definition id="double.def" item="double.sym" type="simple">
 <CMP xml:lang="eng">The doubling function defined by addition</CMP>
 <FMP><OMOBJ><OMBIND>
  <OMS cd="stlc" name="lambda"/>
  <OMBVAR><OMV name="X"/></OMBVAR>
  <OMA><OMS cd="arith1" name="plus"/><OMV name="X"/><OMV name="X"/></OMA>
 </OMBIND></OMOBJ></FMP>
</definition>
```

The CMP (commented mathematical property) element gives an informal characterization of the definition (which is a simple definition for the symbol with the identifier **double.sym** according to the attributes to the **definition** element) and the FMP (formal MP) gives the defining λ-term $\lambda X.(+XX)$ in OPENMATH representation. Note that the question of the semantics of such a term is determined by that of the symbols λ and $+$. These are specified in the MBASE theories given in the **cd** attributes of the OMS elements (the name of the symbol together with the theory establish unique reference in MBASE)

As a consequence of the XML-based approach it is possible to generate other logical formats from OMDOC by specifying simple XSL [Dea99] style sheets; in fact the transformation from OMDOC to the input formats of the ΩMEGA [BCF+97] and INKA [HS96] theorem provers is realized this way. It should be an easy exercise for most other concrete input formats. Furthermore one can generate customized OMDOC documents from MBASE, which can then be presented in one of the more standard presentation media (e.g. LaTeX or HTML/MATHML).

Generating OMDOC from a reasoning system is also quite simple in practice, since OMDOC has a relatively simple structure (fully specified in an XML document type definition [Koh00]) that closely follows the term structure of OPEN-MATH (using the OMS, OMV, OMA, OMBIND elements to describe formula trees made up of symbols, variables, applications and abstractions).

4   Conclusion, Evaluation, and Future Work

We have described the MBASE system, a distributed mathematical knowledge base, it can be obtained from http://www.mathweb.org/mbase. This system differs from other repositories of mathematical data such as the ISABELLE [Isa] or PVS [PVS] libraries in that it is an independent system not tied to a particular deduction system and offers inference services (matching, type-computation,...). The data format is not geared towards a particular application.

It is currently used by the ΩMEGA and INKA theorem provers for storing and sharing logical theories including theorems, definitions, tactics and methods. In particular, the MBASE service can be used as an ontology server fixing the semantics of mathematical objects used in protocols for deduction system integration. Furthermore, the MBASE system is used as the basis of an interactive personalized mathematics book (IDA [CCS99]). Here, the structure information contained in the MBASE version of the IDA data can be used to generate individualized sub-documents of IDA on the fly. While in the first case study the logical formulation of mathematical data is in the center of interest, in the second application textual representation plays a much more prominent role. MBASE supports both formats and even fosters their integration.

The current implementation uses the very simple file-based **gdbm** database system. This is sufficient for the amount of data currently available in ΩMEGA, INKA and IDA. Furthermore it offers a very flexible, open and portable programming base. A version of MBASE that uses ORACLE is currently under development.

Here a comparison to the MDB system [Har97] developed at the University of Erlangen is in order. MDB aims at supplying database support for the MIZAR libraries, and is based on an object-oriented extension of ORACLE. Unfortunately, already the first 13 (of more than 300) articles already need 500 MB disc space in ORACLE. Our division of labor that treats logical formulae in the programming language MOZART and relational, text and structural data in a DBMS pays off here. The size of the data base is only one order of magnitude larger than the size of the OMDOC encoding, which is comparable in size to the encodings used e.g. in ΩMEGA, ISABELLE, or PVS. As an example for relative sizes of representations in MBASE we consider the core theory library of ΩMEGA and the IDA text:

Relative sizes of representations in MBASE (MB)			
System	native	OMDOC	MBASE
ΩMEGA	0.61 (POST)	1.5	4.2
IDA	4.2 (LATEX)	5.0	9.3

Even when MBASE implementations based on industrial strength relational data base systems like e.g. ORACLE are available, we believe that the current gdbm-based implementation can still serve as a local development knowledge base and "proxy" system to ease the load on the central MBASE repository servers. Such a local system will probably also be better suited to support the operations necessary for changing definitions and axiomatizations during the development of a theory.

In the current version, we have not yet treated more advanced structuring concepts like theory morphisms, inheritance wrt. signature mappings, etc. that have been developed for structuring the knowledge base (see [KF00]). There remains much to be done in this direction, and we hope to adopt techniques from algebraic specification (see for instance [Hut99]).

References

[BCF+97] C. Benzmüller, L. Cheikhrouhou, D. Fehrer, A. Fiedler, X. Huang, M. Kerber, M. Kohlhase, K. Konrad, E. Melis, A. Meier, W. Schaarschmidt, J. Siekmann, and V. Sorge. ΩMEGA: Towards a mathematical assistant. In William McCune, editor, CADE'97, Springer LNAI 1249, pages 252–255, 1997.

[CC98] Olga Caprotti and Arjeh M. Cohen. The Open Math standard. The Open Math Society, http://www.nag.co.uk/projects/OpenMath/omstd/, 1998.

[CCS99] Arjeh Cohen, Hans Cuypers, and Hans Sterk. Algebra Interactive! Springer, 1999. Interactive Book on CD.

[Duc98] Denys Duchier. The NEGRA tree bank. Private communication, 1998.

[FHJ+99] Andreas Franke, Stephan M. Hess, Christoph G. Jung, Michael Kohlhase, and Volker Sorge. Agent-oriented integration of distributed mathematical services. Journal of Universal Computer Science, 5:156–187, 1999.

[FK99] Andreas Franke and Michael Kohlhase. System description: MATHWEB, an agent-based communication layer for distributed automated theorem proving. In H. Ganzinger, editor, CADE'99, Springer LNAI 1632, pages 217–221, 1999.

[Har97] Michael Hartmeier. Aufbau einer Datenbank f'ur mathematisches Wissen. Master Thesis, Friedrich-Alexander-Universität Erlangen-Nürnberg, 1997.

[HS96] Dieter Hutter and Claus Sengler. INKA - The Next Generation. In M.A. McRobbie and J.K. Slaney, editors, CADE'96, Springer LNAI 1104, pages 288–292, 1996.

[Hut99] Dieter Hutter. Reasoning about theories. Technical report, Deutsches Forschungszentrum für Künstliche Intelligenz (DFKI), 1999.

[Isa] The isabelle online theory library. Internet interface at http://www4.informatik.tu-muenchen.de/~isabelle/library-Isabelle98-1.

[KF00] Michael Kohlhase and Andreas Franke. Mbase: Representing knowledge and context for the integration of mathematical software systems. Journal of Symbolic Comutation, 2000. forthcoming.

[Koh00] Michael Kohlhase. OMDOC: Towards an OPENMATH representation of mathematical documents. Seki Report SR-00-02, Fachbereich Informatik, Universität des Saarlandes, 2000. http://www.mathweb.org/ilo/omdoc.

[PVS] The Pvs libraries. http://pvs.csl.sri.com/libraries.html.

[QED95] The QED manifesto. Internet Report http://www.cybercom.net/~rbjones/rbjpub/logic/quedres00.htm, 1995.

[Smo95] G. Smolka. The Oz programming model. In Jan van Leeuwen, editor, Computer Science Today, Volume 1000 of LNCS, pages 324–343. Springer, 1995.

[Dea99] Stephen DeachExtensible stylesheet language (xsl) specification. W3c working draft, W3C, 1999. Available at http://www.w3.org/TR/WD-xsl.

System Description: TRAMP
Transformation of Machine-Found Proofs into
Natural Deduction Proofs at the Assertion Level

Andreas Meier

Fachbereich Informatik, Universität des Saarlandes
66041 Saarbrücken, Germany
ameier@ags.uni-sb.de
http://www.ags.uni-sb.de/~ameier

Abstract. The TRAMP system transforms the output of several auto-mated theorem provers for first order logic with equality into natural deduction proofs at the assertion level. Through this interface, other systems such as proof presentation systems or interactive deduction systems can access proofs originally produced by any system interfaced by TRAMP only by adapting the assertion level proofs to their own needs.

1 Introduction

Today's theorem proving systems (automatic and interactive ones) have reached a considerable strength. However, it has become clear that no single system is capable of handling all sorts of deduction tasks. Therefore, it is a well-established approach to delegate subgoals to other (specialist) systems such as automated theorem provers (ATPs). Unfortunately, most ATPs use their own particular formalism. These machine-oriented formalisms make the proofs difficult to read. Hence, in order to make use of the results of the ATPs other systems need to adapt the output of an ATP to input, that they can further process. To minimize the transformation efforts it is advisable to use an interface that transforms the machine-found proofs of various formalisms into a uniform format.

Thereby, interactive deduction systems or proof presentation systems need a uniform format that they can easily transform into a presentation comprehensible to humans. Hence, a uniform format suitable for such systems should consist of intuitive steps and should be compact. Some approaches transform the machine-found proofs into natural deduction (ND) proofs [2,12]. But the resulting ND proofs suffer from the problem that they usually consist of a large number of low-level steps which are pure-syntactic manipulations of logical quantifiers and connectives. An approach to enhance these problems is to produce ND proofs at the assertion level [8]. The assertion level allows for human-oriented macro-steps justified by the application of theorems, lemmas, or definitions which are collectively called *assertions*. For instance, the assertion level step

$$\frac{F \subset G \quad c \in F}{c \in G} \text{DEF} \subset$$

derives the conclusion $c \in G$ by an application of the subset definition DEF$\subset$ — formalized by $\forall S_1.\forall S_2.(S_1 \subset S_2 \Leftrightarrow \forall x.(x \in S_1 \Rightarrow x \in S_2))$ — from the premises

D. McAllester (Ed.): CADE-17, LNAI 1831, pp. 460–464, 2000.
© Springer-Verlag Berlin Heidelberg 2000

$c \in F$ and $F \subset G$. A corresponding basic ND proof consists of a whole sequence of basic ND steps. Other work indicates that assertion level proofs are well suited as basic representation level for further human-oriented proof presentation [7].

In the following we describe the TRAMP system which can transform the output of several ATPs for first order logic with equality into ND proofs at the assertion level. Moreover, we give an example of an assertion level proof produced by TRAMP and discuss current applications of TRAMP and potential extensions.

2   The TRAMP System

TRAMP consists of three parts: (1) For each ATP interfaced by TRAMP there is a minor transformation process that can transform a *problem description* (consisting of a set of first order formulas, the assumptions, and one first order formula, the conclusion) into input suitable for this ATP. (2) At the heart is the transformation process that can transform a problem description and the corresponding output of an ATP into an ND proof at the assertion level. (3) A communication shell handles the access of the ATPs by TRAMP and the way other systems can reach TRAMP.

The transformation processes producing the inputs for the ATPs work all in the same manner: compute the clause normal form of the formulas of the problem description; then use these clauses to create an input file for an ATP. In the following we focus on the transformation of proofs at the assertion level and the integration in a networked proof development environment.

2.1   Proof Transformation at the Assertion Level

The transformation process takes as input a problem description and the corresponding output of an ATP. It produces an ND proof at the assertion level. The proof is created through three subprocesses, all embedded into one nutshell:

Structuring and Transforming into Refutation Graphs: The output of the ATP is structured by cutting off lemmas from the main proof. The resulting proof parts of the (remaining) main problem and the lemmas are each transformed into refutation graphs (refutation graphs are ground clause graphs representing refutation proofs [5]).

Transformation at the Assertion Level: Each refutation graph is transformed into an ND proof at the assertion level. These proofs are connected via the lemmas such that we obtain a single ND proof at the assertion level.

Optional Expansion of Assertion Steps: If requested by the user each assertion application can be expanded to a sequence of basic ND steps such that the resulting proof is a basic ND proof.

We enriched this basic transformation procedure with several heuristics, producing especially short and comprehensible proof parts and avoiding indirect parts. By structuring and transforming the output of the ATPs into refutation graphs we obtain an intermediate uniform representation of the different formalisms of the ATPs. To integrate further ATPs we require only appropriate extensions of this subprocess. We prefer refutation graphs as intermediate uniform representation because proofs in many other refutation based formalisms can be transformed easily into refutation graphs (e.g., a transformation algorithm for resolution proofs is described in [5]). Furthermore, the correspondences between the input clauses (which literals are contradictory?) are directly visible. The second subprocess consists of two phases: First, TRAMP decomposes the assumptions, the conclusion, and the refutation graphs until refutation graphs are reached consisting only of a sequence of steps which represent translatable assertion applications (TRAMP identifies assertion applications already in the refutation graphs). Then, in a second phase these refutation graphs are transformed by translating these steps successively into corresponding assertion applications in the ND proof. A detailed description of this algorithm which is an extension of [9] to refutation graphs with equality can be found in [11].

2.2   Integration in a Networked Proof Development Environment

All transformation processes are implemented in Allegro Common Lisp and run in one lisp process. This lisp process runs within a MATHWEB communication shell[6]. MATHWEB is a system for distributed automated theorem proving. Existing tools are equipped with a communication shell and are integrated into a networked proof development environment. Via MATHWEB, TRAMP can be reached by other MATHWEB services and can reach the ATPs which are also available as MATHWEB services.

As input TRAMP accepts problem descriptions in $\mathcal{POST}$ syntax [1]. Moreover, the user can feed TRAMP directly with the corresponding output of an ATP or can instruct TRAMP to access several ATPs to prove the input problem. Currently, TRAMP is able to produce the input and process the output of the ATPs SPASS, BLIKSEM, OTTER, WALDMEISTER[1], PROTEIN, and EQP (see [14] for references). When instructed to access ATPs TRAMP computes the inputs for the chosen ATPs and distributes these inputs via MATHWEB among the ATPs. The distributed ATPs run competitively. When an ATP is finished it sends its output via MATHWEB back to TRAMP which transforms the output into an ND proof at the assertion level. Thus, when instructed to access an ATP, TRAMP behaves for an external user like an ND-ATP.

As output TRAMP produces again $\mathcal{POST}$ syntax. A created ND proof is expressed in the linearized version first used in [2]. Thereby, a *ND line* consists of a finite set of formulas Δ, called the *hypotheses*, a single formula F, called *the conclusion*, and a justification $(\mathcal{R})$. Such a line is denoted as: $L.\Delta \vdash F \ (\mathcal{R})$ where L is a label for this line. Our set of basic ND rules is based on Gentzen's natural deduction calculus NK, but is enriched with further derived rules to obtain

---

[1] Except WALDMEISTER all ATPs interfaced by TRAMP are refutation based. However, TRAMP can transform the output of WALDMEISTER into refutation graphs by deriving a contradiction between the proved theorem $t = t'$ and its negation $t \neq t'$.

better comprehensible and more compact proofs. The justification 'application of assertion L_A on premises $L_{P_1}, \ldots, L_{P_n}$' is written as $(L_A \; L_{P_1} \; \ldots \; L_{P_n})$.

3 An Example

We apply TRAMP on the problem SET001 of the TPTP problem library [13]. The input for TRAMP is the following problem description:

Assumptions: $\forall S_1.\forall S_2.((S_1 = S_2) \Leftrightarrow ((S_1 \subset S_2) \wedge (S_2 \subset S_1)))$
$\qquad\qquad \forall S_1.\forall S_2.\forall x.((x \in S_1) \wedge (S_1 \subset S_2) \Rightarrow (x \in S_2))$
Conclusion: $\forall S_1.\forall S_2.\forall x.((x \in S_1) \wedge (S_1 = S_2) \Rightarrow (x \in S_2))$,

TRAMP computes the input for an ATP and applies the ATP via MATHWEB. Then, it transforms its output into the following refutation graph G:

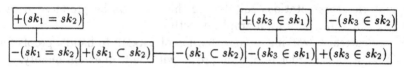

sk_1, sk_2, sk_3 are skolem constants. Afterwards, TRAMP transforms G together with the input problem description into the following assertion level ND proof:

L_1.	L_1	$\vdash \forall S_1.\forall S_2.((S_1 = S_2) \Leftrightarrow ((S_1 \subset S_2) \wedge (S_2 \subset S_1)))$	(Hyp)
L_2.	L_2	$\vdash \forall S_1.\forall S_2.\forall x.((x \in S_1) \wedge (S_1 \subset S_2) \Rightarrow (x \in S_2))$	(Hyp)
L_3.	L_3	$\vdash (c \in F) \wedge (F = G)$	(Hyp)
L_4.	L_3	$\vdash c \in F$	($\wedge E \; L_3$)
L_5.	L_3	$\vdash F = G$	($\wedge E \; L_3$)
L_6.	L_1, L_3	$\vdash F \subset G$	($L_1 \; L_5$)
L_7.	L_1, L_2, L_3	$\vdash c \in G$	($L_2 \; L_6 \; L_4$)
L_8.	L_1, L_2	$\vdash (c \in F) \wedge (F = G) \Rightarrow (c \in G)$	($\Rightarrow I \; L_3 \; L_7$)
L_9.	L_1, L_2	$\vdash \forall x.((x \in F) \wedge (F = G) \Rightarrow (x \in G))$	($\forall I \; L_8$)
L_{10}.	L_1, L_2	$\vdash \forall S_2.\forall x.((x \in F) \wedge (F = S_2) \Rightarrow (x \in S_2))$	($\forall I \; L_9$)
L_{11}.	L_1, L_2	$\vdash \forall S_1.\forall S_2.\forall x.((x \in S_1) \wedge (S_1 = S_2) \Rightarrow (x \in S_2))$	($\forall I \; L_{10}$)

In this proof the lines L_6 and L_7 are justified by assertion applications.

4 Experience, Discussion, and Future Work

We have tested the implementation of TRAMP on about 100 examples from the TPTP problem library which can be proved by the ATPs interfaced by TRAMP. Furthermore, TRAMP is used permanently by the systems PROVERB [10] and ΩMEGA [3]. PROVERB, a proof presentation system, uses TRAMP to obtain assertion level proofs that it can translate into natural language proofs. ΩMEGA, an interactive mathematical assistant system, calls the ATPs via TRAMP on open goals in its proof object. Then ΩMEGA integrates the proofs provided by TRAMP into its own proof.

Our experiments show that, for a problem containing some applicable assertions, the length of the assertion proof is typically about half the length of the basic ND proof that results from expanding the abstract assertion steps (e.g., the assertion proof in Sec. 3 consists of 11 lines whereas the corresponding basic ND proof consists of 21 lines). Although pure equality proofs such as produced by EQP and WALDMEISTER are neither shortened nor abstracted by using TRAMP, we use these systems to push the solvability horizon of ATPs interfaced by TRAMP. Not unexpectedly, transforming proofs to the assertion level requires considerable computational effort for larger proofs. From our experience, this effort is justified for systems that need a human-oriented representation of the machine-found proofs such as interactive deduction systems and proof presentation systems. Such systems can adapt easily the resulting assertion level proofs to their own needs for the following reasons: (1) The resulting assertion level proofs are significantly more compact than basic ND proofs. (2) They contain meaningful steps but hardly indirect parts. (3) Each assertion application can be expanded to a sequence of basic ND steps when a more detailed derivation is needed. Otherwise, for the communication between fully automatic systems other uniform representations suitable for this task can be produced with less effort (e.g., by using refutation graphs directly).

We are currently working on an extension of TRAMP to handle proofs found by LEO [4], a higher order resolution prover with built-in extensionality.

The current version of TRAMP is available at http://www.ags.uni-sb.de/~ameier/tramp.html. Soon there will be also a web interface for TRAMP.

References

1. $\mathcal{POST}$. See at http://www.ags.uni-sb.de/~omega/primer/post.html, 1999.
2. P. B. Andrews. Transforming matings into natural deduction proofs. In *Proc. of CADE-5*, pages 281–292, 1980.
3. C. Benzmüller *et al.* ΩMEGA: Towards a mathematical assistant. In *Proc. of CADE-14*, pages 252–255, 1997.
4. C. Benzmüller and M. Kohlhase. LEO, a higher order theorem prover. In *Proc. of CADE-15*, pages 139–144, 1998.
5. N. Eisinger. *Completeness, confluence, and related properties of clause graph resolution*. PhD thesis, Universität Kaiserslautern, Germany, 1988.
6. A. Franke and M. Kohlhase. MATHWEB, an agent-based communication layer for distributed automated theorem proving. In *Proc. CADE-16*, pages 217–221, 1999.
7. H. Horacek. Presenting proofs in a human-oriented way. In *Proc. of CADE-16*, pages 142–156, 1999.
8. X. Huang. Reconstructing proofs at the assertion level. In *Proc. of CADE-12*, pages 738–752, 1994.
9. X. Huang. Translating machine-generated resolution proofs into ND-proofs at the assertion level. In *Proc. of PRICAI-96*, pages 399–410, 1996.
10. X. Huang and A. Fiedler. Presenting machine-found proofs. In *Proc. of CADE-13*, pages 221–225, 1996.
11. A. Meier. Transformation of machine-found proofs into assertion level proofs. Technical report, 2000. Avaible at http://www.ags.uni-sb.de/~ameier/tramp.html.
12. F. Pfenning. *Proof transformation in higher-order logic*. PhD thesis, CMU, Pittsburgh, Pennsylvania, USA, 1987.
13. G. Sutcliffe *et al.* The TPTP problem library. In *Proc. of CADE-12*, pages 252–266, 1994.
14. G. Sutcliffe and C. Suttner. The results of the CADE-13 ATP system competition. *Journal of Automated Reasoning*, 18(2):259–264, 1997.

On Unification for Bounded Distributive Lattices

Viorica Sofronie-Stokkermans

Max-Planck-Institut für Informatik
Im Stadtwald, D-66123 Saarbrücken, Germany
sofronie@mpi-sb.mpg.de

Abstract. We give a resolution-based procedure for deciding unifiability in the variety of bounded distributive lattices. The main idea is to use a structure-preserving translation to clause form to reduce the problem of testing the satisfiability of a unification problem S to the problem of checking the satisfiability of a set Φ_S of (constrained) clauses. These ideas can be used for unification with free constants and for unification with linear constant restrictions. Complexity issues are also addressed.

1 Introduction

From an algebraic point of view, unification can be seen as solving (systems of) equations in the initial or free algebra of an equational theory. Apart from its theoretical interest, unification is used e.g. in resolution-based theorem proving and in term rewriting to deal with certain equational axioms (such as associativity and commutativity). The unification problem has been thoroughly studied for equationally defined theories characterized by axioms such as associativity, commutativity, distributivity, associativity-commutativity, associativity-commutativity-idempotency; and for several theories related to algebra (Abelian groups, commutative and Boolean rings, semilattices, Boolean algebras, primal algebras, discriminator varieties). For details cf. [5] and the bibliography cited there. The combination of unification algorithms has been studied in [4].

In this paper we present some results on unification in the equational theory of bounded distributive lattices. The study was motivated, on the one hand, by our interest in distributive lattices with operators, and, on the other hand, by the fact that unification problems in semilattice- and lattice-based structures are becoming of increasing interest in computer science (we mention e.g. the results of Baader and Narendran on unification of concept terms in description logics [2]; similar possible applications in set constraints may also be of interest).

It is known that the class D_{01} of bounded distributive lattices has an undecidable first-order theory, but both its universal theory and its positive $\forall\exists$ theory (hence the unification problem with free constants) are decidable. Unification for distributive lattices has only been addressed in a few papers. In [12], Gerhard and Petrich give a criterion for unifiability (with free constants) of *two* terms in the theory of distributive lattices. (We were not able to generalize the argument used in the proof of this result to handle conjunctions of equations.) Then, in the attempt to give a basis set for all unifiers of two terms, they considered

D. McAllester (Ed.): CADE-17, LNAI 1831, pp. 465–481, 2000.
© Springer-Verlag Berlin Heidelberg 2000

terms containing only one of the lattice operations $\vee$ or $\wedge$ and, for more general terms, only particular cases, containing few variables. The results of Ghilardi [13] show that the equational class D of distributive lattices has unification type 0, i.e. there exist D-unification problems with no minimal complete set of unifiers. We are not aware of any other results on unification for distributive lattices, e.g. concerning its complexity. Due to the interaction between operators, neither the ideas used in [19] for distributive unification, nor the results in [4] on the combination of unification algorithms can be applied in this case.

In [20], we gave a resolution-based decision procedure for the universal theory of certain varieties of distributive lattices with operators. The arguments in [20] cannot be used for the positive $\forall\exists$ theory of such varieties without modification. In this paper we further develop the ideas in [20] and show that the use of the Priestley representation for bounded distributive lattices allows us to give an algorithm based on resolution (with constrained clauses) for the unification problem in D_{01}. The algorithm consists of the following steps:

1. *Structure-preserving translation to clause form:* testing the satisfiability of a unification problem S is reduced to the problem of checking the satisfiability of a set Φ_S of clauses. Expressing Φ_S as a set of constrained clauses further simplifies the representation of the problem.

2. *Ordered resolution with selection* for (constrained) clauses is used for testing the satisfiability of Φ_S.

We also show that similar ideas can be used for unification with linear constant restrictions [4]. These results complete and improve the results in [12].

The main advantage of our approach is that the structure-preserving translation to clause form makes it much easier to treat the unification problem for bounded distributive lattices, by using results in resolution theory. As a byproduct, using Prop. 5.6 in [3], our results show that resolution (for ground clauses without equality) can be used for deciding the positive theory of D_{01}.

It seems that many of the results in this paper can be extended without difficulties to other varieties in which the free algebras have a description similar to those in D_{01}. This is the case for many subvarieties of the variety of Ockham algebras (bounded distributive lattices with a lattice antimorphism), such as, e.g., the variety of De Morgan algebras. For the sake of simplicity, in this paper we restrict our attention to the class of bounded distributive lattices only.

The paper is structured as follows. Section 2 contains the background information needed in the paper. Section 3 contains generalities about the unification problem for bounded distributive lattices. In Section 4 we give a resolution-based algorithm for this problem, and an extension to unification with linear constant restrictions. Section 5 contains conclusions and plans for future work.

2   Preliminaries

2.1   Algebra

Let Σ be a signature and $a : \Sigma \to \mathbb{N}$ an arity function. A Σ-algebra is a structure $\mathbf{A} = (A, \{\sigma_A\}_{\sigma \in \Sigma})$, where A is a non-empty set and for every $\sigma \in \Sigma$, $\sigma_A : A^{a(\sigma)} \to A$. We denote by $T_\Sigma(X)$ the term algebra over Σ in the variables X. An *equation* is an expression of the form $t_1 = t_2$ where $t_1, t_2 \in T_\Sigma(X)$. A Σ-algebra $\mathbf{A} = (A, \{\sigma_A\}_{\sigma \in \Sigma})$ *satisfies* an equation $t_1 = t_2$ if t_1 and t_2 become equal for every substitution of elements in A for the variables. An equational class is the class of all algebras that satisfy a set of equations. If E is a set of equations in the signature Σ, then $F_\Sigma^E(X) := T_\Sigma(X)/\equiv_E$ is the free algebra over X in the equational class of all algebras that satisfy E (where $\equiv_E$ is the Σ-congruence on $T_\Sigma(X)$ generated by E). A *system of equations* is a finite set of equations $\mathcal{S} : \{s_1 = t_1, \ldots, s_k = t_k\}$, where $s_i, t_i \in T_\Sigma(X)$ for every $1 \leq i \leq k$. Let $\{y_1, \ldots, y_n\} \subseteq X$ be the set of all variables in $\mathcal{S}$. An algebra $\mathbf{A} = (A, \{\sigma_A\}_{\sigma \in \Sigma})$ satisfies the existential closure, $\exists y_1, \ldots y_n (s_1 = t_1 \land \cdots \land s_k = t_k)$, of $\mathcal{S}$ if there exists a map $h : X \to A$ such that $\overline{h}(s_i) = \overline{h}(t_i)$ for every $1 \leq i \leq k$, where $\overline{h} : T_\Sigma(X) \to \mathbf{A}$ is the unique homomorphism of Σ-algebras that extends h.

2.2   Lattice Theory

For the definition of partially-ordered set and order-filter we refer to [10]. If $\mathbf{X} = (X, \leq)$ is a partially-ordered set, we denote its set of order-filters by $\mathcal{O}(\mathbf{X})$. There is a bijective correspondence between $\mathcal{O}(\mathbf{X})$ and the set of all order-preserving maps from X to the partially-ordered set $\underline{\mathbf{2}} = (\{0, 1\}, \leq)$, where $0 < 1$. A structure $\mathbf{L} = (L, \vee, \wedge)$, where L is a non-empty set and $\vee$ and $\wedge$ are two binary operations on L is a *lattice* if $\vee$ and $\wedge$ are associative, commutative and idempotent and satisfy the absorption laws. A *distributive lattice* is a lattice that satisfies either of the distributive laws. A lattice $\mathbf{L} = (L, \vee, \wedge)$ has a *first element* if there is an element $0 \in L$ such that $0 \leq x$ for every $x \in L$; it has a *last element* if there is an element $1 \in L$ such that $x \leq 1$ for every $x \in L$ (where $x \leq y$ iff $x \vee y = y$). A lattice having both a first and a last element is called *bounded*. In what follows, when we refer to bounded distributive lattices, the first and last element are supposed to be included in the signature. Thus, a bounded distributive lattice is a structure $\mathbf{L} = (L, \vee, \wedge, 0, 1)$, where $(L, \vee, \wedge)$ is a distributive lattice and $0, 1$ are constants such that 0 is first element and 1 last element in $(L, \vee, \wedge)$. We denote the equational class of all bounded distributive lattices by $\mathbf{D}_{01}$. $\mathbf{D}_{01}$ contains e.g. the two-element bounded lattice, $\mathbf{2} = (\{0, 1\}, \vee, \wedge, 0, 1)$, where $0 \vee 1 = 1, 0 \wedge 1 = 0$.

2.3   Priestley Representation

If $\mathbf{L}$ is a bounded distributive lattice, let $D(\mathbf{L}) := \mathsf{Hom}_{\mathbf{D}_{01}}(\mathbf{L}, \mathbf{2})$ be set of all $0,1$-lattice homomorphisms from $\mathbf{L}$ to the two-element bounded distributive lattice. The space $\mathbf{D}(\mathbf{L}) = (D(\mathbf{L}), \leq, \tau)$, where $\leq$ is the pointwise ordering on maps and τ is the topology generated by all sets of the form $X_a = \{h \in D(L) \mid h(a) = 1\}$ and their complements as a subbasis, is called the *Priestley dual* of $\mathbf{L}$. Let $\mathsf{Hom}_P(\mathbf{D(L)}, \underline{2})$ be the lattice of all continuous and order-preserving maps between the ordered topological space $\mathbf{D(L)}$, and the two-element partially-ordered set $\underline{2}$ with the discrete topology. Priestley [18] showed that for every $\mathbf{L} \in \mathsf{D}_{01}$, $\mathbf{L}$ is isomorphic to $\mathsf{Hom}_P(\mathbf{D(L)}, \underline{2})$. In particular, if $\mathbf{L}$ is finite, then τ is the discrete topology, so $\mathbf{L}$ is isomorphic to $(\mathcal{O}(\mathbf{D(L)}), \cup, \cap, \emptyset, D(\mathbf{L}))$.

The dual of a finite distributive lattice is much smaller and less complex than the lattice itself. Therefore, problems concerning finite distributive lattices are likely to become simpler when translated into problems about their duals. We illustrate this by comparing the free algebra in D_{01} over a finite set C, $F_{\mathsf{D}_{01}}(C)$, and its Priestley dual $\mathbf{D}(F_{\mathsf{D}_{01}}(C))$. The theorem below is well-known.

Theorem 1. *Let C be a finite set. The following statements hold:*

(1) The map $p_C : (D(F_{\mathsf{D}_{01}}(C)), \leq) \to (2^C, \leq)$ defined for every $h \in D(F_{\mathsf{D}_{01}}(C))$ by $p_C(h) = h_{|C}$ (the restriction of $h : F_{\mathsf{D}_{01}}(C) \to \mathbf{2}$ to C) is an order-isomorphism, where in both cases the order is defined pointwise.

(2) The map $\eta_C : F_{\mathsf{D}_{01}}(C) \to \mathcal{O}(2^C, \leq)$ defined for every $t \in F_{\mathsf{D}_{01}}(C)$ by $\eta_C(t) = \{f : C \to \{0,1\} \mid \overline{f}(t) = 1\}$ (where for every $f : C \to \{0,1\}$, $\overline{f} : F_{\mathsf{D}_{01}}(C) \to \mathbf{2}$ is the unique extension of f to a 0,1-lattice homomorphism) is a lattice isomorphism. Its inverse is defined for every $U \in \mathcal{O}(2^C, \leq)$ by $\eta_C^{-1}(U) = \bigvee_{f \in U} (\bigwedge_{\{c \mid f(c)=1\}} c)$.

Every member of $F_{\mathsf{D}_{01}}(C)$ can be written as a finite join of finite meets of elements in C. Hence, $F_{\mathsf{D}_{01}}(C)$ is finite, and its number of elements is bounded by $2^{2^{|C|}}$. $|F_{\mathsf{D}_{01}}(C)|$ has been computed only for small values of $|C|$. By Theorem 1(1), $(D(F_{\mathsf{D}_{01}}(C)), \leq)$ is order-isomorphic to $(\mathcal{P}(C), \subseteq)$, hence has $2^{|C|}$ elements.

The main idea of this paper relies on this remark. The relatively simple structure of $D(F_{\mathsf{D}_{01}}(C))$ allows us to define a more efficient method for checking the satisfiability of unification problems with constants compared with methods that use the structure of $F_{\mathsf{D}_{01}}(C)$ and/or equational reasoning.

2.4   Unification

We present the definitions and results on E-unification needed in the paper.

Definition 1. *Let E be an equational theory, Σ its signature, and Δ a signature containing Σ. Let $\mathcal{S} : \{s_1 = t_1, \ldots, s_k = t_k\}$ be a system of equations, where $s_i, t_i \in T_\Delta(Y)$. Then $\mathcal{S}$ defines an E-unification problem over Δ. $\mathcal{S}$ is elementary iff $\Delta \subseteq \Sigma$; $\mathcal{S}$ is an E-unification problem with (free) constants iff $\Delta \backslash \Sigma$ is a set of constant symbols; and $\mathcal{S}$ is an E-unification problem with linear constant restrictions iff it is an E-unification problem with constants and, in addition, a linear ordering $<$ on the variables and free constants occurring in $\mathcal{S}$ is given. In a general E-unification problem $\Delta \backslash \Sigma$ may contain arbitrary function symbols.*

Definition 2. *A unification problem S has a solution w.r.t. E if there is a substitution $\sigma : Y \to T_\Delta(Y)$ such that $\sigma(s_i) \equiv_E \sigma(t_i)$ for every $1 \le i \le k$.*

If S is an E-unification problem with linear constant restrictions, a solution for S is a substitution $\sigma : Y \to T_\Delta(Y)$ with the additional property that for every variable $y \in Y$ and every constant c, if $y < c$ then c does not occur in $\sigma(y)$.

In this context, one can study decidability of unifiability, the existence of unifiers, their classification according to "generality", or the possibility of determining minimal sets of unifiers which are complete, in the sense that all other unifiers are less general. In this paper we focus on testing unifiability. This is sufficient in many applications (e.g. in constraint-based approaches to automated deduction [9, 17, 16]) and is often simpler than computing complete sets of unifiers.

Theorem 2. *For any E-unification problem $S : \{s_1 = t_1, \ldots, s_k = t_k\}$ with free constants in C and variables $Y = \{y_1, \ldots, y_m\}$ the following are equivalent:*

(1) S has a solution w.r.t. E.

(2) The formula $\exists y_1, \ldots, y_m (s_1 = t_1 \wedge \cdots \wedge s_k = t_k)$ is true in $F_{\Sigma \cup C}^E(\emptyset)$.

(3) There exists $h : Y \to F_\Sigma^E(C)$ such that $\overline{h}(s_i) = \overline{h}(t_i)$ for every $1 \le i \le k$, where $\overline{h} : T_\Sigma(Y \cup C) \to F_\Sigma^E(C)$ is the unique extension of h to a homomorphism, such that, for all $c \in C$, $\overline{h}(c) = [c]$ (where $[c]$ is the equivalence class of c in $F_\Sigma^E(C)$).

Proof: (Sketch) The equivalence of (1) and (2) is proved e.g. in [7]. The equivalence of (2) and (3) follows from the fact that $U(F_{\Sigma \cup C}^E(Y))$ is isomorphic to $F_\Sigma^E(Y \cup C)$, where U maps a $\Sigma \cup C$-algebra to a Σ-algebra by forgetting the constants in the signature, i.e. $U((A, \{\sigma_A\}_{\sigma \in \Sigma \cup C})) = (A, \{\sigma_A\}_{\sigma \in \Sigma})$. □

The importance of E-unification with linear constant restrictions is justified by the following theorem.

Theorem 3 ([5, 3]). *Let E be a non-trivial equational theory. The following statements are equivalent:*

(1) The positive theory[1] of E is decidable.

(2) General E-unification is decidable.

(3) E-unification with linear constant restrictions is decidable.

More precisely, as pointed out e.g. in [1], the decision problem for E-unification with linear constant restrictions can be reduced to the decision problem for general unification in linear time. The nondeterministic polynomial algorithm given in [3] can be used to reduce the decision problem for general unification to the decision problem for E-unification with linear constant restrictions.

---

[1] The positive theory of E is the collection of those closed formulae valid in the class of all models of E which are (equivalent to a formula) of the form $(Q_1 x_1) \ldots (Q_m x_m)(\bigvee_{i=1}^q (s_{i1} = t_{i1} \wedge \cdots \wedge s_{in_i} = t_{in_i}))$, where $Q_1, \ldots, Q_m \in \{\exists, \forall\}$.

3   Unification with Constants in D_{01}. General Remarks.

We now study the unification problem with free constants in the equational class D_{01} of bounded distributive lattices. We denote by D_{01} the equational theory of D_{01}. Let $\Sigma = \{\vee, \wedge, 0, 1\}$ be the signature of bounded distributive lattices. The following result is a direct consequence of Theorem 2.

Corollary 1. *For any D_{01}-unification problem $S : \{s_1 = t_1, \ldots, s_k = t_k\}$, with free constants in C and variables Y, the following are equivalent:*

(1) S has a solution w.r.t. D_{01}.

(2) There exists $h : Y \to F_{D_{01}}(C)$ such that $\overline{h}(s_i) = \overline{h}(t_i)$ for every $1 \leq i \leq k$, where $\overline{h} : T_\Sigma(Y \cup C) \to F_{D_{01}}(C)$ is the unique extension of h to a homomorphism, such that $\overline{h}(c) = [c]$ for all $c \in C$.

By Corollary 1 and the fact that $F_{D_{01}}(C)$ is finite for every finite C it follows that D_{01}-unification with free constants is decidable. This problem is co-NP hard: if S contains only one equation and no variables it reduces to the word problem for D_{01}, which has been shown to be co-NP hard [14].

We first present a straightforward (and rather inefficient) method for testing whether a unification problem has a solution. We then show how a simpler case (only one equation) is solved in [12]. In Section 4 we give a simpler method, which allows to test by resolution whether a unification problem has a solution.

The Straightforward Method. Let $S : \{s_1 = t_1, \ldots, s_k = t_k\}$ be a D_{01}-unification problem with free constants in a finite set C and variables in the finite set Y. We can check if S has a solution by checking if there is an instantiation of the variables in S with elements in $F_{D_{01}}(C)$ that satisfies S. There exist at most $(2^{2^{|C|}})^{|Y|}$ such instantiations. For each instantiation $h : Y \to F_{D_{01}}(C)$, one has to check if $h(s_i) \equiv_{D_{01}} h(t_i)$, $1 \leq i \leq k$. There exists an algorithm for disproving the equivalence of two terms which is nondeterministically polynomial in the length of the terms [14]. The elements in $F_{D_{01}}(C)$ can be written as disjunctions of conjunctions of elements in C; the length of such a term is at most $|C| \cdot 2^{|C|}$. Hence, the length of $h(s_i)$ and $h(t_i)$, $1 \leq i \leq k$, can at most be $|Y| \cdot |C| \cdot 2^{|C|} + \max(S)$, where $\max(S)$ is the maximal length of a term occurring in S.

A Special Case. In [12] Gerhard and Petrich present the following criterion for unifiability for one single equation, i.e. for the unification problem $S : \{s = t\}$.

1. Let s' and t' be the disjunctive normal forms of s resp. t.
2. If neither s' or t' has a constant term[2] then s and t are unifiable.
3. If s' or t' has constant terms, let $h : Y \to T_{D_{01}}(C)$ be defined by $h(x) = D$ for every $x \in Y$, where D is the disjunction of all constant terms in s' and t'.
4. If $\overline{h}(s) \equiv_{D_{01}} \overline{h}(t)$ then s and t are unifiable, otherwise they are not unifiable.

---

[2] If $s = \bigvee_i \bigwedge_{j \in I_i} s_j$ is in disjunctive normal form, then a constant term of s is any of the conjunctions $\bigwedge_{j \in I_i} s_j$ in s not containing any variable.

The disjunction D in Step 3 can be determined in polynomial time w.r.t. $\text{length}(s') + \text{length}(t')$. The same holds for the process of replacing every variable in s and t by D. Both the length of D and the length of the result of replacing all variables in s, t by D ($\overline{h}(s)$ resp. $\overline{h}(t)$) is polynomial in $\text{length}(s') + \text{length}(t')$; but may be exponential in $\text{length}(s) + \text{length}(t)$. Hence, the complexity of the criterion above is given by the complexity of Step 1 (computing the disjunctive normal forms of s and t) and Step 4 (solving a word problem). The last problem is co-NP complete [14]; there exists an algorithm for disproving the equivalence of $\overline{h}(s)$ and $\overline{h}(t)$ which is nondeterministically polynomial in $\text{length}(\overline{h}(s)) + \text{length}(\overline{h}(t))$.

4 A Resolution-Based Algorithm

We give a simpler algorithm for the problem of deciding whether a D_{01}-unification problem with free constants has a solution. The algorithm consists of two steps:

1. **Structure-preserving translation to clause form:**
   - Reduce the problem of testing the satisfiability of a unification problem S to checking the satisfiability of a set Φ_S of clauses.
   - Show that Φ_S can be expressed as a set of constrained clauses.

2. **Checking satisfiability by ordered resolution with selection:**
   - Use ordered resolution with selection for constrained sets of clauses to test the satisfiability of Φ_S.

4.1 Structure-Preserving Translation to Clause Form

In this section we reduce the problem of testing the satisfiability of a unification problem S to that of checking the satisfiability of a set of clauses. We do this in two steps: Theorem 4 shows that $F_{D_{01}}(C)$ can be replaced with the lattice of order-filters of $(\mathcal{P}(C), \subseteq)$; Theorem 5 further reduces the problem to that of checking the satisfiability of a set of first-order (ground) clauses.

Theorem 4. *For any D_{01}-unification problem $S : \{s_1 = t_1, \ldots, s_k = t_k\}$ with free constants C and variables Y, the following are equivalent:*

(1) S has a solution w.r.t. D_{01}.

(2) There exists $h : Y \to F_{D_{01}}(C)$ such that $\overline{h}(s_i) = \overline{h}(t_i)$ for every $1 \leq i \leq k$, where $\overline{h} : T_\Sigma(Y \cup C) \to F_{D_{01}}(C)$ is the unique homomorphism that extends h, such that $\overline{h}(c) = [c]$ for all $c \in C$.

(3) There exists $g : Y \to \mathcal{O}(\mathcal{P}(C), \subseteq)$ such that $\overline{g}(s_i) = \overline{g}(t_i)$ for every $1 \leq i \leq k$, where $\overline{g} : T_\Sigma(Y \cup C) \to \mathcal{O}(\mathcal{P}(C), \subseteq)$ is the unique homomorphism that extends g, such that $\overline{g}(c) = \uparrow\{c\} = \{X \subseteq C \mid c \in X\}$ for every $c \in C$.

Proof: (Idea) The equivalence of (1) and (2) follows directly from Corollary 1. The equivalence of (2) and (3) follows from the fact that there exists a 0,1-lattice isomorphism $\eta_C : F_{D_{01}}(C) \to \mathcal{O}(\mathcal{P}(C), \subseteq)$ defined for every $t \in F_{D_{01}}(C)$ by

$\eta_C(t) = \{f^{-1}(1) \cap C \mid f : F_{D_{01}}(C) \to 2 \text{ is a } 0,1\text{-lattice homomorphism}; f(t) = 1\}$
such that for every $c \in C$, $\eta_C([c]) = \uparrow\{c\}$.   □

Theorem 4 justifies a reduction of the problem of checking whether a unification problem with constants S has a solution to the problem of checking the satisfiability of a system of set constraints. This reduction can be then used to give a structure-preserving translation to clause form. Thus, the problem of checking whether a unification problem with constants S has a solution can be reduced to the problem of checking the satisfiability of a set of clauses.

The structure-preserving translation to clause form is inspired by Tseitin's well-known method for transforming quantifier-free formulae to clausal normal form and by the ideas in [20]. The link with set constraints mentioned above also explains the similarities with the structure-preserving translation to clause form presented in [6]. The remarks above are formally expressed by the following theorem.

Theorem 5. *Let $S : \{s_1 = t_1, \ldots, s_k = t_k\}$ be a D_{01}-unification problem with free constants C, and variables $Y = \{y_1, \ldots, y_n\}$. Let $ST(S)$ be the set of all subterms of terms occurring in S. The following are equivalent:*

(1) *There exists $h : \{y_1, \ldots, y_n\} \to \mathcal{O}(\mathcal{P}(C), \subseteq)$ such that $\overline{h}(s_i) = \overline{h}(t_i)$ for every $1 \le i \le k$, where $\overline{h} : T_\Sigma(Y \cup C) \to \mathcal{O}(\mathcal{P}(C), \subseteq)$ is the unique homomorphism that extends h, such that $\overline{h}(c) = \uparrow\{c\}$ for every $c \in C$.*

(2) *There exists a family $\{I_e\}_{e \in ST(S)}$, such that $I_e \subseteq \mathcal{P}(C)$ for all $e \in ST(S)$, and for all $X, X_1, X_2 \subseteq C$ the following hold:*
  - *if $X_1 \in I_y$ and $X_1 \subseteq X_2$ then $X_2 \in I_y$, for every $y \in \{y_1, \ldots, y_n\}$;*
  - *$X \in I_{e_1 \wedge e_2}$ iff $X \in I_{e_1}$ and $X \in I_{e_2}$;*
  - *$X \in I_{e_1 \vee e_2}$ iff $X \in I_{e_1}$ or $X \in I_{e_2}$;*
  - *$I_0 = \emptyset$; $I_1 = C$; and for every $c \in C$, $X \in I_c$ iff $c \in X$;*
  - *$X \in I_{s_i}$ iff $X \in I_{t_i}$ for all $1 \le i \le k$.*

(3) *The conjunction of the following formulae is satisfiable:*

(Her)	$P_y(X_1) \to P_y(X_2)$	for all $X_1 \subseteq X_2 \subseteq C$, $y \in \{y_1, \ldots, y_n\}$
(Ren) (∧n)	$P_{e_1 \wedge e_2}(X) \to P_{e_i}(X)$	for all $X \subseteq C, i = 1, 2$
(∧p)	$P_{e_1}(X) \wedge P_{e_2}(X) \to P_{e_1 \wedge e_2}(X)$	for all $X \subseteq C$
(∨n)	$P_{e_1 \vee e_2}(X) \to P_{e_1}(X) \vee P_{e_2}(X)$	for all $X \subseteq C$
(∨p)	$P_{e_i}(X) \to P_{e_1 \vee e_2}(X)$	for all $X \subseteq C, i = 1, 2$
(1)	$P_1(X)$	for all $X \subseteq C$
(0)	$\neg P_0(X)$	for all $X \subseteq C$
(cp)	$P_c(X)$	for all $X \subseteq C$ with $c \in X$
(cn)	$\neg P_c(X)$	for all $X \subseteq C$ with $c \notin X$
(P)	$P_{s_i}(X) \leftrightarrow P_{t_i}(X)$,	for all $X \subseteq C$, for all $1 \le i \le k$

where each formula in (Her) $\cup$ (Ren) $\cup$ (P) is the conjunction of all formulae obtained by instantiating the variables X, resp. X_1, X_2 with subsets of C satisfying the additional conditions; the indices $e_1 \vee e_2, e_1 \wedge e_2, 0, 1, c$ range over all elements in $ST(S)$; y ranges over all variables in $\{y_1, \ldots, y_n\}$.

Proof: (Sketch) (1) $\Rightarrow$ (2). For every $e \in ST(\mathcal{S})$ let $I_e := \overline{h}(e)$. Since $\overline{h}$ is a 0,1-homomorphism with $\overline{h}(c) = \uparrow\{c\}$, and the lattice operations in $\mathcal{O}(\mathcal{P}(C), \subseteq)$ are union and intersection, the family $\{I_e\}_{e \in ST(\mathcal{S})}$ satisfies the conditions in (2).

(2) $\Rightarrow$ (3) Let $\{I_e\}_{e \in ST(\mathcal{S})}$ be a family satisfying the conditions in (2). Then $(\mathcal{P}(C), \mathcal{I})$, where $\mathcal{I}(P_e) := I_e$ for all $e \in ST(\mathcal{S})$, is a model for (Her)$\cup$(Ren)$\cup$(P).

(3) $\Rightarrow$ (1) Assume that (Her)$\cup$(Ren)$\cup$(P) (a conjunction of ground clauses) is satisfied by the map $\mathcal{I} : \{P_e(X) \mid e \in ST(\mathcal{S}), X \subseteq C\} \to \{0, 1\}$. For every $y \in Y$ let $h(y) := \{X \in \mathcal{P}(C) \mid \mathcal{I}(P_y(X)) = 1\}$. Let $\overline{\overline{h}} : T_\Sigma(Y \cup C) \to \mathcal{O}(\mathcal{P}(C), \subseteq)$ be the unique homomorphism that extends h, such that $\overline{h}(c) = \uparrow\{c\}$ for every $c \in C$. As $\mathcal{I}$ satisfies (Her) $\cup$ (Ren), $\overline{h}(e) = \{X \in \mathcal{P}(C) \mid \mathcal{I}(P_e(X)) = 1\}$ for all $e \in ST(\mathcal{S})$. Since $\mathcal{I}$ satisfies (P), $\overline{h}(s_i) = \overline{h}(t_i)$ for every $1 \leq i \leq k$. $\square$

Corollary 2. *The D_{01}-unification problem $\mathcal{S} : \{s_1 = t_1, \ldots, s_k = t_k\}$ with free constants C has a solution w.r.t. the equational theory of D_{01} iff the set of clauses* (Her) $\cup$ (Ren) $\cup$ (P) *is satisfiable.*

The satisfiability of (Her)$\cup$(Ren)$\cup$(P) can be checked for instance by resolution.

We now give an upper bound on the complexity of deciding the satisfiability of (Her) $\cup$ (Ren) $\cup$ (P), i.e. for deciding the unifiability of $\mathcal{S}$.

Theorem 6. *(1) The problem of deciding whether the D_{01}-unification problem $\mathcal{S}$ has a solution can be solved in at most non-deterministically polynomial time in $|ST(\mathcal{S})|2^{|C|}$ (and in exponential time in $|ST(\mathcal{S})|2^{|C|}$ by using resolution).*

(2) If $\mathcal{S}$ only contains the operation symbols $\wedge, 0, 1$, and, possibly, constants, then the problem can be decided in at most polynomial time in $|ST(\mathcal{S})|2^{|C|}$.

Proof: Note first that the structure-preserving translation to clause form in Theorem 5 is polynomial in $|ST(\mathcal{S})|2^{|C|}$. The size of the conjunction of all formulae in (Her)$\cup$(Ren)$\cup$(P) is also polynomial in $|ST(\mathcal{S})|2^{|C|}$. (1) follows from this and the fact that the number of all distinct literals that can occur in the conjunction of ground clauses (Her)$\cup$(Ren)$\cup$(P) in Theorem 5(3) is bounded by $|ST(\mathcal{S})|2^{|C|}$. To prove (2) note that if only the operators $\wedge, 0, 1$ and, possibly, constants, occur in $\mathcal{S}$, then the clause form of (Her) $\cup$ (Ren) $\cup$ (P) is a set of ground Horn clauses. Dowling and Gallier [11] showed that satisfiability of a set Φ of ground Horn clauses can be proved in linear time w.r.t. the number of clauses in Φ. $\square$

Note. It is not necessary to explicitly add to (Her) $\cup$ (Ren) $\cup$ (P) formulae expressing the order relationship between the elements in $\mathcal{P}(C)$. However, these relationships have to be known when expressing (Her) $\cup$ (Ren) $\cup$ (P) as the conjunction of ground formulae by instantiating the variables with elements in $\mathcal{P}(C)$.

4.2   Translation to Constrained Clause Form

The clause form of the set of formulae (Her)$\cup$(Ren)$\cup$(P) defined in Theorem 5(3) can be naturally expressed by constrained clauses of a special form as follows:

(Her)	$P_y(X_1) \to P_y(X_2)$	$[X_1 \subseteq X_2 \subseteq C], y \in Y$
(Ren) (∧n)	$P_{e_1 \wedge e_2}(X) \to P_{e_i}(X)$	$[X \subseteq C], i = 1,2$
(∧p)	$P_{e_1}(X) \wedge P_{e_2}(X) \to P_{e_1 \wedge e_2}(X)$	$[X \subseteq C]$
(∨n)	$P_{e_1 \vee e_2}(X) \to P_{e_1}(X) \vee P_{e_2}(X)$	$[X \subseteq C]$
(∨p)	$P_{e_i}(X) \to P_{e_1 \vee e_2}(X)$	$[X \subseteq C], i = 1,2$
(1)	$P_1(X)$	$[X \subseteq C]$
(0)	$\neg P_0(X)$	$[X \subseteq C]$
(cp)	$P_c(X)$	$[X \subseteq C, c \in X]$
(cn)	$\neg P_c(X)$	$[X \subseteq C, c \notin X]$
(P)	$P_{s_i}(X) \leftrightarrow P_{t_i}(X),$	$[X \subseteq C]$, for all $1 \leq i \leq k$

Definition 3. *A constrained clause has the form $D[\![\phi]\!]$, where (i) D is a first-order clause with variables $X, X_1, \ldots, X_n, \ldots$ ranging over a countably infinite set V; all predicates occurring in D are unary; and (ii) the constraint ϕ is of the form $\bigwedge_{i \in I_1}(c_i \in X_i) \wedge \bigwedge_{i \in I_2}(c_i \notin X_i) \wedge \bigwedge_{i \in I_3, j \in I_4}(X_i \subseteq X_j) \wedge \bigwedge_{i \in I_1 \cup \cdots \cup I_4}(X_i \subseteq C).$*

Let $\mathcal{S}$ be a D_{01}-unification problem, and $\Phi_{\mathcal{S}}$ the set of constrained clauses associated with $\mathcal{S}$ as explained above. Then $\Phi_{\mathcal{S}}$ can be constructed in polynomial time in the size of $\mathcal{S}$. The size of $\Phi_{\mathcal{S}}$ is polynomial in the size of $\mathcal{S}$.

A substitution of the variables in V is called *ground* when it replaces every variable by an element of $\mathcal{P}(C)$ (this is the Herbrand universe of (Her) $\cup$ (Ren) $\cup$ (P)). A constrained clause $D[\![\phi]\!]$ represents the set $(D[\![\phi]\!])^g = \{D\sigma \mid \sigma$ ground; $\phi\sigma$ true$\}$ of all ground instances of D by instantiations of the variables which satisfy ϕ. We say that a set Φ of constrained clauses is satisfiable if the set of all its ground instances, $\Phi^g = \bigcup_{D[\![\phi]\!] \in N}(D[\![\phi]\!])^g$, is satisfiable.

4.3   A Resolution Calculus for Constrained Clauses

We now formulate a resolution calculus $\mathsf{CRes}^{\succ}_S$ for the type of constrained clauses considered here. The calculus is parameterized by a total ordering $\succ$ on the predicate symbols, and a selection function S that assigns to each constrained clause $D[\![\phi]\!]$ a (possibly empty) multiset of (occurrences) of negative literals, called the selected literals of $D[\![\phi]\!]$. $\mathsf{CRes}^{\succ}_S$ consists of the following rules[3].

Ordered Resolution.

$$\frac{D_1 \vee P_e(X) \, [\![\phi_1(X, X_1, \ldots, X_n)]\!] \quad D_2 \vee \neg P_e(Z) \, [\![\phi_2(Z, Z_1, \ldots, Z_n)]\!]}{D_1 \vee D_2\sigma \, [\![\phi_1(X, X_1, \ldots, X_n) \wedge \phi_2(Z, Z_1, \ldots, Z_n)\sigma]\!]}$$

where $\sigma(Z) = X$ and $\sigma(W) = W$ in rest; P_e is the largest predicate symbol in $D_1 \vee P_e(X)$ and no literal is selected in $D_1 \vee P_e(X)$, and either $\neg P_e(Z)$ is selected in $D_2 \vee \neg P_e(Z)$, or otherwise nothing is selected in $D_2 \vee \neg P_e(Z)$ and P_e is the largest predicate symbol in $D_2 \vee \neg P_e(Z)$.

---

[3] The ordered resolution calculus with selection $\mathsf{Res}^{\succ}_S$ is complete for any well-founded and total ordering $\succ$ on ground literals and any selection function S. Here we consider a less restrictive form of resolution for constrained clauses, in order to simplify the presentation by avoiding the necessity of also handling order constraints (w.r.t. $\succ$).

Positive Factoring.

$$\frac{D \vee P_e(X) \vee P_e(Z) \; [\![\phi(X, Z, X_1, \ldots, X_n)]\!]}{D\sigma \vee P_e(X) \; [\![\phi(X, Z, X_1, \ldots, X_n)\sigma]\!]}$$

where $\sigma(Z) = X$, and $\sigma(W) = W$ in rest; P_e is the largest predicate symbol in $D \vee P_e(X) \vee P_e(Z)$ and nothing is selected in $D \vee P_e(X) \vee P_e(Z)$.

Theorem 7. *Let Φ be a set of constrained clauses, $\succ$ a total order on the predicate symbols, and S a selection function. Φ is unsatisfiable iff the empty clause $\Box[\![\phi]\!]$ (constrained by a satisfiable constraint ϕ) is derivable from Φ in* $\mathsf{CRes}_S^\succ$.

Proof: (Idea) The proof uses the completeness of ordered resolution with selection for ground clauses and a lifting lemma for constrained clauses; the arguments are similar to those in [9]. $\Box$

Example: Decide whether the D_{01}-unification problem $\mathcal{S} : \{y \wedge c = 0, y \vee c = 1\}$ has a solution, where c is a constant and y a variable.
(Note that $\mathcal{S}$ corresponds to the formula: $\forall c \exists y (y \wedge c = 0 \text{ and } y \vee c = 1)$.)

Solution: Let $\succ$ be a total ordering on the predicate symbols, defined such that $P_e \succ P_{e'}$ whenever (i) e' is a subterm of e; or (ii) e is a non-atomic term and c a constant; or (iii) e is a non-atomic term or a constant and x is a variable. Let S be a selection function that selects all negative occurrences of literals except in $\mathsf{Ren}(\vee\mathsf{p}, \wedge\mathsf{p})$, where nothing is selected. By the structure-preserving translation to clause form in Theorem 5(3) we obtain the following set of constrained clauses:

(1)	$\underline{P_y(X)} \to P_y(Y)$	$[\![X \subseteq Y \subseteq C]\!]$
(2)	$\underline{P_{y \wedge c}(X)} \to P_y(X)$	$[\![X \subseteq C]\!]$
(3)	$\underline{P_{y \wedge c}(X)} \to P_c(X)$	$[\![X \subseteq C]\!]$
(4)	$P_y(X) \wedge P_c(X) \to \boxed{P_{y \wedge c}(X)}$	$[\![X \subseteq C]\!]$
(5)	$\underline{P_{y \vee c}(X)} \to \overline{P_y(X) \vee P_c(X)}$	$[\![X \subseteq C]\!]$
(6)	$\underline{P_y(X)} \to \boxed{P_{y \vee c}(X)}$	$[\![X \subseteq C]\!]$
(7)	$\underline{P_c(X)} \to \boxed{P_{y \vee c}(X)}$	$[\![X \subseteq C]\!]$
(8)	$\neg P_0(X)$	$[\![X \subseteq C]\!]$
(9)	$P_1(X)$	$[\![X \subseteq C]\!]$
(10)	$P_c(X)$	$[\![X \subseteq C, c \in X]\!]$
(11)	$\neg P_c(X)$	$[\![X \subseteq C, c \notin X]\!]$
(12)	$\underline{P_{y \wedge c}(X)} \to P_0(X)$	$[\![X \subseteq C]\!]$
(13)	$\underline{P_0(X)} \to P_{y \wedge c}(X)$	$[\![X \subseteq C]\!]$
(14)	$\underline{P_{y \vee c}(X)} \to P_1(X)$	$[\![X \subseteq C]\!]$
(15)	$\underline{P_1(X)} \to P_{y \vee c}(X)$	$[\![X \subseteq C]\!]$

where $C = \{c\}$, the selected literals are underlined and the positive literals containing the maximal predicate symbol are in boxes. All ground inferences of (13) and (14) are redundant (so, (13) and (14) can be considered to be redundant).

We obtain the following deduction of the empty clause $\Box$:

(16) $P_y(X) \wedge P_c(X) \to P_0(X)$ $[\![X \subseteq C]\!]$		(by (12) and (4))
(17) $\overline{P_y(X)} \to \overline{P_0(X)}$	$[\![c \in X, X \subseteq C]\!]$	(by (10) and (16))
(18) $\overline{P_{y \vee c}(X)}$	$[\![X \subseteq C]\!]$	(by (15) and (9))
(19) $P_y(X) \vee \boxed{P_c(X)}$	$[\![X \subseteq C]\!]$	(by (18) and (5))
(20) $P_y(X)$	$[\![c \notin X, X \subseteq C]\!]$	(by (19) and (11))
(21) $P_y(Y)$	$[\![c \notin X, X \subseteq Y \subseteq C]\!]$	(by (20) and (1))
(22) $P_0(X)$	$[\![c \notin X, c \in X, X \subseteq C]\!]$	(by (20) and (17))
(23) $\Box$	$[\![c \notin X, c \in X, X \subseteq C]\!]$	(by (22) and (8))
(24) $P_0(Y)$	$[\![c \notin X, c \in Y, X \subseteq Y \subseteq C]\!]$	(by (21) and (17))
(25) $\Box$	$[\![c \notin X, c \in Y, X \subseteq Y \subseteq C]\!]$	(by (21) and (17))

The constraint in (23) is unsatisfiable, but the constraint in (25) is satisfiable (e.g. by $X = \emptyset$ and $Y = C$).[4] So, the set consisting of the clauses (1)–(15) is unsatisfiable, hence S has no solution.

4.4   Complexity Considerations for Some Special Cases

We now analyze some situations in which deciding D_{01}-unifiability is especially easy. We start by showing that for unification problems of the form $S : \{s = t\}$ the algorithm performs well. We end by analyzing the complexity of unification without constants.

Unification with Free Constants: General Case. Let S be a unification problem. Let $\succ$ be a total ordering on the predicate symbols $\{P_e \mid e \in ST(S)\}$, defined such that $P_e \succ P_{e'}$ whenever (i) e' is a subformula of e; or (ii) e is a non-atomic formula and c a constant; or (iii) e is a non-atomic formula or a constant and y is a variable. Let S be a selection function that (i) selects nothing in $\mathsf{Ren}(\vee \mathsf{p}, \wedge \mathsf{p})$, and (ii) in every other non-positive clause selects the set of all occurrences of negative literals that contain the maximal predicate symbol(s) among those occurring in the negative literals in the clause. Then:

1. no inferences are possible between (Her) and (Ren);
2. all inferences between two clauses in (Ren) generate tautologies;
3. no inferences are possible between (Her) and clauses in (P);
4. inferences between $P_{s_i}(X) \to P_{t_i}(X)$ in (P) and clauses in (Ren) lead to:
   (a) $\bigwedge_{j \in J} P_{e_j^i}(X) \to P_{t_i}(X)$ $[\![X \subseteq C]\!]$, where for every J, $\{e_j^i \mid j \in J\}$ is a multiset of subterms of $ST(s_i)$, containing no repetition of subterms that occur at the same position in s_i;
   (b) $(\bigwedge P_{c_j}(X)) \wedge (\bigwedge P_{x_i}(X)) \to P_{t_i}(X)$ $[\![X \subseteq C, d_i \in X, i \in I]\!]$;
   (c) $\bigwedge P_{x_l}(X) \to P_{t_i}(X)$ $[\![X \subseteq C, d_i \in X, i \in I]\!]$;
   (d) $P_{t_i}(X)$ $[\![X \subseteq C, d_i \in X, i \in I]\!]$.
   There are at most $2^{\mathsf{length}(s_i)}$ such clauses; all constraints are linear in $\mathsf{length}(s_i)$.

---

[4] This shows that inferences with the clause Her in Theorem 5(3) (in particular, with clause (1) for this example) are necessary for the correctness of the method.

5. inferences between $P_{t_i}(X)$ $[\![X \subseteq C, d_i \in X, i \in I]\!]$ and clauses in (Ren) lead to:

   (a') $\bigvee_{j \in J} P_{e_j^i}(X)$ $[\![X \subseteq C, d_i \in X, i \in I]\!]$, where for every J, $\{e_j^i \mid j \in J\}$ is a multiset of subterms of $ST(t_i)$, containing no repetition of subterms that occur at the same position in t_i;

   (b') $(\bigvee P_{c_j}(X)) \vee (\bigwedge P_{x_l}(X))$ $[\![X \subseteq C, d_i \in X, i \in I, d_k' \notin X, k \in K]\!]$;

   (c') $\bigvee P_{x_l}(X)$ $[\![X \subseteq C, d_i \in X, i \in I, d_k' \notin X, k \in K]\!]$;

   (d') $\square$ $[\![X \subseteq C, d_i \in X, i \in I]\!]$.

   or factors thereof. There are at most $2^{\mathsf{length}(t_i)}$ such clauses; all constraints are linear in $\mathsf{length}(s_i) + \mathsf{length}(t_i)$.

6. inferences between clauses of type (c') and (Her) produce clauses of the form.

   (e') $\bigvee P_{x_l}(X_i)$ $[\![X \subseteq C, d_i \in X, i \in I, d_k' \notin X, k \in K, X \subseteq X_i]\!]$;

7. further inferences may be possible between clauses of the form (a)–(c) and clauses of type (d) or (a')–(c') and (e') and resolvents thereof. (These inferences can be further controlled, e.g. by adopting an additional labeling of the predicate symbols P_e that also indicates in which of the terms $t_1, \ldots, t_k, s_1, \ldots, s_k$ and at which position e occurs; and by defining redundancy criteria. Here, we do not enter in further details.)

Corollary 3. *If all clauses generated from* (Her) $\cup$ (Ren) $\cup$ (P) *by* $\mathsf{CRes}_S^{\succ}$ *contain a negative literal then S has a solution.*

Proof: Follows from the remarks above and the fact that if all clauses generated contain a (selected) negative literal then $\square$ cannot be obtained by $\mathsf{CRes}_S^{\succ}$. $\square$

This happens e.g. if for all $1 \leq i \leq k$, the disjunctive normal forms of s_i, t_i do not contain constant terms.

Unification with Free Constants: One Equation. If S consists of only one equation, the last type of inferences can be proved to produce only constrained clauses with the property that all their ground instances are already subsumed by the set of ground instances of clauses of type (d) previously generated, as will be shown in Corollary 4.

Lemma 1. *The satisfiability of a constraint ϕ of size $|\phi|$ can be checked in time linear in $|C| \cdot |\phi|$.*

Proof: (Sketch) Note first that a constraint

$$\phi = \bigwedge_{i \in I_1} (c_i \in X_i) \wedge \bigwedge_{i \in I_2} (d_i \notin X_i) \wedge \bigwedge_{i \in I_3, j \in I_4} (X_i \subseteq X_j) \wedge \bigwedge_{i \in I_1 \cup \cdots \cup I_4} (X_i \subseteq C)$$

is satisfiable iff the set

$$E_\phi = \{P_{X_i}(c_i) \mid i \in I_1\} \cup \{\neg P_{X_j}(d_j) \mid j \in I_2\} \cup \{P_{X_i}(x) \to P_{X_j}(x) \mid i \in I_3, j \in I_4\}$$

of Horn clauses is satisfiable by a Herbrand interpretation. Moreover, E_ϕ is satisfiable iff the set G_ϕ of all its ground instances (which has cardinality $|C| \times |\phi|$) is satisfiable. Since all clauses in G_ϕ are ground Horn clauses, it follows by results in [11] that the satisfiability of G_ϕ can be checked in time linear in the size of G_ϕ i.e. linear in $|C| \times |\phi|$. $\square$

Corollary 4. *Let S : $\{s = t\}$ consist only of one equation. Then the satisfiability of S can be checked in time $2^{\text{length}(s)+\text{length}(t)}(\text{length}(s) + \text{length}(t))$.*

Proof: (Idea) In a first step, by inferences with (Ren), $P_s(X) \to P_t(X)$ generates literals of the form (a)–(c), and, possibly of type (d). Let $P_t(X)[\![\phi_i]\!], i \in I$ be all clauses of type (d) generated this way. Similarly, let $P_s(X)[\![\phi_j']\!], j \in J$ be all clauses of type (d) generated from $P_t(X) \to P_s(X)$. By the remarks at the beginning of this subsection, ϕ_i, ϕ_j contain only constraints of the form $X \subseteq C$ and $c_i \in X$. By inferences between $P_t(X)[\![\phi_i]\!], i \in I$ and $P_t(Z) \to P_s(Z)$ $[\![Z \subseteq C]\!]$ the clause $P_s[\![\phi_i]\!]$ is generated. $P_t[\![\phi_j']\!]$ are generated, for all $j \in J$, in a similar way. The constraint of a clause of type (a')–(d') contains, as a conjunct, one of the constraints ϕ_i or ϕ_j'. Let $D_i \to P_t(X)[\![\psi]\!]$ be one of the clauses in (a)–(c). A resolution with a clause of type (a')–(d') would produce a clause of the form $D_i' \to P_t[\![\psi \wedge \phi_i \wedge \rho]\!]$. The set of ground instances of such a clause is redundant with respect any set of clauses that contains all ground instances of P_t $[\![\phi_i]\!]$, hence (by the proof of Theorem 7) all such clauses can be considered redundant.

Similar arguments concerning redundancy of generated clauses can be used to control the inferences with (Her) and to prove that all resolvents of clauses of type (a')–(d') as well as resolvents of inferences with Her and clauses of type (a)–(d) have the property that all their ground instances are subsumed by ground instances of clauses of type (d). The conclusion follows since $2^{|S|}$ clauses are generated this way and the constraints can be checked in linear time. $\square$

Unification without Constants in D_{01}. If $C = \emptyset$, then $F_{D_{01}}(C)$ is the two-element lattice. By Theorem 2, a D_{01}-unification problem S : $\{s_1 = t_1, \ldots, s_k = t_k\}$ with variables $\{y_1, \ldots, y_n\}$ and no constants has a solution w.r.t. the equational theory of D_{01} iff the existential closure of S, $\exists y_1, \ldots y_n(s_1 = t_1 \wedge \cdots \wedge s_k = t_k)$ is valid in the two-element lattice. The number of clauses corresponding to (Her) $\cup$ (Ren) $\cup$ (P) in Theorem 5 is in this case polynomial in $ST(S)$.

Theorem 8 (Complexity). *Let S be a D_{01}-unification problem without constants. Assume that all terms in S have been simplified by (recursively) applying the following simplification rules[5]: $e \wedge 1 \mapsto e; e \wedge 0 \mapsto 0; e \vee 1 \mapsto 1; e \vee 0 \mapsto e$.*

(1) If $\{0,1\} \not\subseteq ST(S)$, then S always has a solution.

(2) If $\{0,1\} \subseteq ST(S)$ and S consists of only one equation (or else it contains the equation $1 = 0$) then S has no solution.

(3) If S only contains the operators $\wedge, 0, 1$, then the problem of checking whether S has a solution can be solved in polynomial time. The same holds if S only contains the operators $\vee, 0, 1$.

(4) Otherwise, the problem of checking whether S has a solution is NP-complete.

Proof: (Sketch) (1) Assume that $1 \notin ST(S)$. Let Φ_S be the set of clauses associated to S by the structure-preserving translation to clause form in Theorem 5(3).

---

[5] This can be done in polynomial time.

Since no constant occurs in $\mathcal{S}$, all the clauses $\Phi_{\mathcal{S}}$ are non-positive, so $\Phi_{\mathcal{S}}$ is satisfiable (consider a selection function that selects all negative literals in all clauses; then no resolution inference is possible). The case $0 \notin ST(\mathcal{S})$ follows by duality.

(2) is obvious, and (3) follows from the second part of Theorem 6.

(4) The problem of deciding whether a unification problem without constants has a solution is clearly in NP. NP-hardness follows from the fact that the satisfiability problem for Boolean formulae of the form $E = F \wedge \neg G$, where F and G only contain the operators $\vee$ and $\wedge$ (which is NP-complete [15]) can be reduced in polynomial time to the satisfiability of a D_{01}-unification problem with at least two equations $s_i = 0$ and $s_j = 1$, namely $\mathcal{S} : \{F = 1, G = 0\}$. $\square$

4.5   Unification with Linear Constant Restrictions

We now show that the idea used in the method described in Section 4.1 can be adapted to give an algorithm for unification with linear constant restrictions.

We first express the fact that $t \in F_{D_{01}}(C\backslash\{c\})$ by using the isomorphism $\eta_C : F_{D_{01}}(C) \to \mathcal{O}(\mathcal{P}(C), \subseteq)$ defined for every $t \in F_{D_{01}}(C)$ by $\eta_C(t) = \{f^{-1}(1) \cap C \mid f : F_{D_{01}}(C) \to 2$ is a 0,1-lattice homomorphism; $f(t) = 1\}$ (cf. Theorem 1).

Lemma 2. If $t \in F_{D_{01}}(C)$ then there exists $t' \in F_{D_{01}}(C\backslash\{c\})$ such that $t \equiv_{D_{01}} t'$ iff $\eta_C(t) = \emptyset$ or $C\backslash\{c\} \in \eta_C(t)$.

Proof: (Idea) This is a consequence of the fact that for every $t \in F_{D_{01}}(C)$, there exists $t' \in F_{D_{01}}(C\backslash\{c\})$ such that $t \equiv_{D_{01}} t'$ iff for every $X \subseteq C$, if $c \in X$ then either $X \notin \eta_C(t)$ or X is not minimal in $\eta_C(t)$. The proof of the equivalence above uses the way η_D is defined for every D, and the fact that there exists an injective homomorphism $i : F_{D_{01}}(C\backslash\{c\}) \to F_{D_{01}}(C)$, such that for every $U \in \mathcal{O}(\mathcal{P}(C\backslash\{c\}))$, $\eta_C(i(\eta_{C\backslash\{c\}}^{-1}(U)))$ is the order-filter generated by U in $(\mathcal{P}(C), \subseteq)$. $\square$

As in the case of unification with free constants, the remark above justifies a structure-preserving translation to clause form.

Theorem 9. Let $\mathcal{S} : \{s_1 = t_1, \ldots, s_k = t_k\}$ be a D_{01}-unification problem with linear constant restrictions Lcr, constants C and variables Y. The following are equivalent:

(1) S has a solution w.r.t. the equational theory of D_{01}.

(2) The conjunction of the following set of formulae is satisfiable:

(Her)	$P_y(X_1) \to P_y(X_2)$	*for all $X_1 \subseteq X_2 \subseteq C, y \in Y$*
(Ren)	$P_{e_1 \wedge e_2}(X) \leftrightarrow P_{e_1}(X) \wedge P_{e_2}(X)$	*for all $X \subseteq C$*
	$P_{e_1 \vee e_2}(X) \leftrightarrow P_{e_1}(X) \vee P_{e_2}(X)$	*for all $X \subseteq C$*
	$P_1(X)$	*for all $X \subseteq C$*
	$\neg P_0(X)$	*for all $X \subseteq C$*
	$P_c(X)$	*for all $X \subseteq C$ with $c \in X$*
	$\neg P_c(X)$	*for all $X \subseteq C$ with $c \notin X$*
(Lcr)	$P_y(X) \to P_y(C\backslash\{c\})$	*for all $X \subseteq C$ if $y < c \in$ Lcr*
(P)	$P_{s_i}(X) \leftrightarrow P_{t_i}(X)$,	*for all $1 \leq i \leq k$*

Proof: (Sketch) This follows from Definition 2, Lemma 2, and arguments similar to those used for Theorem 5.   □

5 Conclusion

We presented a resolution-based method for deciding unifiability w.r.t. the equational theory of bounded distributive lattices with operators. The method uses the Priestley representation for bounded distributive lattices, in particular the description of the dual $\mathbf{D}(F_{D_{01}}(C))$ of the free lattice in D_{01} over C as $(\mathcal{P}(C), \subseteq)$. This helped us to reduce the problem of checking whether a D_{01}-unification problem S with constants C (and linear constraint restrictions) has a solution, to the problem of checking the satisfiability of a set Φ_S of clauses. Φ_S can be represented both as a finite set of ground clauses, and as a set of constrained clauses; the last representation is much more compact. We formulated a resolution calculus for such constrained clauses and proved its soundness and completeness. The method we give is in general still exponential in $|ST(S)|2^C$. However, in several situations it is more efficient than other existing methods: syntactic information about the terms in S is sometimes reflected by the form of clauses, which allows us to establish better upper bounds for these particular problems. Our algorithm also behaves well for unification problems consisting of only one equation.

It would be interesting to compare our method with more general unification algorithms, e.g. based on rewriting. One such algorithm [8] decides unifiability *over* the free algebra, i.e. in an algebraic extension of the free algebra. We analyzed this more general problem for the equational theory of D_{01}. As part of work in progress, we prove that this reduces to Boolean unifiability (due to the fact that the algebraically closed elements of D_{01} are the Boolean lattices).

Acknowledgments

I thank H. Ganzinger for drawing my attention to the results on unification with linear constant restrictions (cf. e.g. [5]) and on Boolean unification [1], and to the possibility of using constrained clauses (for theorem proving for many-valued logics), an idea which proved to be useful in this paper. I thank J.M. Talbot for the discussion we had on possible applications of unification to set constraints.

References

1. F. Baader. On the complexity of Boolean unification. *Information Processing Letters*, 67(4):215–220, 1998.
2. F. Baader and P. Narendran. Unification of concept terms in description logics. In H. Prade, editor, *Proceedings of ECAI'98*, pages 331–335. Wiley, 1998.
3. F. Baader and K.U. Schulz. Unification in the union of disjoint equational theories: Combining decision procedures. *J. Symbolic Computation*, 21:211–243, 1996.
4. F. Baader and K.U. Schulz. Combination of constraint solvers for free and quasi-free structures. *Theoretical Computer Science*, 192:107–161, 1998.

5. F. Baader and W. Snyder. Unification theory. In J.A. Robinson and A. Voronkov, editors, *Handbook of Automated Reasoning*. Elsevier, 2000. To appear.

6. L. Bachmair, H. Ganzinger, and U. Waldmann. Set constraints are the monadic class. In *Eighth Annual IEEE Symposium on Logic in Computer Science*, pages 75–83, Montreal, Canada, June 19–23, 1993. IEEE Computer Society Press, Los Alamitos, CA, USA.

7. A. Bockmayr. Algebraic and logical aspects of unification. In K.U. Schulz, editor, *Proceedings of Word Equations and Related Topics IWWERT'90, Tübingen, October 1990*, LNCS 572, pages 171–180. Springer Verlag, 1992.

8. A. Bockmayr. Model-theoretic aspects of unification. In K.U. Schulz, editor, *Proceedings of Word Equations and Related Topics IWWERT'90, Tübingen, October 1990*, LNCS 572, pages 181–196. Springer Verlag, 1992.

9. H.J. Bürckert. *A resolution principle for a logic with restricted quantifiers*. LNAI 568. Springer Verlag, 1991.

10. S. Burris and H.P. Sankappanavar. *A Course in Universal Algebra*. Graduate Texts in Mathematics. Springer, 1981.

11. W.F. Dowling and J.H. Gallier. Linear-time algorithms for testing the satisfiability of propositional Horn formulae. *J. Logic Programming*, 1(3):267–284, 1984.

12. J.A. Gerhard and M. Petrich. Unification in free distributive lattices. *Theoretical Computer Science*, 126(2):237–257, 1994.

13. S. Ghilardi. Unification through projectivity. *J. Logic and Computation*, 7(6):733–752, 1997.

14. H.B. Hunt, D.J. Rosenkrantz, and P.A. Bloniarz. On the computational complexity of algebra of lattices. *SIAM Journal of Computation*, 16(1):129–148, 1987.

15. H.B. Hunt and R.E. Stearns. The complexity of very simple Boolean formulas with applications. *SIAM J. Comput.*, 19(1):44–70, 1990.

16. C. Kirchner and H. Kirchner. Constrained equational reasoning. In *UNIF'89 Extended Abstacts of the 3rd Int. Workshop on Unification*, pages 160–171, Pfalza-kademie, Lambrecht, 1989.

17. R. Nieuwenhuis and A. Rubio. Theorem proving with ordering constrained clauses. In D. Kapur, editor, *Proceedings of CADE-11*, LNAI 607, pages 477–491. Springer Verlag, 1992.

18. H.A. Priestley. Ordered topological spaces and the representation of distributive lattices. *Proc. London Math. Soc.*, 3:507–530, 1972.

19. M. Schmidt-Schauß. A decision algorithm for distributive unification. *Theoretical Computer Science*, 208(1–2):111–148, 1998.

20. V. Sofronie-Stokkermans. On the universal theory of varieties of distributive lattices with operators: Some decidability and complexity results. In H. Ganzinger, editor, *Proceedings of CADE-16*, LNAI 1632, pages 157–171. Springer Verlag, 1999.

Reasoning with Individuals for the Description Logic $\mathcal{SHIQ}$

Ian Horrocks[1], Ulrike Sattler[2], and Stephan Tobies[2]

[1] Department of Computer Science, University of Manchester, UK
horrocks@cs.man.ac.uk
[2] LuFg Theoretical Computer Science, RWTH Aachen, Germany
{sattler,tobies}@informatik.rwth-aachen.de

Abstract. While there has been a great deal of work on the development of reasoning algorithms for expressive description logics, in most cases only Tbox reasoning is considered. In this paper we present an algorithm for combined Tbox and Abox reasoning in the $\mathcal{SHIQ}$ description logic. This algorithm is of particular interest as it can be used to decide the problem of (database) conjunctive query containment w.r.t. a schema. Moreover, the realisation of an efficient implementation should be relatively straightforward as it can be based on an existing highly optimised implementation of the Tbox algorithm in the FaCT system.

1 Motivation

A description logic (DL) knowledge base (KB) is made up of two parts, a terminological part (the terminology or Tbox) and an assertional part (the Abox), each part consisting of a set of axioms. The Tbox asserts facts about *concepts* (sets of objects) and *roles* (binary relations), usually in the form of inclusion axioms, while the Abox asserts facts about *individuals* (single objects), usually in the form of instantiation axioms. For example, a Tbox might contain an axiom asserting that Man is subsumed by Animal, while an Abox might contain axioms asserting that both Aristotle and Plato are instances of the concept Man and that the pair ⟨Aristotle, Plato⟩ is an instance of the role Pupil-of.

For logics that include full negation, all common DL reasoning tasks are reducible to deciding KB consistency, i.e., determining if a given KB admits a non-empty interpretation [6]. There has been a great deal of work on the development of reasoning algorithms for expressive DLs [2, 12, 16, 11], but in most cases these consider only Tbox reasoning (i.e., the Abox is assumed to be empty). With expressive DLs, determining consistency of a Tbox can often be reduced to determining the satisfiability of a single concept [2, 23, 3], and—as most DLs enjoy the tree model property (i.e., if a concept has a model, then it has a tree model)—this problem can be decided using a tableau-based decision procedure.

The relative lack of interest in Abox reasoning can also be explained by the fact that many applications only require Tbox reasoning, e.g., ontological engineering [15, 20] and schema integration [10]. Of particular interest in this regard is the DL $\mathcal{SHIQ}$ [18], which is powerful enough to encode the logic $\mathcal{DLR}$ [10], and which can thus be used

D. McAllester (Ed.): CADE-17, LNAI 1831, pp. 482–496, 2000.

for reasoning about conceptual data models, e.g., Entity-Relationship (ER) schemas [9]. Moreover, if we think of the Tbox as a *schema* and the Abox as (possibly incomplete) *data*, then it seems reasonable to assume that realistic Tboxes will be of limited size, whereas realistic Aboxes could be of almost unlimited size. Given the high complexity of reasoning in most DLs [23,7], this suggests that Abox reasoning could lead to severe tractability problems in realistic applications.[1]

However, $\mathcal{SHIQ}$ Abox reasoning is of particular interest as it allows $\mathcal{DLR}$ schema reasoning to be extended to reasoning about conjunctive query containment w.r.t. a schema [8]. This is achieved by using Abox individuals to represent variables and constants in the queries, and to enforce co-references [17]. In this context, the size of the Abox would be quite small (it is bounded by the number of variables occurring in the queries), and should not lead to severe tractability problems.

Moreover, an alternative view of the Abox is that it provides a restricted form of reasoning with *nominals*, i.e., allowing individual names to appear in concepts [22,5, 1]. Unrestricted nominals are very powerful, allowing arbitrary co-references to be enforced and thus leading to the loss of the tree model property. This makes it much harder to prove decidability and to devise decision procedures (the decidability of $\mathcal{SHIQ}$ with unrestricted nominals is still an open problem). An Abox, on the other hand, can be modelled by a *forest*, a set of trees whose root nodes form an arbitrarily connected graph, where number of trees is limited by the number of individual names occurring in the Abox. Even the restricted form of co-referencing provided by an Abox is quite powerful, and can extend the range of applications for the DLs reasoning services.

In this paper we present a tableaux based algorithm for deciding the satisfiability of unrestricted $\mathcal{SHIQ}$ KBs (i.e., ones where the Abox may be non-empty) that extends the existing consistency algorithm for Tboxes [18] by making use of the forest model property. This should make the realisation of an efficient implementation relatively straightforward as it can be based on an existing highly optimised implementation of the Tbox algorithm (e.g., in the FaCT system [14]). A notable feature of the algorithm is that, instead of making a unique name assumption w.r.t. all individuals (an assumption commonly made in DLs [4]), increased flexibility is provided by allowing the Abox to contain axioms explicitly asserting inequalities between pairs of individual names (adding such an axiom for every pair of individual names is obviously equivalent to making a unique name assumption).

2 Preliminaries

In this section, we introduce the DL $\mathcal{SHIQ}$. This includes the definition of syntax, semantics, inference problems (concept subsumption and satisfiability, Abox consistency, and all of these problems with respect to terminologies[2]), and their relationships.

$\mathcal{SHIQ}$ is based on an extension of the well known DL $\mathcal{ALC}$ [24] to include transitively closed primitive roles [21]; we call this logic $\mathcal{S}$ due to its relationship with

---

[1] Although suitably optimised algorithms may make reasoning practicable for quite large Aboxes [13].

[2] We use *terminologies* instead of Tboxes to underline the fact that we allow for general concept inclusions axioms and do not disallow cycles.

the proposition (multi) modal logic $S4_{(m)}$ [23].[3] This basic DL is then extended with inverse roles ($\mathcal{I}$), role hierarchies ($\mathcal{H}$), and qualifying number restrictions ($\mathcal{Q}$).

Definition 1. *Let* **C** *be a set of* concept names *and* **R** *a set of* role names *with a subset* $\mathbf{R}_+ \subseteq \mathbf{R}$ *of* transitive role names. *The set of* roles *is* $\mathbf{R} \cup \{R^- \mid R \in \mathbf{R}\}$. *To avoid considering roles such as R^{--}, we define a function* Inv *on roles such that* $\mathrm{Inv}(R) = R^-$ *if R is a role name, and* $\mathrm{Inv}(R) = S$ *if* $R = S^-$. *We also define a function* Trans *which returns* true *iff R is a transitive role. More precisely,* $\mathrm{Trans}(R) = $ true *iff* $R \in \mathbf{R}_+$ *or* $\mathrm{Inv}(R) \in \mathbf{R}_+$.

A role inclusion axiom *is an expression of the form* $R \sqsubseteq S$, *where R and S are* roles, *each of which can be inverse. A* role hierarchy *is a set of role inclusion axioms. For a role hierarchy $\mathcal{R}$, we define the relation $\stackrel{*}{\sqsubseteq}$ to be the transitive-reflexive closure of $\sqsubseteq$ over $\mathcal{R} \cup \{\mathrm{Inv}(R) \sqsubseteq \mathrm{Inv}(S) \mid R \sqsubseteq S \in \mathcal{R}\}$. A role R is called a* sub-role *(resp. super-role) of a role S if $R \stackrel{*}{\sqsubseteq} S$ (resp. $S \stackrel{*}{\sqsubseteq} R$). A role is* simple *if it is neither transitive nor has any transitive sub-roles.*

The set of $\mathcal{SHIQ}$-concepts is the smallest set such that

- *every concept name is a concept, and,*
- *if C, D are concepts, R is a role, S is a simple role, and n is a nonnegative integer, then* $C \sqcap D$, $C \sqcup D$, $\neg C$, $\forall R.C$, $\exists R.C$, $\geqslant nS.C$, *and* $\leqslant nS.C$ *are also concepts.*

A general concept inclusion axiom *(GCI) is an expression of the form* $C \sqsubseteq D$ *for two $\mathcal{SHIQ}$-concepts C and D. A* terminology *is a set of GCIs.*

Let $\mathbf{I} = \{a, b, c \ldots\}$ *be a set of* individual names. *An* assertion *is of the form* $a : C$, $(a, b) : R$, *or* $a \neq b$ *for* $a, b \in \mathbf{I}$, *a (possibly inverse) role R, and a $\mathcal{SHIQ}$-concept C. An* Abox *is a finite set of assertions.*

Next, we define semantics of $\mathcal{SHIQ}$ and the corresponding inference problems.

Definition 2. *An* interpretation $\mathcal{I} = (\Delta^\mathcal{I}, \cdot^\mathcal{I})$ *consists of a set $\Delta^\mathcal{I}$, called the* domain *of $\mathcal{I}$, and a valuation $\cdot^\mathcal{I}$ which maps every concept to a subset of $\Delta^\mathcal{I}$ and every role to a subset of $\Delta^\mathcal{I} \times \Delta^\mathcal{I}$ such that, for all concepts C, D, roles R, S, and non-negative integers n, the following equations are satisfied, where $\sharp M$ denotes the cardinality of a set M and $(R^\mathcal{I})^+$ the transitive closure of $R^\mathcal{I}$:*

$$R^\mathcal{I} = (R^\mathcal{I})^+ \qquad\qquad \textit{for each role } R \in \mathbf{R}_+$$
$$(R^-)^\mathcal{I} = \{\langle x, y\rangle \mid \langle y, x\rangle \in R^\mathcal{I}\} \qquad \textit{(inverse roles)}$$
$$(C \sqcap D)^\mathcal{I} = C^\mathcal{I} \cap D^\mathcal{I} \qquad\qquad \textit{(conjunction)}$$
$$(C \sqcup D)^\mathcal{I} = C^\mathcal{I} \cup D^\mathcal{I} \qquad\qquad \textit{(disjunction)}$$
$$(\neg C)^\mathcal{I} = \Delta^\mathcal{I} \setminus C^\mathcal{I} \qquad\qquad \textit{(negation)}$$
$$(\exists R.C)^\mathcal{I} = \{x \mid \exists y.\langle x, y\rangle \in R^\mathcal{I} \text{ and } y \in C^\mathcal{I}\} \qquad \textit{(exists restriction)}$$
$$(\forall R.C)^\mathcal{I} = \{x \mid \forall y.\langle x, y\rangle \in R^\mathcal{I} \text{ implies } y \in C^\mathcal{I}\} \qquad \textit{(value restriction)}$$
$$(\geqslant nR.C)^\mathcal{I} = \{x \mid \sharp\{y.\langle x, y\rangle \in R^\mathcal{I} \text{ and } y \in C^\mathcal{I}\} \geqslant n\} \qquad \textit{($\geqslant$-number restriction)}$$
$$(\leqslant nR.C)^\mathcal{I} = \{x \mid \sharp\{y.\langle x, y\rangle \in R^\mathcal{I} \text{ and } y \in C^\mathcal{I}\} \leqslant n\} \qquad \textit{($\leqslant$-number restriction)}$$

An interpretation $\mathcal{I}$ satisfies a role hierarchy $\mathcal{R}$ iff $R^\mathcal{I} \subseteq S^\mathcal{I}$ for each $R \sqsubseteq S$ in $\mathcal{R}$. Such an interpretation is called a model *of $\mathcal{R}$ (written $\mathcal{I} \models \mathcal{R}$).*

---

[3] The logic $\mathcal{S}$ has previously been called $\mathcal{ALC}_{R+}$, but this becomes too cumbersome when adding letters to represent additional features.

An interpretation $\mathcal{I}$ *satisfies a terminology* $\mathcal{T}$ *iff* $C^{\mathcal{I}} \subseteq D^{\mathcal{I}}$ *for each GCI* $C \sqsubseteq D$ *in* $\mathcal{T}$. *Such an interpretation is called a* model *of* $\mathcal{T}$ *(written* $\mathcal{I} \models \mathcal{T}$*)*.

A concept C *is called* satisfiable *with respect to a role hierarchy* $\mathcal{R}$ *and a termi-nology* $\mathcal{T}$ *iff there is a model* $\mathcal{I}$ *of* $\mathcal{R}$ *and* $\mathcal{T}$ *with* $C^{\mathcal{I}} \neq \emptyset$. *A concept* D subsumes *a concept* C *w.r.t.* $\mathcal{R}$ *and* $\mathcal{T}$ *iff* $C^{\mathcal{I}} \subseteq D^{\mathcal{I}}$ *holds for each model* $\mathcal{I}$ *of* $\mathcal{R}$ *and* $\mathcal{T}$. *For an interpretation* $\mathcal{I}$, *an element* $x \in \Delta^{\mathcal{I}}$ *is called an* instance *of a concept* C *iff* $x \in C^{\mathcal{I}}$.

For Aboxes, an interpretation maps, additionally, each individual $a \in \mathbf{I}$ *to some element* $a^{\mathcal{I}} \in \Delta^{\mathcal{I}}$. *An interpretation* $\mathcal{I}$ *satisfies an assertion*

$$a : C \quad \text{iff} \quad a^{\mathcal{I}} \in C^{\mathcal{I}},$$
$$(a, b) : R \quad \text{iff} \quad \langle a^{\mathcal{I}}, b^{\mathcal{I}} \rangle \in R^{\mathcal{I}}, \text{ and}$$
$$a \neq b \quad \text{iff} \quad a^{\mathcal{I}} \neq b^{\mathcal{I}}$$

An Abox $\mathcal{A}$ *is* consistent *w.r.t.* $\mathcal{R}$ *and* $\mathcal{T}$ *iff there is a model* $\mathcal{I}$ *of* $\mathcal{R}$ *and* $\mathcal{T}$ *that satisfies each assertion in* $\mathcal{A}$.

For DLs that are closed under negation, subsumption and (un)satisfiability can be mutu-ally reduced: $C \sqsubseteq D$ iff $C \sqcap \neg D$ is unsatisfiable, and C is unsatisfiable iff $C \sqsubseteq A \sqcap \neg A$ for some concept name A. Moreover, a concept C is satisfiable iff the Abox $\{a : C\}$ is consistent. It is straightforward to extend these reductions to role hierarchies, but ter-minologies deserve special care: In [2, 23, 3], the *internalisation* of GCIs is introduced, a technique that reduces reasoning w.r.t. a (possibly cyclic) terminology to reasoning w.r.t. the empty terminology. For $\mathcal{SHIQ}$, this reduction must be slightly modified. The following Lemma shows how general concept inclusion axioms can be *internalised* us-ing a "universal" role U, that is, a transitive super-role of all roles occurring in $\mathcal{T}$ and their respective inverses.

Lemma 1. *Let* C, D *be concepts,* $\mathcal{A}$ *an Abox,* $\mathcal{T}$ *a terminology, and* $\mathcal{R}$ *a role hierarchy. We define*

$$C_{\mathcal{T}} := \bigsqcap_{C_i \sqsubseteq D_i \in \mathcal{T}} \neg C_i \sqcup D_i.$$

Let U *be a transitive role that does not occur in* $\mathcal{T}$, C, D, $\mathcal{A}$, *or* $\mathcal{R}$. *We set*

$$\mathcal{R}_U := \mathcal{R} \cup \{R \sqsubseteq U, \mathsf{Inv}(R) \sqsubseteq U \mid R \text{ occurs in } \mathcal{T}, C, D, \mathcal{A}, \text{ or } \mathcal{R}\}.$$

- *C is satisfiable w.r.t.* $\mathcal{T}$ *and* $\mathcal{R}$ *iff* $C \sqcap C_{\mathcal{T}} \sqcap \forall U . C_{\mathcal{T}}$ *is satisfiable w.r.t.* $\mathcal{R}_U$.
- *D subsumes* C *with respect to* $\mathcal{T}$ *and* $\mathcal{R}$ *iff* $C \sqcap \neg D \sqcap C_{\mathcal{T}} \sqcap \forall U . C_{\mathcal{T}}$ *is unsatisfiable w.r.t.* $\mathcal{R}_U$.
- $\mathcal{A}$ *is consistent with respect to* $\mathcal{R}$ *and* $\mathcal{T}$ *iff* $\mathcal{A} \cup \{a : C_{\mathcal{T}} \sqcap \forall U . C_{\mathcal{T}} \mid a \text{ occurs in } \mathcal{A}\}$ *is consistent w.r.t.* $\mathcal{R}_U$.

The proof of Lemma 1 is similar to the ones that can be found in [23, 2]. Most importantly, it must be shown that, (a) if a $\mathcal{SHIQ}$-concept C is satisfiable with respect to a terminology $\mathcal{T}$ and a role hierarchy $\mathcal{R}$, then $C, \mathcal{T}$ have a *connected* model, i. e., a model where any two elements are connect by a role path over those roles occuring in C and $\mathcal{T}$, and (b) if y is reachable from x via a role path (possibly involving inverse roles), then $\langle x, y \rangle \in U^{\mathcal{I}}$. These are easy consequences of the semantics and the definition of U.

Theorem 1. *Satisfiability and subsumption of $\mathcal{SHIQ}$-concepts w.r.t. terminologies and role hierarchies are polynomially reducible to (un)satisfiability of $\mathcal{SHIQ}$-concepts w.r.t. role hierarchies, and therefore to consistency of $\mathcal{SHIQ}$-Aboxes w.r.t. role hierarchies.*

Consistency of $\mathcal{SHIQ}$-Aboxes w.r.t. terminologies and role hierarchies is polynomially reducible to consistency of $\mathcal{SHIQ}$-Aboxes w.r.t. role hierarchies.

3   A $\mathcal{SHIQ}$-Abox Tableau Algorithm

With Theorem 1, all standard inference problems for $\mathcal{SHIQ}$-concepts and Aboxes can be reduced to Abox-consistency w.r.t. a role hierarchy. In the following, we present a tableau-based algorithm that decides consistency of $\mathcal{SHIQ}$-Aboxes w.r.t. role hierarchies, and therefore all other $\mathcal{SHIQ}$ inference problems presented.

The algorithm tries to construct, for a $\mathcal{SHIQ}$-Abox $\mathcal{A}$, a tableau for $\mathcal{A}$, that is, an abstraction of a model of $\mathcal{A}$. Given the notion of a tableau, it is then quite straightforward to prove that the algorithm is a decision procedure for Abox consistency.

3.1   A Tableau for Aboxes

In the following, if not stated otherwise, C, D denote $\mathcal{SHIQ}$-concepts, $\mathcal{R}$ a role hierarchy, $\mathcal{A}$ an Abox, $\mathbf{R}_{\mathcal{A}}$ the set of roles occurring in $\mathcal{A}$ and $\mathcal{R}$ together with their inverses, and $\mathbf{I}_{\mathcal{A}}$ is the set of individuals occurring in $\mathcal{A}$.

Without loss of generality, we assume all concepts C occurring in assertions $a : C \in \mathcal{A}$ to be in NNF, that is, negation occurs in front of concept names only. Any $\mathcal{SHIQ}$-concept can easily be transformed into an equivalent one in NNF by pushing negations inwards using a combination of DeMorgan's laws and the following equivalences:

$$\neg(\exists R.C) \equiv (\forall R.\neg C) \qquad \neg(\forall R.C) \equiv (\exists R.\neg C)$$
$$\neg(\leqslant nR.C) \equiv \geqslant (n+1)R.C \quad \neg(\geqslant nR.C) \equiv \leqslant (n-1)R.C \quad \text{where}$$
$$\leqslant (-1)R.C := A \sqcap \neg A \quad \text{for some } A \in \mathbf{C}$$

For a concept C we will denote the NNF of $\neg C$ by $\sim C$. Next, for a concept C, $\mathsf{clos}(C)$ is the smallest set that contains C and is closed under sub-concepts and $\sim$. We use $\mathsf{clos}(\mathcal{A}) := \bigcup_{aC \in \mathcal{A}} \mathsf{clos}(C)$ for the closure $\mathsf{clos}(\mathcal{A})$ of each concept C occurring in $\mathcal{A}$. It is not hard to show that the size of $\mathsf{clos}(\mathcal{A})$ is polynomial in the size of $\mathcal{A}$.

Definition 3. $T = (\mathbf{S}, \mathcal{L}, \mathcal{E}, \mathcal{I})$ *is a tableau for $\mathcal{A}$ w.r.t. $\mathcal{R}$ iff*

- $\mathbf{S}$ *is a non-empty set,*
- $\mathcal{L} : \mathbf{S} \to 2^{\mathsf{clos}(\mathcal{A})}$ *maps each element in $\mathbf{S}$ to a set of concepts,*
- $\mathcal{E} : \mathbf{R}_{\mathcal{A}} \to 2^{\mathbf{S} \times \mathbf{S}}$ *maps each role to a set of pairs of elements in $\mathbf{S}$, and*
- $\mathcal{I} : \mathbf{I}_{\mathcal{A}} \to \mathbf{S}$ *maps individuals occurring in $\mathcal{A}$ to elements in $\mathbf{S}$.*

Furthermore, for all $s, t \in \mathbf{S}$, $C, C_1, C_2 \in \mathsf{clos}(\mathcal{A})$, and $R, S \in \mathbf{R}_{\mathcal{A}}$, T satisfies:

(P1) *if $C \in \mathcal{L}(s)$, then $\neg C \notin \mathcal{L}(s)$,*
(P2) *if $C_1 \sqcap C_2 \in \mathcal{L}(s)$, then $C_1 \in \mathcal{L}(s)$ and $C_2 \in \mathcal{L}(s)$,*

(P3) *if* $C_1 \sqcup C_2 \in \mathcal{L}(s)$, *then* $C_1 \in \mathcal{L}(s)$ *or* $C_2 \in \mathcal{L}(s)$,
(P4) *if* $\forall S.C \in \mathcal{L}(s)$ *and* $\langle s,t \rangle \in \mathcal{E}(S)$, *then* $C \in \mathcal{L}(t)$,
(P5) *if* $\exists S.C \in \mathcal{L}(s)$, *then there is some* $t \in \mathbf{S}$ *such that* $\langle s,t \rangle \in \mathcal{E}(S)$ *and* $C \in \mathcal{L}(t)$,
(P6) *if* $\forall S.C \in \mathcal{L}(s)$ *and* $\langle s,t \rangle \in \mathcal{E}(R)$ *for some* $R \trianglelefteq S$ *with* $\mathsf{Trans}(R)$, *then* $\forall R.C \in \mathcal{L}(t)$,
(P7) $\langle x,y \rangle \in \mathcal{E}(R)$ *iff* $\langle y,x \rangle \in \mathcal{E}(\mathsf{Inv}(R))$,
(P8) *if* $\langle s,t \rangle \in \mathcal{E}(R)$ *and* $R \trianglelefteq S$, *then* $\langle s,t \rangle \in \mathcal{E}(S)$,
(P9) *if* $\leqslant nS.C \in \mathcal{L}(s)$, *then* $\sharp S^T(s,C) \leqslant n$,
(P10) *if* $\geqslant nS.C \in \mathcal{L}(s)$, *then* $\sharp S^T(s,C) \geqslant n$,
(P11) *if* $(\bowtie n\ S\ C) \in \mathcal{L}(s)$ *and* $\langle s,t \rangle \in \mathcal{E}(S)$ *then* $C \in \mathcal{L}(t)$ *or* $\sim C \in \mathcal{L}(t)$,
(P12) *if* $a:C \in \mathcal{A}$, *then* $C \in \mathcal{L}(\mathfrak{J}(a))$,
(P13) *if* $(a,b):R \in \mathcal{A}$, *then* $\langle \mathfrak{J}(a), \mathfrak{J}(b) \rangle \in \mathcal{E}(R)$,
(P14) *if* $a \not\doteq b \in \mathcal{A}$, *then* $\mathfrak{J}(a) \neq \mathfrak{J}(b)$,

where $\bowtie$ *is a place-holder for both* $\leqslant$ *and* $\geqslant$, *and* $S^T(s,C) := \{t \in \mathbf{S} \mid \langle s,t \rangle \in \mathcal{E}(S)\ and\ C \in \mathcal{L}(t)\}$.

Lemma 2. *A $\mathcal{SHIQ}$-Abox $\mathcal{A}$ is consistent w.r.t. $\mathcal{R}$ iff there exists a tableau for $\mathcal{A}$ w.r.t. $\mathcal{R}$.*

Proof: For the *if* direction, if $T = (\mathbf{S}, \mathcal{L}, \mathcal{E}, \mathfrak{J})$ is a tableau for $\mathcal{A}$ w.r.t. $\mathcal{R}$, a model $\mathcal{I} = (\Delta^{\mathcal{I}}, \cdot^{\mathcal{I}})$ of $\mathcal{A}$ and $\mathcal{R}$ can be defined as follows:

$$\Delta^{\mathcal{I}} := \mathbf{S}$$

for concept names A in $\mathrm{clos}(\mathcal{A})$: $A^{\mathcal{I}} := \{s \mid A \in \mathcal{L}(s)\}$

for individual names $a \in \mathbf{I}$: $a^{\mathcal{I}} := \mathfrak{J}(a)$

for role names $R \in \mathcal{R}$: $R^{\mathcal{I}} := \begin{cases} \mathcal{E}(R)^+ & \text{if } \mathsf{Trans}(R) \\ \mathcal{E}(R) \cup \bigcup\limits_{P \trianglelefteq R, P \neq R} P^{\mathcal{I}} & \text{otherwise} \end{cases}$

where $\mathcal{E}(R)^+$ denotes the transitive closure of $\mathcal{E}(R)$. The interpretation of non-transitive roles is recursive in order to correctly interpret those non-transitive roles that have a transitive sub-role. From the definition of $R^{\mathcal{I}}$ and (P8), it follows that, if $\langle s,t \rangle \in S^{\mathcal{I}}$, then either $\langle s,t \rangle \in \mathcal{E}(S)$ or there exists a path $\langle s,s_1 \rangle, \langle s_1,s_2 \rangle, \ldots, \langle s_n,t \rangle \in \mathcal{E}(R)$ for some R with $\mathsf{Trans}(R)$ and $R \trianglelefteq S$.

Due to (P8) and by definition of $\mathcal{I}$, we have that $\mathcal{I}$ is a model of $\mathcal{R}$.

To prove that $\mathcal{I}$ is a model of $\mathcal{A}$, we show that $C \in \mathcal{L}(s)$ implies $s \in C^{\mathcal{I}}$ for any $s \in \mathbf{S}$. Together with (P12), (P13), and the interpretation of individuals and roles, this implies that $\mathcal{I}$ satisfies each assertion in $\mathcal{A}$. This proof can be given by induction on the length $\|C\|$ of a concept C in NNF, where we count neither negation nor integers in number restrictions. The only interesting case is $C = \forall S.E$: let $t \in \mathbf{S}$ with $\langle s,t \rangle \in S^{\mathcal{I}}$. There are two possibilities:

- $\langle s,t \rangle \in \mathcal{E}(S)$. Then (P4) implies $E \in \mathcal{L}(t)$.
- $\langle s,t \rangle \notin \mathcal{E}(S)$. Then there exists a path $\langle s,s_1 \rangle, \langle s_1,s_2 \rangle, \ldots, \langle s_n,t \rangle \in \mathcal{E}(R)$ for some R with $\mathsf{Trans}(R)$ and $R \trianglelefteq S$. Then (P6) implies $\forall R.E \in \mathcal{L}(s_i)$ for all $1 \leq i \leq n$, and (P4) implies $E \in \mathcal{L}(t)$.

In both cases, $t \in E^{\mathcal{I}}$ by induction and hence $s \in C^{\mathcal{I}}$.

For the converse, for $\mathcal{I} = (\Delta^{\mathcal{I}}, \cdot^{\mathcal{I}})$ a model of $\mathcal{A}$ w.r.t. $\mathcal{R}$, we define a tableau $T = (\mathbf{S}, \mathcal{L}, \mathcal{E}, \mathcal{J})$ for $\mathcal{A}$ and $\mathcal{R}$ as follows:

$$\mathbf{S} := \Delta^{\mathcal{I}}, \quad \mathcal{E}(R) := R^{\mathcal{I}}, \quad \mathcal{L}(s) := \{C \in \mathsf{clos}(\mathcal{A}) \mid s \in C^{\mathcal{I}}\}, \quad \text{and} \quad \mathcal{J}(a) = a^{\mathcal{I}}.$$

It is easy to demonstrate that T is a tableau for D.   □

3.2   The Tableau Algorithm

In this section, we present a completion algorithm that tries to construct, for an input Abox $\mathcal{A}$ and a role hierarchy $\mathcal{R}$, a tableau for $\mathcal{A}$ w.r.t. $\mathcal{R}$. We prove that this algorithm constructs a tableau for $\mathcal{A}$ and $\mathcal{R}$ iff there exists a tableau for $\mathcal{A}$ and $\mathcal{R}$, and thus decides consistency of $\mathcal{SHIQ}$ Aboxes w.r.t. role hierarchies.

Since Aboxes might involve several individuals with arbitrary role relationships between them, the completion algorithm works on a *forest* rather than on a *tree*, which is the basic data structure for those completion algorithms deciding satisfiability of a concept. Such a forest is a collection of trees whose root nodes correspond to the individuals present in the input Abox. In the presence of transitive roles, *blocking* is employed to ensure termination of the algorithm. In the additional presence of inverse roles, blocking is *dynamic*, i.e., blocked nodes (and their sub-branches) can be un-blocked and blocked again later. In the additional presence of number restrictions, *pairs* of nodes are blocked rather than single nodes.

Definition 4. *A completion forest $\mathcal{F}$ for a $\mathcal{SHIQ}$ Abox $\mathcal{A}$ is a collection of trees whose distinguished root nodes are possibly connected by edges in an arbitrary way. Moreover, each node x is labelled with a set $\mathcal{L}(x) \subseteq \mathsf{clos}(\mathcal{A})$ and each edge $\langle x, y \rangle$ is labelled with a set $\mathcal{L}(\langle x, y \rangle) \subseteq \mathcal{R}_{\mathcal{A}}$ of (possibly inverse) roles occurring in $\mathcal{A}$. Finally, completion forests come with an explicit inequality relation $\not\doteq$ on nodes and an explicit equality relation $\doteq$ which are implicitly assumed to be symmetric.*

If nodes x and y are connected by an edge $\langle x, y \rangle$ with $R \in \mathcal{L}(\langle x, y \rangle)$ and $R \sqsubseteq S$, then y is called an S-successor of x and x is called an $\mathsf{Inv}(S)$-predecessor of y. If y is an S-successor or an $\mathsf{Inv}(S)$-predecessor of x, then y is called an S-neighbour of x. A node y is a successor (resp. predecessor or neighbour) of y if it is an S-successor (resp. S-predecessor or S-neighbour) of y for some role S. Finally, ancestor is the transitive closure of predecessor.

For a role S, a concept C and a node x in $\mathcal{F}$ we define $S^{\mathcal{F}}(x, C)$ by

$$S^{\mathcal{F}}(x, C) := \{y \mid y \text{ is } S\text{-neighbour of } x \text{ and } C \in \mathcal{L}(y)\}.$$

A node is blocked iff it is not a root node and it is either directly or indirectly blocked. A node x is directly blocked iff none of its ancestors are blocked, and it has ancestors x', y and y' such that

1. *y is not a root node and*
2. *x is a successor of x' and y is a successor of y' and*

3. $\mathcal{L}(x) = \mathcal{L}(y)$ and $\mathcal{L}(x') = \mathcal{L}(y')$ and
4. $\mathcal{L}(\langle x', x \rangle) = \mathcal{L}(\langle y', y \rangle)$.

In this case we will say that y blocks x.

A node y is indirectly blocked iff one of its ancestors is blocked, or it is a successor of a node x and $\mathcal{L}(\langle x, y \rangle) = \emptyset$; the latter condition avoids wasted expansions after an application of the $\leq$-rule.

Given a $\mathcal{SHIQ}$-Abox $\mathcal{A}$ and a role hierarchy $\mathcal{R}$, the algorithm initialises a completion forest $\mathcal{F}_{\mathcal{A}}$ consisting only of root nodes. More precisely, $\mathcal{F}_{\mathcal{A}}$ contains a root node x_0^i for each individual $a_i \in \mathbf{I}_{\mathcal{A}}$ occurring in $\mathcal{A}$, and an edge $\langle x_0^i, x_0^j \rangle$ if $\mathcal{A}$ contains an assertion $(a_i, a_j) : R$ for some R. The labels of these nodes and edges and the relations $\neq$ and $\doteq$ are initialised as follows:

$$\mathcal{L}(x_0^i) := \{C \mid a_i : C \in \mathcal{A}\},$$
$$\mathcal{L}(\langle x_0^i, x_0^j \rangle) := \{R \mid (a_i, a_j) : R \in \mathcal{A}\},$$
$$x_0^i \neq x_0^j \text{ iff } a_i \neq a_j \in \mathcal{A}, \text{ and}$$

the $\doteq$-relation is initialised to be empty. $\mathcal{F}_{\mathcal{A}}$ is then expanded by repeatedly applying the rules from Figure 1.

For a node x, $\mathcal{L}(x)$ is said to contain a clash if, for some concept name $A \in \mathbf{C}$, $\{A, \neg A\} \subseteq \mathcal{L}(x)$, or if there is some concept $\leq nS.C \in \mathcal{L}(x)$ and x has $n + 1$ S-neighbours $y_0, \ldots, y_n$ with $C \in \mathcal{L}(y_i)$ and $y_i \neq y_j$ for all $0 \leq i < j \leq n$. A completion forest is clash-free if none of its nodes contains a clash, and it is complete if no rule from Figure 1 can be applied to it.

For a $\mathcal{SHIQ}$-Abox $\mathcal{A}$, the algorithm starts with the completion forest $\mathcal{F}_{\mathcal{A}}$. It applies the expansion rules in Figure 1, stopping when a clash occurs, and answers "$\mathcal{A}$ is consistent w.r.t. $\mathcal{R}$" iff the completion rules can be applied in such a way that they yield a complete and clash-free completion forest, and "$\mathcal{A}$ and is inconsistent w.r.t. $\mathcal{R}$" otherwise.

Since both the $\leq$-rule and the $\leq_r$-rule are rather complicated, they deserve some more explanation. Both rules deal with the situation where a concept $\leq nR.C \in \mathcal{L}(x)$ requires the identification of two R-neighbours y, z of x that contain C in their labels. Of course, y and z may only be identified if $y \neq z$ is not asserted. If these conditions are met, then one of the two rules can be applied. The $\leq$-rule deals with the case where at least one of the nodes to be identified, namely y, is not a root node, and this can lead to one of two possible situations, both shown in Figure 2. The upper situation occurs when both y and z are successors of x. In this case, we add the label of y to that of z, and the label of the edge $\langle x, y \rangle$ to the label of the edge $\langle x, z \rangle$. Finally, z inherits all inequalities from y, and $\mathcal{L}(\langle x, y \rangle)$ is set to $\emptyset$, thus blocking y and all its successors.

The second situation occurs when both y and z are neighbours of x, but z is the predecessor of x. Again, $\mathcal{L}(y)$ is added to $\mathcal{L}(z)$, but in this case the inverse of $\mathcal{L}(\langle x, y \rangle)$ is added to $\mathcal{L}(\langle z, x \rangle)$, because the edge $\langle x, y \rangle$ was pointing away from x while $\langle z, x \rangle$ points towards it. Again, z inherits the inequalities from y and $\mathcal{L}(\langle x, y \rangle)$ is set to $\emptyset$.

The $\leq_r$ rule handles the identification of two root nodes. An example of the whole procedure is given in the lower part of Figure 2. In this case, special care has to be taken to preserve the relations introduced into the completion forest due to role assertions in

$\sqcap$-rule:	if 1. $C_1 \sqcap C_2 \in \mathcal{L}(x)$, x is not indirectly blocked, and 2. $\{C_1, C_2\} \not\subseteq \mathcal{L}(x)$ then $\mathcal{L}(x) \longrightarrow \mathcal{L}(x) \cup \{C_1, C_2\}$
$\sqcup$-rule:	if 1. $C_1 \sqcup C_2 \in \mathcal{L}(x)$, x is not indirectly blocked, and 2. $\{C_1, C_2\} \cap \mathcal{L}(x) = \emptyset$ then $\mathcal{L}(x) \longrightarrow \mathcal{L}(x) \cup \{E\}$ for some $E \in \{C_1, C_2\}$
$\exists$-rule:	if 1. $\exists S.C \in \mathcal{L}(x)$, x is not blocked, and 2. x has no S-neighbour y with $C \in \mathcal{L}(y)$ then create a new node y with $\mathcal{L}(\langle x, y \rangle) := \{S\}$ and $\mathcal{L}(y) := \{C\}$
$\forall$-rule:	if 1. $\forall S.C \in \mathcal{L}(x)$, x is not indirectly blocked, and 2. there is an S-neighbour y of x with $C \notin \mathcal{L}(y)$ then $\mathcal{L}(y) \longrightarrow \mathcal{L}(y) \cup \{C\}$
$\forall_+$-rule:	if 1. $\forall S.C \in \mathcal{L}(x)$, x is not indirectly blocked, and 2. there is some R with $\mathsf{Trans}(R)$ and $R \trianglelefteq S$, 3. there is an R-neighbour y of x with $\forall R.C \notin \mathcal{L}(y)$ then $\mathcal{L}(y) \longrightarrow \mathcal{L}(y) \cup \{\forall R.C\}$
choose-rule:	if 1. $(\bowtie\ n\ S\ C) \in \mathcal{L}(x)$, x is not indirectly blocked, and 2. there is an S-neighbour y of x with $\{C, \sim C\} \cap \mathcal{L}(y) = \emptyset$ then $\mathcal{L}(y) \longrightarrow \mathcal{L}(y) \cup \{E\}$ for some $E \in \{C, \sim C\}$
$\geqslant$-rule:	if 1. $\geqslant nS.C \in \mathcal{L}(x)$, x is not blocked, and 2. there are no n S-neighbours $y_1, \dots, y_n$ such that $C \in \mathcal{L}(y_i)$ and $y_i \neq y_j$ for $1 \leq i < j \leq n$ then create n new nodes $y_1, \dots, y_n$ with $\mathcal{L}(\langle x, y_i \rangle) = \{S\}$, $\mathcal{L}(y_i) = \{C\}$, and $y_i \neq y_j$ for $1 \leq i < j \leq n$.
$\leqslant$-rule:	if 1. $\leqslant nS.C \in \mathcal{L}(x)$, x is not indirectly blocked, and 2. $\sharp S^{\mathcal{F}}(x, C) > n$, there are S-neighbours y, z of x with not $y \neq z$, y is neither a root node nor an ancestor of z, and $C \in \mathcal{L}(y) \cap \mathcal{L}(z)$, then 1. $\mathcal{L}(z) \longrightarrow \mathcal{L}(z) \cup \mathcal{L}(y)$ and 2. if z is an ancestor of x then $\mathcal{L}(\langle z, x \rangle) \longrightarrow \mathcal{L}(\langle z, x \rangle) \cup \mathsf{Inv}(\mathcal{L}(\langle x, y \rangle))$ else $\mathcal{L}(\langle x, z \rangle) \longrightarrow \mathcal{L}(\langle x, z \rangle) \cup \mathcal{L}(\langle x, y \rangle)$ 3. $\mathcal{L}(\langle x, y \rangle) \longrightarrow \emptyset$ 4. Set $u \neq z$ for all u with $u \neq y$
$\leqslant_r$-rule:	if 1. $\leqslant nS.C \in \mathcal{L}(x)$, and 2. $\sharp S^{\mathcal{F}}(x, C) > n$ and there are two S-neighbours y, z of x which are both root nodes, $C \in \mathcal{L}(y) \cap \mathcal{L}(z)$, and not $y \neq z$ then 1. $\mathcal{L}(z) \longrightarrow \mathcal{L}(z) \cup \mathcal{L}(y)$ and 2. For all edges $\langle y, w \rangle$: i. if the edge $\langle z, w \rangle$ does not exist, create it with $\mathcal{L}(\langle z, w \rangle) := \emptyset$ ii. $\mathcal{L}(\langle z, w \rangle) \longrightarrow \mathcal{L}(\langle z, w \rangle) \cup \mathcal{L}(\langle y, w \rangle)$ 3. For all edges $\langle w, y \rangle$: i. if the edge $\langle w, z \rangle$ does not exist, create it with $\mathcal{L}(\langle w, z \rangle) := \emptyset$ ii. $\mathcal{L}(\langle w, z \rangle) \longrightarrow \mathcal{L}(\langle w, z \rangle) \cup \mathcal{L}(\langle w, y \rangle)$ 4. Set $\mathcal{L}(y) := \emptyset$ and remove all edges to/from y. 5. Set $u \neq z$ for all u with $u \neq y$. 6. Set $y \doteq z$.

Fig. 1. The Expansion Rules for $\mathcal{SHIQ}$-Aboxes.

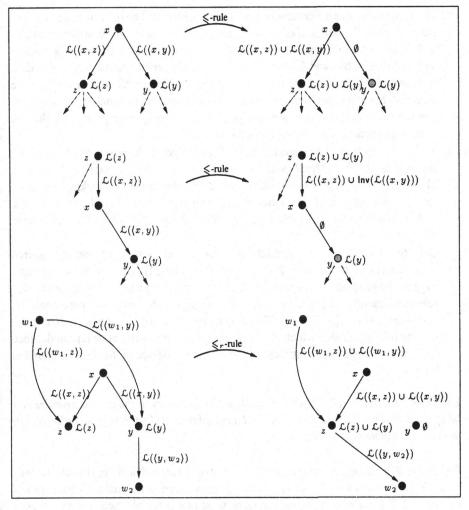

Fig. 2. Effect of the $\leqslant$- and the $\leqslant_r$-rule

the Abox, and to memorise the identification of root nodes (this will be needed in order to construct a tableau from a complete and clash-free completion forest). The $\leqslant_r$ rule includes some additional steps that deal with these issues. Firstly, as well as adding $\mathcal{L}(y)$ to $\mathcal{L}(z)$, the edges (and their respective labels) between y and its neighbours are also added to z. Secondly, $\mathcal{L}(y)$ and all edges going from/to y are removed from the forest. This will not lead to dangling trees, because all neighbours of y became neighbours of z in the previous step. Finally, the identification of y and z is recorded in the $\doteq$ relation.

Lemma 3. *Let $\mathcal{A}$ be a $\mathcal{SHIQ}$-Abox and $\mathcal{R}$ a role hierarchy. The completion algorithm terminates when started for $\mathcal{A}$ and $\mathcal{R}$.*

Proof: Let $m = \sharp\mathsf{clos}(\mathcal{A})$, $n = |\mathbf{R}_{\mathcal{A}}|$, and $n_{\max} := \max\{n \mid {\geqslant}nR.C \in \mathsf{clos}(\mathcal{A})\}$. Termination is a consequence of the following properties of the expansion rules:

1. The expansion rules never remove nodes from the forest. The only rules that remove elements from the labels of edges or nodes are the $\leqslant$- and $\leqslant_r$-rule, which sets them to $\emptyset$. If an edge label is set to $\emptyset$ by the $\leqslant$-rule, the node below this edge is blocked and will remain blocked forever. The $\leqslant_r$-rule only sets the label of a root node x to $\emptyset$, and after this, x's label is never changed again since all edges to/from x are removed. Since no root nodes are generated, this removal may only happen a finite number of times, and the new edges generated by the $\leqslant_r$-rule guarantees that the resulting structure is still a completion forest.

2. Nodes are labelled with subsets of $\text{clos}(\mathcal{A})$ and edges with subsets of $R_\mathcal{A}$, so there are at most 2^{2mn} different possible labellings for a pair of nodes and an edge. Therefore, if a path p is of length at least 2^{2mn}, the pair-wise blocking condition implies the existence of two nodes x, y on p such that y directly blocks y. Since a path on which nodes are blocked cannot become longer, paths are of length at most 2^{2mn}.

3. Only the $\exists$- or the $\geqslant$-rule generate new nodes, and each generation is triggered by a concept of the form $\exists R.C$ or $\geqslant n R.C$ in $\text{clos}(\mathcal{A})$. Each of these concepts triggers the generation of at most $n_{\max}$ successors y_i: note that if the $\leqslant$- or the $\leqslant_r$-rule subsequently causes $\mathcal{L}(\langle x, y_i \rangle)$ to be changed to $\emptyset$, then x will have some R-neighbour z with $\mathcal{L}(z) \supseteq \mathcal{L}(y)$. This, together with the definition of a clash, implies that the rule application which led to the generation of y_i will not be repeated. Since $\text{clos}(\mathcal{A})$ contains a total of at most $m\ \exists R.C$, the out-degree of the forest is bounded by $mn_{\max}n$. □

Lemma 4. *Let $\mathcal{A}$ be a $\mathcal{SHIQ}$-Abox and $\mathcal{R}$ a role hierarchy. If the expansion rules can be applied to $\mathcal{A}$ and $\mathcal{R}$ such that they yield a complete and clash-free completion forest, then $\mathcal{A}$ has a tableau w.r.t. $\mathcal{R}$.*

Proof: Let $\mathcal{F}$ be a complete and clash-free completion forest. The definition of a tableau $T = (\mathbf{S}, \mathcal{L}, \mathcal{E}, \mathcal{J})$ from $\mathcal{F}$ works as follows. Intuitively, an individual in $\mathbf{S}$ corresponds to a *path* in $\mathcal{F}$ from some root node to some node that is not blocked, and which goes only via non-root nodes.

More precisely, a *path* is a sequence of pairs of nodes of $\mathcal{F}$ of the form $p = [\frac{x_0}{x_0'}, \ldots, \frac{x_n}{x_n'}]$. For such a path we define $\text{Tail}(p) := x_n$ and $\text{Tail}'(p) := x_n'$. With $[p|\frac{x_{n+1}}{x_{n+1}'}]$, we denote the path $[\frac{x_0}{x_0'}, \ldots, \frac{x_n}{x_n'}, \frac{x_{n+1}}{x_{n+1}'}]$. The set $\text{Paths}(\mathcal{F})$ is defined inductively as follows:

- For root nodes x_0^i of $\mathcal{F}$, $[\frac{x_0^i}{x_0^i}] \in \text{Paths}(\mathcal{F})$, and
- For a path $p \in \text{Paths}(\mathcal{F})$ and a node z in $\mathcal{F}$:
  - if z is a successor of $\text{Tail}(p)$ and z is neither blocked nor a root node, then $[p|\frac{z}{z}] \in \text{Paths}(\mathcal{F})$, or
  - if, for some node y in $\mathcal{F}$, y is a successor of $\text{Tail}(p)$ and z blocks y, then $[p|\frac{z}{y}] \in \text{Paths}(\mathcal{F})$.

Please note that, since root nodes are never blocked, nor are they blocking other nodes, the only place where they occur in a path is in the first place. Moreover, by construction of Paths($\mathcal{F}$), if $p \in$ Paths($\mathcal{F}$), then Tail(p) is not blocked, Tail(p) = Tail$'(p)$ iff Tail$'(p)$ is not blocked, and $\mathcal{L}$(Tail(p)) = $\mathcal{L}$(Tail$'(p)$).

We define a tableau $T = (\mathbf{S}, \mathcal{L}, \mathcal{E}, \mathcal{J})$ as follows:

$$\mathbf{S} = \text{Paths}(\mathcal{F})$$

$$\mathcal{L}(p) = \mathcal{L}(\text{Tail}(p))$$

$$\mathcal{E}(R) = \{\langle p, [p|\tfrac{x}{x'}]\rangle \in \mathbf{S} \times \mathbf{S} \mid x' \text{ is an } R\text{-successor of Tail}(p)\} \cup$$
$$\{\langle [q|\tfrac{x}{x'}], q\rangle \in \mathbf{S} \times \mathbf{S} \mid x' \text{ is an } \text{Inv}(R)\text{-successor of Tail}(q)\} \cup$$
$$\{\langle [\tfrac{x}{x}], [\tfrac{y}{y}]\rangle \in \mathbf{S} \times \mathbf{S} \mid x, y \text{ are root nodes, and } y \text{ is an } R\text{-neighbour of } x\}$$

$$\mathcal{J}(a_i) = \begin{cases} [\tfrac{x_0^i}{x_0^i}] & \text{if } x_0^i \text{ is a root node in } \mathcal{F} \text{ with } \mathcal{L}(x_0^i) \neq \emptyset \\ [\tfrac{x_0^j}{x_0^j}] & \text{if } \mathcal{L}(x_0^i) = \emptyset, x_0^j \text{ a root node in } \mathcal{F} \text{ with } \mathcal{L}(x_0^j) \neq \emptyset \text{ and } x_0^i \doteq x_0^j \end{cases}$$

Please note that $\mathcal{L}(x) = \emptyset$ implies that x is a root node and that there is another root node y with $\mathcal{L}(y) \neq \emptyset$ and $x \doteq y$. We show that T is a tableau for D.

- T satisfies (**P1**) because $\mathcal{F}$ is clash-free.
- (**P2**) and (**P3**) are satisfied by T because $\mathcal{F}$ is complete.
- For (**P4**), let $p, q \in \mathbf{S}$ with $\forall R.C \in \mathcal{L}(p)$, $\langle p, q\rangle \in \mathcal{E}(R)$. If $q = [p|\tfrac{x}{x'}]$, then x' is an R-successor of Tail(p) and, due to completeness of $\mathcal{F}$, $C \in \mathcal{L}(x') = \mathcal{L}(x) = \mathcal{L}(q)$. If $p = [q|\tfrac{x}{x'}]$, then x' is an Inv(R)-successor of Tail(q) and, due to completeness of $\mathcal{F}$, $C \in \mathcal{L}(\text{Tail}(q)) = \mathcal{L}(q)$. If $p = [\tfrac{x}{x}]$ and $q = [\tfrac{y}{y}]$ for two root nodes x, x, then y is an R-neighbour of x, and completeness of $\mathcal{F}$ yields $C \in \mathcal{L}(y) = \mathcal{L}(q)$. (**P6**) and (**P11**) hold for similar reasons.
- For (**P5**), let $\exists R.C \in \mathcal{L}(p)$ and Tail(p) $= x$. Since x is not blocked and $\mathcal{F}$ complete, x has some R-neighbour y with $C \in \mathcal{L}(y)$.
  - If y is a successor of x, then y can either be a root node or not.
    * If y is not a root node: if y is not blocked, then $q := [p|\tfrac{y}{y}] \in \mathbf{S}$; if y is blocked by some node z, then $q := [p|\tfrac{z}{y}] \in \mathbf{S}$.
    * If y is a root node: since y is a successor of x, x is also a root node. This implies $p = [\tfrac{x}{x}]$ and $q = [\tfrac{y}{y}] \in \mathbf{S}$.
  - x is an Inv(R)-*successor* of y, then either
    * $p = [q|\tfrac{x}{x'}]$ with Tail(q) $= y$.
    * $p = [q|\tfrac{x}{x'}]$ with Tail(q) $= u \neq y$. Since x only has one predecessor, u is not the predecessor of x. This implies $x \neq x'$, x blocks x', and u is the predecessor of x' due to the construction of Paths. Together with the definition of the blocking condition, this implies $\mathcal{L}(\langle u, x'\rangle) = \mathcal{L}(\langle y, x\rangle)$ as well as $\mathcal{L}(u) = \mathcal{L}(y)$ due to the blocking condition.
    * $p = [\tfrac{x}{x}]$ with x being a root node. Hence y is also a root node and $q = [\tfrac{y}{y}]$.
  In any of these cases, $\langle p, q\rangle \in \mathcal{E}(R)$ and $C \in \mathcal{L}(q)$.
- (**P7**) holds because of the symmetric definition of the mapping $\mathcal{E}$.
- (**P8**) is due to the definition of R-neighbours and R-successor.
- Suppose (**P9**) were not satisfied. Hence there is some $p \in \mathbf{S}$ with $(\leqslant nS.C) \in \mathcal{L}(p)$ and $\sharp S^T(p, C) > n$. We will show that this implies $\sharp S^{\mathcal{F}}(\text{Tail}(p), C) > n$, contradicting either clash-freeness or completeness of $\mathcal{F}$. Let $x := \text{Tail}(p)$ and $P := S^T(p, C)$. We distinguish two cases:

- P contains only paths of the form $[p|\frac{y}{y'}]$ and $[\frac{x_0^{i_\ell}}{x_0^{i_\ell}}]$. Then $\sharp P > n$ is impossible since the function Tail' is injective on P: if we assume that there are two distinct paths $q_1, q_2 \in P$ and $\mathsf{Tail}'(q_1) = \mathsf{Tail}'(q_2) = y'$, then this implies that each q_i is of the form $q_i = [p|\frac{y_i}{y'}]$ or $q_i = [\frac{y'}{y'}]$. From $q_1 \neq q_2$, we have that $q_i = [p|\frac{y_i}{y'}]$ holds for some $i \in \{1, 2\}$. Since root nodes occur only in the beginning of paths and $q_1 \neq q_2$, we have $q_1 = [p|(y_1, y')]$ and $q_2 = [p|(y_2, y')]$. If y' is not blocked, then $y_1 = y' = y_2$, contradicting $q_1 \neq q_2$. If y' is blocked in $\mathcal{F}$, then both y_1 and y_2 block y', which implies $y_1 = y_2$, again a contradiction. Hence Tail' is injective on P and thus $\sharp P = \sharp \mathsf{Tail}'(P)$. Moreover, for each $y' \in \mathsf{Tail}'(P)$, y' is an S-successor of x and $C \in \mathcal{L}(y')$. This implies $\sharp S^{\mathcal{F}}(x, C) > n$.

- P contains a path q where $p = [q|\frac{x}{x'}]$. Obviously, P may only contain one such path. As in the previous case, Tail' is an injective function on the set $P' := P \setminus \{q\}$, each $y' \in \mathsf{Tail}'(P')$ is an S-successor of x, and $C \in \mathcal{L}(y')$ for each $y' \in \mathsf{Tail}'(P')$. Let $z := \mathsf{Tail}(q)$. We distinguish two cases:

  * $x = x'$. Hence x is not blocked, and thus x is an $\mathsf{Inv}(S)$-successor of z. Since $\mathsf{Tail}'(P')$ contains only successors of x we have that $z \notin \mathsf{Tail}'(P')$ and, by construction, z is an S-neighbour of x with $C \in \mathcal{L}(z)$.

  * $x \neq x'$. This implies that x' is blocked by x and that x' is an $\mathsf{Inv}(S)$-successor of z. Due to the definition of pairwise-blocking this implies that x is an $\mathsf{Inv}(S)$-successor of some node u with $\mathcal{L}(u) = \mathcal{L}(z)$. Again, $u \notin \mathsf{Tail}'(P')$ and, by construction, u is an S-neighbour of x and $C \in \mathcal{L}(u)$.

- For **(P10)**, let $(\geqslant n S.C) \in \mathcal{L}(p)$. Hence there are n S-neighbours $y_1, \ldots, y_n$ of $x = \mathsf{Tail}(p)$ in $\mathcal{F}$ with $C \in \mathcal{L}(y_i)$. For each y_i there are three possibilities:

  - y_i is an S-successor of x and y_i is not blocked in $\mathcal{F}$. Then $q_i := [p|\frac{y_i}{y_i}]$ or y_i is a root node and $q_i := [\frac{y_i}{y_i}]$ is in **S**.

  - y_i is an S-successor of x and y_i is blocked in $\mathcal{F}$ by some node z. Then $q_i = [p|\frac{z}{y_i}]$ is in **S**. Since the same z may block several of the y_js, it is indeed necessary to include y_i explicitly into the path to make them distinct.

  - x is an $\mathsf{Inv}(S)$-successor of y_i. There may be at most one such y_i if x is not a root node. Hence either $p = [q_i|\frac{x}{x'}]$ with $\mathsf{Tail}(q_i) = y_i$, or $p = [\frac{x}{x}]$ and $q_i = [\frac{y_i}{y_i}]$.

  Hence for each y_i there is a different path q_i in **S** with $S \in \mathcal{L}(\langle p, q_i \rangle)$ and $C \in \mathcal{L}(q_i)$, and thus $\sharp S^T(p, C) \geqslant n$.

- **(P12)** is due to the fact that, when the completion algorithm is started for an Abox $\mathcal{A}$, the initial completion forest $\mathcal{F}_{\mathcal{A}}$ contains, for each individual name a_i occurring in $\mathcal{A}$, a root node x_0^i with $\mathcal{L}(x_0^i) = \{C \in \mathsf{clos}(\mathcal{A}) \mid a_i : C \in \mathcal{A}\}$. The algorithm never blocks root individuals, and, for each root node x_0^i whose label and edges are removed by the $\leqslant_r$-rule, there is another root node x_0^j with $x_0^i \doteq x_0^j$ and $\{C \in \mathsf{clos}(\mathcal{A}) \mid a_i : C \in \mathcal{A}\} \subseteq \mathcal{L}(x_0^j)$. Together with the definition of $\mathfrak{I}$, this yields **(P12)**. **(P13)** is satisfied for similar reasons.

- **(P14)** is satisfied because the $\leqslant_r$-rule does not identify two root nodes x_0^i, y_0^i when $x_0^i \neq y_0^i$ holds. $\square$

Lemma 5. *Let $\mathcal{A}$ be a $\mathcal{SHIQ}$-Abox and $\mathcal{R}$ a role hierarchy. If $\mathcal{A}$ has a tableau w.r.t. $\mathcal{R}$, then the expansion rules can be applied to $\mathcal{A}$ and $\mathcal{R}$ such that they yield a complete and clash-free completion forest.*

Proof: Let $T = (\mathbf{S}, \mathcal{L}, \mathcal{E}, \mathcal{I})$ be a tableau for $\mathcal{A}$ and $\mathcal{R}$. We use T to trigger the application of the expansion rules such that they yield a completion forest $\mathcal{F}$ that is both complete and clash-free. To this purpose, a function π is used which maps the nodes of $\mathcal{F}$ to elements of $\mathbf{S}$. The mapping π is defined as follows:

- For individuals a_i in $\mathcal{A}$, we define $\pi(x_0^i) := \mathcal{I}(a_i)$.
- If $\pi(x) = s$ is already defined, and a successor y of x was generated for $\exists R.C \in \mathcal{L}(x)$, then $\pi(y) = t$ for some $t \in \mathbf{S}$ with $C \in \mathcal{L}(t)$ and $\langle s, t \rangle \in \mathcal{E}(R)$.
- If $\pi(x) = s$ is already defined, and successors y_i of x were generated for $\geqslant nR.C \in \mathcal{L}(x)$, then $\pi(y_i) = t_i$ for n distinct $t_i \in \mathbf{S}$ with $C \in \mathcal{L}(t_i)$ and $\langle s, t_i \rangle \in \mathcal{E}(R)$.

Obviously, the mapping for the initial completion forest for $\mathcal{A}$ and $\mathcal{R}$ satisfies the following conditions:

$$\left.\begin{array}{l} \mathcal{L}(x) \subseteq \mathcal{L}(\pi(x)), \\ \text{if } y \text{ is an } S\text{-neighbour of } x, \text{ then } \langle \pi(x), \pi(y) \rangle \in \mathcal{E}(S), \text{ and} \\ x \neq y \text{ implies } \pi(x) \neq \pi(y). \end{array}\right\} \qquad (*)$$

It can be shown that the following claim holds:

CLAIM: Let $\mathcal{F}$ be generated by the completion algorithm for $\mathcal{A}$ and $\mathcal{R}$ and let π satisfy $(*)$. If an expansion rule is applicable to $\mathcal{F}$, then this rule can be applied such that it yields a completion forest $\mathcal{F}'$ and a (possibly extended) π that satisfy $(*)$.

As a consequence of this claim, (P1), and (P9), if $\mathcal{A}$ and $\mathcal{R}$ have a tableau, then the expansion rules can be applied to $\mathcal{A}$ and $\mathcal{R}$ such that they yield a complete and clash-free completion forest. $\qquad \square$

From Theorem 1, Lemma 2, 3 4, and 5, we thus have the following theorem:

Theorem 2. *The completion algorithm is a decision procedure for the consistency of $\mathcal{SHIQ}$-Aboxes and the satisfiability and subsumption of concepts with respect to role hierarchies and terminologies.*

4 Conclusion

We have presented an algorithm for deciding the satisfiability of $\mathcal{SHIQ}$ KBs where the Abox may be non-empty and where the uniqueness of individual names is not assumed but can be asserted in the Abox. This algorithm is of particular interest as it can be used to decide the problem of conjunctive query containment w.r.t. a schema [17].

An implementation of the $\mathcal{SHIQ}$ Tbox satisfiability algorithm is already available in the FaCT system [14], and is able to reason efficiently with Tboxes derived from realistic ER schemas. This suggests that the algorithm presented here could form the basis of a practical decision procedure for the query containment problem. Work is already underway to test this conjecture by extending the FaCT system with an implementation of the new algorithm.

References

1. C. Areces, P. Blackburn, and M. Marx. A road-map on complexity for hybrid logics. In *Proc. of CSL'99*, number 1683 in *LNCS*, pages 307–321 Springer-Verlag, 1999.
2. F. Baader. Augmenting concept languages by transitive closure of roles: An alternative to terminological cycles. In *Proc. of IJCAI-91*, 1991.
3. F. Baader, H.-J. Bürckert, B. Nebel, W. Nutt, and G. Smolka. On the expressivity of feature logics with negation, functional uncertainty, and sort equations. *Journal of Logic, Language and Information*, 2:1–18, 1993.
4. F. Baader, H.-J. Heinsohn, B. Hollunder, J. Muller, B. Nebel, W. Nutt, and H.-J. Profitlich. Terminological knowledge representation: A proposal for a terminological logic. Technical Memo TM-90-04, DFKI, Saarbrücken, Germany, 1991.
5. P. Blackburn and J. Seligman. What are hybrid languages? In *Advances in Modal Logic*, volume 1, pages 41–62. CSLI Publications, Stanford University, 1998.
6. M. Buchheit, F. M. Donini, and A. Schaerf. Decidable reasoning in terminological knowledge representation systems. *J. of Artificial Intelligence Research*, 1:109–138, 1993.
7. D. Calvanese. Reasoning with inclusion axioms in description logics: Algorithms and complexity. In *Proc. of ECAI'96*, pages 303–307. John Wiley & Sons Ltd., 1996.
8. D. Calvanese, G. De Giacomo, and M. Lenzerini. On the decidability of query containment under constraints. In *Proc. of PODS'98*, pages 149–158. 1998.
9. D. Calvanese, G. De Giacomo, M. Lenzerini, D. Nardi, and R. Rosati. Source integration in data warehousing. In *Proc. of DEXA-98*. IEEE Computer Society Press, 1998.
10. Diego Calvanese, Giuseppe De Giacomo, Maurizio Lenzerini, Daniele Nardi, and Riccardo Rosati. Description logic framework for information integration. In *Proc. of KR-98*, 1998.
11. G. De Giacomo and F. Massacci. Combining deduction and model checking into tableaux and algorithms for converse-PDL. *Information and Computation*, 1998. To appear.
12. Giuseppe De Giacomo and Maurizio Lenzerini. What's in an aggregate: Foundations for description logics with tuples and sets. In *Proc. of IJCAI-95*, 1995.
13. V. Haarslev and R. Möller. An empirical evaluation of optimization strategies for abox reasoning in expressive description logics. In Lambrix et al. [19], pages 115–119..
14. I. Horrocks. FaCT and iFaCT. In Lambrix et al. [19], pages 133–135.
15. I. Horrocks, A. Rector, and C. Goble. A description logic based schema for the classification of medical data. In *Proc. of the 3rd Workshop KRDB'96*. CEUR, June 1996.
16. I. Horrocks and U. Sattler. A description logic with transitive and inverse roles and role hierarchies. *Journal of Logic and Computation*, 9(3):385–410, 1999.
17. I. Horrocks, U. Sattler, S. Tessaris, and S. Tobies. Query containment using a DLR ABox. LTCS-Report 99-15, LuFG Theoretical Computer Science, RWTH Aachen, Germany, 1999.
18. I. Horrocks, U. Sattler, and S. Tobies. Practical reasoning for expressive description logics. In *Proc. of LPAR'99*, number 1705 in *LNAI*, pages 161–180. Springer-Verlag, 1999.
19. P. Lambrix, A. Borgida, M. Lenzerini, R. Möller, and P. Patel-Schneider, editors. *Proc. of the International Workshop on Description Logics (DL'99)*, 1999.
20. E. Mays, R. Weida, R. Dionne, M. Laker, B. White, C. Liang, and F. J. Oles. Scalable and expressive medical terminologies. In *Proc. of the 1996 AMAI Annual Fall Symposium*, 1996.
21. U. Sattler. A concept language extended with different kinds of transitive roles. In *20. Deutsche Jahrestagung für KI*, volume 1137 in *LNAI*. Springer-Verlag, 1996.
22. A. Schaerf. Reasoning with individuals in concept languages. *Data and Knowledge Engineering*, 13(2):141–176, 1994.
23. K. Schild. A correspondence theory for terminological logics: Preliminary report. In J. Mylopoulos, R. Reiter, editors, *Proc. of IJCAI-91*, Sydney, 1991.
24. M. Schmidt-Schauß and G. Smolka. Attributive concept descriptions with complements. *Artificial Intelligence*, 48(1):1–26, 1991.

System Description: Embedding Verification into Microsoft Excel*

Graham Collins[1] and Louise A. Dennis[2]

[1] Department of Computing Science, University of Glasgow, Glasgow G12 8QQ, UK
[2] Division of Informatics, University of Edinburgh
Edinburgh EH1 1HN, UK

Abstract. The aim of the PROSPER project is to allow the embedding of existing verification technology into applications in such a way that the theorem proving is hidden, or presented to the end user in a natural way. This paper describes a system built to test whether the PROSPER toolkit satisfied this aim. The system combines the toolkit with Microsoft Excel, a popular commercial spreadsheet application.

1 Introduction

The PROSPER project is researching and developing a toolkit [1] that allows an expert to easily and flexibly assemble *proof engines* from existing tools to provide embedded formal reasoning support inside applications. The ultimate goal is to make the reasoning and proof support invisible to the end-user—or at least, more realistically, to incorporate it securely within the interface and style of interaction to which they are already accustomed. Several large case studies are taking place within the project to investigate this.

This paper describes a preliminary case study embedding verification into Microsoft Excel without inventing or re-implementing any existing theorem proving techniques or mathematical decision procedures.[1] The primary aim was to show that the technology is effective when applied to real, standard applications not designed by project members. In addition we were interested in investigating a "lightweight" theorem proving approach where only a small amount of theorem proving functionality is added but it is completely hidden from the user.

This paper begins with a brief overview of the PROSPER toolkit (§2) and Excel (§3) followed by a discussion of the system developed.

2 Extending Applications with Custom Proof Engines

A central part of PROSPER's vision is the idea of a proof engine—a custom built verification engine which can be operated by another program through an Application Programming Interface (API). A proof engine can be built by a system

---

* Work funded ESPRIT Framework IV Grant LTR 26241
[1] The case study is available from http://www.collins-peak.net/p-excel/

D. McAllester (Ed.): CADE-17, LNAI 1831, pp. 497–501, 2000.
© Springer-Verlag Berlin Heidelberg 2000

developer using the toolkit provided by the project. A proof engine is based upon the functionality of a theorem prover with additional capabilities provided by 'plugins' formed from existing, off-the-shelf, tools. The toolkit includes a set of libraries based on a language-independent specification, the PROSPER Integration Interface (PII), for communication between components of a final system. The theorem prover's command language is treated as a kind of scripting or glue language for managing plugin components and orchestrating the proofs.

The PII consists of several parts. There is a datatype for communication of data between components of a system which includes the language of higher order logic used by the HOL system[2] and so any formula expressible in higher order logic can be passed between components. There is support for installing procedures in an API and calling them remotely. There are also parts for managing low level communication, which are largely invisible to an application developer. The PII is currently implemented in ML, C, Java, Python, λProlog and ADA.

Proof engines are constructed on top of a small subset of HOL, called the *Core Proof Engine*. This consists of theorems, inference rules for higher order logic and an ML implementation of the PII. A developer can write extensions to the Core Proof Engine and place them in an API to form a custom proof engine. When incorporating a proof engine into an application the developer calls the customised API through the PII.

3   Microsoft Excel

Excel is a spreadsheet package marketed by Microsoft [4]. Its basic constituents are rows and columns of cells into which either values or formulae may be entered. Formulae refer to other cells, which may contain either values or further formulae.

Users of Excel are likely to have no interest in using or guiding mathematical proof, but they do want to know that they have entered formulae correctly. They therefore have an interest in 'sanity checking functions' that they can use to reassure themselves of correctness. This made Excel suited as a case study since the users have a notion of formulae and correctness, all that needs to be hidden is the proof. Another advantage is that Excel was designed to allow new functionality to be added and although its developers were not concerned with verification there is support for calling external tools.

As a simple example, the authors undertook to incorporate a sanity checking function into Excel. We chose to implement an equality checking function which would take two cells containing formulae and attempt to determine whether these formulae were equal for all possible values of the cells to which they refer.

Simplifying assumptions were made for the case study. The most important were that cell values were only natural numbers or booleans and that only a small subset of the functions available in Excel (some simple arithmetical and logical functions) appeared in formulae. Given these assumptions, less than 150 lines of code were needed to produce a prototype. This prototype handled only a small range of formulae decidable by linear arithmetic or propositional logic decision procedures, but it demonstrated the basic functionality.

4 Architecture

The main difficulty in the system was that Excel is Windows based and expects Microsoft's Component Object Model (COM) to be used for communication between processes, whereas the PROSPER toolkit had been developed for UNIX machines[2] and uses sockets for communications between components.

Several possible solutions to this problem were considered including implementing the PII in Visual Basic and using internet sockets to let Excel communicate with a proof engine. We did not take this approach because our aim was to show that theorem proving technology can be incorporated into applications in as natural a way as possible. For Excel this meant making the functionality of the PROSPER tools available as a COM server.

The PROSPER COM server was implemented in Python, a dynamically typed, object oriented scripting language which supports both COM and sockets. The server consists of two parts, the python implementation of the PII and the additional code described below which is specific to this example.

The remaining decision was where to convert Excel's formulae, which we access as strings, into terms. This requires some type inference but is simple to do and could have been written in either the Python or Visual Basic components. This was done in Python since it was the preferred language of the authors.

From the Excel side the Python component is a COM server which makes available a small number of functions that Excel can call. The use of a UNIX based theorem prover is not visible to Excel. From the proof engine side the Python component behaves like any other application calling the proof engine using the PII. The use of Excel is not visible to the theorem prover.

A view of the current (2 operating system) architecture is shown below.

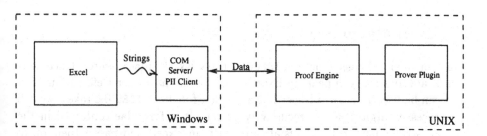

5 Custom Proof Engine

The initial custom proof procedure is very simple-minded. It uses a linear arithmetic decision procedure provided by HOL and a propositional logic plugin (based on Prover Technology's proof tool [6,5]) to decide the truth of formulae. While the approach is not especially robust, it is strong enough to handle many formulae.

---

[2] It is expected that a future version will be ported to Windows.

The additional code required to create this custom proof procedure is very small (approx. 45 extra lines of ML were needed). All the verification code used already existed either in HOL or the plugin, the new code concentrated on gluing together the decision procedures and deciding which should be used.

A proof engine which could handle a wider range of formulae would require more work. It is possible that more decision procedures could be used to provide this, for instance we could exploit HOL's simplifier. Alternatively it might prove necessary to implement some specialised theorem proving algorithms. This would also be possible using the PROSPER toolkit.

6   Python COM Component

The main piece of code developed for this system is the Python implementation of the PII. This was simple to write, partly since the structure is similar to the existing Java PII, and partly because this is the sort of application for which Python was designed. The code makes use of dynamic typing and other features of the language to provide a compact and natural implementation of the PII. Although written for this one application, the Python implementation makes available the objects of the PII, and hence the functionality of the PROSPER tools to any language that supports COM.

In addition to the PII implementation the COM component contains some additional code specific to this example. This first parses the strings to logical terms. This assumes that the semantics of the operators is the same in Excel and HOL. The terms are then passed on to the proof engine. It returns the result of the proof attempt as true, false, or 'unable to decide', which is displayed in the cell containing the ISEQUAL formula. This result can be used by other cells and will be automatically recomputed if necessary.

7   Excel Macro

We wrote a visual basic function, ISEQUAL, using Excel's macro editor. Once written, it automatically appears in Excel's function list as a User Defined Function and can be used in a spreadsheet like any other function. ISEQUAL takes two cell references as arguments. It recursively extracts the formulae contained in the cells as strings (support for this already exists in Excel) and passes them on to the Python object. The macro consists of about 30 lines of Visual Basic code.

8   Conclusions

There are numerous Add-Ins to Excel many of which, unsurprisingly, extend its mathematical ability. The Maple 6 Add-In provides computer algebra techniques to Excel spreadsheets. Interval Solver [3] extends Excel with Interval Constraint Solving to allow spreadsheet users to reason with incomplete and uncertain information. We believe that theorem proving could also have a role to play in this field. We have demonstrated that the PROSPER approach provides a framework in which this could be done.

We were surprised and pleased with the ease that a very basic prototype of verification support for Excel could be produced. It took two programmers, neither of whom had any experience with Visual Basic, Python or COM only 48 hours to get to the point where Excel was able to prove the commutativity of plus. While this may seem uninteresting, the reordering of the mathematical operators in large formulae is exactly the kind of lightweight sanity check that may appeal to users. Extending the system to handle more arithmetic and logical operators was easy and the system has been tested on a range of linear arithmetic and spreadsheet style examples. The system could be extended further and more complex and interesting proof strategies could be programmed.

The system is a proof of concept of the claim made by the PROSPER project that their toolkit would enable the embedding of verification into applications not designed with it specifically in mind. The only significant piece of new code is the Python port of the PII which is a general purpose component that could be used for other systems.

Adding even limited theorem proving functionality by programming a procedure from scratch instead of using existing tools would have taken much longer, as would interfacing to a theorem prover without using the PROSPER tools.

The use of two operating systems is not ideal but could be removed if the PROSPER tools were ported to Windows. The current setup would be reasonable in a networked setting with many copies of Excel accessing one proof engine.

The embedding of verification into Excel also serves as an example of the concepts of "lightweight" theorem proving and the "invisible" use of verification. Here all the infrastructure is invisible to the user who simply gets an extra function available in Excel.

References

1. L. A. Dennis, G. Collins, M. Norrish, R. Boulton, K. Slind, G. Robinson, M. Gordon and T. Melham, The PROSPER Toolkit, *TACAS 2000*, to appear. 2000.
2. M. J. C. Gordon and T. F. Melham (eds), *Introduction to HOL: A theorem proving environment for higher order logic*, Cambridge University Press, 1993.
3. E. Hyvönen and S. De Pascale, A New Basis for Spreadsheet Computing: Interval Solver^{TM} for Microsoft Excel. *Proceedings of 16th National Conference on Artificial Intelligence and 11th Innovative Applications of Artificial Intelligence Conference (AAAI/IAAI-99)*, AAAI Press / The MIT Press, pp. 799–806, 1999.
4. Microsoft Corporation, *Microsoft Excel*, http://www.microsoft.com/excel.
5. M. Sheeran and G. Stålmarck, A tutorial on Stålmarck's proof procedure for propositional logic. *The Second International Conference on Formal Methods in Computer-Aided Design*, Lecture Notes in Computer Science 1522, Springer-Verlag, pp. 82–99, 1998.
6. G. Stålmarck and M. Säflund, Modelling and Verifying Systems and Software in Propositional Logic. *Proceedings of SAFECOMP '90*, Pergamon Press, pp. 31–36, 1990.

System Description: Interactive Proof Critics in XBarnacle

Mike Jackson[1] and Helen Lowe[2]

[1] Department of Electronic and Electrical Engineering, University of Edinburgh, Scotland
Michael.Jackson@ee.ed.ac.uk
[2] Department of Computer Studies, Glasgow Caledonian University
Cowcaddens Road, Glasgow G4 0BA, Scotland
H.Lowe@gcal.ac.uk

1 Introduction

Proof critics [2] extend the power of a theorem prover by, for example, allowing lemmas to be postulated and proved in the course of a proof. However, extending the automated theorem prover CLAM by adding critics also increases the search space. XBarnacle [5] was developed to make the process of interacting with a semi-automatic theorem prover more tractable for the non-expert user. We have now substantially amplified and extended XBarnacle so that it makes the work of expert users more efficient as they interact with proof critics. Of course, we have also made cosmetic improvements to aid navigability and to bridge the *gulf of evaluation* [1] which proves such an obstacle in making theorem provers more accessible, and even their expert users more efficient.

2 System Requirements

In building the new version of XBarnacle, we were able to build on experience with several systems, each with strengths and weaknesses. An obvious meta-requirement was to try to incorporate the strengths of each whilst avoiding the weaknesses.

- Clam version 3.2 could patch proofs by applying critics under certain patterns of failure in the pre-conditions of methods. However, this greatly increases the search space, making interaction necessary.
- XClam [3] was a graphical interface to Clam version 3.2 but, like its parent system had no persistent representation of the partial plan, making undoing and re-planning nodes impossible; it also used non-hierarchical planning, making navigation difficult.
- The version of XBarnacle (based on clam version 2.2) reported in [5] allowed the user to interact with the proof tree in the course of a proof but could not patch proofs automatically.

D. McAllester (Ed.): CADE-17, LNAI, pp. 502-506, 2000.

– Another version of XBarnacle [4] incorporated critics in the "flat" structure of Clam 3.2. This made proof trees very rapidly become large and unnavigable in practice if not in theory.. Co-incidentally, during the evaluation users requested *passive* as opposed to *active* critiquing (so that the decision to apply a critic was the user's rather than the program's).

The proof engine of the new version of XBarnacle was therefore to be an amalgam of CLAM version 2.6, which has no critics but a hierarchical method set; and CLAM version 3.2, which incorporates critics in a flat method structure. In addition, we had found that a common request on the original XBarnacle system reported in [5] was to be able to "open up" the high-level proof steps to see the individual smaller steps within. This facility of hierarchical tree browsing has therefore been provided in the current system, not merely to help less experienced users see how the steps are performed, but so that expert users can interact directly with the step requiring the use of critics.

3   Hierarchical Tree Browsing

With previous versions of XBarnacle only the top level goals were retained and displayed. The intermediate goals arising in sub-plans were not stored. For example, if the method used several rewrite rules only the final result of rewriting was shown, not the intermediate steps. We made modifications so that all of these steps are retained and can be displayed on request. This helps to bridge the gulf of evaluation for users who may find that the granularity is too high, and that the system is making too big a leap from one step to another. It also facilitates the use of critics. Experienced users can open up nodes until they see a likely looking sub-goal for which they know a lemma will also certainly be proposed by a critic.

A data structure in Prolog to facilitate hierarchical tree browsing was provided as a set of *tree_nodes*. The Tcl/Tk side also has a *tree_node* array. This array is indexed by the canvas in which the sub-plan will be drawn if the user requests it. It indicates whether any of the (sub-)goals may be critiqued, in other words, whether a critic might be applicable.

4   Critics

XBarnacle is a co-operative system and allows the user to critique nodes in the proof plan. Interactive proof critics will then propose ways in which the proof might be improved, for example by the addition of some wave rules or generalising a goal and the user is free to accept or reject the proposed patches. All nodes that may be critiqued with the loaded critics will be marked by XBarnacle using a distinctive icon.

XBarnacle allows the user to critique either a specific goal, in which case critics are tested for applicability at that goal only; or an entire sub-plan, in which case every goal that is in the sub-plan (and recursively sub-plans of the sub-plan etc.) will be critiqued (i.e. the system will test for applicability of critics). The type of critiquing supported on a node will be indicated by a shaded mark on the node. The user can then indicate that they wish to critique the goal, possibly by using hierarchical tree browsing, or the sub-plan.

After having selected whether to critique a goal or a sub-plan XBarnacle will display a list of the critics that may be used to critique the goal or sub-plan. Currently these are lemma calculation, lemma speculation, generalization, and inductiiob revision. The user can select one or more of these critics to try to apply to the goal/sub-plan. For example, if the user thinks that the original goal must be generalized before the theorem is provable, they will choose generalization. If there seems to be a missing lemma, lemma calucation and/or lemma speculation will be chosen.

Once the critic is selected, XBarnacle will see if the chosen critics can propose some patches for the selected goal, or goals in a sub-plan.

If the user has no idea which to choose, they can simply leave the choice to XBarnacle. Instead of choosing a selected set of critics to apply as described above, pressing the *Critique with All Critics* button causes XBarnacle to critique the goal/sub-plan using all the allowable critics.

View the Proposed Patches is then facilitated. There will be a short delay as CLAM tests whether the critics propose any patches. Critiquing a sub-plan may take a while since every goal in the sub-plan hierarchy must be critiqued.

If any of the chosen critics propose patches for the goal/sub-plan goals being critiqued then XBarnacle will display a *Proposed Patches* window with a list of the possible patches, each patch including the name of the critic that proposed that patch..

The user can apply a patch or retrieve a number of types of information about a patch. Some patches may need to be customized. All of these actions are specific to the currently selected patch.

Clicking on a patch in the list of patches displayed selects that patch. Actions such as *customization, apply patch, view locations, explanations* and *view patches as wave rules*, will now be specific to the selected patch until the user selects another one.

To apply a selected patch we just press the *Apply* button and XBarnacle will attempt to apply the selected patch. There will be a delay as XBarnacle plans the node where the patch is to be applied and attempts to apply the patch. If the attempt to apply the patch fails then XBarnacle will just apply an applicable method instead.

Two types of explanation may be viewed for each patch proposed in the possible patches window. All are more suitable for users familiar with proof planning and rippling than general or novice users.

- The *Why critic applicable?* button displays an explanation as to why the method associated with the critic failed, in terms of the pattern of precondition failure of the method, and some extra information;
- The *What critic does?* button gives a general explanation as to what the critic will do to patch the proof.

Patches intended to be used as wave rules, which, at present, are the patches proposed by the lemma calculation and lemma speculation critics, may be viewed as wave rules, shaded graphically.

Some patches need to be customized by the user before they can be applied. The user is expected to provide some information. On selecting a customizable patch the *Customize...* button in the *Proposed Patches* window will become active. On pressing this button XBarnacle will display the *Customize* window, displaying the patch. The customizable parts of the patch (higher order meta-terms) will be displayed so that the patch can be customized by editing these to provide the necessary instantiations. The instantiations may use any of the variables listed in the *Customize* window, any of the functions loaded into XBarnacle, and any of the standard constants and operators. Infix versions of common functions, such as + for *plus* are also accepted.

If the user has instantiated all the meta-terms in the patch then they can try to apply it. This is done by pressing the *Apply* button. XBarnacle will first perform checks to ensure that the patch contains no uninstantiated meta-terms, there are no unknown function or other symbols, the term is syntactically correct and that there are no type violations.

5 Obtaining the System

The interface side of XBarnacle is written in Tcl 8.0 and Tk 8.0. These may be downloaded free from http://www.scriptics.com/software/8.0.html. The proof engine is written in SICSTUS Prolog 3.5, obtainable from http://www.sics.se/isl/sicstus.html. XBarnacle is available at http://members.xoom.com/helen_lowe/XBarnacle.tar. After extracting the files, go to the tk_tcl/make subdirectory and edit the Makefile as described in the file at http://members.xoom.com/helen_lowe/readme.txt. Follow the instructions in that file to make and run the executable.

The doc/ sub-directory contains substantial user and developer information. The user guide includes a short tutorial. The developer manual contains details of the architecture, the interplay between Prolog and tcl/tk, data structures used to hold information to be displayed and facilitate navigation, and the design rationale behind some of the design decisions taken.

6 Further Work

The system was built for, and evaluated by, expert users. As HCI practitioners, we focus strongly on the user and the task for which the system is designed. One maxim is "Speak the User's Language" [6]. The users in this case were all members of, or close associates of, the Mathematical Reasoning Group in the Division of Informatics at the University of Edinburgh. Their interests were in theorem proving per se. Redesigning the system so that, for example, it supports users of proof tools, interested primarily in program development, is not just a simple question of rewording explanations, although that could be easily done. However, much of what we have learned can be carried through to other users and other tasks. A hierarchical display of the proof seems fairly generally desirable, merely the level and granularity differing between novices and experts. There is no conclusive evidence that passive as opposed to active critiquing is desired by all users, but these could easily be provided as alternatives, a feature customizable by the user.

References

1. E. L. Hutchins, Hollan, J. D. and Norman D. Direct manipulation interfaces. In *User Centred System Design* (Norman, D. and Draper, S., eds.), pp. 87–124. Hillsdale, NJ: Lawrence Erlbaum Associates, 1986.
2. Ireland, A., and Bundy, A. Productive Use of Failure in Inductive Proof. Special edition of *Journal of Automated Reasoning* on Inductive Proof **16** 1996.
3. Ireland, A., Jackson, M., and Reid, G. Interactive Proof Critics, Formal Aspects of Computing **11**, 302-325, Springer-Verlag, 1999..
4. Jackson, M. Interacting with Semi-automated Theorem Provers via Interactive Proof Critics PhD Thesis, School of Computing, Napier University, 1999.
5. Lowe, H., and Duncan, D. XBarnacle: Making Theorem Provers More Accessible. Proceedings of the Fourteenth conference on Automated Deduction, Townsville, Australia, pp 108–121, 1997.
6. Molich, R. and Nielsen, J. Improving a human-computer dilogue, Communications of the ACM, 33(3), 338-348, 1990.

Tutorial: Meta-logical Frameworks

Carsten Schürmann

Department of Computer Science, Carnegie Mellon University
Pittsburgh, PA 15213, USA
carsten@cs.cmu.edu

The logical framework LF [HHP93] is a meta-language for specifying formal languages and related algorithms. It is typically used to represent programming languages, type systems and logics, such as operational semantics, compilers, natural deduction, sequent calculi, etc. For a survey on logical frameworks consult [Pfe99]. LF derives its expressive power from dependent types together with higher-order representation techniques which directly support common concepts in deductive systems such as variable binding, capture-avoiding substitutions, parametric and hypothetical judgments and substitution properties.

Meta-logical frameworks on the other hand extend logical frameworks by the ability to formalize and in many cases to mechanize the task of reasoning about those languages and their algorithms. Their purpose is to improve the designers productivity by automating tedious and error-prone aspects to their task.

Twelf is such a meta-logical framework. It extends the logical framework LF by a meta-logic $\mathcal{M}_2^+$ [Sch00]. Twelf is designed as a formal tool to express and reason about properties of systems represented in LF while taking advantage of the expressive power of the underlying logical framework. For example, Twelf has been successfully employed to derive various properties of programming languages, type systems and logics, such as for example type preservation and progress of various operational semantics, the consistency of logics, and the admissibility of new inference rules. Other results include automatic proofs of the Church-Rosser theorem, cut-elimination for various logics, soundness and completeness of uniform proof search and resolution.

Unlike theorem provers that are based on standard principles requiring first-order encodings, Twelf derives its deductive power from marrying higher-order representation techniques with the technique of inductive reasoning. Therefore, in these special domains its automated reasoning power far exceeds that of any other theorem prover.

Twelf is written in Standard ML and runs under SML of New Jersey and MLWorks on Unix and Window platforms. The current version is distributed with a complete manual [PS98], example suites, a tutorial in the form of on-line lecture notes [Pfe00], and an Emacs interface. Source and binary distributions are accessible via the Twelf home page http://www.twelf.org.

D. McAllester (Ed.): CADE-17, LNAI 1831, pp. 507–508, 2000.
© Springer-Verlag Berlin Heidelberg 2000

References

[HHP93] Robert Harper, Furio Honsell, and Gordon Plotkin. A framework for defining logics. *Journal of the Association for Computing Machinery*, 40(1):143–184, January 1993.

[Pfe99] Frank Pfenning. Logical frameworks. In Alan Robinson and Andrei Voronkov, editors, *Handbook of Automated Reasoning*. Elsevier Science Publishers, 1999. In preparation.

[Pfe00] Frank Pfenning. *Computation and Deduction*. Cambridge University Press, 2000. In preparation. Draft from April 1997 available electronically.

[PS98] Frank Pfenning and Carsten Schürmann. *Twelf User's Guide*, 1.2 edition, September 1998. Available as Technical Report CMU-CS-98-173, Carnegie Mellon University.

[Sch00] Carsten Schürmann. *Automating the Meta-Theory of Deductive Systems*. PhD thesis, Carnegie-Mellon University, 2000. forthcoming.

Tutorial: Automated Deduction and Natural Language Understanding

Stephen Pulman

University of Cambridge Computer Laboratory
and SRI International, Cambridge
sgp@cl.cam.ac.uk

The purpose of this tutorial is to introduce the Automated Deduction community to a growing area in which their expertise can be applied to a novel set of problems. No knowledge of natural language processing will be assumed. (If you have time for some background reading, James Allen's 'Natural Language Understanding', 2nd edition, Addison Wesley, 1995 can be recommended).

In this tutorial I will describe and illustrate some of the areas in which NLU interacts with theorem proving, and say what our problems are. My hope is that out of this you will get some interesting new problems to work on, and that we will eventually get answers to some of our questions.

The topics to be covered include:

Semantic Assembly. At some point, all interesting natural language processing applications need to relate sentences to a meaning representation of some kind. In our case, the target representation is usually first order logic, augmented with some higher order constructs. We choose first order logic because that is what it is easiest to mechanise inferences for, but natural language is intrinsically higher order. I will describe some of the typical problems that arise in choosing logical representations for English constructs that are (a) capable of being derived compositionally from parsed sentences and (b) capable of supporting the necessary inferences. Usually this is a balancing act between expressiveness and efficiency of inference. What we need from you is some guidance about what higher order constructs are 'safe' in that they can be mechanised with reasonable efficiency, and some lessons in how to transform or compile the rest into something that can be used in practical systems.

Underspecified Representations. Unfortunately, sentences usually contain context-dependent constructs (pronouns, etc.) whose interpretation will depend on the circumstances in which the sentence is produced. This means that semantic interpretation has to take place in two stages: one relatively compositional phase in which the meanings of the words and their syntactic configuration are used to build a 'quasi-logical form', and a second stage in which inferences are made from the context to flesh out the quasi-logical form to something that is evaluable independently. We have two problems here: firstly, the 'quasi-logical' forms only have a semantics indirectly and so we need inference mechanisms for non-standard logics, or at least non-standard, linguistically oriented representations. Secondly, the context may

D. McAllester (Ed.): CADE-17, LNAI 1831, pp. 509–510, 2000.
© Springer-Verlag Berlin Heidelberg 2000

not be fully represented either and so we often need to make conditional or abductive inferences leading to conclusions of the form 'P if Q holds'. Again, many of the constructs we are dealing with are higher order, and typically, the search spaces can be very large. ¡/p¿

Disambiguation. Most sentences are ambiguous. We can use various methods for choosing the most likely reading. Many readings can be eliminated because they are inconsistent with the model that has been built up by previous sentences. I will describe current applications of model building and model checking for this purpose. Our problems here include (a) reasoning efficiently with large numbers of axioms (b) efficient reasoning with equality. Yet again, many of the constructs we are dealing with are higher order.

Some web sites which offer demos of NLP systems using theorem provers in the ways described above:

- `http://www.coli.uni-sb.de/~bos/doris`: Johan Bos's DORIS system
- `http://www.cs.rochester.edu/research/epilog`: Len Schubert's EPILOG demo
- `http://ubatuba.ccl.umist.ac.uk`: Allan Ramsay's PARASITE system

Tutorial: Using TPS for Higher-Order Theorem Proving and ETPS for Teaching Logic

Peter B. Andrews and Chad E. Brown

Department of Mathematical Sciences, Carnegie Mellon University
Pittsburgh, PA 15213, USA
Peter.Andrews@cmu.edu, cebrown@andrew.cmu.edu

TPS is an automated theorem proving system which can be used to prove theorems of first- or higher-order logic automatically, interactively, or in a combination of these modes of operation. Proofs in TPS are presented in natural deduction style. ETPS is a program which was obtained from TPS by deleting all the facilities for proving theorems automatically. ETPS can be used by students to learn how to prove theorems interactively. The objective of the tutorial is to teach participants how to make effective use of TPS and ETPS.

Information about TPS, including manuals and information about obtaining the system, can be found at http://gtps.math.cmu.edu/tps.html.

ETPS is intended to be used as a teaching tool in logic courses. ETPS can be used effectively in a course which is concerned purely with first-order logic as well as one which also deals with higher-order logic. ETPS gives students immediate feedback for both correct and incorrect actions, and makes it easy to display selected parts of proofs, as well as modify and rearrange them. Proofs, and the active lines of the proof, are displayed in proof windows which are automatically updated as the proof is constructed interactively. ETPS enables students to construct rigorous proofs of more difficult theorems than they otherwise find tractable. ETPS checks proofs automatically, and creates records of the theorems proved by each student which can be automatically transferred to the teacher's grade file.

The logical language of TPS is Church's type theory, a formulation of higher-order logic in which theorems of mathematics can be expressed very naturally. The notation of this language is displayed on the screen and in printed proofs. Definitions are handled elegantly by λ-notation. The tutorial presupposes familiarity with first-order logic, but not with higher-order logic. The tutorial includes an introduction to the notation of type theory, examples showing how to express theorems of mathematics (including those involving inductive definitions) in this language, and lessons on how to write theorems and definitions in TPS and put them into a TPS library.

The facilities for constructing natural deduction proofs and an editor for wffs are common to TPS and ETPS. In addition, TPS has tactics for applying natural deduction rules of inference semi-automatically, and automatic procedures for constructing complete proofs or filling in gaps in partially completed natural

* The development of TPS and ETPS was supported by the National Science Foundation under grant CCR-9732312 and previous grants.

D. McAllester (Ed.): CADE-17, LNAI 1831, pp. 511–512, 2000.

deduction proofs. TPS searches for proofs in automatic mode by first searching for an expansion proof, and then translating this into a natural deduction proof. TPS has a number of search procedures, and there are many flags which control the behavior of TPS and set bounds for the many dimensions of proof search in higher-order logic.

TPS is designed to be a research tool as well as a theorem proving system. It has facilities for working on unification problems and mating searches, displaying wffs in vertical path diagrams, printing proofs in various styles including tex, and translating back and forth between natural deduction proofs and expansion proofs. TPS has library facilities, online help, and extensive documentation (some of which is produced automatically).

The tutorial provides opportunities for hands-on experience with TPS and ETPS, and discussion of how to treat examples provided by the participants.

References

1. Peter B. Andrews. *An Introduction to Mathematical Logic and Type Theory: To Truth Through Proof.* Academic Press, 1986.
2. Peter B. Andrews. On Connections and Higher-Order Logic. *Journal of Automated Reasoning*, 5:257–291, 1989.
3. Peter B. Andrews, Matthew Bishop, Sunil Issar, Dan Nesmith, Frank Pfenning, and Hongwei Xi. TPS: A Theorem Proving System for Classical Type Theory. *Journal of Automated Reasoning*, 16:321–353, 1996.
4. Matthew Bishop. A Breadth-First Strategy for Mating Search. In Harald Ganzinger, editor, *Proceedings of the 16th International Conference on Automated Deduction*, volume 1632 of *Lecture Notes in Artificial Intelligence*, pages 359–373, Trento, Italy, 1999. Springer-Verlag.
5. Matthew Bishop and Peter B. Andrews. Selectively Instantiating Definitions. In Claude Kirchner and Hélène Kirchner, editors, *Proceedings of the 15th International Conference on Automated Deduction*, volume 1421 of *Lecture Notes in Artificial Intelligence*, pages 365–380, Lindau, Germany, 1998. Springer-Verlag.
6. Alonzo Church. A Formulation of the Simple Theory of Types. *Journal of Symbolic Logic*, 5:56–68, 1940.
7. Douglas Goldson, Steve Reeves, and Richard Bornat. A Review of Several Programs for the Teaching of Logic. *The Computer Journal*, 36:373–386, 1993.
8. Sunil Issar. Path-Focused Duplication: A Search Procedure for General Matings. In *AAAI-90. Proceedings of the Eighth National Conference on Artificial Intelligence*, volume 1, pages 221–226. AAAI Press/The MIT Press, 1990.
9. Dale A. Miller. A Compact Representation of Proofs. *Studia Logica*, 46(4):347–370, 1987.
10. Frank Pfenning and Dan Nesmith. Presenting Intuitive Deductions via Symmetric Simplification. In M. E. Stickel, editor, *Proceedings of the 10th International Conference on Automated Deduction*, volume 449 of *Lecture Notes in Artificial Intelligence*, pages 336–350, Kaiserslautern, Germany, 1990. Springer-Verlag.

Workshop: Model Computation – Principles, Algorithms, Applications

Peter Baumgartner, Chris Fermueller, Nicolas Peltier, and Hantao Zhang

Computing models of first-order or propositional logic specifications is the complementary problem of refutational theorem proving. A deduction system capable of producing models significantly extends the functionality of purely refutational systems by providing the user with useful information in case that no refutation exists.

Ideally, any theorem prover that terminates without finding a refutational proof should be able to output (information on) countermodels. Characterizing classes of inputs for which termination can be guaranteed, defining appropriate formalisms for representing such models, and providing algorithms for working with the resulting model representations (e.g., evaluating clauses, testing equivalence, etc.) is a great challenge in automated deduction.

Computing models is becoming an increasingly important topic in automated deduction. This is due to the potential application areas like disproving conjectures in classical theorem proving and software verification; discourse representation in natural language, deductive databases, product configuration, hardware verification, model-based diagnosis, planning, model checking etc.

Some of these methods currently rely heavily on first-order logic with finite-domain or propositional logic (e.g. model checking). On the other side, methods for computing models for first-order specifications have been emerging recently by linking fields like term rewriting, term schematizations, and constraint evaluation and their potential is worth to be explored much further.

The workshop is therefore emphasizing model construction principles for the first-order case, although also highly welcoming contributions concentrating on finite-domain, propositional logics, and more expressive logics such as higher-order and modal logics. More specifically, the goal of the workshop is to discuss (non-exclusively) research on the following issues:

- Theoretical background, such as representation formalisms for models and their properties (like expressivity and complexity).
- Calculi and respective procedures to compute models, implementations, experiments and performance issues.
- Applications and related topics, such as finding the appropriate formulation, application problems, and problem sets.

The goal of the workshop is to bring together researchers working on these and related topics. As an outcome, the workshop would help to identify important problems in model computation, concentrate our efforts coming from different directions to attack these problems, get new insights by mutually learning from the various aspects of model computation, and stimulate further research.

Workshop home page: http://www.uni-koblenz.de/~peter/CADE17-WS-MODELS/

D. McAllester (Ed.): CADE-17, LNAI 1831, pp. 513–513, 2000.
© Springer-Verlag Berlin Heidelberg 2000

Workshop: Automation of Proof by Mathematical Induction

Carsten Schürmann

Carnegie Mellon University, Pittsburgh, USA

Mathematical induction is required for reasoning about objects or events containing repetition, e.g. computer programs with recursion or iteration, electronic circuits with feedback loops or parameterized components, and properties that hold for all time forward. It is thus a vital ingredient of formal methods techniques for synthesizing, verifying and transforming software and hardware. The automation of proof by induction strengthens the capabilities of mechanical assistants, it reduces the need for designers to be skilled in mathematical proof techniques, and it improves productivity by automating tedious and error-prone aspects of formal system development. This workshop is organized around four sessions.

Inductive Theorem Proving and Formal Methods: Formal system development is becoming a mature and established discipline and induction is one of the key techniques for dealing with abstract concepts. The aim of this session is to bring together the merits of inductive theorem proving techniques and formal methods in industrial application scenarios.

Higher-Order Inductive Theorem Proving: Higher-order logics provide a rich framework for expressing and reasoning about formal specifications. The importance of mechanizing formal arguments within higher-order logics is reflected by the sustained growth in popularity of verification environments such as HOL, Isabelle, Nuprl, and PVS. The aim of this session is to discuss recent advances of automated reasoning techniques within the context of higher-order logics.

Integrating Inductive and High-Performance Theorem Provers: Many first-order theorem provers are based on tableaux, matrix, and resolution techniques and their implementations are in general very efficient and highly specialized. The aim of this session is to elaborate how to integrate inductive theorem provers with other existing theorem proving technology.

Meeting the Challenges: We are interested in problems which demonstrate the unique merits of inductive theorem proving techniques. Submitted challenge problems will be displayed on the homepage prior to the workshop. Researchers are invited to submit solutions or counter challenges. The aim of this workshop session is to debate the relative merits of challenges and their solutions.

The workshop homepage is located at www.cs.cmu.edu/~carsten/apmi00 and the workshop committee consists of Carsten Schürmann, Andrew Ireland, Deepak Kapur, Christoph Kreitz, and Toby Walsh.

D. McAllester (Ed.): CADE-17, LNAI 1831, pp. 514–514, 2000.
© Springer-Verlag Berlin Heidelberg 2000

Workshop: Type-Theoretic Languages: Proof-Search and Semantics

Didier Galmiche

LORIA - UHP, Nancy, France

Much recent work has been devoted to type theory and its applications to proof- and program- development in various logical settings. The focus of this workshop is on proof-search, with a specific interest on semantic aspects of, and semantics approaches to, type-theoretic languages and their underlying logics (e.g., classical, intuitionistic, linear, substructural). Such languages can be seen as logical frameworks for representing proofs and in some cases formalize connections between proofs and programs that support program-synthesis.

The theory of proof-search has developed mostly along proof-theoretic lines but using many type-theoretic techniques. The utility of type-theoretic methods suggests that semantic methods of the kind found to be valuable in the semantics of programming languages should be useful in tackling the main outstanding difficulty in the theory of proof-search, i.e., the representation of intermediate stages in the search for a proof. The objective of the workshop is to provide a forum for discussion between, on the one hand, researchers interested in all aspects of proof-search in type theory, logical frameworks and their underlying (e.g., classical, intuitionistic, substructural) logics and, on the other, researchers interested in the semantics of computation.

Topics of interest, in this context, include but are not restricted to the following: Foundations of proof-search in type-theoretic languages (sequent calculi, natural deduction, logical frameworks, etc.); Systems, methods and techniques related to proof construction or to counter-models generation (tableaux, matrix, resolution, semantic techniques, proof plans, etc.); Decision procedures, strategies, complexity results; Logic programming as search-based computation, integration of model-theoretic semantics, semantics foundations for search spaces; Computational models based on structures as games and realizability; Proof synthesis vs program synthesis and applications, equational theories and rewriting; Applications of proof-theoretic and semantics techniques to the design and implementation of theorem provers.

Programme Committee:

D. Galmiche, *LORIA - UHP, Nancy, France.*
P. Lincoln, *SRI, Stanford, U.S.A.*
F. Pfenning, *CMU, Pittsburgh, U.S.A.*
D. Pym, *Queen Mary and Westfield College, London, U.K.*
J. Smith, *Chalmers University, Göteborg, Sweden.*

D. McAllester (Ed.): CADE-17, LNAI 1831, pp. 515–515, 2000.
© Springer-Verlag Berlin Heidelberg 2000

Workshop: Automated Deduction in Education

Erica Melis

Universität des Saarlandes, FB Informatik, Germany

One of the potential real-world applications of deduction systems is in mathematics education. Patrick Suppes' education system is an early pioneer in this regard, for example, and while the potential has been mentioned in discussions at previous CADE conferences, currently there is renewed interest in this topic as well as several activities and projects within the CADE community.

In an intelligent tutor system a deduction component might be used, e.g. to provide the expert model, to provide potential models of erroneous reasoning, as a basis for topic sequencing, as a basis for automated diagnosis. However, typically, a mathematics education system will not or not only include a deduction system, because the need for explanation will dominate the requirements for correctness in theorem proving in an educational context. That is, the power of automated deduction has to be combined with appropriate interfaces, user models, theory construction, and explanation functionalities before a system can be didactively effective. Though extensive production-quality systems are still in the future, some of the knowledge and the knowledge representation that is currently used in automated and interactive theorem-proving systems can be employed for educational needs as well.

A purpose of this workshop, in this application area of automated and interactive theorem proving, is to establish more communication between current education projects in the CADE community, to exchange ideas and opinions, and to make available the experience of education systems from other AI-communities. We explicitly encourage the submission of project descriptions.

We plan to focus the workshop on the following topics and questions

- How best can automated and interactive theorem-proving systems contribute to mathematics education?
- What are the proof-presentation and explanation needs for such teaching?
- What sort of integration of specialized reasoning systems (e.g., computer algebra systems) should we expect?
- How do we generate good examples and counter examples in various subjects?
- What is the role of knowledge-based theorem proving for mathematics education?
- What are the human-factors requirements for good systems?
- How do we evaluate the educational success of such systems?

Further information can be available at
http://www.ags.uni-sb.de/~melis/cade00ws.htm

D. McAllester (Ed.): CADE-17, LNAI 1831, pp. 516–516, 2000.
© Springer-Verlag Berlin Heidelberg 2000

Workshop: The Role of Automated Deduction in Mathematics

Simon Colton, Volker Sorge, and Ursula Martin

The purpose of this workshop is to discuss the role of automated deduction in all areas of mathematics. This will include looking at the interaction between automated deduction programs and other computational systems which have been developed over recent years to automate different areas of mathematical activity. Such systems include computer algebra packages, tutoring programs, mathematical discovery systems and systems developed to help present and archive mathematical theories. The workshop will also include discussions of the use of automated theorem proving in the wider mathematical community. Presentations which detail the employment of automated deduction techniques in any area of mathematical research have been encouraged.

With initiatives such as the Calculemus project, automated deduction is increasingly being seen not as an isolated area of research, but as part of an integrated attack on the problem of automating mathematics. We are interested in the interaction of automated theorem proving programs with (i) computer algebra (CA) packages (ii) constraint solvers (iii) model generators (iv) tutoring systems (v) interactive textbooks (vi) theory formation programs and (vi) mathematical databases. In all these fields automated deduction is either already used or could be fruitfully employed to enhance the power and reliability of existing systems. Particular ongoing projects include the use of deduction to certify CA systems, and also to enhance CA systems. Other projects include the incorporation of deduction into mathematical tutoring systems and interactive mathematical textbooks and the use of theory formation to help in automated theorem proving. The interaction between these programs could be in terms of improving automated deduction or in terms of using automated deduction to improve the techniques employed in the other system.

The workshop is intended to inspire the use of automated deduction within other fields of mathematics as well as the incorporation of techniques from other fields into automated deduction. We intend to provide a forum for discussion between researchers from the field of automated deduction and researchers from particular domains of mathematics. In particular, the workshop will address mathematical results proved in part by automated deduction techniques as well as theorems which can potentially be proved with automated techniques. An original goal of automated theorem proving was its application to mathematics, whether by proving established results, enhancing calculation techniques or facilitating discovery of new results. There is still much scope for the use of automated deduction to add to mathematics and we hope to explore these possibilities in the workshop.

The workshop home page is located at:
http://www.dai.ed.ac.uk/~simonco/conferences/CADE00

D. McAllester (Ed.): CADE-17, LNAI 1831, pp. 517–517, 2000.
© Springer-Verlag Berlin Heidelberg 2000

Author Index

Allen, Stuart F. 170
Andrews, Peter B. 164, 511
Appel, Andrew W. 7
Audemard, Gilles 302
Bachmair, Leo 64, 220
Barrett, Clark W. 79
Baumgartner, Peter 200, 513
Belinfante, Johan G.F. 132
Benhamou, Belaid 302
Bezem, Marc 148
Bishop, Matthew 164
Borralleras, Cristina 346
Brown, Chad E. 164, 511
Brown, Marianne 411
Bustan, Doran 255
Chatalic, Philippe 449
Collins, Graham 497
Colton, Simon 517
Constable, Robert L. 170
Degtyarev, Anatoli 365
Dennis, Louise A. 497
Dill, David L. 79
Eaton, Rich 170
Emerson, E. Allen 236
Farmer, William M. 115
Fermueller, Chris 513
Ferreira, Maria 346
Franke, Andreas 455
Fujita, Hiroshi 184
Galmiche, Didier 515
Genet, Thomas 271
Giesl, Jürgen 309
Gillard, Guillaume 417
Giunchiglia, Enrico 291
Grumberg, Orna 255
Harrison, John 1
Hasegawa, Ryuzo 184
Hendriks, Dimitri 148
Henocque, Laurent 302
Horrocks, Ian 482
Horton, Joseph D. 385
Hustadt, Ullrich 433
Jackson, Mike 502
Kahlon, Vineet 236

Kammüller, Florian 99
Kapur, Deepak 324
Kautz, Henry 183
Klay, Francis 271
Kohlhase, Michael 455
Koshimura, Miyuki 184
Kreitz, Christoph 170
Lee, Peter 25
Lorigo, Lori 170
Lowe, Helen 502
Martin, Ursula 517
McCune, William 401
Meier, Andreas 460
Melis, Erica 516
Michael, Neophytos G. 7
Middeldorp, Aart 309
Necula, George C. 25
Nivelle, Hans de 148
Patel-Schneider, Peter 297
Peltier, Nicolas 513
Pulman, Stephen 509
Rubio, Albert 346
Ruess, Harald 220
Sattler, Ulrike 482
Schmidt, Renate A. 433
Schürmann, Carsten 507, 514
Seger, Carl-Johan 235
Shumsky, Olga 401
Simon, Laurent 449
Sinz, Carsten 177
Slind, Konrad 45
Sofronie-Stokkermans, Viorica ... 465
Sorge, Volker 517
Spencer, Bruce 385
Stump, Aaron 79
Subramaniam, Mahadavan 324
Sutcliffe, Geoff 406, 411
Tacchella, Armando 291
Tiwari, Ashish 64, 220
Tobies, Stephan 482
Voronkov, Andrei 365
Zhang, Hantao 513

Author Index

Lecture Notes in Artificial Intelligence (LNAI)

Vol. 1688: P. Bouquet, L. Serafini, P. Brézillon, M. Benerecetti, F. Castellani (Eds.), Modeling and Using Context. Proceedings, 1999. XII, 528 pages. 1999.

Vol. 1692: V. Matoušek, P. Mautner, J. Ocelíková, P. Sojka (Eds.), Text, Speech, and Dialogue. Proceedings, 1999. XI, 396 pages. 1999.

Vol. 1695: P. Barahona, J.J. Alferes (Eds.), Progress in Artificial Intelligence. Proceedings, 1999. XI, 385 pages. 1999.

Vol. 1699: S. Albayrak (Ed.), Intelligent Agents for Telecommunication Applications. Proceedings, 1999. IX, 191 pages. 1999.

Vol. 1701: W. Burgard, T. Christaller, A.B. Cremers (Eds.), KI-99: Advances in Artificial Intelligence. Proceedings, 1999. XI, 311 pages. 1999.

Vol. 1704: Jan M. Żytkow, J. Rauch (Eds.), Principles of Data Mining and Knowledge Discovery. Proceedings, 1999. XIV, 593 pages. 1999.

Vol. 1705: H. Ganzinger, D. McAllester, A. Voronkov (Eds.), Logic for Programming and Automated Reasoning. Proceedings, 1999. XII, 397 pages. 1999.

Vol. 1711: N. Zhong, A. Skowron, S. Ohsuga (Eds.), New Directions in Rough Sets, Data Mining, and Granular-Soft Computing. Proceedings, 1999. XIV, 558 pages. 1999.

Vol. 1712: H. Boley, A Tight, Practical Integration of Relations and Functions. XI, 169 pages. 1999.

Vol. 1714: M.T. Pazienza (Eds.), Information Extraction. IX, 165 pages. 1999.

Vol. 1715: P. Perner, M. Petrou (Eds.), Machine Learning and Data Mining in Pattern Recognition. Proceedings, 1999. VIII, 217 pages. 1999.

Vol. 1720: O. Watanabe, T. Yokomori (Eds.), Algorithmic Learning Theory. Proceedings, 1999. XI, 365 pages. 1999.

Vol. 1721: S. Arikawa, K. Furukawa (Eds.), Discovery Science. Proceedings, 1999. XI, 374 pages. 1999.

Vol. 1724: H.I. Christensen, H. Bunke, H. Noltemeier (Eds.), Sensor Based Intelligent Robots. Proceedings, 1998. VIII, 327 pages. 1999.

Vol. 1730: M. Gelfond, N. Leone, G. Pfeifer (Eds.), Logic Programming and Nonmonotonic Reasoning. Proceedings, 1999. XI, 391 pages. 1999.

Vol. 1733: H. Nakashima, C. Zhang (Eds.), Approaches to Intelligent Agents. Proceedings, 1999. XII, 241 pages. 1999.

Vol. 1735: J.W. Amtrup, Incremental Speech Translation. XV, 200 pages. 1999.

Vol. 1739: A. Braffort, R. Gherbi, S. Gibet, J. Richardson, D. Teil (Eds.), Gesture-Based Communication in Human-Computer Interaction. Proceedings, 1999. XI, 333 pages. 1999.

Vol. 1744: S. Staab, Grading Knowledge: Extracting Degree Information from Texts. X, 187 pages. 1999.

Vol. 1747: N. Foo (Ed.), Adavanced Topics in Artificial Intelligence. Proceedings, 1999. XV, 500 pages. 1999.

Vol. 1757: N.R. Jennings, Y. Lespérance (Eds.), Intelligent Agents VI. Proceedings, 1999. XII, 380 pages. 2000.

Vol. 1759: M.J. Zaki, C.-T. Ho (Eds.), Large-Scale Parallel Data Mining. VIII, 261 pages. 2000.

Vol. 1760: J.-J. Ch. Meyer, P.-Y. Schobbens (Eds.), Formal Models of Agents. Poceedings. VIII, 253 pages. 1999.

Vol. 1761: R. Caferra, G. Salzer (Eds.), Automated Deduction in Classical and Non-Classical Logics. Proceedings. VIII, 299 pages. 2000.

Vol. 1771: P. Lambrix, Part-Whole Reasoning in an Object-Centered Framework. XII, 195 pages. 2000.

Vol. 1772: M. Beetz, Concurrent Reactive Plans. XVI, 213 pages. 2000.

Vol. 1775: M. Thielscher, Challenges for Action Theories. XIII, 138 pages. 2000.

Vol. 1778: S. Wermter, R. Sun (Eds.), Hybrid Neural Systems. IX, 403 pages. 2000.

Vol. 1792: E. Lamma, P. Mello (Eds.), AI*IA 99: Advances in Artificial Intelligence. Proceedings, 1999. XI, 392 pages. 2000.

Vol. 1793: O. Cairo, L.E. Sucar, F.J. Cantu (Eds.), MICAI 2000: Advances in Artificial Intelligence. Proceedings, 2000. XIV, 750 pages. 2000.

Vol. 1794: H. Kirchner, C. Ringeissen (Eds.), Frontiers of Combining Systems. Proceedings, 2000. X, 291 pages. 2000.

Vol. 1805: T. Terano, H. Liu, A.L.P. Chen (Eds.), Knowledge Discovery and Data Mining. Proceedings, 2000. XIV, 460 pages. 2000.

Vol. 1810: R. López de Mántaras, E. Plaza (Eds.), Machine Learning: ECML 2000. Proceedings, 2000. XII, 460 pages. 2000.

Vol. 1821: R. Loganantharaj, G. Palm, M. Ali (Eds.), Intelligent Problem Solving. Proceedings, 2000. XVII, 751 pages. 2000.

Vol. 1822: H.H. Hamilton, Advances in Artificial Intelligence. Proceedings, 2000. XII, 450 pages. 2000.

Vol. 1831: D. McAllester (Ed.), Automated Deduction – CADE-17. Proceedings, 2000. XIII, 519 pages. 2000.

Vol. 1835: D. N. Christodoulakis (Ed.), Natural Language Processing – NLP 2000. Proceedings, 2000. XII, 438 pages. 2000.

Vol. 1849: C. Freksa, W. Brauer, C. Habel, K.F. Wender (Eds.), Spatial Cognition II. XI, 420 pages. 2000.

Lecture Notes in Computer Science

Vol. 1787: J. Song (Ed.), Indormation Security and Cryptology – ICISC'99. Proceedings, 1999. XI, 279 pages. 2000.

Vol. 1789: B. Wangler, L. Bergman (Eds.), Advanced Information Systems Engineering. Proceedings, 2000. XII, 524 pages. 2000.

Vol. 1790: N. Lynch, B.H. Krogh (Eds.), Hybrid Systems: Computation and Control. Proceedings, 2000. XII, 465 pages. 2000.

Vol. 1792: E. Lamma, P. Mello (Eds.), AI*IA 99: Advances in Artificial Intelligence. Proceedings, 1999. XI, 392 pages. 2000. (Subseries LNAI).

Vol. 1793: O. Cairo, L.E. Sucar, F.J. Cantu (Eds.), MICAI 2000: Advances in Artificial Intelligence. Proceedings, 2000. XIV, 750 pages. 2000. (Subseries LNAI).

Vol. 1794: H. Kirchner, C. Ringeissen (Eds.), Frontiers of Combining Systems. Proceedings, 2000. X, 291 pages. 2000. (Subseries LNAI).

Vol. 1795: J. Sventek, G. Coulson (Eds.), Middleware 2000. Proceedings, 2000. XI, 436 pages. 2000.

Vol. 1796: B. Christianson, B. Crispo, J.A. Malcolm, M. Roe (Eds.), Security Protocols. Proceedings, 1999. XII, 229 pages. 2000.

Vol. 1800: J. Rolim et al. (Eds.), Parallel and Distributed Processing. Proceedings, 2000. XXIII, 1311 pages. 2000.

Vol. 1801: J. Miller, A. Thompson, P. Thomson, T.C. Fogarty (Eds.), Evolvable Systems: From Biology to Hardware. Proceedings, 2000. X, 286 pages. 2000.

Vol. 1802: R. Poli, W. Banzhaf, W.B. Langdon, J. Miller, P. Nordin, T.C. Fogarty (Eds.), Genetic Programming. Proceedings, 2000. X, 361 pages. 2000.

Vol. 1803: S. Cagnoni et al. (Eds.), Real-World Applications of Evolutionary Computing. Proceedings, 2000. XII, 396 pages. 2000.

Vol. 1805: T. Terano, H. Liu, A.L.P. Chen (Eds.), Knowledge Discovery and Data Mining. Proceedings, 2000. XIV, 460 pages. 2000. (Subseries LNAI).

Vol. 1806: W. van der Aalst, J. Desel, A. Oberweis (Eds.), Business Process Management. VIII, 391 pages. 2000.

Vol. 1807: B. Preneel (Ed.), Advances in Cryptology – EUROCRYPT 2000. Proceedings, 2000. XVIII, 608 pages. 2000.

Vol. 1811: S.W. Lee, H.. Bülthoff, T. Poggio (Eds.), Biologically Motivated Computer Vision. Proceedings, 2000. XIV, 656 pages. 2000.

Vol. 1815: G. Pujolle, H. Perros, S. Fdida, U. Körner, I. Stavrakakis (Eds.), Networking 2000 – Broadband Communications, High Performance Networking, and Performance of Communication Networks. Proceedings, 2000. XX, 981 pages. 2000.

Vol. 1816: T. Rus (Ed.), Algebraic Methodology and Software Technology. Proceedings, 2000. XI, 545 pages. 2000.

Vol. 1817: A. Bossi (Ed.), Logic-Based Program Synthesis and Transformation. Proceedings, 1999. VIII, 313 pages. 2000.

Vol. 1818: C.G. Omidyar (Ed.), Mobile and Wireless Communications Networks. Proceedings, 2000. VIII, 187 pages. 2000.

Vol. 1819: W. Jonker (Ed.), Databases in Telecommunications. Proceedings, 1999. X, 208 pages. 2000.

Vol. 1821: R. Loganantharaj, G. Palm, M. Ali (Eds.), Intelligent Problem Solving. Proceedings, 2000. XVII, 751 pages. 2000. (Subseries LNAI).

Vol. 1822: H.H. Hamilton, Advances in Artificial Intelligence. Proceedings, 2000. XII, 450 pages. 2000. (Subseries LNAI).

Vol. 1823: M. Bubak, H. Afsarmanesh, R. Williams, B. Hertzberger (Eds.), High Performance Computing and Networking. Proceedings, 2000. XVIII, 719 pages. 2000.

Vol. 1824: J. Palsberg (Ed.), Static Analysis. Proceedings, 2000. VIII, 433 pages. 2000.

Vol. 1830: P. Kropf, G. Babin, J. Plaice, H. Unger (Eds.), Distributed Communities on the Web. Proceedings, 2000. X, 203 pages. 2000.

Vol. 1831: D. McAllester (Ed.), Automated Deduction – CADE-17. Proceedings, 2000. XIII, 519 pages. 2000. (Subseries LNAI).

Vol. 1835: D. N. Christodoulakis (Ed.), Natural Language Processing – NLP 2000. Proceedings, 2000. XII, 438 pages. 2000. (Subseries LNAI).

Vol. 1839: G. Gauthier, C. Frasson, K. VanLehn (Eds.), Intelligent Tutoring Systems. Proceedings, 2000. XIX, 675 pages. 2000.

Vol. 1842: D. Vernon (Ed.), Computer Vision – ECCV 2000. Part I. Proceedings, 2000. XVIII, 953 pages. 2000.

Vol. 1843: D. Vernon (Ed.), Computer Vision – ECCV 2000. Part II. Proceedings, 2000. XVIII, 881 pages. 2000.

Vol. 1845: H.B. Keller, E. Plöderer (Eds.), Reliable Software Technologies Ada-Europe 2000. Proceedings, 2000. XIII, 304 pages. 2000.

Vol. 1846: H. Lu, A. Zhou (Eds.), Web-Age Information Management. Proceedings, 2000. XIII, 462 pages. 2000.

Vol. 1848: R. Giancarlo, D. Sankoff (Eds.), Combinatorial Pattern Matching. Proceedings, 2000. XI, 423 pages. 2000.

Vol. 1849: C. Freksa, W. Brauer, C. Habel, K.F. Wender (Eds.), Spatial Cognition II. XI, 420 pages. 2000. (Subseries LNAI).

Vol. 1850: E. Bertino (Ed.), ECOOP 2000 – Object-Oriented Programming. Proceedings, 2000. XIII, 493 pages. 2000.